CONVECTION HEAT TRANSFER

CONVECTION HEAT TRANSFER

VEDAT S. ARPACI

UNIVERSITY OF MICHIGAN

POUL S. LARSEN

TECHNICAL UNIVERSITY OF DENMARK,
FORMERLY UNIVERSITY OF MICHIGAN

PRENTICE-HALL, INC.

Englewood Cliffs, New Jersey, 07632

Library of Congress Cataloging in Publication Data

Arpaci, Vedat S. (date)
Convection heat transfer

Includes bibliographies and index.
1. Heat—Convection. I. Larsen, Poul Scheel

II. Title
QC327.A76 1984 536′.25 83–15966
ISBN 0-13-172346-4

Printed in the United States of America

10 9 8 7 6 5 4 3 2 1

Editorial/production: Nicholas C. Romanelli
Manufacturing buyer: Anthony Caruso
Cover design: Edsal Enterprises

ISBN 0-13-172346-4

Prentice-Hall International, *London*
Prentice-Hall of Australia Pty. Limited, *Sydney*
Editora Prentice-Hall do Brasil, Ltda., *Rio de Janeiro*
Prentice Hall Canada, Inc., *Toronto*
Prentice-Hall of India Private Limited, *New Delhi*
Prentice-Hall of Japan, Inc., *Tokyo*
Prentice-Hall of Southeast Asia Pte. Ltd., *Singapore*
Whitehall Books Limited, *Wellington, New Zealand*

To Nigâr and Myriam

CONTENTS

PREFACE

In this text we present a coherent account of what we have learned about convection heat transfer during two decades with our students and colleagues. Our philosophy is to show how to develop an approach to *formulating* and *solving* convection problems. This philosophy follows essentially that of the text on *Conduction Heat Transfer* written by one of us some years ago (Arpaci, 1966).

In Part I, following an introduction on the foundations of heat transfer, the concepts needed from mechanics and thermodynamics are reviewed (Chapter 1) and a GENERAL APPROACH to formulation is developed in terms of general principles, local equilibrium, constitutive relations, and the governing equations (Chapter 2). Novel aspects of Part I are the development of differential and integral forms of the general principles in terms of a specific property b, and the disclosure of the need for local equilibrium.

In Part II an INDIVIDUAL APPROACH to integral and differential formulations is developed in terms of six steps and applied to parallel flows (Chapter 3), and to nearly parallel (boundary layer) flows (Chapter 4). Part II reflects the philosophy of the text. Our claim is that physical insight, combined with analytical simplicity, provide the shortcuts leading directly to the core of complex problems and to straightforward solutions: we repudiate any form of sophistication for its own sake. Noticeable features of Part II are the concepts of *penetration depth* and *volumetric rise* and their application to a number of problems. We have stressed the fact (not only in this part but throughout the entire text) that the integral approach is an important tool for engineering problems.

Part III is devoted to SOLUTION procedures most commonly applied to nonlinear convection problems. Chapter 5 deals with similarity transformations, Chapter 6 with periodic convection, and Chapter 7 with computational methods. Unusual aspects of Part III are the application of dimensional analysis to similarity transformations, a clear illustration of the physical significance of ac and dc components of a response to periodic disturbances, and a comparison between the finite difference and finite element methods.

Although the first three parts of this text deal either with general developments or laminar flows, we have no intention of losing sight of the fact (between the foundations of convection and mathematical intricacies of laminar problems) that the actual flows are turbulent. However, because of its complicated nature, it is impractical to learn turbulence simultaneously with other aspects of convection. Therefore, the first three parts should be interpreted as being the logical steps for learning fundamentals (in terms of simple, but to a large extent unrealistic, laminar flows) and Part IV as being the application of these fundamentals (combined with a turbulence model) to actual flows.

The first two chapters of Part IV aim at developing an understanding of TURBULENCE. Chapter 8 deals with (instability) origin and (random or statistical) foundations of turbulence. Since neither stability nor statistical methods are fully capable of handling turbulence, this chapter is intended as an appreciation of the origin and foundations of turbulence. Chapter 9 introduces the integral and micro scales of turbulence and investigates in terms of these scales the equilibrium range of kinetic energy and thermal spectra. Chapter 10 concerns flow prediction and Chapter 11 considers heat transfer prediction. Our aim in these two chapters is to obtain velocity and temperature distributions, as well as the heat transfer, for convection problems. Finally, Chapter 12 discusses the organization of experimental data in terms of DIMENSIONAL ANALYSIS, and interprets (by utilizing the scales of Chapter 9) the hitherto assumed empirical heat transfer CORRELATIONS.

Aside from the classical ones needed for the sake of completeness, the great majority of examples worked in the text and the problems left to the student as homework are our own inventions. In general, the examples are designed to supplement and extend the text. At the risk of being criticised for incompleteness, but with the hope of guiding students to future discoveries, we incorporate some advanced but incomplete examples and problems into the text.

With a few exceptions, the engineering background required of the reader are the customary undergraduate courses in fluid mechanics, thermodynamics, heat transfer, and advanced calculus. Cylindrical and spherical coordinates are intentionally omitted due to space considerations. The reader should have no difficulties in formulating Cartesian examples in other coordinates.

We were tempted to include a wider range of subjects (such as rheology, mass transfer, hydromagnetics, separated, high-speed, and rarefied flows), but we avoided doing so in the belief that it is better to consider a smaller number of topics in depth than attempt to survey the entire field superficially.

In a text combining a number of diverse subjects, the nomenclature creates some problems. We have tried hard to keep a uniform symbol system throughout the text. However, for customary reasons, we have been led to use h for both enthalpy and heat transfer coefficient and u for x-velocity and internal energy. Also, when manipulating with dimensionless numbers, we have used one-letter symbols, and when working with correlations we have employed two-letter symbols.

This text is the result of a series of revisions of the material originally prepared for use in a beginning graduate course on heat transfer in the Mechanical Engineering Department of the University of Michigan. Over the years, because of its basic content, the course also attracted students from aeronautical, chemical, and nuclear engineering departments.

The material included in the text is more than can be covered in a one-semester course. This was done purposely to trigger the curiosity of students who are interested in advancing beyond the minimum on the subject and also to leave some flexibility to instructors in the selection of materials.

For a one-semester graduate course on convection at a typical engineering school in the United States the following outline is suggested:

—Omit Chapters 1 and 2.
—Concentrate on Chapters 3 and 4. Omit those parts of Chapter 3 dealing with the three classical solution methods (separation of variables, Laplace transforms, and variational calculus). These methods apply to linear convection, as well as conduction, but they are not convenient for nonlinear convection. They are intended for the reader who has more than usual interest in analytical tools. Read Section 4.6. Inspecting the figures of this section, develop some appreciation for the changing behavior of nonparallel flows, depending on the Reynolds number or the Rayleigh number.
—Study Chapter 5 to learn the most common and classical solution method of convection problems.

For the remainder of the course two avenues may be followed, consisting of either Chapters 6, 8, 9, and 12, or chapters 7, 10, and 11. In the *qualitative* (*intuitive*) approach comprising the first set of chapters:

—Study Chapter 6 to learn the limitation of perturbation methods for the solution of turbulence.
—Read Chapter 8 for an appreciation of the origin and foundations of turbulence.
—Concentrate on Chapter 9 to learn the scales of turbulence.
—Study Chapter 12 to learn the foundations of heat transfer correlations in terms of these scales.

In the *computational* approach comprising the second set of chapters:

—Study Chapter 7 to learn the numerical solution of laminar convection.
—Study Chapters 10 and 11 to learn the one- and two-equation turbulence models.

We are grateful to Professor E. M. Sparrow of the University of Minnesota for his support, and to Professor M. M. Chen of the University of Illinois and Professor S. Korpela of the Ohio State University for their constructive criticism of the manuscript.

Ann Arbor, Michigan
Lyngby, Denmark

VEDAT S. ARPACI
POUL S. LARSEN

Part I

GENERAL FORMULATION

Chapter 1

INTRODUCTION. CONCEPTS

A thermal science like heat transfer, as well as gas dynamics, always starts with mechanics and thermodynamics but also requires information about the behavior (constitution) of the working fluid. Gas dynamics, for example, considers the equation of state for a gas in addition to the general principles of mechanics and thermodynamics. Heat transfer, on the other hand, requires two constitutive relations by which its two basic modes, diffusion (conduction) and radiation, are distinguished.

Phenomenologically speaking, diffusion is the experimental recognition of heat transfer from a point of higher temperature to an adjacent point of lower temperature in a medium. At the microscopic level, the mechanism of diffusion is visualized as and hypothesized by a model for the exchange of energy between adjacent particles. Consequently, diffusion is local and, being directional, is irreversible, and it can happen only through matter. On the other hand, radiation is the experimental recognition of electromagnetic waves, and the energy carried (heat transferred) by these waves. At the microscopic level, the mechanism of radiation is visualized as the transport of energy by radiation particles (radiation quanta, photons). Consequently, acting at a distance, the radiation is global, and is reversible when it happens through vacuum (Fig. 1.1).

From a conceptual viewpoint, convection is not a basic mode of heat transfer but, rather, is the diffusion and/or radiation in moving media. Therefore, fluid mechanics

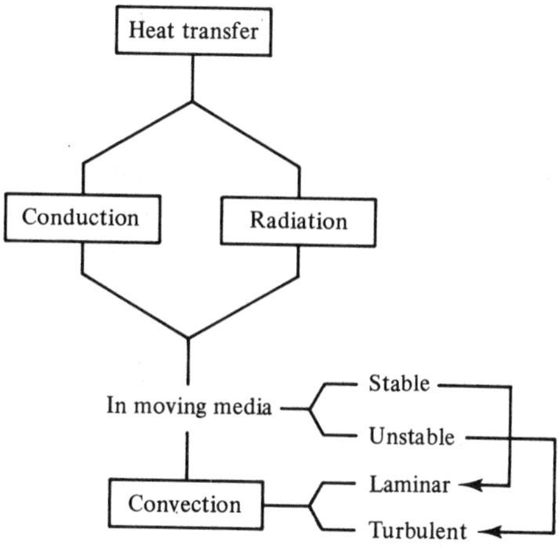

Fig. 1.1 Two modes of heat transfer and convection

plays an important role in convection. For customary reasons only, we shall refer to the diffusion of heat in moving (or stationary) rigid media as conduction, and to the diffusion and/or radiation of heat in moving deformable media as convection. Conduction has already been treated by Arpaci (1966); convection is the subject matter of this text; convection with radiation, however, requires separate treatment, and will not be considered here.

1.1 *CONTINUUM APPROACH VERSUS PARTICLE APPROACH*

In the preceding section we prescribed the modes of heat transfer from the phenomenological and particle viewpoints. The contrast between these views is reflected in the two alternative approaches to heat transfer. In the phenomenological approach, the medium is assumed to be a continuum. That is, the mean free path of particles is small compared with other dimensions existing in the medium, such that a statistical average (or global description) is possible. In the microscopic approach, the medium is assumed to be made of a very large number of particles, and either a statistical average of the particle behavior is not possible or it is possible but not essential. Actually, a very general and logical description of a medium that consists of a spatially distributed particle structure would be one in which the general principles are written for each separate particle. Solving the many-particle system in time and space and then relating a desired continuum concept to particle behavior would obviously produce the result to be obtained from the continuum approach.

The reason for not always starting with the particle approach, apart from the mathematical difficulties and from the uncertainties related to intermolecular forces, is that behavior of particles may not be of particular interest. On the contrary, as in most cases of engineering, the problem is to determine how the medium behaves as a whole, for example, how the velocity and temperature of a medium vary. Here the convenience of using the continuum concept becomes clear. Obviously, there are cases in which it is advantageous to use one of the foregoing approaches in preference to the other. A large number of problems exists, however, for which both approaches can be used conveniently, the choice depending on previous experience or personal conviction. From the viewpoint of physical interpretation the only difference is that the averaging process of the particle structure is undertaken before or after the analysis, depending on the approach used. That is, the statistics either precedes or follows the thermomechanics.

In this text, our interest lies not in the individual behavior of particles, but in their mean effects in space and time. In short, we are interested in a continuum approach to convection heat transfer.

1.2 *LAMINAR AND TURBULENT CONVECTION*

Convection heat transfer, as a diffusion process in a moving medium, can best be classified according to the celebrated Reynolds experiment sketched in Fig. 1.2. Accordingly, the natural appearance of convection is either stable (laminar) or unstable (turbulent). In Part II we shall learn the foundations of convection in terms of simple problems of laminar convection which may exist, if they do, rather infre-

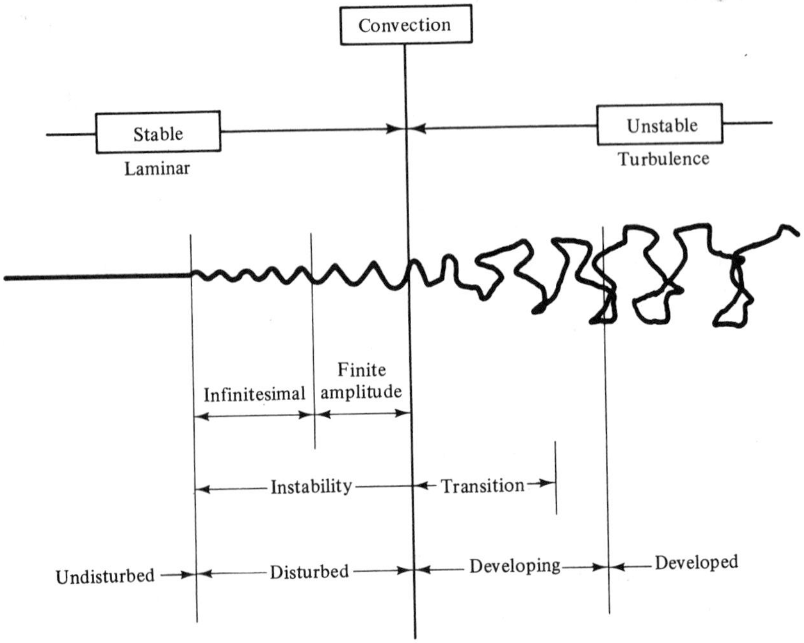

Fig. 1.2 The Reynolds experiment; structure of convection

quently. In Part IV we develop models for turbulence and, in terms of these models, investigate the turbulent problems of convection which are encountered most frequently.

An inspection of the different forms of motion assumed for the classical branches of thermal science locates the place of convection among these branches, as shown in Fig. 1.3. Thus, convection is usually known to deal with incompressible viscous flows, while gas dynamics is usually known to deal with inviscid compressible flows.

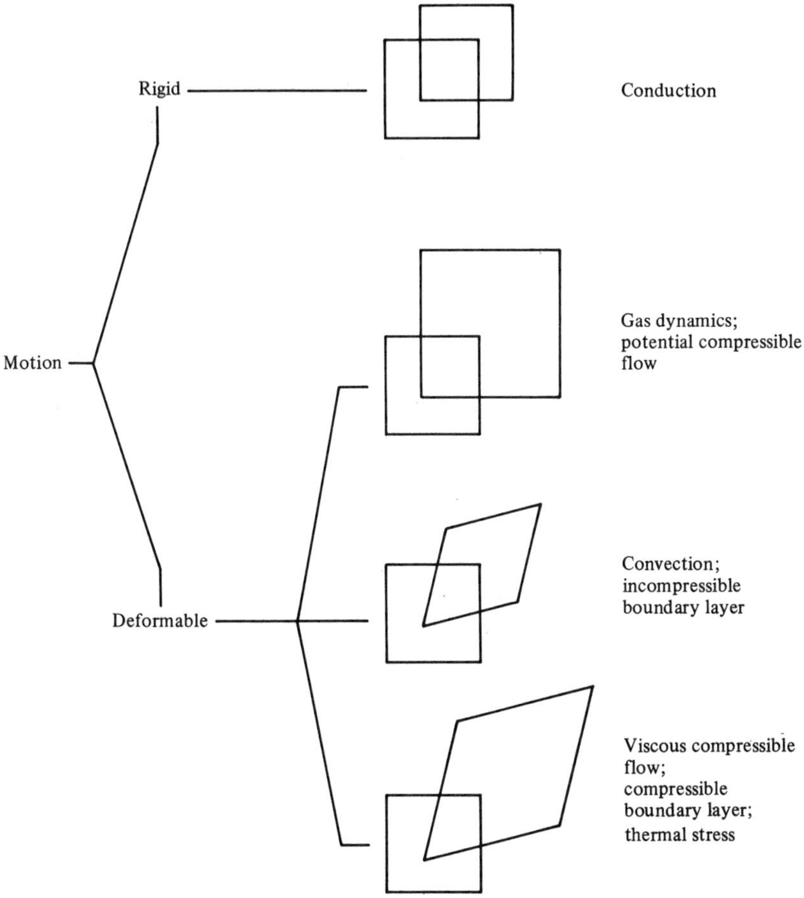

Fig. 1.3 Customary branches of thermal science

Convection heat transfer is usually encountered along a solid–fluid interface (Fig. 1.4). Thus, conduction in a solid and convection in a fluid simultaneously occur in nature, rather than occurring separately. However, because of our inability to solve this coupled actual problem, we cut the problem along the interface and artificially separate it into a conduction problem and a convection problem. Then we try

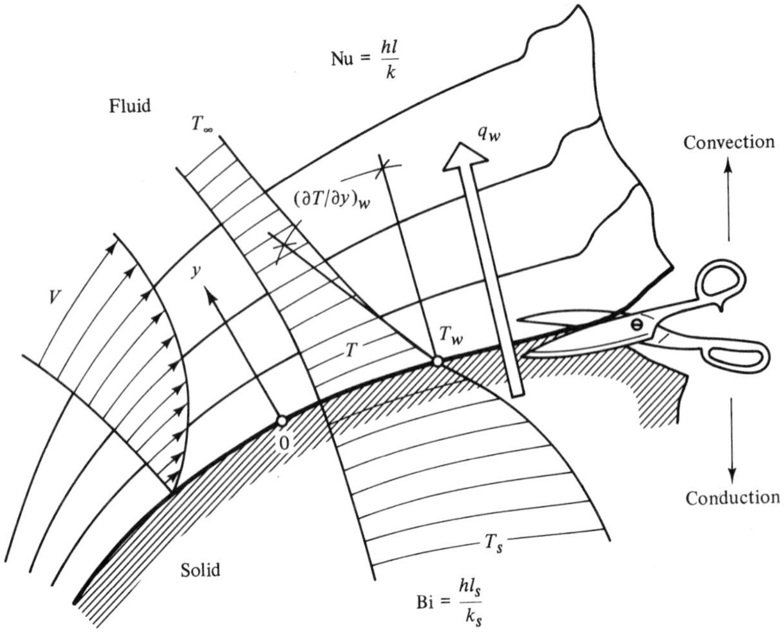

Fig. 1.4 Separation of an actual thermal problem into a conduction problem and a convection problem

to solve each problem separately after replacing the real interface boundary conditions with simpler but somewhat artificial boundary conditions. The present text deals with the convection part of the actual problem.

Defining convection as heat transfer through a solid–fluid interface, this heat transfer per unit area and time is expressed in terms of a heat transfer coefficient h, according to Newton's law of cooling,

$$q_W = h(T_W - T_\infty), \tag{1.2.1}$$

where $T_W - T_\infty$ denotes the difference between interface and ambient temperatures. The heat flux q_W may be also expressed in terms of the thermal conductivity k of the fluid, according to the Fourier law of conduction, as

$$q_W = -k\left(\frac{\partial T}{\partial y}\right)_W. \tag{1.2.2}$$

Combining Eqs. (1.2.1) and (1.2.2) gives

$$-k\left(\frac{\partial T}{\partial y}\right)_W = h(T_W - T_\infty), \tag{1.2.3}$$

or, in terms of a characteristic length l for the fluid domain, as

$$\text{Nu} = \frac{hl}{k} = \frac{\partial}{\partial(y/l)}\left(\frac{T_W - T}{T_W - T_\infty}\right)_W, \tag{1.2.4}$$

where Nu is the *Nusselt number* and y/l the dimensionless distance normal to the interface. Thus, the convection heat transfer through an interface is related to the evaluation of the dimensionless wall gradient of the fluid temperature. The prime objective of this text is the evaluation, by a number of methods, of this temperature gradient, or the Nusselt number.

Clearly, q_W may also be expressed by conduction in the solid, which leads to the definition of the *Biot number*,

$$\text{Bi} = \frac{hl_s}{k_s}, \tag{1.2.5}$$

where the subscript s refers to the solid domain. However, the conceptual difference between the definition of Nusselt and Biot numbers should be noted. In conduction problems, h and T_∞ are given, and Eq. (1.2.1) is employed as a boundary condition. Because of their complexity, convection problems are usually solved in terms of simpler boundary conditions, and Eq. (1.2.3) is employed to evaluate h.

A continuum science such as convection heat transfer is based on the definition of concepts and the statement of natural laws in terms of these concepts. Accordingly, the next section deals with the definition of a number of appropriate concepts.

1.3 RATE OF CHANGE OF A PROPERTY AT A POINT (THREE TIME DERIVATIVES)

In Section 1.1 we learned that convection is *conduction* and/or *radiation* in a *moving* continuum. This definition demonstrates the involvement of unsteady convection problems with a rate of change of a property (such as mass, momentum, moment of momentum, internal energy, entropy, etc.) in this continuum. Actually, we may encounter three different rates of change of a property. Let us demonstrate the physical significance of these rates in terms of the concentration b of some contaminant* in a river to be defined as the limit of the number of particles ΔB in a small volume $\Delta\mathcal{V}$,

$$b = \lim_{\Delta\mathcal{V}\to\Delta\mathcal{V}_0} \frac{\Delta B}{\Delta\mathcal{V}},$$

where $\Delta\mathcal{V}_0$ represents the size of a *physical point depending on the size of contaminant particles.* For mathematical convenience, we may let $\Delta\mathcal{V}_0 \longrightarrow 0$ and reduce the physical point to the mathematical point, and more conveniently define the concentration as

$$b = \lim_{\Delta\mathcal{V}\to 0} \frac{\Delta B}{\Delta\mathcal{V}} = \frac{dB}{d\mathcal{V}}.$$

*Property b may be any passive scalar associated with the fluid, such as the concentration of a chemical compound or the energy (temperature).

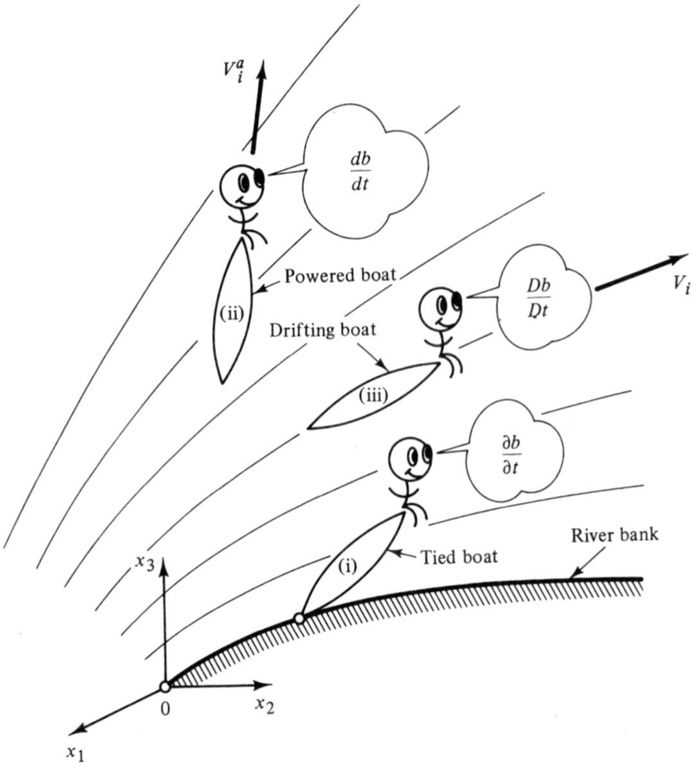

Fig. 1.5 Three time derivatives

Now consider the concentration with respect to an observer in a motorboat on the river. As shown in Fig. 1.5, we may encounter three cases: (i) the boat is tied to the river bank; (ii) the (powered) boat moves with absolute velocity $V_i^a = dx_i^a/dt$ relative to the river bank, x_i^a denoting the position of the boat with respect to a reference frame attached to the river bank; and (iii) the (powerless) boat drifts downstream with the river velocity $V_i = dx_i/dt$, x_i denoting a point on the river coinciding with the boat.

Let us try to express mathematically the rate of change of the concentration relative to the observer in each of the foregoing cases. A differential change in b may readily be written, in terms of the Cartesian tensor notation, for example, as

$$db = \frac{\partial b}{\partial t}\,dt + \frac{\partial b}{\partial x_i}\,dx_i,$$

and the *total derivative* as

$$\frac{db}{dt} = \frac{\partial b}{\partial t} + \left(\frac{dx_i}{dt}\right)\frac{\partial b}{\partial x_i}.$$

In this text no formal commitment will be made to the use of Cartesian tensor or symbolic vector notations. Early developments will sometimes be carried out in terms of one or both of these notations. However, the later developments will be discussed only in terms of the most convenient notation. The reader is expected to be familiar with the vector and tensor calculus (see, for example, Prager 1961).

In the boat tied to the river bank, case (i) and $dx_i/dt = 0$, the observer realizes the *partial derivative* of the concentration with respect to time,

$$\frac{\partial b}{\partial t}.$$

In the cruising boat, case (ii) and $dx_i/dt = V_i^a$, the observer realizes the total derivative based on an arbitrary (absolute) velocity V_i^a,

$$\frac{db}{dt} = \frac{\partial b}{\partial t} + V_i^a \frac{\partial b}{\partial x_i}.$$

In the drifting boat, case (iii) and $dx_i/dt = V_i$, the observer realizes the total derivative based on the river velocity V_i. This particular total derivative is referred to as the *substantial derivative*, the *material derivative*, or the *derivative following the motion*, and is denoted by

$$\frac{Db}{Dt} = \frac{\partial b}{\partial t} + V_i \frac{\partial b}{\partial x_i}. \tag{1.3.1}$$

For a steady process (in the coordinate system attached to the river bank), $\partial b/\partial t = 0$ in each of the foregoing three cases.

A special but indispensable case of Eq. (1.3.1) is the substantial derivative of velocity, the acceleration,

$$a_j = \frac{DV_j}{Dt} = \frac{\partial V_j}{\partial t} + V_i \frac{\partial V_j}{\partial x_i}, \tag{1.3.2}$$

or, in vector notation,

$$\boldsymbol{a} = \frac{D\boldsymbol{V}}{Dt} = \frac{\partial \boldsymbol{V}}{\partial t} + \boldsymbol{V} \cdot \nabla \boldsymbol{V}, \tag{1.3.3}$$

where the two terms on the right represent the *local* and *convective* accelerations, respectively. Furthermore, the second term may be rearranged to give

$$\boldsymbol{V} \cdot \nabla \boldsymbol{V} = \nabla\left(\frac{V^2}{2}\right) - \boldsymbol{V} \times (\nabla \times \boldsymbol{V}). \tag{1.3.4}$$

The relation given by Eq. (1.3.4) may readily be established by the vector identity

$$\boldsymbol{\alpha} \times (\boldsymbol{\beta} \times \boldsymbol{\gamma}) = (\boldsymbol{\alpha} \cdot \boldsymbol{\gamma})\boldsymbol{\beta} - (\boldsymbol{\alpha} \cdot \boldsymbol{\beta})\boldsymbol{\gamma}.$$

Assuming that

$$\boldsymbol{V} \cdot \nabla \boldsymbol{V} = (\boldsymbol{V} \cdot \boldsymbol{e}_{(i)}) \frac{\partial \boldsymbol{V}}{\partial x_i} = (\boldsymbol{\alpha} \cdot \boldsymbol{\beta})\boldsymbol{\gamma},$$

we have

$$(\boldsymbol{V} \cdot \boldsymbol{e}_{(i)}) \frac{\partial \boldsymbol{V}}{\partial x_i} = \left(\boldsymbol{V} \cdot \frac{\partial \boldsymbol{V}}{\partial x_i}\right)\boldsymbol{e}_{(i)} - \boldsymbol{V} \times \left(\boldsymbol{e}_{(i)} \times \frac{\partial \boldsymbol{V}}{\partial x_i}\right),$$

or

$$(\boldsymbol{V} \cdot \boldsymbol{e}_{(i)}) \frac{\partial \boldsymbol{V}}{\partial x_i} = \boldsymbol{e}_{(i)} \frac{\partial}{\partial x_i}\left(\frac{V^2}{2}\right) - \boldsymbol{V} \times \left(\boldsymbol{e}_{(i)} \frac{\partial}{\partial x_i} \times \boldsymbol{V}\right),$$

which, in view of $\nabla \equiv e_{(i)}\, \partial/\partial x_i$, is identical to Eq. (1.3.4) The same result may also be obtained by the tensor calculus. Adding $\pm V_j\, \partial V_j/\partial x_i$, we have for

$$(V \cdot \nabla V)_i = V_j \frac{\partial V_i}{\partial x_j} \pm V_j \frac{\partial V_j}{\partial x_i}$$

$$= \frac{\partial}{\partial x_i}\left(\frac{V^2}{2}\right) + V_j\left(\frac{\partial V_i}{\partial x_j} - \frac{\partial V_j}{\partial x_i}\right),$$

where the second term may be rearranged as

$$V_j\left(\frac{\partial V_i}{\partial x_j} - \frac{\partial V_j}{\partial x_i}\right) = -V_j(\delta_{ip}\delta_{jq} - \delta_{iq}\delta_{jp})\frac{\partial V_q}{\partial x_p},$$

or

$$-V_j \epsilon_{ijk} \epsilon_{kpq} \frac{\partial V_q}{\partial x_p},$$

or, in terms of $\epsilon_{kpq}\, \partial V_q/\partial x_p = (\nabla \times V)_k$, as

$$-\epsilon_{ijk} V_j (\nabla \times V)_k = -[V \times (\nabla \times V)]_i.$$

Introducing the usual definition of vorticity, $\boldsymbol{\omega} = \nabla \times V$, the acceleration given by Eq. (1.3.3) may be rearranged in terms of Eq. (1.3.4) as

$$\boldsymbol{a} = \frac{DV}{Dt} = \frac{\partial V}{\partial t} + \nabla\left(\frac{V^2}{2}\right) - V \times \boldsymbol{\omega}. \tag{1.3.5}$$

This form is useful for the interpretation of the momentum balance, for example in potential flow for which $\boldsymbol{\omega} = \mathbf{0}$ (Section 2.4.4).

1.4 *RATE OF CHANGE OF A PROPERTY WITHIN A VOLUME (THE REYNOLDS TRANSPORT THEOREM)*

We know from mechanics and thermodynamics that the simple original forms of the general principles are in terms of a *system* (of constant mass). Yet it is not convenient to use these forms of the general principles for convection (or for other disciplines associated with moving continua). Clearly, it becomes difficult to identify the boundaries of a moving system for any appreciable length of time. This fact suggests the transformation of general principles for a system to those for a *control volume* when dealing with a formulation of moving continua. The formula needed for this transformation, the Reynolds transport theorem, is the subject matter of this section.

Let us explain first the physical significance of this theorem by returning to the observer in the boat, considered in the preceding section. Suppose that the observer is able to determine the property B within a given volume $\mathcal{V}$, say bounded by a fish net thrown into the river. Since fluid and the associated property flow freely through the net, and the net may deform, it represents a deformable control volume surface. On the other hand, the bounding surface of a system, defined to enclose a constant mass of fluid, must be moving and deforming with the fluid. The Reynolds transport theorem gives a relation between the time rate of change within the system volume $\mathcal{V}_s$ and that within the control volume $\mathcal{V}$.

Now we are ready to establish mathematically the Reynolds transport theorem in terms of the foregoing situation. Let the instantaneous volume of the net be $\mathcal{V}(x_i, t)$ and that of the system be $\mathcal{V}_s(x_i, t)$. At time t let the system $\mathcal{V}_s$ coincide with the control volume $\mathcal{V}$. At time $t + \Delta t$ the new positions of the system and control volume are, as shown in Fig. 1.6, $\mathcal{V}_s(x_i, t + \Delta t)$ and $\mathcal{V}(x_i, t + \Delta t)$, respectively. Clearly, the boundaries of $\mathcal{V}_s$ are determined solely by the motion of the river, while the boundaries of $\mathcal{V}$ are controlled by the observer, that is, are determined by the problem in question.

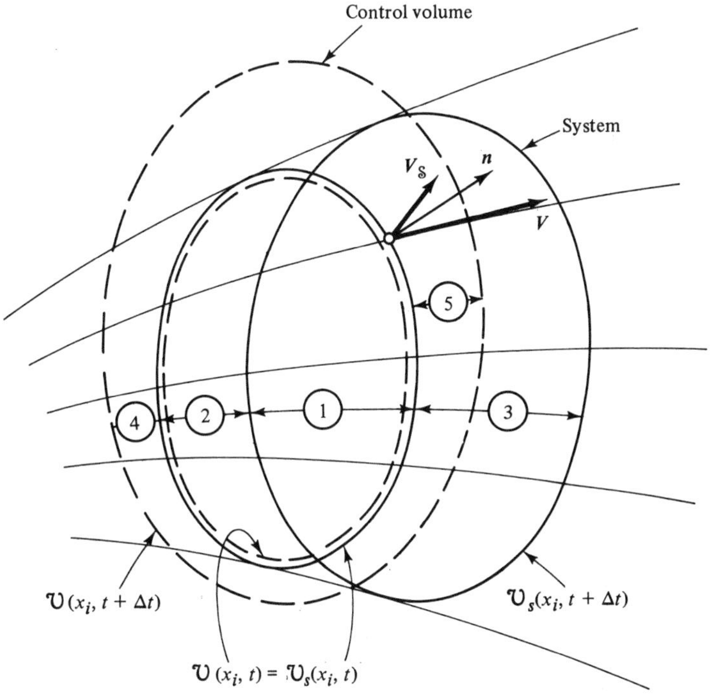

Fig. 1.6 The Reynolds transport theorem

In terms of the spatial regions numbered in Fig. 1.6 and over the time interval Δt, the change of property B within the system is

$$\Delta B_s = B(1 + 3) - B(1 + 2), \tag{1.4.1}$$

and that within the control volume is

$$\Delta B = B(1 + 2 + 4 + 5) - B(1 + 2). \tag{1.4.2}$$

Adding $\pm B(2 + 4 + 5)$ to Eq. (1.4.1), and rearranging the result as

$$\Delta B_s = B(1 + 2 + 4 + 5) - B(1 + 2) + B(3 - 2) - B(4 + 5),$$

we may express ΔB_s in terms of ΔB as

$$\Delta B_s = \Delta B + [B(3) - B(5)] - [B(2) + B(4)]. \tag{1.4.3}$$

Dividing Eq. (1.4.3) by Δt, and letting $\Delta t \longrightarrow 0$ in the usual manner, we obtain

$$\frac{DB_s}{Dt} = \frac{dB}{dt} + \lim_{\Delta t \to 0} \frac{[B(3) - B(5)] - [B(2) + B(4)]}{\Delta t}, \tag{1.4.4}$$

where the limit term shows the net flow of the property B through the control volume. This term, as well as the rate terms, may conveniently be expressed in terms of b. The rate terms are related to b by $B = \int_{\mathcal{V}} b \, d\mathcal{V}$. For the limit term, consider the differential control surface $d\mathcal{S}$ with outward normal vector $\boldsymbol{n}$ (Fig. 1.7). Over the time inverval

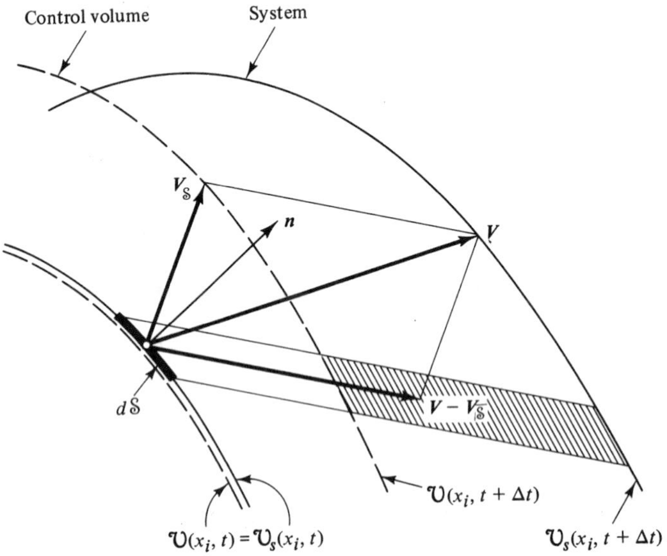

Fig. 1.7 Flow of a property through a deforming control volume

Δt the fluid volume flowing through $d\mathcal{S}$ is $(\boldsymbol{V} - \boldsymbol{V}_{\mathcal{S}}) \cdot \boldsymbol{n} \, d\mathcal{S} \, \Delta t$, and the amount of property b in this volume $b(\boldsymbol{V} - \boldsymbol{V}_{\mathcal{S}}) \cdot \boldsymbol{n} \, d\mathcal{S} \, \Delta t$. Dividing by Δt, letting $\Delta t \longrightarrow 0$, and integrating over the entire control surface gives the second term in Eq. (1.4.4), hence

$$\frac{D}{Dt}\int_{\mathcal{V}_s} b \, d\mathcal{V} = \frac{d}{dt}\int_{\mathcal{V}} b \, d\mathcal{V} + \int_{\mathcal{S}} b(\boldsymbol{V} - \boldsymbol{V}_{\mathcal{S}}) \cdot \boldsymbol{n} \, d\mathcal{S}, \tag{1.4.5}$$

the *Reynolds first transport theorem*. This theorem states that the rate of change of the property b within system $\mathcal{V}_s$ equals the sum of the rate of change of property b within the control volume $\mathcal{V}$ and the net outflow of b through control surface $\mathcal{S}$.

For a control volume fixed in space $\mathcal{V} =$ constant, $\boldsymbol{V}_{\mathcal{S}} = 0$, $d/dt \equiv \partial/\partial t$ and the order of differentiation and integration is interchangeable, and Eq. (1.4.5) reduces to

$$\frac{D}{Dt}\int_{\mathcal{V}_s} b\, d\mathcal{V} = \int_{\mathcal{V}} \frac{\partial b}{\partial t}\, d\mathcal{V} + \int_{\mathcal{S}} b\boldsymbol{V}\cdot\boldsymbol{n}\, d\mathcal{S}, \tag{1.4.6}$$

the *Reynolds second transport theorem*. This theorem is general, however, as it applies to deforming as well as fixed control volumes. To demonstrate this fact, let us set $\boldsymbol{V} = 0$ in Fig. 1.6 and show the result by Fig. 1.8. Noting that for a fixed system $D/Dt \equiv \partial/\partial t$ and the order of differentiation and integration is interchangeable, and

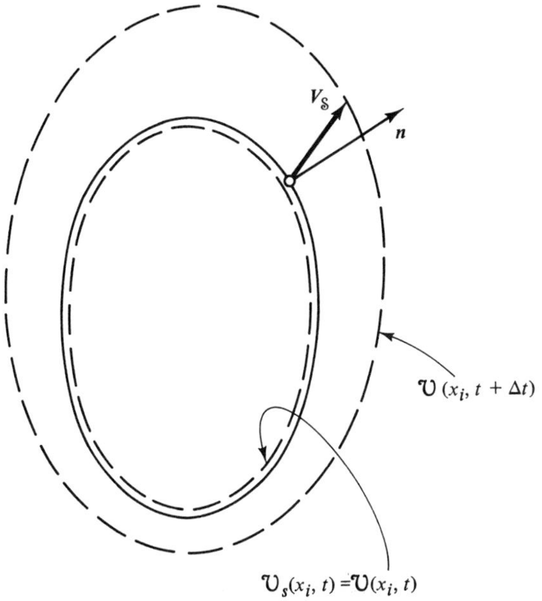

Fig. 1.8 The Leibnitz rule

that $\mathcal{V}_s = \mathcal{V}$ at time t, we have from Eq. (1.4.5)

$$\frac{d}{dt}\int_{\mathcal{V}} b\, d\mathcal{V} = \int_{\mathcal{V}} \frac{\partial b}{\partial t}\, d\mathcal{V} + \int_{\mathcal{S}} b\boldsymbol{V}_{\mathcal{S}}\cdot\boldsymbol{n}\, d\mathcal{S}, \tag{1.4.7}$$

the *Leibnitz rule*, which states that the rate of change of a volume integral over a deforming volume equals the sum of the rate of change of the integrand for fixed $\mathcal{V}$ and the spacewise change of the integrand for fixed time. Combination of Eqs. (1.4.5) and (1.4.7) readily yields Eq. (1.4.6).

Note that the Reynolds theorems stated by Eqs. (1.4.5) and (1.4.6) are in terms of properties with specific values defined per unit volume. Examples are the (mass) density, constituent concentration of a mixture, electric charge density, and so on. The mechanical and thermal properties associated with convection, such as the momentum, moment of momentum, internal energy, and entropy, however, have specific values defined per unit mass. Therefore, suitable forms for convection of the Reynolds first and second theorems, obtained from Eqs. (1.4.5) and (1.4.6) replacing

b by ρb, are

$$\frac{D}{Dt}\int_{\mathcal{V}_s} \rho b \, d\mathcal{V} = \frac{d}{dt}\int_{\mathcal{V}} \rho b \, d\mathcal{V} + \int_{\mathcal{S}} \rho b(\boldsymbol{V} - \boldsymbol{V}_{\mathcal{S}}) \cdot \boldsymbol{n} \, d\mathcal{S} \tag{1.4.8}$$

and

$$\frac{D}{Dt}\int_{\mathcal{V}_s} \rho b \, d\mathcal{V} = \int_{\mathcal{V}} \frac{\partial}{\partial t}(\rho b) \, d\mathcal{V} + \int_{\mathcal{S}} \rho b \boldsymbol{V} \cdot \boldsymbol{n} \, d\mathcal{S}. \tag{1.4.9}$$

For an alternative mathematical form to Eq. (1.4.9), first the surface integral may be converted by the Gauss divergence theorem to a volume integral. This gives us

$$\frac{D}{Dt}\int_{\mathcal{V}_s} \rho b \, d\mathcal{V} = \int_{\mathcal{V}} \left[\frac{\partial}{\partial t}(\rho b) + \nabla \cdot (\rho b \boldsymbol{V})\right] d\mathcal{V}. \tag{1.4.10}$$

Next, invoking the conservation of mass by letting $b = 1$ in Eq. (1.4.10), and noting that $M_s = \int_{\mathcal{V}_s} \rho \, d\mathcal{V} =$ constant, yields

$$0 = \int_{\mathcal{V}} \left[\frac{\partial \rho}{\partial t} + \nabla \cdot (\rho \boldsymbol{V})\right] d\mathcal{V}. \tag{1.4.11}$$

The difference between Eqs. (1.4.10) and (1.4.11) is

$$\frac{D}{Dt}\int_{\mathcal{V}_s} \rho b \, d\mathcal{V} = \int_{\mathcal{V}} \rho \frac{Db}{Dt} \, d\mathcal{V}, \tag{1.4.12}$$

the *Reynolds third transport theorem*.

> Note the physical meaning of the Gauss divergence theorem, equating the surface integral of a vector field with the volume integral of its divergence. It tells us that, for example, $\nabla \cdot (\rho b \boldsymbol{V})$ is the net outflow (a scalar) of the convected property ρb per unit volume.

In the formulation of the general principles for convection, we shall employ the first transport theorem given by Eq. (1.4.8) for deforming integral control volumes, the second transport theorem given by Eq. (1.4.9) for fixed integral control volumes, and the third transport theorem for a (fixed) differential control volume.*

1.5 RATE OF DISPLACEMENT (VELOCITY NEAR A POINT)

We learned in mechanics that the general motion of *rigid* continua is *helicoidal*, being composed of a translation and a rotation. For the formulation of convection, however, we need to know the general motion of *deformable* continua, to which we devote this section.

Let us clarify first the physical and geometric aspects of the section by considering

*Since the limit of a differential control volume represents a typical point of continua, a deforming differential control volume is meaningless.

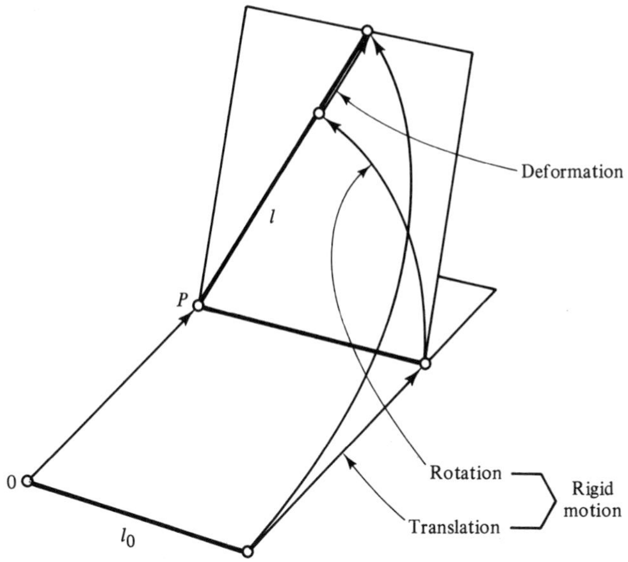

Fig. 1.9 Deformable motion

a fluid particle which, over a time interval Δt, travels a distance l_0 from a typical point O. Let another fluid particle travel, over the same time interval, a distance l from a neighboring point, say P. As shown in Fig. 1.9, l may be obtained from l_0 by a translation, a rotation, and a deformation. This fact is expressed mathematically by expanding the velocity of P, say V_i, into a McLaurin series about the velocity of O, say $(V_i)_0$, as

$$V_i = (V_i)_0 + x_j\left(\frac{\partial V_i}{\partial x_j}\right)_0, \tag{1.5.1}$$

or

$$\boldsymbol{V} = \boldsymbol{V}_0 + \boldsymbol{x} \cdot (\nabla \boldsymbol{V})_0. \tag{1.5.2}$$

Note that the material derivative of Eq. (1.5.1) or Eq. (1.5.2) gives the acceleration defined previously by Eq. (1.3.2) or Eq. (1.3.3).

Let us split the rate of displacement tensor, $\partial V_i/\partial x_j$, into a symmetric tensor and an antisymmetric tensor. Then Eq. (1.5.1) may be rearranged as

$$V_i = (V_i)_0 + x_j(r_{ij})_0 + x_j(S_{ij})_0, \tag{1.5.3}$$

which becomes, after multiplication by Δt, the mathematical statement of Fig. 1.9. Here

$$r_{ij} = \frac{1}{2}\left(\frac{\partial V_i}{\partial x_j} - \frac{\partial V_j}{\partial x_i}\right) \tag{1.5.4}$$

denotes the rate of twist (*spin* or *rotation*) tensor and

$$S_{ij} = \frac{1}{2}\left(\frac{\partial V_i}{\partial x_j} + \frac{\partial V_j}{\partial x_i}\right) \tag{1.5.5}$$

the *rate of deformation* tensor.

A second-order tensor like A_{ij} may be divided into a symmetric part and an anti-symmetric part by adding and subtracting one-half of its transpose, A_{ji},

$$A_{ij} = \tfrac{1}{2}(A_{ij} + A_{ji}) + \tfrac{1}{2}(A_{ij} - A_{ji}) = A^s_{ij} + A^a_{ij}.$$

Since $A^s_{ij} = A^s_{ji}$, a symmetric tensor is composed of six different elements. On the other hand, since $A^a_{ij} = -A^a_{ji}$ and diagonal elements are zero, an antisymmetric tensor is composed of only three different elements, which are the components of a pseudo (or axial) vector

$$r_k = \tfrac{1}{2}\epsilon_{kpq}A_{pq}.$$

It follows then that $A^a_{ij} = \epsilon_{ijk}r_k$.

Furthermore, a symmetric tensor like A^s_{ij} may be split into a deviatoric part α_{ij} and an isotropic part a as

$$A^s_{ij} = \alpha_{ij} + a\delta_{ij}; \qquad a = \frac{A^s_{kk}}{3}.$$

A_{kk} is the trace of A_{ij}.

The geometric as well as physical interpretation of r_{ij} and S_{ij} may best be given by a planar or volumetric (differential) element rather than by a line element as considered in Fig. 1.9. For the sake of simplicity, let us restrict ourselves to plane geometry, and consider the rate of displacement of the differential element shown in Fig. 1.10(b). Clearly, the extension to three-dimensional geometry would readily follow from cyclic permutations.

Defining the rotation as the *rate of twist of the diagonal*, we have from Fig. 1.10(a),

$$\varpi_3 = \frac{D\theta_3}{Dt} = \frac{D}{Dt}\left[\frac{1}{2}(\gamma_1 + \gamma_2)\right],$$

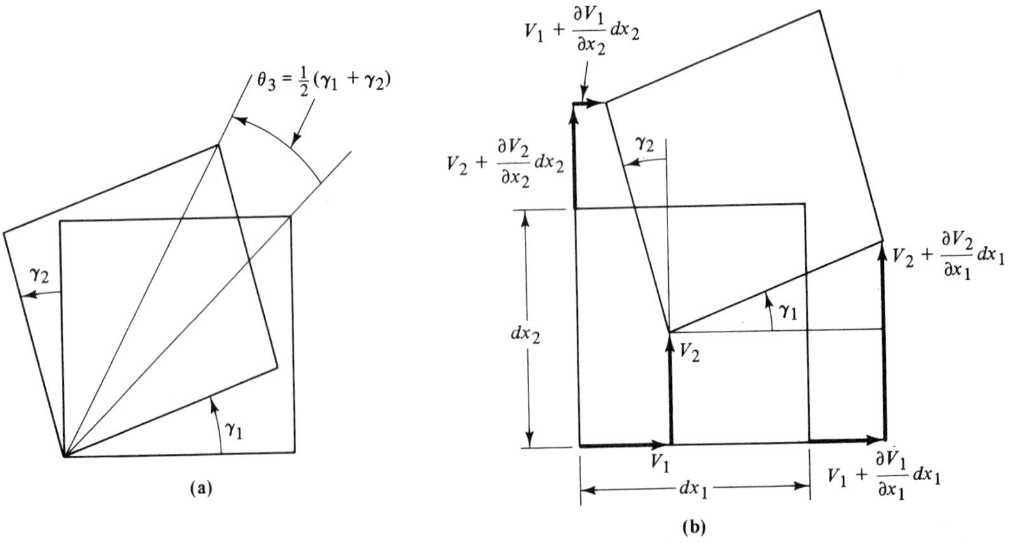

Fig. 1.10 Rate of displacement

which, in view of $D\gamma_1/Dt = \partial V_2/\partial x_1$ and $D\gamma_2/Dt = -\partial V_1/\partial x_2$ (Fig. 1.10(b)), may be written as

$$\varpi_3 = \frac{1}{2}\left(\frac{\partial V_2}{\partial x_1} - \frac{\partial V_1}{\partial x_2}\right),$$

denoting the component of the angular velocity vector normal to the (x_1, x_2)-plane. The angular velocity vector is axial, is related to the rotation tensor by

$$r_{ij} = -\epsilon_{ijk}\varpi_k, \tag{1.5.6}$$

and is equal to one-half of the vorticity vector, $\varpi_k = \omega_k/2$. The second right-hand-side term of Eq. (1.5.3) may then be rearranged as

$$x_j r_{ij} = -\epsilon_{ijk} x_j \varpi_k = (\boldsymbol{\varpi} \times \mathbf{r})_i,$$

and Eq. (1.5.3) becomes

$$\mathbf{V} = \underbrace{\mathbf{V}_0}_{\text{translation}} + \underbrace{\boldsymbol{\varpi}_0 \times \mathbf{r}}_{\text{rotation}} + \underbrace{\mathbf{r} \cdot \mathbf{S}_0}_{\text{deformation}}. \tag{1.5.7}$$

(translation + rotation: rigid motion; rotation + deformation: deformable motion)

Next, the rate of deformation may be defined as the *rotation of a side of the differential element relative to the rotation of its diagonal* (Fig. 1.10a). This gives us

$$\frac{D}{Dt}\left[\gamma_1 - \frac{1}{2}(\gamma_1 + \gamma_2)\right] = -\frac{D}{Dt}\left[\gamma_2 - \frac{1}{2}(\gamma_1 + \gamma_2)\right] = \frac{D}{Dt}\left[\frac{1}{2}(\gamma_1 - \gamma_2)\right],$$

or

$$\frac{D}{Dt}\left[\frac{1}{2}(\gamma_1 - \gamma_2)\right] = \frac{1}{2}\left(\frac{\partial V_2}{\partial x_1} + \frac{\partial V_1}{\partial x_2}\right),$$

which is identical to one component of the rate of deformation tensor, S_{ij}.

Furthermore, the rate of deformation tensor may be split into a *deviatoric* tensor and an *isotropic* tensor, respectively,

$$S_{ij} = s_{ij} + s\delta_{ij}. \tag{1.5.8}$$

The importance of these tensors will become clear in Section 2.3 when we develop the constitutive relations. Meanwhile we concentrate our attention on the tensors themselves.

The isotropic tensor is defined to be the mean of the trace of deformation rate,

$$s = \frac{1}{3}\frac{\partial V_k}{\partial x_k}. \tag{1.5.9}$$

Since the differential conservation of mass* (obtained, for example, by equating the integrand of Eq. 1.4.11 to zero) may be rearranged, in terms of the specific volume

---

*The differential conservation of mass will be elaborated on in Section 2.1.2.

$v = 1/\rho$, as

$$\frac{\partial V_k}{\partial x_k} = \frac{1}{v}\frac{Dv}{Dt}, \tag{1.5.10}$$

the isotropic tensor denotes the *rate of volumetric expansion* or the *rate of dilatation.*

The deviatoric tensor, on the other hand, obtained from Eqs. (1.5.5), (1.5.8), and (1.5.9),

$$s_{ij} = \frac{1}{2}\left(\frac{\partial V_i}{\partial x_j} + \frac{\partial V_j}{\partial x_i}\right) - \frac{1}{3}\frac{\partial V_k}{\partial x_k}\delta_{ij} \tag{1.5.11}$$

shows the *rate of distortion.* Figure 1.11 explains the rate of deformation as the combination of the rate of dilatation and the rate of distortion.

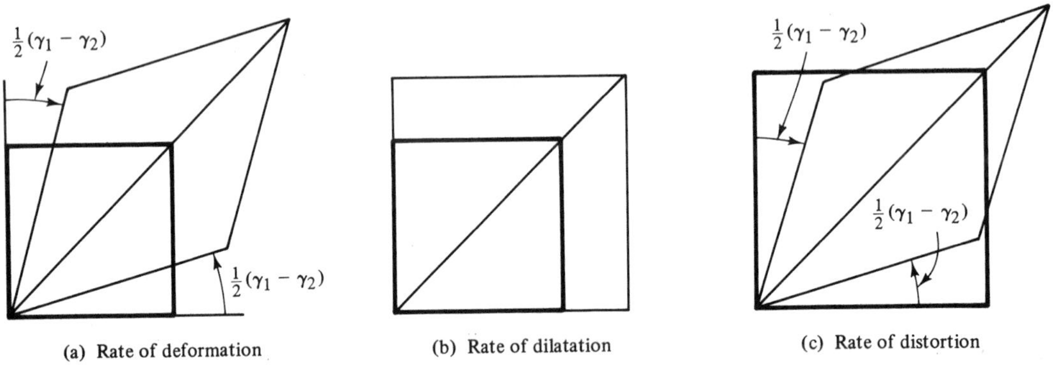

Fig. 1.11 Rate of deformation and its components

1.6 STATE OF STRESS (VOLUME AND SURFACE FORCES)

The last concept to be needed in the development of general principles is the *state of stress at a point*. Before proceeding to this need, however, let us talk briefly about the types of force and review the concept of *stress at a point.*

The forces acting on continua may be distinguished as *volume* (or *body*) forces and *surface* forces. Volume forces are long-range forces and are characterized by slow decrease with increase of distance between interacting parts. Such forces penetrate into continua, and act on all parts of continua. Gravity is the most common example, but two other kinds, electromagnetic and centrifugal forces, are also encountered quite frequently. We shall symbolize these forces acting at time t on a differential system $d\mathcal{V}$ with position vector $\boldsymbol{x}$ by

$$\boldsymbol{f}(\boldsymbol{x}, t)\rho \, d\mathcal{V}, \tag{1.6.1}$$

ρ denoting the local density. Surface forces, on the other hand, are short-range forces of intrinsic (or molecular) origin which decrease very rapidly with increase of distance between interacting parts. They are negligible unless the interacting parts are in direct contact.

Consider now a system (or control volume) separated from a continuum. Let $\Delta\mathcal{S}$ be a differential surface of this system with an outward normal $\boldsymbol{n}$. Assume that the continuum surrounding the system exerts a surface force $\Delta\boldsymbol{F}$ on $\Delta\mathcal{S}$. Then the limit

$$\boldsymbol{S}_{(n)} = \lim_{\Delta\mathcal{S}\to 0} \frac{\Delta \boldsymbol{F}}{\Delta\mathcal{S}} \tag{1.6.2}$$

defines the stress (vector) at a point (Fig. 1.12(a)). However, the differential general principles are stated for a point which is obtained as the limit of a differential volume rather than that of a differential area. Therefore, we need to know the stress condition at a point as $\Delta\mathcal{V} \longrightarrow 0$. This gives the state of stress at a point.

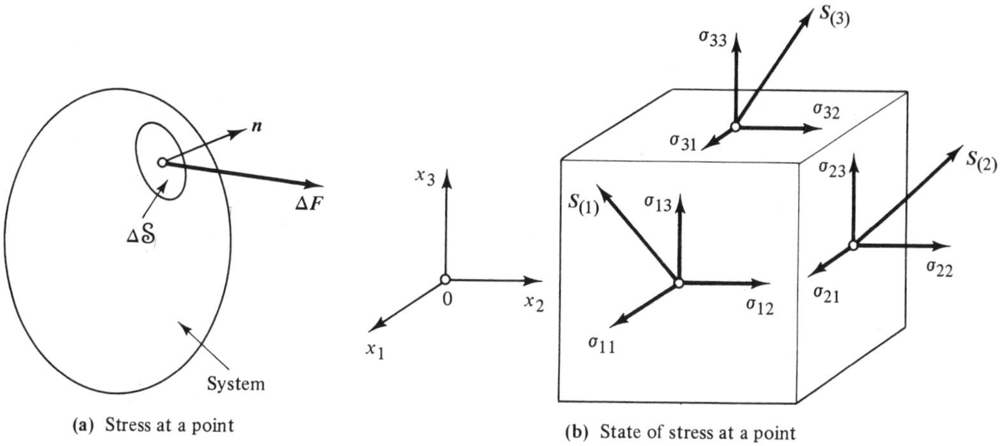

Fig. 1.12 Stress and state of stress

Consider, for example, the Cartesian differential volume shown in Fig. 1.12(b). Assume that the stress (vectors) acting on three surfaces of this volume are $\boldsymbol{S}_{(1)}$, $\boldsymbol{S}_{(2)}$, and $\boldsymbol{S}_{(3)}$. Let σ_{11}, σ_{12}, and σ_{13} be the components of the vector $\boldsymbol{S}_{(1)}$. Similarly, denote the components of the vectors $\boldsymbol{S}_{(2)}$ and $\boldsymbol{S}_{(3)}$ by σ_{21}, σ_{22}, σ_{23} and σ_{31}, σ_{32}, σ_{33}, respectively. σ_{11} is called the *normal stress* for the surface normal to x_1-axis. The components σ_{12} and σ_{13} are tangential to the surface and are called *shear stresses*. Similarly, σ_{22} and σ_{33} are the normal stresses for surfaces normal to the x_2- and x_3-axes, respectively, and σ_{21}, σ_{23} and σ_{31}, σ_{32} are the corresponding shear stresses. The tensor σ_{ij}, defined by the foregoing nine stress components, prescribes the state of stress at a point. Clearly,

$$\boldsymbol{S}_{(j)i} = e_j\sigma_{ij}, \tag{1.6.3}$$

where e_j denotes the unit vector in x_j.

The stress tensor, in a manner similar to the rate of displacement tensor, may be split into a symmetric tensor and an antisymmetric tensor as

$$\sigma_{ij} = \sigma_{ij}^s + \sigma_{ij}^a. \tag{1.6.4}$$

Furthermore, the symmetric part may be divided into a deviatoric tensor and an isotropic tensor as

$$\sigma_{ij}^s = \tau_{ij} - p\delta_{ij}; \qquad p = -\tfrac{1}{3}\sigma_{kk}^s. \tag{1.6.5}$$

Assuming c_k to be the axial vector associated with the antisymmetric tensor,

$$c_k = \tfrac{1}{2}\epsilon_{kpq}\sigma_{pq}, \tag{1.6.6}$$

combining Eqs. (1.6.4), (1.6.5), and (1.6.6) the stress tensor may be rearranged as

$$\sigma_{ij} = \tau_{ij} - p\delta_{ij} + \epsilon_{ijk}c_k. \tag{1.6.7}$$

The right-hand-side terms of Eq. (1.6.7) represent the stress components responsible for distortion, dilatation, and rotation of continua, respectively.

REFERENCES

ARIS, R., 1962, *Vectors, Tensors, and the Basic Equations of Fluid Dynamics*, Prentice-Hall, Englewood Cliffs, N. J.

ARPACI, V. S., 1966, *Conduction Heat Transfer*, Addison-Wesley, Reading, Mass.

BATCHELOR, G. K., 1967, *Fluid Dynamics*, Cambridge University Press, New York.

BIRD, R. B., STEWART, W. E., & LIGHTFOOT, E. N., 1960, *Transport Phenomena*, Wiley, New York.

HILDEBRAND, F. B., 1962, *Advanced Calculus for Applications*, Prentice-Hall, Englewood Cliffs, N. J.

JEFFREYS, H., 1957, *Cartesian Tensors*, Cambridge University Press, New York.

LAMB, H., 1945, *Hydrodynamics*, Dover, New York.

PRAGER, W., 1961, *Introduction to Mechanics of Continua*, Ginn, Lexington, Mass.

PRANDTL, L., & TIETJENS, O. G., 1957, *Fundamentals of Hydro-and Aerodynamics*, Dover, New York.

Chapter 2

DEVELOPMENT OF GOVERNING EQUATIONS

The continuum approach to convection, as well as to other branches of engineering science, is based on the *definition of concepts* and the *statement of natural laws* in terms of these concepts. Having acquired the appropriate concepts in Chapter 1, we are ready for the statement of these laws.

The natural laws of the universe can neither be proved nor disproved but are arrived at inductively on the basis of *phenomenological evidence* collected from a variety of *observations*. As the progress of man continues, the present statements of natural laws will continue to be refined and generalized. For the time being, however, we shall refer to these statements as adequate descriptions of nature, and employ them in the formulation of contemporary problems of engineering. Nature provides these laws with two distinct features; those *independent of the continuum* they apply to are called the *general principles*, and those *dependent on the continuum* they apply to are called the *constitutive relations*. Sections 2.1 and 2.3 are devoted to these laws. Furthermore, in dealing with thermal problems, we are often forced to seek an additional equation, called the *thermodynamic relation*. It expresses the constraining assumption of *local equilibrium* and is developed in Section 2.2. The appropriate combination of the general principles with constitutive relations and the assumption of local equilibrium gives the equations governing our problems. These equations will emerge in Section 2.4.

2.1 GENERAL PRINCIPLES

A general principle is independent of the behavior of the continua to which it applies. Examples are the conservation of mass, balance of linear momentum, balance of the moment of total momentum, conservation of total energy, increase of entropy, conservation of electric charge, the Lorentz force, the Ampere circuit law, the Faraday induction law, and so on. However, being interested primarily in the thermomechanics of single-component continua, we devote the present section only to the first five of these principles. Although the foundation of general principles rests on observations and, consequently, is *phenomenological*, alternative forms are *derived* from them by appropriate combinations; derived principles often become convenient in the formulation of our problems. For later reference, phenomenological and derived principles are summarized in Table 2.1.

TABLE 2.1 General principles

Phenomenological	Derived
1. Conservation of mass	
2. Balance of linear momentum	(a) Balance of moment of linear momentum (b) Balance of linear mechanical energy
3. Balance of moment of total momentum (moment of linear + angular momenta)	(a) Balance of angular momentum (b) Balance of angular mechanical energy
4. Conservation of total energy (mechanical + thermal energies)	(a) Balance of thermal energy
5. Increase of entropy	(a) Balance of entropy (production of entropy)

The (spacewise) statement of a general principle is in terms of a (finite or infinitesimal) system or control volume. This statement can be expressed by differential, integral, or variational means. The foregoing classifications are summarized in Fig. 2.1. The lumped formulation, because of its oversimplification of convection problems, and the difference and variational formulations, because of their special nature, will not be discussed in this chapter. Finite difference and finite element formulations, indispensable for numerical solutions, will be introduced in Chapter 7.

We state next the various forms of general principles outlined in Table 2.1 by means of a "master equation."

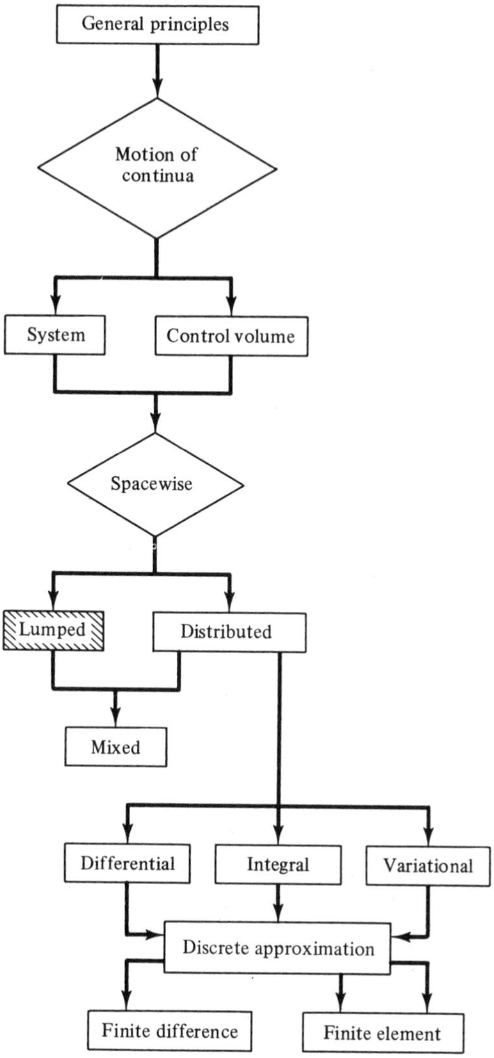

Fig. 2.1 Forms of general principles

2.1.1 A Unified Approach

Let a general *balance* principle for a *lumped system* be expressible by the master equation

$$\frac{DB_s}{Dt} = \Phi + \Gamma, \tag{2.1.1}$$

which states the balance of the time rate of change of a property, B_s, with the combination of flux (surface) terms, Φ, and source (volumetric) terms, Γ. For $\Gamma = 0$ Eq. (2.1.1) reduces to a general *conservation* principle. Equation (2.1.1) applies also to a component of a vector field.

For an *integral system*, Eq. (2.1.1) may be explicitly written as

$$\frac{D}{Dt}\int_{\mathcal{V}_s} \rho b \, d\mathcal{V} = -\int_{\mathcal{S}} \boldsymbol{n} \cdot \boldsymbol{\varphi} \, d\mathcal{S} + \int_{\mathcal{V}_s} \gamma''' \, d\mathcal{V}, \tag{2.1.2}$$

where b, $-\boldsymbol{n} \cdot \boldsymbol{\varphi}$, and γ''', respectively, denote B per unit mass, Φ per unit area, and Γ per unit volume.

For an *integral control volume*, employing the first and second transport theorems (see Eqs. 1.4.8 and 1.4.9), Eq. (2.1.2) may be rearranged as

$$\left.\begin{aligned} &\frac{d}{dt}\int_{\mathcal{V}} \rho b \, d\mathcal{V} + \int_{\mathcal{S}} \boldsymbol{n} \cdot \rho(\boldsymbol{V} - \boldsymbol{V}_{\mathcal{S}}) b \, d\mathcal{S} \\ &\int_{\mathcal{V}} \frac{\partial}{\partial t}(\rho b) \, d\mathcal{V} + \int_{\mathcal{S}} \boldsymbol{n} \cdot \rho \boldsymbol{V} b \, d\mathcal{S} \end{aligned}\right\} = -\int_{\mathcal{S}} \boldsymbol{n} \cdot \boldsymbol{\varphi} \, d\mathcal{S} + \int_{\mathcal{V}} \gamma''' \, d\mathcal{V}. \tag{2.1.3a, b}$$

The meanings of *flow* and *flux* are clearly distinguished in this text. *Flow* is employed for the convection (motion) of a mass-dependent property through a surface. *Flux*, on the other hand, is used for the diffusion of a property through a surface. In Eq. (2.1.3b), for example, $\rho \boldsymbol{V} b$ is the flow and $\boldsymbol{\varphi}$ the flux of B.

Rearranging the left-hand side of Eq. (2.1.3b) by the Gauss divergence theorem (see Eq. 1.4.10), converting the first term of the right-hand side to a volume integral, and eliminating the volume integral operations, we have for a *differential control volume*,

$$\frac{\partial}{\partial t}(\rho b) + \nabla \cdot (\rho \boldsymbol{V} b) = -\nabla \cdot \boldsymbol{\varphi} + \gamma''', \tag{2.1.4}$$

or, expanding the left-hand side, eliminating the conservation of mass (Eq. 2.1.4 for $b = 1$, $\varphi = 0$, and $\gamma''' = 0$),

$$\rho \frac{Db}{Dt} = -\nabla \cdot \boldsymbol{\varphi} + \gamma'''. \tag{2.1.5}$$

Actually, employing the third transport theorem (see Eq. 1.4.12), Eq. (2.1.2) may readily be rearranged as Eq. (2.1.5), by converting the first term on the right-hand side to a volume integral and then eliminating the volume integral operations.

A mixed (three-dimensional) formulation can be obtained by any combination of lumped, differential, and integral formulations. Consequently, the form of a mixed formulation is not unique, and it depends on the particular combination assumed. For the sake of demonstration, here we consider the simplest possible mixed formulation in terms of a *radially lumped-axially differential control volume* attached to a flexible tube of variable cross section shown in Fig. 2.2. The interpretation of Eq. (2.1.1) in terms of this control volume yields

$$\frac{\partial}{\partial t}(\rho A b) + \frac{\partial}{\partial n}(\rho A V_n b) + \rho C V_c b = -\frac{\partial}{\partial n}(A \varphi_n) - C \varphi_c + A \gamma''', \tag{2.1.6}$$

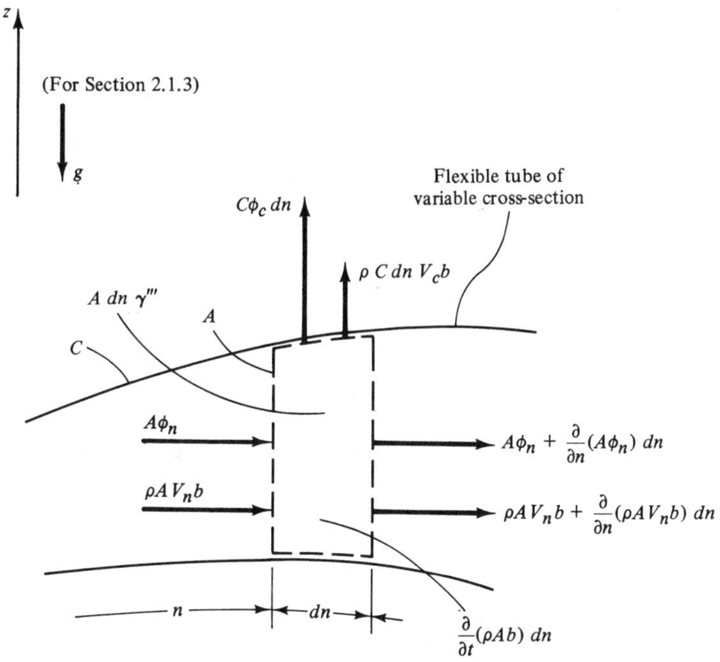

Fig. 2.2 Radially lumped-axially differential control volume

or, subtracting the conservation of mass (Eq. 2.1.6 for $b = 1$, $\varphi_n = 0$, $\varphi_c = 0$, and $\gamma''' = 0$),

$$\rho A \frac{Db}{Dt} = -\frac{\partial}{\partial n}(A\varphi_n) - C\varphi_c + A\gamma''', \tag{2.1.7}$$

where the special form of the material derivative is $D/Dt \equiv \partial/\partial t + V\,\partial/\partial n$.

Having established the master equations for integral and differential formulations and indicated to several possibilities for the mixed formulation, we proceed now to the explicit statement of the general principles of thermomechanics.

2.1.2 *Integral and Differential Formulations*

We shall follow Table 2.1 in the statement of the integral and differential forms of the general principles.

1. *Conservation of mass.* Let $b = 1$, $\boldsymbol{\varphi} = 0$, and $\gamma''' = 0$; then Eq. (2.1.3b) gives the *conservation of mass for an integral control volume*,

$$\int_{\mathcal{V}} \frac{\partial \rho}{\partial t}\, d\mathcal{V} + \int_{\mathcal{S}} \boldsymbol{n} \cdot \rho \boldsymbol{V}\, d\mathcal{S} = 0. \tag{2.1.8}$$

Next, referring to Eq. (2.1.4), we obtain the *conservation of mass for a differential control volume*,

$$\frac{\partial \rho}{\partial t} + \nabla \cdot (\rho V) = 0, \tag{2.1.9}$$

which may be rearranged, after expanding the second term and employing the definition of material derivative, as

$$\frac{D\rho}{Dt} + \rho \nabla \cdot V = 0. \tag{2.1.10}$$

For steady flows $\partial/\partial t \equiv 0$, Eq. (2.1.9) gives $\nabla \cdot (\rho V) = 0$. For incompressible flows $\rho =$ constant, $D\rho/Dt = 0$, Eq. (2.1.10) yields $\nabla \cdot V = 0$.

2. *Balance of linear momentum.* Employing Eq. (2.1.3b) for one component of velocity, three stresses and one component of the body force (all coinciding with the assumed component of velocity) and returning to the symbolic form amount to $b \longrightarrow V$, $\varphi \longrightarrow -\sigma$, and $\gamma''' \longrightarrow \rho f$; then Eq. (2.1.3b) gives the *balance of linear momentum for an integral control volume*,

$$\int_{\mathcal{V}} \frac{\partial}{\partial t}(\rho V)\, d\mathcal{V} + \int_{\mathcal{S}} n \cdot \rho VV\, d\mathcal{S} = \int_{\mathcal{S}} n \cdot \sigma\, d\mathcal{S} + \int_{\mathcal{V}} \rho f\, d\mathcal{V}, \tag{2.1.11}$$

where ρVV is related to the flow of momentum, σ denotes the state of stress defined in Section 1.6, and f represents the body force per unit mass. Among the body forces, the gravitational force is well known; the electromagnetic forces and the inertial forces associated with accelerating coordinates are discussed in texts on fluid mechanics.

The flow of momentum involves the special tensor,

$$VV \equiv V_i V_j \equiv \begin{vmatrix} V_1V_1 & V_1V_2 & V_1V_3 \\ V_2V_1 & V_2V_2 & V_2V_3 \\ V_3V_1 & V_3V_2 & V_3V_3 \end{vmatrix},$$

which is called a *dyad*. The physical meaning of this tensor is explained in Fig. 2.3 by the first two elements of its first column. Also, the elements of each row form the vector components $\rho V_1 V$, $\rho V_2 V$, and $\rho V_3 V$ of the tensor. For example, $\rho V_1 V$ represents the flow ρV_1 of momentum V through a surface with normal $e_{(1)}$. Then the vector

$$n \cdot \rho VV \equiv n_i \rho V_i V_j \equiv n_1 \rho V_1 V + n_2 \rho V_2 V + n_3 \rho V_3 V$$

gives the flow of momentum through a surface with normal n.

Similarly, in terms of the vector components of the stress tensor (see Fig. 1.12), we have

$$n \cdot \sigma \equiv n_i \sigma_{ij} \equiv n_1 S_{(1)} + n_2 S_{(2)} + n_3 S_{(3)}, \tag{2.1.12}$$

which denotes the stress vector $S_{(n)}$ on a surface with normal n.

Next, referring to Eq. (2.1.5) and assuming again that $b \longrightarrow V$, $\varphi \longrightarrow -\sigma$, and $\gamma''' \longrightarrow \rho f$, we arrive at the *balance of linear momentum for a differential control volume*

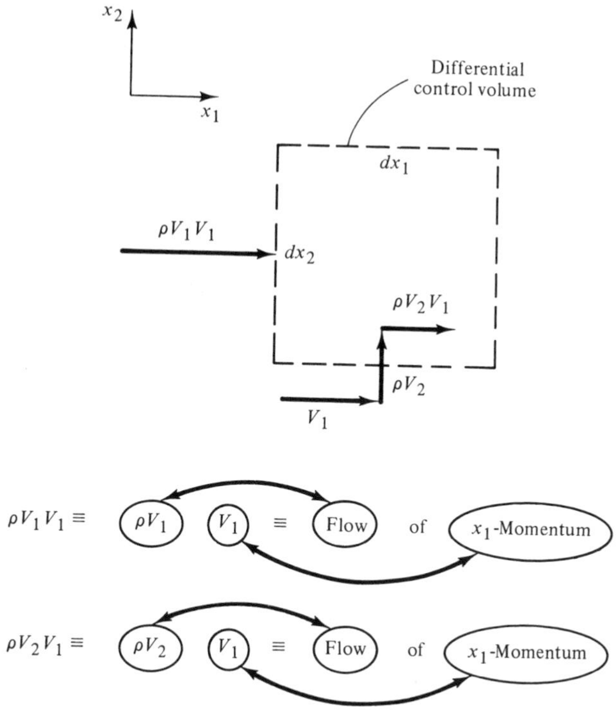

Fig. 2.3 Two components of momentum flow

$$\rho \frac{DV}{Dt} = \nabla \cdot \boldsymbol{\sigma} + \rho f, \tag{2.1.13}$$

where $\nabla \cdot \boldsymbol{\sigma}$ denotes the net change of stress over the differential control volume which is equal to the resulting surface force per unit volume.

In terms of the change in stress vectors (see, for example, Fig. 2.4 for the change in $S_{(1)}$), we have

$$\nabla \cdot \boldsymbol{\sigma} = \frac{\partial S_{(1)}}{\partial x_1} + \frac{\partial S_{(2)}}{\partial x_2} + \frac{\partial S_{(3)}}{\partial x_3}, \tag{2.1.14}$$

which may be expanded, by the help of Eq. (1.6.3), as

$$\begin{aligned} \nabla \cdot \boldsymbol{\sigma} &= \left(\frac{\partial \sigma_{11}}{\partial x_1} + \frac{\partial \sigma_{21}}{\partial x_2} + \frac{\partial \sigma_{31}}{\partial x_3}\right) e_{(1)} \\ &+ \left(\frac{\partial \sigma_{12}}{\partial x_1} + \frac{\partial \sigma_{22}}{\partial x_2} + \frac{\partial \sigma_{32}}{\partial x_3}\right) e_{(2)} \\ &+ \left(\frac{\partial \sigma_{13}}{\partial x_1} + \frac{\partial \sigma_{23}}{\partial x_2} + \frac{\partial \sigma_{33}}{\partial x_3}\right) e_{(3)}. \end{aligned} \tag{2.1.15}$$

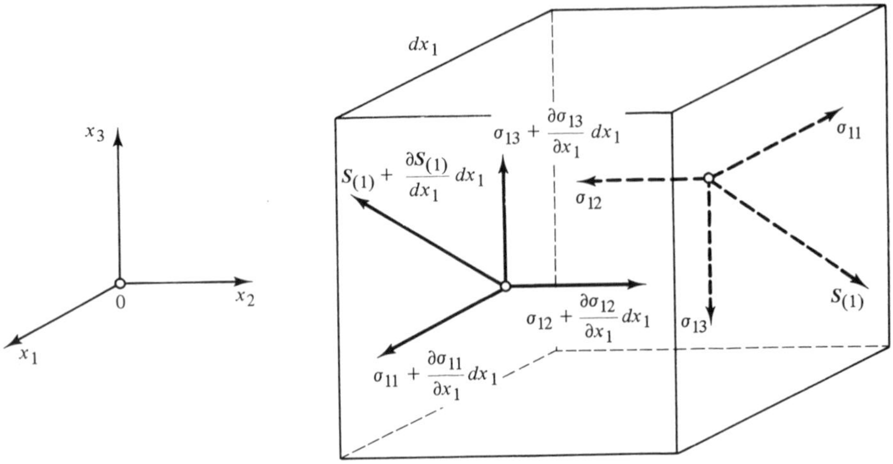

Fig. 2.4 Change of the stress vector $\boldsymbol{S}_{(1)}$

Assuming the balance of linear momentum as observed and *phenomenological* (recall Table 2.1), we *derive* from it the following principles:

2(a). ***Balance of the moment of linear momentum.*** From the definition of the moment of a vector with respect to a point, the cross product of the space vector $\boldsymbol{r}$ with Eq. (2.1.13) yields

$$\boldsymbol{r} \times \left(\rho \frac{D\boldsymbol{V}}{Dt}\right) = \boldsymbol{r} \times (\boldsymbol{\nabla} \cdot \boldsymbol{\sigma} + \rho \boldsymbol{f}). \tag{2.1.16}$$

2(b). ***Balance of mechanical energy* of linear momentum.*** From the definitions of kinetic energy and power (work rate), the dot product of Eq. (2.1.13) with the velocity $\boldsymbol{V}$ gives

$$\rho \frac{D}{Dt}\left(\frac{V^2}{2}\right) = (\boldsymbol{\nabla} \cdot \boldsymbol{\sigma} + \rho \boldsymbol{f}) \cdot \boldsymbol{V}, \tag{2.1.17}$$

which states that the *rate of kinetic energy is balanced with the displacement power of the net surface and body forces.*

For *conservative* and *steady* body forces whose potential is denoted Ψ, $\boldsymbol{f} = -\boldsymbol{\nabla}\Psi$ and $\boldsymbol{f} \cdot \boldsymbol{V} = -D\Psi/Dt$. Then Eq. (2.1.17) may be rearranged as

$$\rho \frac{D}{Dt}\left(\frac{V^2}{2} + \Psi\right) = (\boldsymbol{\nabla} \cdot \boldsymbol{\sigma}) \cdot \boldsymbol{V},$$

which, for steady motion of an incompressible and inviscid fluid, reduces to the Bernoulli equation along a streamline,

$$\frac{p}{\rho} + \frac{V^2}{2} + \Psi = \text{constant},$$

*This balance is also referred to as the balance of kinetic energy.

and for unsteady motion of a rigid body to

$$\frac{V^2}{2} + \Psi = \text{constant.}$$

In Section 2.4.4 we elaborate further on the Bernoulli equation.

3. ***Balance of the moment of total momentum*** (moment of linear momentum + angular momentum). The statement of this principle is considerably simplified for nonpolar continua for which the rate of intrinsic angular momentum, surface, and body couples are negligible or absent.* While polar effects have no significance for thermal problems to be considered in this text, they may become important, for example, for dielectric media subject to an electric field.

For nonpolar continua, letting $b \longrightarrow \boldsymbol{L} = \boldsymbol{r} \times \boldsymbol{V}$, $\boldsymbol{\varphi} \longrightarrow -\boldsymbol{\mu}$, and $\gamma''' \longrightarrow \rho\boldsymbol{r} \times \boldsymbol{f}$ in Eq. (2.1.3b), we obtain the *balance of the moment of total momentum for an integral control volume,*

$$\int_{\mathcal{V}} \frac{\partial}{\partial t}(\rho \boldsymbol{L})\, d\mathcal{V} + \int_{\mathcal{S}} \boldsymbol{n} \cdot \rho \boldsymbol{V}\boldsymbol{L}\, d\mathcal{S} = \int_{\mathcal{S}} \boldsymbol{n} \cdot \boldsymbol{\mu}\, d\mathcal{S} + \int_{\mathcal{V}} \rho \boldsymbol{r} \times \boldsymbol{f}\, d\mathcal{V}. \qquad (2.1.18)$$

The meaning of $\rho\boldsymbol{VL}$ is clear in view of the discussion on $\rho\boldsymbol{VV}$ following the statement of Eq. (2.1.11), and $\boldsymbol{n} \cdot \boldsymbol{\mu} = \boldsymbol{r} \times (\boldsymbol{n} \cdot \boldsymbol{\sigma})$ is the moment of stress.

Introduce the moments of stress vectors, say $\boldsymbol{M}_{(1)}$, $\boldsymbol{M}_{(2)}$, and $\boldsymbol{M}_{(3)}$, where $\boldsymbol{M}_{(1)}$, for example, is

$$\boldsymbol{M}_{(1)} = \boldsymbol{r} \times \boldsymbol{S}_{(1)} \equiv \epsilon_{lmj} x_m \sigma_{1j} \equiv \begin{vmatrix} \boldsymbol{e}_{(1)} & \boldsymbol{e}_{(2)} & \boldsymbol{e}_{(3)} \\ x_1 & x_2 & x_3 \\ \sigma_{11} & \sigma_{12} & \sigma_{13} \end{vmatrix},$$

or, explicitly,

$$\boldsymbol{M}_{(1)} = (x_2\sigma_{13} - x_3\sigma_{12})\boldsymbol{e}_{(1)} + (x_3\sigma_{11} - x_1\sigma_{13})\boldsymbol{e}_{(2)} + (x_1\sigma_{12} - x_2\sigma_{11})\boldsymbol{e}_{(3)}.$$

Then, by definition,

$$\boldsymbol{\mu} \equiv \epsilon_{lmj} x_m \sigma_{ij} \equiv \begin{vmatrix} x_2\sigma_{13} - x_3\sigma_{12} & x_3\sigma_{11} - x_1\sigma_{13} & x_1\sigma_{12} - x_2\sigma_{11} \\ x_2\sigma_{23} - x_3\sigma_{22} & x_3\sigma_{21} - x_1\sigma_{23} & x_1\sigma_{22} - x_2\sigma_{21} \\ x_2\sigma_{33} - x_3\sigma_{32} & x_3\sigma_{31} - x_1\sigma_{33} & x_1\sigma_{32} - x_2\sigma_{31} \end{vmatrix},$$

which implies that $\boldsymbol{\mu} = -\boldsymbol{\sigma} \times \boldsymbol{r}$.

From the balance of moments (in a manner similar to that of stress vectors shown in Fig. 1.12),

$$\boldsymbol{n} \cdot \boldsymbol{\mu} \equiv n_i \epsilon_{lmj} x_m \sigma_{ij} \equiv n_1 \boldsymbol{M}_{(1)} + n_2 \boldsymbol{M}_{(2)} + n_3 \boldsymbol{M}_{(3)},$$

which is the moment of stress vector $\boldsymbol{M}_{(n)} = \boldsymbol{r} \times \boldsymbol{S}_{(n)}$ on a surface with normal $\boldsymbol{n}$.

Next, referring to Eq. (2.1.5), and assuming again that $b \longrightarrow \boldsymbol{L}$, $\boldsymbol{\varphi} \longrightarrow -\boldsymbol{\mu}$, and $\gamma''' \longrightarrow \rho\boldsymbol{r} \times \boldsymbol{f}$, we obtain the *balance of the moment of total momentum for a differential control volume,*

*For polar continua see, for example, Dahler & Scriven (1963) and DeGroot & Mazur (1962).

$$\rho \frac{DL}{Dt} = \nabla \cdot \boldsymbol{\mu} + \rho r \times f, \tag{2.1.19}$$

where the meaning of $\nabla \cdot \boldsymbol{\mu}$ is clear in view of the discussion on $\nabla \cdot \boldsymbol{\sigma}$ following Eq. (2.1.13).

From the foregoing phenomenological principle we derive, in a manner similar to those derived from the balance of linear momentum, the following principles:

3(a). ***Balance of angular momentum.*** Subtracting Eq. (2.1.16) from Eq. (2.1.19), and noting that $DL/Dt = r \times DV/Dt$, gives

$$0 = \nabla \cdot \boldsymbol{\mu} - r \times (\nabla \cdot \boldsymbol{\sigma}),$$

which, in indicial notation, is

$$0 = \frac{\partial}{\partial x_i}(\epsilon_{lmj} x_m \sigma_{ij}) - \epsilon_{lmj} x_m \frac{\partial \sigma_{ij}}{\partial x_i}.$$

This equation, in view of $\partial x_m/\partial x_i = \delta_{im}$ and $\epsilon_{lmj}\delta_{im} = \epsilon_{lij}$, reduces to

$$\epsilon_{lij}\sigma_{ij} = 0, \tag{2.1.20}$$

or

$$\sigma_{ij} = \sigma_{ji}. \tag{2.1.21}$$

That is, for nonpolar continua the balance of angular momentum requires stress to be a *symmetric tensor*. The state of stress can then be prescribed by six rather than nine components.

3(b). ***Balance of mechanical energy of angular momentum.*** The dot product of Eq. (2.1.20) with the angular velocity $\bar{\omega}_l$ gives

$$\bar{\omega}_l \epsilon_{lij} \sigma_{ij} = 0, \tag{2.1.22}$$

which states that the rate of mechanical energy associated with angular momentum vanishes for nonpolar continua.

4. ***Conservation of total (mechanical + thermal) energy.*** As mentioned in Section 2.1.1, a conservation principle does not have any source terms. However, when the conservation of total energy involves electromagnetic, chemical, and nuclear effects, the thermomechanical energy is only part of the total energy and can no longer be conserved. In many practical situations the foregoing energy contributions may be represented by a source term γ''' to be added to the thermomechanical energy balance.

Let $b = u + V^2/2$, $\boldsymbol{\varphi} = \boldsymbol{q} - \boldsymbol{\sigma} \cdot \boldsymbol{V}$, and $\gamma''' = \rho \boldsymbol{f} \cdot \boldsymbol{V} + u'''$. Then Eq. (2.1.3b) yields the *balance of thermomechanical energy for an integral control volume*,

$$\int_{\mathcal{V}} \frac{\partial}{\partial t}\left[\rho\left(u + \frac{V^2}{2}\right)\right] d\mathcal{V} + \int_{\mathcal{S}} \boldsymbol{n} \cdot \rho \boldsymbol{V}\left(u + \frac{V^2}{2}\right) d\mathcal{S}$$
$$= -\int_{\mathcal{S}} \boldsymbol{n} \cdot \boldsymbol{q}\, d\mathcal{S} + \int_{\mathcal{S}} \boldsymbol{n} \cdot (\boldsymbol{\sigma} \cdot \boldsymbol{V})\, d\mathcal{S} + \int_{\mathcal{V}} \rho \boldsymbol{f} \cdot \boldsymbol{V}\, d\mathcal{V} + \int_{\mathcal{V}} u'''\, d\mathcal{V}. \tag{2.1.23}$$

Here u denotes the internal energy, $\boldsymbol{q}$ the heat flux* and u''' the energy generation, including all contributions of electromagnetic, nuclear and chemical origin; $\rho\boldsymbol{f} \cdot \boldsymbol{V}$ and $\boldsymbol{n} \cdot (\boldsymbol{\sigma} \cdot \boldsymbol{V})$ are displacement powers by body forces and stress,† respectively.

In terms of stress vectors, $\boldsymbol{S}_{(1)} \cdot \boldsymbol{V}$ denotes the displacement power by stress on a surface with normal $\boldsymbol{e}_{(1)}$, and

$$\boldsymbol{n} \cdot (\boldsymbol{\sigma} \cdot \boldsymbol{V}) \equiv \boldsymbol{S}_{(n)} \cdot \boldsymbol{V} \equiv n_i \sigma_{ij} V_j,$$

the displacement power by stress on a surface with normal $\boldsymbol{n}$.

Employing the stagnation (internal) energy $e = u + V^2/2$, dividing the stress into its isotropic and deviatoric components (see Eq. 1.6.5), $\boldsymbol{\sigma} = -p\boldsymbol{\delta} + \boldsymbol{\tau}$, and introducing the definitions of enthalpy, $h = u + p/\rho$, and stagnation enthalpy, h^0,

$$h^0 = h + \frac{V^2}{2} = u + \frac{p}{\rho} + \frac{V^2}{2}, \tag{2.1.24}$$

Eq. (2.1.23) may readily be given the alternative form,

$$\int_{\mathcal{V}} \frac{\partial}{\partial t}(\rho e)\, d\mathcal{V} + \int_{\mathcal{S}} \boldsymbol{n} \cdot \rho \boldsymbol{V} h^0\, d\mathcal{S} = -\int_{\mathcal{S}} \boldsymbol{n} \cdot \boldsymbol{q}\, d\mathcal{S} + \int_{\mathcal{S}} \boldsymbol{n} \cdot (\boldsymbol{\tau} \cdot \boldsymbol{V})\, d\mathcal{S} + \int_{\mathcal{V}} \rho \boldsymbol{f} \cdot \boldsymbol{V}\, d\mathcal{V} + \int_{\mathcal{V}} u'''\, d\mathcal{V}. \tag{2.1.25}$$

In arriving at Eq. (2.1.25) we used

$$\int_{\mathcal{S}} \boldsymbol{n} \cdot (-p\boldsymbol{\delta} \cdot \boldsymbol{V})\, d\mathcal{S} = -\int_{\mathcal{S}} \boldsymbol{n} \cdot \rho \boldsymbol{V}\left(\frac{p}{\rho}\right) d\mathcal{S},$$

which may be interpreted as the flux of displacement power by pressure and the flow of "pressure energy," respectively.

Next, introducing $b = u + V^2/2$, $\boldsymbol{\varphi} = \boldsymbol{q} - \boldsymbol{\sigma} \cdot \boldsymbol{V}$, and $\gamma''' = \rho \boldsymbol{f} \cdot \boldsymbol{V} + u'''$ into Eq. (2.1.5), we obtain the *balance of thermomechanical energy for a differential control volume*,

$$\rho \frac{D}{Dt}\left(u + \frac{V^2}{2}\right) = -\nabla \cdot \boldsymbol{q} + \nabla \cdot (\boldsymbol{\sigma} \cdot \boldsymbol{V}) + \rho \boldsymbol{f} \cdot \boldsymbol{V} + u''', \tag{2.1.26}$$

where $\nabla \cdot \boldsymbol{q}$ and $\nabla \cdot (\boldsymbol{\sigma} \cdot \boldsymbol{V})$, respectively, show the net change in heat flux and displacement power by stress over the differential control volume.

In terms of stress vectors,

$$\nabla \cdot (\boldsymbol{\sigma} \cdot \boldsymbol{V}) = \frac{\partial}{\partial x_1}(\boldsymbol{S}_{(1)} \cdot \boldsymbol{V}) + \frac{\partial}{\partial x_2}(\boldsymbol{S}_{(2)} \cdot \boldsymbol{V}) + \frac{\partial}{\partial x_3}(\boldsymbol{S}_{(3)} \cdot \boldsymbol{V}),$$

---

*$\boldsymbol{n} \cdot \boldsymbol{q}$ denotes a positive flux of heat *from* the control volume.

†According to the definition of stress given in Section 1.6, $\boldsymbol{n} \cdot (\boldsymbol{\sigma} \cdot \boldsymbol{V})$ denotes a positive flux of power *to* the control volume.

or, explicitly,

$$\nabla \cdot (\boldsymbol{\sigma} \cdot \boldsymbol{V}) \equiv \frac{\partial}{\partial x_i}(\sigma_{ij}V_j) \equiv \frac{\partial}{\partial x_1}(\sigma_{11}V_1 + \sigma_{12}V_2 + \sigma_{13}V_3) + \frac{\partial}{\partial x_2}(\sigma_{21}V_1 + \sigma_{22}V_2 + \sigma_{23}V_3) + \frac{\partial}{\partial x_3}(\sigma_{31}V_1 + \sigma_{32}V_2 + \sigma_{33}V_3).$$

This expression may be rearranged as

$$\frac{\partial}{\partial x_i}(\sigma_{ij}V_j) = \frac{\partial \sigma_{ij}}{\partial x_i}V_j + \sigma_{ij}\frac{\partial V_j}{\partial x_i}, \tag{2.1.27}$$

where

$$\frac{\partial \sigma_{ij}}{\partial x_i}V_j \equiv \left(\frac{\partial \sigma_{11}}{\partial x_1} + \frac{\partial \sigma_{21}}{\partial x_2} + \frac{\partial \sigma_{31}}{\partial x_3}\right)V_1 + \left(\frac{\partial \sigma_{12}}{\partial x_1} + \frac{\partial \sigma_{22}}{\partial x_2} + \frac{\partial \sigma_{32}}{\partial x_3}\right)V_2 + \left(\frac{\partial \sigma_{13}}{\partial x_1} + \frac{\partial \sigma_{23}}{\partial x_2} + \frac{\partial \sigma_{33}}{\partial x_3}\right)V_3$$

and

$$\sigma_{ij}\frac{\partial V_j}{\partial x_i} \equiv \sigma_{11}\frac{\partial V_1}{\partial x_1} + \sigma_{12}\frac{\partial V_2}{\partial x_1} + \sigma_{13}\frac{\partial V_3}{\partial x_1} + \sigma_{21}\frac{\partial V_1}{\partial x_2} + \sigma_{22}\frac{\partial V_2}{\partial x_2} + \sigma_{23}\frac{\partial V_3}{\partial x_2} + \sigma_{31}\frac{\partial V_1}{\partial x_3} + \sigma_{32}\frac{\partial V_2}{\partial x_3} + \sigma_{33}\frac{\partial V_3}{\partial x_3}.$$

In symbolic notation,

$$\nabla \cdot (\boldsymbol{\sigma} \cdot \boldsymbol{V}) = (\nabla \cdot \boldsymbol{\sigma}) \cdot \boldsymbol{V} + \boldsymbol{\sigma} : \nabla \boldsymbol{V}, \tag{2.1.28}$$

where $(\nabla \cdot \boldsymbol{\sigma}) \cdot \boldsymbol{V}$ denotes the displacement power by net stress over the differential control volume, and $\boldsymbol{\sigma} : \nabla \boldsymbol{V}$ denotes the (reversible and irreversible) conversion of the mechanical energy into thermal energy. The irreversible part of this conversion is the *dissipation* associated with viscous and plastic deformations.

From the foregoing phenomenological principle we derive the following principle:

4(a). *Balance of thermal energy.* Subtracting the mechanical energy given by Eq. (2.1.17) from the thermomechanical energy stated by Eq. (2.1.26), and noting Eq. (2.1.28), we obtain the *balance of thermal energy for a differential control volume,*

$$\rho\frac{Du}{Dt} = -\nabla \cdot \boldsymbol{q} + \boldsymbol{\sigma} : \nabla \boldsymbol{V} + u'''. \tag{2.1.29}$$

For nonpolar continua stress becomes symmetric (see Eq. 2.1.21), and Eq. (2.1.29) reduces to

$$\rho\frac{Du}{Dt} = -\nabla \cdot \boldsymbol{q} + \boldsymbol{\sigma} : \boldsymbol{S} + u'''. \tag{2.1.30}$$

Dividing stress and rate of deformation into deviatoric and isotropic parts, using Eqs. (1.5.8)–(1.5.11) and (1.6.5),

$$\boldsymbol{\sigma} : \boldsymbol{S} = \boldsymbol{\tau} : \boldsymbol{s} - p\nabla \cdot \boldsymbol{V} = \boldsymbol{\tau} : \boldsymbol{s} - p\rho \frac{D(1/\rho)}{Dt}, \tag{2.1.31}$$

Eq. (2.1.30) becomes

$$\rho \frac{Du}{Dt} = -\nabla \cdot \boldsymbol{q} + \boldsymbol{\tau} : \boldsymbol{s} - p\nabla \cdot \boldsymbol{V} + u'''. \tag{2.1.32}$$

Furthermore, by introducing the definition $h = u + p/\rho$, it may be expressed in terms of the enthalpy

$$\rho \frac{Dh}{Dt} - \frac{Dp}{Dt} = -\nabla \cdot \boldsymbol{q} + \boldsymbol{\tau} : \boldsymbol{s} + u'''. \tag{2.1.33}$$

In Section 2.2 we shall return to Eqs. (2.1.32) and (2.1.33) in the development of the thermodynamic relation.

5. *Increase of entropy.* Contrary to other principles nature provides this one in the form of an inequality,

$$\frac{DB_s}{Dt} \geq \Phi, \tag{2.1.34}$$

which replaces Eq. (2.1.1). Treating also Eq. (2.1.3b) as an inequality with $\gamma''' = 0$, and letting $b = s$ and $\boldsymbol{\varphi} = \boldsymbol{q}/T$, we obtain the *increase of entropy for an integral control volume*,

$$\int_{\mathcal{V}} \frac{\partial}{\partial t}(\rho s)\, d\mathcal{V} + \int_{\mathcal{S}} \boldsymbol{n} \cdot \rho \boldsymbol{V} s\, d\mathcal{S} \geq -\int_{\mathcal{S}} \boldsymbol{n} \cdot \left(\frac{\boldsymbol{q}}{T}\right) d\mathcal{S}, \tag{2.1.35}$$

where s denotes the entropy and $\boldsymbol{q}/T$ the entropy flux. Applying the same procedure, this time to Eq. (2.1.5), we have the *increase of entropy for a differential control volume*,

$$\rho \frac{Ds}{Dt} \geq -\nabla \cdot \left(\frac{\boldsymbol{q}}{T}\right). \tag{2.1.36}$$

5(a). *Entropy production.* Actually, introducing the missing source term into Eq. (2.1.34), the principle of the increase of entropy may be now interpreted as a "balance principle." The source term corresponds to the generation of entropy, and is called the *entropy production*, s'''. Letting $b = s$, $\boldsymbol{\varphi} = \boldsymbol{q}/T$, and $\gamma''' = s'''$ in Eq. (2.1.3b), we get the *entropy production for an integral control volume*,

$$\int_{\mathcal{V}} \frac{\partial}{\partial t}(\rho s)\, d\mathcal{V} + \int_{\mathcal{S}} \boldsymbol{n} \cdot \rho \boldsymbol{V} s\, d\mathcal{S} = -\int_{\mathcal{S}} \boldsymbol{n} \cdot \left(\frac{\boldsymbol{q}}{T}\right) d\mathcal{S} + \int_{\mathcal{V}} s'''\, d\mathcal{V}. \tag{2.1.37}$$

Next, referring to Eq. (2.1.5), we have the *entropy production for a differential control volume*,

$$\rho \frac{Ds}{Dt} = -\nabla \cdot \left(\frac{\boldsymbol{q}}{T}\right) + s'''. \tag{2.1.38}$$

Clearly, s''' is *a measure* of the irreversibility per unit volume. The evaluation of this quantity is one of the objectives of the irreversible thermodynamics.

Having established the integral and differential forms of general principles, we proceed to the mixed forms of these principles. However, as we mentioned in the introduction of the present chapter, several forms may be given for this formulation, depending on the particular combination of the assumed lumped, integral, and differential directions. Here we confine ourselves to the simplest possible (radially lumped-axially differential) form of the mixed formulation developed already in terms of Fig. 2.2 and stated by Eqs. (2.1.6) and (2.1.7).

2.1.3 Mixed Formulation: An Example

Consider a compressible inviscid fluid flowing unsteadily through a flexible tube of variable cross section. We wish to formulate the problem which would give the averaged fluid velocity and other variables along the tube.

Inserting $b = 1$, $\varphi_n = 0$, $V_n = V$, $\varphi_c = 0$, $V_c = 0$ and $\gamma''' = 0$ into Eq. (2.1.6), we obtain the *conservation of mass for the mixed* (radially lumped-axially differential) *control volume* shown in Fig. 2.2,

$$\frac{\partial}{\partial t}(\rho A) + \frac{\partial}{\partial n}(\rho A V) = 0. \tag{2.1.39}$$

Letting $b = V$, $\varphi_n = p$, $C\varphi_c = -p\,\partial A/\partial n$, and $\gamma''' = -\rho g\,\partial z/\partial n$, we have from Eq. (2.1.7) the *balance of momentum*,

$$\rho A\frac{DV}{Dt} = -A\frac{\partial p}{\partial n} - A\rho g\frac{\partial z}{\partial n}. \tag{2.1.40}$$

Multiplying Eq. (2.1.40) by the velocity V, we get the *balance of mechanical energy*,

$$\rho A\frac{D}{Dt}\left(\frac{V^2}{2}\right) = -AV\frac{\partial p}{\partial n} - A\rho g V\frac{\partial z}{\partial n}. \tag{2.1.41}$$

Again the use of Eq. (2.1.7), this time with $b = u + V^2/2$, $\varphi_n = q_n + pV$, $\varphi_c = q_w$, and $\gamma''' = -\rho g V\,\partial z/\partial n$, yields the *balance of thermomechanical energy*,

$$\rho A\frac{D}{Dt}\left(u + \frac{V^2}{2}\right) = -\frac{\partial}{\partial n}(Aq_n) - Cq_w - \frac{\partial}{\partial n}(pAV) - A\rho g V\frac{\partial z}{\partial n}. \tag{2.1.42}$$

Subtracting the mechanical energy given by Eq. (2.1.41) from the thermomechanical energy stated by Eq. (2.1.42) results in the *balance of thermal energy*,

$$\rho A\frac{Du}{Dt} = -\frac{\partial}{\partial n}(Aq_n) - Cq_w - p\frac{\partial}{\partial n}(AV). \tag{2.1.43}$$

Finally, letting $b = s$, $\varphi_n = q_n/T$, $\varphi_c = q_w/T$, and $\gamma''' = s'''$ in Eq. (2.1.7), we have the *entropy production*

$$\rho A\frac{Ds}{Dt} = -\frac{\partial}{\partial n}\left(\frac{Aq_n}{T}\right) - \frac{Cq_w}{T} + As'''. \tag{2.1.44}$$

In Section 2.2.5 we evaluate s''' for this example.

2.2 THE THERMODYNAMIC RELATION

In Section 2.1 we developed the general principles appropriate for thermomechanics. Let us see what happens when we apply these principles to particular problems.

Consider first the motion of an incompressible and inviscid fluid which can be established in terms of four scalar variables, say $\boldsymbol{V}$ and p. Clearly, the use of the conservation of mass and the balance of momentum suffices for a complete formulation in terms of the *general principles alone*.

The motion of an incompressible but viscous fluid, on the other hand, requires 10 variables, say $\boldsymbol{V}$, p, and $\boldsymbol{\tau}$. Now we are faced with the need of six additional conditions which are provided by the constitutive relations for viscous stress $\boldsymbol{\tau}$. Problems in this category demand simultaneous consideration of the *general principles and constitutive relations*.

Consider finally the motion and temperature of a compressible and viscous fluid which can be prescribed by 16 variables, say ρ, $\boldsymbol{V}$, p, $\boldsymbol{\tau}$, u, $\boldsymbol{q}$, and T. General principles readily yield five equations; constitutive relations provide nine more equations (six for viscous stress and three for heat diffusion). The remaining two conditions are constitutive, say a (p, ρ, T)-relation and a (u, ρ, T)-relation. Conceptually, however, these relations are derived from one general equation of state with the help of the thermodynamic relation. The latter plays the role of a compatibility condition resting on the assumption of local thermodynamic equilibrium. Problems in this category require the simultaneous use of *general principles*, *constitutive equations*, and the *thermodynamic relation*.

In this section, assuming *local equilibrium* for problems which are in *global nonequilibrium*, we derive the thermodynamic relation, which is also referred to as the *Gibbs relation*. Although the departure from local equilibrium may be important for rapid processes, such as the propagation of shock and ultrasonic waves, it turns out to be negligible for thermal problems to be considered in this text.

2.2.1 The Gibbs Relation

First, recall the balance of thermal energy and the production of entropy,

$$\rho \frac{Du}{Dt} = -\nabla \cdot \boldsymbol{q} + \boldsymbol{\sigma} : \boldsymbol{S} + u''', \tag{2.1.30}$$

$$\rho \frac{Ds}{Dt} = -\nabla \cdot \left(\frac{\boldsymbol{q}}{T}\right) + s'''. \tag{2.1.38}$$

Note that these principles are assumed, on the basis of phenomenological evidence, to govern the actual nonequilibrium processes but there is no a priori restriction that u, s, and T should be identical to their counterparts in thermodynamics. However, the assumption of local equilibrium secures this identity.

Next, consider a reversible process resulting in the same rate of changes. Then Eqs. (2.1.30) and (2.1.38) become

$$\rho \frac{Du}{Dt} = -(\nabla \cdot \boldsymbol{q})^R + (\boldsymbol{\sigma} : \boldsymbol{S})^R,$$

$$\rho \frac{Ds}{Dt} = -\frac{1}{T}(\nabla \cdot \boldsymbol{q})^R.$$

Elimination of $(\nabla \cdot \boldsymbol{q})^R$ between these equations yields the *Gibbs thermodynamic relation*,

$$\frac{Du}{Dt} = T\frac{Ds}{Dt} + \frac{1}{\rho}(\boldsymbol{\sigma} : \boldsymbol{S})^R, \tag{2.2.1}$$

which establishes a relation among the thermodynamic properties.

Specifically, consider a compressible viscous fluid and divide the total deformation power into its deviatoric and isotropic parts, as given by Eq. (2.1.31). Deviatoric stresses are viscous and yield the irreversible deformation power while the isotropic dilatation is reversible. Therefore,

$$(\boldsymbol{\sigma} : \boldsymbol{S})^I = \boldsymbol{\tau} : \boldsymbol{s}; \qquad (\boldsymbol{\sigma} : \boldsymbol{S})^R = -p\rho\frac{Dv}{Dt}, \tag{2.2.2}$$

the superscripts I and R denoting the irreversible and reversible parts, and the Gibbs relation becomes

$$\frac{Du}{Dt} = T\frac{Ds}{Dt} - p\frac{Dv}{Dt}. \tag{2.2.3}$$

Note that Eq. (2.2.3) is the rate form of the well-known relation of classical thermodynamics for simple compressible media, $du = T\,ds - p\,dv$. For more complex media, such as viscoelastic fluids, part of the deviatoric stress may be elastic and reversible, and Eq. (2.2.3) then includes a term of the form $(\boldsymbol{\tau} : \boldsymbol{s})^R$.

2.2.2 *Other Thermodynamic Relations*

Let us review other thermodynamic relations which are derived from the Gibbs relation. These, as well as the Gibbs relation, will be used in expressing the balance of thermal energy in terms of temperature.

Confining attention to the compressible viscous fluid Eq. (2.2.3) may be written for a differential change

$$du = T\,ds - p\,dv, \tag{2.2.4}$$

which suggests a general *equation of state* of the form

$$u = u(s, v). \tag{2.2.5}$$

Differentiating Eq. (2.2.5) and comparing the result with Eq. (2.2.4) yields

$$T = \left(\frac{\partial u}{\partial s}\right)_v; \qquad -p = \left(\frac{\partial u}{\partial v}\right)_s.$$

Here the subscripts indicate which variables are to be kept constant during differentiation. Next, differentiating T with respect to v, and p with respect to s, gives

$$\left(\frac{\partial T}{\partial v}\right)_s = -\left(\frac{\partial p}{\partial s}\right)_v, \tag{2.2.6}$$

one of the well-known *Maxwell relations*.

Let us introduce now the definitions of enthalpy h, the Helmholtz function a, and the Gibbs function g,

$$h = u + pv, \tag{2.2.7}$$

$$a = u - Ts, \tag{2.2.8}$$

$$g = h - Ts. \tag{2.2.9}$$

The Gibbs relation may be given alternative forms in terms of each one of these properties by taking their differential and using Eq. (2.2.4),

$$dh = T\,ds + v\,dp, \tag{2.2.10}$$

$$da = -s\,dT - p\,dv, \tag{2.2.11}$$

$$dg = -s\,dT + v\,dp. \tag{2.2.12}$$

Corresponding to each one of these equations, a Maxwell relation may be derived following the approach leading to Eq. (2.2.6). For example, comparing the differential of $a(T, v)$ to Eq. (2.2.11) and cross-differentiating the coefficients results in

$$-\left(\frac{\partial s}{\partial v}\right)_T = -\left(\frac{\partial p}{\partial T}\right)_v. \tag{2.2.13}$$

Similarly, starting from $g(T, p)$ and using Eq. (2.2.12) gives

$$-\left(\frac{\partial s}{\partial p}\right)_T = \left(\frac{\partial v}{\partial T}\right)_p. \tag{2.2.14}$$

We are ready now to express the thermal energy in terms of temperature.

2.2.3 *Balance of Thermal Energy in Terms of Temperature*

In the application of general principles to individual problems we shall need the balance of energy in terms of temperature, as well as other measurable quantities.

Let u now be a function of T and v, and replace Eq. (2.2.5) with

$$u = u(T, v). \tag{2.2.15}$$

Differentiating this relation,

$$du = \left(\frac{\partial u}{\partial T}\right)_v dT + \left(\frac{\partial u}{\partial v}\right)_T dv, \tag{2.2.16}$$

where the coefficient of the first term on the right-hand side defines the *specific heat at constant volume*,

$$c_v = \left(\frac{\partial u}{\partial T}\right)_v. \tag{2.2.17}$$

To evaluate the coefficient of the second term, we first divide Eq. (2.2.4) by dv at constant T,

$$\left(\frac{\partial u}{\partial v}\right)_T = T\left(\frac{\partial s}{\partial v}\right)_T - p,$$

and then eliminate entropy by the use of Eq. (2.2.13),

$$\left(\frac{\partial u}{\partial v}\right)_T = T\left(\frac{\partial p}{\partial T}\right)_v - p. \tag{2.2.18}$$

Inserting Eqs. (2.2.17) and (2.2.18) into Eq. (2.2.16) gives

$$du = c_v\, dT + \left[T\left(\frac{\partial p}{\partial T}\right)_v - p\right] dv. \tag{2.2.19}$$

This expression involves only changes in temperature and reversible deformation. Consequently, rather than a general equation of state, say Eq. (2.2.15), it suffices to know the specific heat and the (p, v, T)-relation in order to evaluate a change in internal energy.

Finally, inserting Eq. (2.2.19) into Eq. (2.1.30),we get the *balance of thermal energy in terms of temperature* and other measurable quantities,

$$\rho c_v \frac{DT}{Dt} + \rho\left[T\left(\frac{\partial p}{\partial T}\right)_v - p\right]\frac{Dv}{Dt} = -\nabla \cdot \boldsymbol{q} + \boldsymbol{\sigma} : \boldsymbol{S} + u'''. \tag{2.2.20}$$

Dividing deformation power on the right-hand side of this equation into reversible and irreversible parts using Eq. (2.2.2) (or inserting Eq. 2.2.19 into Eq. 2.1.32) gives

$$\rho c_v \frac{DT}{Dt} + \rho T\left(\frac{\partial p}{\partial T}\right)_v \frac{Dv}{Dt} = -\nabla \cdot \boldsymbol{q} + \boldsymbol{\tau} : \boldsymbol{s} + u'''. \tag{2.2.21}$$

The second term on the left-hand side of Eq. (2.2.21) is the thermomechanical coupling, expressing the change of temperature due to reversible deformation. The terms on the right-hand side show, respectively, the contribution of heat transfer, irreversible deformation (dissipation), and energy generation to the change of temperature.

Repeating the foregoing development in terms of $h = h(T, p)$ and using Eqs. (2.2.10) and (2.2.14), we may obtain

$$dh = c_p\, dT + \left[-T\left(\frac{\partial v}{\partial T}\right)_p + v\right] dp, \tag{2.2.22}$$

where the *specific heat at constant pressure* is defined by

$$c_p = \left(\frac{\partial h}{\partial T}\right)_p. \tag{2.2.23}$$

Inserting Eq. (2.2.22) into Eq. (2.1.33) we get the balance of thermal energy, now expressed in terms of c_p,

$$\rho c_p \frac{DT}{Dt} - \rho T\left(\frac{\partial v}{\partial T}\right)_p \frac{Dp}{Dt} = -\nabla \cdot \boldsymbol{q} + \boldsymbol{\tau} : \boldsymbol{s} + u'''. \tag{2.2.24}$$

2.2.4 *Entropy Production*

The Gibbs relation also becomes useful in evaluating the entropy production. Employing Eq. (2.2.1) to eliminate Du/Dt and Ds/Dt between Eqs. (2.1.30) and (2.1.38), the entropy production may be obtained as

$$s''' = \boldsymbol{q} \cdot \nabla\left(\frac{1}{T}\right) + \frac{1}{T}(\boldsymbol{\tau} : \boldsymbol{s})^I + \frac{u'''}{T}. \tag{2.2.25}$$

The three terms on the right-hand side are measures for the *degree of irreversibility* associated with heat transfer, irreversible deformation, and energy generation, respectively.

As provided by nature, the entropy production, and each term contributing to it, are positive or zero. Equation (2.2.25) is the starting point when the theory of irreversible thermodynamics is applied to obtain constitutive relations for irreversible fluxes, such as heat flux and irreversible stress. Finally, we reestablish the thermodynamic relation in terms of the mixed (radially lumped-axially differential) formulation.

2.2.5 *Mixed Formulation. An Example*

Let us reconsider the mixed control volume shown in Fig. 2.2. For a reversible process, Eqs. (2.1.43) and (2.1.44) are

$$\rho A \frac{Du}{Dt} = -\frac{\partial}{\partial n}(Aq_n^R) - Cq_w - p\frac{\partial}{\partial n}(AV),$$

$$\rho A \frac{Ds}{Dt} = -\frac{1}{T}\frac{\partial}{\partial n}(Aq_n^R) - \frac{1}{T}Cq_w.$$

Elimination of the heat fluxes between these equations yields

$$\frac{Du}{Dt} = T\frac{Ds}{Dt} - \frac{p}{\rho A}\frac{\partial}{\partial n}(AV). \tag{2.2.26}$$

Here the last term may be rearranged, by the use of the conservation of mass given by Eq. (2.1.39), as

$$\frac{\partial}{\partial n}(AV) = -\frac{A}{\rho}\frac{D\rho}{Dt} = A\rho\frac{Dv}{Dt},$$

and we immediately recover Eq. (2.2.3).

To express the balance of thermal energy in terms of temperature, we insert Eq. (2.2.19) into Eq. (2.1.43):

$$\rho A\left\{c_v\frac{DT}{Dt} + \left[T\left(\frac{\partial p}{\partial T}\right)_v - p\right]\frac{Dv}{Dt}\right\} = -\frac{\partial}{\partial n}(Aq_n) - Cq_w - p\frac{\partial}{\partial n}(AV).$$

Using conservation of mass, the foregoing equation may be reduced to

$$\rho A c_v \frac{DT}{Dt} + T\left(\frac{\partial p}{\partial T}\right)_v \frac{\partial}{\partial n}(AV) = -\frac{\partial}{\partial n}(Aq_n) - Cq_w, \tag{2.2.27}$$

which is the balance of thermal energy in terms of temperature.

Alternatively, expressing Eq. (2.1.43) in terms of the enthalpy, $h = u + pv$,

$$A\left(\rho \frac{Dh}{Dt} - \frac{Dp}{Dt}\right) = -\frac{\partial}{\partial n}(Aq_n) - Cq_w, \tag{2.2.28}$$

and using Eq. (2.2.22), gives

$$\rho A\left[c_p \frac{DT}{Dt} - T\left(\frac{\partial v}{\partial T}\right)_p \frac{Dp}{Dt}\right] = -\frac{\partial}{\partial n}(Aq_n) - Cq_w. \tag{2.2.29}$$

Finally, eliminating Du/Dt and Ds/Dt (given respectively by Eqs. 2.1.43 and 2.1.44) in Eq. (2.2.3), we obtain the entropy production

$$s''' = q_n \frac{\partial}{\partial n}\left(\frac{1}{T}\right), \tag{2.2.30}$$

which shows the irreversibility associated with heat transfer in the direction of flow. The result should be expected since we assumed a radially lumped-axially distributed temperature. This assumption makes the peripheral heat flux reversible. Also, both the energy generation and irreversible deformation (friction) were ignored from the start.

2.3 *CONSTITUTIVE RELATIONS*

In Section 2.1 we stated general principles as the natural laws *independent of* the continua they apply to. Next we learned in Section 2.2 that constitutive relations for *state*, *momentum flux* (*stress*), and *heat flux*, as well as a *thermodynamic relation*, are needed for a complete formulation of convection problems. Having established the thermodynamic relation, we now turn to the constitution.

Constitutive relations are the natural laws that depend on the continua they apply to, describing their intrinsic (*inherent*) *behavior*. In the investigation of this behavior one follows either a *particle* (molecular) approach based on hypothesis about particle interaction, or a *continuum* (macroscopic) approach based on phenomenology, both to be verified by experimental observation. In this text we confine ourselves to the continuum view, which can be characterized by the *local* and *directional* behavior of matter (Fig. 2.5). The local behavior, on the assumption of local equilibrium, is prescribed by the equation of state. The directional behavior, on the other hand, is prescribed by fluxes of momentum (stress) and energy, expressing the response of matter to global nonequilibrium.

To represent intrinsic behavior of matter, the constitutive relations for moving

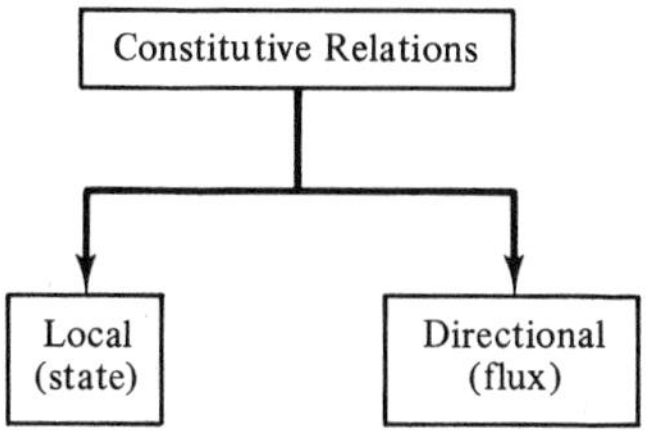

Fig. 2.5 Types of constitutive relations

and deforming continua need to be formulated in the local frame of convected coordinates. Later these relations are transformed to fixed coordinates in which the general principles are most conveniently stated. State properties such as energy, density, temperature, and pressure are scalars and relations among them are unchanged from one frame to the other. The directional behavior involves vectors, such as heat flux and temperature gradient, or tensors, such as stress and deformation rate, which require appropriate coordinate transformations.

We begin the discussion of constitutive relations by stress because continua are conveniently classified according to their response to deformation. Also, we need to identify state properties associated with the reversible work of deformation for explicit forms of the equation of state and the thermodynamic relation. Confining attention to isotropic fluids, we concentrate on the linear viscous fluid because of its importance. For this fluid model we later discuss the constitutive relations for heat flux and state.

2.3.1 *Stress and Pressure*

We learned the state of stress σ_{ij} in Section 1.6, and in Section 2.1 we defined nonpolar fluids by the symmetry of this state,

$$\sigma_{ij} = \sigma_{ji}, \tag{2.1.21}$$

which implies the absence of intrinsic angular momentum, surface and body couples. Throughout this text we deal with *isotropic* and *homogeneous* nonpolar fluids. The intrinsic behavior of an isotropic fluid has no *directional preference*, and that of a homogeneous fluid has no *local dependence*. Here, intrinsic behavior implies uniform structure rather than uniform property.

The *pressure* of a fluid is defined as the *mean normal compressive stress*

$$p = -\tfrac{1}{3}\sigma_{kk}. \tag{2.3.1}$$

Accordingly, a symmetric stress is written in terms of its deviatoric and isotropic components as

$$\sigma_{ij} = \tau_{ij} - p\delta_{ij}. \tag{2.3.2}$$

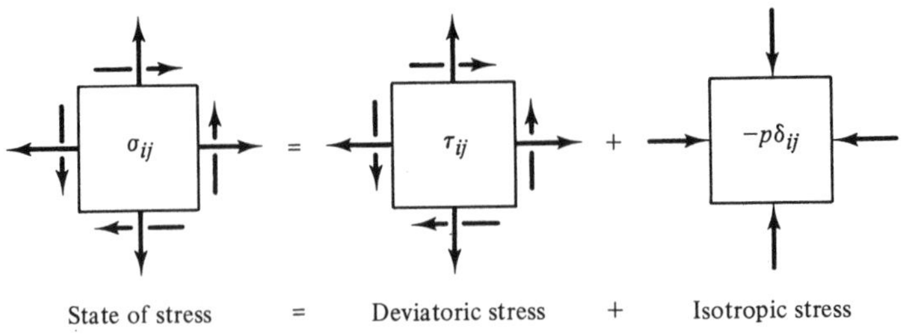

Fig. 2.6 Symmetric state of stress divided into deviatoric and isotropic components

Deviatoric stress tends to distort and isotropic stress (negative pressure) tends to dilate the fluid (Fig. 2.6). The rate of such distortion and dilatation of the fluid is described by the deviatoric and isotropic components of deformation rate (recall Fig. 1.11).

For a fluid at rest and free of shear,* deviatoric stress is zero and the state of stress is isotropic, $\sigma_{ij} = -p_s\delta_{ij}$, where p_s denotes the familiar *hydrostatic pressure*. Clearly, p_s is identical to the thermodynamic pressure of equilibrium. For a fluid in motion, on the other hand, nonisotropic normal stress may arise and the pressure defined by Eq. (2.3.1) takes on a new meaning. Only by the assumption of local equilibrium can the mean compressive stress be interpreted as the thermodynamic pressure, as we show in Section 2.3.2.

Following these general remarks we proceed now to the stress of viscous fluids.

2.3.2 *Stress in Linear Viscous Fluids*

In this section we obtain the constitutive relation for viscous stress from simple physical reasoning based on the observations of pure shearing and pure dilating motions.†

Consider the one-dimensional flow of steady shearing motion between parallel plates (Fig. 2.7). For gases and many liquids (including water) it is well known that shear stress is proportional to the velocity gradient,

$$\tau = \mu \frac{du}{dy}. \tag{2.3.3}$$

This equation is the *Newton law of friction* and it defines the coefficient of (dynamic) *viscosity*. Fluids responding to shear deformation according to Eq. (2.3.3) are called *Newtonian*. Here μ may depend on the thermodynamic state but *not* on the rate of

---

*Certain viscoplastic fluids at rest can sustain shear stress; hence we shall not commit ourselves to the traditional definition of fluids "as media that deform continuously under the action of shear forces."

†See Prandtl & Tietjens (1957) for a similar derivation of deviatoric stress. Alternative approaches are the analogy to the linear elastic solid (see, for example, Schlichting 1968) and the mathematical approach based on tensor calculus (see, for example, Prager 1961 or Aris 1962).

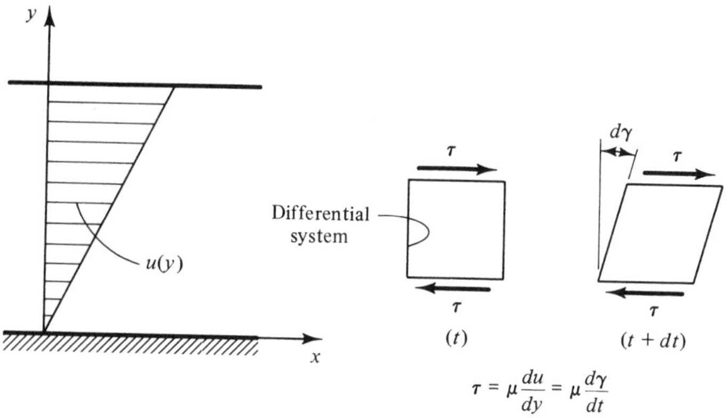

Fig. 2.7 Steady shearing motion of Newton fluid

deformation. For any other response we speak of *non-Newtonian* fluids, which include nonlinear viscous fluids as well as those showing elastic and plastic behavior. An example of a simple nonlinear viscous fluid is that subscribing to the power-law model, which for shearing motion is

$$\tau = K\left|\frac{du}{dy}\right|^{n-1}\frac{du}{dy}, \tag{2.3.4}$$

where K and n are constants. Referring to Eq. (2.3.3), one may define the *apparent viscosity* as $\tau/(du/dy)$, which depends on shear rate for the nonlinear fluid. Many real fluids are non-Newtonian, such as certain oils, polymer solutions, paints, toothpaste, molten glass, and plastics.

Now, in terms of Fig. 2.7 we may express Eq. (2.3.3) as

$$\tau = \mu\frac{du}{dy} = \mu\frac{d\gamma}{dt} = 2\mu\frac{d(\frac{1}{2}\gamma)}{dt},$$

where the rate of change of $\frac{1}{2}\gamma = \frac{1}{2}(\gamma_1 - \gamma_2)$ exactly equals the rate of deformation S_{21} (recall Section 1.5). For two-dimensional shearing motion (Fig. 2.8), we therefore obtain $\sigma_{21} = 2\mu S_{21}$, and generalizing to three-dimensional shearing motion, recalling the symmetry of σ_{ij},

$$\begin{aligned}
\sigma_{12} = \sigma_{21} &= \mu\left(\frac{\partial V_1}{\partial x_2} + \frac{\partial V_2}{\partial x_1}\right) = 2\mu S_{12} \\
\sigma_{23} = \sigma_{32} &= \mu\left(\frac{\partial V_2}{\partial x_3} + \frac{\partial V_3}{\partial x_2}\right) = 2\mu S_{23} \\
\sigma_{31} = \sigma_{13} &= \mu\left(\frac{\partial V_3}{\partial x_1} + \frac{\partial V_1}{\partial x_3}\right) = 2\mu S_{31}.
\end{aligned} \tag{2.3.5}$$

Next, to relate normal stress to the rate of deformation for the same motion, consider the square drawn with a dashed line inside the original differential system (Fig. 2.8). The equilibrium of triangular elements shown in the figure requires that only normal stresses act on the dashed square, whose normals are therefore the principal directions

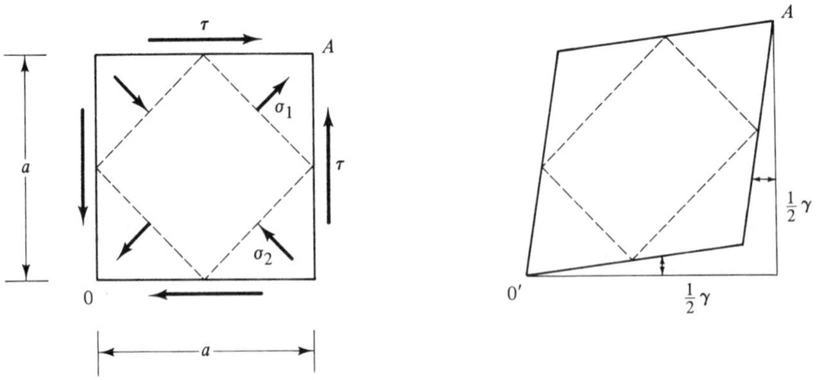

Fig. 2.8 Differential system in shearing motion

of pure tension, $\sigma_1 = \tau$, and pure compression, $\sigma_2 = -\tau$. Hence, we have

$$\sigma_1 - \sigma_2 = 2\tau. \tag{2.3.6}$$

The normal strains ε_1 and ε_2 associated with the deformation of the dashed square into a rectangle equals the relative elongation of the diagonals of the differential system in Fig. 2.8, that is,

$$\varepsilon_1 = \frac{O'A' - OA}{OA} = \frac{\sqrt{2}\,a(\cos\gamma/2 + \sin\gamma/2) - \sqrt{2}\,a}{\sqrt{2}\,a} \simeq \frac{1}{2}\gamma,$$

and similarly, $\varepsilon_2 = -\frac{1}{2}\gamma$, or

$$\varepsilon_1 - \varepsilon_2 = \gamma. \tag{2.3.7}$$

Using the previous result, $\tau = \mu\, d\gamma/dt$, to eliminate γ between Eqs. (2.3.6) and (2.3.7), we obtain

$$\sigma_1 - \sigma_2 = 2\mu \frac{d}{dt}(\varepsilon_1 - \varepsilon_2),$$

which for a three-dimensional shearing motion is generalized to*

$$\begin{aligned}\sigma_{11} - \sigma_{22} &= 2\mu(S_{11} - S_{22})\\ \sigma_{22} - \sigma_{33} &= 2\mu(S_{22} - S_{33})\\ \sigma_{33} - \sigma_{11} &= 2\mu(S_{33} - S_{11}).\end{aligned}$$

The difference between two of these equations in cyclic order gives, after some rearrangement,

$$\begin{aligned}\sigma_{11} &= 2\mu(S_{11} - \tfrac{1}{3}S_{kk}) + \tfrac{1}{3}\sigma_{kk}\\ \sigma_{22} &= 2\mu(S_{22} - \tfrac{1}{3}S_{kk}) + \tfrac{1}{3}\sigma_{kk}\\ \sigma_{33} &= 2\mu(S_{33} - \tfrac{1}{3}S_{kk}) + \tfrac{1}{3}\sigma_{kk}.\end{aligned} \tag{2.3.8}$$

Noting that the terms in parentheses in Eq. (2.3.8) are the diagonal elements of the

*Here we switch to the more conventional notation S_{ij} from $d\varepsilon_{ij}/dt$.

deviatoric rate of deformation s_{ij} defined by Eq. (1.5.11), we may combine Eqs. (2.3.5) and (2.3.8) to a single (tensor) relation

$$\sigma_{ij} = 2\mu s_{ij} + \tfrac{1}{3}\sigma_{kk}\delta_{ij}. \tag{2.3.9}$$

This result shows, when compared with Eq. (2.3.2), that *viscous deviatoric stress is proportional to deviatoric rate of deformation*. Note that s_{ij} is the rate of distortion explained in Section 1.5 and deviatoric stress τ_{ij} is the part of stress that tends to distort the differential system (Fig. 2.6).

Having exhausted the information available from pure shearing motion, we now consider pure dilating motion (of a compressible fluid). Because of isotropy, the deviatoric components of stress and rate of deformation vanish. If the dilating motion were arrested, the isotropic compressive stress would reduce to the static pressure (which also is the thermodynamic pressure)

$$\tfrac{1}{3}\sigma_{kk} = -p. \tag{2.3.1}$$

One may expect the same relation to hold for slow (quasi-steady) expansion or compression; in fact, we shall *assume* that Eq. (2.3.1) holds for all situations encountered, a postulate that is consistent with the assumption of *local equilibrium*. Consequently, the complete constitutive relation for stress in a *Newtonian* (linear viscous) fluid becomes

$$\sigma_{ij} = \tau_{ij} - p\delta_{ij}, \tag{2.3.10}$$

where deviatoric stress is viscous and irreversible,

$$\sigma^{I}_{ij} = \tau_{ij} = 2\mu s_{ij}, \tag{2.3.11}$$

with

$$s_{ij} = \frac{1}{2}\left(\frac{\partial V_i}{\partial x_j} + \frac{\partial V_j}{\partial x_i}\right) - \frac{1}{3}\frac{\partial V_k}{\partial x_k}\delta_{ij}, \tag{1.5.11}$$

while isotropic stress is reversible,

$$\sigma^{R}_{ij} = \tfrac{1}{3}\sigma_{kk} = -p\delta_{ij}, \tag{2.3.12}$$

and is a local state property, $p(v, T)$, to be discussed in Section 2.3.4. Note that for an incompressible fluid p is a dynamic variable and has no thermodynamic significance.

A more general linear fluid model based only on the assumption that stress (in excess of isotropic pressure p_1) is proportional to rate of deformation leads to (see, for example, Prager 1961)

$$\sigma_{ij} = 2\mu S_{ij} + \lambda S_{kk}\delta_{ij} - p_1\delta_{ij}, \tag{2.3.13}$$

or, after collecting deviatoric and isotropic parts, to

$$\sigma_{ij} = 2\mu s_{ij} + (\kappa S_{kk} - p_1)\delta_{ij}, \tag{2.3.14}$$

where λ denotes the *second viscosity* and $\kappa = \tfrac{2}{3}\mu + \lambda$ the *bulk viscosity*. Equation (2.3.14) reduces to Eqs. (2.3.10)–(2.3.12) on the assumption of zero bulk viscosity, $\kappa = 0$, known as the *Stokes hypothesis*. This statement is consistent with the assumption of local equilibrium implying that $p_1 = p$ is the thermodynamic pressure. For an incompressible fluid, however, $S_{kk} = \tfrac{1}{3}\,\partial V_k/\partial x_k = 0$ and we recover the same result without any assumption.

Equation (2.3.14) is analogous to Hooke's law of isothermal elastic solids, obtained replacing s_{ij} by ϵ_{ij}, μ by elastic shear modulus μ_e, κ by elastic bulk modulus κ_e, and equating p_1 to zero.

> From observation of processes involving rapid volume changes, say in ultrasonics, it may be deduced that the pressure $p_m = -\frac{1}{3}\sigma_{kk}$ deviates from the thermodynamic pressure. The latter may be defined for a nonequilibrium process as the equilibrium pressure corresponding to instantaneous local values of specific volume and energy. The aforementioned discrepancy between pressures is explained by a bulk viscous contribution to normal stress which—for linear fluids—is taken to be proportional to the isotropic rate of deformation as indicated by Eq. (2.3.14),
>
> $$\tfrac{1}{3}\sigma_{kk} = -p_m = \kappa S_{kk} - p,$$
>
> or, by the use of conservation of mass,
>
> $$\frac{1}{3v}\frac{Dv}{Dt} = \frac{p - p_m}{\kappa}. \tag{2.3.15}$$
>
> A formal derivation of Eq. (2.3.15) for gases and an evaluation of the bulk viscosity κ (which is not a material constant, but depends also on the rate of volume change) belongs to statistical mechanics. It suffices to realize that bulk viscosity is a manifestation of departure from local equilibrium related to molecular relaxation processes, and is therefore zero for monatomic gases at low density. It remains negligible also for other gases and for liquids whenever the time characterizing global processes is large compared to that of relaxation processes. Clearly, the assumption of local equilibrium, implying that pressure may well depend on specific volume *but not on its rate of change*, leads us to specify that $\kappa = 0$.

Rewriting Eq. (2.3.11) for an incompressible fluid in one-dimensional shearing flow as

$$\tau_{21} = \nu \frac{\partial \rho V_1}{\partial x_2}, \tag{2.3.16}$$

where $\nu = \mu/\rho$ is the *kinematic viscosity*, suggests the diffusion nature of viscous (irreversible) stress. Accordingly, τ_{ij} represents the negative diffusion flux of x_j-momentum density in the x_i-direction and ν is interpreted as the *momentum diffusivity*.

2.3.3 Heat Flux

The heat flux represents the transfer of thermal energy due to temperature differences, from regions of higher temperature to regions of lower temperature. Two mechanisms for this transfer are distinguished, *diffusion* (conduction) and *radiation*. The interaction of thermal radiation with fluids will not be discussed here.

The diffusion heat flux q_i is the intrinsic (molecular) response of matter to nonuniform temperatures characterizing global nonequilibrium. For isotropic fluids the heat flux is proportional to the temperature gradient as expressed by the *Fourier law of conduction*,

$$q_i = -k \frac{\partial T}{\partial x_i}, \tag{2.3.17}$$

where k denotes the *thermal conductivity*, which may be a function of local state. The negative sign in Eq. (2.3.17) indicates that thermal energy is transferred in the direction of decreasing temperature in accordance with the increase of entropy.

For an incompressible fluid having constant specific heat, $\rho\, du = \rho c\, dT = d(\rho c T)$ (see Eq. 2.3.23), we may rewrite Eq. (2.3.17) as

$$q_i = -a \frac{\partial \rho c T}{\partial x_i}, \tag{2.3.18}$$

where $a = k/\rho c$ denotes the *thermal diffusivity*. Accordingly, the diffusion flux of internal energy in a given direction is proportional to the gradient of internal energy density in this direction.

2.3.4 Equation of State

The equation of state describes the local behavior of the continuum in terms of state properties at a point. For *compressible viscous fluids* reversible deformation is associated only with volume change and we have

$$c_v = c_v(T, v); \qquad p = p(T, v). \tag{2.3.19}$$

Specifically, *gases at low density* are adequately described by the equation of *ideal gas*

$$pv = RT, \tag{2.3.20}$$

where R denotes the gas constant and T the absolute temperature. Inserting Eq. (2.3.20) into Eqs. (2.2.19) and (2.2.22) gives

$$du = c_v\, dT; \qquad dh = c_p\, dT, \tag{2.3.21}$$

where $h = u + pv$ denotes the enthalpy. Consequently, specific heats depend only on temperature and are related by $c_p(T) - c_v(T) = R$. For *dense gases* Eq. (2.3.20) is replaced by more elaborate equations of state (van der Walls, Beattie–Bridgeman, or other virial equations) and specific heats then depend on specific volume (or pressure) in addition to temperature. For *liquids* there is no simple algebraic equation of state. For a small change of state, for example, it may be sufficient to integrate a differential equation of state, such as

$$d\rho = \left(\frac{\partial \rho}{\partial T}\right)_p dT + \left(\frac{\partial \rho}{\partial p}\right)_T dp \tag{2.3.22}$$

with the assumption of constant *coefficient of thermal expansion* $\beta = -(1/\rho)(\partial\rho/\partial T)_p$ and constant *isothermal compressibility* $\beta_T = (1/\rho)(\partial\rho/\partial p)_T$. Also, the specific heats of liquids depend on temperature and pressure in a complex way, but for small changes in state far from the critical point they may be assumed constant. We shall use these approximations for both liquids and gases when dealing with problems of natural convection (see Section 2.4.2). For *incompressible viscous fluids* there is no reversible deformation and Eq. (2.2.19) becomes

$$du = c\, dT, \tag{2.3.23}$$

where $c = c(T)$ is the (only) specific heat. Although pressure is now a dynamical variable without any thermodynamic significance, the enthalpy, $h = u + p/\rho$, remains a useful concept for which we have $(\partial h/\partial T)_p = c(T)$.

2.4 GOVERNING EQUATIONS

In the preceding three sections we have developed the general principles, local equilibrium, and constitutive relations. Here we combine the results for the purpose of obtaining the differential formulation of convection.

2.4.1 Viscous Flows

Our starting point, arranged in Table 2.2, is the conservation of mass, balance of linear momentum, balance of thermal energy, and constitutive relations for a viscous Newton–Fourier fluid taken from previous sections. In Eq. (2.1.13) we have utilized

TABLE 2.2 General formulation of a Newton–Fourier fluid

General principles:

$$\frac{\partial \rho}{\partial t} + \frac{\partial \rho V_k}{\partial x_k} = 0 \tag{2.1.9}$$

$$\rho \frac{DV_j}{Dt} = \rho f_j + \frac{\partial \tau_{ij}}{\partial x_i} - \frac{\partial p}{\partial x_j} \tag{2.1.13}$$

$$\rho c_v \frac{DT}{Dt} + \rho T \left(\frac{\partial p}{\partial T}\right)_v \frac{Dv}{Dt} = -\frac{\partial q_k}{\partial x_k} + \tau_{ij} s_{ij} + u''' \tag{2.2.21}$$

$$\rho c_p \frac{DT}{Dt} - \rho T \left(\frac{\partial v}{\partial T}\right)_p \frac{Dp}{Dt} = -\frac{\partial q_k}{\partial x_k} + \tau_{ij} s_{ij} + u''' \tag{2.2.24}$$

Constitutive relations:

$$\tau_{ij} = 2\mu s_{ij}; \qquad s_{ij} = \frac{1}{2}\left(\frac{\partial V_i}{\partial x_j} + \frac{\partial V_j}{\partial x_i}\right) - \frac{1}{3}\frac{\partial V_k}{\partial x_k}\delta_{ij} \tag{2.3.11}$$

$$q_k = -k \frac{\partial T}{\partial x_k} \tag{2.3.17}$$

$$c_v = c_v(T, v) \quad \text{or} \quad c_p = c_p(T, p) \quad \text{and} \quad p = p(T, v) \tag{2.3.19}$$

$\sigma_{ij} = \tau_{ij} - p\delta_{ij}$, and we state both Eqs. (2.2.21) and (2.2.24), which are the two alternative forms of the balance of thermal energy, involving respectively c_v and c_p. Inserting constitutive relations into balance equations, we obtain the governing equations. The explicit Cartesian form of these equations is written in Table 2.3. The *dissipation function* Φ_v, given by Eq. (2.4.6), is defined according to $\tau_{ij}s_{ij} = 2\mu s_{ij}s_{ij} = \mu\Phi_v$. Equations (2.4.2)–(2.4.4) are the celebrated Navier–Stokes equations.

Specifying initial and boundary conditions for velocity and temperature, the equations of Table 2.3 determine the time-dependent, three-dimensional velocity and temperature fields of convection. However, the five resulting partial differential equations are both nonlinear and coupled (through property variation). Note that for

TABLE 2.3 Variable property convection

Conservation of mass:

$$\frac{\partial \rho}{\partial t} + \frac{\partial \rho u}{\partial x} + \frac{\partial \rho v}{\partial y} + \frac{\partial \rho w}{\partial z} = 0 \tag{2.4.1}$$

Balance of momentum (Navier–Stokes equations):

$$\begin{aligned} \rho\left(\frac{\partial u}{\partial t} + u\frac{\partial u}{\partial x} + v\frac{\partial u}{\partial y} + w\frac{\partial u}{\partial z}\right) &= \rho f_x - \frac{\partial p}{\partial x} + \frac{\partial}{\partial x}\left[\frac{4}{3}\mu\frac{\partial u}{\partial x} - \frac{2}{3}\mu\left(\frac{\partial v}{\partial y} + \frac{\partial w}{\partial z}\right)\right] \\ &\quad + \frac{\partial}{\partial y}\left[\mu\left(\frac{\partial u}{\partial y} + \frac{\partial v}{\partial x}\right)\right] + \frac{\partial}{\partial z}\left[\mu\left(\frac{\partial w}{\partial x} + \frac{\partial u}{\partial z}\right)\right] \end{aligned} \tag{2.4.2}$$

$$\begin{aligned} \rho\left(\frac{\partial v}{\partial t} + u\frac{\partial v}{\partial x} + v\frac{\partial v}{\partial y} + w\frac{\partial v}{\partial z}\right) &= \rho f_y - \frac{\partial p}{\partial y} + \frac{\partial}{\partial x}\left[\mu\left(\frac{\partial u}{\partial y} + \frac{\partial v}{\partial x}\right)\right] \\ &\quad + \frac{\partial}{\partial y}\left[\frac{4}{3}\mu\frac{\partial v}{\partial y} - \frac{2}{3}\mu\left(\frac{\partial u}{\partial x} + \frac{\partial w}{\partial z}\right)\right] + \frac{\partial}{\partial z}\left[\mu\left(\frac{\partial w}{\partial y} + \frac{\partial v}{\partial z}\right)\right] \end{aligned} \tag{2.4.3}$$

$$\begin{aligned} \rho\left(\frac{\partial w}{\partial t} + u\frac{\partial w}{\partial x} + v\frac{\partial w}{\partial y} + w\frac{\partial w}{\partial z}\right) &= \rho f_z - \frac{\partial p}{\partial z} + \frac{\partial}{\partial x}\left[\mu\left(\frac{\partial w}{\partial x} + \frac{\partial u}{\partial z}\right)\right] \\ &\quad + \frac{\partial}{\partial y}\left[\mu\left(\frac{\partial v}{\partial z} + \frac{\partial w}{\partial y}\right)\right] + \frac{\partial}{\partial z}\left[\frac{4}{3}\mu\frac{\partial w}{\partial z} - \frac{2}{3}\mu\left(\frac{\partial u}{\partial x} + \frac{\partial v}{\partial y}\right)\right] \end{aligned} \tag{2.4.4}$$

Balance of thermal energy:

$$\left.\begin{aligned} &\rho c_v\left(\frac{\partial T}{\partial t} + u\frac{\partial T}{\partial x} + v\frac{\partial T}{\partial y} + w\frac{\partial T}{\partial z}\right) + T\left(\frac{\partial p}{\partial T}\right)_v\left(\frac{\partial u}{\partial x} + \frac{\partial v}{\partial y} + \frac{\partial w}{\partial z}\right) \\ &\rho c_p\left(\frac{\partial T}{\partial t} + u\frac{\partial T}{\partial x} + v\frac{\partial T}{\partial y} + w\frac{\partial T}{\partial z}\right) - \rho T\left(\frac{\partial v}{\partial T}\right)_p\left(\frac{\partial p}{\partial t} + u\frac{\partial p}{\partial x} + v\frac{\partial p}{\partial y} + w\frac{\partial p}{\partial z}\right) \end{aligned}\right\}$$

$$= \frac{\partial}{\partial x}\left(k\frac{\partial T}{\partial x}\right) + \frac{\partial}{\partial y}\left(k\frac{\partial T}{\partial y}\right) + \frac{\partial}{\partial z}\left(k\frac{\partial T}{\partial z}\right) + \mu\Phi_v + u''' \tag{2.4.5a, b}$$

Dissipation function:

$$\begin{aligned} \Phi_v = 2\left[\left(\frac{\partial u}{\partial x}\right)^2 + \left(\frac{\partial v}{\partial y}\right)^2 + \left(\frac{\partial w}{\partial z}\right)^2\right] + \left(\frac{\partial u}{\partial y} + \frac{\partial v}{\partial x}\right)^2 \\ + \left(\frac{\partial w}{\partial y} + \frac{\partial v}{\partial z}\right)^2 + \left(\frac{\partial w}{\partial x} + \frac{\partial u}{\partial z}\right)^2 - \frac{2}{3}\left(\frac{\partial u}{\partial x} + \frac{\partial v}{\partial y} + \frac{\partial w}{\partial z}\right)^2 \end{aligned} \tag{2.4.6}$$

Equation of state:

$$c_v = c_v(T, v) \quad \text{or} \quad c_p = c_p(T, p) \quad \text{and} \quad p = p(T, v) \qquad (v = 1/\rho) \tag{2.4.7}$$

liquids, ρ is essentially constant and the second term on the left of Eqs. (2.4.5a) and (2.4.5b) is zero. This implies that $c_v = c_p$, say c (recall Section 2.3.4). For gases, on the other hand, the second term on the left of Eqs. (2.4.5a) or (2.4.5b) vanishes, respectively, for an essentially incompressible flow or for an essentially constant pressure flow.

Many problems of convection may be treated as *incompressible, constant-property* flows. Introducing $\rho \simeq$ constant, the kinematic viscosity $\nu = \mu/\rho$ and the thermal diffusivity $a = k/\rho c_v$, the equations of Table 2.3, now selecting Eq. (2.4.5a), reduce to those of Table 2.4.* Moreover, assuming a body force independent of temperature,

TABLE 2.4 Incompressible, constant-property convection

Conservation of mass:

$$\frac{\partial u}{\partial x} + \frac{\partial v}{\partial y} + \frac{\partial w}{\partial z} = 0 \tag{2.4.8}$$

Balance of momentum (Navier–Stokes equations):

$$\frac{\partial u}{\partial t} + u\frac{\partial u}{\partial x} + v\frac{\partial u}{\partial y} + w\frac{\partial u}{\partial z} = f_x - \frac{1}{\rho}\frac{\partial p}{\partial x} + \nu\left(\frac{\partial^2 u}{\partial x^2} + \frac{\partial^2 u}{\partial y^2} + \frac{\partial^2 u}{\partial z^2}\right) \tag{2.4.9}$$

$$\frac{\partial v}{\partial t} + u\frac{\partial v}{\partial x} + v\frac{\partial v}{\partial y} + w\frac{\partial v}{\partial z} = f_y - \frac{1}{\rho}\frac{\partial p}{\partial y} + \nu\left(\frac{\partial^2 v}{\partial x^2} + \frac{\partial^2 v}{\partial y^2} + \frac{\partial^2 v}{\partial z^2}\right) \tag{2.4.10}$$

$$\frac{\partial w}{\partial t} + u\frac{\partial w}{\partial x} + v\frac{\partial w}{\partial y} + w\frac{\partial w}{\partial z} = f_z - \frac{1}{\rho}\frac{\partial p}{\partial z} + \nu\left(\frac{\partial^2 w}{\partial x^2} + \frac{\partial^2 w}{\partial y^2} + \frac{\partial^2 w}{\partial z^2}\right) \tag{2.4.11}$$

Balance of thermal energy:

$$\frac{\partial T}{\partial t} + u\frac{\partial T}{\partial x} + v\frac{\partial T}{\partial y} + w\frac{\partial T}{\partial z} = a\left(\frac{\partial^2 T}{\partial x^2} + \frac{\partial^2 T}{\partial y^2} + \frac{\partial^2 T}{\partial z^2}\right) + \frac{\nu}{c_v}\Phi_v + \frac{u'''}{\rho c_v} \tag{2.4.12}$$

Dissipation function:

$$\Phi_v = 2\left[\left(\frac{\partial u}{\partial x}\right)^2 + \left(\frac{\partial v}{\partial y}\right)^2 + \left(\frac{\partial w}{\partial z}\right)^2\right] + \left(\frac{\partial u}{\partial y} + \frac{\partial v}{\partial x}\right)^2 + \left(\frac{\partial w}{\partial y} + \frac{\partial v}{\partial z}\right)^2 + \left(\frac{\partial w}{\partial x} + \frac{\partial u}{\partial z}\right)^2 \tag{2.4.13}$$

the fluid motion (Eqs. 2.4.8–2.4.11) may be solved separately. Further simplifications are possible for limiting values of dimensionless numbers which are discussed next.†

On dimensional grounds Eqs. (2.4.9)–(2.4.11) yield, with the distinction between flow and flux stated in Section 2.1.1,

$$\frac{\text{momentum flow}}{\text{momentum flux}} = \frac{\text{inertia force}}{\text{viscous force}} \sim \frac{VV}{\nu V/l} = \frac{Vl}{\nu} = \text{Re}, \tag{2.4.14}$$

*For the derivation of the corresponding differential formulation in cylindrical or spherical coordinates, see Appendix C, and for orthogonal curvelinear coordinates, see Hughes & Gaylord (1964).

†For a list of the dimensionless numbers of convection, see Appendix D.

where Re denotes the *Reynolds number*, V being a characteristic velocity, and l a length of the problem being considered. Introducing also a characteristic temperature difference ΔT, we obtain, in a similar way, from Eq. (2.4.12)

$$\frac{\text{energy flow}}{\text{energy flux}} = \frac{\text{convection}}{\text{diffusion}} \sim \frac{V\,\Delta T}{a\,\Delta T/l} = \frac{Vl}{a} = \text{Pe}, \tag{2.4.15}$$

where Pe denotes the *Péclet number*. The ratio of Eqs. (2.4.15) and (2.4.14), eliminating the effects of flow and geometry, yields

$$\frac{\text{(flux/flow) of momentum}}{\text{(flux/flow) of energy}} \sim \frac{\nu}{a} = \text{Pr}, \tag{2.4.16}$$

where Pr denotes the *Prandtl number*, a fluid property. Finally, from Eq. (2.4.12), using Eq. (2.4.13),

$$\frac{\text{energy dissipation}}{\text{change of energy flux}} \sim \frac{(\nu/c_v)(V/l)^2}{a\,\Delta T/l^2} = \frac{V^2}{c_v\,\Delta T}\,\frac{\nu}{a} = \text{EcPr}, \tag{2.4.17}$$

where $\text{Ec} = V^2/(c_v\,\Delta T)$ denotes the *Eckert number*.

The Reynolds number is an important parameter for viscous flow. Its magnitude determines whether the flow will be laminar or turbulent. Furthermore, since it is a measure of the ratio of inertia to viscous forces, it determines, for a given geometry, the resulting flow pattern. The governing equations simplify in the limit of large Reynolds number to the boundary layer equations (Chapter 4) and in the limit of small Reynolds number to those of creeping flow. For zero viscosity Eqs. (2.4.9)–(2.4.11) reduce to the Euler equations of inviscid flow. The Prandtl number, in addition to being a fluid property, is a measure of the diffusion of momentum relative to that of heat. Simplified governing equations for liquid metals and viscous oils correspond to small and large values of the Prandtl number. The magnitude of the product EcPr is a measure of the importance of the viscous dissipation. It is small, hence may be neglected, for most convection problems not involving the flow of viscous oils. In the following section, allowing density to depend on temperature in the body force term, we derive the governing equations of natural convection.

2.4.2 *Buoyancy-Driven Flows*

As observed by the Greeks, a body immersed in a fluid experiences a buoyant force equal to the weight of the displaced fluid (the Archimedes principle); when its density is more or less than that of the surrounding fluid, the body sinks or rises, respectively. The buoyant force is the driving mechanism of natural convection; so, when the effect of temperature on fluid density is taken into account, lighter parts of a heated fluid flow upward and the heavier ones downward relative to a reference state in equilibrium.

The most frequently encountered buoyancy is associated with gravity. It is seen in the heating of rooms from convectors or space heaters, in providing the draft in chimneys, in cooling products in refrigerators or cooling houses, in the cooling of transistors and transformers, as well as human beings and animals standing in a quiescent atmosphere. On a large scale, buoyancy contributes to driving the atmospheric circulation of the earth. Other sources of natural convection are centrifugal forces, which provide the internal cooling of turbine blades, and inertial forces, which affect cryogenic liquids in accelerating rockets.

Here we formulate the natural convection in terms of a body force $\boldsymbol{f} = -\boldsymbol{g}$ which may be constant (gravity) or variable. For a given temperature distribution, the motion of the fluid is governed by

$$\rho \frac{DV_j}{Dt} = -\rho g_j - \frac{\partial p}{\partial x_j} + \frac{\partial \tau_{ij}}{\partial x_i}, \qquad (2.1.13)$$

where p denotes the pressure defined in Section 2.3.1. Now, if the temperature were to approach a uniform value, say T_0, convection would stop and Eq. (2.1.13) reduce to

$$0 = -\rho_0 g_j - \frac{\partial p_0}{\partial x_j}, \qquad (2.4.18)$$

where p_0 is the hydrostatic pressure corresponding to density ρ_0 and temperature T_0. Subtracting Eq. (2.4.18) from Eq. (2.1.13) gives

$$\rho \frac{DV_j}{Dt} = -(\rho - \rho_0)g_j - \frac{\partial(p - p_0)}{\partial x_j} + \frac{\partial \tau_{ij}}{\partial x_i}. \qquad (2.4.19)$$

For buoyancy-driven flows, $p - p_0$ is negligibly small. However, for combined (forced and buoyancy-driven) flows the pressure term becomes appreciable.

For most problems of natural convection the density variations are mainly caused by the thermal expansion of the fluid and can be expressed as

$$\rho = \rho_0[1 - \beta(T - T_0)], \qquad (2.4.20)$$

where $\beta = -(1/\rho)(\partial\rho/\partial T)_p$ denotes the coefficient of thermal expansion, which is positive for most fluids. It is negative for water in the range 0 to 4°C.

Integration of Eq. (2.3.22) for small departures of density from an equilibrium state $\rho_0 = \rho(p_0, T_0)$ gives

$$\rho = \rho_0[1 - \beta(T - T_0) + \beta_T(p - p_0)], \qquad (2.4.21)$$

where $\beta_T = (1/\rho)(\partial\rho/\partial p)_T$ denotes the isothermal compressibility. In the absence of forced convection, $p - p_0 = 0$ for horizontal temperature differences in the gravity field. For vertical temperature differences the actual $p - p_0$ depends on the solution of the problem, and cannot a priori be determined. However, since the natural convection is characterized by slow motion and small rates of deformation, here p may be assumed hydrostatic for all practical purposes. Then $p - p_0 \sim \rho g H$, where H is a characteristic vertical length, and

$$\frac{\text{compressibility}}{\text{thermal expansion}} \sim \frac{\beta_T \rho g H}{\beta\, \Delta T}.$$

For water (from Appendix E: $\beta \sim 2 \times 10^{-4}\ °\text{C}^{-1}$, $\beta_T \sim 5 \times 10^{-10}\ \text{Pa}^{-1}$) at $\Delta T \sim$ 10°C and $H = 1$ m, this ratio is $\sim 2.5 \times 10^{-3}$ and the effect of pressure on density change can be safely neglected. This fact also holds for other fluids. For ideal gases, employing Eq. (2.3.20), the foregoing ratio may be written as $gH/R\,\Delta T$. For atmospheric air ($R = 286$ J/kg K) at $\Delta T = 10$°C and $H = 1$ m, this ratio becomes

$\sim 3.5 \times 10^{-3}$, and again the effect of compressibility can be neglected. However, this condition may not be satisfied for thick (atmospheric) layers subjected to small temperature differences.

Furthermore, for most fluids $\beta \sim 10^{-3}$ to $10^{-4}\ ^\circ\mathrm{C}^{-1}$, and for small variations in temperature (say 10°C) the variations in density are at most 1%. We may therefore treat density as a constant in all terms in the equations governing natural convection, except in the buoyancy term. This is the well-known *Boussinesq approximation* (see Eshghy & Morrison 1966 for a study which excludes this approximation). For small temperature variations, we may also ignore other property variations. Then, inserting Eq. (2.4.20) into the buoyancy term of Eq. (2.4.19), letting $\rho = \rho_0$ everywhere else, the equations of Table 2.2 reduce to those given in Table 2.5, where $\nu = \mu/\rho_0$ and $a = k/\rho_0 c_p$. (The reader may readily write the explicit Cartesian forms of these equations by referring to Table 2.4.)

TABLE 2.5 Constant-property buoyancy-driven convection

$$\frac{\partial V_k}{\partial x_k} = 0 \tag{2.4.22}$$

$$\frac{DV_j}{Dt} = \beta(T - T_0)g_j - \frac{1}{\rho_0}\frac{\partial(p - p_0)}{\partial x_j} + \nu \frac{\partial^2 V_j}{\partial x_k\, \partial x_k} \tag{2.4.23}$$

$$\frac{DT}{Dt} = a\frac{\partial^2 T}{\partial x_k\, \partial x_k} + \frac{\nu}{c_p}\Phi_v + \frac{u'''}{\rho_0 c_p} \tag{2.4.24}$$

As expected, the retention of property variation in the buoyancy term couples the momentum and energy balances (except for one-dimensional parallel flows, Section 3.4), and the equations must be solved simultaneously.

For dimensionless numbers associated with the natural convection, first, from the balance of thermal energy, Eq. (2.4.24),

$$\frac{\text{energy dissipation}}{\text{change of energy flux}} \sim \frac{(\nu/c_p)(V/l)^2}{a\,\Delta T/l^2} = \frac{V^2}{c_p\,\Delta T}\frac{\nu}{a} = \mathrm{EcPr}, \tag{2.4.25}$$

which, as Eq. (2.4.17) for forced convection, is the product of the Eckert and Prandtl numbers. As may be shown, this product is small, and the dissipation may be ignored, in all problems of natural convection. Furthermore, from Eq. (2.4.24),

$$\frac{\text{induced energy flow}}{\text{energy flux}} \sim \frac{V\,\Delta T}{a\,\Delta T/l} = \frac{V}{a/l}.$$

Assuming this ratio to be of order unity* gives

$$V \sim \frac{a}{l}. \tag{2.4.26}$$

*In Chapter 12 we elaborate the dimensional analysis in terms of two-length scales associated with flow and flux terms and will then be able to estimate the magnitude of this ratio.

Next, to account for the inherent coupling between energy and momentum balances, we employ this velocity in Eq. (2.4.23) to obtain

$$\frac{\text{buoyancy force}}{\text{change of momentum flux}} \sim \frac{g\beta\,\Delta T}{\nu V/l^2} = \frac{g\beta\,\Delta T\, l^3}{\nu a} = \text{Ra}, \tag{2.4.27}$$

where Ra denotes the *Rayleigh number*, the counterpart to the Reynolds number of the forced convection. The magnitude of the Rayleigh number determines whether the flow will be laminar or turbulent and, for a given geometry, the resulting flow pattern. Finally, for mixed (buoyancy and forced) convection, again using Eq. (2.4.23),

$$\frac{\text{buoyancy force}}{\text{change of momentum flow}} \sim \frac{g\beta\,\Delta T}{VV/l} = \frac{g\beta\,\Delta T\, l}{V^2} = \text{Ri}, \tag{2.4.28}$$

where Ri denotes the *Richardson number*. This number, although independent of the viscosity, is traditionally expressed as $\text{Ri} = \text{Gr}/\text{Re}^2$, where $\text{Gr} = g\beta\,\Delta T\, l^3/\nu^2$ denotes the *Grashof number* and $\text{Re} = Vl/\nu$, the Reynolds number.

Having completed the formulation of the governing equations of convection we consider next the vorticity and the stream function as dependent variables in place of velocity and pressure in the equations governing the fluid motion.

2.4.3 Vorticity Transport

An alternative approach studying fluid mechanics is to assume the flow as a process of vorticity transport. This approach adds a new dimension to our understanding of many inviscid and viscous flow phenomena (see Lighthill 1963), and it will be particularly useful to our study of the turbulent flow (Part IV).

Consider an *incompressible, constant-property* fluid including the effect of buoyancy. Using vector notation and rewriting the left-hand side of Eq. (2.4.23) by Eq. (1.3.5), the momentum balance becomes

$$\frac{\partial \boldsymbol{V}}{\partial t} + \nabla\left(\frac{V^2}{2}\right) - \boldsymbol{V} \times \boldsymbol{\omega} = \beta \boldsymbol{g}(T - T_0) - \frac{\nabla(p - p_0)}{\rho_0} + \nu \nabla^2 \boldsymbol{V}. \tag{2.4.29}$$

The curl of this equation gives

$$\frac{\partial \boldsymbol{\omega}}{\partial t} - \nabla \times (\boldsymbol{V} \times \boldsymbol{\omega}) = \beta \nabla \times [\boldsymbol{g}(T - T_0)] + \nu \nabla^2 \boldsymbol{\omega}. \tag{2.4.30}$$

Note that linear operators are commutative and that the curl of a gradient of a scalar is identically zero.* Hence, pressure does not appear in the governing equations of motion. This is one advantage of studying vorticity transport. Applying the vector identity

$$\nabla \times (\boldsymbol{A} \times \boldsymbol{B}) = \boldsymbol{B} \cdot \nabla \boldsymbol{A} - \boldsymbol{A} \cdot \nabla \boldsymbol{B} - \boldsymbol{B}(\nabla \cdot \boldsymbol{A}) + \boldsymbol{A}(\nabla \cdot \boldsymbol{B})$$

to the second term on the left of Eq. (2.4.30), with $\boldsymbol{A} = \boldsymbol{V}$ and $\boldsymbol{B} = \boldsymbol{\omega} = \nabla \times \boldsymbol{V}$, noting

*The contribution from any body force of form $\boldsymbol{f} = -\nabla\Psi$, if included in Eq. (2.4.29), would vanish in this operation.

$\nabla \cdot V = 0$ by Eq. (2.4.22) and $\nabla \cdot (\nabla \times V) = 0$ by identity, we obtain the *vorticity transport equation*

$$\frac{D\boldsymbol{\omega}}{Dt} = \boldsymbol{\omega} \cdot \nabla V + \nu \nabla^2 \boldsymbol{\omega} + \beta \nabla \times [g(T - T_0)], \tag{2.4.31}$$

or, in indical notation,

$$\underbrace{\frac{\partial \omega_l}{\partial t} + V_k \frac{\partial \omega_l}{\partial x_k}}_{\text{change}} = \underbrace{\omega_k \frac{\partial V_l}{\partial x_k}}_{\text{stretching}} + \underbrace{\nu \frac{\partial^2 \omega_l}{\partial x_k \, \partial x_k}}_{\text{diffusion}} + \underbrace{\beta \epsilon_{lij} \frac{\partial}{\partial x_i}[g_j(T - T_0)]}_{\text{buoyancy source}}. \tag{2.4.32}$$

Accordingly, the material change of a given component of vorticity is composed of the vortex stretching, viscous diffusion, and buoyancy effects. Buoyancy contributes a volume source of vorticity perpendicular to the plane formed by the temperature gradient ∇T and the body force g. Stretching can involve contributions from all components of vorticity for a three-dimensional straining fluid motion. Replacing the rate-of-displacement tensor ∇V by $S + r$ from Eqs. (1.5.4) and (1.5.5) and recalling Eq. (1.5.6), the stretching term reduces to $\boldsymbol{\omega} \cdot S$. This is readily interpreted as the rates involving the rotation of deformation, or—equivalently—the deformation of rotation. To be specific, consider the explicit contributions to the changes in the x-component of vorticity,*

$$\omega_x \frac{\partial u}{\partial x} + \omega_y \frac{\partial u}{\partial y} + \omega_z \frac{\partial u}{\partial z},$$

where

$$\omega_x = \frac{\partial w}{\partial y} - \frac{\partial v}{\partial z}; \qquad \omega_y = \frac{\partial u}{\partial z} - \frac{\partial w}{\partial x}; \qquad \omega_z = \frac{\partial v}{\partial x} - \frac{\partial u}{\partial y}.$$

The term $\omega_x \, \partial u/\partial x$ would contibute to increase ω_x, for example, in a swirling flow ($\omega_x > 0$) along a channel having a decreasing area in the direction of flow ($\partial u/\partial x > 0$, by the conservation of mass). The contributions from the terms $\omega_y \, \partial u/\partial y + \omega_z \, \partial u/\partial z$, which may be reduced to

$$-\frac{\partial w}{\partial x}\frac{\partial u}{\partial y} + \frac{\partial v}{\partial x}\frac{\partial u}{\partial z},$$

are more difficult to visualize. They involve shear deformation of rotation perpendicular to x. Terms such as these are effective, for example, in the development of secondary flows (streamwise vorticity) in curved channels. These secondary flows may be alternatively explained by the fact that fluid particles near the core of the flow have higher velocities than those near the wall, hence are subject to a larger centrifugal acceleration in the bend.

The vortex stretching may also be viewed as the deformation of a *vortex tube.* A vortex tube is formed by the *vortex lines* through a closed path C (Fig. 2.9(a)), where a vortex line is defined as a curve having $\boldsymbol{\omega}$ as tangent at every point. The identity $\nabla \cdot \boldsymbol{\omega} = 0$ expresses that the net vector flux of vorticity leaving a unit volume is

*Here u, v, w denote the x, y, z-component, respectively, of velocity.

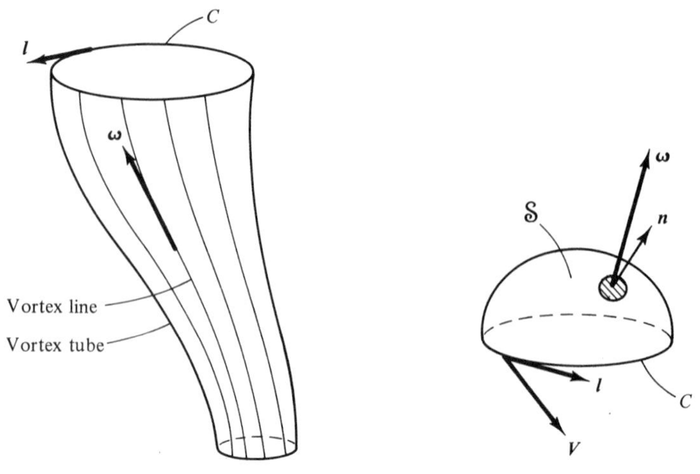

(a) Vortex lines through path C forming a vortex tube

(b) The Stokes theorem

Fig. 2.9 Vortex tube

zero. Integrating this identity over the surface of a portion of a vortex tube (Fig. 2.9(a)) shows that at any instant the mean intensity of vorticity of the vortex tube varies inversely with its cross-sectional area. Therefore, a nonuniform fluid motion that reduces the area of a straight vortex tube, by stretching it, causes an increase of mean vorticity within the tube. Also, a nonuniform fluid motion that bends an originally straight vortex tube contributes through deformation to vorticity components in directions other than that of the original tube axis.

In fluid mechanics we define the circulation Γ_C along the closed path C and show, by the Stokes theorem, that it equals the total vorticity in the vortex tube (Fig. 2.9(b)),

$$\Gamma_C = \oint_C \boldsymbol{V} \cdot d\boldsymbol{l} = \int_{\mathcal{S}} (\nabla \times \boldsymbol{V}) \cdot \boldsymbol{n}\, d\mathcal{S} = \int_{\mathcal{S}} \boldsymbol{\omega} \cdot \boldsymbol{n}\, d\mathcal{S}. \tag{2.4.33}$$

Furthermore, according to the Kelvin theorem, for an inviscid fluid without buoyancy effects, the circulation about any closed path C moving with the fluid remains constant,

$$\frac{D\Gamma_C}{Dt} = 0. \tag{2.4.34}$$

Actually, projection of Eq. (2.4.31) on a vortex line, $\boldsymbol{e} = \boldsymbol{\omega}/|\boldsymbol{\omega}|$, ignoring the last two terms in the equation, gives the Helmholtz theorem, $D\omega_e/Dt = 0$, which may be integrated to give Eq. (2.4.34).

These theorems describe the permanency of vortex lines and vortex tubes in inviscid flows without sources of vorticity. The vortex tube cannot begin or end in the bulk of the fluid. It can translate, bend, stretch, shorten, or otherwise deform while its circulation remains constant. Specifically, if once zero, $\boldsymbol{\omega}$ remains zero, which is the Lagrange theorem. Finally, the cross-sectional reduction of a vortex tube to zero, keeping its circulation Γ_C constant, leads to the idealized line vortex, a singularity used both in classical potential theory (see, for example, Milne-Thomson 1967 or Robertson 1965) and in modern discrete vortex methods (see, for example, Chow 1979 for an introduction, and Chorin 1973 and 1978 for viscous flow simulations).

For *two-dimensional flow*, say parallel to the (x, y)-plane, we have $\omega_x = \omega_y = 0$, $w = 0$, so vortex stretching dissappears, and Eq. (2.4.31) reduces to the z-component, say $\omega_z = \omega$,

$$\frac{D\omega}{Dt} = \nu \nabla^2\omega + \beta\left[\frac{\partial g_y (T - T_0)}{\partial x} - \frac{\partial g_x (T - T_0)}{\partial y}\right]. \tag{2.4.35}$$

Furthermore, we may replace the two dependent variables u and v of a two-dimensional flow by a single variable, the *stream function* ψ, defined by

$$u = \frac{\partial \psi}{\partial y}, \qquad v = -\frac{\partial \psi}{\partial x}, \tag{2.4.36}$$

so that the conservation of mass is satisfied identically.

The equation $\psi(x, y, t) =$ constant describes, in the (x, y)-plane, a *streamline* to which the velocity is tangent at every point and time. Integrating Eq. (2.4.36) along a path joining, say, points A and B,

$$\psi_B - \psi_A = \int (u\, dy - v\, dx), \tag{2.4.37}$$

shows that the change in the stream function between two points is equal to the volume flow crossing the path connecting these points. Hence, sketching streamlines corresponding to equidistant values of ψ gives a useful picture of a flow, not only in terms of the orientation of the velocity vector, being everywhere tangent to streamlines, but also in terms of the magnitude of the velocity, being everywhere inversely proportional to the spacing between streamlines. In steady flow, streamlines are fixed in space and coincide with path lines followed by fluid particles.

Introducing Eqs. (2.4.36) into $\omega = \partial v/\partial x - \partial u/\partial y$ gives the *ω-equation*,

$$\nabla^2\psi + \omega = 0, \tag{2.4.38}$$

and the equations of Table 2.5 reduce—for two-dimensional convection—to those of Table 2.6. In this way, after eliminating pressure, three governing equations in usual (primitive) variables u, v, p are reduced to two equations in alternative variables ψ, ω. From a solution in terms of ψ, ω we may obtain variables u, v by integrating Eqs. (2.4.36) and (if needed) p by integrating the divergence of the momentum balance which, by use of the conservation of mass, may be written as

$$2\left[\left(\frac{\partial^2\psi}{\partial x\, \partial y}\right)^2 - \frac{\partial^2\psi}{\partial x^2}\frac{\partial^2\psi}{\partial y^2}\right] = -\frac{1}{\rho_0}\left(\frac{\partial^2 p}{\partial x^2} + \frac{\partial^2 p}{\partial y^2}\right) + \beta\left[\frac{\partial g_x (T - T_0)}{\partial x} + \frac{\partial g_y (T - T_0)}{\partial y}\right]. \tag{2.4.42}$$

Instead of the (ψ, ω)-formulation, which often proves convenient in computational convention (Chapter 7), we may eliminate ω between Eqs. (2.4.39) and (2.4.40)

TABLE 2.6 (ψ, ω)-Formulation of two-dimensional incompressible, constant-property convection

$$\frac{\partial^2 \psi}{\partial x^2} + \frac{\partial^2 \psi}{\partial y^2} + \omega = 0 \tag{2.4.39}$$

$$\frac{\partial \omega}{\partial t} + \frac{\partial \psi}{\partial y}\frac{\partial \omega}{\partial x} - \frac{\partial \psi}{\partial x}\frac{\partial \omega}{\partial y} = \nu\left(\frac{\partial^2 \omega}{\partial x^2} + \frac{\partial^2 \omega}{\partial y^2}\right) + \beta\left[\frac{\partial g_y(T - T_0)}{\partial x} - \frac{\partial g_x(T - T_0)}{\partial y}\right] \tag{2.4.40}$$

$$\frac{\partial T}{\partial t} + \frac{\partial \psi}{\partial y}\frac{\partial T}{\partial x} - \frac{\partial \psi}{\partial x}\frac{\partial T}{\partial y} = a\left(\frac{\partial^2 T}{\partial x^2} + \frac{\partial^2 T}{\partial y^2}\right) + \frac{\nu}{c_p}\Phi_v + \frac{u'''}{\rho c_p} \tag{2.4.41}$$

to obtain the *equation for the stream function*,

$$\frac{\partial}{\partial t}(\nabla^2 \psi) + \frac{\partial \psi}{\partial y}\frac{\partial \nabla^2 \psi}{\partial x} - \frac{\partial \psi}{\partial x}\frac{\partial \nabla^2 \psi}{\partial y} = \nu \nabla^4 \psi - \beta\left[\frac{\partial g_y (T - T_0)}{\partial x} - \frac{\partial g_x (T - T_0)}{\partial y}\right], \tag{2.4.43}$$

where the Laplacian and biharmonic operators are

$$\nabla^2 \psi = \frac{\partial^2 \psi}{\partial x^2} + \frac{\partial^2 \psi}{\partial y^2}; \qquad \nabla^4 \psi = \frac{\partial^4 \psi}{\partial x^4} + 2\frac{\partial^4 \psi}{\partial x^2 \partial y^2} + \frac{\partial^4 \psi}{\partial y^4}. \tag{2.4.44}$$

Equation (2.4.43), involving a single dependent variable, is a convenient form for the solution of some two-dimensional viscous flows.

Having completed our study of vorticity transport, we now turn to flows of negligible vorticity.

2.4.4 Potential Flows

In some regions of convective flows, for example, in regions far from a body in a uniform flow, the vorticity is negligible. Then the flow may be treated as irrorational,

$$\boldsymbol{\omega} = \nabla \times \boldsymbol{V} = 0. \tag{2.4.45}$$

This condition implies the existence of a scalar function φ, the *velocity potential*, defined by

$$\boldsymbol{V} = \nabla \varphi. \tag{2.4.46}$$

Conversely, a potential flow, prescribed by Eq. (2.4.46), is irrotational.

In a manner similar to streamlines, we may define *equipotential lines* by the equation $\varphi(x, y, z, t) = \text{constant}$. For two-dimensional flow equipotential lines are everywhere (except at points for which $\boldsymbol{V} = 0$) orthogonal to streamlines, $\nabla \varphi \cdot \nabla \psi = 0$, which follows from Eqs. (2.4.46) and (2.4.36).

Now, reexamine the momentum balance for incompressible, constant property flow, Eq. (2.4.29). Assume buoyancy effects to be absent,* body froces to be conservative,† $\boldsymbol{f} = -\nabla\Psi$, and use the vector identity

$$\nabla^2 \boldsymbol{V} = \nabla(\nabla \cdot \boldsymbol{V}) - \nabla \times \boldsymbol{\omega} = -\nabla \times \boldsymbol{\omega}.$$

Then Eq. (2.4.29) may be rearranged as

$$\frac{\partial \boldsymbol{V}}{\partial t} + \nabla\left(\frac{V^2}{2} + \frac{p}{\rho} + \Psi\right) = \boldsymbol{V} \times \boldsymbol{\omega} - \nu \nabla \times \boldsymbol{\omega}. \tag{2.4.47}$$

For inviscid flow ($\nu = 0$) this reduces to the *Euler equations*

$$\frac{\partial \boldsymbol{V}}{\partial t} + \nabla\left(\frac{V^2}{2} + \frac{p}{\rho} + \Psi\right) = \boldsymbol{V} \times \boldsymbol{\omega}, \tag{2.4.48}$$

and for irrotational flow to the (unsteady) *Bernoulli equation*

$$\frac{\partial \varphi}{\partial t} + \frac{1}{2}(\nabla\varphi)^2 + \frac{p}{\rho} + \Psi = \text{constant}. \tag{2.4.49}$$

Finally, inserting Eq. (2.4.46) into the conservation of mass, $\nabla \cdot \boldsymbol{V} = 0$, gives

$$\nabla^2 \varphi = 0. \tag{2.4.50}$$

The solution of a potential flow (see, for example, Milne-Thomson 1967, Robertson 1965 and Chow 1979) amounts to determine φ from Eq. (2.4.50). Then the pressure distribution may be calculated from Eq. (2.4.49), which holds throughout a potential flow. Note that for inviscid flow, even if rotational, Eq. (2.4.49) holds along a streamline, which is seen by projecting Eq. (2.4.48) onto $\boldsymbol{l} = \boldsymbol{V}/|\boldsymbol{V}|$.

It is important to note that the potential flow theory may apply to certain regions of high-Reynolds-number flows of real fluids which are in fact viscous. Consider, for example, a body (say a cylinder of diameter D) which moves with velocity U through a fluid at rest, or—equivalently—uniform flow past a body at rest. Because of the smallness of the kinematic viscosity of fluids like air or water (Appendix E) the Reynolds number for most flows of practical importance is so large (say 10^3 or 10^4 for air or water at the even moderate values of $U = 1$ m/s, $D = 0.01$ m) that viscous effects should be negligible. This is, in fact, the case for all flow regions away from and upstream of the body. However, close to the surfaces (in the boundary layer and in possible separated flow regions) as well as downstream (in the wake) of the body, potential flow theory fails. It fails because a real fluid must satisfy the no-slip boundary condition at the surface, no matter how small the viscosity of the fluid. This boundary condition cannot be satisfied by the inviscid equations.

Here, a final remark on the vorticity transport may be in order. Consider a body set in motion through a fluid which is initially at rest, hence initially at zero vorticity everywhere. Once motion has started the vorticity generated at surfaces diffuses into

*If present, they constitute a volume source of vorticity (Section 2.4.3).

†Such as gravity, for which the body force potential Ψ equals gz, when $\boldsymbol{f} = (0, 0, -g)$.

the fluid, which sweeps it downsteam. The greater the Reynolds number, the more convection will dominate diffusion; consequently, the thinner will be the layer near the forward facing surfaces of the body to which vorticity (hence viscous effects) will be confined. Behind the body the vorticity fills the entire wake region. Note that the foregoing discussion on the external flow applies equally well to the entrance region of the flow in a channel, which is a fundamental problem of convection.

REFERENCES AND SUPPLEMENTARY READING

General principles

ARPACI, V. S., 1966, *Conduction Heat Transfer*, Addison-Wesley, Reading, Mass.

BIRD, R. B., STEWART, W. E., & LIGHTFOOT, E. N., 1960, *Transport Phenomena*, Wiley, New York.

PRANDTL, L., & TIETJENS, O. G., 1957, *Fundamentals of Hydro- and Aerodynamics*, Dover, New York.

SHAPIRO, A. H., 1953, *The Dynamics and Thermodynamics of Compressible Flow*, Ronald Press, New York.

Continuum Mechanics

ARIS, R., 1962, *Vectors, Tensors and the Basic Equations of Fluid Dynamics*, Prentice-Hall, Englewood Cliffs, N.J.

DAHLER, J. S., & SCRIVEN, L. E., 1963, Proc. R. Soc. A, **275**, 504.

ERINGEN, A. C., 1967 *Mechanics of Continua*, Wiley, New York.

PRAGER, W., 1961, *Introduction to Mechanics of Continua*, Ginn, Lexington, Mass.

TRUESDELL, C., & TOUPIN, R., 1960, The Classical Field Theories, in *Handbuch der Physik*, Vol. III, Pt. 1 (ed. S. Flügge), Springer-Verlag, New York.

Fluid Mechanics

BATCHELOR, G. K., 1967, *Fluid Dynamics*, Cambridge University Press, New York.

CHORIN, A. J., 1973, J. Fluid Mech., **57**, 785–796.

CHORIN, A. J., 1978, J. Comp. Phys., **27**, 428.

CHOW, C.-Y., 1979, *An Introduction to Computational Fluid Mechanics*, Wiley, New York.

HUGHES, W. F., & GAYLORD, E. W., 1964, *Basic Equations of Engineering Science*, Schaum McGraw-Hill, New York.

LAMB, H., 1932, *Hydrodynamics*, Dover, New York.

LANDAU, L. D., & LIFSHITZ, E. M., 1959, *Fluid Mechanics*, Pergamon Press, Elmsford, N.Y.

LIGHTHILL, M. J., 1963, Part II of *Laminar Boundary Layers* (ed. L. Rosenhead), Oxford University Press, New York.

MILNE-THOMSON, L. M., 1967, *Theoretical Hydrodynamics*, 5th ed., Macmillan, New York.

ROBERTSON, J. M., 1965, *Hydrodynamics in Theory and Applications*, Prentice-Hall, Englewood Cliffs, N.J.

ROSENHEAD, L., 1963, *Laminar Boundary Layers*, Oxford University Press, New York.

SCHLICHTING, H., 1968, *Boundary Layer Theory*, 6th ed., McGraw-Hill, New York.

WHITE, F. M., 1974, *Viscous Fluid Flow*, McGraw-Hill, New York.

Natural Convection

ESHGHY, S., & MORRISON, JR., F. A., 1966, Proc. R. Soc. A, **293**, 395–407.

GEBHART, B., 1973, Natural Convection Flows and Stability, *Advances in Heat Transfer* (ed. T. F. Irvine, Jr., & J. P. Hartnett), Vol. 9, pp. 273–348, Academic Press, New York.

OSTRACH, S., 1972, Natural Convection in Enclosures, *Advances in Heat Transfer* (ed. T. F. Irvine, Jr., & J. P. Hartnett), Vol. 8, pp. 171–227, Academic Press, New York.

TURNER, J. S., 1973, *Buoyancy Effects in Fluids*, Cambridge University Press, New York.

Thermodynamics

CALLEN, H. B., 1960, *Thermodynamics*, Wiley, New York.

DEGROOT, S. R., & MAZUR, P., 1962, *Non-equilibrium Thermodynamics*, North-Holland, Amsterdam.

HAASE R., 1969, *Thermodynamics of Irreversible Processes*, Addison-Wesley, Reading, Mass.

HATSOPOULOS, G. N., & KEENAN, J. H., 1965, *Principles of General Thermodynamics*, Wiley, New York.

PENNER, S. S., 1968, *Thermodynamics*, Addison-Wesley, Reading, Mass.

REIF, F., 1965, *Statistical and Thermal Physics*, McGraw-Hill, New York.

SONNTAG, R. E., & VAN WYLEN, G. J., 1965, *Fundamentals of Classical Thermodynamics*, Wiley, New York.

VINCENTI, W. G., & KRUGER, C. H., 1967, *Introduction to Physical Gas Dynamics*, Wiley, New York.

Constitutive Relations and Thermophysical Properties

BATCHELOR, G. K., 1974, Annu. Rev. Fluid Mech., **6**, 227–255.

BIRD, R. B., WARNER, H. R., & EVANS, D. C., 1971, Adv. Polym. Phys., **8**, 1.

BRETSZNAJDER, S., 1972, *Prediction of Transport and Other Physical Properties of Fluids*, Pergamon Press, Elmsford, N.Y.

COLEMAN, B. D., & NOLL, W., 1971, Rev. Mod. Phys., **33**, 239.

FREDRICKSON, A. G., 1964, *Principles and Applications of Rheology*, Prentice-Hall, Englewood Cliffs, N.J.

GREEN, A. E., & RIVLIN, R. S., 1960, Arch. Rat. Mech. Anal., **4**, 387.

JAUMANN, G., 1911, Sitzungsber. Akad. Wiss. Wien, IIa, **120**, 385.

METZNER, A. B., & WHITE, J. L., 1965, AIChE J., **11**, 989.

OLDROYD, J. G., 1950, Proc. R. Soc. A, **200**, 523.

OLDROYD, J. G., 1958, Proc. R. Soc. A, **245**, 278.

REID, R. C., & SHERWOOD, T. K., 1966, *The Properties of Gases and Liquids*, 2nd ed., McGraw-Hill, New York.

RIVLIN, R. S., & ERICKSEN, J. L., 1955, J. Rat. Mech. Anal., **4**, 323.

WILLIAMS, M. C., & BIRD, R. B., 1962, AIChE J., **8**, 378.

PROBLEMS

2-1 Formulate the mass conservation and the momentum balance for the motion in a shallow water layer of varying thickness.

2-2 Repeat the steps of Section 2.2.2 for a medium having reversible components of deviatoric stress τ_{ij}^R. Specifically, derive the equivalents of Eqs. (2.2.4), (2.2.19), (2.2.21), and (2.2.24).

2-3 Sketch, in a τ_{yx} versus $\partial u/\partial y$ plot, the rheological behavior of fluids that are Newtonian, pseudo-plastic ($n < 1$ in Eq. 2.3.4), and dilatant ($n > 1$ in Eq. 2.3.4).

2-4 Consider fully developed flow between parallel plates emerging to form a free, plane jet. Determine the ratio of downstream jet width to plate spacing for:
(a) A Newton fluid.
(b) A power-law fluid.

2-5 Derive the equations of Table 2.3 from those of Table 2.2.

2-6 Derive Eq. (2.4.43).

2-7 Show that the curl of a gradient of a scalar, as well as the divergence of the curl of a vector, is identically zero.

2-8 Discuss the differences between Eqs. (2.2.21) and (2.2.24), say for liquids and for ideal gases. In the latter case, noting the significant difference between c_v and c_p, which specific heat should be used in the Prandtl number of a convection problem?

2-9 Derive the Bernoulli equation along a streamline.

2-10 Enumerate the assumptions required to derive the (steady) Bernoulli equation, both from the balance of total energy and from the balance of linear momentum. Discuss these assumptions and verify that they are mutually compatible.

2-11 Verify that the velocity vector of two-dimensional flow is tangent to a streamline.

2-12 Verify that $\psi = 2xy$ describes the potential flow normal to a plane along the x-axis, having a stagnation point at the origin. Determine the velocity potential $\varphi(x, y)$ and show that equipotential lines are orthogonal to streamlines, except at the stagnation point.

2-13 Write the explicit contributions from the vorticity stretching term to changes in, say, the x-component of vorticity. Discuss longitudinal deformation (swirl flow in contracting channel) and shear deformation (development of secondary flow in curved channel, such as a river bend).

Part II

INDIVIDUAL FORMULATION (LAMINAR FLOWS)

In Part I we described the general formulation, starting with the integral form of general laws and concluding with the differential form of these laws for Newton–Fourier fluids. We might now be expected to deduce the formulation of each specific problem from the general formulation. This, of course, is possible and often done, but it is not always convenient, especially if a formulation to be lumped in one or more directions is desired (the last statement will be clarified by the problems of Fig. II(c)). The application of the general formulation to a specific problem is a mathematical deduction process which usually overshadows the physical aspects of the problem. By contrast, we may consider an *individual approach* which treats each problem separately with special emphasis on the physical significance of each term involved in its formulation. To illustrate how the formulation of a specific problem can be deduced from a general formulation or can be developed individually, we consider the following three problems requiring the one-dimensional Cartesian formulation of the conservation of mass.

The first problem deals with a parallel flow of infinite extent. Equating the rate of change of mass to the net flow of mass through boundaries of the differential control volume shown in Fig. II(a), the individual approach yields

$$\frac{\partial \rho}{\partial t} + \frac{\partial}{\partial x}(\rho u) = 0. \tag{II.1}$$

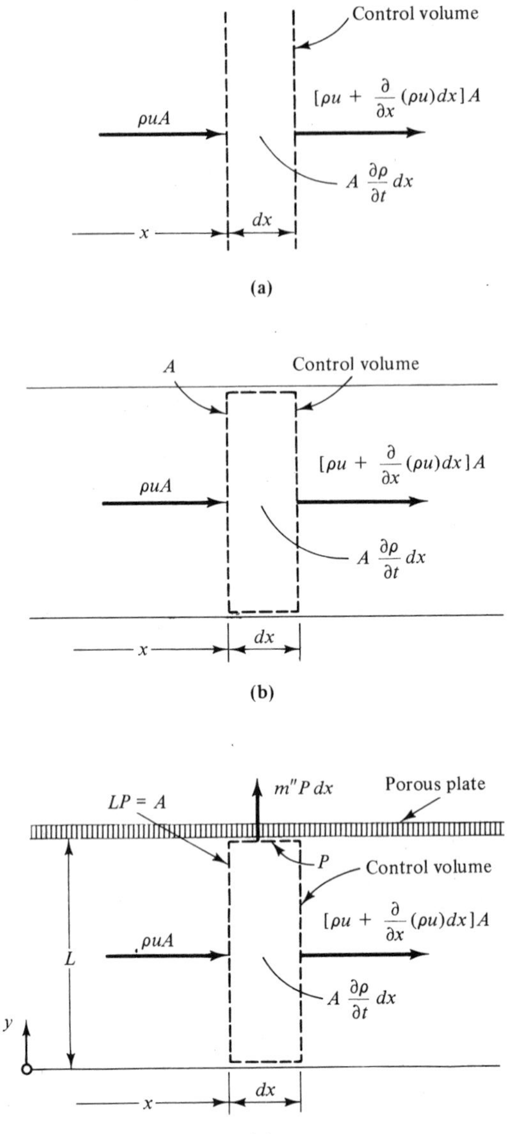

Fig. II Conservation of mass: individual formulation

The general formulation, reduced to the one-dimensional Cartesian form of Eq. (2.4.1), readily gives the same result.

Next, consider the flow between two parallel plates separated a distance L apart. Following the individual approach in terms of the mixed (transversally lumped and longitudinally differential) control volume of Fig. II(b), or employing again the one-dimensional Cartesian form of Eq. (2.4.1), we end up with the result of the preceding problem, Eq. (II.1).

Finally, assume one of the plates of the foregoing problem to be porous. By the individual treatment, applying the conservation of mass to the mixed control volume shown in Fig. II(c), we obtain

$$\frac{\partial \rho}{\partial t} + \frac{\partial}{\partial x}(\rho u) + \frac{m''}{L} = 0, \tag{II.2}$$

where m'' denotes the mass flow through unit area of the porous plate. By contrast, saying that we were asked for the one-dimensional formulation of the problem and hastily write the one-dimensional form of Eq. (2.4.1), again we arrive in Eq. (II.1) which discounts the mass flow through the porous plate. This difficulty, however, may be circumvented by considering instead the two-dimensional form of Eq. (2.4.1),

$$\frac{\partial \rho^*}{\partial t} + \frac{\partial}{\partial x}(\rho^* u^*) + \frac{\partial}{\partial y}(\rho^* v^*) = 0, \tag{II.3}$$

where ρ^*, u^*, and v^* now depend on y as well as x and t. Averaging Eq. (II.3) transversally, that is, multiplying each term by dy and integrating the result over L, yields

$$L\frac{\partial \rho}{\partial t} + L\frac{\partial}{\partial x}(\rho u) + \rho^* v^* |_L = 0, \tag{II.4}$$

which, in view of $m'' = \rho^* v^* |_L$, is identical to Eq. (II.2). Here the transversal averaging of a quantity, say $\rho^* u^*$, is defined as

$$\rho(x, t)u(x, t) = \frac{1}{L}\int_0^L \rho^*(x, y, t)u^*(x, y, t)\, dy.$$

Two methods of formulation demonstrated in terms of the foregoing three problems may now be summarized as follows: Whenever the general formulation is available, the formulation of a specific problem may be deduced from the interpretation of the general formulation in terms of the coordinates of the specific problem. By contrast, the individual approach involves the application of the steps leading to the general formulation directly to the specific problem. For one- or multidimensional problems to be formulated including all dimensions of the problem (such as the first two of the foregoing examples), the deduction from the general formulation proves to be shorter than the individual approach. However, for multidimensional problems which we may wish to formulate in fewer dimensions, that is, which we may wish to lump in one or more directions (such as the third example), the deduction process, requiring an averaging, becomes lengthy and inconvenient.

The foregoing discussion suggests that, for many engineering applications, the individual approach be the preferred method of formulation. For convenience and later reference, this method of formulation is summarized by the following *six steps*:

1. *Define a system or control volume.* This step includes the selection of (a) a lumped,* distributed, or mixed formulation; (b) a coordinate system for the latter two; and (c) a system or control volume in terms of (a) and (b).

*The lumped formulation often oversimplifies the convection problems.

2. *Construct general principles individually for (1).* The differential form of these principles depends on the direction but not the origin of the coordinates, whereas the integral form of them depends on the origin as well as the direction of coordinates. Although the differential form is local, namely, it is written for a point of the problem, the lumped and integral forms are global, that is, they are stated for the entire problem.
3. *Construct the thermodynamic relation by combining the principles of the balance of thermal energy and the increase of entropy with the assumption of local equilibrium.* This relation helps to express the balance of thermal energy in terms of temperature and the entropy production in terms of irreversibilities (related to the diffusion of momentum and heat).
4. *State constitutive relations for (2).* These relations are either local or directional. The local relations (referred to as the equations of state) are independent of coordinates, whereas the directional relations (referred to as the laws of diffusion) depend on the direction but not the origin of coordinates.
5. *Obtain governing equations from (2) to (4).* These equations, depending on our choice for (1), are lumped, distributed, or mixed. They involve the desired quantities (dependent variables) as the only unknowns. The governing equations, except for their flow terms, are independent of the origin and direction of coordinates.
6. *Specify the initial and/or boundary conditions for (5).* The initial condition is global and is valid only for an instant (which is most commonly assumed to be the start of the problem), whereas the boundary conditions are local and are valid for all time. Both conditions depend on the origin as well as the direction of coordinates.

Having molded the individual formulation into six steps, we proceed now to a variety of illustrative examples to be formulated in light of these steps. Chapter 3 deals with problems associated with parallel flows and Chapter 4 with those of nearly parallel (boundary layer) flows. Although the emphasis in these chapters will be placed on formulation, we shall also carry the solution whenever possible. Throughout, we confine attention to incompressible flows of constant property Newton–Fourier fluids. Moreover, we consider examples in the Cartesian geometry. Other cases are left to problems. The six steps of formulation are explicitly emphasized in early examples. Later, for brevity, they are used tacidly.

Chapter 3

PARALLEL FLOWS

This chapter is devoted to a number of examples associated with laminar incompressible parallel flows. First. however, we study some common features of these flows.

3.1 TWO-DIMENSIONAL PARALLEL FLOWS

Consider an unsteady, incompressible parallel flow of infinite extent. We wish to describe this flow, and the related temperature field, following the six steps of the individual approach outlined in the introduction to Part II. Here we are concerned only with the differential formulation, leaving the integral formulation to specific problems to be discussed in the following sections.

Let the flow be in the x-direction. Then, by definition, $u = u(y, z, t)$, $v = w = 0$, and $\partial/\partial x \equiv 0$ except for $-\partial p/\partial x$. Assume the temperature field to be three-dimensional, $T = T(x, y, z, t)$. In terms of the Cartesian coordinates and the differential control volume shown in Fig. 3.1, the general principles are:

The conservation of mass (Fig. 3.1)

$$\frac{\partial u}{\partial x} = 0. \qquad (3.1.1)$$

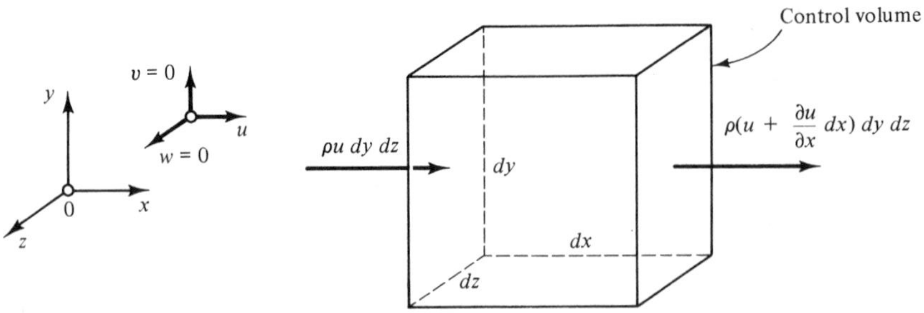

Fig. 3.1 Coordinate system, velocity components, control volume, and conservation of mass

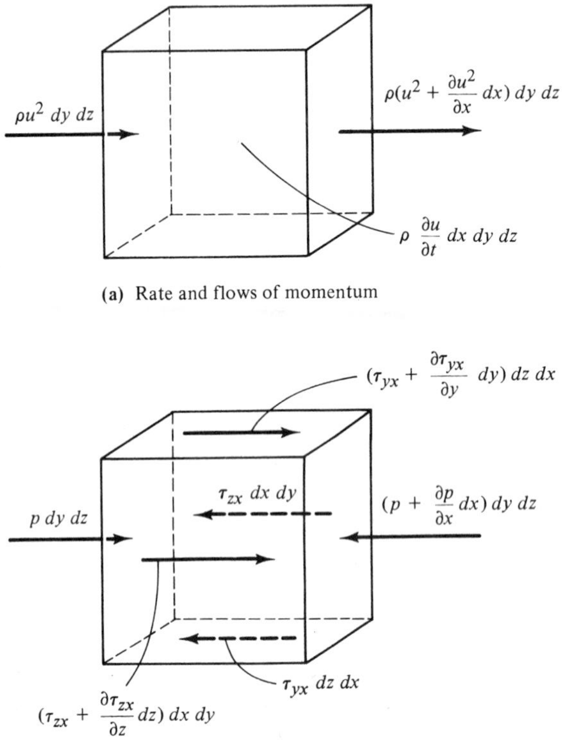

(a) Rate and flows of momentum

(b) Fluxes of momentum

Fig. 3.2 Balance of momentum

The balance of momentum* (Fig. 3.2)

$$\rho \frac{\partial u}{\partial t} = -\frac{\partial p}{\partial x} + \frac{\partial \tau_{yx}}{\partial y} + \frac{\partial \tau_{zx}}{\partial z}, \tag{3.1.2}$$

*Note that the change in momentum flow, not the momentum flow itself, is zero.

where p and τ_{yx}, τ_{zx} denote the pressure and the components of shear stress, respectively. The body force ρf_x is here ignored* but its effects will be considered in Section 3.4.

The balance of mechanical energy, obtained multiplying Eq. (3.1.2) by u,

$$\rho \frac{\partial}{\partial t}\left(\frac{1}{2}u^2\right) = \left(-\frac{\partial p}{\partial x} + \frac{\partial \tau_{yx}}{\partial y} + \frac{\partial \tau_{zx}}{\partial z}\right)u, \tag{3.1.3}$$

states the balance of kinetic energy with the power of net stress.

The conservation of total (mechanical + thermal) energy (Fig. 3.3)

$$\rho \frac{De}{Dt} = -\nabla \cdot \boldsymbol{q} - \frac{\partial}{\partial x}(pu) + \frac{\partial}{\partial y}(\tau_{yx}u) + \frac{\partial}{\partial z}(\tau_{zx}u) + u''', \tag{3.1.4}$$

where $e = u_0 + u^2/2$ denotes the stagnation (internal) energy per unit mass, u_0 the internal energy per unit mass, $\boldsymbol{q}$ the heat flux, and u''' the energy generation per unit volume. (Which terms in Eq. 3.1.4 are identically zero?)

The balance of thermal energy, obtained subtracting Eq. (3.1.3) from Eq. (3.1.4),

$$\rho \frac{Du_0}{Dt} = -\nabla \cdot \boldsymbol{q} + \tau_{yx}\frac{\partial u}{\partial y} + \tau_{zx}\frac{\partial u}{\partial z} + u''', \tag{3.1.5}$$

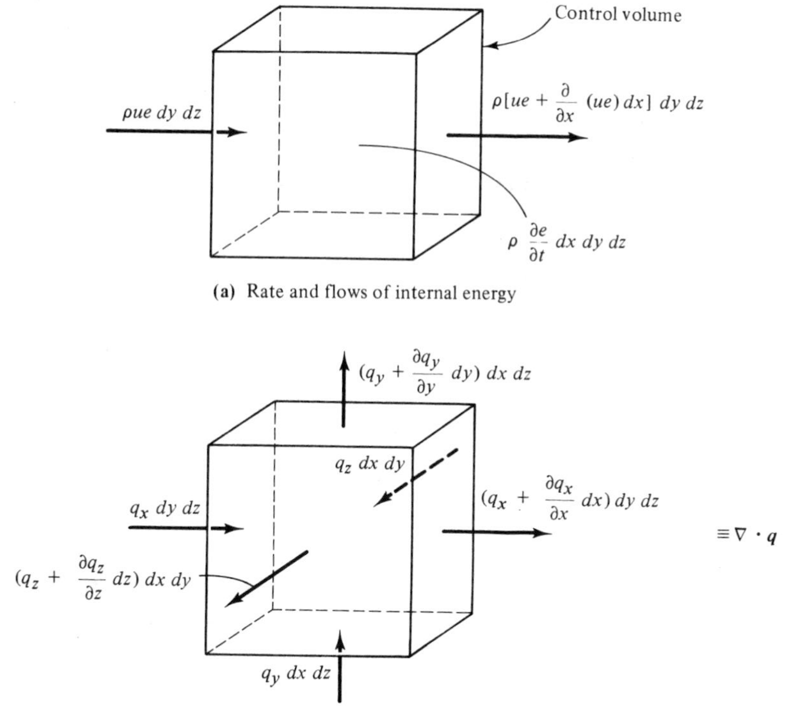

(a) Rate and flows of internal energy

(b) Fluxes of heat

Fig. 3.3 Conservation of total energy

*This point is elaborated in the discussion following Eq. (4.1.28).

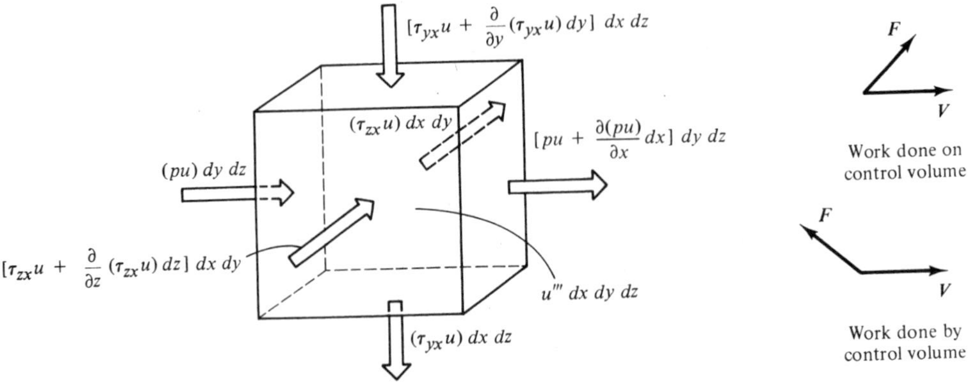

(c) Fluxes of power

Fig. 3.3 (cont.)

where the terms related to shear stress denote the irreversible conversion of mechanical energy into thermal energy. (What happened to the reversible conversion?)

The entropy production (Fig. 3.4)

$$\rho \frac{Ds}{Dt} = -\nabla \cdot \left(\frac{\mathbf{q}}{T}\right) + s''', \tag{3.1.6}$$

where s denotes the entropy per unit mass, $\boldsymbol{q}/T$ the entropy flux, and s''' the entropy production per unit volume. Having stated the general principle, we proceed now to the thermodynamic relation.

Invoking the local equilibrium and assuming the transfer of heat to be reversible, Eqs. (3.1.5) and (3.1.6) may be reduced to

$$\rho \frac{Du_0}{Dt} = -(\nabla \cdot \boldsymbol{q})^R, \tag{3.1.5a}$$

$$\rho \frac{Ds}{Dt} = -\frac{1}{T}(\nabla \cdot \boldsymbol{q})^R. \tag{3.1.6a}$$

Eliminating between Eqs. (3.1.5a) and (3.1.6a) the terms related to heat flux gives the thermodynamic relation,

$$\rho \frac{Du_0}{Dt} = \rho T \frac{Ds}{Dt}. \tag{3.1.7}$$

In Chapter 2 we learned how this relation becomes indispensable in the process of expressing (1) the entropy production in terms of irreversibilities, and (2) the internal energy of compressible fluids in terms of the temperature.

Inserting Eqs. (3.1.5) and (3.1.6) into Eq. (3.1.7) and rearranging yields the entropy production in parallel flows,

$$s''' = -\frac{1}{T}\left(\frac{\boldsymbol{q}}{T} \cdot \nabla T\right) + \frac{1}{T}\left(\tau_{yx}\frac{\partial u}{\partial y} + \tau_{zx}\frac{\partial u}{\partial z}\right) + \frac{u'''}{T}, \tag{3.1.8}$$

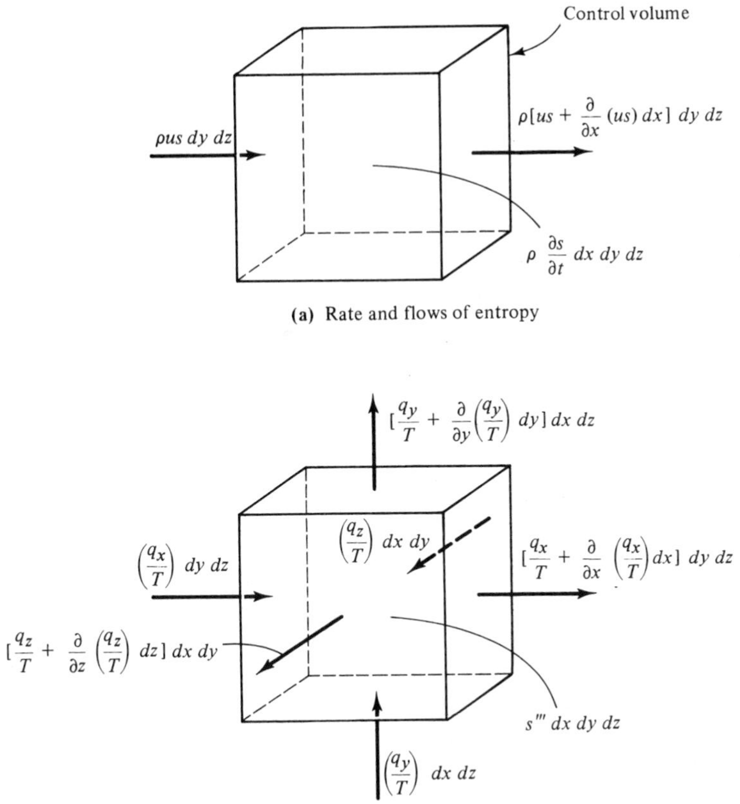

(a) Rate and flows of entropy

(b) Fluxes and production of entopy

Fig. 3.4 Entropy production

where the first and second terms denote irreversibilities associated with the diffusion of heat and momentum,* respectively, and the third one the irreversibility associated with other effects that are interpreted as an energy generation.

So far, following the first three steps of the individual approach, we are able to carry the differential formulation of parallel flows for any media. To proceed further, that is, to express the formulation of these flows only in terms of u and T, we must specify the constitution of media. Under the assumption of local equilibrium, $u_0 = u_0(T)$ for incompressible media, and

$$du_0 = \left(\frac{\partial u_0}{\partial T}\right)_v dT = c_v \, dT, \tag{3.1.9}$$

where c_v denotes the usual definition of the specific heat at constant volume. Inserting

*The irreversibility related to the diffusion of momentum is frequently referred to as the dissipation.

Eq. (3.1.9) into Eq. (3.1.5), we may rearrange the thermal energy as

$$\rho c_v \frac{DT}{Dt} = -\nabla \cdot q + \left(\tau_{yx}\frac{\partial u}{\partial y} + \tau_{zx}\frac{\partial u}{\partial z}\right) + u'''. \tag{3.1.10}$$

Consider, for example, a Newton–Fourier fluid. Then the components of the shear stress not vanishing for parallel flows become

$$\tau_{yx} = \mu\frac{\partial u}{\partial y}, \qquad \tau_{zx} = \mu\frac{\partial u}{\partial z}, \tag{3.1.11}$$

and the heat flux

$$q = -k\,\nabla T. \tag{3.1.12}$$

Inserting Eqs. (3.1.11) and (3.1.12) appropriately into Eqs. (3.1.2) and (3.1.10), we obtain the equations governing parallel flows of a Newton–Fourier fluid,

$$\frac{1}{\nu}\frac{\partial u}{\partial t} = -\frac{1}{\mu}\frac{\partial p}{\partial x} + \left(\frac{\partial^2 u}{\partial y^2} + \frac{\partial^2 u}{\partial z^2}\right), \tag{3.1.13}$$

$$\frac{1}{a}\frac{DT}{Dt} = \nabla^2 T + \frac{\mu}{k}\left[\left(\frac{\partial u}{\partial y}\right)^2 + \left(\frac{\partial u}{\partial z}\right)^2\right] + \frac{u'''}{k}, \tag{3.1.14}$$

where $\nu = \mu/\rho$ denotes the momentum diffusivity (kinematic viscosity) and $a = k/\rho c_v$ the heat diffusivity, which are both assumed to be constant.

The initial and boundary conditions to be imposed on Eqs. (3.1.13) and (3.1.14) are left to specific problems to be considered in the following sections.

For parallel flows of a Newton–Fourier fluid, the entropy production becomes, after inserting Eqs. (3.1.11) and (3.1.12) into Eq. (3.1.8),

$$s''' = \frac{k}{T^2}(\nabla T)^2 + \frac{\mu}{T}\left[\left(\frac{\partial u}{\partial y}\right)^2 + \left(\frac{\partial u}{\partial z}\right)^2\right] + \frac{u'''}{T}, \tag{3.1.15}$$

which shows a positive contribution from each source of irreversibility. This result, in view of the principle of the increase of entropy, is expected. Having completed our discussion on the differential formulation of parallel flows, we proceed to a number of illustrative examples related to these flows.

3.2 "VOLUMETRIC RISE" VERSUS "PENETRATION DEPTH"

In this section we lay the foundations of integral formulation in terms of the concepts of *volumetric rise* and *penetration depth* as they are applied to three illustrative examples. The differential formulation of these examples may be directly taken from the preceeding section. However, the solution techniques available for differential formulation, such as the separation of variables, transform calculus, variational calculus, and so on, are the subject matter of texts on *boundary value problems*, and will not be explored in this text. However, whenever appropriate, solutions obtained by these techniques are given.

EXAMPLE 3.2.1 Consider a stagnant viscous fluid between two parallel horizontal plates separated a distance $2l$ apart. This fluid is suddenly pressurized by a constant pressure gradient. We wish to determine the unsteady velocity distribution of the fluid.

The differential formulation of the problem, its governing equation being the one-dimensional form of Eq. (3.1.13), is

$$\frac{1}{\nu}\frac{\partial u}{\partial t} = -\frac{1}{\mu}\frac{\partial p}{\partial x} + \frac{\partial^2 u}{\partial y^2}; \qquad u(y, 0) = 0,$$
$$\frac{\partial u(0, t)}{\partial y} = 0, \qquad u(l, t) = 0. \tag{3.2.1}$$

(Describe the physics of the conduction problem which leads to the same formulation.)

The foregoing formulation, as it stands, cannot be solved by the separation of variables because of the nonhomogeneity of the governing equation. However, the superposition

$$u(y, t) = \bar{u}(y) + u'(y, t)$$

eliminates this difficulty. Now, $\bar{u}$ and u' satisfy

$$\begin{aligned} 0 &= -\frac{1}{\mu}\frac{dp}{dx} + \frac{d^2\bar{u}}{dy^2}, & \frac{1}{\nu}\frac{\partial u'}{\partial t} &= \frac{\partial^2 u'}{\partial y^2}, \\ & & u'(y, 0) &= -\bar{u}(x), \\ 0 &= \frac{d\bar{u}(0)}{dy}, & \frac{\partial u'(0, t)}{\partial y} &= 0, \\ 0 &= \bar{u}(l), & u'(l, t) &= 0. \end{aligned} \tag{3.2.2}$$

The solution of the steady problem is the well-known Poiseuille flow,

$$\bar{u}(y) = \frac{-dp/dx}{2\mu}(l^2 - y^2). \tag{3.2.3}$$

The solution of the unsteady problem, obtained by the separation of variables, is

$$u'(y, t) = \sum_{n=0}^{\infty} a_n e^{-\lambda_n^2 \nu t} \cos \lambda_n y, \tag{3.2.4}$$

where $\lambda_n l = (2n + 1)\pi/2$, $n = 0, 1, 2, \ldots$, and

$$a_n = -2\frac{(-dp/dx)}{\mu}\frac{(-1)^n l^2}{(\lambda_n l)^3}.$$

The details of this solution are left to the interested reader (see also Ex. 5-4 of Arpaci 1966). Combining Eqs. (3.2.3) and (3.2.4), we obtain the unsteady velocity distribution,

$$\frac{u(y, t)}{(-dp/dx)l^2/\mu} = \frac{1}{2}\left[1 - \left(\frac{y}{l}\right)^2\right] - 2\sum_{n=0}^{\infty}\frac{(-1)^n}{(\lambda_n l)^3} e^{-\lambda_n^2 \nu t} \cos \lambda_n y. \tag{3.2.5}$$

A more elegant approach, which leads to the same result, is the use of transform calculus (see Ex. 7.12 of Arpaci 1966, or see Özişik 1968).

We proceed now to the main objective of this example, that is, the integral formulation of the unsteady Poiseuille flow and its solution by approximate profiles.

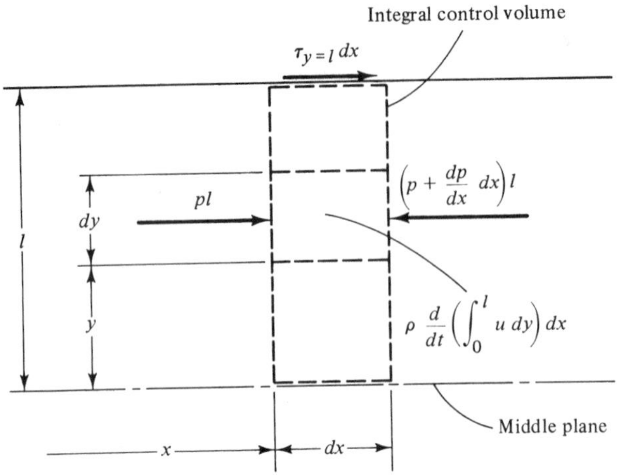

Fig. 3.5 Unsteady Poiseuille flow, Ex. 3.2.1

Accordingly, consider the integral control volume* shown in Fig. 3.5. The balance of momentum for this control volume, evaluating the rate of momentum and pressure terms from the integration of corresponding terms of the differential control volume, results in

$$\rho \frac{d}{dt}\int_0^l u\,dy = \left(-\frac{dp}{dx}\right)l + \tau_{y=l}. \tag{3.2.6}$$

Assuming a Newtonian fluid,

$$\tau_{y=l} = \mu\left(\frac{\partial u}{\partial y}\right)_{y=l}, \tag{3.2.7}$$

inserting Eq. (3.2.7) into Eq. (3.2.6) yields the governing equation

$$\frac{1}{\nu}\frac{d}{dt}\int_0^l u\,dy = \frac{1}{\mu}\left(-\frac{dp}{dx}\right)l + \left(\frac{\partial u}{\partial y}\right)_{y=l} \tag{3.2.8}$$

subject to the initial and boundary conditions identical to those of the differential formulation,

$$u(y, 0) = 0; \qquad \frac{\partial u(0, t)}{\partial y} = 0, \qquad u(l, t) = 0. \tag{3.2.9}$$

We show next how the foregoing formulation can be utilized in the process of getting an approximate solution to the unsteady Poiseuille flow. Assuming the solution of this flow to have the form $u(y, t) = U(y)\,\delta(t)$, and observing the fact that the scaled-down steady velocity approximates the unsteady velocity, that is, $U(y) = \bar{u}(y)$, we propose that

$$u(y, t) = \frac{-dp/dx}{2\mu}(l^2 - y^2)\,\delta(t), \tag{3.2.10}$$

---

*Is this not a mixed control volume?

where $\delta(t)$ denotes the unknown scale factor. Note that Eq. (3.2.10) satisfies the boundary conditions of Eq. (3.2.9). Inserting Eq. (3.2.10) into Eq. (3.2.8) results in

$$\frac{d\delta}{dt} + \frac{3\nu}{l^2}(\delta - 1) = 0; \qquad \delta(0) = 0, \tag{3.2.11}$$

where the initial condition is obtained from the initial condition of Eq. (3.2.9). The solution of Eq. (3.2.11) is

$$\delta(t) = 1 - \exp\left(-3\frac{\nu t}{l^2}\right). \tag{3.2.12}$$

Inserting Eq. (3.2.12) into Eq. (3.2.10), we obtain an approximate solution for the unsteady Poiseuille flow,

$$\frac{u(y, t)}{(-dp/dx)l^2/\mu} = \frac{1}{2}\left[1 - \left(\frac{y}{l}\right)^2\right]\left[1 - \exp\left(-3\frac{\nu t}{l^2}\right)\right]. \tag{3.2.13}$$

We shall comment, following the next example, on the classification and improvement of this solution.

So far, in terms of Eqs. (3.2.5) and (3.2.13), we have obtained exact and approximate solutions to our problem. However, flow development being completed—for all practical purposes—over a finite time interval, say t_0, suggests yet another approximate velocity profile, such as

$$u(y, t) = -\frac{dp/dx}{2\mu}(l^2 - y^2)\left(2\frac{t}{t_0} - \frac{t^2}{t_0^2}\right).$$

Inserting this profile into Eq. (3.2.8), now also integrating over the time $(0, t_0)$ gives $t_0 = l^2/\nu$. Numerical values of t_0 for four typical fluids (Appendix E) in a narrow channel ($l = 10$ mm) are given in Table 3.1. (Discuss the experimental setup that would provide the fully developed liquid metal flow, say for a steady mean velocity of 1.0 m/s.)

TABLE 3.1 Approximate times for flow development between parallel plates spaced $2l = 20$ mm apart, Ex. 3.2.1

Fluid	Light oil	Air	Water	Mercury
$\nu \times 10^6$ (m²/s)	84	15.7	1.0	0.093
$t_0 \simeq l^2/\nu$ (s)	1.2	6.4	100	1080

The use of a finite "penetration depth" in the time domain for a problem which mathematically penetrates to infinity is explored further in the following example.

EXAMPLE 3.2.2 Reconsider the stagnant fluid of Ex. 3.2.1. This time let the distance between plates be l, and let one of the plates, say the upper one, suddenly assume a uniform velocity, U. We wish to determine the unsteady velocity distribution of the fluid.

The differential formulation of the problem, its governing equation being the one-dimensional form of Eq. (3.1.13) with $\partial p/\partial x = 0$, is

$$\frac{1}{\nu}\frac{\partial u}{\partial t} = \frac{\partial^2 u}{\partial y^2}; \qquad u(y, 0) = 0, \qquad u(0, t) = 0, \qquad u(l, t) = U. \tag{3.2.14}$$

(Describe the physics of the conduction problem which leads to the same formulation.)

The foregoing formulation, as it stands, cannot be solved by the separation of variables because of the nonhomogeneity of boundary condition for the upper plate. However, the superposition

$$u(y, t) = \bar{u}(y) + u'(y, t)$$

eliminates this difficulty. Now $\bar{u}$ and u' satisfy

$$\begin{aligned} 0 &= \frac{d^2\bar{u}}{dy^2}, & \frac{1}{\nu}\frac{\partial u'}{\partial t} &= \frac{\partial^2 u'}{\partial y^2}, \\ & & u'(y, 0) &= -\bar{u}(y), \\ 0 &= \bar{u}(0), & u'(0, t) &= 0, \\ U &= \bar{u}(l), & u'(l, t) &= 0. \end{aligned} \tag{3.2.15}$$

The solution of the steady problem is the well-known Couette flow,

$$\bar{u}(y) = U\frac{y}{l}. \tag{3.2.16}$$

The solution of the unsteady problem, obtained by the separation of variables, is

$$u'(y, t) = \sum_{n=1}^{\infty} a_n e^{-\lambda_n^2 \nu t} \sin \lambda_n y, \tag{3.2.17}$$

where $\lambda_n l = n\pi$, $n = 1, 2, 3, \ldots$, and

$$a_n = \frac{2U}{\lambda_n l}(-1)^n.$$

The details of this solution is left to the interested reader. Combining Eqs. (3.2.16) and (3.2.17), we obtain the unsteady velocity distribution,

$$\frac{u(y, t)}{U} = \frac{y}{l} + 2\sum_{n=1}^{\infty} \frac{(-1)^n}{\lambda_n l} e^{-\lambda_n^2 \nu t} \sin \lambda_n y. \tag{3.2.18}$$

The use of transform calculus leads also to the same answer.

We proceed now to the main objective of this example, that is, the integral formulation of the unsteady Couette flow and its solution by approximate profiles. First, observe the important fact that the momentum flux diffuses from the moving upper plate downward, and in a finite *penetration time*, say t_0, it reaches the lower plate. This suggests for $t < t_0$ a *penetration depth* (boundary layer), say δ_0, below which the effect of momentum diffusion is negligible. Consequently, it is convenient to investigate the problem in two domains, say the first domain and the second domain, corresponding to $0 \leq t \leq t_0$ and $t \geq t_0$, respectively.

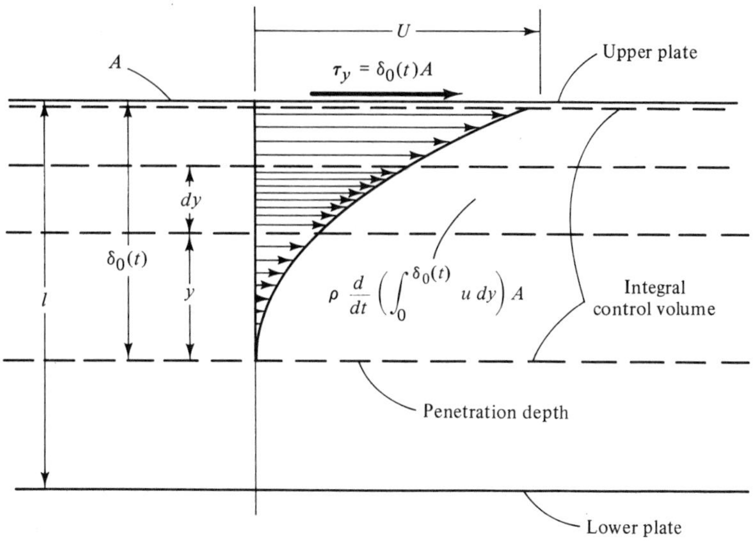

Fig. 3.6 Unsteady Couette flow—first domain, Ex. 3.2.2

For the first domain, the balance of momentum applied to the integral control volume shown in Fig. 3.6 yields, after the assumption of Newtonian fluid,

$$\frac{1}{\nu}\frac{d}{dt}\int_0^{\delta_0(t)} u\,dy = \left(\frac{\partial u}{\partial y}\right)_{y=\delta_0(t)} \tag{3.2.19}$$

subject to

$$u(y, 0) = 0; \qquad u(0, t) = 0, \qquad u(\delta_0, t) = U, \tag{3.2.20}$$

where $u(0, t)$ specifies the velocity at the penetration depth rather than that of the lower plate. The y-coordinate measured from the penetration depth upward proves convenient in the construction of the approximate velocity profile.

Since, as $t \longrightarrow t_0$, the unsteady velocity of this domain approaches the unknown initial velocity of the second domain, the idea of constructing unsteady profiles by scaling the steady solution of the problem fails. Instead, we consider simple functions (such as polynomials) and require that they satisfy the boundary conditions of the problem. In terms of polynomials, for example, the parabola satisfying the boundary conditions of Eq. (3.2.20) is readily found to be

$$u(y, t) = U\left(\frac{y}{\delta_0}\right)^2. \tag{3.2.21}$$

Inserting Eq. (3.2.21) into Eq. (3.2.19) results in the simple but *nonlinear* differential equation,

$$d\delta_0^2 = 12\nu\,dt \tag{3.2.22}$$

subject to the initial condition of Eq. (3.2.20),

$$\delta_0(0) = 0. \tag{3.2.23}$$

The solution of Eq. (3.2.22) which satisfies Eq. (3.2.23) is

$$\delta_0(t) = (12\nu t)^{1/2}. \tag{3.2.24}$$

Inserting Eq. (3.2.24) into Eq. (3.2.21), we obtain an approximate solution for the unsteady Couette flow,

$$\frac{u(y, t)}{U} = \frac{y^2}{12\nu t}, \qquad 0 \leq t \leq t_0. \tag{3.2.25}$$

This velocity (or the momentum flux related to it) reaches the lower plate as $\delta_0 \longrightarrow l$. Then, from Eq. (3.2.24), we have the *penetration time*,

$$t_0 = \frac{l^2}{12\nu}. \tag{3.2.26}$$

We proceed next to the second domain. The balance of momentum gives, in terms of the integral control volume shown in Fig. 3.7 and for a Newtonian fluid,

$$\frac{l^2}{\nu}\frac{d}{dt}\int_0^1 u\, d\eta = \left(\frac{\partial u}{\partial \eta}\right)_{\eta=1} - \left(\frac{\partial u}{\partial \eta}\right)_{\eta=0}, \tag{3.2.27}$$

where $\eta = y/l$. For an approximate profile, consider the parabola

$$u = a_0 + a_1\eta + a_2\eta^2,$$

which, after satisfying the boundary conditions of Eq. (3.2.14) and introducing $\delta_1(t) = a_2/U$, becomes

$$\frac{u(\eta, t)}{U} = \eta - \eta(1 - \eta)\,\delta_1(t). \tag{3.2.28}$$

[Sketch and discuss the components of this profile. Is there any relation, for $t = t_0$,

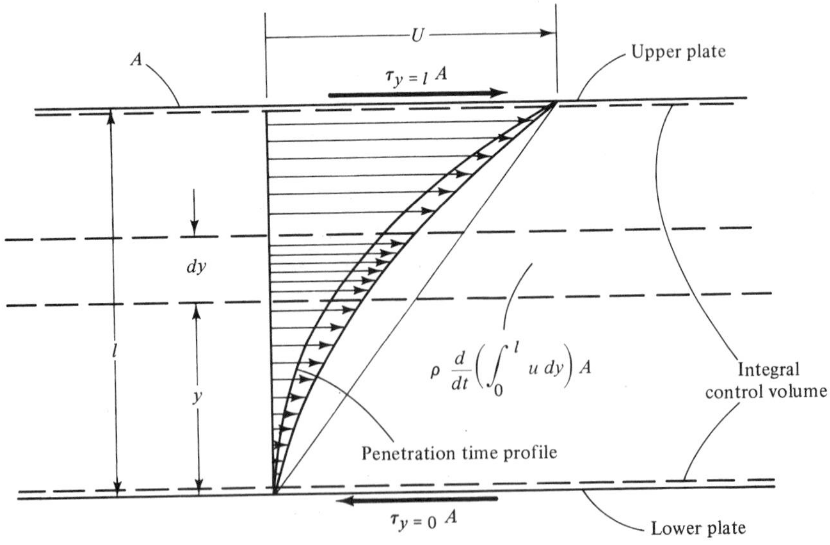

Fig. 3.7 Unsteady Couette flow—second domain, Ex. 3.2.2

between Eqs. (3.2.28) and (3.2.21)? What happens to Eq. (3.2.28) as $t \longrightarrow \infty$?] Inserting Eq. (3.2.28) into Eq. (3.2.27) yields

$$\frac{d\delta_1}{dt} + \frac{12\nu}{l^2}\delta_1 = 0; \qquad \delta_1(t_0) = 1. \tag{3.2.29}$$

The solution of Eq. (3.2.29) is

$$\delta_1(t) = \exp\left[-12\frac{\nu(t - t_0)}{l^2}\right]. \tag{3.2.30}$$

Inserting Eq. (3.2.30) into Eq. (3.2.28) gives an approximate solution for the unsteady Couette flow,

$$\frac{u(\eta, t)}{U} = \eta - \eta(1 - \eta)\exp\left[-12\frac{\nu(t - t_0)}{l^2}\right], \qquad t \geq t_0. \tag{3.2.31}$$

Note that for $t = t_0$, $\delta_0 = l$, $\delta_1 = 1$ and profiles of the first and second domains coincide; also, as $t \longrightarrow \infty$, $\delta_1 \longrightarrow 0$ and the second domain profile approaches the steady solution of the problem.

Clearly, we have developed the integral formulation of the Poiseuille flow and that of Couette flow by different approaches. Hereafter, we shall conveniently refer to problems to be formulated following the approach used for the Poiseuille flow as *volumetric rise problems* and to those to be formulated following the approach used for the Couette flow as *penetration depth* (*boundary layer*) *problems*. An external cause for volumetric rise problems, such as a pressure gradient or an energy generation, instantaneously affects the entire geometry of the problem. On the other hand, an external cause for penetration depth problems, such as a change in a flow and/or a thermal boundary condition, penetrates from boundaries inward and, in early stages of the problem, affects only a part of the geometry.

Volumetric rise problems are one-domain problems in time. Their unsteady profiles can be constructed from scaling their steady solution. Penetration depth problems in finite spatial domains are two-domain problems in time. Their unsteady profiles *cannot* be constructed from scaling their steady solution; they are constructed from simple functions satisfying boundary conditions of the problem; the profiles constructed for the first and second domains must coincide at the penetration time.

Since our main objective is to distinguish the two forms of integral formulation, and to be prepared to use the concept of penetration depth in *nonlinear* boundary layer problems, we shall not proceed to the profile improvement of the foregoing *linear* problems. This improvement can be accomplished either by considering higher-order profiles or by the use of variational calculus; for an informal study, see Arpaci (1966); for a formal study, see Finlayson (1972) and the review article by Finlayson & Scriven (1966).

EXAMPLE 3.2.3 Reconsider Ex. 3.2.1. Now assume that the z-direction is finite, say $2L$. We wish to determine the steady velocity distribution and pressure drop of the fluid flowing in a channel of rectangular cross section.

The differential formulation of the problem, its governing equation obtained from Eq. (3.1.13) by letting $\partial/\partial t \equiv 0$ and $\partial p/\partial x = dp/dx$, is

$$0 = -\frac{1}{\mu}\frac{dp}{dx} + \left(\frac{\partial^2 u}{\partial y^2} + \frac{\partial^2 u}{\partial z^2}\right), \tag{3.2.32}$$

$$\frac{\partial u(0, z)}{\partial y} = 0, \qquad u(l, z) = 0,$$

$$\frac{\partial u(y, 0)}{\partial z} = 0, \qquad u(y, L) = 0.$$

(Describe the physics of the conduction problem which leads to the same formulation.)

The foregoing formulation, as it stands, cannot be solved by the separation of variables because of the nonhomogeneity of the governing equation. However, either one of the superpositions $u(y, z) = \bar{u}(y) + u'(y, z)$, $u(y, z) = \bar{u}(z) + u'(y, z)$ eliminates this difficulty. Here $\bar{u}(y)$ and u', for example, satisfy

$$\begin{aligned} 0 &= -\frac{1}{\mu}\frac{dp}{dx} + \frac{d^2\bar{u}}{dy^2}, & 0 &= \frac{\partial^2 u'}{\partial y^2} + \frac{\partial^2 u'}{\partial z^2}, \\ 0 &= \frac{d\bar{u}(0)}{dy}, & 0 &= \frac{\partial u'(0, z)}{\partial y}, \\ 0 &= \bar{u}(l), & 0 &= u'(l, z), \\ & & 0 &= \frac{\partial u'(y, 0)}{\partial z}, \\ & & -\bar{u}(y) &= u'(y, L). \end{aligned} \tag{3.2.33}$$

The complete solution of the problem is

$$\frac{u(y, z)}{(-dp/dx)l^2/\mu} = \frac{1}{2}\left[1 - \left(\frac{y}{l}\right)^2\right] - 2\sum_{n=1}^{\infty}\frac{(-1)^n}{(\lambda_n l)^3}\left(\frac{\cosh \lambda_n z}{\cosh \lambda_n L}\right)\cos \lambda_n y, \tag{3.2.34}$$

with $\lambda_n l = (2n + 1)\pi/2$, $n = 0, 1, 2, \ldots$. The details of this solution are left to the interested reader (see also Ex. 4.10 of Arpaci 1966).

We proceed now to the main objective of this example, that is, the integral formulation of the steady channel flow and its solution by approximate profiles. Clearly, the problem is one of the volumetric rise. Then, in terms of the integral control volume shown in Fig. 3.8(a), the balance of momentum yields, for a Newtonian fluid,

$$0 = \int_0^L \left[\frac{1}{\mu}\left(-\frac{dp}{dx}\right)l + \left(\frac{\partial u}{\partial y}\right)_{y=l} + \int_0^l \left(\frac{\partial^2 u}{\partial z^2}\right) dy\right] dz, \tag{3.2.35}$$

or, in terms of the integral control volume shown in Fig. 3.8b,

$$0 = \frac{1}{\mu}\left(-\frac{dp}{dx}\right)lL + \int_0^L \left(\frac{\partial u}{\partial y}\right)_{y=l} dz + \int_0^l \left(\frac{\partial u}{\partial z}\right)_{z=L} dy. \tag{3.2.36}$$

(What is the relation between Eqs. 3.2.35 and 3.2.36?) Depending on the particular problem under consideration, one of these formulations may be more convenient to use than the other.

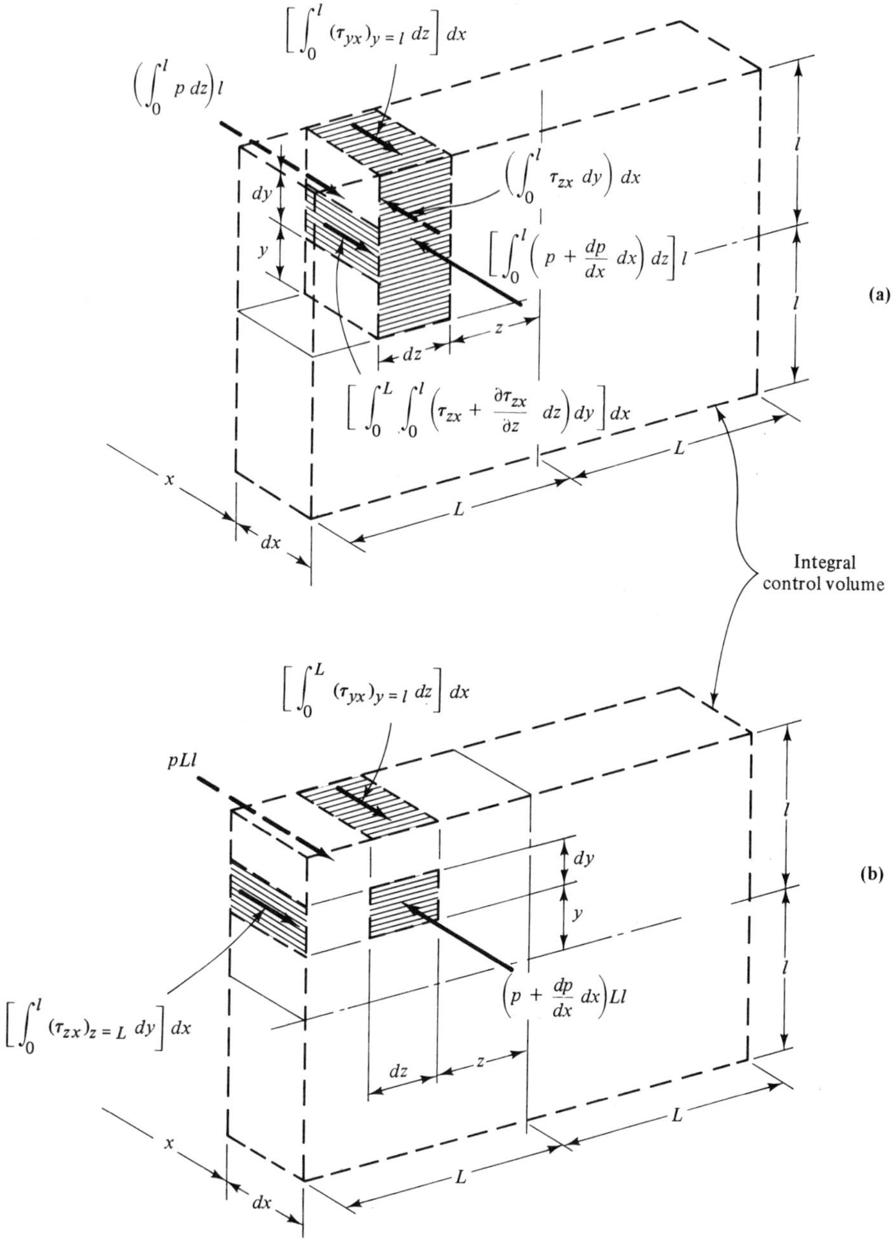

Fig. 3.8 Steady channel flow, Ex. 3.2.3

Let $L > l$ and assume that the problem has a middle domain in which the velocity distribution becomes one-dimensional. Then, expressing two-dimensional effects in terms of the scale factor $\delta(z)$, an approximate velocity profile may be written as

$$u(y, z) = \frac{-dp/dx}{2\mu}(l^2 - y^2)\,\delta(z). \tag{3.2.37}$$

(Comment on the conceptual similarity of Eqs. 3.2.37 and 3.2.10.) Inserting Eq. (3.2.37) into Eq. (3.2.35) results in

$$\int_0^L \left(1 - \delta + \frac{l^2}{3}\frac{d^2\delta}{dz^2}\right) dz = 0. \tag{3.2.38}$$

Since Eq. (3.2.38) is true for an arbitrary length L, the integrand itself must vanish everywhere in the interval $(0, L)$. Consequently,

$$\frac{d^2\delta}{dz^2} + \frac{3}{l^2}(\delta - 1) = 0; \qquad \frac{d\delta(0)}{dz} = 0, \qquad \delta(L) = 0, \tag{3.2.39}$$

where the boundary conditions are obtained from the corresponding conditions in Eq. (3.2.32). The solution of Eq. (3.2.39) is

$$\delta(z) = 1 - \frac{\cosh(\sqrt{3}/l)z}{\cosh(\sqrt{3}/l)L}. \tag{3.2.40}$$

Inserting Eq. (3.2.40) into Eq. (3.2.37), we obtain an approximate solution for steady channel flow,

$$\frac{u(y, z)}{(-dp/dx)l^2/\mu} = \frac{1}{2}\left[1 - \left(\frac{y}{l}\right)^2\right]\left(1 - \frac{\cosh(\sqrt{3}/l)z}{\cosh(\sqrt{3}/l)L}\right). \tag{3.2.41}$$

For improvements on this solution by the use of higher-order profiles and/or the variational calculus, see Exs. 4.11 and 8.6 of Arpaci (1966).

Next, returning to the fact that the two-dimensional flow in the channel may have a one-dimensional middle domain, we wish to determine the condition under which this domain exists. Clearly, our volumetric rise problem now becomes a penetration depth problem. Let Δ be the penetration depth of the two-dimensional flow. Then the balance of momentum for the integral control volume shown in Fig. 3.9, also available from Eq. (3.2.36) by letting $z \to \zeta$ and $L \to \Delta$, gives

$$0 = \frac{1}{\mu}\left(-\frac{dp}{dx}\right) l\Delta + \int_0^\Delta \left(\frac{\partial u}{\partial y}\right)_{y=l} d\zeta + \int_0^l \left(\frac{\partial u}{\partial \zeta}\right)_{\zeta=\Delta} dy. \tag{3.2.42}$$

Specifying now δ of Eq. (3.2.37) in terms of Δ, the approximate velocity profile may be assumed to be

$$u(y, \zeta) = \frac{-dp/dx}{2\mu}(l^2 - y^2)\left(1 - \frac{\zeta^2}{\Delta^2}\right). \tag{3.2.43}$$

Inserting Eq. (3.2.43) into Eq. (3.2.42) yields

$$\Delta = \sqrt{2}\, l. \tag{3.2.44}$$

Thus, whenever $L > \Delta$, the two-dimensional flow in the channel becomes one-dimensional over a middle domain of width $2(L - \sqrt{2}\, l)$.

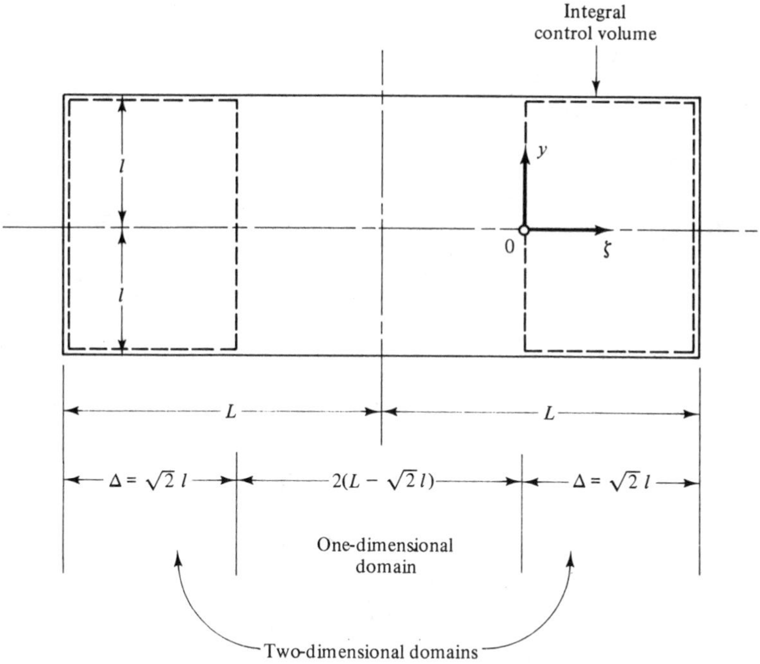

Fig. 3.9 One- and two-dimensional domains of parallel channel flow, Ex. 3.2.3

So far in this section we solved three key problems by the integral approach. The first problem was an example of volumetric rise, the second of penetration depth. In the third we showed how a volumetric rise problem can be converted to a penetration depth problem by the selection of different profiles. Since each of these problems readily admits an exact solution (either by the separation of variables or by the transform calculus), we discussed briefly the exact solution of each problem in terms of the separation of variables, which is the simplest of the exact methods of solution. These problems, however, only demonstrate the use of the integral approach as a convenient method for obtaining an approximate solution to problems which are also solvable by exact methods. Next we shall illustrate, in terms of a volumetric rise problem, that the integral approach is not merely convenient but is indispensable when considering problems whose exact solutions are not readily available.

EXAMPLE 3.2.4 Reconsider Ex. 3.2.3. Now assume the fluid flowing in a channel whose cross section is bounded by the curves $y = \pm f(z)$ and the straight lines $z = \text{a}$ and $z = \text{b}$ as shown in Fig. 3.10(a). We wish to determine the steady velocity of the fluid.

The integral formulation of the problem, obtained in a manner similar to the development of Eq. (3.2.35), is

$$\int_a^b \left[-\frac{2}{\mu}\left(\frac{dp}{dx}\right) f + \frac{\partial u}{\partial z}\bigg|_{-f}^{f} + \int_{-f}^{f} \left(\frac{\partial^2 u}{\partial y^2}\right) dz \right] dy = 0. \qquad (3.2.45)$$

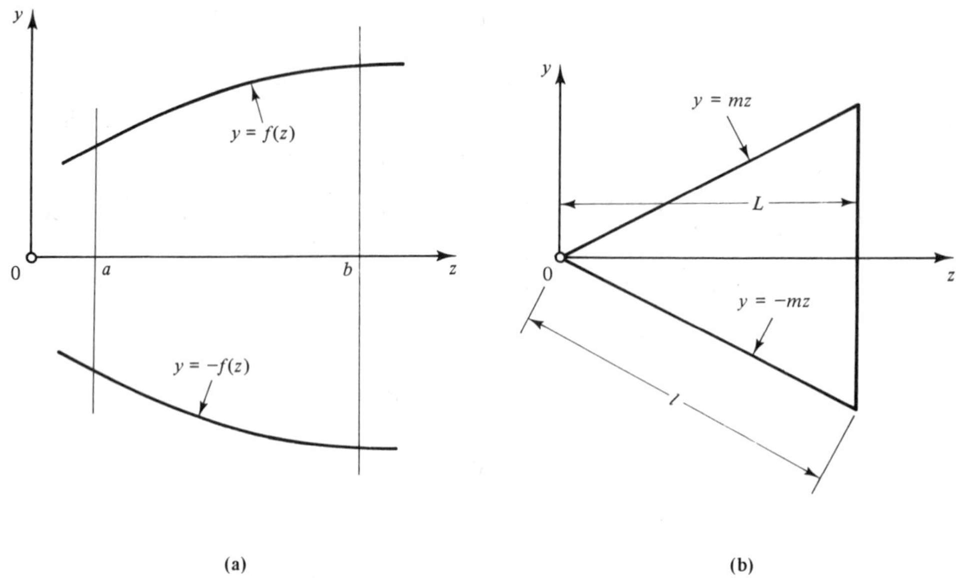

Fig. 3.10 Channel cross sections, Ex. 3.2.4

Another formulation, which yields more accurate approximate solutions, may be given by the use of the variational calculus (see Sec. 8-9 of Arpaci 1966).

Inserting the approximate velocity profile

$$u(y, z) = \frac{-dp/dx}{2\mu}[f^2(z) - y^2]\,\delta(z), \tag{3.2.46}$$

constructed by the generalization of Eq. (3.2.37), into Eq. (3.2.45) gives, after some rearrangement,

$$\left(\frac{1}{3}f^2\right)\frac{d^2\delta}{dz^2} + (2ff')\frac{d\delta}{dz} + (f'^2 + ff'' - 1)\,\delta = -1, \tag{3.2.47}$$

subject to

$$\delta(a) = 0 \qquad \text{and} \qquad \delta(b) = 0. \tag{3.2.48}$$

The special case corresponding to $f(z) = l$, with $\delta(-L) = 0$ [or $d\delta(0)/dz = 0$] and $\delta(L) = 0$, reduces Eq. (3.2.47) to Eq. (3.2.39), which gives the flow in a rectangular channel considered in Ex. 3.2.3.

An important case that prescribes the flow in a triangular channel (Fig. 3.10b) leads, after inserting $f(z) = mz$ into Eq. (3.2.47), to the well-known equidimensional (or Euler or Cauchy) equation,

$$z^2\frac{d^2\delta}{dz^2} + 6z\frac{d\delta}{dz} + 3\left(1 - \frac{1}{m^2}\right)\delta = -\frac{3}{m^2}, \tag{3.2.49}$$

with

$$\delta(0) = 0 \qquad \text{and} \qquad \delta(L) = 0. \tag{3.2.50}$$

The general solution of Eq. (3.2.49) is readily found to be

$$\delta = C_1 e^{r_1 z} + C_2 e^{r_2 z} + \frac{1}{1 - m^2}, \tag{3.2.51}$$

where r_1 and r_2, resulting from the insertion of $\delta = z^r$ into the homogeneous part of Eq. (3.2.49), satisfy

$$r^2 + 5r + 3\left(1 - \frac{1}{m^2}\right) = 0. \tag{3.2.52}$$

Furthermore, $m = 1/\sqrt{3}$ for an equilateral triangle, and Eq. (3.2.52) reduces to

$$r^2 + 5r - 6 = 0,$$

which readily gives $r_1 = 1$ and $r_2 = -6$. Then Eq. (3.2.51) yields, with $\delta(0) = 0$ and $\delta(L = \sqrt{3}\,l/2) = 0$,

$$\delta = \frac{3}{2} - z\frac{\sqrt{3}}{l}. \tag{3.2.53}$$

Finally, inserting $f(z) = mz = z/\sqrt{3}$ and Eq. (3.2.53) into Eq. (3.2.46), we obtain

$$u(y, z) = \frac{-dp/dx}{2\mu}\left(\frac{z^2}{3} - y^2\right)\left(\frac{3}{2} - z\frac{\sqrt{3}}{l}\right), \tag{3.2.54}$$

the steady flow of a Newtonian fluid in an equilateral triangular channel.

For flow in a channel it is customary to define a dimensionless *friction factor* in terms of the pressure gradient associated with friction,

$$f_m = \frac{(-dp/dx)D_e}{\frac{1}{2}\rho U_m^2}, \tag{3.2.55}$$

where U_m is the mean velocity and

$$D_e = 4\frac{\text{cross-sectional flow area}}{\text{wetted perimeter}} = \frac{4A}{P_w} \tag{3.2.56}$$

defines the hydraulic or *equivalent diameter* (so that it equals the diameter of a pipe of circular cross section). Equation (3.2.55) defines the Moody (or Weissbach) friction factor. The alternative definition, $f = c_f = \tau_w/(\frac{1}{2}\rho U_m^2)$, called the Fanning friction factor, is also used. Note that $f_m = 4f$.

For a channel of rectangular cross section, for example, integrating Eq. (3.2.41) over the flow area gives

$$\frac{U_m}{(-dp/dx)l^2/\mu} = \frac{1}{3}\left(1 - \frac{l}{L\sqrt{3}}\tanh\frac{L\sqrt{3}}{l}\right).$$

Eliminating the pressure gradient and introducing $D_e = 4l/(1 + l/L)$, Eq. (3.2.55) may be rearranged as

$$f_m = \frac{96}{\text{Re}}\left[\left(1 + \frac{l}{L}\right)^2\left(1 - \frac{l}{L\sqrt{3}}\tanh\frac{L\sqrt{3}}{l}\right)\right]^{-1}, \tag{3.2.57}$$

where $\text{Re} = \rho U_m D_e/\mu$ denotes the Reynolds number based on mean velocity and equivalent diameter. Figure 3.11 compares results from Eq. (3.2.57) to those of the exact solution, Eq. (3.2.34).

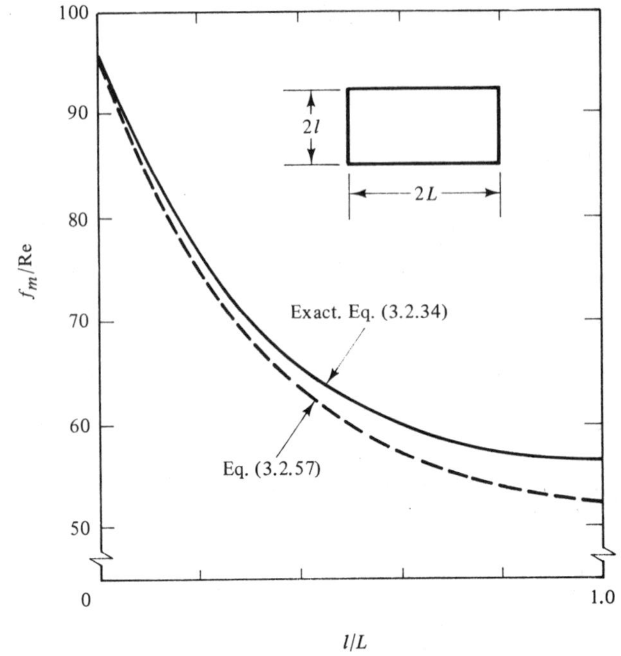

Fig. 3.11 Friction factor for developed flow in rectangular channel, Ex. 3.2.3

For a channel of triangular cross section, now using Eq. (3.2.54) and $L = l\sqrt{3}/2$, $D_e = l\sqrt{3}$ gives

$$\frac{U_m}{(-dp/dx)l^2/\mu} = \frac{1}{80}; \qquad f_m = \frac{160}{3}\frac{1}{\text{Re}}. \tag{3.2.58}$$

For a pipe of circular cross section (see Problem 3-26) the result is $f_m = 64/\text{Re}$.

The lack of geometric similarity between cross sections leads to different constants in the expressions for the friction factor of laminar flows. These differences turn out to become negligible, however, for turbulent flows. Equation (3.2.32) has been solved for many shapes of channel cross sections (see the review by Berker 1963). All these laminar flow solutions become invalid above Re $\sim$ 2300, where the flows usually experience transition to turbulence.

EXAMPLE 3.2.5 Reconsider Ex. 3.2.1, the sudden start of Poiseuille flow between parallel plates spaced a distance $2l$ apart. We wish to interpret and solve this problem in terms of vorticity.

For the parallel flow $u = u(y, t)$, only the z-component of the vorticity $\boldsymbol{\omega} = \nabla \times \boldsymbol{V}$ is nonzero and it reduces to $\omega = -\partial u/\partial y$. It equals twice the instantaneous rate of rotation of a fluid element in the (x, y)-plane. Differentiating Eq. (3.2.1) with respect to y yields a diffusion equation in ω,

$$\frac{1}{\nu}\frac{\partial \omega}{\partial t} = \frac{\partial^2 \omega}{\partial y^2}. \tag{3.2.59}$$

However, the boundary conditions for ω are different from those for u. From Eq. (3.2.1), rearranged as $(1/\nu)\partial u/\partial t = -(1/\mu)dp/dx - \partial\omega/\partial y$, we have for these boundary conditions,

$$\frac{\partial\omega(0, t)}{\partial y} = -\frac{1}{\mu}\frac{dp}{dx}; \qquad \frac{\partial\omega(2l, t)}{\partial y} = -\frac{1}{\mu}\frac{dp}{dx} \quad \text{or} \quad \omega(l, t) = 0, \tag{3.2.60}$$

where the more convenient last condition follows from the antisymmetry of the problem, implying zero vorticity at the middle plane. The initial condition is $\omega(y, 0) = 0$. Clearly, in terms of the vorticity, we now have a penetration depth problem resulting from a suddenly imposed and constant vorticity flux at the wall.

For the first time domain, $0 < t \leq t_1$, the profile

$$\omega(y, t) = \frac{1}{2\mu}\frac{dp}{dx}\delta\left(1 - \frac{y}{\delta}\right)^2, \tag{3.2.61}$$

satisfies Eq. (3.2.60), where y is measured upward from the lower plate (Fig. 3.12).

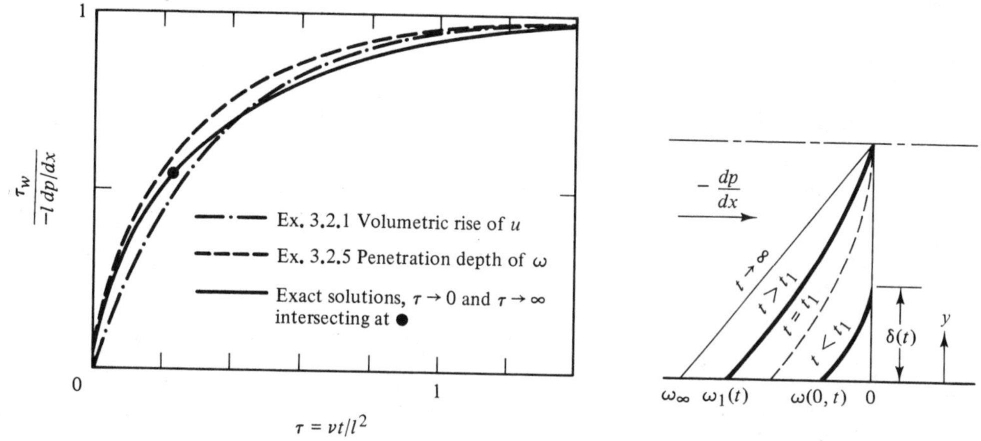

Fig. 3.12 Vorticity in unsteady Poiseuille flow, Ex. 3.2.5

Inserting Eq. (3.2.61) into Eq. (3.2.59) and integrating from $y = 0$ to δ yields

$$\frac{2\delta}{3}\frac{d\delta}{dt} = 2,$$

subject to $\delta(0) = 0$. Thus

$$\delta(t) = (6\nu t)^{1/2}. \tag{3.2.62}$$

With this result Eq. (3.2.61) gives the approximate vorticity distribution until $t = t_1 = l^2/6\nu$, the penetration time to the middle plane. At this time the vorticity at the lower plate has increased to

$$\omega(0, t_1) = \omega_1(t_1) = \frac{l}{2\mu}\frac{dp}{dx} = \frac{\omega_\infty}{2}, \tag{3.2.63}$$

where ω_∞ denotes the developed steady value.

For the second time domain $t \geq t_1$, the profile in terms of $\eta = y/l$,

$$\omega(y, t) = \omega_\infty(1 - \eta) - \omega_1(t)(1 - \eta^2), \tag{3.2.64}$$

satisfies Eq. (3.2.60), $\omega_1(t)$ being a scale parameter to be determined. Inserting Eq. (3.2.64) into Eq. (3.2.59) and integrating from $\eta = 0$ to 1 yields

$$-\frac{2}{3\nu}\frac{d\omega_1}{dt} = \frac{2\omega_1}{l^2},$$

subject to Eq. (3.2.63). Thus

$$\omega_1(t) = \frac{\omega_\infty}{2}\exp\left[-\frac{3\nu(t-t_1)}{l^2}\right], \tag{3.2.65}$$

and the approximate vorticity distribution for $t \geq t_1$ is given by Eq. (3.2.64).

Figure 3.12 shows the time dependent wall shear, being proportional to the vorticity at the lower plate, $\tau_w = \mu\, \partial u(0, t)/\partial y = -\mu\omega(0, t)$, as obtained from Eqs. (3.2.61) and (3.2.64). Also shown in the figure is the approximate solution obtained in Ex. 3.2.1 by the consideration of the volumetric rise of velocity, and the asymptotes of the exact solution for small and large times, respectively (see Problem 3-12). Note the higher accuracy of the small time vorticity solution, which apparently is a better approximation to the problem.

The foregoing example shows how an unsteady problem formulated as a volumetric rise of velocity may also be formulated as a penetration depth of vorticity. Mathematically, this difference results from the consideration of the velocity gradient instead of the velocity itself. Physically, when the bulk of the fluid is subjected to a sudden pressure gradient (or body force) we expect it to accelerate with velocity $U(t) = [-(1/\rho)dp/dx]t$, while viscous boundary layers build up from the bounding walls. In fact, the unsteady flat plate boundary layer arising from the free stream velocity $U(t)$ gives identically the small time vorticity solution of Eq. (3.2.61) (see the related Problem 4-6).

Now, consider the sudden start of Couette flow (Ex. 3.2.2 and Problem 3-13). The initial step change in wall velocity suggests a penetration depth in terms of velocity. In terms of vorticity, the problem is governed by Eq. (3.2.59) and (in the absence of a pressure gradient, so that $(1/\nu)\partial u/\partial t = -\partial\omega/\partial y$) subject to zero vorticity flux at walls. The initial condition, however, is unusual. It amounts to a sudden release of a finite amount of circulation at the accelerated plate, hence $\omega(y, 0)$ is a delta function. This vorticity penetrates by diffusion into and fills the space between plates, and ultimately reaches a uniform distribution as $t \longrightarrow \infty$, $\omega(y, \infty) = -U/l$.

Having illustrated the concept of vorticity diffusion, and some boundary conditions associated with vorticity, we return to the velocity and temperature distributions of parallel flows, now under the added effect of viscous dissipation.

3.3 DISSIPATION EFFECTS

For parallel flows, the irreversible conversion of mechanical power by viscous stress into internal energy is given by

$$\mu\left[\left(\frac{\partial u}{\partial y}\right)^2 + \left(\frac{\partial u}{\partial z}\right)^2\right]$$

from Eq. (3.1.14). In Section 2.4.1 we have stated that this volumetric source in the balance of thermal energy may be neglected in most problems of convection. The exception is the flow of viscous oils at high shear rates. Such flows exist in oil-lubricated journal bearings, which involve substantial dissipated power. This power is removed partly by convection to bounding surfaces and partly by the net enthalpy flow of oil through the bearing.

Leaving temperature dependence of properties as well as the development of flow and temperature to problems, here we consider Couette flow with a superposed pressure gradient.

EXAMPLE 3.3.1 Reconsider the steady part of Ex. 3.2.2, now including a pressure gradient. The upper plate is insulated and the lower plate is kept at temperature T_0. We wish to determine the distributions of velocity and temperature, and the wall heat flux which gives the dissipated power.

The differential formulation, using the steady, one-dimensional forms of Eqs. (3.1.13) and (3.1.14) and referring to Fig. 3.13, is

$$0 = -\frac{1}{\mu}\frac{dp}{dx} + \frac{d^2u}{dy^2}; \qquad 0 = \frac{d^2T}{dy^2} + \frac{\mu}{k}\left(\frac{du}{dy}\right)^2;$$
$$u(0) = 0, \quad u(l) = U; \qquad T(0) = T_0, \quad \frac{dT(l)}{dy} = 0. \tag{3.3.1}$$

Integration readily gives the velocity distribution

$$u(\eta) = U\eta + 6U_p\eta(1-\eta), \tag{3.3.2}$$

where $\eta = y/l$, and $U_p = (l^2/12\mu)(-dp/dx)$ denotes the mean velocity of Poiseuille flow if it occurred separately. As expected for a linear problem, Eq. (3.3.2) is the superposition of Couette flow and Poiseuille flow.

Substituting Eq. (3.3.2) into the second of Eq. (3.3.1), twice integration gives the temperature distribution

$$\frac{2k}{\mu U^2}[T(\eta) - T_0] = \eta(2-\eta) - 4\frac{U_p}{U}\eta^2(3-2\eta) + 12\left(\frac{U_p}{U}\right)^2\eta(2 - 3\eta + 4\eta^2 - 2\eta^3). \tag{3.3.3}$$

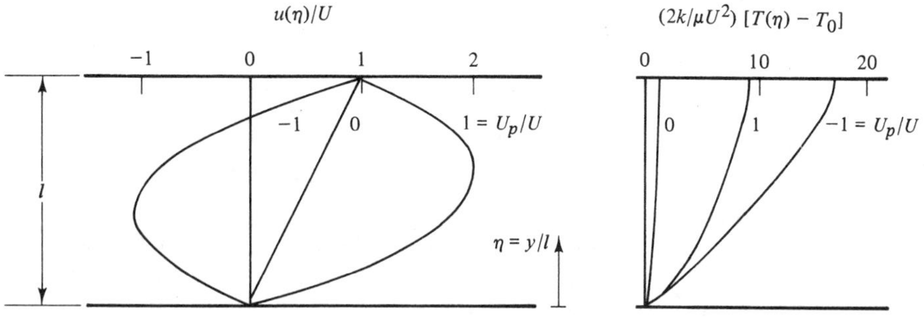

Fig. 3.13 Couette flow with pressure gradient and viscous dissipation; upper moving plate insulated, Ex. 3.3.1

The temperature rise at the upper adiabatic plate is

$$\Delta T = T(l) - T_0 = \frac{\mu U^2}{2k}\left[1 - 4\frac{U_p}{U} + 12\left(\frac{U_p}{U}\right)^2\right]. \tag{3.3.4}$$

The heat flux at the lower plate is

$$q_0 = k\frac{dT(0)}{dy} = \frac{\mu}{l}(U^2 + 12U_p^2). \tag{3.3.5}$$

For a typical bearing flow, ignoring curvature and taking $U_p = 0$, $U = 10$ m/s, $l = 10^{-3}$ m, $\mu = 0.0722$ kg/m s, $\mu/k \sim 2°\text{C s}^2/\text{m}^2$ for light oils, we find $\Delta T \sim 100°\text{C}$ and a heat flux of $q_0 \sim 7220$ W/m^2.

Recalling Eq. (2.4.25), the importance of viscous dissipation relative to change in energy flux is measured by

$$\text{EcPr} = \frac{\mu U^2}{k\,\Delta T}.$$

Clearly, this quantity is of order unity, as seen from Eq. (3.3.4), when the temperature distribution results solely from dissipation.

3.4 *BUOYANCY-DRIVEN FLOWS*

When the effect of temperature on fluid density is taken into account, lighter parts of a heated fluid in the gravity field flow upward and the heavier ones downward relative to a reference state in equilibrium, say at $\rho_0 = \rho(T_0, p_0)$. In Section 2.4.2 we learned that the effect of pressure on density is negligible, that we can put $\rho \simeq \rho_0$ in all terms except the buoyancy term (the Boussinesq approximation), and that the viscous dissipation can be safely ignored in problems of convection.

Consider now one-dimensional parallel flow between vertical plates, say x measured vertical upward and y across, and the resulting governing equations, Eqs. (2.4.22)–(2.4.24), reduce to

$$\frac{\partial u}{\partial x} = 0, \tag{3.4.1}$$

$$\frac{\partial u}{\partial t} = -\frac{1}{\rho}\frac{d}{dx}(p - p_0) + g\beta(T - T_0) + \nu\frac{\partial^2 u}{\partial y^2}, \tag{3.4.2}$$

$$\frac{\partial T}{\partial t} = a\frac{\partial^2 T}{\partial y^2}. \tag{3.4.3}$$

Here $\beta = -(1/\rho)(\partial\rho/\partial T)_p$ denotes the coefficient of thermal expansion and g the acceleration of gravity (for plates inclined an angle θ with the vertical, g is replaced by $g\cos\theta$). For these parallel flows the unknown T_0 is determined by Eq. (3.4.1) such that the buoyancy-induced motion is free of a vertical pressure gradient. Any gradient in pressure $p - p_0$ in Eq. (3.4.2) must then be associated with external sources, inducing additional fluid motion. Note that, contrary to the multidimensional case, Eq. (3.4.3) is not coupled to Eq. (3.4.2) and can be solved separately. Moreover, Eq. (3.4.2) is linear.

The following examples illustrate the concept of mixed (forced and buoyancy driven) flow, the penetration depth problem of transient approach to steady convection, and the concept of vorticity transport, all in a slot of infinite height. These idealized problems represent regions removed from "end zones" of real slots, which are, of course, of finite height.

EXAMPLE 3.4.1 Consider a Newtonian fluid confined in a closed vertical slot (Fig. 3.14). The vertical side walls of the slot are spaced a distance l apart and are kept at uniform temperatures T_1 and T_2. One wall moves downward with constant velocity U. We wish to determine the steady motion of the fluid in the slot (see Problem 3-16 for a related practical problem associated with falling films).

The purpose of this example is to show that the explicit pressure term is related only to the forced flow, and, because of its linearity, the forced and buoyancy-driven parts of the problem can be solved separately.

The momentum formulation, taken from Eq. (3.4.2), is

$$0 = -\frac{1}{\mu}\frac{d}{dx}(p - p_0) + \frac{g\beta}{\nu}(T - T_0) + \frac{d^2u}{dy^2} \tag{3.4.4}$$

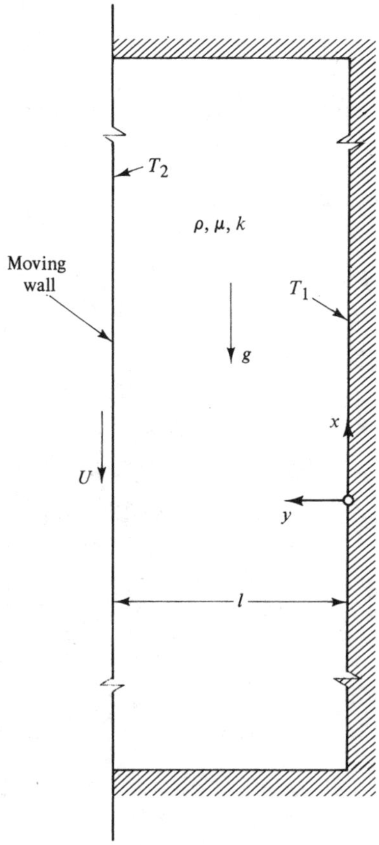

Fig. 3.14 Forced and buoyancy-driven flows combined, Ex. 3.4.1

subject to boundary conditions

$$u(0) = 0, \qquad u(l) = -U, \tag{3.4.5}$$

and the conservation of mass

$$\int_0^l u(y)\,dy = 0. \tag{3.4.6}$$

The temperature distribution

$$T = T_1 + \frac{y}{l}\Delta T, \qquad \Delta T = T_2 - T_1, \tag{3.4.7}$$

is readily obtained from Eq. (3.4.3).

Note that the pressure gradient explicitly involved in Eq. (3.4.4) is induced by the flow related to the moving side wall. Then, considering the problem as the superposition of this forced (isothermal) flow, u_1, and the buoyancy-driven flow, u_2, we have

$$0 = -\frac{1}{\mu}\frac{d}{dx}(p - p_0) + \frac{d^2u_1}{dy^2}, \qquad 0 = \frac{g\beta}{\nu}\left[(T_1 - T_0) + \frac{y}{l}\Delta T\right] + \frac{d^2u_2}{dy^2},$$

$$u_1(0) = 0, \qquad u_2(0) = 0,$$

$$u_1(l) = -U, \qquad u_2(l) = 0,$$

$$\int_0^l u_1(y)\,dy = 0, \qquad \int_0^l u_2(y)\,dy = 0,$$

where the latter formulation already incorporates Eq. (3.4.7).

The solution of the forced flow, after twice integration and the use of boundary conditions, is

$$u_1(y) = -\frac{y}{l}U - \frac{1}{2}\left(\frac{l^2}{\mu}\right)\frac{d}{dx}(p - p_0)\left(1 - \frac{y}{l}\right)\frac{y}{l}, \tag{3.4.8}$$

which combines the Couette flow with the Poiseuille flow. (Is this an expected result?) Inserting Eq. (3.4.8) into $\int_0^l u_1(y)\,dy = 0$, we get the magnitude of the induced pressure gradient,

$$-\frac{d}{dx}(p - p_0) = 6\frac{\mu U}{l^2}. \tag{3.4.9}$$

Combining Eqs. (3.4.8) and (3.4.9) determines the forced flow,

$$u_1(y) = U\left[2\left(\frac{y}{l}\right) - 3\left(\frac{y}{l}\right)^2\right]. \tag{3.4.10}$$

The solution of the buoyancy-driven flow, after integrating twice, is

$$u_2(y) = \frac{g\beta\,\Delta T l^2}{\nu}\left\{\frac{1}{6}\left[\frac{y}{l} - \left(\frac{y}{l}\right)^3\right] - \frac{1}{2}\left(\frac{T_0 - T_1}{\Delta T}\right)\left[\frac{y}{l} - \left(\frac{y}{l}\right)^2\right]\right\}, \tag{3.4.11}$$

which yields, with the conservation of mass,

$$T_0 - T_1 = \tfrac{1}{2}\Delta T, \tag{3.4.12}$$

where T_0 is the temperature relative to which the buoyancy occurs. Combining Eqs. (3.4.11) and (3.4.12) determines the buoyancy-driven flow,

$$u_2(y) = -\frac{g\beta\,\Delta T l^2}{\nu}\left[\frac{y}{l} - 3\left(\frac{y}{l}\right)^2 + 2\left(\frac{y}{l}\right)^3\right]. \tag{3.4.13}$$

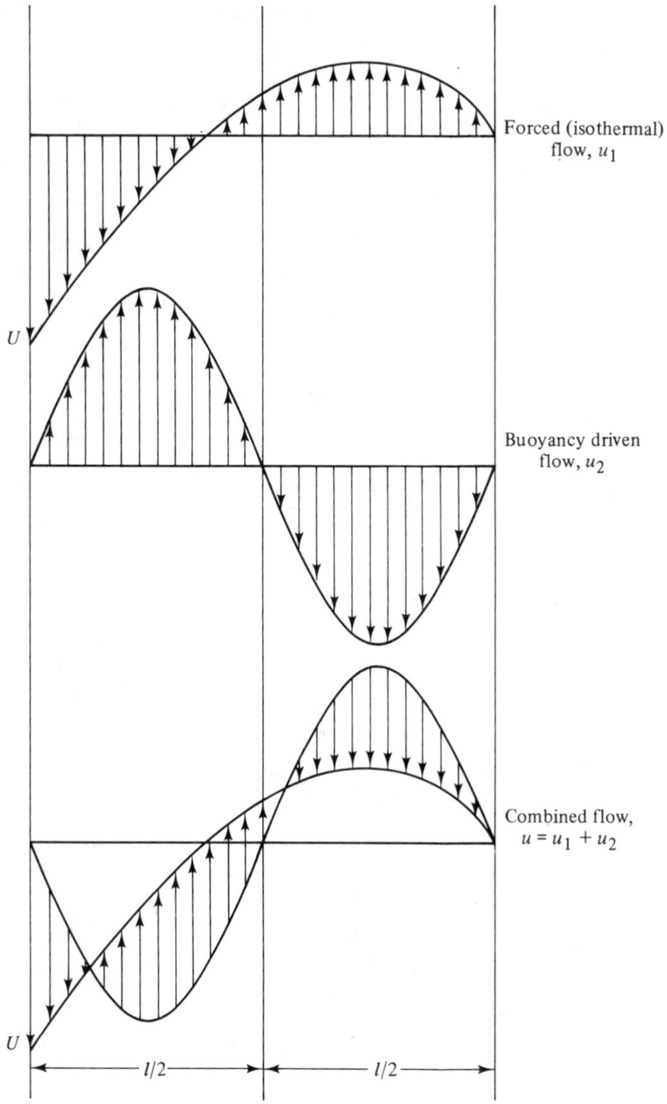

Fig. 3.15 Forced flow, buoyant flow, and combined flow, Ex. 3.4.1

The superposition of Eqs. (3.4.10) and (3.4.13) gives the combined (forced and buoyancy-driven) flow (Fig. 3.15). Here the relative contribution of the forced and buoyancy-driven flows to the combined flow is measured by the ratio

$$g\beta \frac{\Delta T l^2}{\nu U} = \frac{\text{Gr}}{\text{Re}}, \tag{3.4.14}$$

where $\text{Gr} = g\beta\,\Delta T l^3/\nu^2$ is the Grashof number and $\text{Re} = Ul/\nu$ is the Reynolds number.

Having explained the origin of the explicit pressure gradient in combined flows, we proceed now to a buoyancy-driven unsteady flow.

EXAMPLE 3.4.2 Consider a stagnant Newtonian fluid confined in a closed vertical slot (Fig. 3.16). The vertical side walls of the slot are separated by a distance $2l$. From an initial isothermal condition at T_0, the temperature of one of the side walls is suddenly raised to $T_0 + \Delta T$ and the other is lowered to $T_0 - \Delta T$. We wish to determine the unsteady motion of the fluid in the slot for a Prandtl number of unity (see Problem 3-17 for $\Pr \neq 1$).

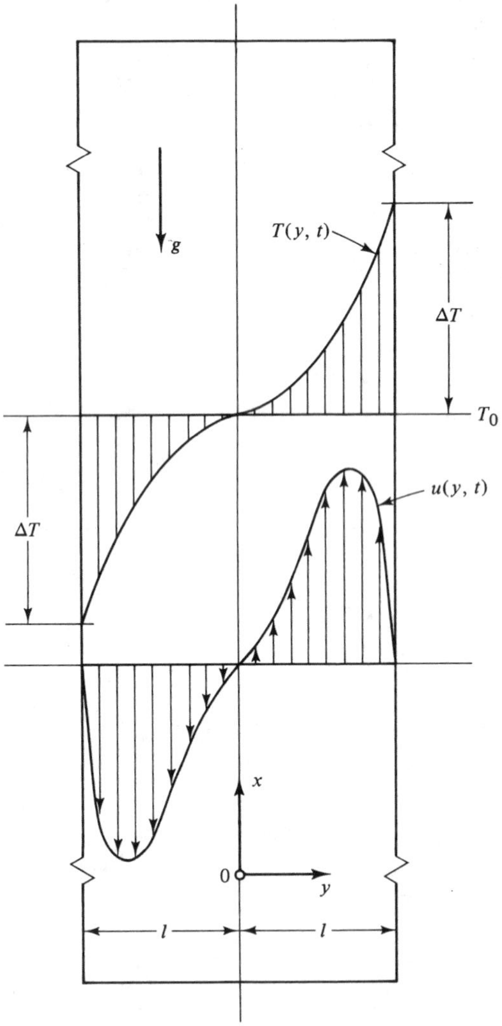

Fig. 3.16 Unsteady natural convection in a vertical slot with sudden change in wall temperature, Ex. 3.4.2

The differential formulation of the problem for the right one-half of the slot is

$$\frac{1}{\nu}\frac{\partial u}{\partial t} = \frac{g\beta}{\nu}(T - T_0) + \frac{\partial^2 u}{\partial y^2} \tag{3.4.15}$$

subject to

$$u(y, 0) = 0; \qquad u(0, t) = 0, \qquad u(l, t) = 0,$$

and

$$\frac{1}{a}\frac{\partial T}{\partial t} = \frac{\partial^2 T}{\partial y^2} \tag{3.4.16}$$

subject to

$$T(y, 0) = T_0; \qquad T(0, t) = T_0, \qquad T(l, t) = T_0 + \Delta T,$$

and the condition

$$\int_{-l}^{l} u(y, t)\, dy = 0 \tag{3.4.17}$$

requires asymmetric temperature and velocity distributions for the left one-half.

The temperature problem readily gives, either by the use of the separation of variables or by the Laplace transforms,

$$\frac{T(y, t) - T_0}{\Delta T} = \frac{y}{l} + 2\sum_{n=1}^{\infty}\frac{(-1)^n}{\lambda_n l}e^{-a\lambda_n^2 t}\sin \lambda_n y, \tag{3.4.18}$$

where $\lambda_n l = n\pi$, $n = 1, 2, 3, \ldots$. However, the velocity problem is somewhat involved. While the use of the separation of variables, requiring the consideration of the Duhamel integral, leads to lengthy calculations, the use of Laplace transforms presents no difficulties.

Thus, taking the transform of both the temperature and velocity problems, solving the resulting ordinary differential equations with the assumption that $\nu = a$, and satisfying the transformed boundary conditions, we have for the temperature

$$\bar{T}(y, p) - \frac{T_0}{p} = \Delta T\frac{\sinh qy}{p \sinh ql},$$

and in terms of this temperature, we have for the velocity

$$\frac{\bar{u}(y, p)}{g\beta\, \Delta T/2\nu} = \frac{l \cosh ql(\sinh qy/\sinh ql) - y \cosh qy}{pq \sinh ql}, \tag{3.4.19}$$

where $q = (p/\nu)^{1/2}$ and p is the transform variable. Hereafter ν denotes the common value of $\nu = a$. The inversion of Eq. (3.4.19) requires the evaluation of residues associated with the poles of each of its two terms

$$\frac{u(y, t)}{g\beta\, \Delta T/2\nu} = \text{Res}_1\,(0) + \sum_{n=1}^{\infty}\text{Res}_1\,(-\nu\lambda_n^2) - \left[\text{Res}_2\,(0) + \sum_{n=1}^{\infty}\text{Res}_2\,(-\nu\lambda_n^2)\right].$$

Omitting the lengthy details (see Problem 3-18), the unsteady velocity of the natural convection becomes

$$\frac{u(y, t)}{g\beta\, \Delta T l^2/2\nu} = \frac{1}{3}\left(\frac{y}{l}\right)\left[1 - \left(\frac{y}{l}\right)^2\right] + 2\sum_{n=1}^{\infty}\frac{(-1)^n}{(\lambda_n l)^2}e^{-\nu\lambda_n^2 t}\left\{\left[2(\lambda_n l)\frac{\nu t}{l^2} + \frac{3}{\lambda_n l}\right]\sin \lambda_n y - \left(\frac{y}{l}\right)\cos \lambda_n y\right\}, \tag{3.4.20}$$

where $\lambda_n l = n\pi$, $n = 1, 2, 3, \ldots$. This solution, although uniformly convergent, converges slowly for small values of time, and, in a sense, corresponds to the solution of the second time domain of an integral formulation. Thus, a complementary solution which rapidly converges for small values of time, and, in a sense, corresponds to the solution of the first time domain of an integral formulation, is needed. To get this solution, let Eq. (3.4.19) be rearranged and expanded in terms of exponentials as

$$\frac{\bar{u}(y, p)}{g\beta\,\Delta T/2\nu} = \frac{1}{pq}\sum_{n=0}^{\infty}\{[(n+1)l - y]e^{-q[(2n+1)l-y]} - [(n+1)l + y]e^{-q[(2n+1)l+y]} + (n+1)l(e^{-q[(2n+3)l-y]} - e^{-q[(2n+3)l+y]})\},$$

which is inverted to give the complementary solution to the unsteady velocity,

$$\begin{aligned}\frac{u(y, t)}{g\beta\,\Delta T l^2/2\nu} = \sum_{n=0}^{\infty}\Bigg(&\left[(n+1) - \left(\frac{y}{l}\right)\right]\Bigg\{2\sqrt{\frac{\nu t}{\pi l^2}}e^{-[(2n+1)-(y/l)]^2 l^2/4\nu t} \\ &- \left[(2n+1) - \left(\frac{y}{l}\right)\right]\operatorname{erfc}\left[\frac{(2n+1)-(y/l)}{2\sqrt{\nu t/l^2}}\right]\Bigg\} \\ &- \left[(n+1) + \left(\frac{y}{l}\right)\right]\Bigg\{2\sqrt{\frac{\nu t}{\pi l^2}}e^{-[(2n+1)+(y/l)]^2 l^2/4\nu t} \\ &- \left[(2n+1) + \left(\frac{y}{l}\right)\right]\operatorname{erfc}\left[\frac{(2n+1)+(y/l)}{2\sqrt{\nu t/l^2}}\right]\Bigg\} \\ &+ (n+1)\Bigg\{2\sqrt{\frac{\nu t}{\pi l^2}}e^{-[(2n+3)-(y/l)]^2 l^2/4\nu t} \\ &- [(2n+3) - (y/l)]\operatorname{erfc}\left[\frac{(2n+3)-(y/l)}{2\sqrt{\nu t/l^2}}\right] \\ &\quad - 2\sqrt{\frac{\nu t}{\pi l^2}}e^{-[(2n+3)+(y/l)]^2 l^2/4\nu t} \\ &+ [(2n+3) + (y/l)]\operatorname{erfc}\left[\frac{(2n+3)+(y/l)}{2\sqrt{\nu t/l^2}}\right]\Bigg\}\Bigg).\end{aligned} \tag{3.4.21}$$

The foregoing solutions are algebraically somewhat involved, both to derive and to apply. Our aim being the solution of engineering problems suggests an alternative but simpler solution to be based on the integral formulation, which we consider next.

Clearly, the problem is one of penetration depth, and must be solved in terms of two time domains. However, the temperature problem is simple and needs no elaboration (see Ex. 3.2.2). For the right one-half of the slot, when $t \leq t_0$,

$$\frac{T(y, t) - T_0}{\Delta T} = \left(\frac{y}{\Delta}\right)^2, \tag{3.4.22}$$

where $\Delta = (12at)^{1/2}$ and $t_0 = l^2/12a$, and when $t > t_0$,

$$\frac{T(y, t) - T_0}{\Delta T} = \eta - \eta(1 - \eta)\exp(\tau_0 - \tau), \tag{3.4.23}$$

where $\eta = y/l$ and $\tau = 12at/l^2$. Asymmetric profiles hold for the left one-half of the slot.

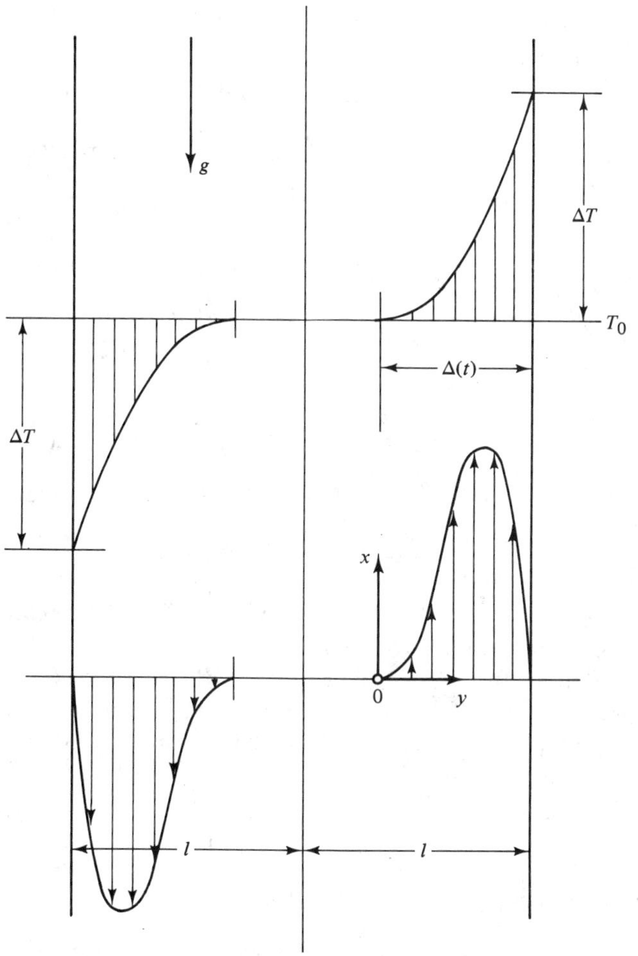

Fig. 3.17 First time domain; temperature and velocity profiles, Ex. 3.4.2

For the first time domain of the velocity problem, assuming because of Pr = 1 the penetration depth of the velocity profile to be equal to that of the temperature profile, we have from Fig. 3.17,

$$u(y, t) = U(t)\left(1 - \frac{y}{\Delta}\right)\left(\frac{y}{\Delta}\right)^2, \tag{3.4.24}$$

where $U(t)$ is the scale factor to be determined according to the integral formulation

$$\frac{1}{\nu}\frac{d}{dt}\int_0^\Delta u\,dy = \frac{g\beta}{\nu}\int_0^\Delta (T - T_0)\,dy + \left(\frac{\partial u}{\partial y}\right)_{y=\Delta} \tag{3.4.25}$$

Inserting Eqs. (3.4.22) and (3.4.24) into Eq. (3.4.25), and rearranging the result, we have

$$\frac{d}{dt}(U\Delta) = 8(g\beta\,\Delta T)\Delta - 12\nu\left(\frac{U}{\Delta}\right). \tag{3.4.26}$$

In terms of $\Delta = (12at)^{1/2}$ and $\nu = a$, Eq. (3.4.26) is reduced to

$$t\frac{dU}{dt} + \frac{3}{2}U = 4(g\beta\,\Delta T)t,$$

whose integration subject to $U(0) = 0$ yields

$$U(t) = \tfrac{8}{5}(g\beta\,\Delta T)t. \tag{3.4.27}$$

Substituting Eq. (3.4.27) into Eq. (3.4.24), and with some rearrangement, we get when $t \leq t_0$,

$$\frac{U(y, t)}{g\beta(\Delta T)\Delta^2/2\nu} = \frac{4 \times 4}{5}\left(\frac{at}{\Delta^2}\right)\left(1 - \frac{y}{\Delta}\right)\left(\frac{y}{\Delta}\right)^2. \tag{3.4.28}$$

For the second time domain, Fig. 3.18 suggests the profile

$$\frac{u(y, t)}{g\beta\,\Delta T l^2/2\nu} = \tfrac{1}{3}\eta(1 - \eta^2)[1 - \delta(\tau)] + \tfrac{16}{5}(1 - \eta)\eta^2\,\delta(\tau), \tag{3.4.29}$$

where $\eta = y/l$, $\tau = 12at/l^2$, and $\delta(\tau)$ is a scale factor expected to be exponentially decreasing with time. Note that the first term in Eq. (3.4.29) is the steady solution valid both for the left and right halves of the slot. However, the second term, chosen to satisfy Eq. (3.4.28) at the end of the first time domain, is valid only for the right half of the slot. Now, inserting Eqs. (3.4.23) and (3.4.29) into the integral formulation of the second domain,

$$\frac{1}{\nu}\frac{d}{dt}\int_0^l u\,dy = \frac{g\beta}{\nu}\int_0^l (T - T_0)\,dy + \left.\frac{\partial u}{\partial y}\right|_{y=0}^{y=l} \tag{3.4.30}$$

and rearranging the result, using $\nu = a$, we have

$$\frac{d\delta}{dt} + (\tau - \tau_0) = -\tfrac{5}{33}\exp(\tau_0 - \tau),$$

whose solution for $\delta(\tau_0) = 1$ is

$$\delta(\tau) = [1 - \tfrac{5}{33}(\tau - \tau_0)]\exp(\tau_0 - \tau). \tag{3.4.31}$$

Introducing Eq. (3.4.31) into Eq. (3.4.29) gives the velocity distribution for the second domain.

Finally, we examine the steady solution of this example in terms of the vorticity. Noting that $\omega = -\partial u/\partial y$ and differentiating Eq. (3.4.29) for $\tau \to \infty$ gives

$$\omega(y) = \frac{g\beta\,\Delta T l}{2\nu}\left[\left(\frac{y}{l}\right)^2 - \frac{1}{3}\right].$$

This distribution is readily seen to be the solution of the equation of vorticity transport

$$\frac{\partial^2\omega}{\partial y^2} = \frac{g\beta\,\Delta T}{\nu l},$$

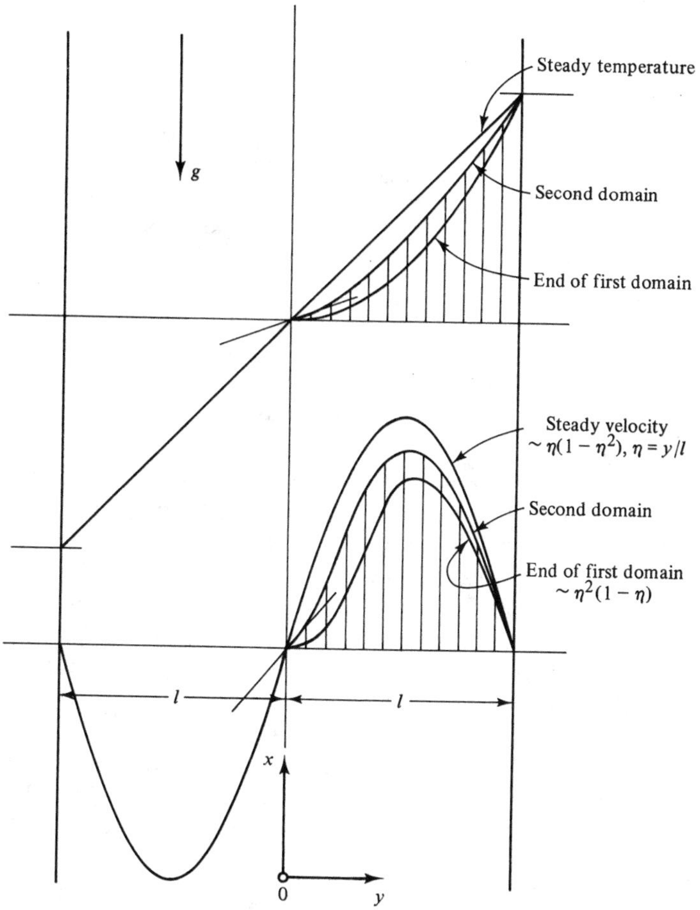

Fig. 3.18 Second time domain; temperature and velocity profiles, Ex. 3.4.2

obtained either by reducing Eq. (2.4.31) or by differentiating the steady part of Eq. (3.4.15) with respect to y and inserting $\partial T/\partial y = \Delta T/l$. Note that the constant temperature gradient provides a uniform source of vorticity, yielding a parabolic distribution as vorticity diffuses to the walls.

Here we terminate our study of forced and buoyancy-driven parallel flows, and proceed to thermal boundary layers associated with forced flows.

3.5 *THERMAL BOUNDARY LAYERS*

As we already stated in Chapter 1, convection studies are ultimately concerned with heat transfer through a solid–fluid interface. Because of its importance, here it is worth repeating and somewhat elaborating this fact. The heat transfer, say q_w, is conveniently expressed in terms of the coefficient of heat transfer h defined according to

$$q_w = h(T_w - T_\infty), \tag{3.5.1}$$

where T_w and T_∞ denote the interface and ambient temperatures, respectively. Also, expressing q_w by means of conduction in the fluid (Fig. 1.4), Eq. (3.5.1) becomes

$$-k\left(\frac{\partial T}{\partial y}\right)_w = h(T_w - T_\infty),$$

which may be rearranged in the dimensionless form

$$\text{Nu} = \frac{hl}{k} = -\frac{[\partial T/\partial(y/l)]_w}{T_w - T_\infty}, \tag{3.5.2}$$

where Nu is the *Nusselt number*, k and l are the thermal conductivity of and a characteristic length for the fluid, and y is the curvilinear coordinate in the direction of heat transfer. Consequently, the evaluation of q_w is reduced to that of the coefficient of heat transfer, which in turn is proportional to the gradient of the fluid temperature normal to the interface. Here we evaluate this gradient from an exact or approximate solution of the fluid temperature. The boundary layer (or penetration depth) concept provides a convenient tool for an approximate solution, and is emphasized in this section. Later we shall study two other methods of evaluating the coefficient of heat transfer. (1) The analogy between heat transfer and momentum transfer, to be supplemented by wall friction measurements, is considered in Chapter 4 for laminar flow and in Chapter 11 for turbulent flow. (2) Dimensional analysis, employing also the foregoing analogy and to be supplemented by some heat transfer measurements, is considered in Chapter 12 for turbulent flow for which it is particularly suitable.

Clearly, q_w may also be expressed by means of conduction in the solid (Fig. 1.4). Then we have

$$-k_s\left(\frac{\partial T_s}{\partial y}\right)_w = h(T_w - T_\infty),$$

or, in the dimensionless form,

$$\text{Bi} = \frac{hl_s}{k_s} = -\frac{[\partial T_s/\partial(y/l_s)]_w}{T_w - T_\infty}, \tag{3.5.3}$$

where Bi is the *Biot number* k_s and l_s are the thermal conductivity of and a characteristic length for the solid. When the geometry of the fluid is finite, T_∞ of Eqs. (3.5.2) and (3.5.3) is conveniently replaced by a mean temperature for the fluid, say the *bulk temperature* for incompressible flow,

$$T_b = \frac{1}{AU}\int_A u(r, t)T(r, t)\, dA,$$

where A is the area transversal to motion and U the transversally averaged velocity of the fluid.

Note the conceptual difference in the use of Eqs. (3.5.2) and (3.5.3). In conduction problems h and T_∞ (or T_b) are given, and Eq. (3.5.3) is employed as a boundary condition; on the other hand, because of their complexity, convection problems are first solved in terms of simpler boundary conditions unrelated to h (such as specified temperature or specified heat flux); later Eq. (3.5.2) is utilized for the evaluation of h.

Returning to the evaluation of h from the solution of thermal problems associated with incompressible flows, let us first comment on the velocities of these problems. For the incompressible flow of a constant-property fluid, the momentum equation becomes decoupled from the energy equation and it can be solved separately. Thus, from our background on isothermal incompressible flows (recall the penetration depth of Ex. 3.2.2), we have for the thickness of the momentum (velocity) boundary layer

$$\delta \sim \nu^n,$$

where n is a fractional number, being equal to $\frac{1}{2}$ in problems discussed so far. Also, for the thickness of the analogous thermal boundary layers, we have

$$\Delta \sim a^n.$$

Then the thickness ratio of the momentum and thermal boundary layers is

$$\frac{\delta}{\Delta} \sim \mathrm{Pr}^n,$$

Pr denoting the *Prandtl number*. The range of this number for various fluids (see Fig. 3.19) indicates that the thickness ratio of boundary layers varies between the limits of

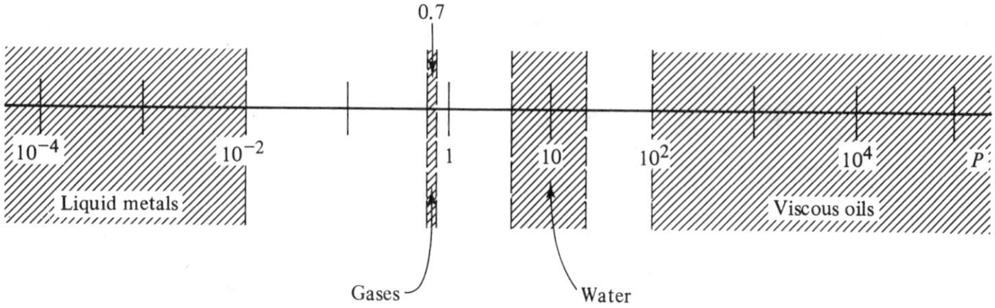

Fig. 3.19 Range of Prandtl number for various fluids

0 and ∞. This ratio usually assumes the values $\delta/\Delta \ll 1$ for liquid metals, $\delta/\Delta \sim 1$ for gases, $\delta/\Delta > 1$ for water, and $\delta/\Delta \gg 1$ for viscous oils. Thus, for liquid metals, the limit $\delta/\Delta \longrightarrow 0$ of $\delta/\Delta \ll 1$ suggests that the momentum boundary layer be ignored relative to the thermal boundary layer. For gases, being slightly less than unity, δ/Δ is very closely approximated when thicknesses of both layers are assumed identical. For water, δ/Δ is somewhat greater than unity, and there appears to be no simplifying limit for this case. For viscous oils, the limit $\delta/\Delta \longrightarrow \infty$ of $\delta/\Delta \gg 1$ suggests that the development of the momentum boundary layer, being much faster than that of the thermal boundary layer, be ignored and the momentum layer be assumed fully developed relative to the thermal layer. For various fluids, the thickness ratio of momentum and thermal layers and their limits are sketched in Fig. 3.20. Leaving to

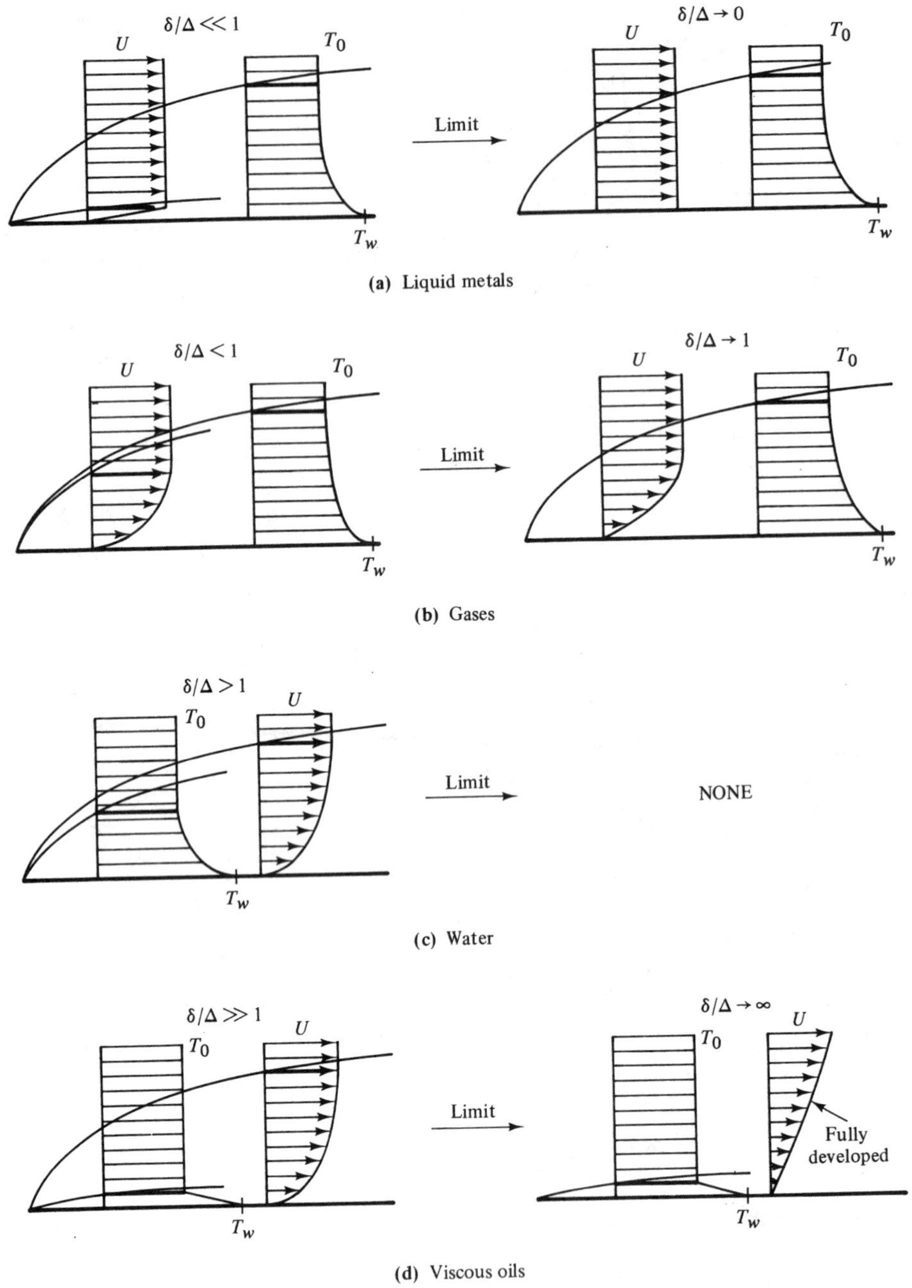

Fig. 3.20 Thickness ratio of momentum and thermal boundary layers for various fluids

Chapter 4 the study of thermal, as well as momentum, boundary layers, we devote this section to examples on parallel flows prescribed by the limit $\delta/\Delta \longrightarrow 0$ for liquid metals and by $\delta/\Delta \longrightarrow \infty$ for viscous oils.

EXAMPLE 3.5.1 A liquid metal flows steadily between two parallel plates separated a distance l apart. The upper plate is insulated, while the lower plate is kept at uniform temperature T_w. The inlet temperature of the liquid is T_∞ and, since $\delta/\Delta \longrightarrow 0$, the velocity of the liquid may be assumed uniform, say U. We wish to determine the temperature of the liquid.

After neglecting the effect of axial conduction,* the differential formulation of the problem with respect to the coordinate system attached at the inlet to the upper plate is

$$\frac{U}{a}\frac{\partial T}{\partial x} = \frac{\partial^2 T}{\partial y^2}; \qquad T(0, y) = T_0,$$

$$\frac{\partial T(x, 0)}{\partial y} = 0, \qquad T(x, l) = T_w.$$

The solution, obtained either by the use of separation of variables or Laplace transforms, is

$$\frac{T - T_w}{T_0 - T_w} = 2\sum_{n=0}^{\infty} \frac{(-1)^n}{(\lambda_n l)} e^{-\lambda_n^2 ax/U} \cos \lambda_n y, \tag{3.5.4}$$

with $\lambda_n l = (2n + 1)\pi/2$, $n = 0, 1, 2, \ldots$. (See Ex. 4.5 of Arpaci 1966. Reformulate the present problem with respect to an observer moving with velocity U. See Ex. 5.2 of the same text.) Although uniformly convergent, the solution given by Eq. (3.5.4) converges slowly for small values of x, and, in a sense, corresponds to the solution of the second domain of an integral formulation. Thus, a complementary solution which converges rapidly for small values of x, and, in a sense, corresponds to the solution of the first domain of an integral formulation, is needed. To get this solution, the Laplace transform of the differential formulation may be rearranged in terms of exponentials before inversion, which then gives

$$\frac{T - T_w}{T_0 - T_w} = 1 - \sum_{n=0}^{\infty} (-1)^n \left\{\operatorname{erfc}\left[\frac{(2n+1) - y/l}{2(ax/Ul^2)^{1/2}}\right] - \operatorname{erfc}\left[\frac{(2n+1) + y/l}{2(ax/Ul^2)^{1/2}}\right]\right\}. \tag{3.5.5}$$

This solution, although obtained with the aim of having rapid convergence for small values of x, turns out to be uniformly convergent for all values of x. (Example 7.23 of Arpaci 1966 gives the solution of our problem relative to an observer moving with velocity U. From this solution, replacing t with x/U, we readily obtain Eq. 3.5.5.)

We now proceed to the main objective of this example, that is, to the integral formulation and its solution by approximate profiles. Clearly, the problem is one of boundary layer (penetration depth), and must be solved in terms of two space domains. For the first domain of a thermal boundary layer, now measuring y from the lower plate (Fig. 3.21) gives

$$\rho c \frac{d}{dx}\int_0^\Delta u(T - T_0)\, dy = -k\left(\frac{\partial T}{\partial y}\right)_{y=0} \tag{3.5.6}$$

---

*For the effect of axial conduction on this problem, see Ex. 4.7 of Arpaci 1966.

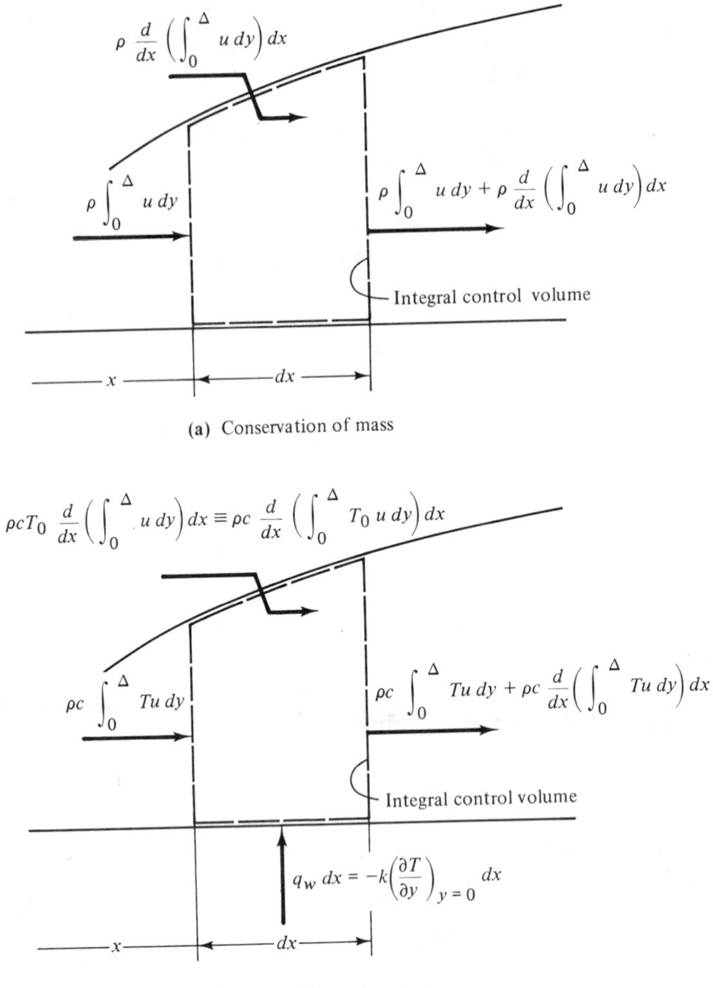

Fig. 3.21 First domain thermal boundary layer, Ex. 3.5.1

subject to

$$T(0, y) = T_0, \qquad T(x, 0) = T_w, \qquad T(x, \Delta) = T_0, \tag{3.5.7}$$

and

$$\frac{\partial T(x, \Delta)}{\partial y} = 0. \tag{3.5.8}$$

Note that the differential solution inherently satisfies but does not require the explicit use of Eq. (3.5.8). However, a first-order profile, defined to approximate the apparent physics of the problem, explicitly needs the use of Eq. (3.5.8).

Since the first domain profile of boundary layer problems cannot be obtained by scaling the asymptotic (fully developed) solution of these, we start with simple functions (such as polynomials) and require them to satisfy the last two conditions of Eq. (3.5.7), and Eq. (3.5.8). This gives a first-order profile,

$$\frac{T - T_w}{T_0 - T_w} = 2\left(\frac{y}{\Delta}\right) - \left(\frac{y}{\Delta}\right)^2. \tag{3.5.9}$$

Inserting Eq. (3.5.9) into Eq. (3.5.6) yields

$$d\Delta^2 = 12\left(\frac{a}{U}\right) dx, \qquad \Delta(0) = 0$$

and

$$\Delta(x) = \left(\frac{12ax}{U}\right)^{1/2}. \tag{3.5.10}$$

[Which condition of Eq. 3.5.7 is related to $\Delta(0) = 0$? How?] Introducing Eq. (3.5.10) into Eq. (3.5.9) gives a first-order approximation for the temperature of the liquid metal in the first domain,

$$\frac{T - T_w}{T_0 - T_w} = 2\frac{y}{(12ax/U)^{1/2}} - \frac{y^2}{12ax/U}, \qquad 0 \le x \le x_0. \tag{3.5.11}$$

The thermal boundary layer reaches the upper plate as $\Delta = l$. Then, from Eq. (3.5.10), we have the *penetration distance* of the first domain (Fig. 3.22),

$$x_0 = \frac{Ul^2}{12a}. \tag{3.5.12}$$

For the second domain, the balance of thermal energy yields, in terms of the integral control volume shown in Fig. 3.22, or, after letting $\Delta = l$ in Eq. (3.5.6),

$$\frac{l^2}{a}\frac{d}{dx}\int_0^1 u(T_0 - T)\, d\eta = \left(\frac{\partial T}{\partial \eta}\right)_{\eta=0}, \tag{3.5.13}$$

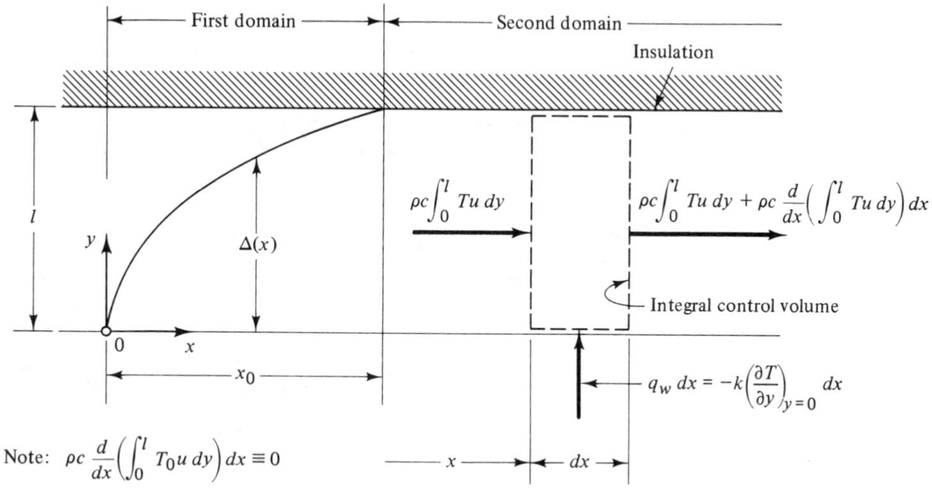

Fig. 3.22 Penetration distance of first domain; control volume for second domain, Ex. 3.5.1

where $\eta = y/l$. A first-order profile for this domain may be constructed by scaling the terminal profile of the first domain. Accordingly,

$$\frac{T - T_w}{T_0 - T_w} = (2\eta - \eta^2)\,\Delta_1(x), \tag{3.5.14}$$

$\Delta_1(x)$ being the scale factor. Inserting Eq. (3.5.14) into Eq. (3.5.13) results in

$$\frac{d\Delta_1}{dx} + 3\left(\frac{a}{Ul^2}\right)\Delta_1 = 0 \qquad \text{with} \qquad \Delta_1(x_0) = 1. \tag{3.5.15}$$

Introducing the solution of Eq. (3.5.15),

$$\Delta_1(x) = \exp\left[\frac{-3a(x - x_0)}{Ul^2}\right],$$

into Eq. (3.5.14) yields a first-order approximation for the temperature of the liquid metal in the second domain,

$$\frac{T - T_w}{T_0 - T_w} = (2\eta - \eta^2)\exp\left[\frac{-3a(x - x_0)}{Ul^2}\right], \qquad x \geq x_0. \tag{3.5.16}$$

Clearly, Eq. (3.5.16) shows how the temperature of the second domain *volumetrically collapses* to $T_w(< T_0)$. If we wish to define a *collapse distance* for the second domain, our problem now becomes a penetration depth problem in x. Then, from the balance of thermal energy for the integral control volume shown in Fig. 3.23, we have

$$\frac{Ul^2}{a}\int_0^1 (T_{\xi=0} - T_w)\,d\eta = \int_0^{\xi_1}\left(\frac{\partial T}{\partial \eta}\right)_{\eta=0} d\xi, \tag{3.5.17}$$

where $\xi_1 = x_1 - x_0$ is the collapse distance. A first-order profile involving this distance is

$$\frac{T - T_w}{T_0 - T_w} = (2\eta - \eta^2)\left(1 - \frac{\xi}{\xi_1}\right)^2. \tag{3.5.18}$$

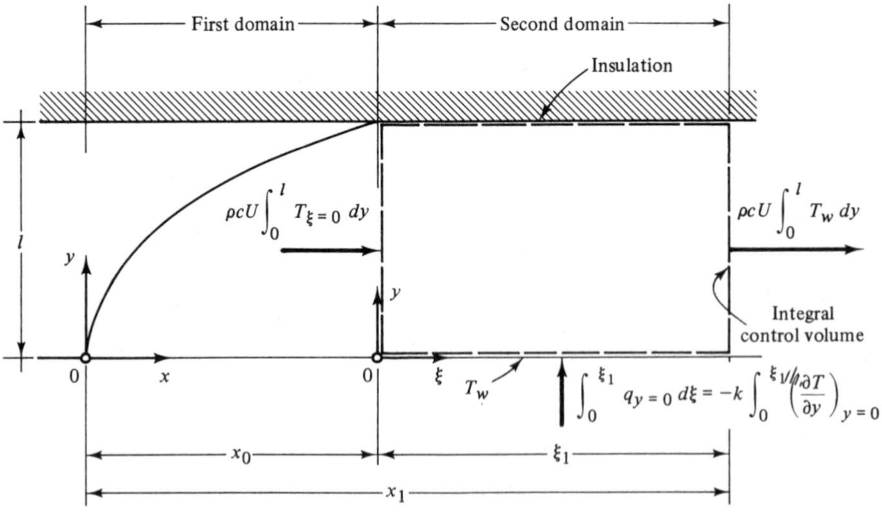

Fig. 3.23 Collapse distance ξ_1 for second domain, Ex. 3.5.1

Inserting Eq. (3.5.18) into Eq. (3.5.17) results in the approximate collapse distance for the second domain,

$$\xi_1 = \frac{Ul^2}{a}.$$

As already mentioned in the beginning of this section, convection studies are ultimately concerned with the evaluation of the heat transfer coefficient along a solid–fluid interface. We now evaluate this coefficient for the first domain of the present problem.

Selecting $T_w - T_0$ as the appropriate temperature difference Eq. (3.5.2), evaluated in terms of Eqs. (3.5.9) and (3.5.10), gives the Nusselt number of the thermal boundary layer in uniform flow along the lower isothermal plate

$$\text{Nu}_x = \frac{hx}{k} = -\frac{[\partial T/\partial(y/\Delta)]_{y=0}}{T_w - T_0}\frac{x}{\Delta} = \frac{1}{\sqrt{3}}\left(\frac{Ux}{a}\right)^{1/2},$$

or

$$\text{Nu}_x = \frac{1}{\sqrt{3}}\text{Pe}_x^{1/2}, \tag{3.5.19}$$

where $\text{Pe}_x = Ux/a$ denotes the *Péclet number*.

For the second domain, see Problem 3-22. Here, employing Eqs. (3.5.16) and (3.5.18), the evaluation of the heat transfer coefficient requires use of the appropriate bulk temperature, to be illustrated later, in Ex. 3.5.5.

EXAMPLE 3.5.2 Resolve Ex. 3.5.1 after assuming the lower plate insulated for $x \geq x_0$.

This actual problem may best be solved by means of a superposition. To help the superposition, consider a second problem for $x \geq x_0$; let the upper plate of Ex. 3.5.1 remain insulated, while the lower plate is subjected to a uniform heat flux, say q'', and the inlet temperature of the liquid is T_w. Later, in a third problem, the value of q'' is to be adjusted to cancel the heat loss from the lower plate. The superposition of the second and third problems with that of Ex. 3.5.1 is identical to the actual problem to be solved.

The integral formulation of the second problem is (Fig. 3.24)

$$\rho c U \frac{d}{dx}\int_0^{\Delta_2}(T - T_w)\,dY = q''. \tag{3.5.20}$$

Satisfying the apparent physics of the problem, a first-order profile in Y measured from the boundary layer downward is

$$T - T_w = \frac{q''Y^2}{2k\Delta_2}, \tag{3.5.21}$$

where Δ_2 satisfies, after inserting Eq. (3.5.21) into Eq. (3.5.20),

$$d\Delta^2 = 6\left(\frac{a}{U}\right)dx, \qquad \Delta_2(0) = 0,$$

which yields

$$\Delta_2(x) = \left(\frac{6ax}{U}\right)^{1/2}. \tag{3.5.22}$$

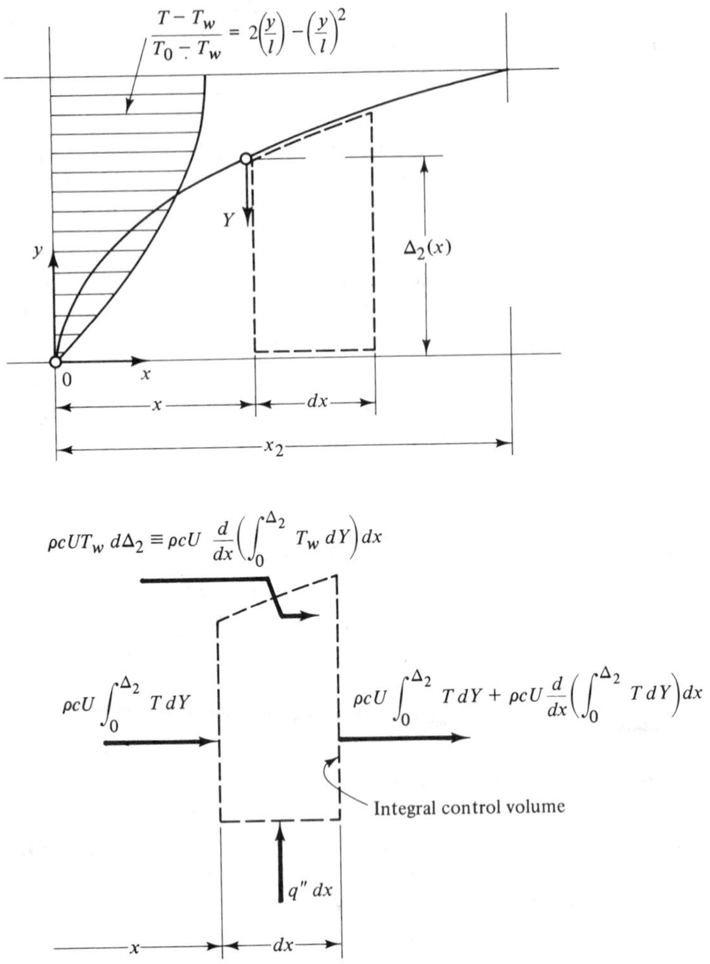

Fig. 3.24 Second problem of second domain, Ex. 3.5.2

For $\Delta_2 = l$, we have a new penetration distance from Eq. (3.5.22),

$$x_2 = \frac{Ul^2}{6a}. \tag{3.5.23}$$

Now, employing $q_{y=0}$ of Ex. 3.5.1, we obtain the specific value of q'' from

$$q_{y=0} + q'' = 0,$$

which gives

$$+k\left(\frac{\partial T}{\partial y}\right)_{y=0} = q'' = 2k\frac{T_0 - T_w}{l}. \tag{3.5.24}$$

Introducing Eq. (3.5.24) into Eq. (3.5.21) gives

$$\frac{T - T_w}{T_0 - T_w} = \frac{Y^2}{l\Delta_2}. \tag{3.5.25}$$

Clearly, the inlet temperature at $x = x_0$, that is, Eq. (3.5.9) for $\Delta = l$ [or Eq. (3.5.16) for $x = x_0$] plus Eq. (3.5.25) satisfy the condition of no heat flux at the lower plate. However, the second problem with q'' adds energy to the liquid through the lower plate, which is supposed to be insulated. To eliminate this energy, consider a third problem for $x \geq x_0$; let the upper plate remain insulated, while the previous heat flux, q'', is extracted from the lower plate, and the inlet temperature of the liquid is T_w.

The lumped formulation of the third problem is (Fig. 3.25)

$$\rho c U l \frac{dT}{dx} = -q'', \qquad T(0) = T_w$$

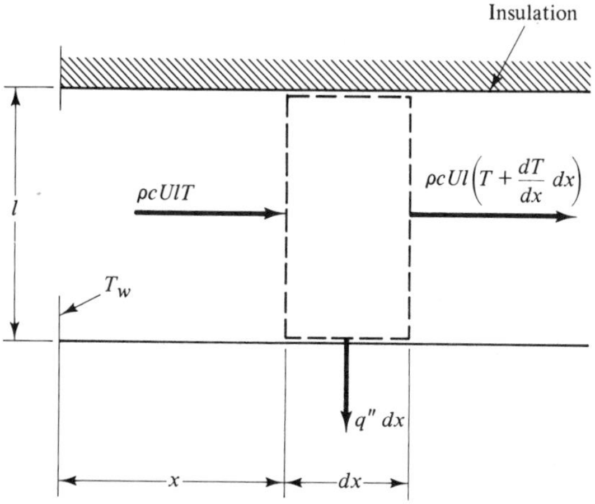

Fig. 3.25 Third problem of second domain, Ex. 3.5.2

which yields

$$T = T_w - \frac{q''x}{\rho c U l},$$

or, in view of Eq. (3.5.24),

$$\frac{T - T_w}{T_0 - T_w} = -2\frac{ax}{Ul^2}. \tag{3.5.26}$$

Finally, the superposition of Eq. (3.5.9) for $\Delta = l$ [or Eq. (3.5.16) for $x = x_0$] with Eqs. (3.5.25) and (3.5.26) gives for $x \geq x_0$ a first-order approximation for the actual temperature of the liquid metal,

$$\frac{T - T_w}{T_0 - T_w} = 2\left(\frac{y}{l}\right) - \left(\frac{y}{l}\right)^2 + \frac{Y^2}{l\Delta_2} - 2\frac{ax}{Ul^2}, \qquad \Delta_2 = \left(\frac{6ax}{U}\right)^{1/2}. \tag{3.5.27}$$

The three profiles of this temperature are separately shown in Fig. 3.26(a) and the result of their superposition is shown shaded in Fig. 3.26(b). (What happened to the second domain of the second problem? What is the distance over which the inlet temperature, $(T - T_w)/(T_0 - T_w) = 2(y/l) - (y/l)^2$, becomes uniform? Compare the mean value of this temperature with the value of Eq. (3.5.27) for $\Delta_2 = l$, say at

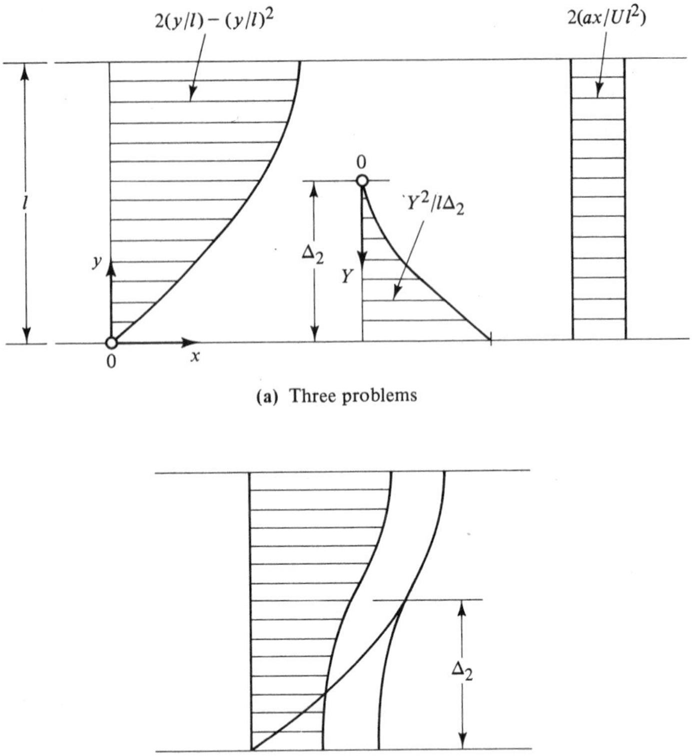

(a) Three problems

(b) Actual problem

Fig. 3.26 Superposition of three temperatures, Ex. 3.5.2

$x = x_2$. Why are these values identical?) The upper and lower plate temperatures, obtained by letting $y/l = 1,\ Y = 0$ and $y = 0,\ Y = \Delta_2 = (6ax/U)^{1/2}$, respectively, in Eq. (3.5.27),

$$\frac{T - T_w}{T_0 - T_w} = 1 - \frac{2}{\mathrm{Gz}_x}, \tag{3.5.28}$$

$$\frac{T - T_w}{T_0 - T_w} = \left(\frac{6}{\mathrm{Gz}_x}\right)^{1/2} - \frac{2}{\mathrm{Gz}_x}, \tag{3.5.29}$$

$1/\mathrm{Gz}_x = (x/l)/\mathrm{Pe}$, $\mathrm{Pe} = Ul/a$, Gz_x being the *Graetz number* and Pe the Péclet number, are plotted against $1/\mathrm{Gz}_x$ in Fig. 3.27. (Why is the upper plate temperature linear?)

EXAMPLE 3.5.3 Reconsider the first domain of Ex. 3.5.1. Assume that the lower plate now has a finite thickness l_w, and a finite thermal conductivity k_w. The lower surface of the plate is kept isothermal (say by dropwise condensation of a vapor at saturation temperature T_s). The fluid is inviscid and flows past the plate with a uniform velocity U and a uniform temperature T_0. We wish to determine the effect of the plate thickness on the problem.

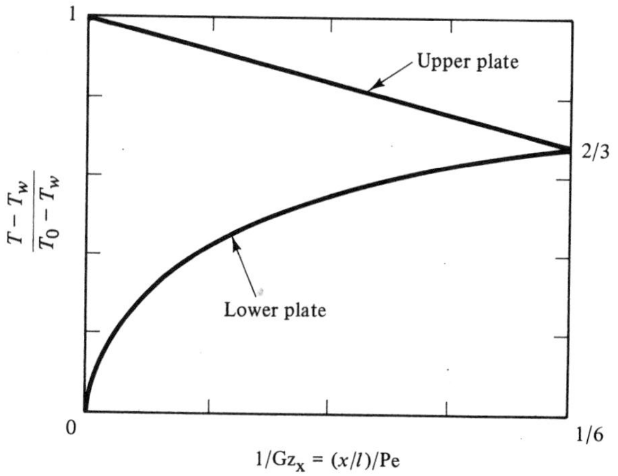

Fig. 3.27 Upper and lower plate temperatures, Ex. 3.5.2

The complexity of convection problems usually prevents the plate thickness from being included into the formulation of these problems. Here, in terms of the present example, we investigate the influence of this thickness by evaluating the variation of the interface temperature, and, consequently, show how good is the assumption of constant wall temperature made for some of these problems.

Our problem involves steady developing temperatures in two domains, coupled along the interface (Fig. 3.28). For the fluid, the balance of thermal energy gives (see Eq. 3.5.6)

$$\rho c U \frac{d}{dx}\int_0^{\Delta}(T - T_0)\,dy = q_w = -k\left(\frac{\partial T}{\partial y}\right)_{y=0}. \tag{3.5.30}$$

For the plate, neglecting the conduction along the plate,

$$q_w = -k\left(\frac{\partial T}{\partial y}\right)_{y=0} = \frac{k_w}{l_w}(T_s - T_w). \tag{3.5.31}$$

In terms of the appropriate parabolic profile—Eq. (3.5.9) now with T_w assumed x-dependent—Eqs. (3.5.30) and (3.5.31) yield

$$\frac{d}{dx}[\Delta(T_w - T_0)] = 6\frac{a}{U\Delta}(T_w - T_0) \tag{3.5.32}$$

and

$$2\frac{k}{\Delta}(T_w - T_0) = \frac{k_w}{l_w}(T_s - T_w). \tag{3.5.33}$$

Rearranging Eq. (3.5.33), we have

$$\frac{T_w(x) - T_0}{T_s - T_0} = \frac{\tilde{\Delta}}{1 + \tilde{\Delta}} \tag{3.5.34}$$

and

$$\frac{T_s - T_w(x)}{T_s - T_0} = \frac{1}{1 + \tilde{\Delta}}, \tag{3.5.35}$$

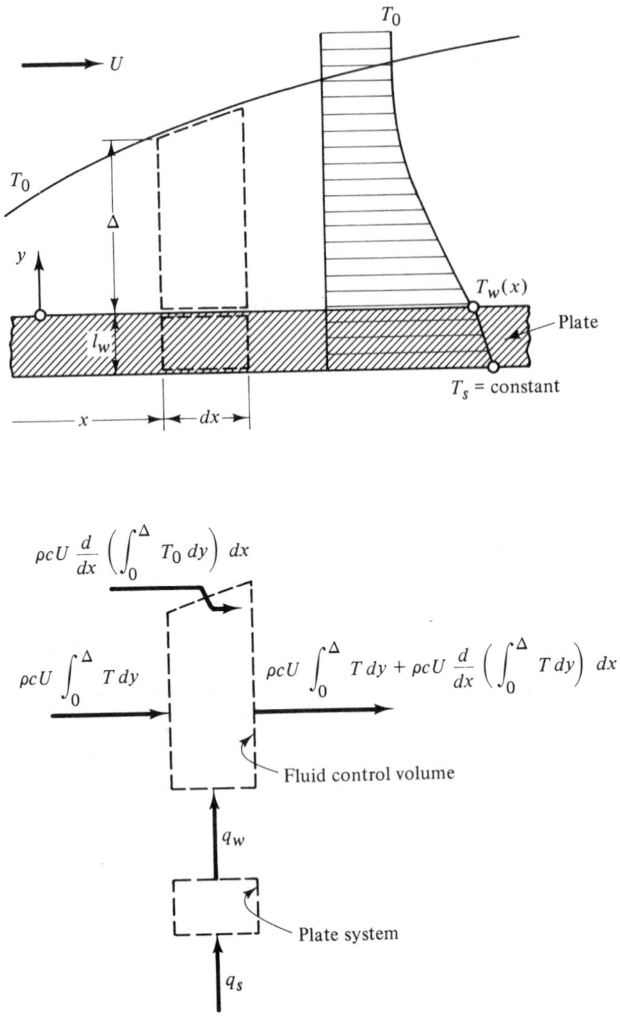

Fig. 3.28 Slug flow past plate of finite thickness, Ex. 3.5.3

where $\tilde{\Delta} = \frac{1}{2}(k_w/k)\Delta/l_w$. Now, inserting Eq. (3.5.34) into Eq. (3.5.32) results in

$$\frac{d}{dx}\left(\frac{\tilde{\Delta}^2}{1+\tilde{\Delta}}\right) = \frac{3}{2}\,\frac{(k_w/k)^2}{(1+\tilde{\Delta})l_w\,\mathrm{Pe}}, \tag{3.5.36}$$

where $\mathrm{Pe} = Ul_w/a$ is the Péclet number. Integration of Eq. (3.5.36) subject to $\tilde{\Delta}(0) = 0$ gives

$$\tfrac{1}{2}\tilde{\Delta}^2 + \tilde{\Delta} - \ln(1+\tilde{\Delta}) = \frac{\frac{3}{2}(k_w/k)^2}{\mathrm{Gz}_x}, \tag{3.5.37}$$

where $1/\mathrm{Gz}_x = (x/l_w)/\mathrm{Pe}$. Finding $1/\mathrm{Gz}_x$ for a given $\tilde{\Delta}$ is a straightforward numerical matter. Then, with known $\tilde{\Delta} = \tilde{\Delta}[(k_w/k)^2/\mathrm{Gz}_x]$ distribution, Eq. (3.5.35) plotted versus

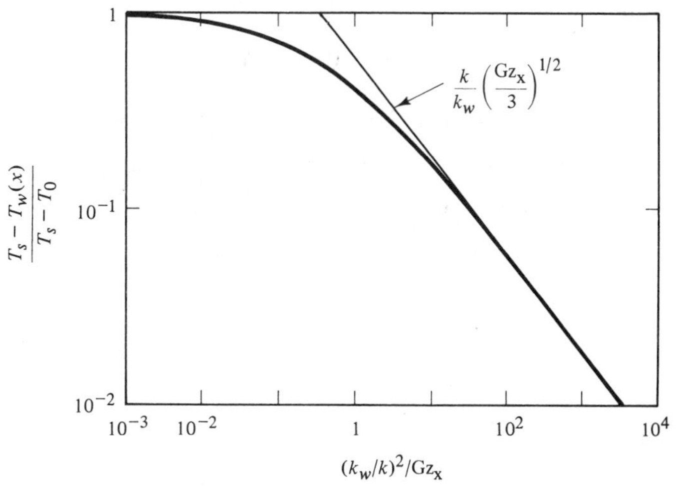

Fig. 3.29 Effect of finite plate thickness, Ex. 3.5.3

$(k_w/k)^2/\mathrm{Gz_x}$ shows in Fig. 3.29 the effect of finite plate thickness. $(T_s - T_w)/(T_s - T_0)$ decreases with increasing $(k_w/k)^2/\mathrm{Gz_x}$, as expected. More specifically, for example, at $x = 30$ mm downstream of a 3 m/s flow, the dimensionless temperature drop across a 3-mm-thick steel plate is quite appreciable, being 15% for air and 33% for water. However, the temperature drop across a 3-mm-thick copper plate is negligible, being 0.3% for air and 2% for water.

EXAMPLE 3.5.4 Consider an incompressible viscous fluid between two parallel plates, one of which, say the upper one, is moving with a uniform velocity, U. The upper plate is insulated, while the lower one is kept at uniform temperature T_w. The distance between the plates is l. The inlet temperature of the fluid is T_0. We wish to determine the temperature of the fluid.

The differential formulation of the problem in terms of $\theta = T - T_w$ ($\theta_0 = T_0 - T_w$), and with respect to the coordinate system attached at the inlet to the lower plate, is

$$\frac{U}{a}\left(\frac{y}{l}\right)\frac{\partial \theta}{\partial x} = \frac{\partial^2 \theta}{\partial y^2}; \qquad \theta(0, y) = \theta_0,$$
$$\theta(x, 0) = 0, \qquad \frac{\partial \theta(x, l)}{\partial y} = 0. \tag{3.5.38}$$

Here the velocity is that of the steady Couette flow given by Eq. (3.2.16).

The solution of Eq. (3.5.38), conveniently obtained by considering the separation of variables, and before the use of the nonhomogeneous boundary condition related to the inlet temperature, is

$$T - T_w = \sum_{n=0}^{\infty} a_n e^{-(al/U)\lambda_n^2 x}\, \varphi_n(y). \tag{3.5.39}$$

Here, obtained from the characteristic-value problem,

$$\frac{d^2Y}{dy^2} + \lambda^2 yY = 0; \qquad Y(0) = 0, \qquad \frac{dY(l)}{dy} = 0,$$

the characteristic functions* are (see, for example, Table 3.1 on p. 138 of Arpaci 1966)

$$\varphi_n(y) = y^{1/2}\,\mathrm{J}_{1/3}\left(\tfrac{2}{3}\lambda_n y^{3/2}\right), \tag{3.5.40}$$

and the characteristic values satisfy the roots of

$$\left.\frac{d\varphi_n(y)}{dy}\right|_{y=l} = \frac{d}{dy}\left[y^{1/2}\,\mathrm{J}_{1/3}\left(\tfrac{2}{3}\lambda_n y^{3/2}\right)\right]_{y=l} = 0. \tag{3.5.41}$$

Finally, the use of the inlet temperature, yielding

$$a_n = \theta_0 \frac{\int_0^l y\varphi_n\,dy}{\int_0^l y\varphi_n^2\,dy}, \tag{3.5.42}$$

completes the solution of the differential formulation. However, the integrals appearing in Eq. (3.5.42) are algebraically involved and need some attention. This involvement may be circumvented by employing (the differential equation of) the characteristic-value problem, which helps to express these integrals in terms of the derivatives of the characteristic functions. Accordingly, integrating over $(0, l)$ the characteristic-value problem, now written in terms of the characteristic function φ_n, we readily have, for the numerator of Eq. (3.5.42),

$$\int_0^l y\varphi_n\,dy = -\frac{1}{\lambda_n^2}\left.\left(\frac{d\varphi_n}{dy}\right)\right|_0^l. \tag{3.5.43}$$

The denominator, on the other hand, requires some elaboration. First, multiplying by φ_n the characteristic-value problem already expressed in φ_n, integrating the result over $(0, l)$, and rearranging by an integration by parts, yields

$$\left.\varphi_n \frac{d\varphi_n}{dy}\right|_0^l - \int_0^l \left(\frac{d\varphi_n}{dy}\right)^2 dy + \lambda_n^2 \int_0^l y\varphi_n^2\,dy = 0,$$

whose differential with respect to λ_n is

$$\begin{aligned}\left.\left(\frac{d\varphi_n}{d\lambda_n}\right)\left(\frac{d\varphi_n}{dy}\right)\right|_0^l + \left.\varphi_n \frac{d}{d\lambda_n}\left(\frac{d\varphi_n}{dy}\right)\right|_0^l - 2\int_0^l \left(\frac{d\varphi_n}{dy}\right)\frac{d}{d\lambda_n}\left(\frac{d\varphi_n}{dy}\right) dy \\ + 2\lambda_n \int_0^l y\varphi_n^2\,dy + \lambda_n^2 \frac{d}{d\lambda_n}\int_0^l y\varphi_n^2\,dy = 0.\end{aligned} \tag{3.5.44}$$

Second, multiplying the characteristic-value problem by $d\varphi_n/d\lambda_n$, integrating the result over $(0, l)$, and rearranging by an integration by parts, results in

$$\left.\left(\frac{d\varphi_n}{d\lambda_n}\right)\left(\frac{d\varphi_n}{dy}\right)\right|_0^l - \int_0^l \left(\frac{d\varphi_n}{dy}\right)\frac{d}{dy}\left(\frac{d\varphi_n}{d\lambda_n}\right) dy + \tfrac{1}{2}\lambda_n^2 \frac{d}{d\lambda_n}\int_0^l y\varphi_n^2\,dy = 0. \tag{3.5.45}$$

---

*As $y \longrightarrow 0$, behaving like $y^{-5/6}$, the other particular solution, $y^{-1/2}\,\mathrm{J}_{-1/3}\left(\tfrac{2}{3}\lambda y^{3/2}\right)$, becomes infinite.

Now, multiplying Eq. (3.5.45) by 2, then subtracting it from Eq. (3.5.44), gives, after noting that $(d/dy)d/d\lambda_n = (d/d\lambda_n)d/dy$,

$$\int_0^l y\varphi_n^2\,dy = \frac{1}{2\lambda_n}\left[\left(\frac{d\varphi_n}{dy}\right)\left(\frac{d\varphi_n}{d\lambda_n}\right)\Big|_0^l - \varphi_n\frac{d}{d\lambda_n}\left(\frac{d\varphi_n}{dy}\right)\Big|_0^l\right]. \tag{3.5.46}$$

Equations (3.5.43) and (3.5.46) introduced into Eq. (3.5.42) yield

$$a_n = \frac{-2\theta_0\left(\dfrac{d\varphi_n}{dy}\right)\Big|_0^l}{\lambda_n\left[\left(\dfrac{d\varphi_n}{d\lambda_n}\right)\left(\dfrac{d\varphi_n}{dy}\right)\Big|_0^l - \varphi_n\dfrac{d}{d\lambda_n}\left(\dfrac{d\varphi_n}{dy}\right)\Big|_0^l\right]}. \tag{3.5.47}$$

Consequently, inserting Eq. (3.5.47) into Eq. (3.5.39), and noting that $\varphi_n(0) = 0$ and $d\varphi_n(l)/dy = 0$, leads to a convenient form of the differential solution,

$$\frac{\theta(x, y)}{\theta_0} = -2\sum_{n=0}^{\infty}\frac{\left(\dfrac{d\varphi_n}{dy}\right)_{y=0}\exp\left(-\dfrac{al}{U}\lambda_n^2 x\right)\varphi_n(y)}{\lambda_n\left[\left(\dfrac{d\varphi_n}{d\lambda_n}\right)\left(\dfrac{d\varphi_n}{dy}\right)\Big|_{y=0} + \varphi_n\dfrac{d}{d\lambda_n}\left(\dfrac{d\varphi_n}{dy}\right)\Big|_{y=l}\right]}. \tag{3.5.48}$$

Leaving to the interested reader a numerical study of Eq. (3.5.48) in terms of $y^{1/2}\,J_{1/3}\,(\frac{2}{3}\lambda_n y^{3/2}) = \varphi_n(y)$, here we consider an alternative but simpler solution to be based on the integral formulation.

Clearly, the integral formulations given by Eqs. (3.5.6) and (3.5.13), and already employed, respectively, with the first and second domains of Ex. 3.5.1, apply equally to the present problem. Then, in terms of the first-order profiles, Eqs. (3.5.9) and (3.5.14), and the present velocity, $u = Uy/l$, we have, for the first domain,

$$\frac{T - T_w}{T_0 - T_w} = 2\left(\frac{y}{\Delta}\right) - \left(\frac{y}{\Delta}\right)^2, \qquad \Delta^3 = 36\left(\frac{al}{U}\right)x, \tag{3.5.49}$$

and the penetration distance $x_0 = Ul^2/36a$, and for the second domain,

$$\frac{T - T_w}{T_0 - T_w} = (2\eta - \eta^2)\exp\left[-\frac{24}{5}\frac{a(x - x_0)}{Ul^2}\right]. \tag{3.5.50}$$

(Why is the penetration distance shorter and the exponential decay of the second domain steeper than those of Ex. 3.5.1?)

So far, the only interesting result is that

$$\Delta \sim x^{1/3}$$

rather than $\Delta \sim x^{1/2}$ obtained in preceding examples on slug flow. Furthermore, we explore now an important question, the accuracy of the first-order profiles, in terms of the present example. As a matter of numerical experience, we know these profiles to yield answers that are no longer closely but rather grossly approximate when used beyond unsteady conduction problems and steady slug flow problems. Consequently, higher-order profiles to be considered with the variational calculus or with a method of weighted residuals are needed.*

*Refer, for an informal study, to Sec. 3.10 and to Chap. 8 of Arpaci (1966), or, for a formal study, to Collatz (1948), Ames (1965), Finlayson & Scriven (1966), or Finlayson (1972).

Here we employ the simplest and the most common procedure for boundary layers. This procedure is based on satisfying at appropriate places the differential formulation and, if necessary, its space derivatives normal to the direction of flow. Because of the well-known smoothing nature of the process of integration, the use of variational calculus or a method of weighted residuals may be expected to yield answers more accurate than those to be obtained by the use of the differential formulation at a point. However, satisfying the differential formulation at a point at which a maximum accuracy is desired, we usually get comparable accuracy with reasonably less algebra.

Returning to the example, in the process of constructing a first-order profile, we employed two conditions at the edge of the boundary layer,

$$T(x, \Delta) = T_0, \qquad \frac{\partial T(x, \Delta)}{\partial y} = 0, \tag{3.5.51}$$

and only one condition at the wall,

$$T(x, 0) = T_w \tag{3.5.52}$$

(see Eqs. 3.5.7 and 3.5.8). The profile thus obtained behaves more accurately near the edge of the boundary layer because it satisfies more conditions there. However, because of our strong interest in the heat transfer coefficient, which is porportional to the temperature gradient at the wall, we need our profiles to behave most accurately near the wall. A way of readily accomplishing this task is to consider the differential formulation

$$\frac{U}{a}\left(\frac{y}{l}\right)\frac{\partial T}{\partial x} = \frac{\partial^2 T}{\partial y^2},$$

which, at the wall, satisfies

$$\frac{\partial^2 T(x, 0)}{\partial y^2} = 0. \tag{3.5.53}$$

A second-order profile, constructed from a polynomial satisfying Eqs. (3.5.51)–(3.5.53), is

$$\frac{T - T_w}{T_0 - T_w} = \frac{3}{2}\left(\frac{y}{\Delta}\right) - \frac{1}{2}\left(\frac{y}{\Delta}\right)^3. \tag{3.5.54}$$

Inserting Eq. (3.5.54) into Eq. (3.5.6), noting that $u = Uy/l$, results in

$$d\Delta^3 = \frac{45}{2}\left(\frac{al}{U}\right) dx, \qquad \Delta(0) = 0,$$

with

$$\Delta(x) = \left(\frac{45}{2}\frac{alx}{U}\right)^{1/3}, \tag{3.5.55}$$

and, after letting $\Delta = l$, the penetration distance

$$x_0 = \frac{2}{45}\frac{Ul^2}{a}. \tag{3.5.56}$$

Also, obtained by the same procedure, a second-order profile for the second domain is

$$\frac{T - T_w}{T_0 - T_w} = \left(\frac{3}{2}\eta - \frac{1}{2}\eta^3\right)\Delta_1(x) \tag{3.5.57}$$

with $\eta = y/l$ and $\Delta_1(x)$ as the scale factor. Introducing Eq. (3.5.57) into Eq. (3.5.13), noting that $u = Uy/l$, results in

$$\frac{d\Delta_1}{dx} + \frac{15}{4}\left(\frac{a}{Ul^2}\right)\Delta_1 = 0, \qquad \Delta_1(x_0) = 1. \tag{3.5.58}$$

The solution of Eq. (3.5.58),

$$\Delta_1(x) = \exp\left[-\frac{15}{4}\frac{a(x - x_0)}{Ul^2}\right],$$

inserted into Eq. (3.5.57), yields a second-order profile for the temperature of the fluid in the second domain

$$\frac{T - T_w}{T_0 - T_w} = \left(\frac{3}{2}\eta - \frac{1}{2}\eta^3\right)\exp\left[-\frac{15}{4}\frac{a(x - x_0)}{Ul^2}\right]. \tag{3.5.59}$$

The considerable numerical difference between the first- and second-order profiles given by Eqs. (3.5.49) and (3.5.54) for the first domain, and by Eqs. (3.5.50) and (3.5.59) for the second domain should be mentioned. However, on the basis of our numerical experience, the consideration of profiles higher than order 2 appears to be unnecessary. For example, calculating the Nusselt number for the first time domain, employing Eq. (3.5.2) and the temperature difference $T_w - T_0$, gives

$$\mathrm{Nu}_x = \frac{hx}{k} = C\left(\frac{Ux}{a}\frac{x}{l}\right)^{1/3},$$

where $C = (\frac{2}{9})^{1/3} = 0.606$ for the first-order profile, Eq. (3.5.49), and $C = (\frac{3}{20})^{1/3} = 0.531$ for the second-order profile, Eq. (3.5.54). For the exact solution, $C = 0.538$ (see Ex. 5.3.2).

We proceed now to the last example, the well-known Graetz problem.

EXAMPLE 3.5.5 A viscous oil flows steadily between two parallel plates separated a distance $2l$ apart and kept at uniform temperature T_w. The inlet temperature of the oil is T_0, and since $\delta/\Delta \longrightarrow \infty$, the velocity of the oil may be assumed fully developed. We wish to determine the temperature of the oil.

The differential formulation of the problem in terms of $\theta = T - T_w$ ($\theta_0 = T_0 - T_w$) and with respect to the coordinate system attached at the inlet to the middle plane is

$$\frac{3}{2}\,\mathrm{Pe}\,(1 - y^2)\frac{\partial\theta}{\partial x} = \frac{\partial^2\theta}{\partial y^2}; \qquad \theta(0, y) = \theta_0,$$
$$\frac{\partial\theta(x, 0)}{\partial y} = 0, \qquad \theta(x, 1) = 0, \tag{3.5.60}$$

where $\mathrm{Pe} = U_m l/a$ is the Péclet number, x and y are dimensionless (representing x/l and y/l), $\frac{3}{2}U_m(1 - y^2)$ is the dimensionless form of the Poiseuille velocity given by Eq.

(3.2.3), and $U_m = (-dp/dx)l^2/3\mu$ is the mean velocity. We shall consider three methods of solution. In the first two, to be based on the separation of variables and the variational formulation, respectively, the dimensionless formulation will become convenient in the evaluation of the heat transfer coefficient. In the third method, to be based on the integral formulation, we employ dimensional variables.

(1) The solution of Eq. (3.5.60), readily obtained by considering the separation of variables, and before the use of the nonhomogeneous boundary condition related to the inlet temperature, is

$$T - T_w = \sum_{n=0}^{\infty} a_n e^{-(2/3)\lambda_n^2 x/\mathrm{Pe}} G_0^{(1)}(\lambda_n y). \tag{3.5.61}$$

Here, obtained from the characteristic-value problem,

$$\frac{d^2Y}{dy^2} + \lambda^2(1 - y^2)Y = 0; \qquad \frac{dY(0)}{dy} = 0, \qquad Y(1) = 0, \tag{3.5.62}$$

the characteristic functions are the Graetz functions corresponding to the even particular solution, and the characteristic values satisfy the roots of

$$G_0^{(1)}(\lambda_n) = 0. \tag{3.5.63}$$

Finally, the use of the inlet temperature, yielding

$$a_n = \theta_0 \frac{\int_0^1 (1 - y^2) G_0^{(1)}(\lambda_n y)\, dy}{\int_0^1 (1 - y^2)[G_0^{(1)}(\lambda_n y)]^2\, dy}, \tag{3.5.64}$$

completes the solution of the differential formulation. The integrals appearing in Eq. (3.5.64) may be evaluated in a manner identical to those appearing in Eq. (3.5.42) of Ex. 3.5.4. The result is

$$a_n = \frac{-2\theta_0}{\lambda_n dG_0^{(1)}(\lambda_n y)/d\lambda_n |_{y=1}}. \tag{3.5.65}$$

(For details, see Ex. 4-6 of Arpaci 1966.) Inserting Eq. (3.5.65) into Eq. (3.5.61) leads to a convenient form of the differential solution,

$$\frac{\theta(x, y)}{\theta_0} = -2 \sum_{n=0}^{\infty} \frac{e^{-(2/3)\lambda_n^2 x/\mathrm{Pe}} G_0^{(1)}(\lambda_n y)}{\lambda_n\, dG_0^{(1)}(\lambda_n y)/d\lambda_n |_{y=1}}. \tag{3.5.66}$$

The dimensionless heat transfer coefficient, the Nusselt number, obtained from Eq. (3.5.2) after replacing T_∞ with the bulk temperature T_b, is

$$\mathrm{Nu} = \frac{hl}{k} = -\frac{\partial\theta/\partial y |_{y=1}}{\theta_b}, \tag{3.5.67}$$

where

$$\theta_b = \tfrac{3}{2} \int_0^1 (1 - y^2)\theta(x, y)\, dy.$$

Now, employing Eq. (3.5.66), we have

$$\left.\frac{d\theta}{dy}\right|_{y=1} = -2 \sum_{n=0}^{\infty} \left.\frac{dG_0^{(1)}(\lambda_n y)/dy}{\lambda_n\, dG_0^{(1)}(\lambda_n y)/d\lambda_n}\right|_{y=1} e^{-(2/3)\lambda_n^2 x/\mathrm{Pe}}$$

and

$$\theta_b = 3 \sum_{n=0}^{\infty} \frac{dG_0^{(1)}(\lambda_n y)/dy}{\lambda_n^3 \, dG_0^{(1)}(\lambda_n y)/d\lambda_n}\bigg|_{y=1} e^{-(2/3)\lambda_n^2 x/\mathrm{Pe}}.$$

Finally, inserting these into Eq. (3.5.67) yields the Nusselt modulus,

$$\mathrm{Nu} = \frac{2}{3} \frac{\displaystyle\sum_{n=0}^{\infty} \frac{dG_0^{(1)}(\lambda_n y)/dy}{\lambda_n \, dG_0^{(1)}(\lambda_n y)/d\lambda_n}\bigg|_{y=1} e^{-(2/3)\lambda_n^2 x/\mathrm{Pe}}}{\displaystyle\sum_{n=0}^{\infty} \frac{dG_0^{(1)}(\lambda_n y)/dy}{\lambda_n^3 \, dG_0^{(1)}(\lambda_n y)/d\lambda_n}\bigg|_{y=1} e^{-(2/3)\lambda_n^2 x/\mathrm{Pe}}}. \tag{3.5.68}$$

In Fig. 3.30 Nu versus $1/\mathrm{Gz}_x = x/\mathrm{Pe}$ is plotted, Gz_x being the Graetz number. (Can you explain in simple terms why $\mathrm{Nu} \longrightarrow \infty$ as $1/\mathrm{Gz}_x \longrightarrow 0$? See Section 3.6 for a simpler analysis yielding only the limit $\mathrm{Nu} = 1.885$ as $1/\mathrm{Gz}_x \longrightarrow \infty$.)

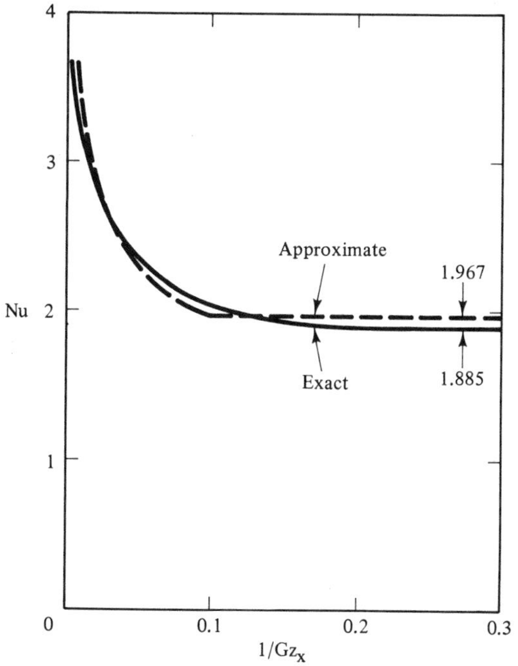

Fig. 3.30 Exact and approximate values of Nu versus $1/\mathrm{Gz}_x$, Ex. 3.5.5 (exact curve adapted from Prins, Mulder and Schenk 1951; used by permission)

Graetz functions have not been explored to the extent of some mathematical functions such as Bessel functions. So unless they are programmed for a digital computer, Eqs. (3.5.66) and (3.5.68) are not convenient for numerical computations. This suggests the use of alternative procedures more convenient for these computations. The most elegant method, leading to an optimum solution in terms of the selected profiles, is the variational calculus. Let us now apply this method, not to the whole problem, but to the part of the problem leading to computational difficulties. This part is the characteristic-value problem given by Eq. (3.5.62). In the solution of Eq. (3.5.62), the variational calculus allows the consideration of functions simpler than Graetz functions.

(2) Now, assume Eq. (3.5.62) to be the *Euler equation* of the desired variational formulation.* Then we have

$$\int_0^1 \left[\frac{d^2Y}{dy^2} + \lambda^2(1 - y^2)Y\right]\delta Y\, dy = 0. \tag{3.5.69}$$

An integration by parts, after the use of boundary conditions which happens to be *natural* in the variational sense, leads to the variational formulation,

$$\delta I = 0, \tag{3.5.70}$$

where the *functional* is

$$I = \frac{1}{2}\int_0^1 \left[\left(\frac{dY}{dy}\right)^2 + \lambda^2(1 - y^2)Y^2\right]. \tag{3.5.71}$$

The following procedure† is identical to the use of Eq. (3.5.69) (see, for example, the Ritz method in Sec. 8.4 of Arpaci 1966). Consider a characteristic-value problem with boundary conditions identical to those of Eq. (3.5.62) but satisfying characteristic functions simpler than, yet grossly behaving like, the Graetz functions of Eq. (3.5.62). For example, the characteristic-value problem,

$$\frac{d^2Y}{dy^2} + \Lambda^2 Y = 0; \qquad \frac{dY(0)}{dy} = 0, \qquad Y(1) = 0, \tag{3.5.72}$$

leading to circular (trigonometric) functions,

$$Y = \sum_{n=0}^{\infty} a_n \cos \Lambda_n y, \qquad \Lambda_n = (2n + 1)\frac{\pi}{2}, \qquad n = 0, 1, 2, \ldots, \tag{3.5.73}$$

satisfies these requirements. Inserting Eq. (3.5.73) into Eq. (3.5.62), multiplying each term by $\cos \Lambda_m y\, dy$, and integrating the result over (0, 1) yields a homogeneous set of algebraic equations of infinite order involving the unknown characteristic values of Eq. (3.5.62). The determinant of this set (represented by a typical term) is‡

$$\| \tfrac{1}{2}(\lambda_n^2 - \Lambda_m^2)\, \delta_{nm} - \lambda_n^2 P(n, m) \| = 0, \tag{3.5.74}$$

where δ_{nm} is the Kronecker delta, and

$$P(n, m) = \int_0^1 y^2 \cos \Lambda_n y \cos \Lambda_m y\, dy.$$

---

*See foot note p. 115.

†An alternative procedure based on the orthogonalization of the Stodola method has been extensively used by Chandrasekhar (1961). This procedure is no longer variational when used in the absence of $\delta I = 0$, and it is then referred to as the Galerkin method.

‡Note that any two terms of Eq. (3.5.73) corresponding to different values of Λ_n are orthogonal with respect to Eq. (3.5.72) but *not* with respect to Eq. (3.5.62).

Expressing the product of cosines in terms of the sum of cosines, and twice integrating by parts, yields

$$P(n, m) = \frac{(-1)^{n+m}}{\pi^2}\left[\frac{1}{(n-m)^2} - \frac{1}{(n+m-1)^2}\right], \qquad n \neq m. \tag{3.5.75}$$

The same steps, after letting $n = m$, gives

$$P(n, n) = \frac{1}{6} - \frac{1}{\pi^2(2n+1)^2}, \qquad n = m. \tag{3.5.76}$$

From the first diagonal term of Eq. (3.5.74),

$$\tfrac{1}{2}(\lambda_0^2 - \Lambda_0^2) - \lambda_0^2 P(0, 0) = 0,$$

we have a first approximation for the first characteristic value of Eq. (3.5.62),

$$\lambda_0^2 = \frac{\Lambda_0^2}{1 - 2P(0, 0)}. \tag{3.5.77}$$

Inserting $P(0, 0) = \frac{1}{6} - 1/\pi^2$, obtained from Eq. (3.5.76) for $n = 0$, into Eq. (3.5.77), we have

$$\lambda_0^2 = \frac{\Lambda_0^2}{\frac{2}{3} + 2/\pi^2},$$

or, noting that $\Lambda_0 = \pi/2$,

$$\lambda_0 \cong 1.6847, \tag{3.5.78}$$

whose exact value, obtained from Eq. (3.5.63), is

$$\lambda_0 = 1.6816.$$

However, for characteristic values beyond the first one, the order of the determinant needed for accuracy must be increased. So the foregoing method, although rather convenient for the first characteristic value, gets algebraically involved and less accurate for higher characteristic values. Procedures devised to eliminate some of these difficulties have been extensively studied (see Wilkinson 1965, or Fröberg 1969 and the references cited therein). Separately, in place of the circular functions of the characteristic-value problem given by Eq. (3.5.72), we may consider Legendre, Chebyshev, or other polynomials associated with appropriate characteristic-value problems. The use of polynomials appears to considerably improve the convergence (see Fox & Parker 1968, or, Orszag 1971a, b). A complementary asymptotic method, the so-called WKBJ procedure (see Morse & Feshbach 1953), yields more accurate answers for higher characteristic values than those for the lower ones.

The search for a way to understand the Graetz problem, with no reference to the mathematics so far employed, leads us to consider an approximate solution of this problem by the integral method.

(3) Here, for the reasons given in Ex. 3.5.4, we limit ourselves to second-order (cubic) profiles. Thus, for the first domain, inserting into Eq. (3.5.6) the Poiseuille velocity $u = (3U_m/2l^2)(2l - y)y$ and Eq. (3.5.54) (now measuring dimensional y

from the lower plate), we have

$$\Delta d\left(\frac{1}{5}\Delta^2 - \frac{1}{24}\frac{\Delta^3}{l}\right) = \left(\frac{al}{U_m}\right) dx, \qquad \Delta(0) = 0,$$

which integrates to

$$\Delta^3\left(1 - \frac{15}{64}\frac{\Delta}{l}\right) = \frac{15}{2}\left(\frac{al}{U_m}\right)x. \tag{3.5.79}$$

Now, for each Δ, we numerically find x. For small values of x, noting that $\Delta/l \ll 1$ and neglecting the second left-hand term in parentheses of Eq. (3.5.79), we have

$$\Delta(x) \cong \left(\frac{15}{2}\frac{alx}{U_m}\right)^{1/3}, \qquad x \longrightarrow 0. \tag{3.5.80}$$

For $\Delta = l$, Eq. (3.5.79) gives the penetration distance,

$$x_0 = \tfrac{2}{15}\left(\tfrac{49}{64}\right)\frac{U_m l^2}{a} \tag{3.5.81}$$

at which the upper and lower boundary layers coincide. (What physical approximation is made when the second left-hand term in parentheses of Eq. 3.5.79 is neglected? Why is Eq. 3.5.80 not identical to Eq. 3.5.55?)

For the second domain, introducing Eq. (3.5.57) and $u = (3U_m/2l^2)(2l - y)y$ into Eq. (3.5.13) results in

$$\frac{d\Delta_1}{dx} + \frac{120}{61}\left(\frac{a}{U_m l^2}\right)\Delta_1 = 0, \qquad \Delta_1(x_0) = 1. \tag{3.5.82}$$

The solution of Eq. (3.5.82),

$$\Delta_1(x) = \exp\left[-\frac{120}{61}\frac{a(x - x_0)}{U_m l^2}\right],$$

inserted into Eq. (3.5.57), yields

$$\frac{T - T_w}{T_0 - T_w} = \left(\frac{3}{2}\eta - \frac{1}{2}\eta^3\right)\exp\left[-\frac{120}{61}\frac{a(x - x_0)}{U_m l^2}\right]. \tag{3.5.83}$$

Now we proceed to the evaluation of the Nusselt number,

$$\mathrm{Nu} = -\frac{l(\partial T/\partial y)_{y=0}}{T_b - T_w}. \tag{3.5.84}$$

For the first domain,

$$l\left(\frac{\partial T}{\partial y}\right)_{y=0} = \frac{3}{2}\left(\frac{l}{\Delta}\right)(T_0 - T_w) \tag{3.5.85}$$

and the bulk temperature is calculated from

$$U_m l(T_b - T_w) = \int_0^\Delta u(T - T_w)\,dy + (T_0 - T_w)\int_\Delta^l u\,dy,$$

which yields

$$T_b - T_w = \frac{3}{2}\left[\frac{2}{3} - \frac{1}{5}\left(\frac{\Delta}{l}\right)^2 + \frac{1}{24}\left(\frac{\Delta}{l}\right)^3\right](T_0 - T_w). \tag{3.5.86}$$

Introducing Eqs. (3.5.85) and (3.5.86) into Eq. (3.5.84), we have

$$\mathrm{Nu} = \frac{1}{(\Delta/l)[\frac{2}{3} - \frac{1}{5}(\Delta/l)^2 + \frac{1}{24}(\Delta/l)^3]}, \qquad \frac{\Delta}{l} \leq 1. \tag{3.5.87}$$

For the second domain,

$$l\left(\frac{\partial T}{\partial y}\right)_{y=0} = \frac{3}{2} \exp\left[-\frac{120}{61}\frac{a(x - x_0)}{U_m l^2}\right] \tag{3.5.88}$$

and

$$U_m(T_b - T_w) = \int_0^1 u(T - T_w)\, dy,$$

which gives

$$T_b - T_w = \frac{3}{2}\left(\frac{61}{120}\right) \exp\left[-\frac{120}{61}\frac{a(x - x_0)}{U_m l^2}\right]. \tag{3.5.89}$$

Inserting Eqs. (3.5.88) and (3.5.89) into Eq. (3.5.84), we get

$$\mathrm{Nu} = \frac{120}{61} = 1.967.$$

In Fig. 3.30 the exact and approximate Nu versus $1/\mathrm{Gz}_x = x/\mathrm{Pe}$ are plotted. The convenience of the integral method should be once more emphasized because it yields sufficiently accurate results by a solution procedure much simpler than any procedure leading to the exact solution.

The fact that Nu is constant for $x \geq x_0$ suggests the concept of fully developed heat transfer, which will be explored next.

3.6 *FULLY DEVELOPED HEAT TRANSFER*

In Section 3.5 we studied the spatially developing (steady) temperature distributions in parallel flows in terms of thermal boundary layers. We also learned (in Ex. 3.5.5) that, following an entrance length, the Nusselt number downstream approaches a constant value for flow in a channel having constant wall temperature.

By *definition*, fully developed heat transfer exists when the heat transfer coefficient is independent of position, say x, along the channel; that is, using Eqs. (3.5.1) and (3.5.2),

$$h = \frac{q_w}{T_w - T_b} = -\frac{k}{T_w - T_b}\frac{\partial T}{\partial y}\bigg|_w \neq f(x). \tag{3.6.1}$$

Here y denotes the coordinate normal to the wall and, for a constant-property fluid, the bulk temperature is given by

$$T_b = \frac{1}{U_m A}\int_A uT\, dA, \tag{3.6.2}$$

and the mean velocity by

$$U_m = \frac{1}{A}\int_A u\, dA, \tag{3.6.3}$$

where A denotes the cross-sectional area of the flow. Note that u and U_m are independent of x for parallel (developed) flow considered here, but T_b may depend on x.

Since T_w and T_b are independent of y, Eq. (3.6.1) may be rearranged as

$$\text{Nu} = \frac{hl}{k} = \frac{\partial}{\partial(y/l)}\left(\frac{T_w - T}{T_w - T_b}\right)\bigg|_w \neq f(x), \tag{3.6.4}$$

implying the existence of a fully developed dimensionless temperature θ:

$$\theta = \frac{T_w - T}{T_w - T_b} \neq f(x) \qquad \text{or} \qquad \frac{\partial \theta}{\partial x} = 0,$$

or equivalently,

$$\frac{\partial T_w}{\partial x} - \frac{\partial T}{\partial x} = \frac{T_w - T}{T_w - T_b}\left(\frac{\partial T_w}{\partial x} - \frac{\partial T_b}{\partial x}\right). \tag{3.6.5}$$

Let us consider two standard boundary conditions that lead to fully developed heat transfer. First, for the case of constant wall heat flux q_w, since h is constant, Eq. (3.6.1) gives $T_w - T_b =$ constant, or $\partial T_w/\partial x = \partial T_b/\partial x$, and using Eq. (3.6.5) we obtain

$$\frac{\partial T}{\partial x} = \frac{\partial T_w}{\partial x} = \frac{\partial T_b}{\partial x} \qquad (q_w = \text{constant}). \tag{3.6.6}$$

Next, for the case of constant wall temperature T_w, Eq. (3.6.5) gives directly

$$\frac{\partial T}{\partial x} = \left(\frac{T_w - T}{T_w - T_b}\right)\frac{\partial T_b}{\partial x} \qquad (T_w = \text{constant}). \tag{3.6.7}$$

The boundary condition of constant heat flux applies to a wall subject to electrical resistance heating or to a radiant flux. That of constant temperature is approached when the outer wall is subject to phase change, such as condensation or boiling (recall Ex. 3.5.3). In many situations neither of these conditions are satisfied, yet calculation of heat exchangers often employ results based on a constant heat transfer coefficient.

The following examples treat the case of constant heat flux, for which the temperature distribution can be calculated explicitly, and the case of constant wall temperature, for which an iterative procedure is needed.

EXAMPLE 3.6.1 Consider the fully developed steady flow of a viscous fluid between two parallel plates spaced a distance $2l$ apart and each subjected to a constant heat flux q_w. We wish to determine the downstream fully developed temperature distribution and the heat transfer coefficient.

In terms of $\eta = y/l$ measured from the centerline the differential formulation, with $\partial T/\partial x$ replaced by $\partial T_b/\partial x \neq f(x)$ from Eq. (3.6.6), is

$$\frac{ul^2}{a}\frac{\partial T_b}{\partial x} = \frac{\partial^2 T}{\partial \eta^2} \tag{3.6.8}$$

$$\frac{\partial T(0)}{\partial \eta} = 0; \qquad \frac{\partial T(1)}{\partial \eta} = \frac{q_w l}{k}.$$

Integrating Eq. (3.6.8) once, using the first boundary condition, gives

$$\frac{dT}{d\eta} = \frac{l^2}{a}\frac{\partial T_b}{\partial x}\int_0^\eta u(\eta')\,d\eta' \tag{3.6.9}$$

which, using the second boundary condition and Eq. (3.6.3), yields

$$\frac{q_w l}{k} = \frac{U_m l^2}{a} \frac{\partial T_b}{\partial x}. \tag{3.6.10}$$

Eliminating $\partial T_b/\partial x$ between Eqs. (3.6.9) and (3.6.10), the result may be integrated to

$$T_w - T = \frac{q_w l}{k} \int_\eta^1 \int_0^{\eta'} \frac{u(\eta'')}{U_m} \, d\eta'' \, d\eta'.$$

Using $q_w = h(T_w - T_b)$ and Eq. (3.6.2), or

$$T_w - T_b = \int_0^1 \frac{u(\eta)}{U_m}(T_w - T) \, d\eta,$$

we finally obtain the dimensionless temperature distribution

$$\frac{T_w - T}{T_w - T_b} = \frac{\int_\eta^1 \int_0^{\eta'} (u/U_m) \, d\eta'' \, d\eta'}{\int_0^1 (u/U_m)\left[\int_\eta^1 \int_0^{\eta'} (u/U_m) \, d\eta'' \, d\eta'\right] d\eta}, \tag{3.6.11}$$

and the Nusselt number, based on l,

$$\mathrm{Nu} = \left\{\int_0^1 \frac{u}{U_m}\left[\int_\eta^1 \int_0^{\eta'} \frac{u}{U_m} \, d\eta'' \, d\eta'\right] d\eta\right\}^{-1}. \tag{3.6.12}$$

For the parabolic velocity distribution of Poiseuille flow,

$$\frac{u}{U_m} = \frac{3}{2}(1 - \eta^2), \tag{3.6.13}$$

Eqs. (3.6.11) and (3.6.12) become

$$\frac{T_w - T}{T_w - T_b} = \frac{35}{136}(5 - 6\eta^2 + \eta^4), \tag{3.6.14}$$

$$\mathrm{Nu} = \frac{35}{17} \simeq 2.06. \tag{3.6.15}$$

Following a similar procedure for flow in a pipe of diameter D yields Nu = 4.36 (see Problem 3-26).

Equations (3.6.10) and (3.6.6) show that the fully developed temperature at each η increases linearly with x. Also, being governed by diffusion through parallel fluid layers moving at different velocities the heat transfer is independent of the Reynolds number and depends only on the velocity distribution. For the present example, Eq. (3.6.12) shows that the Nusselt number increases from the pure conduction limit (Nu = 1) to a maximum (Nu = 3) as the velocity distribution changes from being zero (except of a Dirac delta function at $\eta = 0$) to slug flow ($u/U_m = 1$). While the former case is hypothetical, the latter case of slug flow is a good approximation to a liquid metal flow for which fully developed heat transfer may be established before any significant flow development has occurred.

These considerations illustrate in simple terms the convective heat transfer in parallel flows. Clearly, a further increase in heat transfer may be expected if there exists a velocity component away from the wall at some distance from it. This is, in fact, the case for developing (nearly parallel) flows, considered in Chapter 4.

EXAMPLE 3.6.2 Reconsider Ex. 3.6.1, now with isothermal walls kept at T_w. Assuming the fluid to enter the channel with a temperature different from T_w, we seek the downstream developed temperature distribution and the heat transfer coefficient.

Substituting $\partial T/\partial x$ from Eq. (3.6.7) into the differential formulation, we obtain

$$\frac{ul^2}{a}\theta(\eta)\frac{\partial T_b}{\partial x} = \frac{\partial^2 T}{\partial \eta^2} \tag{3.6.16}$$

$$\frac{\partial T(0)}{\partial \eta} = 0; \qquad T(1) = T_w,$$

where $\theta = (T_w - T)/(T_w - T_b)$ is the yet unknown dimensionless temperature distribution. Proceeding formally as in Ex. 3.6.1, we integrate Eq. (3.6.16) twice:

$$\frac{\partial T}{\partial \eta} = \frac{l^2}{a}\frac{\partial T_b}{\partial x}\int_0^\eta u(\eta')\theta(\eta')\,d\eta' \tag{3.6.17}$$

$$T_w - T(\eta) = \frac{l^2}{a}\frac{\partial T_b}{\partial x}\int_\eta^1\int_0^{\eta'} u(\eta'')\theta(\eta'')\,d\eta''\,d\eta'. \tag{3.6.18}$$

Using Eq. (3.6.2) to evaluate the bulk temperature difference from Eq. (3.6.18),

$$T_w - T_b = \frac{U_m l^2}{a}\frac{\partial T_b}{\partial x}\int_0^1 \frac{u}{U_m}\left[\int_\eta^1\int_0^{\eta'}\frac{u}{U_m}\theta\,d\eta''\,d\eta'\right]d\eta, \tag{3.6.19}$$

we eliminate $\partial T_b/\partial x$ in Eq. (3.6.17) to obtain the dimensionless temperature distribution

$$\frac{T_w - T}{T_w - T_b} = \theta(\eta) = \frac{\displaystyle\int_\eta^1\int_0^{\eta'}\frac{u}{U_m}\theta\,d\eta''\,d\eta'}{\displaystyle\int_0^1\frac{u}{U_m}\left[\int_\eta^1\int_0^{\eta'}\frac{u}{U_m}\theta\,d\eta''\,d\eta'\right]d\eta}. \tag{3.6.20}$$

From the ratio of Eq. (3.6.17), evaluated at the wall ($\eta = 1$), and Eq. (3.6.19), we obtain the heat transfer coefficient, or according to Eq. (3.6.4), the Nusselt number

$$\mathrm{Nu} = \frac{\displaystyle\int_0^1 (u/U_m)\theta\,d\eta}{\displaystyle\int_0^1 (u/U_m)\left[\int_\eta^1\int_0^{\eta'}(u/U_m)\theta\,d\eta''\,d\eta'\right]d\eta}. \tag{3.6.21}$$

The velocity $u(\eta)/U_m$ being known, $\theta(\eta)$ is found by solving Eq. (3.6.20), say, by successive approximations. Assume, for example, a polynomial profile satisfying the boundary conditions, such as $\theta = c(1 - \eta^2)$. Substituting it into the right-hand side of Eq. (3.6.20) and performing the integrations yields an improved profile. The process is repeated until the Nusselt number, evaluated from Eq. (3.6.21) for each profile, converges to a limiting value. Experience shows that a few iterations are sufficient. Following this procedure we find, using Eq. (3.6.13), that

$$\mathrm{Nu} = 1.885. \tag{3.6.22}$$

Actually, the integral solution to the second domain in Ex. 3.5.5, leading to Nu $= 1.967$, is the first-order approximation, identically obtained by inserting into Eq. (3.6.21) the cubic profile $\theta(\eta) = c(1 - \frac{3}{2}\eta^2 + \frac{1}{2}\eta^3)$, which is Eq. (3.5.57) for $\eta = y/l$ measured from the centerline. It gives a Nusselt number within 4% of the exact value, so any iterative improvement of the profile by use of Eq. (3.6.20) is hardly warranted. Following a similar procedure for a pipe of diameter D yields Nu $= 3.66$ (see Problem 3-27).

The analysis presented in Exs. 3.6.1 and 3.6.2 may be carried out to yield fully developed heat transfer for flow in channels of arbitrary cross section subject to constant temperature or heat flux at the wall. Table 3.2 for constant wall temperature

TABLE 3.2 Nusselt number of fully developed heat transfer in parallel channel flow at constant wall temperature

Cross Section		D_e	Nu $= \bar{h}D_e/k$
Triangular	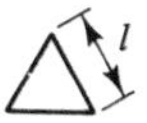	$l\sqrt{3}$	2.35
Square	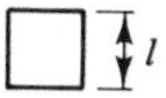	l	2.98
Circular	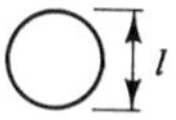	l	3.66
Planar	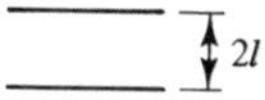	$2l$	3.77

shows typical results, expressed in customary terms of Nu $= \bar{h}D_e/k$, where $\bar{h}$ is the peripherally averaged heat transfer coefficient, and $D_e = 4A/P_h$ denotes the equivalent heated diameter, four times the flow area A divided by the heated perimeter P_h. When all wetted surfaces are heated, D_e equals the equivalent diameter defined by Eq. (3.2.56).

For cross sections with corners (e.g., rectangular or triangular channels) subject to constant flux corner temperatures become unbounded. (See Clark & Kays 1953 for solutions based on isothermal perifery, but constant total heating. The resulting Nusselt numbers are representative of the constant flux case and are larger than those of constant wall temperature.) The case of constant wall temperature, however, presents no problem. It leads to peripheral variations in local heat flux and heat transfer coefficient, both being zero at corners. The results in Table 3.2 are accordingly based on the peripherally averaged heat transfer coefficient. As noted for the friction factor (Section 3.2), the lack of geometrical similarity between cross sections leads to different constants in the Nusselt number for different geometries for the case of laminar flow. Again, this proves not to be the case for turbulent flows, hence the utility of the concept of equivalent diameter. Having completed our study of parallel flows, we turn now to nearly parallel flows.

REFERENCES

AMES, W. F., 1965, *Nonlinear Partial Differential Equations of Engineering*, Academic Press, New York.

ARPACI, V. S., 1966, *Conduction Heat Transfer*, Addison-Wesley, Reading, Mass.

BERKER, R., 1963, *Handbuch der Physik*, Vol. VIII, Pt. 2, pp. 1–384, Springer-Verlag, New York.

CARSLAW, H. S., & JAEGER, J. C., 1959, *Conduction of Heat in Solids*, Oxford University Press, New York.

CHANDRASEKHAR, S., 1961, *Hydrodynamic and Hydromagnetic Stability*, Oxford University Press, New York.

CLARK, S. H., & KAYS, W. M., 1953, Trans. ASME, **75**, 859–866.

COLLATZ, L., 1948, *Eigenwertprobleme und ihre Numerische Behandlung*, Chelsea, New York.

FINLAYSON, B. A., 1972, *The Method of Weighted Residuals and Variational Principles*, Academic Press, New York.

FINLAYSON, B. A., & SCRIVEN, L. E., 1966, Appl. Mech. Rev., **19**, 735–748.

FOX, L., & PARKER, I. B., 1968, *Chebyshev Polynomials in Numerical Analysis*, Oxford University Press, New York.

FRÖBERG, C.-E., 1969, *Introduction to Numerical Analysis*, 2nd ed., Addison-Wesley, Reading, Mass.

MORSE, P. M., & FESHBACH, H., 1953, *Methods of Theoretical Physics*, Vol. II, McGraw-Hill, New York.

ORSZAG, S. A., 1971a, J. Fluid Mech., **49**, 75–112.

ORSZAG, S. A., 1971b, J. Fluid Mech., **50**, 689–703.

ÖZISIK, M. N., 1980, *Heat Conduction*, Wiley-Intersience.

PRINS, J. A., MULDER, J., & SCHENK, J., 1951, Appl. Sci. Res., **A2**, 431.

WILKINSON, J. H., 1965, *The Algebraic Eigenvalue Problem*, Oxford University Press, New York.

PROBLEMS

3-1 A viscous liquid layer of thickness l rests initially on a flat plate of infinite extent. The kinematic viscosity of the fluid is ν. The plate suddenly assumes a uniform velocity U.

(a) Find the velocity variation in the fluid by the integral method.

(b) Devise a simple experimental procedure for determining the kinematic viscosity of the fluid.

(c) How long would it take the velocity of the upper surface of the fluid to reach its steady value?

3-2 Consider a steady Couette flow between two parallel plates separated a distance l apart. Assume that the plate with the velocity U suddenly stops. Determine the unsteady flow velocity by integral means.

3-3 Resolve part (a) of Problem 3-1 by including the effect of a thick air layer above the liquid.

3-4 Consider a steady Couette–Poiseuille flow between two parallel plates separated a distance l apart. One of the plates moves with a constant velocity U. The fluid is subject to a pressure gradient $-dp/dx$ in the direction opposite to the moving plate. Find the velocity distribution, the flow rate, and the condition for zero flow rate.

3-5 Let a large lake be approximated by a pool of water with a uniform thickness l. The kinematic viscosity of the water is ν. Assume that a sudden wind generates a uniform velocity V on the upper surface of the lake (Fig. 3P-5).

(a) Find the steady velocity distribution in the lake.

(b) Find the unsteady velocity variation in the lake.

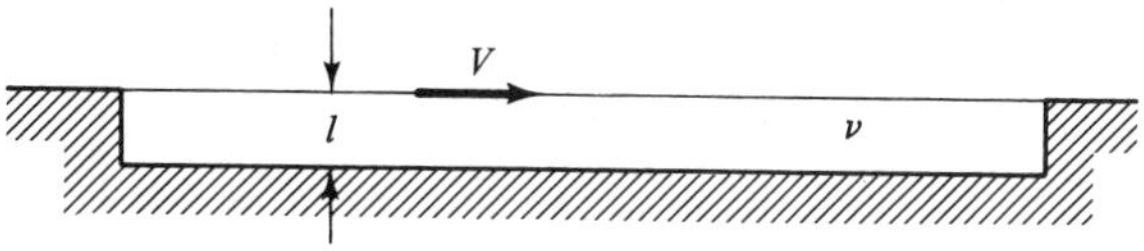

Fig. 3P-5

3-6 Consider an oil bearing with a uniform gap h. The shaft rotates with an angular velocity ω. The oil enters the bearing at a pressure Δp above atmospheric pressure (Fig. 3P-6). Find the steady velocity of the oil.

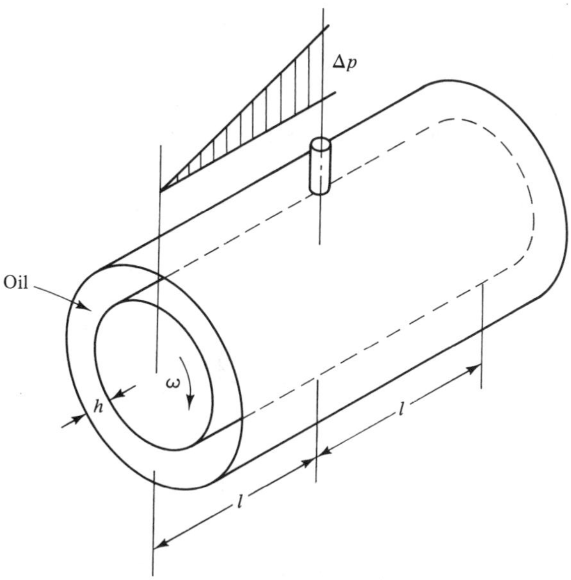

Fig. 3P-6

3-7 Consider an infinitely long flat plate of thickness L and density ρ_0 vertically dividing a viscous fluid of infinite extent. A force F in the vertical direction is suddenly applied to the plate (Fig. 3P-7(a)).

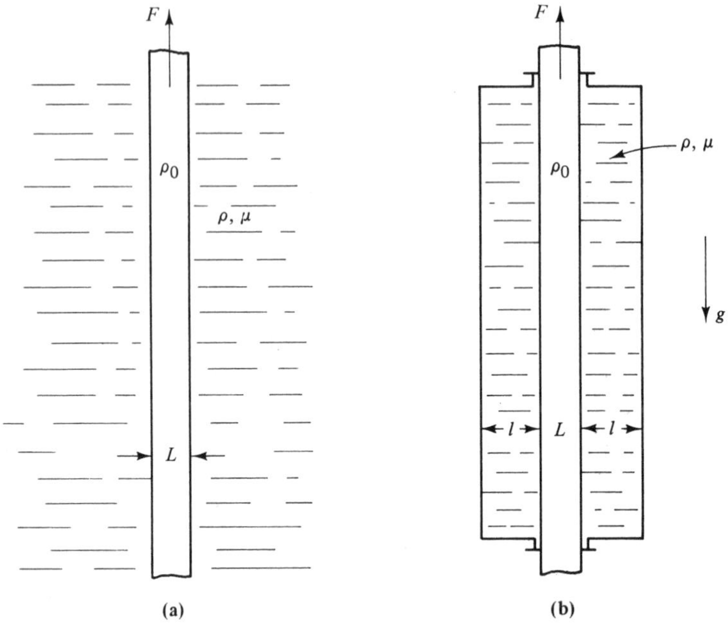

Fig. 3P-7

(a) Neglecting the plate inertia, find the velocity of the fluid.
(b) Assume that the plate suddenly starts falling under the influence of its own weight. Find the velocity of the plate and the fluid.
(c) Repeat part (a) for a slot (fluid of thickness l on both sides of the plate). This problem may apply to a withdrawal of control rods in a liquid-cooled nuclear reactor (Fig. 3P-7(b)).

3-8 A vertical belt moves upward in a viscous fluid with a velocity U. Find the critical value of the velocity at which an upward steady flow of liquid exists in a slot formed by the belt and a vertical plate placed at a distance from the belt (Fig. 3P-8). Solve by integral means the transient motion following a sudden stop of the belt.

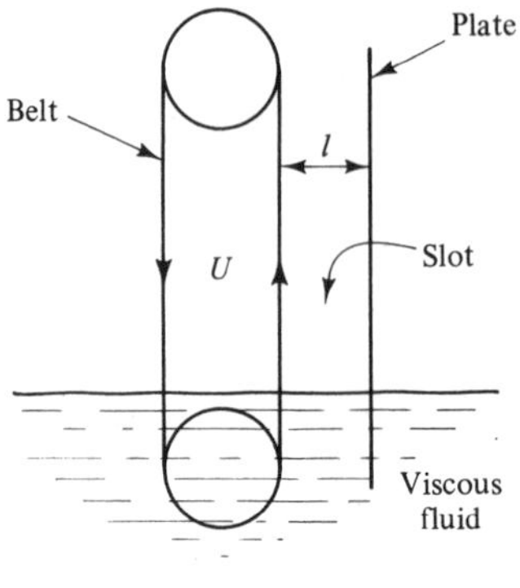

Fig. 3P-8

3-9 Consider a stagnant Newton fluid in a channel of cross section $2l \times h$. The upper boundary suddenly assumes a uniform velocity U. Determine the resulting unsteady velocity of the fluid (Fig. 3P-9).

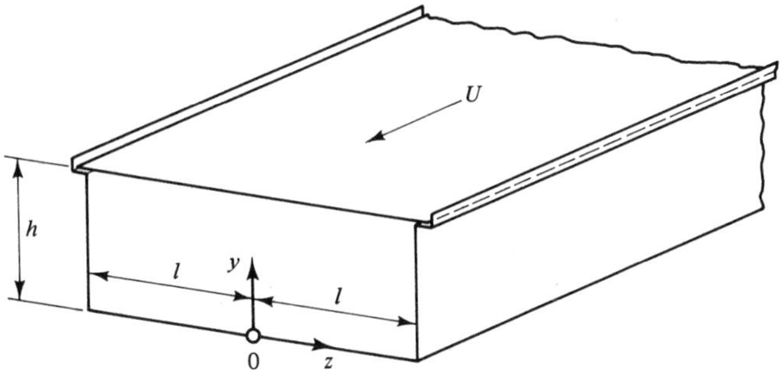

Fig. 3P-9

3-10 Consider a Newton fluid in a channel whose cross section is shown in Fig. 3P-10. The fluid is suddenly subjected to a constant pressure gradient.
(a) Find the steady velocity.
(b) Find the unsteady velocity.

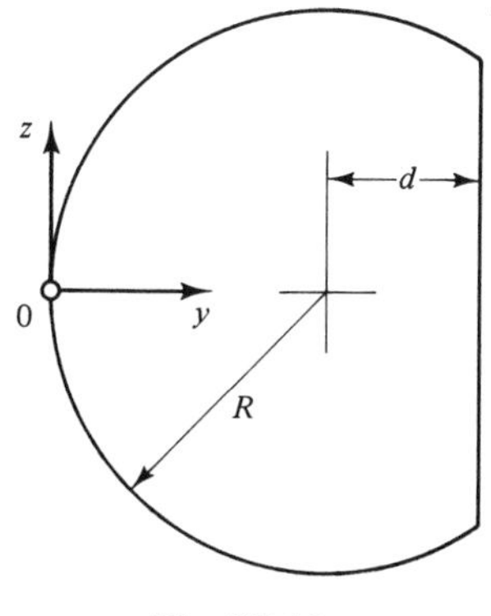

Fig. 3P-10

3-11 Obtain a solution to the second domain of Ex. 3.2.2, now considered as a penetration problem in time (recall Ex. 3.2.1). Is it surprising that the time for this development is on the order of four times shorter than that for Ex. 3.2.1?

3-12 Compare solutions of Exs. 3.2.1 and 3.2.5 to the exact solution, say in terms of the shear stress on the lower plate (Fig. 3.12). Evaluate also the approximate finite development time.

3-13 Interpret and resolve Ex. 3.2.2 as a problem of vorticity diffusion.

3-14 A Newton fluid flows steadily between parallel porous plates subject to a uniform injection velocity v_0 at one wall and uniform suction v_0 at the other wall. Determine the developed velocity distributions as function of $\text{Re}_0 = 2lv_0/\nu$.

3-15 Reconsider Ex. 3.3.1 including the temperature dependence on viscosity, assumed to be of the linear form

$$\mu(T) = \mu_0[1 + \epsilon(T - T_0)].$$

3-16 A liquid film flows steadily down one wall of a vertical slot having unequal wall temperatures T_1 and T_2 (Fig. 3P-16). Accounting for the buoyancy, determine the velocity distributions in the liquid and gas layers.

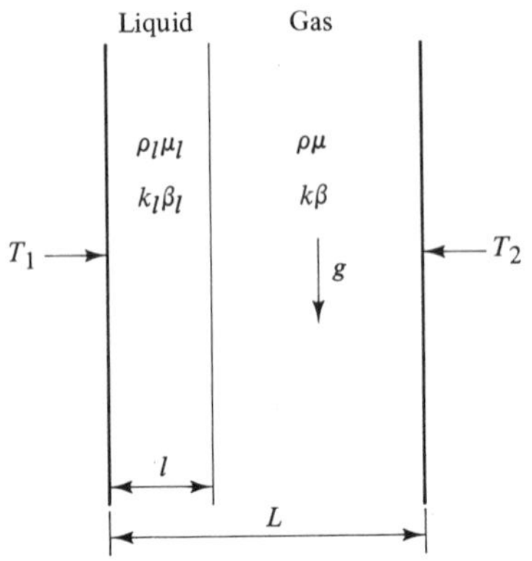

Fig. 3P-16

3-17 Resolve Ex. 3.4.2 for a Prandtl number different from unity. Discuss the limiting solutions for very large and very small Prandtl numbers.

3-18 Using Laplace transforms, derive the solutions given in Ex. 3.4.2.

3-19 Consider a stagnant Newton fluid in a closed vertical slot whose walls are spaced a distance $2l$ apart. From an initial isothermal condition at temperature T_w, the uniform internal energy u''' is suddenly generated in the fluid. Assuming the temperature of the side walls to remain at T_w, we wish to determine the unsteady motion of the fluid in the slot. Consider separately the cases of a Prandtl number of unity and different from unity, and consider both solutions obtained by the Laplace transforms and by integral means.

3-20 Consider an infinitely long and closed slot of a rectangular cross section $l \times L$. Two vertical surfaces of this slot facing each other are insulated while the other two are kept at uniform temperatures T_1 and T_2. Find the steady buoyancy-driven flow.

3-21 Reconsider the slot of Problem 3-20. Let $T_1 = T_2$. The uniform internal energy u''' is generated in the fluid. Find the steady buoyancy-driven flow.

3-22 Evaluate the Nusselt number for the second domain of Ex. 3.5.1 employing the wall-to-bulk temperature difference.

3-23 Find, by the integral method, the steady two-dimensional temperature distribution in a fluid flowing with a uniform velocity U between two parallel plates subject to each of the three boundary conditions shown in Fig. 3P-23.

3-24 A flat plate of thickness l at uniform temperature T_0 moves with a constant velocity U. The upper surface of the plate is insulated while the lower surface over a distance L ($>x_0$ = penetration distance) is subject to a uniform heat flux q'', and is insulated beyond this distance (Fig. 3P-24). Find the steady temperature distribution in the plate by integral means.

3-25 Determine the fully developed heat transfer for a liquid metal flowing steadily between parallel plates subject to a constant heat flux. Show that Eq. (3.6.12) yields Nu = 3 for this case.

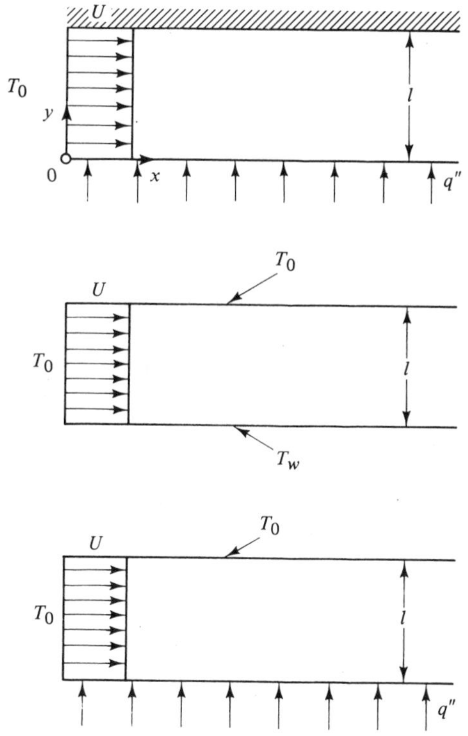

Fig. 3P-23

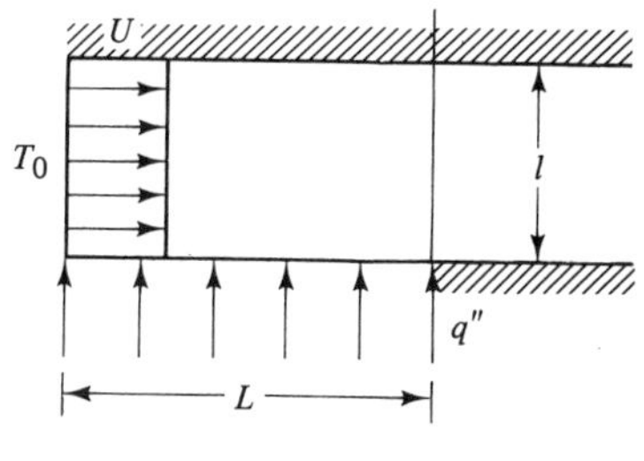

Fig. 3P-24

3-26 Determine the friction factor and resolve Ex. 3.6.1 for flow in a pipe of diameter D to show that $\text{Nu} = \frac{48}{11} = 4.36$.

3-27 Resolve Ex. 3.6.2 for flow in a pipe of diameter D and show that $\text{Nu} = 3.66$.

3-28 Determine the temperature distribution and the peripherally averaged Nusselt number for fully developed heat transfer in a channel of rectangular cross section and with isothermal walls (see Clark & Kays 1953).

3-29 The black energy-absorbing plate of a solar collector is made from corrugated aluminum plates forming lens-shaped channels for the cooling fluid (Fig. 3P-29). Determine the friction factor and the Nusselt number for the developed flow, assuming isothermal walls.

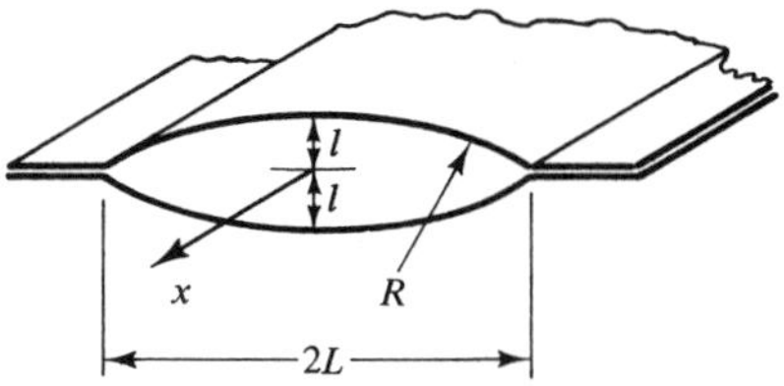

Fig. 3P-29

3-30 Show that any exponentially varying heat flux along a channel leads to a state of fully developed heat transfer downstream. Note that Exs. 3.6.1 and 3.6.2 are special cases.

3-31 Using the representative thermophysical properties given in Appendix E, order the four typical fluids according to their values of the following properties: ρ, μ, ν, k, a, Pr, Gr, and Ra.

3-32 A liquid film of thickness l flows under the influence of gravity down an inclined surface. Upstream the fluid is isothermal, while downstream it is subjected to a uniform heat flux (Fig. 3P-32). Determine the downstream heat transfer depending on the angle of inclination.

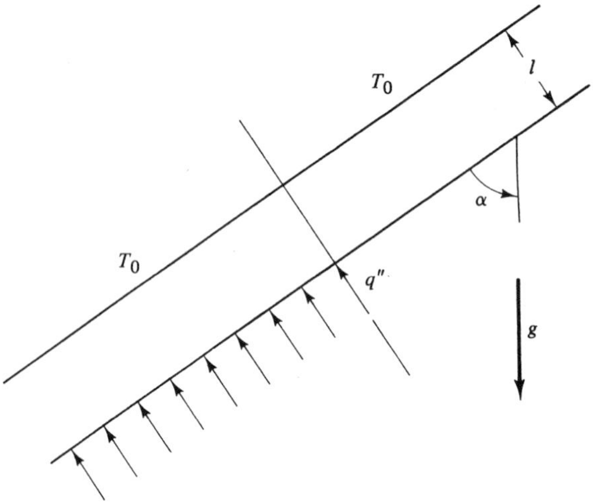

Fig. 3P-32

Chapter 4

NEARLY PARALLEL (BOUNDARY LAYER) FLOWS

This chapter is devoted to boundary layers associated with two-dimensional incompressible flows. The first section deals with the differential formulation governing momentum and thermal boundary layers, and with general aspects of integral formulations. Later sections are assigned to examples, each illustrating a fundamental aspect of boundary layers. Throughout the chapter emphasis is placed on individual integral formulations, requiring solution by approximating functions. Differential formulations, leading to solution via similarity transformation, are postponed until Chapter 5.

4.1 NEARLY PARALLEL FLOWS

In Section 3.5 we discussed the steady thermal boundary layer (a penetration problem in space). We learned how energy diffuses from the heated wall into the moving fluid and is then convected downstream. For large Péclet numbers the temperature distribution is confined to a thin thermal layer whose thickness for slug flow is

$$\frac{\Delta}{x} \sim \left(\frac{a}{Ux}\right)^{1/2} = \mathrm{Pe}_x^{-1/2}, \tag{4.1.1}$$

where Pe_x denotes the Péclet number.

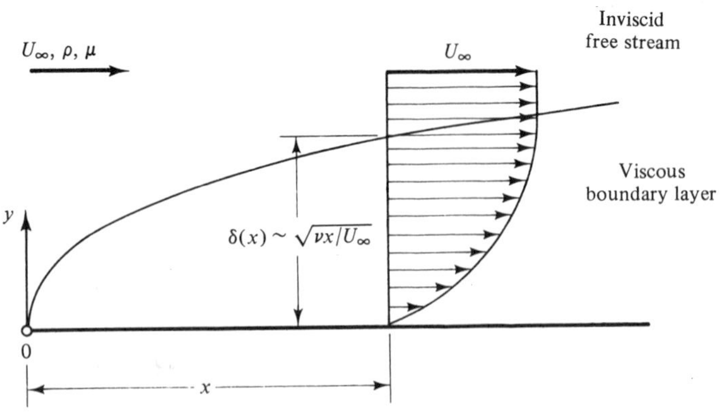

Fig. 4.1 Steady (nearly parallel) boundary layer flow over flat plate

Much the same situation exists in uniform flow past a body, say a thin flat plate (Fig. 4.1). Here momentum diffuses from the fluid to the wall and the resulting momentum deficiency is convected downstream to form a thin momentum layer for large Reynolds numbers. The redistribution of momentum associated with this process alters the parallel flow to a nearly parallel (two-dimensional) flow. The two problems appear to have some common features. For example, we expect—and later find in Section 4.2—that the momentum layer is of thickness

$$\frac{\delta}{x} \sim \left(\frac{\nu}{Ux}\right)^{1/2} = \mathrm{Re}_x^{-1/2}, \tag{4.1.2}$$

where Re_x denotes the Reynolds number and ν is the momentum diffusivity in analogy to the thermal diffusivity a.

In this section, employing the concept of nearly parallel flow, the formulation of boundary layers is developed by an order-of-magnitude approach (for the original approach of Prandtl, refer to Schlichting 1968, and for a formal derivation by the method of singular perturbation, see van Dyke 1964). First, recall the differential formulation for two-dimensional flow of an incompressible, constant property Newton–Fourier fluid, obtained from Table 2.4,

$$\boxed{\frac{\partial u}{\partial x} + \frac{\partial v}{\partial y} = 0} \tag{4.1.3}$$

$$\frac{\partial \rho u}{\partial t} + \frac{\partial}{\partial x}(\rho uu - \tau_{xx}) + \frac{\partial}{\partial y}(\rho vu - \tau_{yx}) = \rho f_x - \frac{\partial p}{\partial x} \tag{4.1.4}$$

$$\frac{\partial \rho v}{\partial t} + \frac{\partial}{\partial x}(\rho uv - \tau_{xy}) + \frac{\partial}{\partial y}(\rho vv - \tau_{yy}) = \rho f_y - \frac{\partial p}{\partial y} \tag{4.1.5}$$

$$\frac{\partial \rho cT}{\partial t} + \frac{\partial}{\partial x}(\rho cuT + q_x) + \frac{\partial}{\partial y}(\rho cvT + q_y) = \mu\Phi_v + u''', \tag{4.1.6}$$

where

$$\tau_{xx} = 2\mu \frac{\partial u}{\partial x}, \qquad \tau_{yx} = \tau_{xy} = \mu\left(\frac{\partial u}{\partial y} + \frac{\partial v}{\partial x}\right), \qquad \tau_{yy} = 2\mu \frac{\partial v}{\partial y}$$
$$q_x = -k\frac{\partial T}{\partial x}, \qquad q_y = -k\frac{\partial T}{\partial y}, \tag{4.1.7}$$

$$\Phi_v = 2\left(\frac{\partial u}{\partial x}\right)^2 + \left(\frac{\partial u}{\partial y} + \frac{\partial v}{\partial x}\right)^2 + 2\left(\frac{\partial v}{\partial y}\right)^2. \tag{4.1.8}$$

Here we have used Eq. (4.1.3) to restore the divergence form of convective terms of momentum and energy in Eqs. (4.1.4)–(4.1.6), and furthermore grouped diffusion terms with corresponding convective terms for ready comparison.

Now, examine this formulation on the assumption of nearly parallel flow along a wall over which a momentum layer develops. From the conservation of mass, Eq. (4.1.3), an order-of-magnitude estimate yields

$$\frac{V}{U} \sim \frac{\delta}{x}, \tag{4.1.9}$$

which shows that transverse velocity v is small compared to axial velocity u because of Eq. (4.1.2).

From the differential balance of x-momentum, Eq. (4.1.4) and Fig. 4.2, the ratio of the axial transport of x-momentum by convection (flow) to that by diffusion (flux) may be estimated, after using Eq. (4.1.7), as

$$\frac{\text{axial momentum flow}}{\text{axial momentum flux}} = \frac{\rho uu}{\tau_{xx}} \sim \frac{\rho UU}{\mu\, U/x} = \frac{Ux}{\nu} = \text{Re}_x. \tag{4.1.10}$$

Hence, for large values of Re_x, axial flux is negligible compared to axial flow and may be ignored in the balance of axial momentum. Turning next to the transversal transport

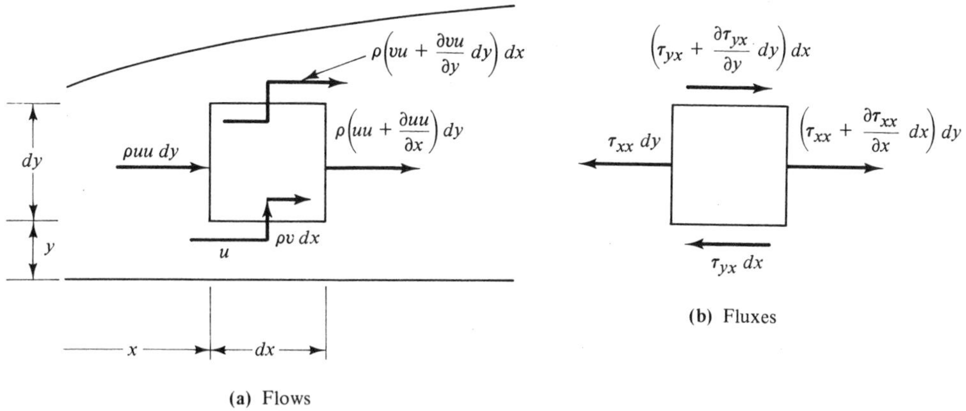

Fig. 4.2 Axial and transversal flows and fluxes of x-momentum

of axial momentum, we first evaluate τ_{yx}, given by Eq. (4.1.7), and using Eq. (4.1.9) obtain

$$\tau_{yx} = \mu\left(\frac{\partial u}{\partial y} + \frac{\partial v}{\partial x}\right) \sim \mu\left(\frac{U}{\delta} + \frac{V}{x}\right) \sim \frac{\mu U}{\delta}\left(1 + \frac{\delta^2}{x^2}\right),$$

or, neglecting the second term in view of Eq. (4.1.2),

$$\tau_{yx} \simeq \mu \frac{\partial u}{\partial x}. \tag{4.1.11}$$

The ratio of transverse flow and flux of x-momentum, referring to Eq. (4.1.4) and Fig. 4.2. and using Eqs. (4.1.9) and (4.1.11), then yields

$$\frac{\text{transverse momentum flow}}{\text{transverse momentum flux}} = \frac{\rho v u}{\tau_{yx}} \sim \frac{\rho(U\delta/x)U}{\mu\, U/\delta} = \frac{Ux}{\nu}\frac{\delta^2}{x^2} \sim 1, \tag{4.1.12}$$

or

$$\frac{\delta}{x} \sim \mathrm{Re}_x^{-1/2}, \tag{4.1.2}$$

which was already conjectured by analogy to thermal boundary layers. Equation (4.1.12) must be of order unity or less; otherwise, we recover the inviscid Euler equations that do not satisfy the no-slip boundary condition on walls and there would be no viscous boundary layer. Consequently, retaining transverse flow and flux as being equally important in Eq. (4.1.4), using Eq. (4.1.11) for the transverse flux, ignoring axial flux and rewriting flow terms by Eq. (4.1.3), we obtain the boundary layer approximation of the axial momentum balance,

$$\boxed{\frac{\partial u}{\partial t} + u\frac{\partial u}{\partial x} + v\frac{\partial u}{\partial y} = f_x - \frac{1}{\rho}\frac{\partial p}{\partial x} + \nu\frac{\partial^2 u}{\partial y^2}.} \tag{4.1.13}$$

The differential balance of transversal y-momentum, Eq. (4.1.5), immediately reduces to

$$\boxed{0 = -\frac{\partial p}{\partial y},} \tag{4.1.14}$$

since by Eqs. (4.1.9) and (4.1.2),

$$\frac{\text{transverse momentum}}{\text{axial momentum}} \sim \frac{\rho v}{\rho u} \sim \frac{1}{\mathrm{Re}_x^{1/2}}. \tag{4.1.15}$$

Thus, the transversal momentum is negligible compared to the axial momentum for large values of Re_x.

Equations (4.1.3), (4.1.13), and (4.1.14) comprise the *Prandtl* (1904) *boundary layer approximation for a Newton fluid.*

Outside the boundary layer, for $\delta/x > \mathrm{Re}_x^{-1/2}$ and Re_x large, the transversal flux of axial momentum may also be neglected, and Eq. (4.1.13) reduces to the x-component of the inviscid Euler equations, Eq. (2.4.48). This equation, evaluated at the outer edge of the boundary layer $y = \delta$, or in view of its negligible thickness at $y = 0$, gives the free-stream condition

$$\boxed{\frac{\partial U}{\partial t} + U\frac{\partial U}{\partial x} = f_x - \frac{1}{\rho}\frac{\partial p}{\partial x}} \tag{4.1.16}$$

to be imposed on the boundary layer formulation. For the solution of a boundary layer problem, first U of Eq. (4.1.16) is obtained by solving the inviscid free-stream flow (see the discussion in Section 2.4.4). Then Eqs. (4.1.3) and (4.1.13) are solved to give the viscous flow near the boundary.

In summary, *the viscous boundary layer is nearly parallel and streamwise diffusion of momentum is negligible. The pressure gradient across the boundary layer is negligible while along the boundary layer it is given by that of the inviscid free stream.* These features, as well as the preceding boundary layer equations, apply to flows over any two-dimensional curved surface (Fig. 4.3) as long as the radius of curvature is large compared with the boundary layer thickness and as long as the flow follows the surface.* For three-dimensional boundary layers, we refer the reader to Rosenhead (1963) and Schlichting (1968).

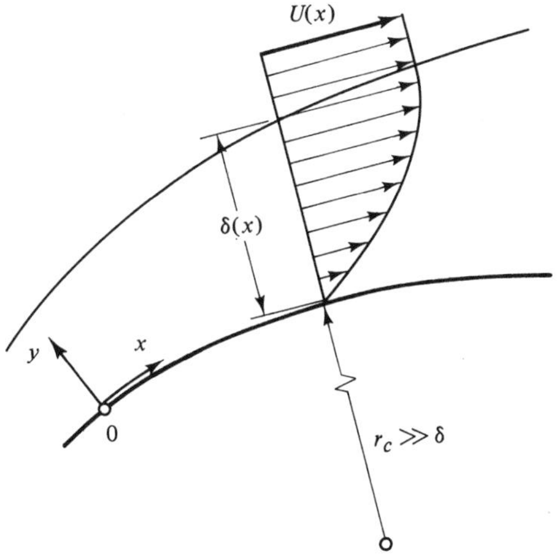

Fig. 4.3 Boundary layer flow over curved surface

Turning now to the balance of thermal energy, Eq. (4.1.6), we examine the ratio of flow and flux in the axial transport of energy in the energy layer of thickness Δ. Employing q_x from Eq. (4.1.7), we obtain

$$\frac{\text{axial energy flow}}{\text{axial energy flux}} = \frac{\rho c u T}{q_x} \sim \frac{\rho c U T}{k\,T/x} = \frac{Ux}{a} = \text{Pe}_x. \tag{4.1.17}$$

*For significant positive pressure gradients the deceleration of the flow leads to a considerable increase in the boundary layer thickness and—ultimately—to separation with reversed flow near the wall. Here, simple boundary layer theory is not valid.

Hence, for large values of the Péclet number Pe_x, axial flux is negligible compared to axial flow and may be ignored. Now, in the thermal layer Eq. (4.1.3) gives

$$\frac{V}{U} \sim \frac{\Delta}{x};$$

hence evaluation of the ratio of transversal flow and flux of energy yields

$$\frac{\text{transverse energy flow}}{\text{transverse energy flux}} = \frac{\rho c v T}{q_y} \sim \frac{\rho c (U\Delta/x)T}{kT/\Delta} = \frac{Ux}{a}\frac{\Delta^2}{x^2} \sim 1, \tag{4.1.18}$$

or

$$\frac{\Delta}{x} \sim \mathrm{Pe}_x^{-1/2}. \tag{4.1.1}$$

Equation (4.1.18) must be of order unity or less to retain the highest derivative, thus to satisfy thermal boundary conditions at both free stream and wall. Comparing Eqs. (4.1.2) and (4.1.1) shows the ratio of momentum to thermal layers to be of the order $\delta/\Delta \sim \mathrm{Pr}^{1/2}$. In general, we shall find that

$$\frac{\delta}{\Delta} \sim \mathrm{Pr}^n, \qquad n \text{ fractional.} \tag{4.1.19}$$

Finally, evaluation of the terms in the dissipation function Φ_v of Eq. (4.1.8) readily shows $\mu(\partial u/\partial y)^2$ to be the only significant term, and

$$\frac{\text{energy dissipation}}{\text{change of energy flux}} = \frac{\mu\Phi_v}{\partial q_y/\partial y} \sim \frac{\mu U^2/\Delta^2}{k\,\Delta T/\Delta^2} = \frac{U^2}{c\,\Delta T}\frac{\nu}{a} = \mathrm{Ec\,Pr}, \tag{4.1.20}$$

where $\mathrm{Ec} = U/c\,\Delta T$ denotes the Eckert number. Dissipation is important and must be included if Ec Pr is of order unity or greater, as for viscous oils (recall the discussions in Sections 2.4 and 3.3). With the foregoing results, Eq. (4.1.6) reduces to the thermal energy balance for boundary layers,

$$\boxed{\frac{\partial T}{\partial t} + u\frac{\partial T}{\partial x} + v\frac{\partial T}{\partial y} = a\frac{\partial^2 T}{\partial y^2} + \frac{\mu}{\rho c}\left(\frac{\partial u}{\partial y}\right)^2 + \frac{u'''}{\rho c}.} \tag{4.1.21}$$

In summary, the thermal boundary layer approximation implies that *streamwise diffusion of thermal energy is negligible compared to convection, and that viscous dissipation contributes through transverse velocity gradient only provided that Ec Pr is of order unity or greater*. For water at 20°C, for example, $c = 4.2$ kJ/kg °C, $\mathrm{Pr} \sim 7$ and a temperature difference of 10°C requires a velocity of

$$U_0 = \left(\frac{c\,\Delta T}{\mathrm{Pr}}\right)^{1/2} = \left(\frac{4200 \times 10}{7}\right)^{1/2} \sim 77 \text{ m/s}$$

for a significant contribution. On the other hand, for a very viscous oil, say $c = 2.2$ kJ/kg °C, $\mathrm{Pr} \sim 10^4$, viscous dissipation must be accounted for even at moderate velocities of $U_0 \sim 1.5$ m/s. For such fluids viscosity depends significantly on the temperature and it may become necessary to include variable properties into the formulation.

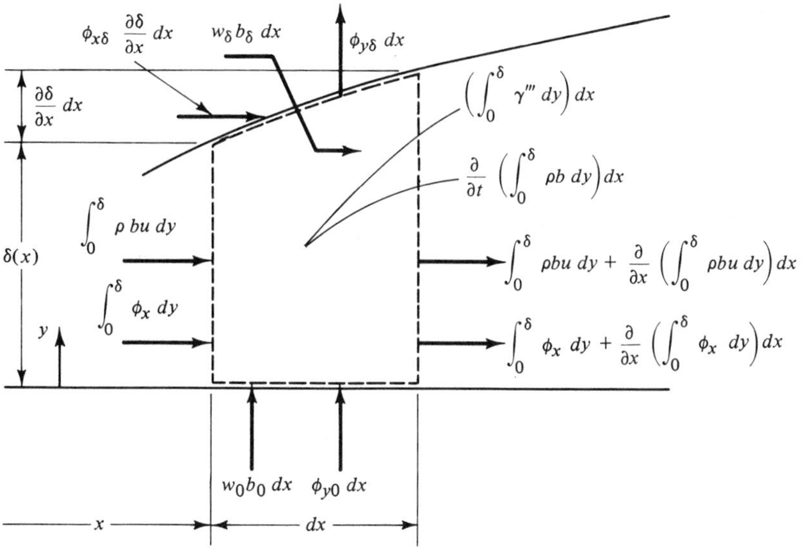

Fig. 4.4 General balance principle for integral control volume of boundary layer

Next, we proceed to the general aspects of the corresponding integral formulations. Our starting point is the master equation, Eq. (2.1.3a), which states a general balance principle involving property b per unit mass. Applying Eq. (2.1.3a) to the differential-integral control volume shown in Fig. 4.4, we obtain for two-dimensional boundary layers

$$\begin{aligned}\frac{\partial}{\partial t}\int_0^\delta \rho b\,dy + \frac{\partial}{\partial x}\int_0^\delta \rho bu\,dy - w_\delta b_\delta - w_0 b_0 \\ = \varphi_{x\delta}\frac{\partial\delta}{\partial x} - \frac{\partial}{\partial x}\int_0^\delta \varphi_x\,dy + \varphi_{y0} - \varphi_{y\delta} + \int_0^\delta \gamma'''\,dy,\end{aligned} \qquad (4.1.22)$$

where w_δ and $w_0 = \rho v_0$ denote mass flow per unit area entering the boundary layer from the free stream and through the (porous) surface, respectively. The first two terms on the right of Eq. (4.1.22) may be combined by the Leibnitz rule to a single term

$$\varphi_{x\delta}\frac{\partial\delta}{\partial x} - \frac{\partial}{\partial x}\int_0^\delta \varphi_x\,dy = -\int_0^\delta \frac{\partial\varphi_x}{\partial x}\,dy,$$

which represents the axial (streamwise) fluxes of b. The diffusive contributions to these fluxes may usually be ignored according to the boundary layer approximation. Transverse diffusion fluxes of b are φ_{y0} at the wall and $\varphi_{y\delta}$ at the free stream, where $\varphi_{y\delta}$ is usually zero by the physics of the problem. The production of b per unit volume is γ'''.

We eliminate the unknown mass flow w_δ from Eq. (4.1.22) by first deriving the *conservation of mass* for $b = 1$, $\varphi = 0$, $\gamma''' = 0$:

$$\boxed{\frac{\partial}{\partial t}\int_0^\delta \rho\, dy + \frac{\partial}{\partial x}\int_0^\delta \rho u\, dy - w_\delta - w_0 = 0,} \tag{4.1.23}$$

whose steady form is

$$\frac{d}{dx}\int_0^\delta \rho u\, dy - w_\delta - w_0 = 0. \tag{4.1.24}$$

Observe that the integral of Eq. (4.1.24) depends only on x, since $u = u(x, y)$ and $\delta = \delta(x)$. Multiplying Eq. (4.1.23) by b_δ and subtracting the result from Eq. (4.1.22) gives

$$\boxed{\begin{aligned}&\frac{\partial}{\partial t}\int_0^\delta \rho b\, dy - b_\delta\frac{\partial}{\partial t}\int_0^\delta \rho\, dy + \frac{\partial}{\partial x}\int_0^\delta \rho b u\, dy - b_\delta \frac{\partial}{\partial x}\int_0^\delta \rho u\, dy - w_0(b_0 - b_\delta)\\ &\quad = -\int_0^\delta \frac{\partial \varphi_x}{\partial x}\, dy + \varphi_{y0} - \varphi_{y\delta} + \int_0^\delta \gamma'''\, dy,\end{aligned}} \tag{4.1.25}$$

which for $b_\delta =$ constant simplifies to

$$\begin{aligned}&\frac{\partial}{\partial t}\int_0^\delta \rho(b - b_\delta)\, dy + \frac{\partial}{\partial x}\int_0^\delta \rho u(b - b_\delta)\, dy - w_0(b_0 - b_\delta)\\ &\quad = -\int_0^\delta \frac{\partial \varphi_x}{\partial x}\, dy + \varphi_{y0} - \varphi_{y\delta} + \int_0^\delta \gamma'''\, dy.\end{aligned} \tag{4.1.26}$$

To obtain the *balance of momentum* in x, let $b = u$, $b_\delta = U$, $b_0 = 0$, $\varphi_x = -\tau_{xx} + p$, $\varphi_y = -\tau_{yx}$, $\gamma''' = \rho f_x$, and Eq. (4.1.25) yields

$$\begin{aligned}&\frac{\partial}{\partial t}\int_0^\delta \rho u\, dy - U\frac{\partial}{\partial t}\int_0^\delta \rho\, dy + \frac{\partial}{\partial x}\int_0^\delta \rho u u\, dy - U\frac{\partial}{\partial x}\int_0^\delta \rho u\, dy + w_0 U\\ &\quad = \int_0^\delta \frac{\partial}{\partial x}(\tau_{xx} - p)\, dy + \tau_{yx}|_{y=\delta} - \tau_{yx}|_{y=0} + \int_0^\delta \rho f_x\, dy.\end{aligned} \tag{4.1.27}$$

Ignoring axial diffusion τ_{xx} and the variation in pressure and body force across the boundary layer and assuming that $\tau_{yx}|_{y=\delta} = 0$, we obtain for *steady flow*

$$\frac{d}{dx}\int_0^\delta \rho u u\, dy - U\frac{d}{dx}\int_0^\delta \rho u\, dy + w_0 U = \delta\Big(\rho f_x - \frac{dp}{dx}\Big) - \tau_{yx}|_{y=0}, \tag{4.1.28}$$

where the pressure gradient is given by Eq. (4.1.16). Further simplifications arise for constant free stream and for an impermeable surface.

It is worthwhile to mention that whenever a constant density fluid is subjected to a constant body force, say gravity, the term ρf_x represents a constant hydrostatic pressure gradient. This term may be combined with the pressure gradient in Eqs. (4.1.13), (4.1.16), and (4.1.28) to form the (modified) pressure gradient, which is associated with the fluid motion alone. Thus, for problems that are independent of gravity we ignore the body force term on the tacit assumption that either $f_x = 0$ or that pressure is measured relative to the hydrostatic distribution. Note that the foregoing remarks do not apply to gravity flow of liquid films (Ex. 4.4.2) or to buoyancy-driven flows (Sections 3.4 and 4.3).

To obtain the *balance of thermal energy*, we consider the thermal boundary layer of thickness Δ and replace δ by Δ in Eqs. (4.1.22)–(4.1.26). Note that the conservation of mass for the thermal boundary layer is needed to eliminate the mass flow entering the boundary layer. Recalling the steps of Section 2.1.2 leading to Eq. (2.1.29), let $b = u_0$ denote the internal energy per unit mass, $\varphi_x = q_x$ axial diffusion, $\varphi_y = q_y$ transverse diffusion, and $\gamma''' = -p\nabla \cdot V + \mu\Phi_v + u'''$ the sum of deformation power by pressure, viscous dissipation, and internal energy generation; then Eq. (4.1.25) becomes

$$\frac{\partial}{\partial t}\int_0^\Delta \rho u_0\, dy - u_{0,\Delta}\frac{\partial}{\partial t}\int_0^\Delta \rho\, dy + \frac{\partial}{\partial x}\int_0^\Delta \rho u_0 u\, dy - u_{0,\Delta}\frac{\partial}{\partial x}\int_0^\Delta \rho u\, dy - w_0(u_{0,0} - u_{0,\Delta})$$
$$= \int_0^\Delta \frac{\partial q_x}{\partial x}\, dy + q_y|_{y=0} - q_y|_{y=\Delta} + \int_0^\Delta (-p\nabla \cdot V + \mu\Phi_v + u''')\, dy. \tag{4.1.29}$$

We next ignore axial diffusion q_x and reduce viscous dissipation to $\mu\Phi_v = \tau_{xy}\, \partial u/\partial y$ because of nearly parallel flow according to the steps of Eqs. (4.1.17)–(4.1.20). Then, considering an *incompressible fluid*, $\nabla \cdot V = 0$, we have by Eq. (2.3.23), $du_0 = c\, dT$, which for constant specific heat gives $u_0 - u_{0,\Delta} = c(T - T_\Delta)$. Furthermore, for constant free-stream temperature and for *steady problems*, Eq. (4.1.29) reduces to

$$\frac{d}{dx}\int_0^\Delta \rho\, cu(T - T_\Delta)\, dy - w_0 c(T_0 - T_\Delta) = q_y|_{y=0} + \int_0^\Delta \left(\tau_{yx}\frac{\partial u}{\partial y} + u'''\right) dy, \tag{4.1.30}$$

where $q_y|_{y=\Delta} = 0$ has been assumed. For an essentially incompressible flow of an ideal gas (low velocity compared to that of sound), ρ may be assumed constant (see Section 2.4.1) and Eq. (4.1.30) applies with $c = c_p$.

The completion of integral formulations, which includes constitutive relations and boundary conditions, is left to the individual examples in the following sections.

4.2 *MOMENTUM BOUNDARY LAYERS*

In this section we study the velocity distribution and wall friction associated with steady isothermal boundary layers. Starting with the flat plate in a Newton fluid at uniform velocity, we later investigate the effect of pressure gradient and the formation of a free surface.

EXAMPLE 4.2.1 Consider the steady, uniform flow over a thin flat plate of length l (Fig. 4.1). The fluid is incompressible and Newtonian. We wish to find the drag force per unit width of the plate.

Assuming the plate to be thin enough that the flow far from it is not significantly changed, the inviscid free-stream velocity is uniform, U_∞, and Eq. (4.1.16) yields

$$0 = \rho f_x - \frac{dp}{dx},$$

implying a pressure gradient that is hydrostatic for $f_x \neq 0$, or zero for $f_x = 0$. In either case, for $f_x =$ constant, the term in the integral balance of momentum for the boundary layer, Eq. (4.1.28), is zero. Introducing $w_0 = 0$ (in the absence of wall injection) and the constitutive relation for a Newton fluid in nearly parallel flow, $\tau_{yx} = \mu(\partial u/\partial y + \partial v/\partial x) \simeq \mu\, \partial u/\partial y$, Eq. (4.1.28) may be rearranged as

$$\frac{d}{dx}\int_0^\delta u(U_\infty - u)\, dy = \nu\left(\frac{\partial u}{\partial y}\right)_{y=0}. \tag{4.2.1}$$

The boundary conditions are

$$u(x, 0) = 0; \qquad u(x, \delta) = U_\infty; \qquad u(0, y) = U_\infty; \qquad \frac{\partial u(x, \delta)}{\partial y} = 0. \tag{4.2.2}$$

For a solution by approximating functions, recall our discussion of the construction of first- and second-order temperature profiles in Exs. 3.5.1 and 3.5.4. Following the same reasoning and being interested in high accuracy for friction at the wall, we consider the second-order profile, which, in addition to Eqs. (4.2.2), also satisfies the differential formulation, Eq. (4.1.13), at the wall:

$$\frac{\partial^2 u(x, 0)}{\partial y^2} = 0. \tag{4.2.3}$$

This yields the cubic velocity profile

$$\frac{u}{U_\infty} = \frac{3}{2}\left(\frac{y}{\delta}\right) - \frac{1}{2}\left(\frac{y}{\delta}\right)^3. \tag{4.2.4}$$

Inserting Eq. (4.2.4) into Eq. (4.2.1) gives

$$d\,\delta^2 = \frac{280}{13}\left(\frac{\nu}{U_\infty}\right) dx; \qquad \delta(0) = 0$$

and

$$\delta(x) = \left(\frac{280}{13}\frac{\nu x}{U_\infty}\right)^{1/2}. \tag{4.2.5}$$

Note the relationsip $\delta/x \sim \mathrm{Re}_x^{-1/2}$, where $\mathrm{Re}_x = U_\infty x/\nu$ is the local Reynolds number. Introducing Eq. (4.2.5) into Eq. (4.2.4) gives the approximate velocity in the boundary layer.

In terms of this example we may explain the inherent limitation of the boundary layer approximation. Clearly, the limit $\mathrm{Re}_x \longrightarrow \infty$ cannot be upheld for $x \longrightarrow 0$, no matter how small the viscosity. Near the leading edge the flow is not nearly parallel, axial diffusion cannot be ignored, and the approximation fails. The boundary layer formulation may therefore also be thought of as the asymptotic formulation valid downstream.

To evaluate the drag force, the shear stress at the wall,

$$\tau_w = \mu\left(\frac{\partial u}{\partial y}\right)_{y=0},$$

is integrated over the length l of the plate to give the friction force per unit width of one side of the plate,

$$F_w = \int_0^l \tau_w \, dx.$$

Inserting Eq. (4.2.5) into (4.2.4) and the result into τ_w yields

$$\tau_w = \frac{3}{2}\left(\frac{13}{280}\right)^{1/2} \mu U_\infty \left(\frac{U_\infty}{\nu x}\right)^{1/2},$$

or, in terms of the *local coefficient of skin friction*,

$$c_x = \frac{\tau_w}{\rho U_\infty^2/2} = 3\left(\frac{13}{280}\right)^{1/2} \mathrm{Re}_x^{-1/2}, \tag{4.2.6}$$

and the drag force becomes

$$F_w = 3\left(\frac{13}{280}\right)^{1/2} (\rho\mu l)^{1/2} U_\infty^{3/2}.$$

Note that the friction force is proportional to the power $\frac{3}{2}$ of velocity, whereas in creeping motion with negligible inertial effects this proportionality is to the first power of velocity. Also, since the integrated friction increases with the power $\frac{1}{2}$ of length, the local contribution to the total friction decreases from upstream toward downstream. In other words, the thicker the boundary layer, the smaller is the shear stress.

For flow past a body it is customary to define the dimensionless *drag coefficient*

$$C_D = \frac{F_D}{\frac{1}{2}\rho U_\infty^2 A}, \tag{4.2.7}$$

where F_D denotes the total drag force and A the wetted surface area (or the area perpendicular to the flow for blunt bodies). The drag equals the skin friction, since no pressure forces contribute to it. Hence $F_D = 2F_w$ and with $A = 2l$ Eq. (4.2.7) becomes

$$C_D = 6\left(\frac{13}{280}\right)^{1/2} \mathrm{Re}^{-1/2}, \tag{4.2.8}$$

where $\mathrm{Re} = U_\infty l/\nu$. The form of Eq. (4.2.8) is identical to that of the exact solution, Eq. (5.3.22), derived in Chapter 5, the coefficient $6(13/280)^{1/2} = 1.293$ being 2.7% smaller than the exact value 1.328. The good accuracy in regard to wall shear may in part be attributed to the use of the second-order profile which satisfies the differential formulation at the wall. Experience shows that the use of still higher-order profiles satisfying additional differential conditions may not improve the accuracy of the present problem (see Finlayson 1972, Chap. 4.2). The integral formulation and solution by polynomial profiles is known as the von Kármán-Pohlhausen procedure and was developed originally for boundary layers with arbitrary pressure gradient (see Rosenhead 1963 and Schlichting 1968 for a discussion of this and related classical methods). In the context of methods of weighted residuals this procedure corresponds to the moment of the differential balance of momentum. That is, we satisfy the integral (over boundary layer thickness) of the balance of momentum multiplied by unit weighting function, after replacing the dependent variable with the approximating function. Similarly, the moment weighted with respect to a Dirac delta function of argument $y = 0$ would give the differential formulation evaluated at the wall.

For improved accuracy of the present problem, methods of several moments employing different continuous weighting functions, or several integral subdomains employing one weighting function, should be used (Finlayson 1972). An elaboration of these procedures is beyond the scope of this text. However, the following example is an exception. There we compare solutions obtained by satisfying one and two moments, respectively, employing weighting functions unity in one case and unity and velocity u in the other case.

EXAMPLE 4.2.2 Consider the steady flow of an incompressible Newtonian fluid between two parallel plates spaced a distance $2l$ apart (Fig. 4.5). We wish to find the velocity distribution and pressure drop in the entrance region.

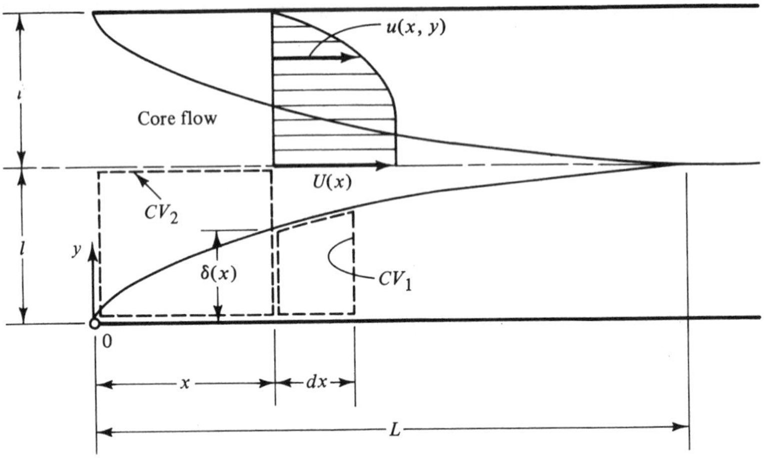

Fig. 4.5 Entrance flow between parallel plates, Ex. 4.2.2

The entrance flow is a special case of nearly parallel flow subject to a pressure gradient. From a uniform velocity of the inlet, boundary layers develop along the walls until they merge at the centerline to form a fully developed (parabolic) velocity distribution. The reduction of velocity in the boundary layers causes an acceleration of the core flow because of the conservation of mass. For a boundary layer formulation the pressure gradient is therefore unknown a priori, and is a part of the solution of the entire problem.

The flow development in a channel or pipe has been studied extensively because of its importance to viscometric flows, heat exchanger design, and so on, which require information such as pressure drop and heat transfer in the entrance region. Schiller (1922) first solved the problem using the integral boundary layer formulation treating the core as potential flow. The problem has since been considered by means of various series expansions and perturbation procedures (see Schlichting 1968 and reference cited therein). On the other hand, Cambell & Slattery (1963) employed the integral balance of both momentum and mechanical energy to eliminate the pressure gradient. Mohanty & Asthana (1978) reconsidered the integral formulation, dividing the development into an inlet region and a filled region, using higher-order profiles.

(1) Here, in the context of integral formulations we begin with the conservation of mass and balance of momentum for control volume CV_1 of Fig. 4.5, leading again to Eq. (4.1.27), which for a Newton fluid becomes

$$\frac{d}{dx}\int_0^\delta uu\,dy - U\frac{d}{dx}\int_0^\delta u\,dy = -\frac{\delta}{\rho}\frac{dp}{dx} - \nu\left(\frac{\partial u}{\partial y}\right)_{y=0}, \tag{4.2.9}$$

The velocity U in the core flow is related to the boundary layer development by the conservation of mass for the whole channel, or in terms of the control volume indicated by CV_2 in Fig. 4.5,

$$\int_0^\delta u\,dy + U(l-\delta) - U_0 l = 0. \tag{4.2.10}$$

Next following Schiller and consistent with the boundary layer approximation, we assume the core flow to be inviscid, so that we can employ Eq. (4.1.16),

$$U\frac{dU}{dx} = -\frac{1}{\rho}\frac{dp}{dx}, \tag{4.2.11}$$

which relates the pressure gradient to the acceleration of the core flow. Although this approximation is reasonable near the entrance, it is questionable downstream, where the boundary layers are not thin but approach the centerline. In view of this approximation we retain ourselves to the first-order profile

$$\frac{u}{U} = \begin{cases} 2\left(\frac{y}{\delta}\right) - \left(\frac{y}{\delta}\right)^2; & 0 \leq y \leq \delta, \\ 1 \quad ; & \delta \leq y = l, \end{cases} \tag{4.2.12}$$

which satisfies the boundary conditions

$$u(x,0) = 0; \qquad u(x,\delta) = U; \qquad \frac{\partial u(x,\delta)}{\partial y} = 0,$$

and which coincides with the fully developed (parabolic) velocity for $\delta = l$. It should be noted that also satisfying Eq. (4.1.13) at the wall would introduce the variable pressure gradient into the resulting (second-order) cubic profile. As in the formalized von Kármán-Pohlhausen procedure, the formulation then requires a numerical integration in x.

After inserting Eq. (4.2.12) into Eq. (4.2.9) and employing Eq. (4.2.11), we carry out the integrations in y. Also introducing dimensionless variables

$$\tilde{\delta} = \frac{\delta}{l}; \qquad \tilde{U} = \frac{U}{U_0}; \qquad \tilde{x} = \left(\frac{x}{l}\right) \mathrm{Re}^{-1},$$

where $\mathrm{Re} = U_0 l/\nu$ is the Reynolds number based on the channel half-width, Eqs. (4.2.9) and (4.2.10) become

$$\frac{8}{15}\frac{d}{d\tilde{x}}(\tilde{U}^2\tilde{\delta}) - \frac{2}{3}\tilde{U}\frac{d}{d\tilde{x}}(\tilde{U}\tilde{\delta}) = \tilde{\delta}\tilde{U}\frac{d\tilde{U}}{d\tilde{x}} - 2\frac{\tilde{U}}{\tilde{\delta}}, \tag{4.2.13}$$

$$\left(1 - \frac{1}{3}\tilde{\delta}\right)\tilde{U} = 1. \tag{4.2.14}$$

Carrying out the differentiations in Eq. (4.2.13) and inserting, from Eq. (4.2.14),

$$\tilde{U} = \frac{1}{1 - \frac{1}{3}\tilde{\delta}}; \qquad \frac{d\tilde{U}}{d\tilde{x}} = \frac{\frac{1}{3}}{(1 - \frac{1}{3}\tilde{\delta})^2}\frac{d\tilde{\delta}}{d\tilde{x}},$$

we eliminate $\tilde{U}$ to obtain, after some rearrangement,

$$\left[7 - \frac{16}{1 - \frac{1}{3}\tilde{\delta}} + \frac{9}{(1 - \frac{1}{3}\tilde{\delta})^2}\right]\frac{d\tilde{\delta}}{d\tilde{x}} = 10,$$

which, subject to $\tilde{\delta}(0) = 0$, may be integrated to yield

$$7\tilde{\delta} + 48\ln(1 - \tfrac{1}{3}\tilde{\delta}) + \frac{9\tilde{\delta}}{1 - \frac{1}{3}\tilde{\delta}} = 10\tilde{x}. \tag{4.2.15}$$

This equation implicitly determines $\tilde{\delta}(\tilde{x})$, and evaluating $\tilde{U}(\tilde{x})$ from Eq. (4.2.14), we can calculate the velocity distribution from Eq. (4.2.12). Note that expanding the terms of Eq. (4.2.15) for small values of $\tilde{\delta}$, we obtain

$$7\tilde{\delta} + 48\left[-\left(\frac{\tilde{\delta}}{3}\right) - \frac{1}{2}\left(\frac{\tilde{\delta}}{3}\right)^2 - \cdots\right] + 9\tilde{\delta}\left[1 + \left(\frac{\tilde{\delta}}{3}\right) + \cdots\right] = 10\tilde{x},$$

or

$$\tilde{\delta}(\tilde{x}) \simeq (30\tilde{x})^{1/2}; \qquad \tilde{x} \longrightarrow 0,$$

which is the approximate solution of the flat-plate boundary layer for a parabolic velocity profile, consistent with a vanishing pressure gradient at $x = 0$.

Although Eq. (4.2.15) is only approximately valid near the entrance, it correctly yields the developed velocity profile which is established at *finite* entrance length L, obtained from Eq. (4.2.15) for $\tilde{\delta} = 1$,

$$\frac{L}{l} = \frac{1}{10}\left[\frac{41}{2} - 48\ln\left(\frac{3}{2}\right)\right]\mathrm{Re} = 0.104\mathrm{Re}. \tag{4.2.16}$$

Next, the pressure gradient in the entrance region, available from Eq. (4.2.11), is

$$-\frac{d\tilde{p}}{dx} = \frac{2/3}{(1 - \tilde{\delta}/3)^3}\frac{d\tilde{\delta}}{d\tilde{x}} = \frac{10}{\tilde{\delta}(1 - \tilde{\delta}/3)(1 + 7\tilde{\delta}/6)}; \qquad \tilde{\delta} \leq 1, \tag{4.2.17}$$

where $\tilde{p} = p/(\rho U_0^2/2)$. Integrating Eq. (4.2.10) and substituting $\tilde{U} = \tilde{U}(\tilde{\delta})$ from Eq. (4.2.14) gives the pressure drop

$$\tilde{p}(0) - \tilde{p}(\tilde{x}) = \tilde{U}^2 - 1 = \frac{(\tilde{\delta}/3)(2 - \tilde{\delta}/3)}{(1 - \tilde{\delta}/3)^2}; \qquad \tilde{\delta} \leq 1, \tag{4.2.18}$$

whose value for the full entrance length is

$$\tilde{p}(0) - \tilde{p}(L) = \tfrac{5}{4}.$$

It is of some interest to show that the pressure drop due to momentum change (acceleration) amounts to 32% and that due to wall friction to 68%. Also, the pressure gradient predicted by Eq. (4.2.17) at $\tilde{\delta} = 1$ is $\frac{90}{13} = 6.92$, which differs from $-d\tilde{p}/d\tilde{x} = 6$, valid for fully developed flow (recall Eq. 3.2.57 for $l/L = 0$, and note that $-d\tilde{p}/d\tilde{x} = f_m(l/D_e)^2\text{Re}$, where f_m is defined by Eq. 3.2.55).

The preceding solution predicts a flow development faster than the actual one, as well as discontinuities in axial gradient of velocity and pressure at $x = L$. As already mentioned, the use of Eq. (4.2.11) is suspect downstream and we should try to improve the estimate of the pressure gradient. For the present nearly parallel flow the pressure is a *global* parameter affecting the whole flow at any axial position. Therefore, retaining Eq. (4.2.11) and employing higher-order profiles obtained by satisfying differential conditions at *local* points of the flow may not be adequate. Instead, we employ an additional moment which weights the effect of pressure over the whole flow. The next higher-order moment with a clear physical meaning is the integral balance of mechanical energy.

(2) For the formulation employing two moments, consider the control volume CV_3 shown in Fig. 4.6. The balance of momentum,

$$\frac{d}{dx}\int_0^l uu\,dy = -\frac{l}{\rho}\frac{dp}{dx} - \nu\left(\frac{\partial u}{\partial y}\right)_{y=0}, \tag{4.2.19}$$

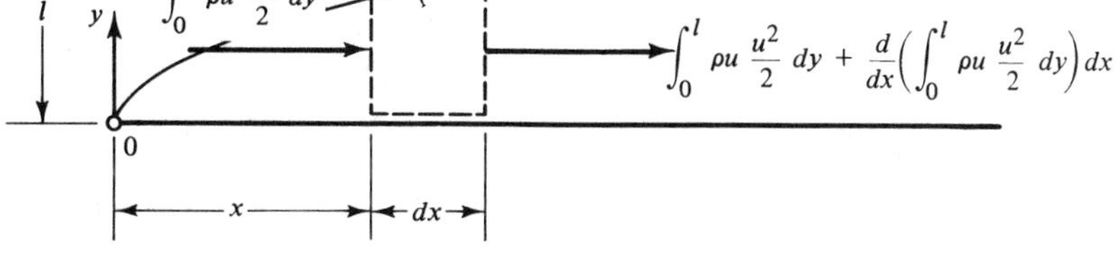

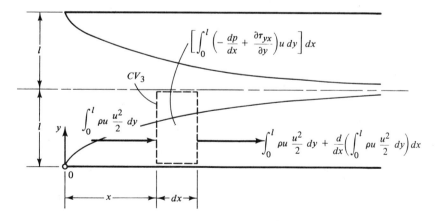

Fig. 4.6 Balance of mechanical energy, Ex. 4.2.2

employing Eq. (4.2.12), and use of the previously introduced dimensionless variables, results in

$$\frac{d}{d\tilde{x}}\left[\tilde{U}^2\left(1 - \frac{7}{15}\tilde{\delta}\right)\right] = -\frac{1}{2}\frac{d\tilde{p}}{d\tilde{x}} - \frac{2\tilde{U}}{\tilde{\delta}}. \tag{4.2.20}$$

The balance of mechanical energy (Fig. 4.6), with the assumption of nearly parallel flow (ignoring axial diffusion and the power by velocity component in y), gives

$$\frac{d}{dx}\int_0^l \rho u\frac{u^2}{2}\,dy = -\int_0^l \frac{dp}{dx}u\,dy + \int_0^l \frac{\partial \tau_{yx}}{\partial y}u\,dy. \tag{4.2.21}$$

Rearranging the term in the last integral of Eq. (4.2.21) as (recall Eq. 2.1.28)

$$\frac{\partial \tau_{yx}}{\partial y}u = \frac{\partial \tau_{yx}u}{\partial y} - \tau_{yx}\frac{\partial u}{\partial y},$$

we note that the integral of the first of these terms on the right is zero because $u(x, 0) = 0$ and $\tau_{yx}(x, l) = 0$. Then, employing $\tau_{yx} = \mu\,\partial u/\partial y$, Eqs. (4.2.12) and (4.2.14) and dimensionless variables, we reduce Eq. (4.2.21) to

$$\frac{1}{2}\frac{d}{d\tilde{x}}\left[U^3\left(1 - \frac{19}{35}\tilde{\delta}\right)\right] = -\frac{1}{2}\frac{d\tilde{p}}{d\tilde{x}} - \frac{4}{3}\frac{\tilde{U}^2}{\tilde{\delta}}. \tag{4.2.22}$$

Eliminating $d\tilde{p}/d\tilde{x}$ between Eqs. (4.2.20) and (4.2.22) and then $\tilde{U}$ by Eq. (4.2.14), we obtain, after some algebra,

$$\left[7 - \frac{411}{28}\frac{1}{1 - \tilde{\delta}/3} + \frac{17}{28}\frac{1}{1 - \tilde{\delta}} + \frac{99}{14}\frac{1}{(1 - \tilde{\delta}/3)^2}\right]\frac{d\tilde{\delta}}{d\tilde{x}} = 10,$$

which, subject to $\tilde{\delta}(0) = 0$, integrates to

$$7\tilde{\delta} + \frac{1233}{28}\ln\left(1 - \frac{\tilde{\delta}}{3}\right) - \frac{17}{28}\ln(1 - \tilde{\delta}) + \frac{99}{14}\frac{\tilde{\delta}}{1 - \tilde{\delta}/3} = 10\tilde{x}. \tag{4.2.23}$$

Expanding Eq. (4.2.23) for small values of $\tilde{\delta}$ gives

$$\tilde{\delta}(\tilde{x}) \simeq \left(\frac{140}{3}\tilde{x}\right)^{1/2}; \qquad \tilde{x} \longrightarrow 0,$$

which, near the entrance, shows faster boundary layer growth than that of the previous solution. Downstream, however, the growth is slower and the solution tends asymptotically toward that of the fully developed flow. Figure 4.7 shows the centerline velocity $\tilde{U}(\tilde{x})$ determined from Eq. (4.2.14) using Eqs. (4.2.15) and (4.2.23), respectively, for $\tilde{\delta}(\tilde{x})$. For comparison the figure also shows results from a numerical integration of the full Navier–Stokes equations at Re = 75 (Gosmann et al. 1969).

Having considered two classical boundary layer flows, we examine now the formation of a free surface.

EXAMPLE 4.2.3 Consider a liquid film on a moving belt leaving a slot as it changes from upstream Couette flow to downstream uniform flow (Fig. 4.8). We seek the film thickness $\delta(x)$ and the velocity distribution for a constant-property Newton fluid. Gravity and surface

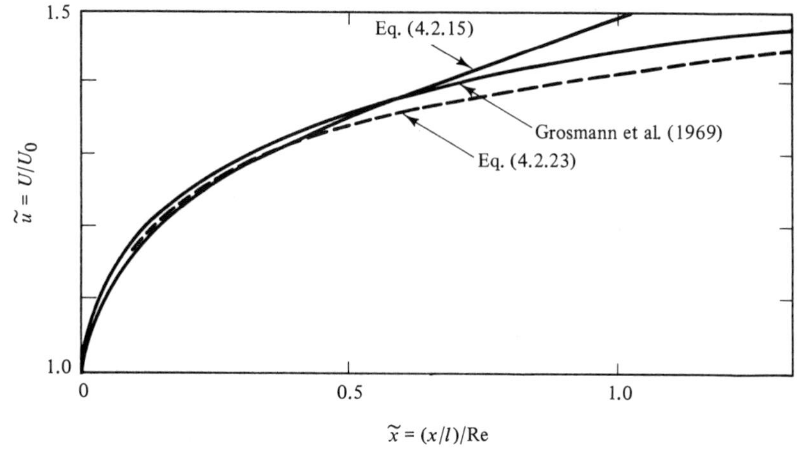

Fig. 4.7 Centerline velocity $\tilde{U}(\tilde{x})$ for entrance flow between parallel plates, Ex. 4.2.2

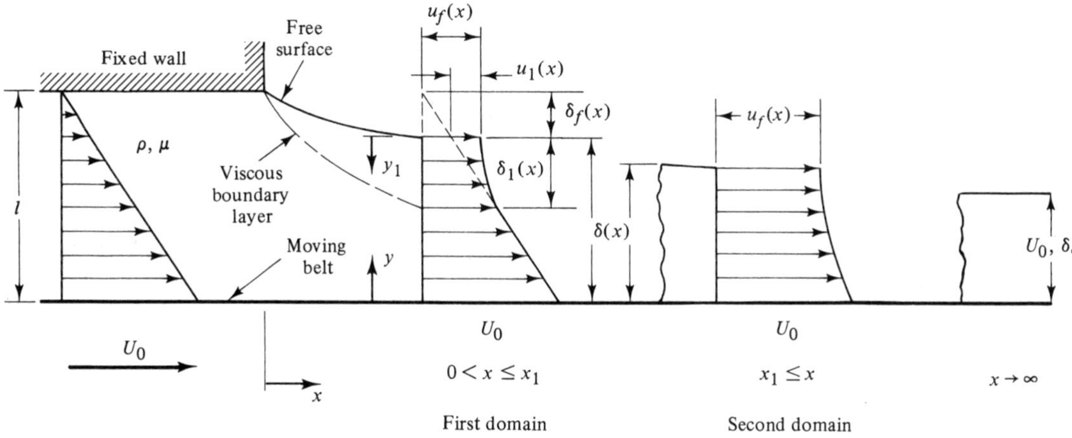

Fig. 4.8 Formation of liquid film on moving belt, Ex. 4.2.3

tension are ignored and the surrounding air is considered inviscid and at uniform pressure.*

After the film of initial thickness l leaves the slot at $x = 0$ with fully developed velocity distribution

$$u(0, y) = U_0\left(1 - \frac{y}{l}\right), \tag{4.2.24}$$

*For a discussion of liquid films including capillary waves, see Levich (1962).

the free upper surface satisfies a boundary condition of zero shear

$$\tau_{yx}(x, \delta) = \mu \frac{\partial u(x, \delta)}{\partial y} = 0; \qquad x > 0. \tag{4.2.25}$$

The momentum diffusing to the free surface therefore begins to accumulate in this region, causing an acceleration of the fluid. This causes, in turn, by the conservation of mass, a reduction of film thickness $\delta_f(x)$ which advances until the whole film assumes a uniform velocity U_0 and thickness

$$\delta_\infty = \tfrac{1}{2} l. \tag{4.2.26}$$

The problem involves a free surface to be determined, a nonuniform upstream condition, and a step change in boundary condition. A differential analysis appears to be impractical. Consequently, we employ an integral formulation, which requires consideration of two space domains.

For the *first domain*, $0 < x \leq x_1$, a suitable profile is readily constructed as the sum of the upstream linear profile and a parabolic profile penetrating to depth $\delta_1(x)$ from the free surface (see Fig. 4.8)

$$\begin{aligned} &0 \leq y \leq l - \delta_f - \delta_1: \qquad u(x, y) = U_0\left(1 - \frac{y}{l}\right), \\ &\left.\begin{matrix} l - \delta_f - \delta_1 \leq y \leq l - \delta_f \\ 0 \leq y_1 \leq \delta_1 \end{matrix}\right\}: \quad u(x, y) = U_0\left(1 - \frac{y}{l}\right) + u_1(x)\left(1 - \frac{y_1}{\delta_1}\right)^2, \end{aligned} \tag{4.2.27}$$

where $y_1 = l - \delta_f - y$, and $\delta_f(x)$, $\delta_1(x)$, and $u_1(x)$ remain to be determined. Satisfying Eq. (4.2.25) gives

$$u_1(x) = \tfrac{1}{2} U_0 \frac{\delta_1}{l}. \tag{4.2.28}$$

The two other unknowns, $\delta_f(x)$ and $\delta_1(x)$, are determined by considering an integral control volume in both y and x.

First, the conservation of mass

$$\int_0^\delta \rho u \, dy - \left(\int_0^l \rho u \, dy\right)_{x=0} = 0,$$

or

$$\int_0^\delta \rho u \, dy - \tfrac{1}{2} \rho U_0 l = 0, \tag{4.2.29}$$

with Eqs. (4.2.27) and (4.2.28), gives after some algebra

$$\delta_1(x) = \sqrt{3}\, \delta_f(x). \tag{4.2.30}$$

Next, the balance of momentum in x becomes

$$\int_0^\delta \rho u u \, dy - \left(\int_0^\delta \rho u u \, dy\right)_{x=0} = -\int_0^x \tau_{yx}(x', 0) \, dx'. \tag{4.2.31}$$

Since shear stress is known both at the free surface, $\tau_{yx}(x, \delta) = 0$, and at the moving belt,

$$\tau_{yx}(x, 0) = \mu \frac{\partial u(x, 0)}{\partial y} = -\frac{\mu U_0}{l},$$

we can employ an integral rather than a differential control volume in Eq. (4.2.31). Also introducing the velocity profiles and carrying out the integrations in Eq. (4.2.31), we obtain, after some lengthy but simple algebra,

$$\frac{\delta_f(x)}{l} = \left(\frac{15}{6\sqrt{3} + 10} \frac{x/l}{\text{Re}}\right)^{1/3}, \tag{4.2.32}$$

where $\text{Re} = U_0 l/\nu$ denotes the Reynolds number for the upstream Couette flow.

The film thickness $\delta(x) = l - \delta_f(x)$ and velocity distribution are then known in this domain, which extends to $x = x_1$, where we have $\delta_f + \delta_1 = l$, and by Eq. (4.2.30),

$$\frac{\delta_f(x_1)}{l} = (\sqrt{3} + 1)^{-1} \tag{4.2.33}$$

and Eq. (4.2.32) gives

$$\frac{x_1/l}{\text{Re}} = \frac{1}{15}(6\sqrt{3} + 10)(\sqrt{3} + 1)^{-3} = \frac{1}{15}. \tag{4.2.34}$$

For the *second domain*, $x_1 < x$, continuing the approximating function from the preceding domain as a uniform velocity U_0 minus a paralbola, we select the two-parameter profile (see Fig. 4.8)

$$u(x, y) = U_0 - [U_0 - u_f(x)]\left(\frac{y}{\delta}\right)\left(2 - \frac{y}{\delta}\right). \tag{4.2.35}$$

Clearly, by this choice Eqs. (4.2.27) and (4.2.35) are identical for $\delta(x_1) = \delta_1(x_1)$.

Conservation of mass, Eq. (4.2.29), relates the unknown parameters as

$$U_0 - u_f(x) = \frac{\frac{3}{2}U_0(\delta - \delta_\infty)}{\delta}, \tag{4.2.36}$$

and the parameter remaining, say $\delta(x)$, is determined from the balance of momentum

$$\frac{d}{dx}\int_0^\delta \rho uu\, dy = -\tau_{yx}|_{y=0}. \tag{4.2.37}$$

Inserting

$$\tau_{yx}(x, 0) = -\frac{3\mu U_0(\delta - \delta_\infty)}{\delta^2}, \tag{4.2.38}$$

and employing Eqs. (4.2.35), (4.2.36), and (4.2.38) in Eq. (4.2.37), gives

$$\frac{\tilde{\delta}^2 - 6}{\tilde{\delta} - 1}\frac{d\tilde{\delta}}{d(x/l)} = \frac{60}{\text{Re}}, \tag{4.2.39}$$

where $\tilde{\delta} = \delta/\delta_\infty = 2\delta/l$. Integrating Eq. (4.2.39) subject to $\delta(x_1) = \sqrt{3}\,\delta_f(x_1)$, where $\delta_f(x_1)$ is given by Eq. (4.2.33), yields the implicit expression for film thickness

$$\tfrac{1}{2}[\tilde{\delta}^2 - \tilde{\delta}^2(x_1)] + [\tilde{\delta} - \tilde{\delta}(x_1)] - 5\ln\frac{\tilde{\delta} - 1}{\tilde{\delta}(x_1) - 1} = \frac{60}{\text{Re}}\frac{x - x_1}{l}. \tag{4.2.40}$$

Having completed the solution, we may now estimate the length L for development of the constant film thickness δ_∞. From Eq. (4.2.34) the first domain terminates at $(x_1/l)/\text{Re} = 0.067$, where $\delta(x_1)/l = 0.634$. From Eq. (4.2.40) $[(L - x_1)/l]/\text{Re} \sim 0.26$, for example, when the film thickness is within 1% of its asymptotic value, that is, for $\delta/l = 0.505$. For an alternative approach to an estimate of a *finite* development length for the second domain, see Ex. 3.5.1.

Certain features of the present example appear in the formation of liquid jets (Problem 4-5) and in the analysis of continuous casting processes prior to solidification. Cooling the film from above or below introduces boundary layers of temperature and solidification fronts which advance into the film (see Problems 4-18 and 4-19). To learn to solve realistic problems of this kind, we turn now to thermal boundary layers in nearly parallel flows.

4.3 *MOMENTUM AND THERMAL BOUNDARY LAYERS*

Having studied the momentum boundary layers, we now proceed to the associated thermal boundary layers, and successively consider forced and buoyancy-driven flows, including separately the effects of viscous dissipation and wall suction.

EXAMPLE 4.3.1 Reconsider Ex. 4.2.1 of steady uniform flow over a flat plate. The fluid is at uniform temperature T_∞ and the plate is isothermal at T_w. We wish to find the local heat transfer along the plate (Fig. 4.9).

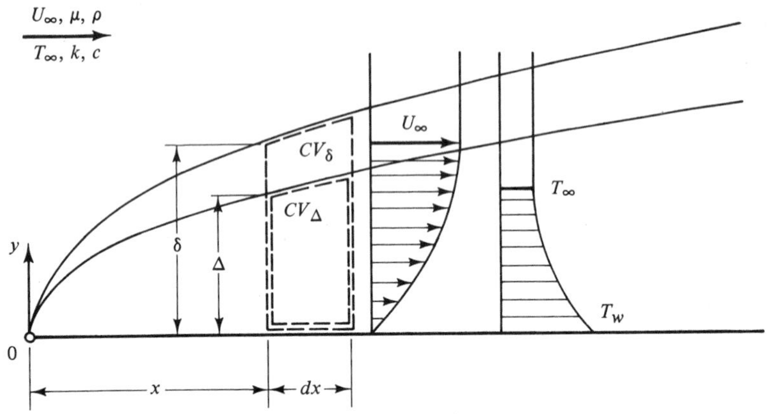

Fig. 4.9 Momentum and thermal boundary layers in uniform flow over isothermal flat plate, Pr ≥ 1, Ex. 4.3.1

The integral balance of momentum and thermal energy for control volumes CV_δ and CV_Δ shown in Fig. 4.9 lead to Eqs. (4.1.28) and (4.1.30), respectively, with $U = U_\infty$, $v_0 = 0$, $\tau_{yx} = \mu\, \partial u/\partial y$ and $T_\Delta = T_\infty$, $q_y = -k\, \partial T/\partial y$, and $u''' = 0$, that is,

$$\frac{d}{dx}\int_0^\delta u(U_\infty - u)\, dy = \nu\left(\frac{\partial u}{\partial y}\right)_{y=0}, \tag{4.3.1}$$

$$\frac{d}{dx}\int_0^{\Delta} u(T - T_\infty)\,dy = -a\left(\frac{\partial T}{\partial y}\right)_{y=0} + \frac{\nu}{c}\int_0^{\Delta}\left(\frac{\partial u}{\partial y}\right)^2 dy, \qquad (4.3.2)$$

subject to boundary conditions

$$u(0, y) = U_\infty; \qquad u(x, 0) = 0; \qquad u(x, \delta) = U_\infty; \qquad \frac{\partial u(x, \delta)}{\partial y} = 0, \qquad (4.3.3)$$

$$T(0, y) = T_\infty; \qquad T(x, 0) = T_w; \qquad T(x, \Delta) = T_\infty; \qquad \frac{\partial T(x, \Delta)}{\partial y} = 0. \qquad (4.3.4)$$

Clearly, the momentum boundary layer of a constant-property fluid is independent of the thermal boundary layer and its formulation and solution become identical to those already obtained in Ex. 4.2.1. The temperature, on the other hand, depends on the velocity distribution.

Now, for solutions by approximating functions we employ second-order profiles for velocity and temperature satisfying the differential formulations at the wall, Eqs. (4.1.13) and (4.1.21), with dissipation neglected,

$$\frac{\partial^2 u(x, \delta)}{\partial y^2} = 0; \qquad \frac{\partial^2 T(x, \Delta)}{\partial y^2} = 0,$$

in addition to Eqs. (4.3.3) and (4.3.4). The resulting profiles (already obtained as Eqs. 4.2.4 and 3.5.54) are

$$\frac{u}{U_\infty} = \frac{3}{2}\left(\frac{y}{\delta}\right) - \frac{1}{2}\left(\frac{y}{\delta}\right)^3, \qquad (4.3.5)$$

$$\frac{T - T_\infty}{T_w - T_\infty} = 1 - \frac{3}{2}\left(\frac{y}{\Delta}\right) + \frac{1}{2}\left(\frac{y}{\Delta}\right)^3. \qquad (4.3.6)$$

Recalling Fig. 3.20 and $\delta/\Delta \sim$ Pr (n positive and fractional) the cases Pr > 1 and Pr $<$ 1 need be considered separately because of the integral in Eq. (4.3.2).

Heat transfer for $P \geq 1$.* Noting that $\Delta < \delta$ and defining $\zeta = y/\Delta$ and $\chi = \Delta/\delta$, we substitute Eqs. (4.3.5) and (4.3.6) into Eq. (4.3.2) and divide the resulting equation by $U_\infty(T_w - T_\infty)$ to obtain

$$\frac{d}{dx}\left[\Delta\int_0^1\left(\frac{3}{2}\chi\zeta - \frac{1}{2}\chi^3\zeta^3\right)\left(1 - \frac{3}{2}\zeta + \frac{1}{2}\zeta^3\right)d\zeta\right]$$

$$= \frac{3}{2}\frac{a}{U_\infty\Delta} + \frac{\nu U_\infty \chi^2}{c(T_w - T_\infty)\Delta}\int_0^1 \frac{9}{4}(1 - \chi^2\zeta^2)^2\,d\zeta,$$

or

$$\frac{d}{dx}\left[\Delta\chi\left(\frac{3}{20} - \frac{3}{280}\chi^2\right)\right] = \frac{3}{2}\frac{a}{U_\infty\Delta} + \frac{9}{4}\frac{\nu U_\infty}{c(T_w - T_\infty)\Delta}\chi^2\left(1 - \frac{3}{2}\chi^2 + \frac{1}{5}\chi^4\right), \qquad (4.3.7)$$

*P and Pr are used interchangeably for the Prandtl number, and E and Ec for the Eckert number.

which is subject to $\Delta(0) = 0$, in view of $T(0, y) = T_\infty$. From the momentum balance, substituting Eq. (4.3.5) into Eq. (4.3.1), we obtain the solution already derived in Ex. 4.2.1,

$$\delta(x) = \left(\frac{280}{13}\frac{\nu x}{U_\infty}\right)^{1/2}. \tag{4.3.8}$$

Now, if χ were assumed to be a constant, integration of Eq. (4.3.7) would give

$$\frac{(14 - \chi^2)\chi}{13}\Delta^2 = \frac{280}{13}\frac{\nu x}{U_\infty}\left[\frac{1}{\mathrm{P}} + \frac{3}{2}\mathrm{E}\chi^2\left(1 - \frac{3}{2}\chi^2 + \frac{1}{5}\chi^4\right)\right], \tag{4.3.9}$$

where $E = U_\infty^2/c(T_w - T_\infty)$ denotes the Eckert number. Equation (4.3.9) implies that $\Delta \sim x^{1/2}$ and hence, in view of Eq. (4.3.8), confirms the assumption. In fact, substituting Eq. (4.3.8) into Eq. (4.3.9) yields an algebraic equation

$$\frac{(14 - \chi^2)\chi^3\mathrm{P}}{13} = 1 + \tfrac{3}{2}\mathrm{PE}\chi^2(1 - \tfrac{3}{2}\chi^2 + \tfrac{1}{5}\chi^4), \tag{4.3.10}$$

from which the relation $\chi = \chi(\mathrm{P}, \mathrm{E})$ is readily evaluated numerically with the result shown in Fig. 4.10. Next, the wall heat flux, employing Eq. (4.3.6),

$$q_w = -k\left(\frac{\partial T}{\partial y}\right)_{y=0} = \frac{\tfrac{3}{2}k(T_w - T_\infty)}{\Delta},$$

introduced into Eq. (3.5.2), yields the local Nusselt number

$$\mathrm{Nu}_x = \frac{hx}{k} = \frac{3}{2}\frac{x}{\Delta} = \frac{3}{2\chi(\mathrm{P}, \mathrm{E})}\frac{x}{\delta(x)}, \tag{4.3.11}$$

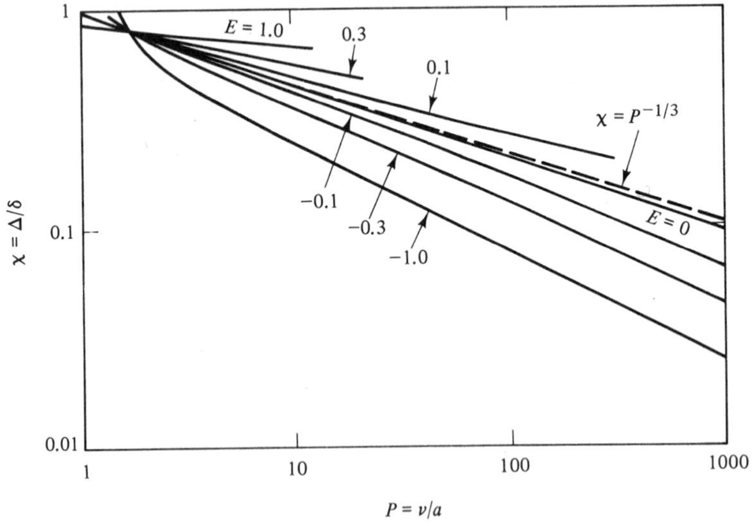

Fig. 4.10 Ratio of thermal and momentum boundary layers, $\Delta/\delta = \chi(\mathrm{P}, \mathrm{E})$, from Eq. (4.3.10); integral solution for $\mathrm{P} \geq 1$, Ex. 4.3.1

or, after using Eq. (4.3.8),

$$\mathrm{Nu_x} = \left(\frac{117}{1120}\right)^{1/2} \frac{\mathrm{Re}_x^{1/2}}{\chi(\mathrm{P}, \mathrm{E})}. \tag{4.3.12}$$

For most applications we can ignore the dissipation and Eq. (4.3.10) for $\mathrm{E} = 0$ gives as special cases, $\chi = 1$ for $\mathrm{P} = 1$, $\chi \simeq \mathrm{P}^{-1/3}$ for $\mathrm{P} \gtrsim 1$ (obtained by taking $\chi^2 = 1$ in the factor $14 - \chi^2$ on the left of Eq. 4.3.10), and $\chi = (14\mathrm{P}/13)^{-1/3}$ for $\mathrm{P} \longrightarrow \infty$. As it turns out, the approximation $\chi = \mathrm{P}^{-1/3}$ proves to be satisfactory for the whole range of Prandtl numbers, and Eq. (3.4.12) may be written as

$$\mathrm{Nu_x} = \frac{3}{4}\sqrt{\frac{13}{70}}\,\mathrm{Re}_x^{1/2}\,\mathrm{P}^{1/3} \qquad (\mathrm{P} \geq 1, \mathrm{E} = 0). \tag{4.3.13}$$

The effect of viscous dissipation need be included only when $|\mathrm{PE}| > 1$ as also evidenced by Fig. 4.10. Then, for heat transfer from fluid to wall ($\mathrm{E} < 0$), dissipation thins the thermal boundary layer, yielding a higher heat transfer coefficient than in the absence of dissipation. These trends are reversed when the heat transfer is from wall to fluid ($\mathrm{E} > 0$). The results shown in Fig. 4.10 should be accepted with some reservation since they are based on a temperature profile which satisfies the differential formulation at the wall only in the absence of dissipation (see Problem 4-8 for an improved profile and see Ex. 5.3.4 for the exact solution).

Heat transfer for P < 1. Noting that $\Delta > \delta$ and employing u from Eq. (4.3.5) for $0 \leq y \leq \delta$ and $u = U_\infty$ for $\delta \leq y \leq \Delta$ in the integral on the left on Eq. (4.3.2), we proceed as in the preceding case. Leaving out the details (see Problem 4-9), dropping the term of order Δ^{-4} arising from the integral and ignoring the dissipation here, we obtain in place of Eq. (4.3.10),

$$\chi = \frac{\frac{1}{2}[(52/35 - 3\mathrm{P}/5)^{1/2} + \mathrm{P}^{1/2}]}{\mathrm{P}^{1/2}}. \tag{4.3.14}$$

Employing this result in Eq. (4.3.12) gives

$$\mathrm{Nu_x} = \frac{3}{4\sqrt{2}} \frac{\mathrm{Re}_x^{1/2}\,\mathrm{P}^{1/2}}{(1 - 21\mathrm{P}/52)^{1/2} + (35\mathrm{P}/52)^{1/2}} \qquad (\mathrm{P} < 1, \mathrm{E} = 0). \tag{4.3.15}$$

Heat transfer for P ⟶ ∞ and P ⟶ 0. Considering the limits of large and small Prandtl numbers, respectively, from Eqs. (4.3.13) and (4.3.15), we obtain, in the absence of viscous dissipation, $\mathrm{E} = 0$,

$$\mathrm{Nu_x}\,\mathrm{Re}_x^{-1/2} = \begin{cases} \dfrac{3}{4}\sqrt{\dfrac{13}{70}}\,\mathrm{P}^{1/3} = 0.332\mathrm{P}^{1/3} & (\mathrm{P} \longrightarrow \infty) \\ \dfrac{3}{4\sqrt{2}}\,\mathrm{P}^{1/2} = 0.531\mathrm{P}^{1/2} & (\mathrm{P} \longrightarrow 0). \end{cases} \tag{4.3.16}$$

The corresponding exact solutions, discussed in Ex. 5.3.4, are of the same form as Eq. (4.3.16), the numerical values of coefficients being 0.339 and 0.564, respectively. The utility of asymptotic solutions, such as Eq. (4.3.16), is apparent from the loga-

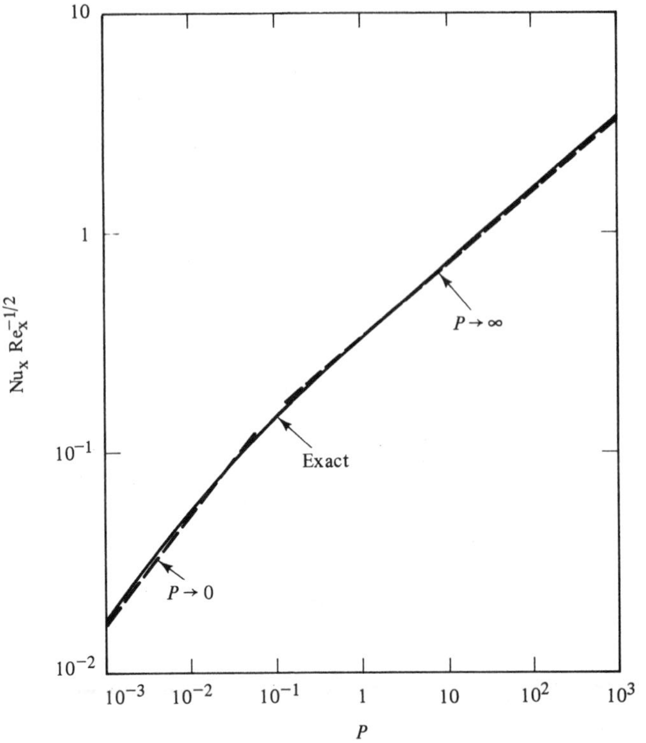

Fig. 4.11 Heat transfer from isothermal flat plate in uniform flow; comparison between approximate asymptotic solution for $P \longrightarrow \infty$ and $P \longrightarrow 0$ (Eq. 4.3.16) and exact solution (Eqs. 5.3.51 and 5.3.55), Ex. 4.3.1

rithmic plot of $\mathrm{Nu_x Re_x^{-1/2}}$ versus P shown in Fig. 4.11. For comparison the exact solution for all P is also shown in the figure.

Now, let us show how we may obtain the foregoing asymptotic solutions without calculating first the complete solutions. In the first case, recalling Fig. 3.20(d) for $P \longrightarrow \infty$, the thermal boundary layer is confined to a region so close to the wall that the velocity distribution may be approximated by shear flow

$$u(x, y) \simeq \left(\frac{\partial u}{\partial y}\right)_{y=0} y = \frac{\tau_w(x)}{\mu} y,$$

where from Eq. (4.3.5), $\tau_w(x)/\mu = 3U_\infty/2\delta(x)$. This approximation to the velocity corresponds to ignoring the second term in Eq. (4.3.5), and hence the term of order χ^3 on the left of Eq. (4.3.7). Ignoring also the dissipation Eq. (4.3.7) integrated with $\Delta(0) = 0$ yields

$$\Delta^3 = \frac{15a}{U_\infty} \delta^{3/2} \int_0^x \delta^{-1/2}\, dx', \tag{4.3.17}$$

which indicates a certain universality of the solution because $\delta(x)$ has not been specified yet. In fact, within the accuracy of the assumed velocity profile, the result is valid for arbitrary boundary layer flow provided that the temperatures of wall and free stream are constant (see Ex. 5.3.5 for a similarity transformation to this problem). Finally, employing Eq. (4.3.8) in Eq. (4.3.17), we obtain $\Delta/\delta = (14\text{P}/13)^{-1/3}$, which is the limit for $\text{P} \longrightarrow \infty$ of Eq. (4.3.10). Replacing $(14/13)^{-1/3}$ by unity, we recover the first of Eqs. (4.3.16). In the second case, recalling Fig. 3.20(a) for $\text{P} \longrightarrow 0$, the thermal boundary layer reaches far outside the momentum boundary layer and the velocity distribution in the balance of thermal energy may be approximated by that of the free stream, $u = U_\infty$. Again ignoring the dissipation we obtain in place of Eq. (4.3.7),

$$\frac{3}{8}\frac{d\Delta}{dx} = \frac{3}{2}\frac{a}{U_\infty \Delta},$$

which is integrated to $\Delta = (8ax/U_\infty)^{1/2}$ and Eq. (4.3.11) then yields exactly the second of Eqs. (4.3.16).

The Reynolds analogy. The idea that there may exist, under certain conditions, a simple proportionality between the heat transfer and the friction near a surface in forced convection was introduced by Reynolds (1874). The idea is based on the analogy between the processes governing transport of thermal energy and momentum by the mechanism of molecular diffusion (as well as that of "mechanical mixing," that is, turbulence, to be discussed in Chapter 10).

The ratio of heat to momentum flux in the direction normal to the surface at an arbitrary point of the flow,

$$\frac{\text{energy flux}}{\text{momentum flux}} = \frac{q_y}{\tau_{yx}} = \frac{-k\,\partial T/\partial y}{\mu\,\partial u/\partial y},$$

may be rearranged in the dimensionless form

$$\frac{q_y/\rho c(T_w - T_\infty)}{\tau_{yx}/\rho U_\infty} = \frac{1}{\text{P}} \frac{\dfrac{\partial}{\partial y}\left(\dfrac{T_\infty - T}{T_\infty - T_w}\right)}{\dfrac{\partial}{\partial y}\left(\dfrac{u}{U_\infty}\right)}. \tag{4.3.18}$$

Evaluated at the wall, introducing the definitions

$$h = \frac{q_w}{(T_w - T_\infty)}; \qquad \text{Nu}_x = \frac{hx}{k}; \qquad c_x = \frac{\tau_w}{\frac{1}{2}\rho U_\infty^2}$$

on the left of Eq. (4.3.18), we obtain

$$\frac{2}{c_x}\frac{\text{Nu}_x}{\text{Re}_x\text{P}} = \frac{\text{St}_x}{c_x/2} = \frac{1}{\text{P}}\left(\frac{d(T_\infty - T)/(T_\infty - T_w)}{du/U_\infty}\right)_{y=0}, \tag{4.3.19}$$

where $\text{St}_x = \text{Nu}_x/\text{Re}_x\text{P} = h/\rho c U_\infty$ denotes the local Stanton number. Whenever the right of Eq. (4.3.19) is independent of position x along the surface, the heat flux is proportional to the friction. That is, given the local coefficient of skin friction c_x, analytically or experimentally, we may calculate the local Nusselt number without actually solving the thermal problem.

The foregoing condition is satisfied for the present problem for P = 1, $\chi = \Delta/\delta = 1$ and in the absence of viscous dissipation, as seen from the integral formulation. In fact, Eqs. (4.3.1) and (4.3.2) and Eqs. (4.3.3) and (4.3.4), respectively, are identical when expressed in terms of u/U_∞ and $(T_\infty - T)/(T_\infty - T_w)$, implying that

$$\frac{u(x, y)}{U_\infty} = \frac{T_\infty - T(x, y)}{T_\infty - T_w} \tag{4.3.20}$$

at any point of the flow. At the wall Eq. (4.3.19) then yields

$$\text{Nu}_x = \frac{c_x}{2}\text{Re}_x \qquad (\text{P} = 1, \text{E} = 0). \tag{4.3.21}$$

Clearly, inserting Eq. (4.2.6) into Eq. (4.3.21) gives Eq. (4.3.12) for $\chi = 1$.

Let us extend the Reynolds analogy to include the Prandtl number effect. Employing the profiles Eqs. (4.3.5) and (4.3.6) in Eq. (4.3.18), we obtain

$$\frac{q_y/\rho c(T_w - T_\infty)}{\tau_{yx}/\rho U_\infty} = \frac{1}{\text{P}\,\chi^3}\frac{\chi^2 - (y/\delta)^2}{1 - (y/\delta)^2}, \tag{4.3.22}$$

which shows that there is only proportionality between heat flux and shear stress throughout the flow for $\chi = 1$. However, in this formulation the proportionality persists at the wall for all values of χ, yielding

$$\text{Nu}_x = \frac{c_x}{2}\text{Re}_x\,\chi^{-3}.$$

Introducing furthermore the approximation $\chi \simeq \text{P}^{-1/3}$, we obtain the generalization of the Reynolds analogy

$$\text{Nu}_x = \frac{c_x}{2}\text{Re}_x\text{P}^{1/3} \qquad (\text{P} \gtrsim 1, \text{E} = 0), \tag{4.3.23}$$

which suggests the Colburn factor

$$\text{j}_\text{H} = \frac{\text{Nu}_x}{\text{Re}_x\text{P}^{1/3}}, \tag{4.3.24}$$

a widely used parameter group for the correlation of heat transfer in both laminar and turbulent flow, and Eq. (4.3.23) simply states that $\text{j}_\text{H} = \frac{1}{2}c_x$. We shall pursue further the foregoing ideas in Chapter 11.

EXAMPLE 4.3.2 Reconsider Ex. 4.3.1 of steady uniform flow over the isothermal flat plate, but let the plate be porous and subject to a uniform suction $v(x, 0) = -v_0$ (Fig. 4.12). We wish to find the skin friction and heat transfer.

First, we treat the fully developed flow which is established far downstream from the leading edge and which is a particular solution independent of x. For $u = u(y)$, Eq. (4.1.3) subject to $v(0) = -v_0$ gives $v(y) = -v_0$, independent of y, and Eq. (4.1.13) reduces to

$$-v_0\frac{du}{dy} = \nu\frac{d^2u}{dy^2},$$

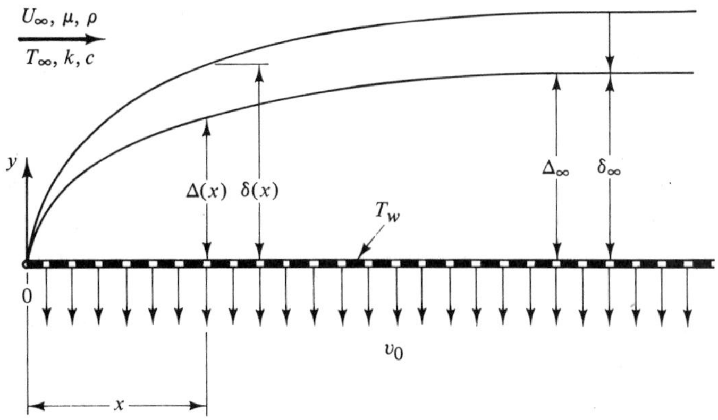

Fig. 4.12 Boundary layers in uniform flow over isothermal flat plate with uniform suction, Ex. 4.3.2

with the solution

$$u(y) = U_\infty\left[1 - \exp\left(-\frac{v_0 y}{\nu}\right)\right], \tag{4.3.25}$$

which assumes that $v_0 > 0$ to satisfy $u(\infty) = U_\infty$. The skin friction,

$$c_\infty = \frac{\tau_w}{\rho U_\infty^2/2} = \frac{2v_0}{U_\infty}, \tag{4.3.26}$$

turns out to be independent of viscosity because the sum of diffusion flux τ_{yx} and convection flow $\rho u v_0$ of x-momentum toward the wall is the same constant value $\rho U_\infty v_0$ for all y. For the temperature, say $T - T_w = \theta(y)$, Eq. (4.1.21) with $u(y)$ inserted from Eq. (4.3.25) reduces to

$$-v_0\frac{d\theta}{dy} = a\frac{d^2\theta}{dy^2} + \frac{\mu}{\rho c}\left(\frac{U_\infty v_0}{\nu}\right)^2 \exp\left(-\frac{2v_0 y}{\nu}\right).$$

This equation, subject to $\theta(0) = 0$ and $\theta(\infty) = \theta_\infty$, has the solution

$$\theta = \theta_\infty\left\{1 - \exp\left(-\frac{\mathrm{P}v_0 y}{\nu}\right) - \frac{\mathrm{EP}}{2(\mathrm{P}-2)}\left[\exp\left(-\frac{2v_0 y}{\nu}\right) - \exp\left(-\frac{\mathrm{P}v_0 y}{\nu}\right)\right]\right\}; \quad \mathrm{P} \neq 2, \tag{4.3.27}$$

which in the limit for $\mathrm{P} \longrightarrow 2$ becomes

$$\theta = \theta_\infty\left[1 - \left(1 + \frac{\mathrm{E}v_0 y}{\nu}\right)\exp\left(-\frac{2v_0 y}{\nu}\right)\right]; \quad \mathrm{P} = 2, \tag{4.3.28}$$

where $\mathrm{E} = U_\infty^2/c(T_w - T_\infty)$. The coefficient of heat transfer becomes independent of the thermal conductivity (explain why),

$$h = \frac{-k(\partial T/\partial y)_{y=0}}{T_w - T_\infty} = (1 - \tfrac{1}{2}\mathrm{E})\rho c v_0. \tag{4.3.29}$$

In the absence of a geometrical length we employ the length scale a/v_0 in terms of which the Nusselt number becomes $\text{Nu}_\infty = h(a/v_0)/k = 1 - \frac{1}{2}\text{E}$. The case $\text{E} = 2$ corresponds to an adiabatic wall, $h = 0$. Physically, the dissipation increases the fluid temperature near the wall to a level approaching T_w, corresponding to thermal insulation, and only the dissipated energy is transferred to the ambient. At still greater dissipation, $\text{E} > 2$, there is heat transfer to the wall even though it is at a temperature above that of the free stream (see Ex. 5.3.4 for the impermeable plate). These phenomena are usually important only for gas dynamics, where compressibility should also be considered.

Let us make a further comment on the foregoing particular solution. Assuming a velocity profile of the form $u/U_\infty = f(y/\delta)$, the integral conservation of mass for a steady boundary layer, Eq. (4.1.23), becomes

$$c_1 \frac{d\delta}{dx} = w_\delta + w_0, \tag{4.3.30}$$

where $c_1 = \int_0^1 f(y/\delta)\, d(y/\delta)$ is a positive constant and the mass flow into the boundary layer is w_δ from the free stream and w_0 from the porous wall (see Fig. 4.4). For the impermeable plate ($w_0 = 0$) we learned in Ex. 4.2.1 that $\delta \sim x^{1/2}$, so mass w_δ flows continuously into the viscous boundary layer as it grows in thickness. For the porous plate, blowing ($w_0 > 0$), according to Eq. (4.3.30), aids the boundary layer growth while suction ($w_0 < 0$) impedes growth. Uniform suction, for the flat plate in a uniform flow, yields the downstream particular solution corresponding to $w_\delta = -w_0 = \rho v_0$, implying a constant boundary layer thickness. Note also that the wall shear stress equals the free-stream momentum flow $\rho U_\infty v_0$ drawn through the surface, and, ignoring dissipation, the heat flux equals the enthalpy transport of free-stream fluid through the surface, $\rho c \theta_\infty v_0$.

Let us turn now to the integral formulation of the complete problem to answer approximately the practical question relating to at what distance from the leading edge the foregoing asymptotic solution is valid. We expect that the complexity of the problem precludes analytical solution of the differential formulation.

Referring to Fig. 4.4 and Eq. (4.1.28), we obtain for the present case

$$\frac{d}{dx}\int_0^\delta u(U_\infty - u)\, dy + v_0 U_\infty = \nu\left(\frac{\partial u}{\partial y}\right)_{y=0}. \tag{4.3.31}$$

Note, that writing the right-hand side as τ_w/ρ, the particular solution which is independent of x, gives us immediately the skin friction of Eq. (4.3.26). For a solution to the full equation by an approximating function, we insert the parabolic profile

$$\frac{u}{U_\infty} = 2\left(\frac{y}{\delta}\right) - \left(\frac{y}{\delta}\right)^2, \tag{4.3.32}$$

satisfying the boundary conditions

$$u(x, 0) = 0; \qquad u(x, \delta) = U_\infty; \qquad \frac{\partial u(x, \delta)}{\partial y} = 0,$$

into Eq. (4.3.31), which then becomes

$$\frac{2}{15}U_\infty^2\frac{d\delta}{dx} + v_0 U_\infty = \frac{2\nu U_\infty}{\delta}. \tag{4.3.33}$$

This result may be rearranged in terms of the asymptotic solution

$$\delta_\infty = \frac{2\nu}{v_0} \tag{4.3.34}$$

as

$$\frac{\delta}{\delta_\infty - \delta}\frac{d\delta}{dx} = \frac{15}{2}\frac{v_0}{U_\infty}. \tag{4.3.35}$$

Integrating Eq. (4.3.35) with $\delta(0) = 0$ gives

$$-\ln\left(1 - \frac{\delta}{\delta_\infty}\right) - \frac{\delta}{\delta_\infty} = \frac{15}{2}\frac{v_0 x}{U_\infty \delta_\infty}, \tag{4.3.36}$$

which determines the boundary layer development $\delta = \delta(x)$. Expanding the left of Eq. (4.3.36) for $\delta/\delta_\infty \ll 1$ shows $\delta \sim x^{1/2}$ for $x \longrightarrow 0$, while deletion of the second term on the left of Eq. (4.3.36) indicates $\delta \longrightarrow \delta_\infty$ exponentially for $x \longrightarrow \infty$. However, it may be useful to evaluate a *finite* development length L by an alternative approximation.

Considering the problem to be one of penetration in both y- and x-directions (recall Ex. 3.2.1), we now specify the complete velocity distribution in terms of Eq. (4.3.32) and an approximating function for $\delta(x)$ involving the penetration length L, say

$$\frac{\delta}{\delta_\infty} = \left(\frac{x}{L}\right)^{1/2}. \tag{4.3.37}$$

Inserting Eq. (4.3.37) into Eq. (4.3.33) and integrating from $x = 0$ to $x = L$ yields

$$\frac{L}{\delta_\infty} = \frac{2}{15}\frac{U_\infty}{v_0}. \tag{4.3.38}$$

At this distance from the leading edge Eq. (4.3.36) gives a boundary layer thickness of $\delta/\delta_\infty \simeq 0.87$ as shown in Fig. 4.13, where δ/δ_∞ from Eqs. (4.3.36) and (4.3.37) are plotted versus $v_0 x/U_\infty\delta_\infty$. Next, employing Eq. (4.3.32), we find $\tau_w = 2\mu U_\infty/\delta$ and the local skin friction becomes $c_x = 4\nu/U_\infty\delta(x)$, which over the development length depends on viscosity and profile shape as well as the suction momentum flow.

The development of the thermal boundary layer may be also studied by the integral formulation. However, a closed-form solution is not in general possible without further approximation because the ratio of boundary layers $\chi = \Delta/\delta$ is not independent of position as it was in Ex. 4.3.1. The formulation, referring to Fig. 4.4 and Eq. (4.1.30) and employing the temperature $\theta = T - T_w$, becomes

$$\frac{d}{dx}\int_0^\Delta u(\theta_\infty - \theta)\,dy + v_0\theta_\infty = a\left(\frac{\partial\theta}{\partial y}\right)_{y=0} - \frac{\mu}{\rho c}\int_0^\Delta \left(\frac{\partial u}{\partial y}\right)^2 dy. \tag{4.3.39}$$

It is left as an exercise to show that for a parabolic profile, say, $\theta/\theta_\infty = 2(y/\Delta) - (y/\Delta)^2$,

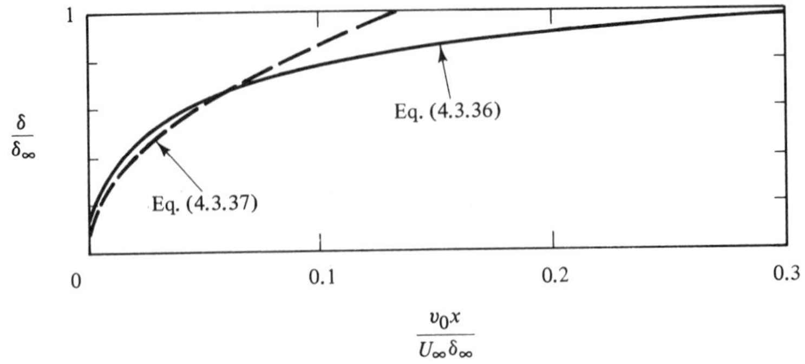

Fig. 4.13 Development length of boundary layer over flat plate with uniform suction, Ex. 4.3.2

the asymptotic boundary layer thickness is given by the implicit relation

$$\chi_\infty \mathrm{P} = 1 + 2\mathrm{EP}\chi_\infty^2(1 - \chi_\infty - \tfrac{1}{3}\chi_\infty^2), \tag{4.3.40}$$

where $\chi_\infty = \Delta_\infty/\delta_\infty$. When dissipation can be ignored, comparing Eqs. (4.3.40) and (4.3.34), we find that $\chi_\infty = \mathrm{P}^{-1}$, and for the case $\mathrm{P} = 1$, Eq. (4.3.39) may be further integrated to yield the trivial results $\chi = 1$ and $\theta/\theta_\infty = u/U_\infty$; hence the Reynolds analogy Eq. (4.3.21) is valid. For a survey of convective heat transfer along surfaces with fluid injection, see Goldstein (1971).

EXAMPLE 4.3.3 Consider a thin vertical plate of uniform temperature T_w placed in a fluid of infinite extent and of temperature $T_\infty < T_w$. We wish to find the heat transfer from the plate.

Initially the fluid is at rest, but as fluid layers near the plate are heated they expand, become lighter than layers far from the plate, and rise. Here we consider the steady-state natural convection (buoyancy-driven) flow (Fig. 4.14), which, as shown later, is confined to a boundary layer. The upflow near the plate must be compensated for by a downflow far from it. However, the velocity of this downflow may safely be put equal to zero because of the large extent of the fluid, and in the absence of imposed flow the free stream is at the hydrostatic state $\rho = \rho_\infty(T_\infty, p_\infty)$, where from Eq. (4.1.16),

$$0 = -\rho_\infty g - \frac{dp_\infty}{dx}. \tag{4.3.41}$$

Note that unlike cases of confined parallel flow (see Ex. 3.4.1), the reference temperature T_∞, with respect to which the buoyancy effect is measured, is determined from the onset.

For an integral formulation the balance of momentum, Eq. (4.1.28) with $U = 0$, $w_0 = 0$, $\tau_{xx} = 0$, $\tau_{yx} = \mu\, \partial u/\partial y$, and $f_x = -g$, becomes

$$\frac{d}{dx}\int_0^\delta \rho u u\, dy = \int_0^\delta \left(-\rho g - \frac{\partial p}{\partial x}\right) dy - \mu\left(\frac{\partial u}{\partial y}\right)_{y=0}, \tag{4.3.42}$$

where $\partial p/\partial x = dp_\infty/dx$ of Eq. (4.3.41) because the change in pressure is negligible over the boundary layer according to Eq. (4.1.14). Then, substituting Eq. (4.3.41) into Eq.

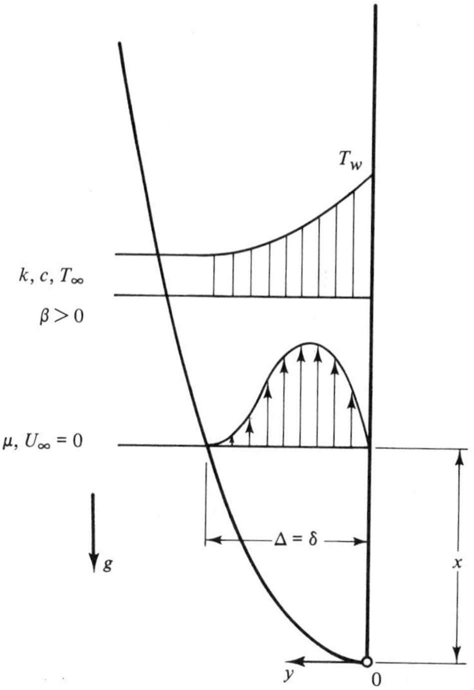

Fig. 4.14 Natural convection boundary layers over isothermal vertical plate, Pr $\sim$ 1, Ex. 4.3.3

(4.3.42) and (recall the Boussinesq approximation introduced in Section 2.4.2) considering ρ to be constant, say ρ_∞, in all terms except the buoyancy term, where its temperature dependency is included by the linear form

$$\rho - \rho_\infty = -\rho_\infty \beta (T - T_\infty), \tag{2.4.20}$$

Eq. (4.3.42) becomes

$$\frac{d}{dx}\int_0^\delta uu\,dy = g\beta \int_0^\delta (T - T_\infty)\,dy - \nu\left(\frac{\partial u}{\partial y}\right)_{y=0}. \tag{4.3.43}$$

Here $\nu = \mu/\rho_\infty$ and β denotes the coefficient of thermal expansion. The balance of thermal energy, Eq. (4.1.30) with $U = 0$, $w_0 = 0$, $q_y = -k\,\partial T/\partial y$, and negligible dissipation, becomes

$$\frac{d}{dx}\int_0^\Delta u(T - T_\infty)\,dy = -a\left(\frac{\partial T}{\partial y}\right)_{y=0}. \tag{4.3.44}$$

Equations (4.3.43) and (4.3.44) are mutually coupled and must be solved simultaneously subject to the boundary conditions

$$\begin{aligned} &u(x, 0) = 0; \qquad u(x, \delta) = 0; \qquad \frac{\partial u(x, \delta)}{\partial y} = 0, \\ &T(x, 0) = T_w; \qquad T(x, \Delta) = T_\infty; \qquad \frac{\partial T(x, \Delta)}{\partial y} = 0. \end{aligned} \tag{4.3.45}$$

Although the motion induced by buoyancy is governed by the penetration of the temperature profile into the fluid, the boundary layer thickness δ of momentum need not equal that of thermal energy Δ (see Fig. 5.6 for exact solutions, and consider Problem 4-13). However, confining attention to a Prandtl number range near unity, we may assume that $\delta = \Delta$.

For a solution by approximating functions we introduce the first-order profiles,* in terms of $\eta = y/\delta(x)$,

$$\begin{aligned} u(x, y) &= u_1 \eta (1 - \eta)^2, \\ T(x, y) - T_\infty &= (T_w - T_\infty)(1 - \eta)^2, \end{aligned} \tag{4.3.46}$$

which satisfy Eqs. (4.3.45). The two parameter functions $\delta(x)$ and $u_1(x)$ are determined by the differential equations resulting from substituting Eq. (4.3.46) into Eqs. (4.3.43) and (4.3.44):

$$\begin{aligned} \frac{1}{105}\frac{d}{dx}(u_1^2\delta) &= \frac{1}{3} g\beta(T_w - T_\infty)\delta - \frac{\nu u_1}{\delta} \\ \frac{1}{30}(T_w - T_\infty)\frac{d}{dx}(u_1\delta) &= \frac{2a(T_w - T_\infty)}{\delta}. \end{aligned} \tag{4.3.47}$$

Despite their nonlinear character these equations are satisfied by solutions of the form

$$u_1(x) = C_1 x^m; \qquad \delta(x) = C_2 x^n. \tag{4.3.48}$$

Substitution yields $m = \frac{1}{2}$, $n = \frac{1}{4}$, and solving for the constants C_1 and C_2 from the resulting algebraic equations gives, finally,

$$u_1(x) = \left(\frac{80}{3}\right)^{1/2}\left(\frac{\mathrm{Gr}_x}{20/21 + \mathrm{Pr}}\right)^{1/2}\frac{\nu}{x}, \tag{4.3.49}$$

$$\frac{\delta(x)}{x} = \left[\frac{240(1 + 20/21\,\mathrm{Pr})}{\mathrm{Pr}\,\mathrm{Gr}_x}\right]^{1/4}, \tag{4.3.50}$$

where $\mathrm{Gr}_x = g\beta(T_w - T_\infty)x^3/\nu^2$ is the local Grashof number and $\mathrm{Pr} = \nu/a$. Equation (4.3.50) shows that $\delta(x) \sim x^{1/4}$ and that the boundary layer approximation is appropriate for large values of Rayleigh number $\mathrm{Ra}_x = \mathrm{Pr}\mathrm{Gr}_x$. Laminar flow is experienced for $\mathrm{Ra}_x < 10^8 \cdots 10^9$.

Using Eq. (4.3.46), the local heat flux,

$$q_w(x) = -k\left(\frac{\partial T}{\partial y}\right)_{y=0} = \frac{2k(T_w - T_\infty)}{\delta(x)},$$

integrated over length l of the plate gives the total heat transfer from one side per unit width,

$$Q = \int_0^l q_w(x)\,dx.$$

Employing Eq. (4.3.50), the local and average Nusselt numbers become

$$\mathrm{Nu}_x = \frac{2x}{\delta(x)} = \left[\frac{\mathrm{Pr}}{15(\frac{20}{21} + \mathrm{Pr})}\right]^{1/4}\mathrm{Ra}_x^{1/4}, \tag{4.3.51}$$

$$\mathrm{Nu}_l = \tfrac{4}{3}\mathrm{Nu}_{x=l}. \tag{4.3.52}$$

*The present formulation and solution was apparently first given by Squire (1938).

For Pr $= 0.733$ of air near room temperature, for example, Eq. (4.3.51) gives $\mathrm{Nu}_x = 0.413\mathrm{Ra}_x^{1/4}$, which is within 7% of the exact solution of Eq. (5.3.62).

The foregoing solution may be improved for Prandtl numbers different from unity by considering the thickness of momentum and thermal boundary layers to be different (Problem 4-13). Because of the lack of similarity between velocity and temperature profiles for the problem, we find (contrary to the results of Ex. 4.3.1) that $\chi = \Delta/\delta$ may be different from unity for Pr $= 1$. Furthermore, χ does not continue to diminish for increasingly large Prandtl numbers (as in Ex. 4.3.1) because the present flow is driven by the buoyancy, which is confined to thickness Δ. It appears that $\chi \longrightarrow$ constant for Pr $\longrightarrow \infty$.

There is a variety of buoyancy driven nearly parallel flows involving modifications of the foregoing problem. For example, the body force may depend on the streamwise coordinate x, such as for the heated horizontal cylinder or the cooling of a turbine blade from fluid circulating in cavities within it. The fluid may have plastic or elastic constitution. The surface may be permeable with blowing or suction applied, leading to features common to those of forced boundary layer flows. Finally, the free stream may be in forced motion, leading to problems of *combined* (forced and buoyancy-driven) *convection*, to be considered next.

EXAMPLE 4.3.4 Reconsider Ex. 4.3.3 of steady natural convection from a vertical isothermal plate at T_w, but let the plate be porous and subject to uniform suction and let the fluid at T_∞ be in uniform upward motion U_∞ (Fig. 4.15). We wish to find the skin friction and heat transfer of this combined (forced and buoyancy driven) flow in the downstream, fully developed region.

In Ex. 4.3.2 we learned that uniform suction applied to the flat plate in uniform forced flow exactly compensates for the boundary layer growth downstream, leading to developed solutions, $u = u(y)$ and $T - T_\infty = \theta(y)$. Anticipating a similar result here we employ a differential formulation (see Problem 4-15 for the integral procedure).

Conservation of mass, in view of $\partial u/\partial x = 0$, again gives $v(y) = -v_0$. Inserting this result into the balance of momentum and thermal energy, Eqs. (2.4.23) and (2.4.24), yields

$$-\frac{v_0}{\nu}\frac{du}{dy} = \frac{\beta g \theta}{\nu} + \frac{d^2u}{dy^2}, \tag{4.3.53}$$

$$-\frac{v_0}{a}\frac{d\theta}{dy} = \frac{d^2\theta}{dy^2}, \tag{4.3.54}$$

subject to

$$u(0) = 0; \qquad u(\infty) = U_\infty; \qquad \theta(0) = \theta_w; \qquad \theta(\infty) = 0. \tag{4.3.55}$$

Because of the simplicity of the problem Eq. (4.3.54) is not coupled to Eq. (4.3.53) and the temperature may be determined first, as for parallel flows (recall Section 3.4). Twice integration of Eq. (4.3.54) subject to Eq. (4.3.55) gives

$$\theta(\eta) = \theta_w e^{-\eta}; \qquad \eta = \frac{v_0 y}{a} > 0. \tag{4.3.56}$$

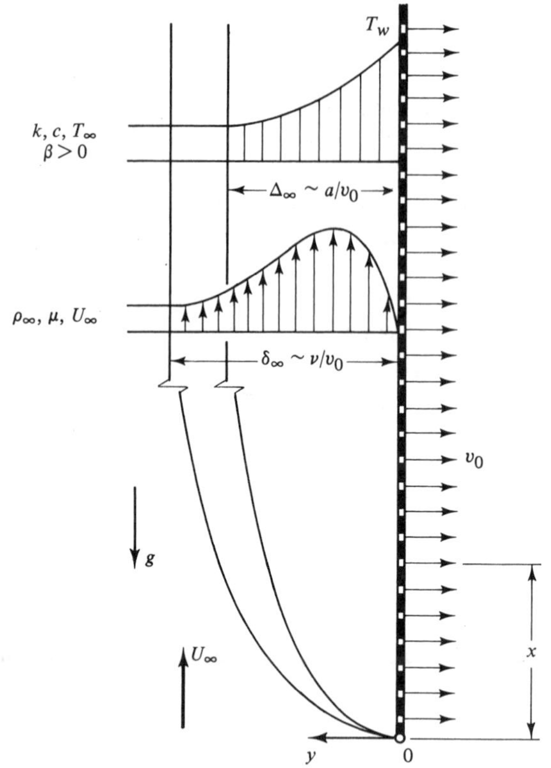

Fig. 4.15 Combined (forced and buoyancy-driven) convection over isothermal vertical plate with uniform suction, Ex. 4.3.4

Inserting this result into Eq. (4.3.53) and rearranging the equation in terms of the variable η as

$$\frac{d^2u}{d\eta^2} + \frac{1}{\mathrm{P}}\frac{du}{d\eta} = -\frac{ag\beta\theta_w}{\mathrm{P}v_0^2}e^{-\eta},$$

subsequent integration subject to Eq. (4.3.55) gives

$$u(\eta) = \frac{ag\beta\theta_w}{v_0^2(\mathrm{P}-1)}(e^{-\eta/\mathrm{P}} - e^{-\eta}) + U_\infty(1 - e^{-\eta/\mathrm{P}}); \qquad \mathrm{P} \neq 1, \qquad (4.3.57)$$

which in the limit for $\mathrm{P} = 1$ becomes

$$u(\eta) = \frac{ag\beta\theta_w}{v_0^2}\eta e^{-\eta} + U_\infty(1 - e^{-\eta}); \qquad \mathrm{P} = 1. \qquad (4.3.58)$$

Evaluating the wall shear $\tau_w = \mu\,\partial u(0)/\partial y$ from Eq. (4.3.57), we may express the coefficient of skin friction, referred to suction velocity, as

$$c_x = \frac{2\tau_w}{\rho v_0^2} = 2\mathrm{PGr} + \frac{2U_\infty}{v_0}, \qquad (4.3.59)$$

where $\mathrm{Gr} = g\beta\theta_w(a/v_0)^3/\nu^2$ is based on the characteristic thermal length a/v_0 (being proportional to the thermal boundary layer thickness). Evidently, the two terms on the right of Eq. (4.3.59) represent the contributions from suction plus buoyancy and suction plus forced flow, respectively. Their ratio may be written in the form which is typical for parallel flows,

$$\frac{\text{buoyancy flow}}{\text{forced flow}} \sim \frac{\mathrm{Gr}}{\mathrm{Re}} \qquad \text{(parallel flow)},$$

where $\mathrm{Re} = U_\infty(a/v_0)/\nu$. For developing boundary layers of combined convection, however, the foregoing ratio becomes

$$\frac{\text{buoyancy flow}}{\text{forced flow}} \sim \frac{\mathrm{Gr_L}}{\mathrm{Re_L^2}} \qquad \text{(nearly parallel flow)},$$

which was derived as Eq. (2.4.28) in Section 2.4.

Finally, the heat flux, employing Eq. (4.3.56), becomes

$$q_w = -k\left(\frac{\partial T}{\partial y}\right)_{y=0} = \frac{k(T_w - T_\infty)v_0}{a}, \tag{4.3.60}$$

implying that $\mathrm{Nu}_\infty = h(a/v_0)/k = 1$, which is identical to Eq. (4.3.29) of forced flow in the absence of dissipation. Thus, heat transfer is independent of the induced and forced flows when these are parallel to the plate, and it depends only on the suction.

This example concludes our preliminary discussion of thermal and momentum boundary layers. In Section 4.4 we include the effect of phase change on these problems.

4.4 *CHANGE OF PHASE*

When part of a continuum has temperatures below or above the temperature at which the continuum, with the absorption or liberation of heat, changes from one phase to another, there exists a boundary between the two phases relative to which the continuum is in motion. With the assumption of local equilibrium the boundary temperature becomes the temperature of phase equilibrium for the prevailing pressure, while other state and transport properties change discontinuously across the interface.

In these problems (including sublimation, melting, freezing, evaporation, and condensation) the location and possible motion of the interface must be determined together with the velocity and temperature in the two phases. Here, before considering specific problems, we formulate the interface conditions which serve as boundary conditions for convection problems of the adjacent phases.

Consider the differential control volume surrounding the phase boundary shown in Fig. 4.16. Let v_1 and v_2 denote velocities in the direction normal to the boundary and $N(t)$ the position of the boundary. Then conservation of mass gives

$$\rho_2\left(v_2 - \frac{dN}{dt}\right) - \rho_1\left(v_1 - \frac{dN}{dt}\right) = 0,$$

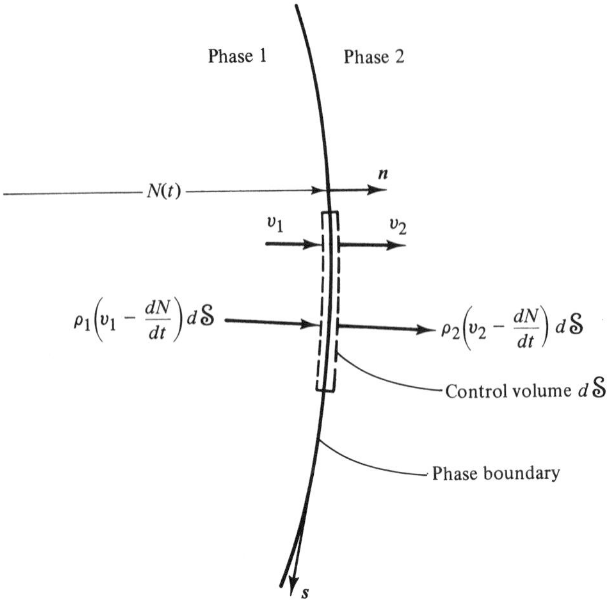

Fig. 4.16 Conservation of mass at phase boundary

or

$$\rho_2 v_2 - \rho_1 v_1 = (\rho_2 - \rho_1)\frac{dN}{dt}, \tag{4.4.1}$$

where dN/dt is the boundary velocity in direction of the normal $\boldsymbol{n}$.

The interface is considered to be massless,* hence without momentum and energy, other than free surface energy which per unit area equals the surface tension σ_s. Then, ignoring normal viscous stress, the balance of momentum in direction of the normal $\boldsymbol{n}$ becomes (Fig. 4.17)

$$\rho_2\left(v_2 - \frac{dN}{dt}\right)v_2 - \rho_1\left(v_1 - \frac{dN}{dt}\right)v_1 = p_1 - p_2 - \frac{\sigma_s}{r_c}, \tag{4.4.2}$$

where r_c denotes the radius of curvature measured from phase 1, $1/r_c = \partial\theta/\partial s$. For a doubled curved surface $1/r_c$ is replaced by $1/r_{c1} + 1/r_{c2}$, where r_{c1} and r_{c2} denote the principal radii of curvature both measured from phase 1. Since momentum fluxes are negligible for slow phase change controlled by heat transfer, any pressure difference across the interface arises from surface curvature. Such pressure differences, however, are significant only for large values of curvature $1/r_c$ (for water at room temperature, for example, the capillary pressure σ_s/r_c and the hydrostatic pressure $\rho g r_c$ are of same magnitude when $r_c \sim 2.5$ mm). Excluding such cases, Eq. (4.4.2) reduces to

$$p_1 = p_2, \tag{4.4.3}$$

*For general surface phenomena, see Scriven (1960), Davies & Rideal (1961), and Slattery (1964).

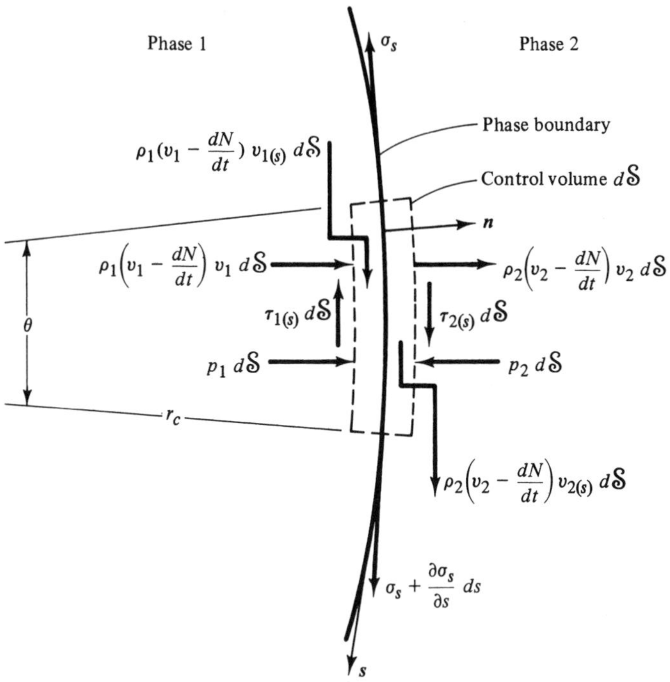

Fig. 4.17 Balance of momentum in $\boldsymbol{n}$ and $\boldsymbol{s}$ at phase boundary

say, p_s, the interface pressure. Next, the balance of momentum in the direction of tangent $\boldsymbol{s}$ yields (Fig. 4.17)

$$\rho_2\left(v_2 - \frac{dN}{dt}\right)v_{2(s)} - \rho_1\left(v_1 - \frac{dN}{dt}\right)v_{1(s)} = \tau_{1(s)} - \tau_{2(s)} + \frac{\partial \sigma_s}{\partial s}, \tag{4.4.4}$$

where the subscript (s) refers to components of velocity and stress in $\boldsymbol{s}$. Also, from the continuum hypothesis, tangential velocity is continuous,

$$v_{1(s)} = v_{2(s)}, \tag{4.4.5}$$

and, in view of Eq. (4.4.1) and for uniform surface tension, shear stress is also continuous,

$$\tau_{1(s)} = \tau_{2(s)}. \tag{4.4.6}$$

Finally, including flux of displacement power by pressure in the enthalpy flow (recall Eq. 2.1.25), the balance of thermal energy becomes (Fig. 4.18)

$$\rho_2\left(v_2 - \frac{dN}{dt}\right)h_2 - \rho_1\left(v_1 - \frac{dN}{dt}\right)h_1 = q_1 - q_2,$$

which by use of Eq. (4.4.1) may be rearranged as

$$\rho_1\left(v_1 - \frac{dN}{dt}\right)h_{12} = q_1 - q_2, \tag{4.4.7}$$

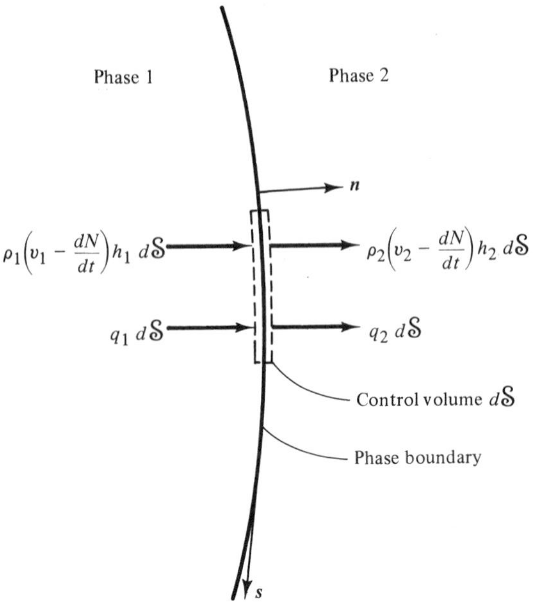

Fig. 4.18 Balance of thermal energy at phase boundary

where $h_{12} = h_2 - h_1$ denotes the latent heat of phase change. For diffusional heat transfer in Fourier continua, Eq. (4.4.7) becomes

$$\rho_1\left(v_1 - \frac{dN}{dt}\right)h_{12} = -k_1\left(\frac{\partial T_1}{\partial n}\right)_s + k_2\left(\frac{\partial T_2}{\partial n}\right)_s, \tag{4.4.8}$$

and for Newton fluids Eq. (4.4.6) gives

$$\mu_1\left(\frac{\partial v_{1(s)}}{\partial n}\right)_s = \mu_2\left(\frac{\partial v_{2(s)}}{\partial n}\right)_s. \tag{4.4.9}$$

In addition to the foregoing boundary conditions, the assumption of local equilibrium implies phase equilibrium at the boundary, yielding a unique relation between temperature and pressure (with the assumptions leading to Eq. 4.4.3)

$$T_s = T_s(p_s). \tag{4.4.10}$$

Recall from thermodyanamics (see, for example, Van Wylen & Sonntag 1972) that phase equilibrium of a pure compressible substance implies equality of temperature, pressure, and the Gibbs function (defined by Eq. 2.2.9). The latter condition reduces the number of independent state properties by one and leads to Eq. (4.4.10). Often pressure is constant along the boundary, which then becomes isothermal. This fact introduces some simplification into the otherwise complex problems of phase change.

Having completed the discussion on interface conditions, we now turn to some examples involving phase change.

EXAMPLE 4.4.1 Consider a liquid initially at temperature T_∞ suddenly brought into contact with a plane wall at constant temperature T_w. Depending on the temperature of the liquid relative to that of melting T_s, and depending on whether the wall is insulated or cooled, several cases may be studied (Fig. 4.19). Here we consider the case of a liquid at the melting point solidifying on an isothermal subcooled wall (Fig. 4.19(d)). We wish to find the thickness of the solidified layer, $\delta(t)$.

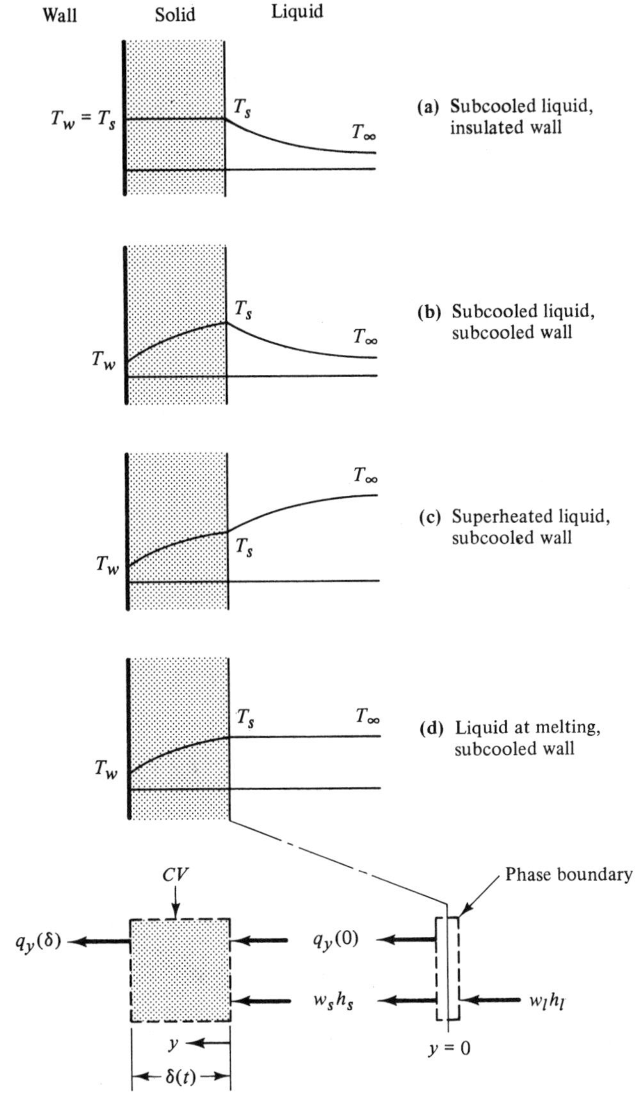

Fig. 4.19 Solidification at plane wall

Clearly, this unsteady problem is one of penetration depth (recall Section 3.2) and it is well suited for integral formulation. In terms of the expanding control volume which encloses the solid (but not the interface) the conservation of mass and balance of thermal energy become

$$\frac{d}{dt}\int_0^\delta \rho dy - w_s = 0, \tag{4.4.11}$$

$$\frac{d}{dt}\int_0^\delta \rho h\, dy - w_s h_s = q_y|_{y=0} - q_y|_{y=\delta}, \tag{4.4.12}$$

where h_s is the enthalpy and w_s the mass flow of solidified material entering the control volume at $y = 0$. Eliminating w_s between these equations and employing the constitutive relations of an incompressible solid, $dh = c\, dT$ and $q_y = -k\, \partial T/\partial y$, and the condition $h_s = h(T_s)$ at $y = 0$, yields

$$\frac{d}{dt}\int_0^\delta (T - T_s)\, dy = -a\left(\frac{dT}{\partial y}\right)_{y=0} + a\left(\frac{\partial T}{\partial y}\right)_{y=\delta}. \tag{4.4.13}$$

Equations (4.4.1) and (4.4.8) for the control volume enclosing the phase boundary give

$$w_s - w_l = 0; \qquad w_s h_{sl} = -k\,\frac{\partial T(0, t)}{\partial y},$$

where $h_{sl} = h_l - h_s$ denotes the heat of fusion. Introducing w_s, obtained by carrying out the integration in Eq. (4.4.11),

$$\rho\frac{d\delta}{dt} = w_s, \tag{4.4.14}$$

we obtain

$$\rho h_{sl}\frac{d\delta}{dt} = -k\left(\frac{\partial T}{\partial y}\right)_{y=0}. \tag{4.4.15}$$

Assuming a second-order polynomial profile for the solid satisfying the boundary conditions

$$T(0, t) = T_s; \qquad T(\delta, t) = T_w, \tag{4.4.16}$$

yields the parabola

$$\frac{T - T_s}{T_w - T_s} = (1 - a_2)\left(\frac{y}{\delta}\right) + a_2\left(\frac{y}{\delta}\right)^2. \tag{4.4.17}$$

Inserting Eq. (4.4.17) into Eqs. (4.4.13) and (4.4.15) gives a set of coupled nonlinear ordinary differential equations in terms of parameters $\delta(t)$ and a_2,

$$\frac{d}{dt}\left[\frac{1}{2}\,\delta\left(1 - \frac{1}{3}a_2\right)\right] = 2a\frac{a_2}{\delta}, \tag{4.4.18}$$

$$\frac{d\delta}{dt} = a\mathrm{J}\frac{1 - a_2}{\delta}, \tag{4.4.19}$$

where* $\mathrm{J} = c(T_s - T_w)/h_{sl}$ denotes the *Jakob number* and $a = k/\rho c$ the thermal diffusivity. The initial condition of Eq. (4.4.16) is $\delta(0) = 0$.

---

*J and Ja are used interchangeably for the Jakob number.

From our experience with penetration problems we expect a solution of the form

$$\delta(t) = 2\lambda\sqrt{at}, \tag{4.4.20}$$

which is in fact suggested by Eq. (4.4.19) provided that a_2 is a constant, whence the phase growth constant λ is given by $\lambda^2 = \frac{1}{2}\text{J}(1 - a_2)$ Inserting Eq. (4.4.20) into Eqs. (4.4.18) and (4.4.19) confirms the foregoing fact, yielding

$$a_2 = 2\left(1 + \frac{3}{\text{J}}\right) - \left[4\left(1 + \frac{3}{\text{J}}\right)^2 - 3\right]^{1/2}$$

and

$$\lambda = [\tfrac{1}{2}(36 + 24\text{J} + \text{J}^2)^{1/2} - \tfrac{1}{2}(6 + \text{J})]^{1/2}. \tag{4.4.21}$$

Given the Jakob number, λ is evaluated from Eq. (4.4.21), the thickness of solidified layer $\delta(t)$ from Eq. (4.4.20), and the approximate solid temperature from (4.4.17). In addition, the liquid velocity v_l, positive toward the wall when $\rho > \rho_l$ (see Eq. 4.4.1), is

$$v_l = \frac{w_l}{\rho_l} = \left(\frac{\rho}{\rho_l} - 1\right)\frac{d\delta}{dt}.$$

The growth constant λ increases with Jakob number from

$$\lambda = \sqrt{\tfrac{1}{2}\text{J}}; \qquad \text{J} \longrightarrow 0, \tag{4.4.22}$$

to a maximum value

$$\lambda = \sqrt{3}; \qquad \text{J} \longrightarrow \infty. \tag{4.4.23}$$

The case of rapid phase growth, $\text{J} \gg 1$, corresponds to negligible heat of fusion compared to sensible heat $c(T_s - T_w)$ and the solution reduces to that of the cooling of a semi-infinite quiescent liquid. The case of slow phase growth, $\text{J} \ll 1$, on the other hand, implies that the temperature distribution in the solid changes so slowly that we may ignore the left-hand side of Eq. (4.4.13). Consistent with this idea, we may employ a linear profile instead of Eq. (4.4.17). Then $a_2 = 0$ and Eq. (4.4.19) yields Eq. (4.4.22), which is the *quasi-steady solution*, to be elaborated in Section 4.5. Physically, the quasi-steady solution ignores the heat removal required to steadily subcool the solidified layer as it grows in thickness.

Let us now compare the foregoing approximate solutions with the exact solution which may be obtained by a similarity analysis or transform calculus applied to the differential formulation,

$$\frac{\partial T}{\partial t} = a\frac{\partial^2 T}{\partial y^2}, \tag{4.2.24}$$

$$T(y, 0) = T_s; \qquad T(0, t) = T_w; \qquad T(\delta, t) = T_s; \qquad \rho h_{sl}\frac{d\delta}{dt} = k\frac{\partial T(\delta, t)}{\partial y},$$

with the origin being now at the wall, $y = 0$. The solution, satisfying the first two boundary conditions, may be written as

$$T - T_w = C\,\text{erf}\left[\frac{y}{2(at)^{1/2}}\right],$$

where C is the integration constant. Employing Eq. (4.4.20) and the third boundary condition gives $C = (T_s - T_w)/\text{erf}\,\lambda$, and the last interface condition yields

$$\sqrt{\pi}\,\lambda\,\exp(\lambda^2)\,\text{erf}\,\lambda = \text{J}. \tag{4.4.25}$$

Figure 4.20 shows $\lambda = \lambda(\text{J})$ of Eqs. (4.4.21), (4.4.22), and (4.4.25). As expected the quasi-steady solution is only valid for $\text{J} < 1$, becoming exact in the limit $\text{J} \to 0$, as seen by expanding Eq. (4.4.25). The integral solution is within 5% of the exact solution for $\text{J} < 50$, but tends toward a constant, Eq. (4.4.23), for large J.

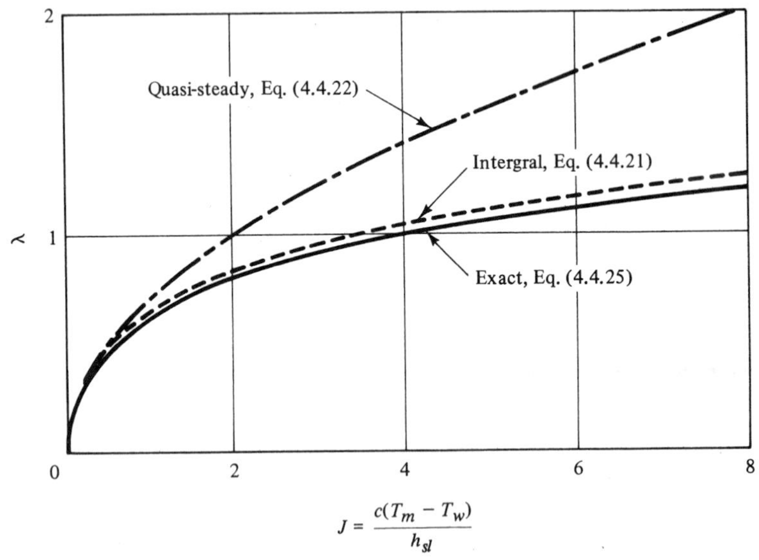

Fig. 4.20 Solidification at plane wall, $\delta(t) = 2\lambda\sqrt{at}$, $\lambda = \lambda(\text{J})$, Ex. 4.4.1

Having introduced phase change by a one-dimensional unsteady problem which involves a stagnant solid with moving interface, we consider now the nearly parallel flow of a steady problem which involves a spatially developing phase boundary.

EXAMPLE 4.4.2 Problems of filmwise condensation (or evaporation) may involve heat transfer to the interface from liquid, vapor, or both, as well as convective flow in either or both phases. We consider the steady condensation of vapor at the saturation temperature T_s to form a liquid film flowing down an inclined isothermal wall at $T_w < T_s$ (Fig. 4.21). Assuming a continuous smooth film starting at $x = 0$, we seek the variation in film thickness $\delta(x)$ and the local Nusselt number.

Clearly, any falling film, because of the continuity of tangential velocity at the interface (Eq. 4.4.5), will create a momentum boundary layer in a surrounding stagnant fluid. This boundary layer is furthermore subject to suction (blowing) when condensation (evaporation) occurs at the interface. We shall include the contribution

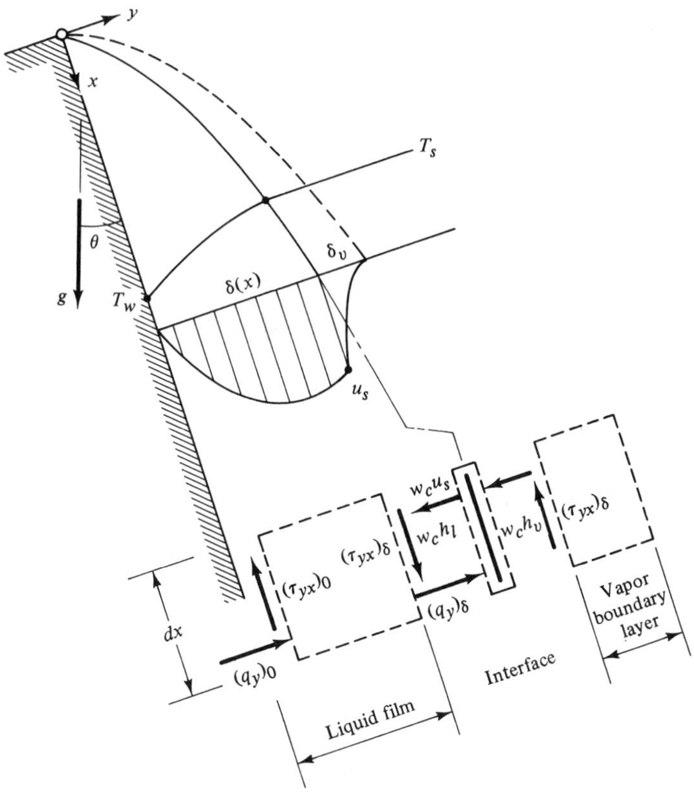

Fig. 4.21 Filmwise condensation of saturated vapor on inclined, isothermal wall, Ex. 4.4.2

of condensation to the interfacial shear stress in the approximate manner suggested by Ex. 4.3.2. We may ignore the change in momentum flow within the vapor boundary layer when $\rho_v \mu_v \ll \rho\mu$, which is assumed here (see the discussion in the related Ex. 4.5.1, or Chen 1961 and Koh, Sparrow & Hartnett 1961 for alternative solutions of the present problem).

The control volume enclosing the liquid film (but excluding the interface) increases in thickness with x as vapor condenses at mass flow w_c. Employing the integral boundary layer formulation, we have the conservation of mass, balance of momentum, and balance of thermal energy (ignoring dissipation)

$$\frac{d}{dx}\int_0^\delta \rho u\, dy - w_c = 0, \tag{4.4.26}$$

$$\frac{d}{dx}\int_0^\delta \rho uu\, dy - w_c\, u(x,\delta) = \rho g\delta\cos\theta - \delta\frac{dp}{dx} - (\tau_{yx})_0 + (\tau_{yx})_\delta, \tag{4.4.27}$$

$$\frac{d}{dx}\int_0^\delta \rho h_l\, dy - w_c\, h_l(x,\delta) = (q_y)_0 - (q_y)_\delta. \tag{4.4.28}$$

For the control volume of the interface, the balance of thermal energy, Eq. (4.4.8), gives

$$-w_c h_{lv} = (q_y)_\delta, \tag{4.4.29}$$

where $h_{lv} = h_v - h_l$ denotes the heat of evaporation. The remaining boundary conditions in y are

$$u(x, 0) = 0; \qquad \mu \frac{\partial u(x, \delta)}{\partial y} = (\tau_{yx})_\delta; \qquad T(x, 0) = T_w; \qquad T(x, \delta) = T_s, \tag{4.4.30}$$

while the upstream condition is $\delta(0) = 0$. To evaluate the interfacial shear stress, we consider also the integral momentum balance of the vapor boundary layer

$$\frac{d}{dx}\int_\delta^{\delta+\delta_v} \rho_v u_v u_v \, dy + w_c u(x, \delta) = \rho_v g \delta_v \cos\theta - \delta_v \frac{dp}{dx} - (\tau_{yx})_\delta. \tag{4.4.31}$$

Now, assuming no motion in the vapor far from the interface the static pressure gradient of the vapor

$$0 = \rho_v g \cos\theta - \frac{dp}{dx} \tag{4.4.32}$$

is imposed on both the boundary layer and the liquid film. Introducing this result into Eq. (4.4.31) and neglecting the first term (on account of the discussion above) gives

$$-w_c u(x, \delta) = (\tau_{yx})_\delta. \tag{4.4.33}$$

Introducing Eqs. (4.4.32) and (4.4.33) into Eq. (4.4.27), the notation $u(x, \delta) = u_s(x)$ and $T(x, \delta) = T_s$ and constitutive relations $\tau_{yx} = \mu\, \partial u/\partial y$, $q_y = -k\, \partial T/\partial y$, and $dh = c\, dT$, and eliminating w_c from Eq. (4.4.28) by use of Eq. (4.4.26), we obtain

$$\frac{d}{dx}\int_0^\delta uu\, dy = \left(1 - \frac{\rho_v}{\rho}\right) g\delta \cos\theta - \nu \left.\frac{\partial u}{\partial y}\right|_{y=0}, \tag{4.4.34}$$

$$\frac{d}{dx}\int_0^\delta u(T - T_s)\, dy = -a \left.\frac{\partial T}{\partial y}\right|_{y=0} + a \left.\frac{\partial T}{\partial y}\right|_{y=\delta}, \tag{4.4.35}$$

where $a = k/\rho c$. Note that the condensing momentum flow cancels the interfacial shear stress in Eq. (4.4.27).

For a solution by approximating functions, we select polynomial profiles in $\eta = y/\delta$ for velocity and temperature, satisfying Eqs. (4.4.29), (4.4.30), and (4.4.33),

$$\frac{u}{u_s} = (1 + b_0)(2\eta - \eta^2) - b_0\eta; \qquad \frac{T - T_s}{T_w - T_s} = 1 - b_1\eta + (b_1 - 1)\eta^2. \tag{4.4.36}$$

Note that from Eq. (4.4.29),

$$\frac{w_c\delta}{\mu} = (2 - b_1)\frac{\text{Ja}}{\text{Pr}},$$

and from Eq. (4.4.33),

$$\frac{w_c\delta}{\mu} = b_0;$$

hence, for $b_0 \neq 0$, b_1 and b_0 are related by

$$b_0 = (2 - b_1)\frac{\text{Ja}}{\text{Pr}}, \tag{4.4.37}$$

where $\text{Ja} = c(T_s - T_w)/h_{lv}$ denotes the Jakob number and $\text{Pr} = \nu/a$, the Prandtl number. Employing Eq. (4.4.36) in Eqs. (4.4.26), (4.4.34), and (4.4.35), carrying out integrations and rearranging, we obtain

$$\frac{d}{dx}(f_0 u_s \delta) = b_0 \frac{\nu}{\delta}, \tag{4.4.38}$$

$$\frac{d}{dx}(f_1 u_s^2 \delta) = \left(1 - \frac{\rho_v}{\rho}\right) g\delta \cos\theta - \left(\frac{u_s \nu}{\delta}\right)(2 + b_0), \tag{4.4.39}$$

$$\frac{d}{dx}(f_2 u_s \delta) = 2\left(\frac{a}{\delta}\right)\left(1 - b_0 \frac{\text{Pr}}{\text{Ja}}\right), \tag{4.4.40}$$

where, having used Eq. (4.4.37) to eliminate b_1,

$$f_0 = \frac{2}{3}\left(1 + \frac{1}{4}b_0\right); \qquad f_1 = \frac{8}{15}\left(1 + \frac{7}{16}b_0 + \frac{1}{16}b_0^2\right);$$
$$f_2 = \frac{2}{15}\left[1 + \frac{3}{8}b_0 + \frac{7}{8}b_0\frac{\text{Pr}}{\text{Ja}} + \frac{1}{4}b_0^2\frac{\text{Pr}}{\text{Ja}}\right]. \tag{4.4.41}$$

Equations (4.4.38)–(4.4.40), subject to $\delta(0) = 0$, determine the three unknowns $\delta(x)$, $u_s(x)$, and b_0. Inspection shows that Eqs. (4.4.38) and (4.4.40) become linearly dependent if b_0 is a constant, implying constant profile shapes (similarity). Eliminating $d(u_s\delta)/dx$ between Eqs. (4.4.38) and (4.4.40) gives

$$2f_0\left(1 - b_0\frac{\text{Pr}}{\text{Ja}}\right) - f_2 b_0 \text{Pr} = 0, \tag{4.4.42}$$

which determines $b_0 = b_0(\text{Pr}, \text{Ja}/\text{Pr})$. Next, on the basis of our experience on Ex. 4.3.3, we try a solution of the form

$$\delta(x) = C_1 x^m; \qquad u_s(x) = C_2 x^n, \tag{4.4.43}$$

which, when inserted into Eqs. (4.4.38) and (4.4.39), gives

$$m = \tfrac{1}{4}; \qquad n = \tfrac{1}{2},$$

while coefficients C_1 and C_2 must satisfy the algebraic equations

$$\tfrac{3}{4} f_0 C_1 C_2 = b_0 \frac{\nu}{C_1},$$

$$\tfrac{5}{4} f_1 C_1 C_2^2 = \left(1 - \frac{\rho_v}{\rho}\right) g \cos\theta\, C_1 - (2 + b_0)\frac{\nu C_2}{C_1}.$$

Inserting the solutions of these equations into Eq. (4.4.43), we find

$$\delta(x) = \left\{\frac{4\nu^2 x}{(1 - \rho_v/\rho) g \cos\theta} \frac{b_0[5f_1 b_0 + 3f_0(2 + b_0)]}{9f_0^2}\right\}^{1/4}, \tag{4.4.44}$$

$$u_s(x) = \left[\frac{4b_0(1 - \rho_v/\rho) g \cos\theta\, x}{5f_1 b_0 + 3f_0(2 + b_0)}\right]^{1/2}, \tag{4.4.45}$$

and the local heat transfer may be expressed as

$$\mathrm{Nu}_x = \frac{hx}{k} = \frac{(\partial T/\partial \eta)_{\eta=0}}{T_s - T_w}\frac{x}{\delta} = \frac{b_1 x}{\delta},$$

or

$$\mathrm{Nu}_x = \left(\frac{\mathrm{Ra}_x}{4\,\mathrm{Ja}}\right)^{1/4} f_3, \tag{4.4.46}$$

$$f_3 = \left\{\frac{9b_1^4 f_0^2}{(2-b_1)[5f_1 b_0 + 3f_0(2+b_0)]}\right\}^{1/4},$$

where $\mathrm{Ra}_x = (\Delta\rho/\rho)g\cos\theta\, x^3/\nu a$, with $\Delta\rho = \rho - \rho_v$, denotes the two-phase Rayleigh number.

Assuming a parabolic velocity and a linear temperature Nusselt (1916) obtained Eq. (4.4.46) with $f_3 = 1$. This corresponds to neglecting the interfacial shear ($b_0 = 0$) and the change in energy of the liquid ($b_1 = 1$). In fact, excepting liquid metals, f_3 is close to unity because $\mathrm{Ja} \ll 1$ for most practical cases (see Problem 4-16 for the case of viscous oils). For liquid metals, on the other hand, the rate of condensation is high and the interfacial shear causes a noticable decrease in the Nusselt number as the parameter Ja/Pr increases. For this case, consider $b_1 = 1$, hence $b_0 = \mathrm{Ja/Pr}$ from Eq. (4.4.37), and

$$f_3 = \left\{\frac{(1+\frac{1}{4}\mathrm{Ja/Pr})^2}{(1+\frac{1}{4}\mathrm{Ja/Pr})(1+\frac{1}{2}\mathrm{Ja/Pr}) + \frac{2}{3}(\mathrm{Ja/Pr})[1+\frac{7}{16}\mathrm{Ja/Pr} + \frac{1}{16}(\mathrm{Ja/Pr})^2]}\right\}^{1/4}. \tag{4.4.47}$$

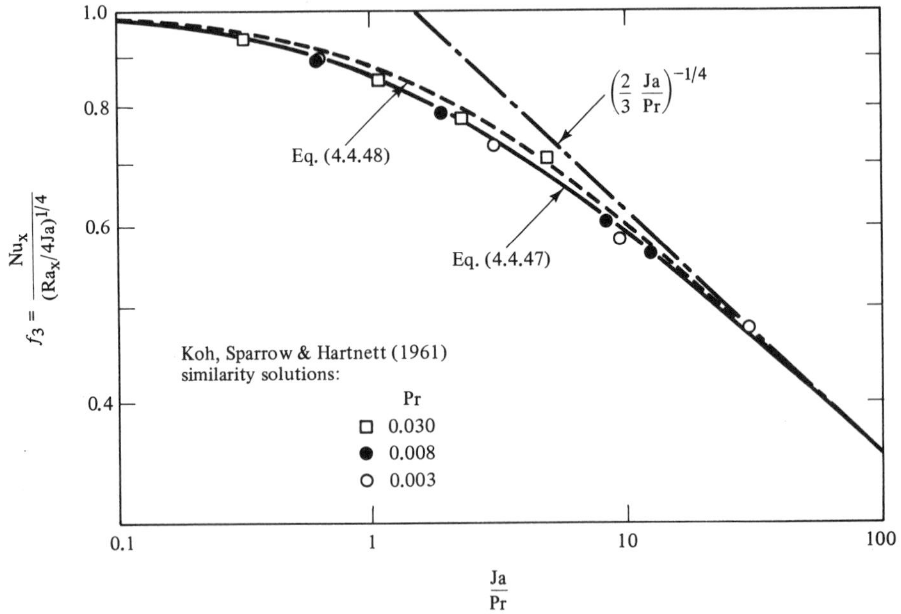

Fig. 4.22 Filmwise condensation, Ex. 4.4.2; departure from the Nusselt theory ($f_3 = 1$) for liquid metals

Noting that $f_3 \longrightarrow 1$ for Ja/Pr $\longrightarrow 0$ and $f_3 \longrightarrow (\frac{2}{3}\text{Ja/Pr})^{-1/4}$ for Ja/Pr $\longrightarrow \infty$, Eq. (4.4.47) may be approximated by

$$f_3 \simeq \left(1 + \frac{2}{3}\frac{\text{Ja}}{\text{Pr}}\right)^{-1/4}. \tag{4.4.48}$$

Figure 4.22 compares Eq. (4.4.47) to results by Koh, Sparrow & Hartnett (1961) for Pr $= 0.03$, 0.008, and 0.003 obtained by a similarity solution.

The present example shows how a strong coupling among the conservation of mass, balance of momentum, and energy arises in problems involving phase change. The coupling results from including the change associated with the development of the film.

4.5 QUASI-STEADY AND QUASI-DEVELOPED FORMULATION

For an *unsteady* problem whose formulation involves more than one general principle a *quasi-steady* formulation suggests that unsteady terms be neglected in at least one, but not all, of the general principles. Similarly, for a steady *developing* two-dimensional problem a *quasi-developed* formulation suggests that one spatial dependence (of developing terms) be neglected in at least one, but not all, of the general principles.

Let us illustrate these ideas by explaining first the nature of the *quasi-steady* approximation in terms of Ex. 4.4.1 for the solidification of a liquid (Fig. 4.19(d)). The integral formulation involves the conservation of mass and the balance of thermal energy, Eqs. (4.4.11) and (4.4.12). Clearly, between two general principles we cannot ignore the unsteady term in the conservation of mass. Such a step would make the problem steady. On the other hand, we may ignore the unsteady term in Eq. (4.4.12) without changing the basic character of the problem. With this assumption the left-hand side of Eq. (4.4.13) vanishes, implying a linear temperature distribution. Hence, $a_2 = 0$ in Eq. (4.4.17), which leads to Eq. (4.4.20) with λ given by Eq. (4.4.22). As already mentioned in Ex. 4.4.1, the quasi-steady solution corresponds to ignoring the sensible heat compared with the latent heat; thus it is the limiting solution for Ja $= c(T_s - T_w)/h_{sl} \longrightarrow 0$. As shown in Fig. 4.20, the solution to the quasi-steady formulation is a satisfactory approximation to the exact solution for Ja < 1.

Next, let us explain the nature of the *quasi-developed* approximation in terms of Ex. 4.4.2 for the filmwise condensation (Fig. 4.21). From the integral formulation, Eqs. (4.4.26)–(4.4.28), we see three possible quasi-developed formulations. To retain the developing aspects of the problem, only the left-hand sides of either Eq. (4.4.27) or Eq. (4.4.28), or both, may be ignored.

In the first case, ignoring the left of Eq. (4.4.27), or its reduced form Eq. (4.4.39), yields the developed film velocity

$$u_s = \frac{(1 - \rho_v/\rho)(g\delta^2/\nu)}{2 + b_0} \cos\theta, \tag{4.5.1}$$

whence Eq. (4.4.38) may be integrated, using also Eq. (4.4.37) and nothing $b_0 \ll 1$, to

$$\delta(x) = \left[\frac{4\nu^2}{(1 - \rho_v/\rho)g \cos\theta} \frac{\mathrm{Ja}}{\mathrm{Pr}}(2 - b_1)\right]^{1/4}, \tag{4.5.2}$$

where b_1 is given by Eq. (4.4.42) as before. Clearly, a developed velocity profile implies a large viscosity and, when compared to a still developing temperature profile, also a large Prandtl number. Consistent with this observation, introducing the approximation

$$b_0 = (2 - b_1)\frac{\mathrm{Ja}}{\mathrm{Pr}} \ll 1 \tag{4.5.3}$$

into Eq. (4.4.44), we correctly recover Eq. (4.5.2).

In the second case we ignore the left of Eq. (4.4.28), or its reduced form Eq. (4.4.40), which yields $b_1 = 1$, implying a linear developed temperature profile. Repeating the algebra of Ex. 4.4.2 then yields Eq. (4.4.44) with $b_1 = 1$, and thus Eq. (4.4.47). This case corresponds to small Prandtl numbers and terms with Ja/Pr cannot be ignored compared to unity even though Ja needs to be small to justify any quasi-developed approximation.

In the third case the foregoing two independent approximations both enter. Thus, $\mathrm{Pr} \sim o(1)$, $\mathrm{Ja} < 1$ and the solution is given by Eq. (4.5.2) for $b_1 = 1$. This leads to the Nusselt solution, $f_3 = 1$ in Eq. (4.4.46).

We proceed to apply the foregoing concepts to some problems whose solutions become practical only after the use of these concepts.

EXAMPLE 4.5.1 Consider steady forced convection film boiling of a subcooled or saturated liquid over an isothermal flat plate. We wish to determine the local skin friction and heat transfer from the plate.

The wall temperature exceeds the saturation temperature of the liquid ($T_w > T_s$) by an amount sufficient to establish and maintain a continuous vapor film (Fig. 4.23).

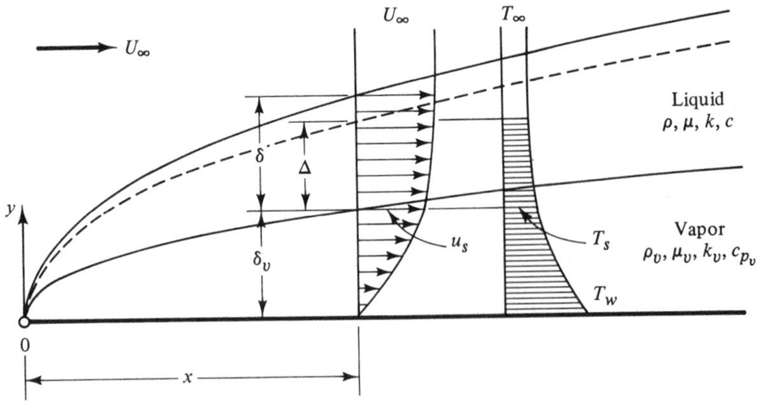

Fig. 4.23 Forced convection film boiling in liquid flow over heated plate, Ex. 4.5.1

The vapor flow is produced by evaporation from the liquid interface, owing to heat transfer from the wall through the vapor layer.*

The complete integral formulation in terms of conservation of mass, balance of momentum, and thermal energy is, for the vapor layer,

$$\frac{d}{dx}\int_0^{\delta_v} \rho_v u_v \, dy - w_s = 0, \tag{4.5.4}$$

$$\frac{d}{dx}\int_0^{\delta_v} \rho_v u_v u_v \, dy - w_s u_s = \tau_{yx}|_{y=\delta_v} - \tau_{yx}|_{y=0}, \tag{4.5.5}$$

$$\frac{d}{dx}\int_0^{\delta_v} \rho_v u_v h_v \, dy - w_s h_{vs} = q_y|_{y=0} - q_y|_{y=\delta_v}, \tag{4.5.6}$$

for the liquid momentum layer,

$$\frac{d}{dx}\int_0^{\delta} \rho u \, dy_1 + w_s - w_\delta = 0, \tag{4.5.7}$$

$$\frac{d}{dx}\int_0^{\delta} \rho u u \, dy_1 + w_s u_s - w_\delta U_\infty = -\tau_{yx}|_{y_1=0}, \tag{4.5.8}$$

and for the liquid thermal layer,

$$\frac{d}{dx}\int_0^{\Delta} \rho u \, dy_1 + w_s - w_\Delta = 0, \tag{4.5.9}$$

$$\frac{d}{dx}\int_0^{\Delta} \rho u h \, dy_1 + w_s h_s - w_\Delta h_\infty = q_y|_{y_1=0}, \tag{4.5.10}$$

where the subscript s refers to interface, the subscript v to vapor, $y_1 = y - \delta_v$, and h denotes the enthalpy. Substituting w_s, w_δ, and w_Δ from Eqs. (4.5.4), (4.5.7), and (4.5.9) into Eqs. (4.5.6), (4.5.8), and (4.5.10) and using $dh = c_p \, dT$ gives, for the vapor layer,

$$\frac{d}{dx}\int_0^{\delta_v} \rho_v c_{pv} u_v (T_v - T_s) \, dy = q_y|_{y=0} - q_y|_{y=\delta_v}, \tag{4.5.11}$$

and for the liquid

$$\frac{d}{dx}\int_0^{\delta} \rho u (U_\infty - u) \, dy_1 + (U_\infty - u_s) w_s = \tau_{yx}|_{y_1=0}, \tag{4.5.12}$$

$$\frac{d}{dx}\int_0^{\Delta} \rho c u (T - T_\infty) \, dy_1 + c(T_s - T_\infty) w_s = q_y|_{y_1=0}, \tag{4.5.13}$$

where w_s is given by Eq. (4.5.4). The boundary conditions are

$$u_v(x, 0) = 0; \qquad u_v(x, \delta_v) = u_s; \qquad T_v(x, 0) = T_w; \qquad T_v(x, \delta_v) = T_s,$$

*Gas injection through a porous flat plate in uniform liquid flow would produce a similar double boundary layer flow at isothermal conditions (see Problem 4-23).

$$u(x,0)=u_s;\qquad u(x,\delta)=U_\infty;\qquad \frac{\partial u(x,\delta)}{\partial y_1}=0$$
$$T(x,0)=T_s;\qquad T(x,\Delta)=T_\infty;\qquad \frac{\partial T(x,\Delta)}{\partial y_1}=0, \tag{4.5.14}$$

and, referring to Eqs. (4.4.9)–(4.4.8), interface conditions are

$$\mu_v\frac{\partial u_v(x,\delta_v)}{\partial y}=\mu\frac{\partial u(x,0)}{\partial y_1}, \tag{4.5.15}$$

$$w_s h_{lv}=-k_v\frac{\partial T_v(x,\delta)}{\partial y}+k\frac{\partial T(x,0)}{\partial y_1} \tag{4.5.16}$$

Because of the complexity of the problem, evidenced by the number of parameters involved, here we limit ourselves to the case of liquid Prandtl number near unity, so that $\Delta\sim\delta$. Furthermore, we consider only the quasi-developed formulation on the assumption that sensible heat of the vapor can be neglected compared with the latent heat of evaporation. This assumption implies a small Jakob number, $\mathrm{J}=c_{p_v}(T_w-T_s)/h_{lv}$. For a vapor Prandtl number of order unity, this approximation is justified for both balance of thermal energy and momentum in view of the Reynolds analogy. Hence, ignoring the left of Eqs. (4.5.5) and (4.5.11), velocity and temperature distributions in the vapor become linear,

$$\frac{u_v(x,y)}{u_s}=\frac{y}{\delta_v(x)};\qquad \frac{T_v(x,y)-T_s}{T_w-T_s}=\frac{y}{\delta_v(x)}. \tag{4.5.17}$$

For the liquid we assume first-order profiles satisfying Eq. (4.5.14),

$$\frac{u(x,y_1)}{U_\infty}=\bar{u}_s+(1-\bar{u}_s)(2\eta-\eta^2);\qquad \frac{T(x,y_1)-T_\infty}{T_s-T_\infty}=(1-\eta)^2, \tag{4.5.18}$$

where $\bar{u}_s=u_s/U_\infty$ and $\eta=y_1/\delta(x)$.

Employing Eqs. (4.5.17) and (4.5.18), the formulation, in terms of Eqs. (4.5.4), (4.5.12), (4.5.13), (4.5.15), and (4.5.16), reduces to

$$w_s=\frac{1}{2}\rho_v U_\infty\frac{d}{dx}(\delta_v\bar{u}_s), \tag{4.5.19}$$

$$\frac{2}{15}\frac{\delta}{1-\bar{u}_s}\frac{d}{dx}\left[\delta(1-\bar{u}_s)\left(1+\frac{3}{2}\bar{u}_s\right)\right]+\frac{1}{2}\frac{\rho_v}{\rho}\delta\frac{d}{dx}(\delta_v\bar{u}_s)=\frac{2\nu}{U_\infty}, \tag{4.5.20}$$

$$\frac{2}{15}\delta\frac{d}{dx}\left[\delta\left(1+\frac{3}{2}\bar{u}_s\right)\right]+\frac{1}{2}\frac{\rho_v}{\rho}\delta\frac{d}{dx}(\delta_v\bar{u}_s)=\frac{2a}{U_\infty}, \tag{4.5.21}$$

$$\frac{\delta_v}{\delta}=\frac{1}{2}\frac{\mu_v}{\mu}\frac{\bar{u}_s}{1-\bar{u}_s}, \tag{4.5.22}$$

$$\frac{\mathrm{P_v}}{2\mathrm{J}}\frac{U_\infty\delta_v}{\nu_v}\frac{d}{dx}(\delta_v\bar{u}_s)=1-2\frac{k}{k_v}\frac{T_s-T_\infty}{T_w-T_\infty}\frac{\delta_v}{\delta}. \tag{4.5.23}$$

Being concerned with the case $\mathrm{P}=\nu/a\simeq 1$, we expect Eqs. (4.5.20) and (4.5.21) to be identical, according to the Reynolds analogy (recall Ex. 4.3.1). Inspection reveals that these equations become identical provided that $\bar{u}_s=$ constant, whence Eq. (4.5.22) shows that δ_v/δ is constant. These conditions are indeed compatible with the entire formulation, admitting solutions of the usual boundary layer form, $\delta_v \sim x^{1/2}$ and $\delta \sim x^{1/2}$. It may be shown (Problem 4-21) that similar solutions arise also for liquid Prandtl numbers different from unity; however, the ratio of boundary layers $\chi = \Delta/\delta$ then depends not only on P (as in Eq. 4.3.14 of Ex. 4.3.1) but on other parameters of the problem as well.

Eliminating δ_v by Eq. (4.5.22) and integrating Eq. (4.5.20) subject to $\delta(0) = 0$ gives

$$\delta(x) = \left(30\frac{\nu x}{U_\infty}\right)^{1/2} \psi_u^{-1/2}, \tag{4.5.24}$$

where

$$\psi_u = 1 + \frac{3}{2}\bar{u}_s + \frac{15}{8}\frac{\rho_v\mu_v}{\rho\mu}\frac{\bar{u}_s^2}{1-\bar{u}_s}, \tag{4.5.25}$$

and where the constant $\bar{u}_s$ is determined from Eq. (4.5.23), which may be rearranged as

$$\frac{15}{8}\frac{\mathrm{P_v}}{\mathrm{J}}\frac{\rho_v\mu_v}{\rho\mu}\frac{\bar{u}_s^3}{(1-\bar{u}_s)^2\psi_u} = 1 - \frac{k\mu_v}{k_v\mu}\frac{T_s - T_\infty}{T_w - T_s}\frac{\bar{u}_s}{1-\bar{u}_s}. \tag{4.5.26}$$

The velocity and temperature profiles as well as the thickness of the vapor layer $\delta_v(x)$ are then known. The shear stress at the wall, noting Eq. (4.5.15), is

$$\tau_w = \mu_v\left(\frac{\partial u_v}{\partial y}\right)_{y=0} = \mu\left(\frac{\partial u}{\partial y_1}\right)_{y_1=0} = \frac{2\mu U_\infty}{\delta(x)}(1-\bar{u}_s),$$

or, inserting Eq. (4.5.24),

$$\tau_w = \left(\frac{2}{15}\right)^{1/2}\mu U_\infty\left(\frac{U_\infty}{\nu x}\right)^{1/2}(1-\bar{u}_s)\psi_u^{1/2}, \tag{4.5.27}$$

and the local coefficient of skin friction becomes

$$c_x = \frac{\tau_w}{\frac{1}{2}\rho U_\infty^2} = \left(\frac{8}{15}\right)^{1/2}\mathrm{Re_x^{-1/2}}(1-\bar{u}_s)\psi_u^{1/2}, \tag{4.5.28}$$

where $\mathrm{Re_x} = U_\infty x/\nu$. For $\bar{u}_s = 0$ this result reduces to Eq. (4.2.6) for the single-phase (liquid) boundary layer, the difference in numerical coefficients resulting from the use of parabolic and cubic profiles, respectively. When $\bar{u}_s$ is close to unity, which will be the case for film boiling of most fluids at states not close to the critical point, the skin friction is significantly reduced. The possibility of reducing the drag by more than an order of magnitude by means of gas injection for bodies submerged in a liquid was pointed out by Cess & Sparrow (1961), who first formulated the present example and derived its similarity solution (see Ex. 5.3.7).

Turning to the heat transfer, the wall heat flux, employing Eqs. (4.5.17) and (4.5.22), becomes

$$q_w = -k_v\left(\frac{\partial T_v}{\partial y}\right)_{y=0} = \frac{k_v(T_w - T_s)}{\delta_v(x)} = \frac{2k_v(T_w - T_s)}{\delta(x)}\frac{\mu}{\mu_v}\frac{1-\bar{u}_s}{\bar{u}_s},$$

and the local Nusselt number, referred to wall superheat and vapor conductivity,

$$\mathrm{Nu_x} = \frac{q_w x}{(T_w - T_s)k_v} = \left(\frac{2}{15}\right)^{1/2}\mathrm{Re_x^{1/2}}\left(\frac{k_v\mu}{k\mu_v}\right)\frac{1-\bar{u}_s}{\bar{u}_s}\psi_u^{1/2}. \tag{4.5.29}$$

The present example illustrates the complexity of the coupled two-domain problems involving momentum and heat transfer in both domains. An inspection of Eq. (4.5.25) and the expressions for friction and heat transfer shows the problem to depend on

$$\frac{\mathrm{P_V}}{\mathrm{J}}, \quad \frac{\rho_v \mu_v}{\rho \mu}, \quad \frac{k \mu_v}{k_v \mu}, \quad \frac{T_s - T_\infty}{T_w - T_s}.$$

In many situations $(\rho_v \mu_v)/(\rho \mu) \ll 1$ ($\sim 3 \times 10^{-5}$, for example, for water at 1 atm, where $\mathrm{P_V/J} \sim 0.1$ for $T_w - T_s = 100°\mathrm{C}$) and we may ignore the last term in Eq. (4.5.25). This follows from the asymptotic behavior of Eq. (4.5.26). As may readily be shown,

$$\bar{u}_s \sim \left(\frac{8}{15} \frac{\mathrm{P_V}}{\mathrm{J}} \frac{\rho_v \mu_v}{\rho \mu}\right)^{-1/3}$$

for $(\rho_v \mu_v)/(\rho \mu) \longrightarrow \infty$, while

$$\bar{u}_s \sim 1 - \left(\frac{3}{4} \frac{\mathrm{P_V}}{\mathrm{J}} \frac{\rho_v \mu_v}{\rho \mu}\right)^{1/2}$$

for $(\rho_v \mu_v)/(\rho \mu) \longrightarrow 0$ for any finite value of $\mathrm{P_V/J}$. Furthermore, for saturated liquid the last term in Eq. (4.5.26) is zero. For this case, we may compare the present approximate solution to the exact solution given by Cess & Sparrow (1961), say in terms of the wall stress,

$$\tau_w = \mu U_\infty \left(\frac{U_\infty}{\nu x}\right)^{1/2} f''(0, \bar{u}_s), \tag{4.5.30}$$

where $f''(0, \bar{u}_s)$ is the dimensionless velocity gradient at the vapor–liquid interface (see Ex. 5.3.7). The present solution, Eq. (4.5.27) with ψ_u from Eq. (4.5.25) where the last term is now neglected, becomes

$$\tau_w = \mu U_\infty \left(\frac{U_\infty}{\nu x}\right)^{1/2} \left(\frac{2}{15}\right)^{1/2} (1 - \bar{u}_s)\left(1 + \frac{3}{2} u_s\right). \tag{4.5.31}$$

The factor $(\frac{2}{15})^{1/2}(1 - \bar{u}_s)(1 + \frac{3}{2}\bar{u}_s)$ is compared to $f''(0, \bar{u}_s)$, taken from the tabulation by Cess & Sparrow (1961), in Fig. 4.24. The overall behavior of the approximate

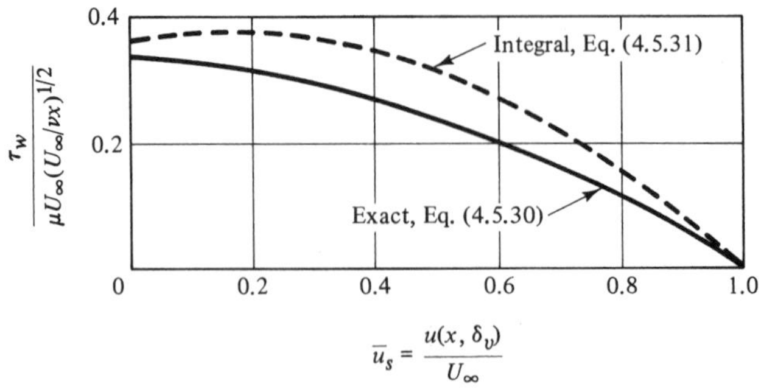

Fig. 4.24 Film-boiling flow over heated plate, exact and integral solutions, Ex. 4.5.1

solution is seen to be satisfactory, but its accuracy is poor as would be expected from the use of first-order profiles.

Problems associated with film boiling play an important role in heat transfer equipment and its safe operation, and they are often aimed at determining the wall superheat required to remove a specified heat flux. This mode of boiling lends itself most readily to theoretical analysis and it has been studied for numerous geometries and flow conditions (for a review, see Jordan 1968 and Kalinin, Berlin & Kostyuk 1975). Here, we consider the problem associated with film boiling of liquid drops.

EXAMPLE 4.5.2 We wish to evaluate the heat transfer coefficient for dropwise film boiling on an isothermal horizontal plate.

Experiments show that a drop of liquid evaporates rapidly when placed on a horizontal surface heated to a few degrees above the saturation temperature; the evaporation is slow, however, if the surface is well above saturation. In the first case the liquid can wet the surface creating nucleate boiling with bubble formation, liquid film evaporation, and high rates of heat transfer. In the second case the drop remains suspended on a poorly conducting vapor film, which prevents direct contact between liquid and hot surface. The latter is the Leidenfrost problem of film boiling and is of concern here.

To attack this difficult problem, involving a complex geometry, unsteadiness, and several regions coupled at boundaries whose locations are not known a priori, we shall make several assumptions without any loss in essential physics.

First, we consider the evaporation to be quasi-steady. The weight of the drop is then balanced by the surface integral of pressure forces. The excess pressure in the vapor film arises from that required to maintain the outward viscous flow of vapor escaping from under the drop. In addition, we shall use a quasi-developed integral formulation for the flow of vapor which is generated from the drop by conduction through the vapor film. Since heat transfer, vapor flow, and hence excess pressure increase with decreasing film thickness, the drop tends to settle at a quasi-steady film thickness to be determined as a part of the solution.

The geometry of an actual drop, which is determined by gravity, surface tension, and pressure distribution, is approximated here by a prismatic slab of rectangular cross section (Fig. 4.25). The liquid is at rest and isothermal at saturation, T_s. The heating surface is isothermal at T_w, which exceeds the Leidenfrost temperature.

The quasi-steady integral formulation for the vapor film is (see Fig. 4.26)

$$\frac{d}{dx}\int_0^\delta \rho u\, dy - w_\delta = 0, \tag{4.5.32}$$

$$\frac{d}{dx}\int_0^\delta \rho uu\, dy - u(x,\delta)w_\delta = -\delta\frac{dp}{dx} + \tau_{yx}|_{y=\delta} - \tau_{yx}|_{y=0}, \tag{4.5.33}$$

$$\frac{d}{dx}\int_0^\delta \rho c_p uT\, dy - c_p T(x,\delta)w_\delta = -q_y|_{y=\delta} + q_y|_{y=0}, \tag{4.5.34}$$

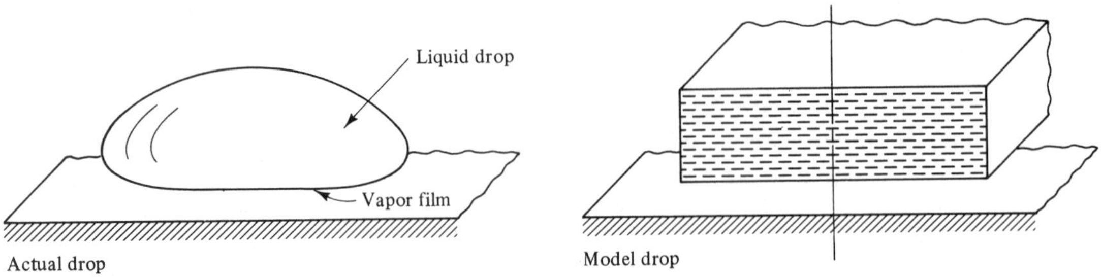

Fig. 4.25 Liquid drop modeling

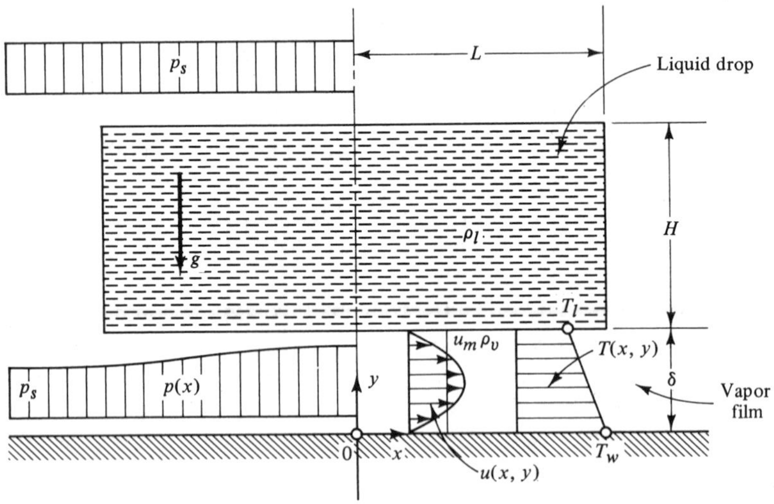

Fig. 4.26 Idealized drop evaporation, Ex. 4.5.2

where w_δ denotes the evaporating mass flow per unit area, and

$$u(x, 0) = 0; \qquad u(x, \delta) = 0; \qquad T(x, 0) = T_w; \qquad T(x, \delta) = T_s, \tag{4.5.35}$$

$$\tau_{yx} = \mu \frac{\partial u}{\partial y}; \qquad q_y = -k \frac{\partial T}{\partial y}. \tag{4.5.36}$$

The balance of thermal energy at the liquid–vapor interface is (recall Eq. 4.4.8)

$$h_{lv} w_\delta = -k \frac{\partial T(x, \delta)}{\partial y}, \tag{4.5.37}$$

where h_{lv} denotes the heat of evaporation.

Now, for a quasi-developed solution, we ignore the left of Eq. (4.5.34), which implies a linear temperature distribution,

$$\frac{T - T_s}{T_w - T_s} = 1 - \frac{y}{\delta}. \tag{4.5.38}$$

For this problem it may not be justified to ignore also the left of Eq. (4.5.33), because the mass flow due to evaporation increases with x. This causes a change in momentum flow which may be significant for the pressure distribution. Rather, the effect on profile shape of vapor injection along the liquid boundary is ignored, which suggests the developed parabolic profile,

$$\frac{u}{u_m} = 6\frac{y}{\delta}\left(1 - \frac{y}{\delta}\right), \tag{4.5.39}$$

where $u_m(x)$ denotes the mean vapor velocity. A quasi-developed formulation of both the momentum and energy balances (as employed in Ex. 4.5.1) is justified only by the Reynolds analogy provided that the Prandtl number is of order unity and the pressure gradient is zero.

Using Eqs. (4.5.38) and (4.5.39), eliminating w_δ by Eq. (4.5.37), we reduce Eqs. (4.5.32) and (4.5.33) to

$$\delta\frac{du_m}{dx} = \frac{aJ}{\delta}, \tag{4.5.40}$$

$$\frac{6}{5}\rho\delta\frac{du_m^2}{dx} = -\delta\frac{dp}{dx} - 12\frac{\mu u_m}{\delta}, \tag{4.5.41}$$

where $\mathrm{J} = c_p(T_w - T_s)/h_{lv}$ denotes the Jakob number. Using symmetry and denoting by p_s the pressure outside the drop, boundary conditions may be stated as

$$u_m(0) = 0; \qquad \frac{dp(0)}{dx} = 0; \qquad p(L) = p_s; \qquad \delta = \text{const.} \tag{4.5.42}$$

Introducing $u_m = a\mathrm{J}x/\delta^2$, obtained from Eq. (4.5.40), into Eq. (4.5.41) and integrating once yields

$$p(x) - p_s = 6\mu a\mathrm{J}\left(\frac{L^2}{\delta^4}\right)\left(1 + \frac{1}{5}\frac{\mathrm{J}}{\mathrm{P}}\right)\left[1 - \left(\frac{x}{L}\right)^2\right], \tag{4.5.43}$$

where $\mathrm{P} = \nu/a$ and the factor $\frac{1}{5}\mathrm{J}/\mathrm{P}$ is a measure for the ratio of inertia to viscous contribution.

Now, the vertical balance of momentum for the drop, ignoring unsteadiness and momentum flow due to evaporation, gives (see Fig. 4.26)

$$0 = \int_0^l [p(x) - p_s]\,dx - (\rho_l - \rho)gHL, \tag{4.5.44}$$

where the subscript l refers to liquid. Inserting Eq. (4.5.43) into Eq. (4.5.44), and defining the equivalent characteristic length $L_e = L^2/H$, yields the film thickness

$$\delta = \left[\frac{a\mu\mathrm{J}L_e}{4g(\rho_l - \rho)}\left(1 + \frac{1}{5}\frac{\mathrm{J}}{\mathrm{P}}\right)\right]^{1/4}. \tag{4.5.45}$$

Introducing the mean instantaneous heat transfer coefficient

$$\bar{h}(t) = \frac{1}{T_w - T_s}\frac{1}{L}\int_0^L q_y(x, 0)\,dy,$$

noting from Eq. (4.5.38) that $\bar{h} = k/\delta$ is constant over the drop surface, and using Eq. (4.5.45), we obtain

$$\bar{h}(t) = \left[\frac{k^4 g(\rho_l - \rho)}{4a\mu\mathrm{J}L_e(1 + \frac{1}{5}\mathrm{J}/\mathrm{P})}\right]^{1/4}, \tag{4.5.46}$$

which may be rearranged as

$$\bar{h}(t) = \frac{k}{L_e}\left[\frac{\text{Ra}}{4\text{J}(1 + \frac{1}{5}\text{J/P})}\right]^{1/4},$$

where $\text{Ra} = (\Delta\rho/\rho)gL_e^3/a\nu$ is the two-phase Rayleigh number, $\Delta\rho = \rho_l - \rho$.

It is left as an exercise to evaluate time for complete evaporation of a drop, for example, assuming L_e to be constant, and to determine the corresponding time-averaged heat transfer coefficient over drop life. Also, the effect of a cylindrical drop shape and the effect of nonuniform film thickness important for smaller drops may be studied (see Problem 4-20).

EXAMPLE 4.5.3 We wish to determine the steady performance of a heat pipe in terms of its total heat transfer capability.

A heat pipe consists of a duct closed at both ends. The inside wall is covered by a layer of porous, capillary material (the wick), which is saturated with the liquid phase of the working fluid. The vapor phase fills the open central part of the duct (Fig. 4.27). Thermal energy is transferred from one end to the other by a process of continuous evaporation from the wick at the evaporator end, flow of vapor to the

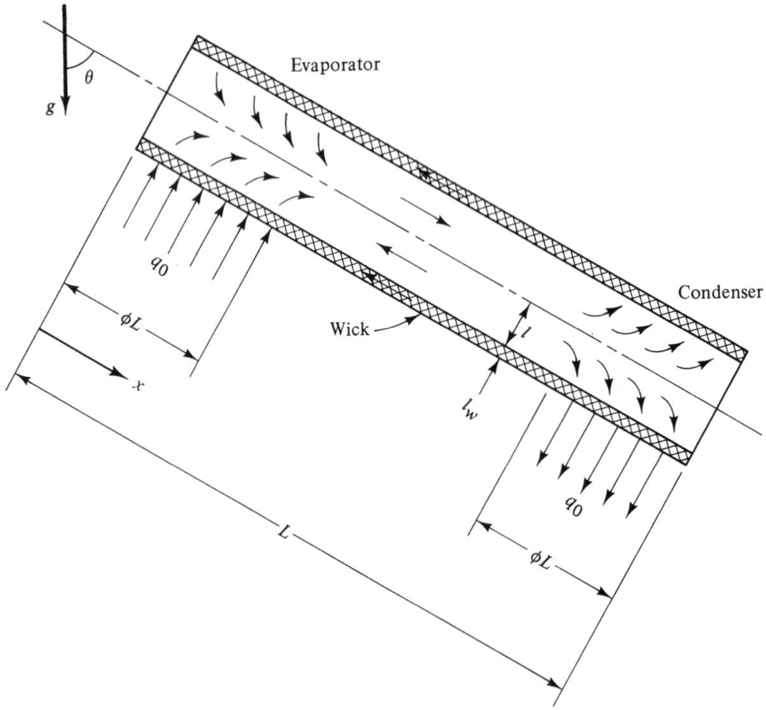

Fig. 4.27 Heat pipe, Ex. 4.5.3

condenser end, recondensation on the wick, and return flow of liquid through capillary action in the wick. The ends of the heat pipe contact, respectively, a heat source and a sink, differing in temperature by a small amount.

Appropriate selection of working fluid, pressure level, and wick porosity according to reservoir temperatures a heat pipe may be designed to transfer a large heat power by moderate space and weight requirements (see, for example, Winter & Barsch 1971 for a survey).

The heat pipe performance depends on the flow of vapor and liquid, requiring the pressure gradient usually to be negative in the vapor, $dp/dx < 0$, and positive in the liquid, $dp_l/dx > 0$. The pressure difference between vapor and liquid is made possible by capillary action of the wick. The radius of curvature of the liquid meniscus along the free wick surface tends to adjust itself to these requirements.

It is apparent that the heat pipe presents a number of interesting problems, each of which may be dealt with in considerable detail. Here we carry a simplified incompressible analysis* of the pressure drop in the vapor and liquid flows, and then couple the two domains by means of the interface heat balance and capillary wick pressure difference to evaluate the total heat transfer.

Pressure gradient in vapor flow. Consider steady flow between parallel plates with uniform injection v_0. For the symmetrical half-width l of the duct, the integral formulation is (see Fig. 4.28)

$$\frac{d}{dx}\int_0^l \rho u \, dy - \rho v_0 = 0, \qquad (4.5.47)$$

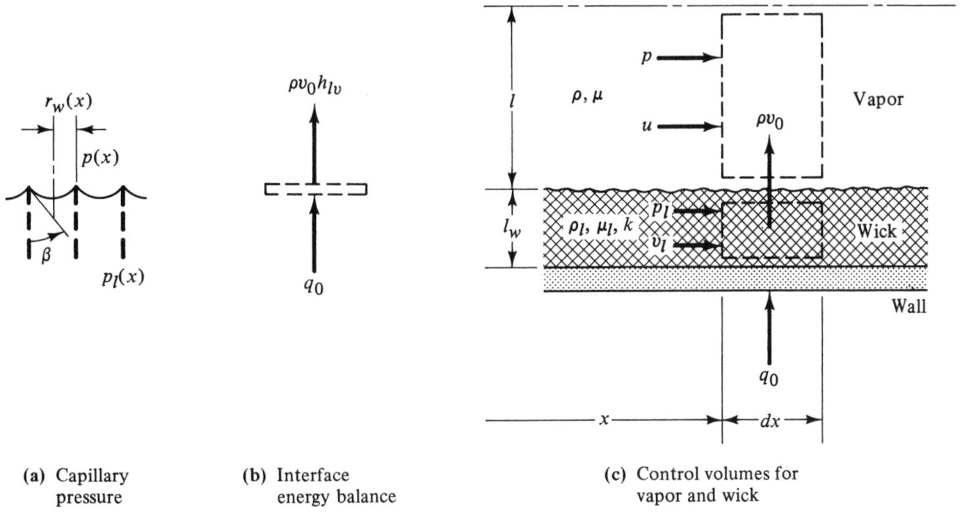

(a) Capillary pressure (b) Interface energy balance (c) Control volumes for vapor and wick

Fig. 4.28 Heat pipe model, Ex. 4.5.3

*The effect of vapor compressibility may be important under certain conditions.

$$\frac{d}{dx}\int_0^l \rho uu\, dy = -l\frac{dp}{dx} + \rho gl\cos\theta - \tau_{yx}|_{y=0}. \tag{4.5.48}$$

Ignoring the effect of injection on the profile, we assume a developed parabolic profile for a quasi-developed analysis

$$\frac{u}{u_m} = \frac{3}{2}\frac{y}{l}\left(2 - \frac{y}{l}\right), \tag{4.5.49}$$

where $u_m(x)$ denotes the mean vapor velocity. Inserting Eq. (4.5.41) into Eq. (4.5.47) gives

$$\frac{du_m}{dx} = \frac{v_0}{l}. \tag{4.5.50}$$

Using this result, $\tau_{yx} = \mu\, \partial u/\partial y$, and Eq. (4.5.49), Eq. (4.5.48) becomes

$$-\frac{dp}{dx} = \frac{3\mu u_m}{l^2}\left(1 + \frac{4}{5}\mathrm{R_e}\right) - \rho g\cos\theta, \tag{4.5.51}$$

where $\mathrm{R_e} = \rho l v_0/\mu$ is the Reynolds number characterizing the rate of vapor injection (evaporation). The term $\frac{4}{5}\mathrm{R_e}$ is a measure of the ratio of momentum to viscous contributions to pressure drop.

For a long heat pipe, ignoring end effects and transitions, Eqs. (4.5.50) and (4.5.51) apply to evaporator $\mathrm{R_e} > 0$, insulated section $\mathrm{R_0} = 0$, and condenser $\mathrm{R_c} < 0$. Assume now for simplicity that evaporator and condenser sections both are of length φL, $0 < \varphi < \frac{1}{2}$, so that the insulated section is of length $(1 - 2\varphi)L$, and $\mathrm{R_e} = -\mathrm{R_c}$. Integration of Eq. (4.5.50) over the three sections subject to

$$u_m(0) = 0; \qquad u_m(L) = 0,$$

gives

$$u_m(x) = \begin{cases} \dfrac{v_0 x}{l}, & 0 \le x \le \varphi L \\ \dfrac{v_0 \varphi L}{l}, & \varphi L \le x \le (1-\varphi)L \\ \dfrac{v_0(L-x)}{l}, & (1-\varphi)L \le x \le L. \end{cases} \tag{4.5.52}$$

Inserting Eq. (4.5.52) into Eq. (4.5.51), we may obtain the pressure distribution, which changes from parabolic to linear and again to parabolic distribution in the three sections. In the present case the contributions from the acceleration and deceleration cancel and the total pressure change becomes

$$p(0) - p(L) = \frac{3\mu v_0 L^2}{l^3}(1-\varphi)\varphi - \rho gL\cos\theta. \tag{4.5.53}$$

Pressure gradient in liquid flow. The flow of liquid in the wick is treated as viscous flow in porous media of permeability K for which (according to the Darcy law) the constitutive relation for Newton fluids is

$$\nabla \cdot \boldsymbol{\tau} \simeq -\frac{\mu_l}{K}V_l. \tag{4.5.54}$$

Because viscous effects completely dominate the flow, inertia terms are ignored from the momentum equation, which reduces to

$$0 = -\nabla p + \rho f - \frac{\mu_l}{K} V_l. \tag{4.5.55}$$

Assuming one-dimensional flow, noting that v_l is in the negative x-direction (see Fig. 4.28), Eq. (4.5.55) becomes

$$\frac{dp_l}{dx} = \rho_l g \cos\theta + \frac{\mu_l}{K} v_l, \tag{4.5.56}$$

where the liquid velocity v_l is related to the vapor injection velocity v_0 by conservation of mass,

$$\frac{dv_l}{dx} = \frac{\rho v_0}{\rho_l l_w}, \tag{4.5.57}$$

l_w denoting the thickness of the wick.

Integrating Eq. (4.5.57) subject to $v_l(0) = v_l(L) = 0$, substituting the result into Eq. (4.5.56), and subsequent integration yields the total pressure change

$$p_l(L) - p_l(0) = \rho_l gL \cos\theta + \frac{\rho\mu_l v_0 L^2}{\rho_l K l_w}\varphi(1-\varphi). \tag{4.5.58}$$

Capillary interface. To maintain the foregoing pressure distributions there must be a pressure difference between liquid and vapor along the wick. Given a capillary radius $r_w(x)$ (see Fig. 4.28) and wetting angle β (measured through the liquid) the radius of curvature of the interface is $r_w(x)/\cos\beta$, and using Eq. (4.4.2), neglecting its terms on the left, we require that for all x

$$p(x) - p_l(x) \leq \frac{2\sigma_s \cos\beta}{r_w(x)}, \tag{4.5.59}$$

where σ_s denotes the surface tension. Assuming, for example, a plane meniscus at $x = L$, $p(L) = p_l(L)$, Eq. (4.5.59) evaluated at $x = 0$ may be written as

$$p(0) - p(L) + p_l(L) - p_l(0) \leq \frac{2\sigma_s \cos\beta}{r_w(0)},$$

or, substituting from Eqs. (4.5.53) and (4.5.58), as

$$\frac{3\mu v_0 L^2}{l^3}\varphi(1-\varphi) + (\rho_l - \rho)gL\cos\theta + \frac{\rho\mu_l v_0 L^2}{\rho_l K l_w}\varphi(1-\varphi) \leq \frac{2\sigma_s\cos\beta}{r_w(0)}. \tag{4.5.60}$$

Given $r_w(0)$, Eq. (4.5.60) determines the maximum injection velocity v_0, and the interface balance of thermal energy,

$$\rho v_0 h_{lv} = q_0, \tag{4.5.61}$$

gives the maximum heat flux q_0, and the heat removal power $\varphi L q_0$ per unit width of the parallel plate heat pipe,

$$\varphi L q_0 = \frac{h_{lv} l^3}{\nu L(1-\varphi)} \frac{2\sigma_s \cos\beta / r_w(0) - (\rho_l - \rho)gL\cos\theta}{3 + (\nu_l l^3)/(\nu K l_w)}. \tag{4.5.62}$$

Note that $\varphi \leq \frac{1}{2}$ and that the capillary pressure must overcome a possible hydrostatic head difference between liquid and vapor for the heat pipe to function. The temperature difference between the ends is determined by the pressure difference $p(0) - p(L)$ from vapor pressure data for the working fluid.

Having completed our study of nearly parallel flows, we comment briefly on nonparallel flows.

4.6 NONPARALLEL (RECIRCULATING) FLOWS

In the preceding sections we learned the convection involving nearly parllel flows. Actually, nonparallel flows are encountered more often in engineering applications, and we therefore comment briefly on these flows. For a nonparallel flow there is no dominant direction of motion, and the diffusion (flux) and convection (flow) of momentum and energy play equal roles, at least in parts of the flow domain. Consequently, the governing equations cannot be simplified as in the preceding sections and solution requires numerical methods, to be introduced in Chapter 7. For these reasons the present section provides only a descriptive introduction to forced and buoyancy-driven nonparallel two-dimensional flows.

As an illustration, reconsider Example 3.4.1, the driven cavity flow including buoyancy effects, but let the height H of the cavity be finite and of the order of the width l (Fig. 4.29). Clearly, end effects associated with the turning of flows now

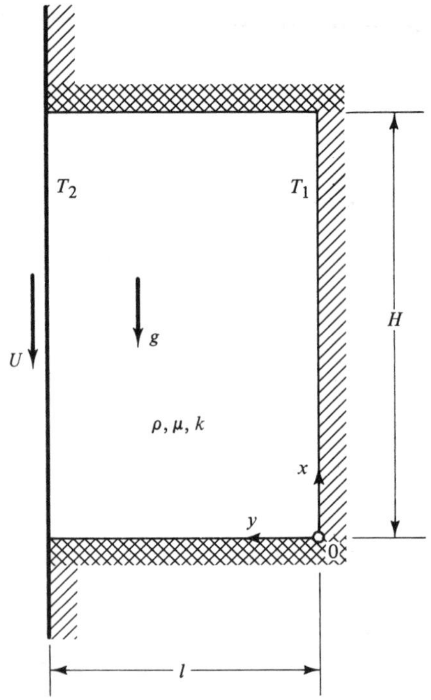

Fig. 4.29 Driven cavity flow with buoyancy effects

influence the central region, which cannot be approximated by the parallel flow expected to prevail in the cavity of large H/l ratio shown in Fig. 3.15. As a result, the friction and heat transfer in end zones amount to significant parts of the totals and cannot be ignored.

Assuming the velocity and temperature distributions to be two-dimensional, properties to be constant, viscous dissipation to be negligible, and accounting for the buoyancy effect through the Boussinesq approximation, the differential formulation of the problem may be written from Table 2.5.

Inspection of this formulation (Problem 4-40) on the basis of our experience from Chapters 3 and 4 shows that the governing equations cannot a priori be simplified by ignoring diffusion compared to convection, or by lumping one direction. In the absence of similarity, the nonlinear terms rule out the application of analytical methods of solution. Also, the approximate methods based on the integral formulation, because of the difficulties of constructing profiles, are not useful.

A measure of the complexity of the problem is the number of governing physical parameters, which may be reduced to four dimensionless numbers,

$$\mathrm{Re} = \frac{Ul}{\nu}; \qquad \mathrm{Ra} = \frac{g\beta(T_2 - T_1)H^3}{\nu a}; \qquad \mathrm{Pr} = \frac{\nu}{a}; \qquad \mathrm{L} = \frac{H}{l}, \tag{4.6.1}$$

affecting the solution. To demonstrate this complexity, consider Figs. 4.30 and 4.31 which show numerical solutions to two simpler problems: the cavity flow without buoyancy, and the buoyancy-driven flow in a cavity. With few exceptions (see, for example, Graebel 1981) the solution of nonparallel convection problems requires numerical methods.

The gross behavior of the driven cavity flow is the recirculating fluid motion portrayed by the closed streamlines in the bulk of the cavity (Fig. 4.30). However, at the corners far from the moving wall small recirculation zones (corner vortices) exist at the Reynolds numbers shown. An additional recirculation zone on the upper fixed wall near the moving wall has developed at Re = 2000. Clearly, these secondary recirculation zones, as well as the somewhat irregular pattern of the bulk motion, would not be intuitively expected on the basis of our experience on parallel and nearly parallel flows. The associated temperature distributions of this flow, portrayed by the isotherms in Fig. 4.30, show how the thermal stratification from being confined to the hot moving wall at low Reynolds numbers is pushed toward the stationary cold walls as the Reynolds number is increased. This development leaves a nearly isothermal large core of rotating fluid at high Reynolds numbers.

The example of buoyancy driven flow in a square cavity (Fig. 4.31) illustrates the effect of increasing Rayleigh number on streamline pattern and isotherms. Flow regimes range from that of conduction (Ra = 100), via that of transition (Ra = 2500) to the boundary layer regime (Ra = 25,000 and 100,000).

Recirculation zones, such as those observed at corners in Fig. 4.30, often occur in nonparallel flows. Such a zone is bounded by a streamline leaving the wall at a separation point and joining it at a reattachment point. The phenomenon of separation in boundary layer flows over aerofoils or in ducts should be well known from fluid mechanics. When subjected to an adverse pressure gradient (decreasing free

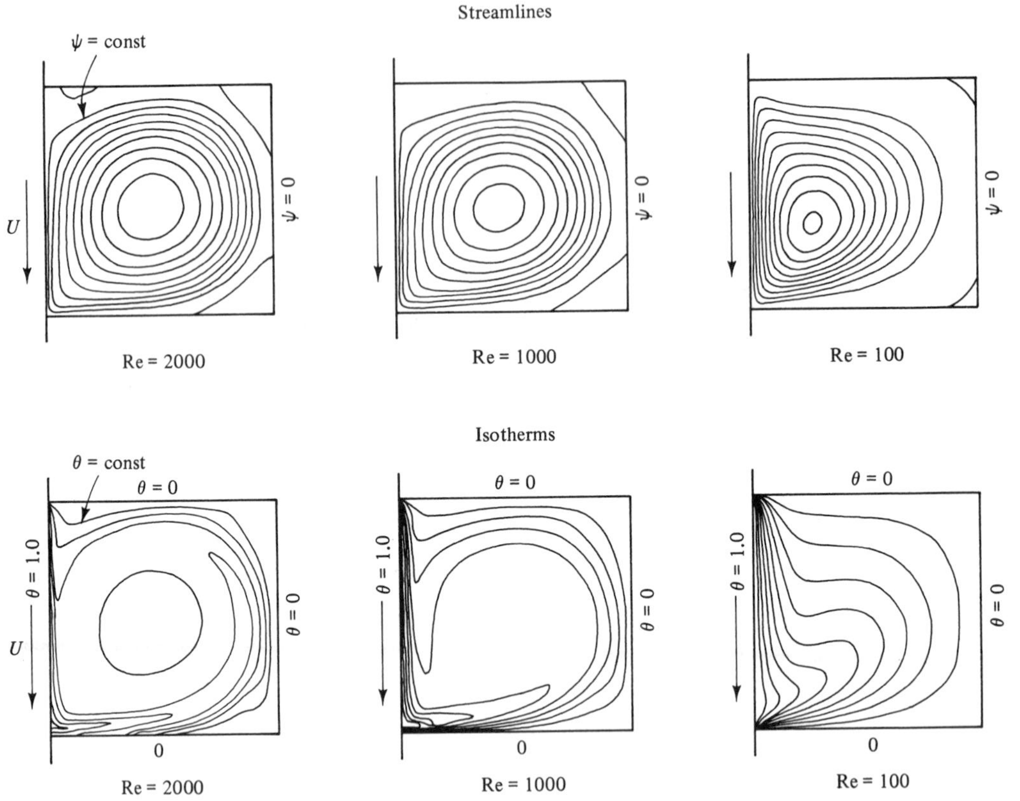

Fig. 4.30 Effect of Reynolds number, Re $= Ul/\nu$, on driven square cavity flow; computed stream lines and isotherms (Pr $= 1$) for $\theta = 1$ on moving wall and $\theta = 0$ on other walls (adapted from Chen, Naseri-Neshat & Ho 1980; used by permission)

stream velocity in the direction of flow), the boundary layer thickness increases as momentum is consumed by both wall shear and pressure gradient, leading at some point to separation. The shear stress becomes zero at the separation point and the flow breaks away from the wall, creating a recirculation zone downstream of this point. If the flow deceleration leading to separation is followed by an acceleration, the flow may reattach, leaving a separation bubble with recirculating fluid. Actually, the boundary layer approximation of nearly parallel flow is applicable only to some distance upstream of the separation point. The rapid boundary layer growth and the recirculation involve nonparallel flows. The foregoing mechanism leading to separation and subsequent reattachment is also at play for flow along a wall toward a corner, because here the pressure reaches its maximum (stagnation) value.

One further distinction between the parallel or nearly parallel flows and the nonparallel flows deserves comment. The former flows are governed by simplified

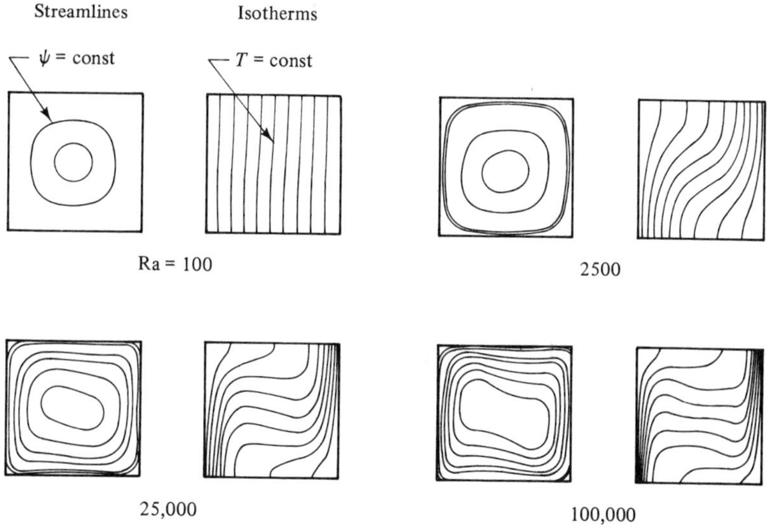

Fig. 4.31 Effect of Rayleigh number, $\text{Ra} = \beta g \Delta T l^3/\nu a$, on buoyancy-driven flow in square cavity, having isothermal side walls and insulated top and bottom, at near-unity Prandtl number, $\text{Pr} = \nu/a = 0.733$; computed streamlines (left) and isotherms (right) (adapted from Roux, Grondin, Bontoux & Gilly 1978; used by permission)

forms of the Navier–Stokes equations (that is, ordinary differential equations or parabolic partial differential equations), whose laminar solutions are unique and independent of the Reynolds number. Nonparallel flows, however, are governed by the full Navier–Stokes equations, whose solutions are more complex and depend on the Reynolds number. This is due to the mathematical complexity of the Navier–Stokes equations, which are nonlinear partial differential equations of the elliptical type. In two dimensions these equations reduce, for $\text{Re} \longrightarrow 0$, to the linear biharmonic equation $\nabla^4\psi = 0$ (Eq. 2.4.43 for $\nu \longrightarrow \infty$), which has a unique solution. For increasing Reynolds number, the nonlinear convective terms play an increasing role and uniqueness of steady solutions cannot be proved. Typically, for $\text{Re} > \text{Re}_1$, say, there may exist more than one steady solution. For still higher values, $\text{Re} > \text{Re}_2$, say, there may exist no steady solution, but only oscillating solutions. For still higher values $\text{Re} > \text{Re}_c$, which is beyond the stability limit of laminar motion, the unsteady solutions become chaotic, and we have a turbulent flow. The well-known visualization studies of flow past a cylinder for increasing values of the Reynolds number clearly illustrate these features.

In summary, it may be conjectured (Ames 1977) that for given boundary conditions a steady solution to the incompressible constant property Navier–Stokes equations may only exist below a certain first Reynolds number Re_1; that several solutions may exist between this first limit and a second limit Re_2; and that no steady solution exists above this second limit.

Since analytical (exact as well as approximate) methods fail, the solution of nonparallel flows requires numerical methods. In general, employing an unsteady formulation, the time evolution of the solution will reveal if a steady solution exists.

REFERENCES

AMES, W. F., 1977, *Numerical Methods for Partial Differential Equations*, 2nd ed., Academic Press, New York.

CAMBELL, J. W. D., & SLATTERY, J. C., 1963, J. Basic Eng. (ASME), **85**, 41.

CESS, R. D., & SPARROW, E. M., 1961, J. Heat Transfer (ASME), **83**, 370, 377.

CHEN, C.-J., NASERI-NESHAT, H., & HO, K.-S., 1981, Numer. Heat Transfer, **4**, 179.

CHEN, M. M., 1961, J. Heat Transfer (ASME), **83**, 48–54.

DAVIES, J. T., & RIDEAL, E. K., 1961, *Interfacial Phenomena*, Academic Press, New York.

FINLAYSON, B. A., 1972, *The Method of Weighted Residuals and Variational Principles*, Academic Press, New York.

GOLDSTEIN, R. J., 1971, *Advances in Heat Transfer*, (ed. T. F. IRVINE, Jr., and J. P. HARNETT), Vol. 7, pp. 321–379, Academic Press, New York.

GOSMANN, A. D., PUN, W. M., RUNCHAL, A. K., SPALDING, D. B., & WOLFSHTEIN, M., 1969, *Heat and Mass Transfer in Recirculating Flows*, Academic Press, New York.

GRAEBEL, W. P., 1981, Int. J. Heat Mass Transfer, **24**, 125.

JORDAN, D. P., 1968, *Advances in Heat Transfer*, (ed. T. F. IRVINE, Jr., and J. P. HARNETT), Vol. 5, pp. 55–128, Academic Press, New York.

KALININ, E. K., BERLIN, I. I., & KOSTYUK, V. V., 1975, *Advances in Heat Transfer*, (ed. T. E. IRVINE, Jr., and J. P. HARNETT), Vol. 11, pp. 51–197, Academic Press, New York.

KOH, J. C. Y., SPARROW, E. M., & HARTNETT, J. P., 1961, Int. J. Heat Mass Transfer, **2**, 69–82.

LEVICH, V. G., 1962, *Physicochemical Hydrodynamics*, Prentice-Hall, Englewood Cliffs, N.J.

METZNER, A. B., & WHITE, J. L., 1965, AIChE J., **11**, 989.

MOHANTY, A. K., & ASTHANA, S. B. L., 1978, J. Fluid Mech., **90**, 433–447.

NUSSELT, W., 1916, Z. Ver. Dtsch. Ing., **60**, 541, 569.

PRANDTL, L., 1904, Proc. 3rd Int. Math. Congr., Heidelberg.

REYNOLDS, O., 1874, Proc. Lit. Philos. Soc. Manchester, **14**.

ROSENHEAD, L., ed., 1963, *Laminar Boundary Layers*, Oxford University Press, New York.

ROUX, B., GRONDIN, J. C., BONTOUX, P., & GILLY, B., 1978, Numer. Heat Transfer, **1**, 331.

SCHILLER, L., 1922, ZAMM, **2**, 96.

SCHLICHTING, H., 1968, *Boundary Layer Theory*, 6th ed., McGraw-Hill, New York.

SCRIVEN, L. E., 1960, Chem. Eng. Sci., **12**, 98.

SLATTERY, J. C., 1964, Chem. Eng. Sci., **19**, 379.

SQUIRE, H. B., 1938, in *Modern Developments in Fluid Dynamics* (ed. S. Goldstein), Oxford University Press, New York.

VAN DYKE, M., 1964, *Perturbation Methods in Fluid Mechanics*, Academic Press, New York.

VAN WYLEN, G. J., & SONNTAG, R. E., 1972, *Classical Thermodynamics*, Wiley, New York.

WHITE, J. L., 1966, AIChE J., **12**, 1019.

WHITE, J. L., & METZNER, A. B., 1965, AIChE J., **11**, 324.

WINTER, E. R. F., & BARSCH, W. O., 1971, *Advances in Heat Transfer*, (ed. T. F. IRVINE, Jr., and J. P. HARNETT), Vol. 7, p. 219, Academic Press, New York.

PROBLEMS

4-1 Reconsider Ex. 4.2.1 of uniform flow past a flat plate (Fig. 4.1). Determine the effect of fluid elasticity on boundary layer growth and drag force for a fluid which for high shear-rates subscribes to

$$\tau_{yx} = K\left(\frac{\partial u}{\partial x}\right)^n; \qquad \tau_{xx} - \tau_{yy} = m\left(\frac{\partial u}{\partial y}\right)^s, \qquad s \leq 2.$$

The streamwise diffusion of momentum may not be negligible for non-Newtonian fluids which develop normal viscous stress due to their viscoelasticity. Furthermore, $\partial\sigma_{yy}/\partial y$ and not the pressure gradient itself vanishes across the boundary layer. In view of this fact, include the normal deviator stresses on the assumption that the Weissenberg number $Ws_x = (m/K)(U_\infty/x)^{s-n}$ is of order unity (White 1966, White & Metzner 1965, and Metzner & White 1965).

4-2 Reconsider Ex. 4.2.3 (see Fig. 4.8). The liquid is at uniform temperature T_0 as it leaves the slot, but is then cooled from below the moving belt by an imposed constant heat flux for $x \geq 0$. Determine the temperature distribution of the liquid. Consider also the limiting cases of $\Pr \longrightarrow 0$ and $\Pr \longrightarrow \infty$.

4-3 Determine the velocity distribution of a smooth liquid film of thickness δ flowing steadily down a plane wall, inclined an angle θ to the vertical. Also, determine the developing momentum boundary layer in the quiescent fluid in contact with the developed liquid film, and discuss its effect on the velocity of the liquid film.

4-4 A viscous liquid layer of thickness l flows under the influence of gravity over an inclined surface. The fluid is sucked with a uniform velocity V from the lower end of the plate (Fig. 4P-4). Evaluate δ and x_0 in terms of first- and second-order profiles.

4-5 A viscous liquid flowing steadily out of a faucet is approximated by the Poiseuille flow from the termination of two parallel plates separated a distance $2l$ apart (Fig. 4P-5).
(a) Describe the change in this Poiseuille profile as x increases.
(b) Find x_0 and δ_0 at which the effect of viscosity is negligible.

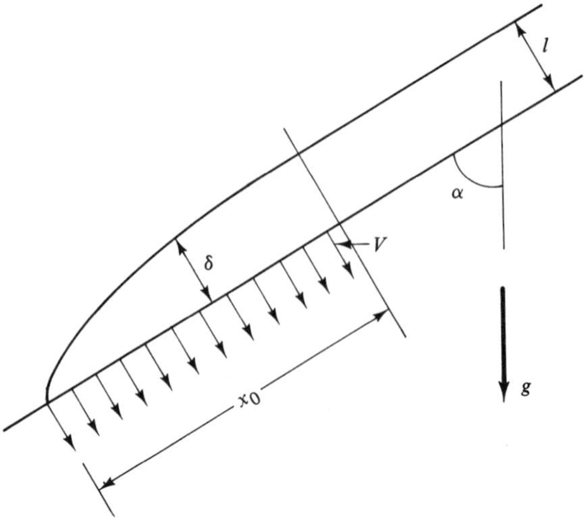

Fig. 4P-4

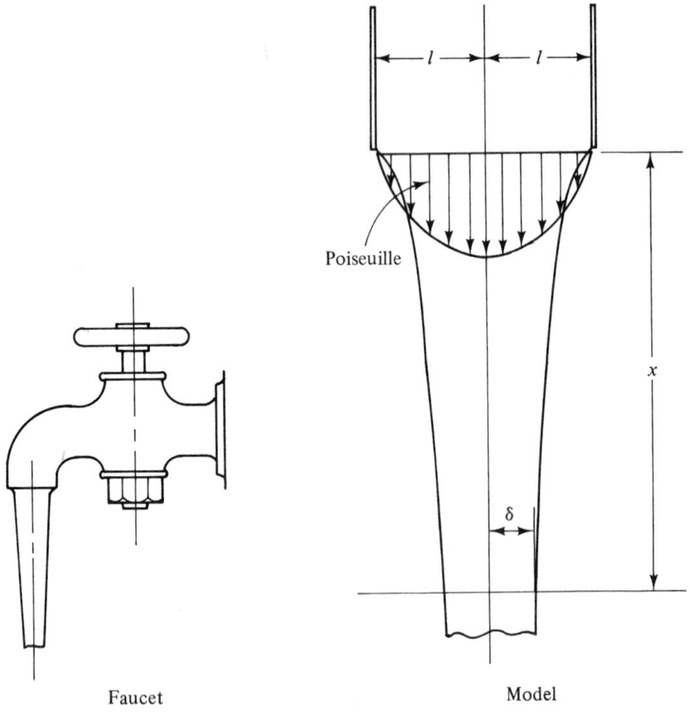

Fig. 4P-5

4-6 A viscous oil, initially at rest, fills a channel of height H formed by two vertical plates spaced a distance $2l$ apart and closed at the bottom by a plate (Fig. 4P-6(a)). At time $t = 0$ the bottom is suddenly removed to permit the liquid to drain from the channel under the influence of gravity. Ignoring the effect of surface tension, determine the velocity distribution, the position of the free surface, and the amount of liquid in the channel as functions of time.

Consider also the second time domain of the draining of a container, defined to start when the bulk liquid has left the container, leaving liquid films on the walls. The photograph in Fig. 4P-6(b) shows, as multiple strobe exposures, the position and width of the horizontal part of the free surface at equidistant times.

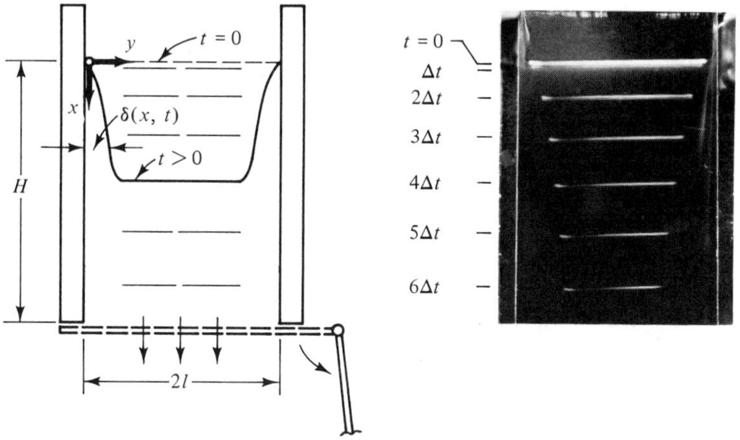

(a) (b)

Fig. 4P-6

4-7 Reconsider Ex. 4.3.1 for q_w = constant.
(a) Evaluate the local heat transfer for $\delta/\Delta \lesseqgtr 1$.
(b) Find the limits of this heat transfer for $\mathrm{Pr} \longrightarrow 0$ and for $\mathrm{Pr} \longrightarrow \infty$.

4-8 Reconsider Ex. 4.3.1 for a viscous oil, improving the temperature profile by including the effect of dissipation.

4-9 Derive Eqs. (4.3.14) and (4.3.15) of Ex. 4.3.1 for the case $\mathrm{Pr} < 1$. Why can the effect of the dissipation be ignored here?

4-10 Reconsider Ex. 4.3.2 for q_w = constant. Evaluate the heat transfer.

4-11 Reconsider Ex. 4.3.3 for q_w = constant. Evaluate the local heat transfer.

4-12 Reconsider Ex. 4.3.3 with an imposed free-stream velocity U_∞.

4-13 Reconsider Ex. 4.3.3, now employing $\delta \neq \Delta$, expecting the thickness of momentum and thermal boundary layers to be different. Again using the integral formulation, which now also involves the variable $\chi = \Delta/\delta$, an additional condition is required: say, the integral balance of mechanical energy (recall Ex. 4.2.2) or the wall compatibility condition obtained by satisfying the differential momentum balance at the wall. Employing the latter condition, obtain the relation $u_1 = (g\beta\,\Delta T/4\nu)\delta^2$. Comparing the integral balance of momentum and thermal energy, obtain for $\chi \leq 1$,

$$\mathrm{Pr} = \frac{40}{21}\left[\chi^3(5 - 4\chi + \chi^2)\left(\frac{4}{3}\chi - 1\right)\right]^{-1},$$

giving $\chi = \chi(\mathrm{Pr})$, and obtain

$$\mathrm{Nu}_x = \left(\frac{5 - 4\chi + \chi^2}{4\chi}\right)^{1/4}\mathrm{Ra}_x^{1/4}.$$

Note that $\chi \neq 1$ for $\mathrm{Pr} = 1$, and $\chi \longrightarrow$ constant for $\mathrm{Pr} \longrightarrow \infty$. Consider separately the case $\chi \geq 1$.

4-14 Redo Problem 4-13 for $\chi \leq 1$ and $\chi \geq 1$, now employing a cubic temperature profile which satisfies the differential balance of thermal energy at the wall (Suryanarayana, N. V., 1977, private communication). Compare the results to those of Problem 4-13 and those of Ex. 4.3.3.

4-15 Reconsidering Ex. 4.3.4, now using the integral procedure, determine the length L from the leading edge at which the developed downstream solution is valid (recall Ex. 4.3.2).

4-16 Reconsider Ex. 4.4.2 for large Prandtl numbers and moderate Jakob numbers. Determine the departure from the Nusselt theory (Chen 1961).

4-17 Study by the integral method the film boiling from a vertical plate at constant temperature.

4-18 Resolve Problem 4-2, now including liquid solidification on the cooled belt for $x \geq 0$. The phase change takes place at $T_s \leq T_0$ and involves the specific enthalpy h_{ls}. For simplicity we ignore the volume change on solidification.

4-19 Reconsider Ex. 4.2.3. The moving belt is cooled by the constant heat flux starting at some length L upstream of the end of the slot. At $x = -L$ (see Fig. 4.8) we have fully developed Couette flow of a liquid at its saturation temperature T_s. The fixed wall is insulated. Determine the length L to ensure a steady process producing a solid sheet, leaving the slot with velocity U_0 on the moving belt at $x = 0$.

4-20 Resolve Ex. 4.5.2 for the idealized drop geometry of a circular cylinder of diameter D and height H (recall Figs. 4.25 and 4.26).

4-21 Resolve Ex. 4.5.1 for $\mathrm{Pr} > 1$.

4-22 A thin liquid film of a constant-property Newton fluid flows steadily down an inclined wall (Fig. 4P-22). The uniform wall temperature T_w exceeds the saturation temperature T_s, causing evaporation from the free liquid surface into the surrounding saturated vapor at T_s without nucleate boiling at the wall. Given a film thickness δ_0 at $x = 0$, we seek the variation in film thickness $\delta(x)$ for $x \geq 0$, and in particular we wish to find the distance L at which the film has completely evaporated, that is, dry-out at $\delta(L) = 0$.

4-23 Determine the velocity field and skin friction for uniform liquid flow past a porous flat plate subject to uniform gas injection.

4-24 A fully developed subcooled liquid enters a heated section of a channel. The imposed constant heat flux q_w is sufficient to sustain nucleate boiling at the wall, giving rise to the formation of a bubble boundary layer (Fig. 4P-24). The recondensation of bubbles in contact with the subcooled flow of the core constitutes the heat transfer to the core. Derive a dif-

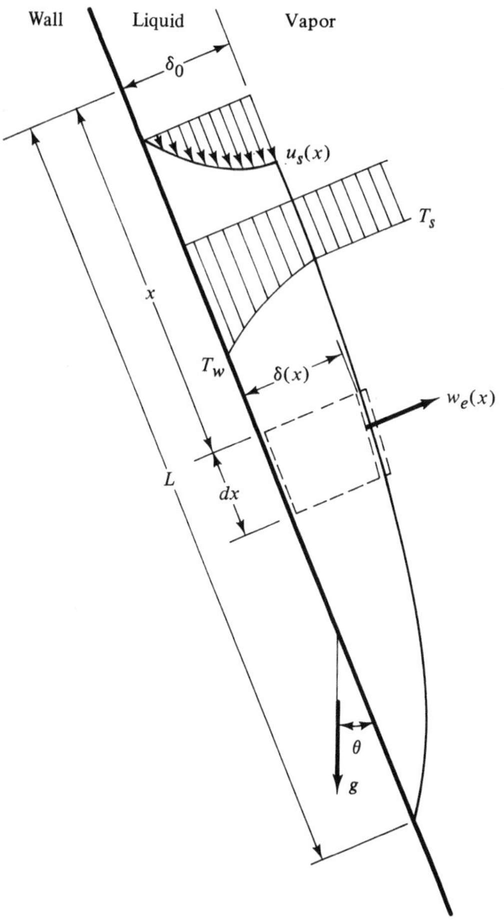

Fig. 4P-22

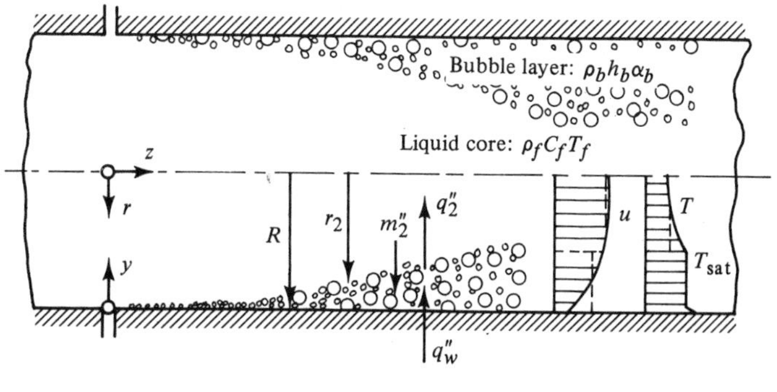

Fig. 4P-24

ferential-integral formulation whose solution yields the developing void fraction (ratio of vapor volume to total volume) in the heated section (Larsen, P. S., & Tong, L. S., 1969, J. Heat Transfer (ASME), **91**, 471–476).

4-25 Resolve Ex. 4.4.2 for an isothermal horizontal cylinder.

4-26 Resolve Ex. 4.3.3 for an isothermal cylinder.

4-27 Show that the integral balance of momentum for steady, incompressible boundary layer flow, employing Eqs. (4.1.23), (4.1.28), and (4.1.16), may be written as

$$\frac{\delta_1}{U}\frac{dU}{dx} + \frac{1}{U^2}\frac{d(U^2\delta_2)}{dx} - \frac{w_0}{\rho U} = \frac{1}{2}f,$$

where

$$\delta_1 = \int_0^\delta \left(1 - \frac{u}{U}\right) dy, \qquad \delta_2 = \int_0^\delta \frac{u}{U}\left(1 - \frac{u}{U}\right) dy, \qquad f = \frac{\tau_w}{\rho U^2/2},$$

or as

$$\frac{d\delta_2}{dx} + (2 + H)\frac{\delta_2}{U}\frac{dU}{dx} - \frac{w_0}{\rho U} = \frac{1}{2}f,$$

where $H = \delta_1/\delta_2$. Study the von Karman–Pohlhausen integral procedure, which is based on a polynominal velocity profile of fourth order (see, for example, Schlichting 1968). The alternative method of Thwaites (Thwaites, B., 1949, Aeronaut. Quant., **1**, 245), which is not based on explicit use of velocity profiles but relies on experimentally determined relations, starts from the following approximation to the integral momentum balance for an impermeable surface,

$$\frac{d}{dx}\left(\frac{\delta_2}{\nu}\right) + b\frac{dU/dx}{U}\left(\frac{\delta_2}{\nu}\right) = \frac{a}{U},$$

where $a \simeq 0.45$ and $b \simeq 6$. This form may be readily integrated to

$$\frac{\delta_2}{\nu} = \frac{0.45}{U^6}\int_0^x U^5\, dx,$$

which yields $\delta_2(x)$ when $U(x)$ is given. Then $\delta_1(x)$ and the shape factor $H(x) = \delta_1/\delta_2$ are determined from an empirical relation $H(\lambda)$, and the local skin friction $\tau_w(x)$ from an empirical relation $l(\lambda)$, where

$$l = \frac{\tau_w \delta_2}{\mu U}, \qquad \lambda = \left(\frac{\delta_2^2}{\nu}\right)\frac{dU}{dx}.$$

The empirical relations may be expressed as (Cebeci, T., & Bradshaw, P., 1977, *Momentum Transfer in Boundary Layers*, Hemisphere/McGraw-Hill, New York)

$$-0.1 \le \lambda \le 0 \quad \begin{cases} H(\lambda) = \dfrac{0.0731}{0.14 + \lambda} + 2.088 \\[2ex] l(\lambda) = 0.22 + 1.402\lambda + \dfrac{0.018\lambda}{0.107 + \lambda} \end{cases}$$

$$0 \le \lambda \le 0.1 \quad \begin{cases} H(\lambda) = 2.61 - 3.75\lambda + 5.24\lambda^2 \\ l(\lambda) = 0.22 + 1.57\lambda - 1.8\lambda^2. \end{cases}$$

Separation of a decelerating boundary layer should occur at $\lambda \simeq -0.090$.

4-28 An amount of a viscous liquid epoxy is poured into a container and spreads to form a thin uniform liquid film of thickness δ_∞. Unavoidable initial nonuniformities in thickness will dissappear with time, owing to the effect of gravity. Using an approximate integral and quasi-steady formulation, show that the film thickness $\delta(x, t)$ for the two-dimensional plane problem (Fig. 4P-28) is governed by

$$\frac{\partial \delta}{\partial t} = \frac{g}{12\nu} \frac{\partial^2 \delta^4}{\partial x^2},$$

subject to $\partial \delta / \partial x = 0$ at $x = 0$ and L.

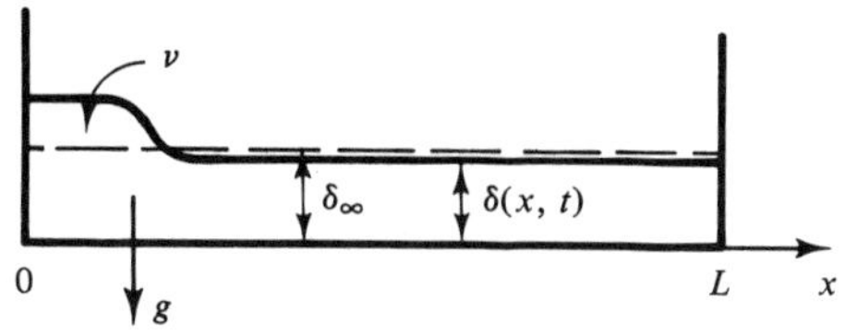

Fig. 4P-28

4-29 Study Problem 4-28 and formulate the problem of the dripping from a thin liquid film of condensate formed on a horizontal ceiling.

4-30 Reconsider Ex. 4.3.3. Study the transient natural convection from the vertical isothermal plate (Ingham, D. B., 1978, Int. J. Heat Mass Transfer, **21**, 67).

4-31 Study the natural convection, steady buoyancy-driven flow in a long, shallow, horizontal cavity, heated from below (Shiralkar, G., Gadgil, A., & Tien, C. L., 1981, Int. J. Heat Mass Transfer, **24**, 1621).

4-32 Consider a shallow rectangular cavity of a liquid-filled (ρ, μ, β) porous medium (permeability K) of width L and height H ($\ll L$). Impermeable side walls are isothermal at T_1 and $T_2 > T_1$, while upper and lower walls are insulated. Formulate and solve the problem giving the buoyancy-driven convective flow through the porous medium and the heat transfer between side walls (Bejan, A., & Tien, C. L., 1978, J. Heat Transfer (ASME), **100**, 191).

4-33 A number of problems involve the squeezing of a liquid film between two surfaces as they move together. Consider a plate of width $2L$ and of infinite depth normal to the plane (Fig. 4P-33). From some initial position $\delta(0) = \delta_0$ the plate is forced to move toward a plane surface parallel to the plate. A viscous fluid fills the gap δ between plate and plane. For two typical problems [specified motion $\delta(t)$ or specified force $F_1(t)$] formulate that of $F_1 =$ constant, determine $\delta(t)$, and find the total viscous dissipation in the liquid film between plate and plane.

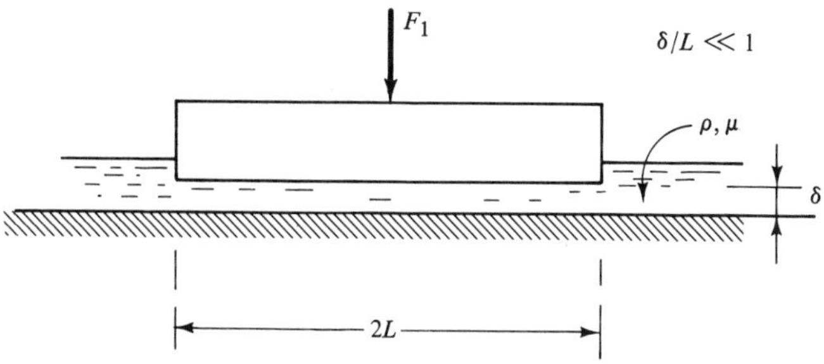

Fig. 4P-33

4-34 Determine the velocity distribution near a plane porous belt moving with constant velocity U_0 and subject to uniform suction v_0.

4-35 In the continuous casting and rolling process, molten metal flows from a crucible as a liquid sheet of uniform velocity V and thickness $2b$ into a cooling zone where heat is transferred to the ambient at T_∞ with the constant heat transfer coefficient h. The metal is pure, solidifying at temperature T_m and having latent heat of solidification h_{sl} (Fig. 4P-35). Ignoring the contraction of the sheet due to the acceleration of gravity, we wish to determine the length L of the cooling zone to ensure complete solidification. Determine the temperature field, solid–liquid interface position, and length L without and with account for solid subcooling (recall Section 4.5).

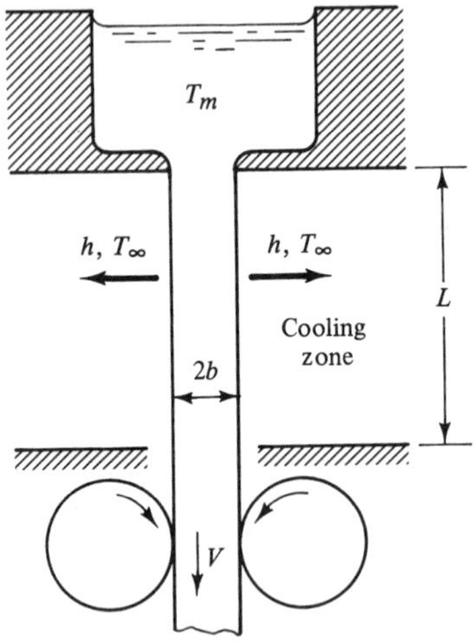

Fig. 4P-35

4-36 In the process of coating wires (say, copper) with a tin film, the wire is drawn from a bath of molten coating material into a cooling zone where the coating film solidifies. Discuss the formulation of the problem leading to determination of the height of the cooling zone to ensure solidification of the coat.

4-37 A plane sheet (jet) of fluid at temperature T_∞ impinges perpendicularly onto an isothermal plate at T_w (Fig. 4P-37). Near the stagnation point S we have a (Hiemenz) stagnation flow (see, for example, Schlichting 1968) for which the inviscous free-stream velocity satisfies (recall Problem 2-12)

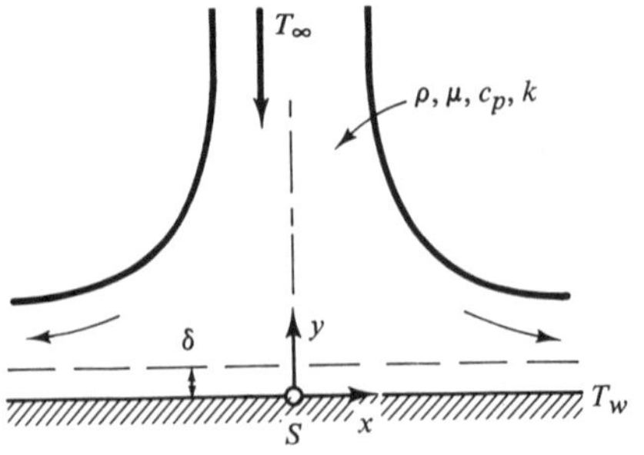

Fig. 4P-37

$$U(x) = U_0 \frac{x}{L},$$

where U_0 and L are reference values of velocity and length. Determine the viscous boundary layer thickness (it becomes a constant), the skin friction, and the heat transfer along the plate.

4-38 Consider the thermal boundary layer in a steady liquid metal flow along a curved isothermal surface kept at T_w. The velocity is specified as $u = U(x)$ and the free-stream temperature is constant, T_∞. Using the integral formulation with a first-order temperature profile, derive the following expression for the local Nusselt number:

$$\text{Nu}_x = \frac{hR}{k} = \frac{c_1 U(x)}{[\int [aU(x)/R]\, d(x/R) + c_2]^{1/2}},$$

where R is a reference length and c_1 and c_2 constants depending on profile and boundary conditions. Next, for a circular cylinder in uniform flow

$$U(x) = 2U_0 \sin\frac{x}{R},$$

calculate the local Nusselt number along the upstream part of the surface, $0 \leq x/R < \pi/2$ (Fig. 4P-38).

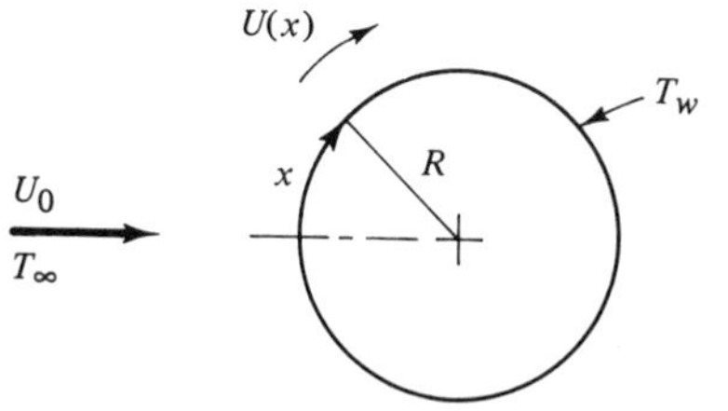

Fig. 4P-38

4-39 Reconsider Ex. 4.3.1 for the case $\text{Pr} \ll 1$ (but not $\text{Pr} = 0$). Show that Eq. (4.3.15) reduces to

$$\text{Nu}_x = \frac{c_1 \text{Pe}_x^{1/2}}{1 + c_2 \text{Pr}^{1/2}}.$$

4-40 Referring to Table 2.5, write the governing equations as well as boundary and initial conditions for two-dimensional driven cavity flow including buoyancy effects (Fig. 4.29). Consider formulations in terms of both u, v, p, T and ψ, ω, T.

4-41 Reconsider Problem 3-8 in the absense of the plate. Note that, given the belt velocity U and liquid film thickness δ_∞, the steady mass flow lifted from the bath can be readily calculated. It is the effect of surface tension in the region of film formation that determines δ_∞ and $\delta(x)$ (Fig. 4P-41). Given U, ρ, μ, σ_s we seek δ_∞ (White, D. A., & Tallmadge, J. A., 1965, Chem. Eng. Sci., **20**, 33).

4-42 A thin flat plate, whose leading edge is initially in contact with the free surface of a large bath of viscous liquid, is released at $t = 0$ to fall into the liquid (Fig. 4P-42). The initial velocity is zero and the plate is assumed to move in its vertical plane. We seek the velocity $V(t)$ of the plate. Note that the plate is submerged partially in a first time domain ($0 \leq t \leq t_1$) and then fully ($t_1 \leq t$), to ultimately reach its terminal velocity V_∞. Consider both lumped and integral solutions. Related problems involve a nonzero initial velocity, an initially partial or full submerged plate, and the heat transfer from plate to liquid.

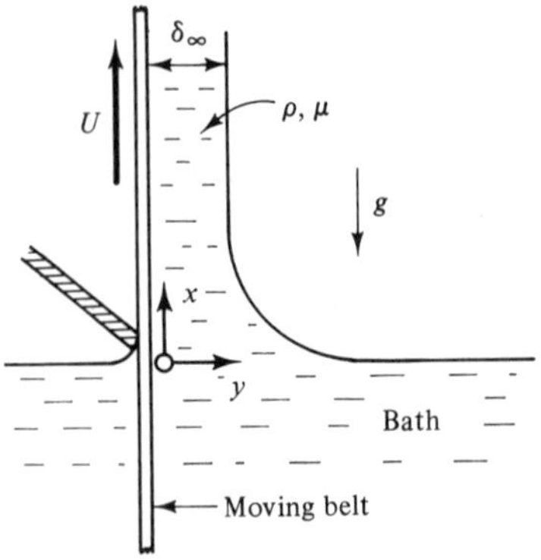

Fig. 4P-41

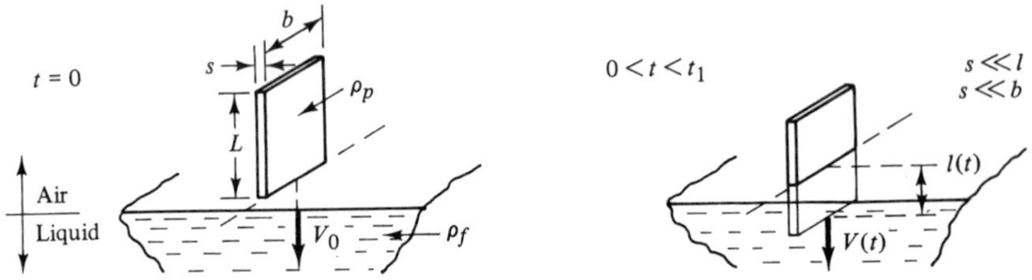

Fig. 4P-42

4-43 A flat plate of thickness L of a porous material (porosity P, specific heat c_s), initially saturated with liquid at its saturation temperature $T(x, 0) = T_s$ and insulated on one side, is suddenly subjected to a constant heat flux q_0 on the other side (Fig. 4P-43). Determine the transient temperature distribution in the dried-out region $0 \leq x \leq X(t)$, specifically the surface temperature $T_w(t)$ and the position of the receding vapor–liquid interface. Discuss various approximations and their physical significance as well as the same problem subject to a step change in surface temperature instead of heat flux (Hilding, W. E., & Chen, U., 1982, Paper CP12, Int. Heat Transfer Conf., Munich).

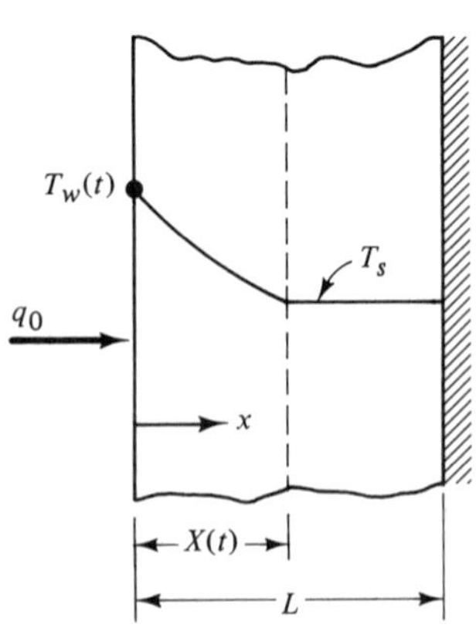

Fig. 4P-43

Part III

SOLUTION

In Part II we learned the individual formulation of laminar forced and natural convection problems associated with parallel flows and boundary layers. The emphasis in Part II was placed on the solution of these problems in terms of an integral formulation coupled with approximate profiles. In Chapter 3 we also discussed the differential formulation of (linear) parallel flow problems, and obtained exact and approximate solutions for these problems. Since most convection problems are nonlinear, in the present part we explore a number of solution techniques particularly suitable to nonlinear problems. Chapter 5 deals with similarity transformations, Chapter 6 with periodic solutions, and Chapter 7 with computational methods.

Chapter 5

SIMILARITY TRANSFORMATION

In this chapter we study the similarity transformations for the solution of convection problems. Although extensively developed for nonlinear (boundary layer) problems, these transformations apply equally to linear problems.

5.1 ORIGIN OF SIMILARITY TRANSFORMATION

Delaying momentarily the method of searching for similarity variables and making similarity transformations with these variables, let us first look for the origin of these variables in terms of an example. For this purpose, reconsider Ex. 3.5.1, now for the semi-infinite domain over a flat plate. The formulation of the new problem in $\theta = T - T_w$ is

$$\frac{U}{a}\frac{\partial \theta}{\partial x} = \frac{\partial^2 \theta}{\partial y^2}, \qquad \theta(0, y) = \theta_0,$$
$$\theta(x, 0) = 0, \qquad \theta(x, \infty) = \theta_0. \tag{5.1.1}$$

A convenient way of solving this problem is by the use of Laplace transforms. Omitting the classical solution procedure, which has no significance for the present discussion, we have

$$\frac{\theta(x, y)}{\theta_0} = \text{erf}\left[\frac{y}{2(ax/U)^{1/2}}\right], \tag{5.1.2}$$

or, in terms of the variable

$$\eta = \frac{y}{(ax/U)^{1/2}}, \tag{5.1.3}$$

$$\frac{\theta(\eta)}{\theta_0} = \text{erf}\left(\frac{\eta}{2}\right). \tag{5.1.4}$$

Now, suppose that we know η one way or another but have difficulty of solving Eq. (5.1.1). Let us see what happens when this equation is expressed in terms of η. Introducing the dimensionless temperature $f = \theta/\theta_0$, and noting that

$$\frac{1}{\theta_0}\frac{\partial \theta}{\partial x} = -\left(\frac{\eta}{2x}\right)\frac{df}{d\eta},$$

$$\frac{1}{\theta_0}\frac{\partial^2 \theta}{\partial y^2} = \left(\frac{U}{ax}\right)\frac{d^2 f}{d\eta^2},$$

Eq. (5.1.1) may be rearranged as

$$\frac{d^2 f}{d\eta^2} + \frac{1}{2}\eta\frac{df}{d\eta} = 0; \qquad f(0) = 0, \qquad f(\infty) = 1. \tag{5.1.5}$$

(What happened to one of the boundary conditions of Eq. 5.1.1?) Thus, in terms of η, the partial differential equation of Eq. (5.1.1) is *reduced* to the ordinary differential equation of Eq. (5.1.5). Clearly, any reduction in the number of independent variables of a problem is an important simplification toward the solution of the problem, whether the problem is to be solved analytically or numerically. Presently, employing $F = df/d\eta$, Eq. (5.1.5) may be readily integrated twice to give

$$\frac{\theta(\eta)}{\theta_0} = \frac{\int_0^{\eta} e^{-z^2/4}\,dz}{\int_0^{\infty} e^{-z^2/4}\,dz}. \tag{5.1.6}$$

This solution can be shown to be identical to Eq. (5.1.4).*

The variable η is known to be the *similarity variable* of Eq. (5.1.1), in terms of η the reduction of Eq. (5.1.1) to Eq. (5.1.5) is a *similarity transformation*, and Eq. (5.1.4) is a *similarity solution*. This solution has the property that two temperature profiles located at different x differ only by an x-dependent scale factor in y. That is, all temperature profiles become identical in a plot for θ/θ_0 versus $\eta = y/(ax/U)^{1/2}$. It is from this fact that we get the name *similarity*.

---

*Note that $\int_0^{\infty} e^{-z^2/4}\,dz = \int_0^{\infty} e^{-\xi}\xi^{(1/2)-1}\,d\xi = \Gamma(\frac{1}{2}) = \sqrt{\pi}$, and the definition of the error function, $\text{erf}\,(z) = (2/\sqrt{\pi})\int_0^{z} e^{-\xi^2}\,d\xi$.

5.2 SIMILARITY VARIABLE VIA DIMENSIONAL ANALYSIS

The quest for a similarity variable, and a transformation leading to a similarity solution in terms of this variable, for problems otherwise difficult to solve, is the concern of this section. Apparently, there are a number of methods that follow different procedures, but all lead to the same answer. Three decades after the original work of Blasius (1908) on the boundary layer along a flat plate, Goldstein (1939) studied the types of potential flows which would yield similarity solution for boundary layers (for later work, see references cited in Schlichting 1968). Following a different approach, Sedov (1943 translation 1959) explored and to a large extent developed the theory of dimensional analysis; he stressed the fact that physical phenomena should rely on the invariant nature of the governing laws relative to the choice of units for physical variables; furthermore, he pointed out the analogy between the theories of dimensional analysis, similarity, and geometric invariants relative to coordinate transformations: then, he proceeded to the solution of many important problems in gas dynamics. Following Sedov, a number of applications of dimensional analysis and similarity have been made to widely different problems in physics, continuum mechanics, and to certain mathematical problems related to the use of group theory in solving differential equations (see the text by Birkhoff 1950 or the review paper by Friedrichs 1955 and the text by Hansen 1964).

Our method of search for a similarity variable rests on the dimensional analysis, and is not new; however, the procedure we follow differs from those given in the literature, including the one by Hellums & Churchill (1964). The outline of the method consists of the following two steps:

1. Make dependent and independent variables dimensionless in terms of the *inherent characteristic properties* or, in the absence of any characteristic property, in terms of *arbitrarily selected reference quantities*.
2. Eliminate all arbitrarily selected reference quantities by successively employing the *mathematical principle* (which states the invariance of the number of dependent and independent variables of a mathematical expression under any transformation) and the *physical principle* (which states the dimensional homogeneity of a physical expression).

Then, *a similarity variable may be found whenever a characteristic property does not inherently exist which would make an independent variable dimensionless.*

Once the similarity variable has been found, the governing (differential) equations, initial and/or boundary conditions, are transformed in terms of this variable. The transformation is successful if the similarity variable remains as the only independent variable. Since the transformation reduces the number of independent variables by one, two of the original initial-boundary conditions must reduce to one.

Let us now apply the aforementioned steps to the example already considered, and *search* for a similarity variable.

EXAMPLE 5.2.1 Reconsider Ex. 3.5.1 for the semi-infinite domain, as already stated by Eq. (5.1.1). We wish to find a similarity variable for the temperature.

Let x and y be made dimensionless in terms of the *arbitrarily selected reference lengths* x_0 and y_0, respectively. Also, placing θ dimensionless in terms of the *characteristic temperature* θ_0, the differential equation of Eq. (5.1.1) may be rearranged as

$$\left(\frac{Uy_0^2}{ax_0}\right)\frac{\partial(\theta/\theta_0)}{\partial(x/x_0)} = \frac{\partial^2(\theta/\theta_0)}{\partial(y/y_0)^2},$$

which implies that

$$\frac{\theta}{\theta_0} = f\left(\frac{x}{x_0}, \frac{y}{y_0}, \frac{Uy_0^2}{ax_0}\right). \tag{5.2.1}$$

Note that the use of θ_0 incorporates the effect of boundaries into Eq. (5.2.1). Now, according to the *mathematical principle*, we are free to transform the variables of Eq. (5.2.1) in any way we like, but only without changing the number of these variables. Among the possible transformations we pick the one to be convenient when we consider the physical principle. Thus, we transform the independent variables in such a way that only one term remains involving one of the independent variables, say x_0. We do this by introducing a new variable in place of Uy_0^2/ax_0, obtained dividing Uy_0^2/ax_0 by x/x_0. The result is

$$\frac{\theta}{\theta_0} = f_1\left(\frac{x}{x_0}, \frac{y}{y_0}, \frac{Uy_0^2}{ax}\right). \tag{5.2.2}$$

According to the *physical principle*, Eq. (5.2.2) assumes significance only when the left- and right-hand sides are dimensionally homogeneous. Since the problem statement clearly indicates to the absence of any characteristic length in the x-direction, the temperature and, consequently, the right side of Eq. (5.2.2) must be independent of x_0. This fact reduces Eq. (5.2.2) to

$$\frac{\theta}{\theta_0} = f_1\left(\frac{y}{y_0}, \frac{Uy_0^2}{ax}\right). \tag{5.2.3}$$

Next, we repeat for y the preceding steps pertaining to x. Thus, reconsider the mathematical principle, and without any reduction, transform independent variables of Eq. (5.2.3)—in a way suitable to later physical interpretation—such that only one term remains depending on y_0. This may be done by introducing a new variable in place of Uy_0^2/ax, obtained multiplying Uy_0^2/ax by $(y/y_0)^2$. The result is

$$\frac{\theta}{\theta_0} = f_2\left(\frac{y}{y_0}, \frac{Uy^2}{ax}\right). \tag{5.2.4}$$

Since there is no characteristic length in the y-direction, according to the physical principle, the temperature and, consequently, the right side of Eq. (5.2.4) must be dimensionally homogeneous, that is, independent of y_0, and Eq. (5.2.4) must reduce to

$$\frac{\theta}{\theta_0} = f_2\left(\frac{Uy^2}{ax}\right), \tag{5.2.5}$$

which may be rearranged as

$$\frac{\theta}{\theta_0} = f(\eta), \tag{5.2.6}$$

where $\eta = y/(ax/U)^{1/2}$. Clearly, our search for a similarity variable turns out to be successful, and we are able to combine two independent variables into one (similarity) variable. It is worth noting that any power of η can be used as the similarity variable. Among these, the one linear in y is conveniently selected to be the similarity variable. For the solution, recall the similarity transformation of Eq. (5.1.1) in terms of η, and the similarity solution of the resulting formulation stated by Eq. (5.1.5). The solution obtained this way is equally valid for the first domain of the original Ex. 3.5.1. However, there exists no similarity solution for the second domain of that example because of the finite geometry, which implies a characteristic length in y.

An alternative method of searching for a similarity variable employs the general idea of separation of variables. The method is based on the assumption that a similarity variable may be written as a product of two functions, each depending on one independent variable,

$$\eta = g(x)h(y). \tag{5.2.7}$$

The simplest possible form of Eq. (5.2.7) in terms of one unknown function (to be determined by the governing equation) is

$$yg(x) \qquad \text{or} \qquad xh(y).$$

However, the experience so far gained on boundary layers suggests the convenient form

$$\eta = \frac{y}{g(x)}. \tag{5.2.8}$$

It is important to note here that the η-variable of similarity solutions corresponds to the $y/\delta(x)$-variable of approximate profiles. The differential formulation specifies $g(x)$ exactly, while the integral solution determines $\delta(x)$ approximately.

Now, $g(x)$ should be evaluated such that

$$\theta(x, y) = \theta_0 f(\eta), \tag{5.2.9}$$

where θ_0 is a reference temperature which merely makes $f(\eta)$ dimensionless. In terms of Eqs. (5.2.8) and (5.2.9),

$$\frac{\partial\theta}{\partial x} = -\theta_0\left(\frac{y}{g^2}\right)\left(\frac{dg}{dx}\right)\frac{df}{d\eta},$$

$$\frac{\partial^2\theta}{\partial y^2} = \left(\frac{\theta_0}{g^2}\right)\frac{d^2 f}{d\eta^2},$$

and, we have from Eq. (5.1.1),

$$\frac{d^2 f}{d\eta^2} + \left(\frac{U}{a}g\frac{dg}{dx}\right)\eta\frac{df}{d\eta} = 0. \tag{5.2.10}$$

Clearly, Eq. (5.2.10) becomes a function of η alone, and identical to Eq. (5.1.5), when

$$2\frac{U}{a}g\frac{dg}{dx} = 1. \tag{5.2.11}$$

Integration of Eq. (5.2.11) leads to

$$g(x) = \left(\frac{ax}{U}\right)^{1/2}, \tag{5.2.12}$$

where the integration constant is set equal to zero so that $\theta(0, y) = \theta_0$ becomes compatible with $\theta(x, \infty) = \theta_0$ (recall the boundary conditions in Eq. 5.1.1). The combination of Eqs. (5.2.8) and (5.2.12) gives the similarity variable obtained previously by dimensional analysis of the same problem.

The comparison of the methods of dimensional analysis and separation of variables points out the somewhat *indirect* nature of the latter because of a priori assumptions needed for the special forms of $\eta(x, y)$ and $\theta(x, y)$. For example, because of its simplicity, the present example is not a serious test for the application of the separation of variables. In fact, if the uniform wall temperature of the present problem were replaced with a uniform heat flux (see Ex. 5.3.1), noting that the new boundary condition depends on x, we now have to assume that

$$\theta(x, y) = A(x)f(\eta),$$

which requires the simultaneous evaluation of $A(x)$ and $g(x)$ (see Problem 5-15). Unlike this approach, dimensional analysis *directly* yields $A(x)$. In view of these remarks, we shall employ only the method of dimensional analysis in the remainder of this chapter.

5.3 SIMILARITY SOLUTION OF BOUNDARY LAYERS

This section, after two transitory examples from Chapter 3, is devoted to similarity solution of the examples which were solved in Chapter 4 by approximate profiles.

EXAMPLE 5.3.1 Reconsider Ex. 5.2.1, after replacing the condition of uniform plate temperature with that of uniform heat flux, say q''. We wish to find a similarity solution for the temperature distribution.

The formulation of the problem in $\theta = T - T_0$ is

$$\frac{U}{a}\frac{\partial \theta}{\partial x} = \frac{\partial^2 \theta}{\partial y^2}, \qquad \theta(0, y) = 0,$$
$$-k\frac{\partial \theta(x, 0)}{\partial y} = q'', \qquad \theta(x, \infty) = 0. \tag{5.3.1}$$

Place x, y, and θ dimensionless in terms of arbitrarily selected reference lengths x_0 and y_0, and a reference temperature θ_R, respectively. Then, the differential equation and the heat flux boundary condition of Eq. (5.3.1) may be rearranged as

$$\left(\frac{Uy_0^2}{ax_0}\right)\frac{\partial(\theta/\theta_R)}{\partial(x/x_0)} = \frac{\partial^2(\theta/\theta_R)}{\partial(y/y_0)^2}, \qquad -\frac{\partial(\theta/\theta_R)}{\partial(y/y_0)} = \frac{q''y_0}{k\theta_R},$$

which implies that

$$\frac{\theta}{\theta_R} = f\left(\frac{x}{x_0}, \frac{y}{y_0}, \frac{Uy_0^2}{ax_0}, \frac{q''y_0}{k\theta_R}\right). \tag{5.3.2}$$

Since the physics of the problem rejects θ_R, x_0, and y_0 as being characteristic properties, let us successively eliminate these arbitrary quantities. Start, for example, with θ_R, and transform Eq. (5.3.2) such that only one term remains depending on θ_R. The result is the mathematical expression

$$\frac{\theta}{q''y_0/k} = f_1\left(\frac{x}{x_0}, \frac{y}{y_0}, \frac{Uy_0^2}{ax_0}, \frac{q''y_0}{k\theta_R}\right),$$

which assumes physical significance in θ only when it is independent of θ_R. Thus,

$$\frac{\theta}{q''y_0/k} = f_1\left(\frac{x}{x_0}, \frac{y}{y_0}, \frac{Uy_0^2}{ax_0}\right).$$

The elimination of x_0 and y_0, on the other hand, follows identically the elimination process of these variables in Ex. 5.2.1. Ultimately, we have

$$\frac{\theta}{q''y/k} = f_3\left(\frac{Uy^2}{ax}\right),$$

which may be rearranged as

$$\frac{\theta}{q''/k} = yf(\eta) \tag{5.3.3}$$

or, introducing $F = \eta f$, as

$$\frac{\theta}{q''/k} = \left(\frac{ax}{U}\right)^{1/2} F(\eta), \tag{5.3.4}$$

where $\eta = y/(ax/U)^{1/2}$ is identical to the similarity variable of the preceding example. Apparently, any change in boundary conditions of a problem affects the relation between the dependent variables of the original and similarity problems but not the similarity variable.

In terms of Eq. (5.3.4), for example, we have

$$\frac{\partial\theta}{\partial x} = \left(\frac{q''}{k}\right)\left[\left(\frac{ax}{U}\right)^{1/2}\frac{F}{2x} - \left(\frac{y}{2x}\right)\frac{dF}{d\eta}\right],$$

$$\frac{\partial\theta}{\partial y} = \left(\frac{q''}{k}\right)\frac{dF}{d\eta},$$

$$\frac{\partial^2\theta}{\partial y^2} = \left(\frac{q''}{k}\right)\left(\frac{ax}{U}\right)^{-1/2}\frac{d^2F}{d\eta^2},$$

by which Eq. (5.3.1) reduces to the similarity formulation,

$$\frac{d^2F}{d\eta^2} + \frac{1}{2}\eta\frac{dF}{d\eta} - \frac{1}{2}F = 0; \qquad -\frac{dF(0)}{d\eta} = 1, \qquad F(\infty) = 0. \tag{5.3.5}$$

Most conveniently obtained by the use of Laplace transforms and the inversion tables, the solution of Eq. (5.3.1) is

$$\frac{\theta(x, \eta)}{q''/k} = 2\left(\frac{ax}{U}\right)^{1/2} i\,\mathrm{erfc}\left(\frac{\eta}{2}\right),$$

where $\eta = y/(ax/U)^{1/2}$. On the other hand, an analytical solution for Eq. (5.3.5) does not appear to be readily available.

Here, only an inductive solution procedure is offered to the interested reader. By definition,

$$i^n \operatorname{erfc} x = \int_x^\infty i^{n-1} \operatorname{erfc} \xi \, d\xi; \qquad n = 1, 2, 3, \dots$$
$$i^1 \operatorname{erfc} x = i \operatorname{erfc} x, \qquad i^0 \operatorname{erfc} x = \operatorname{erfc} x. \tag{a}$$

Then, integrating by parts, we have, successively,

$$i \operatorname{erfc} x = \pi^{-1/2} e^{-x^2} - x \operatorname{erfc} x,$$
$$i^2 \operatorname{erfc} x = \tfrac{1}{4}[(1 + 2x^2) \operatorname{erfc} x - 2\pi^{-1/2} x e^{-x^2}], \tag{b}$$

which may be rearranged, in terms of (b), as

$$i^2 \operatorname{erfc} x = \tfrac{1}{4}(\operatorname{erfc} x - 2xi \operatorname{erfc} x), \tag{c}$$

and so on. The recurrence formula is

$$2ni^n \operatorname{erfc} x = i^{n-2} \operatorname{erfc} x - 2xi^{n-1} \operatorname{erfc} x. \tag{d}$$

It follows from (d) that $y = i^n \operatorname{erfc} x$ satisfies the differential equation

$$\frac{d^2y}{dx^2} + 2x \frac{dy}{dx} - 2ny = 0. \tag{e}$$

Clearly, $y = F$ and $x = \eta/2$ transform Eq. (e) for $n = 1$ to the differential equation of Eq. (5.3.5).

Therefore, an analytical solution to a similarity formulation may or may not be easier to get than that of the corresponding original formulation. However, most of our problems are not suited for analytical treatment, and we are often forced to use a numerical procedure such as power series, finite difference, and so on. When this happens to be the case, the numerical solution of a similarity formulation [associated with ordinary differential equation(s)] is always easier to obtain than that of the corresponding original formulation [associated with partial differential equation(s)]. For example, Eq. (5.3.5) may be replaced by two coupled first-order differential equations,

$$2G' + \eta G = F, \quad F' = G; \qquad -G(0) = 1, \quad F(\infty) = 0,$$

which can readily be solved by a standard procedure such as the Runge–Kutta or the predictor–corrector method (see Section 7.2 and, for example, Fröberg 1970). So far, we considered thermal boundary layers associated with a uniform velocity. Next, we investigate the effect of a distributed velocity on thermal boundary layers.

EXAMPLE 5.3.2 Reconsider the first domain of Ex. 3.5.4. We wish to find a similarity solution for the temperature distribution.

Now, assume the temperature field to extend to the entire half-space over the lower plate. The differential formulation of the problem in $\theta = T - T_w$ is

$$\frac{U}{a}\left(\frac{y}{l}\right)\frac{\partial \theta}{\partial x} = \frac{\partial^2 \theta}{\partial y^2}; \qquad \theta(0, y) = \theta_0,$$
$$\theta(x, 0) = 0, \qquad \theta(x, \infty) = \theta_0. \tag{5.3.6}$$

Here l is a characteristic length for the velocity but not for the temperature. Place x, y, and θ dimensionless in terms of arbitrarily selected reference lengths x_0 and y_0, and the inherent characteristic temperature θ_0, respectively. Then the differential equation of Eq. (5.3.6) may be rearranged as

$$\left(\frac{Uy_0^3}{ax_0 l}\right)\left(\frac{y}{y_0}\right)\frac{\partial(\theta/\theta_0)}{\partial(x/x_0)} = \frac{\partial^2(\theta/\theta_0)}{\partial(y/y_0)^2},$$

which implies that

$$\frac{\theta}{\theta_0} = f_1\left(\frac{x}{x_0}, \frac{y}{y_0}, \frac{Uy_0^3}{ax_0 l}\right).$$

Since the physics of the problem rejects x_0 and y_0 as being characteristic lengths, let us eliminate them as explained in Ex. 5.2.1. Ultimately, we have

$$\frac{\theta}{\theta_0} = f_2\left(\frac{Uy^3}{axl}\right),$$

which may be rearranged as

$$\frac{\theta}{\theta_0} = f(\eta), \tag{5.3.7}$$

where $\eta = y/(axl/U)^{1/3}$ is the similarity variable. Clearly, the only difference between Exs. 5.2.1 and 5.3.2 is in the velocity distribution, and this difference affects the similarity variable.

In terms of Eq. (5.3.7), we have

$$\frac{\partial\theta}{\partial x} = -\theta_0\left(\frac{\eta}{3x}\right)\frac{df}{d\eta},$$

$$\frac{\partial^2\theta}{\partial y^2} = \theta_0\left(\frac{axl}{U}\right)^{-2/3}\frac{d^2 f}{d\eta^2},$$

by which Eq. (5.3.6) reduces to

$$3f'' + \eta^2 f' = 0; \qquad f(0) = 0, \quad f(\infty) = 1. \tag{5.3.8}$$

The analytical solution of Eq. (5.3.8) happens to be simple. In terms of $F = f'$, this solution is readily found to be*

$$\frac{\theta(\eta)}{\theta_0} = \frac{\int_0^\eta e^{-z^3/9}\,dz}{\int_0^\infty e^{-z^3/9}\,dz}, \tag{5.3.9}$$

which was originally obtained by Levêque (1928) as an approximation to the entrance region of the Graetz problem. Note the simplicity and numerical convenience of Eq. (5.3.9) compared with the solution given by Eq. (3.5.48), which is based on the method of characteristic values. Note also the comparable simplicity of Eqs. (3.5.49) and (5.3.9) but the greater accuracy of the latter.

*Expressing its denominator in terms of the Γ-function, Eq. (5.3.9) may be rearranged as $\theta(\eta)/\theta_0 = \int_0^\eta e^{-z^3/9}\,dz/[9^{1/3}\Gamma(\frac{4}{3})]$, where $1/[9^{1/3}\Gamma(\frac{4}{3})] = 0.5384$.

We proceed now to the main objective of the chapter, the similarity solution of momentum as well as thermal boundary layers.

EXAMPLE 5.3.3 Reconsider the flow over a flat plate, Ex. 4.2.1. We wish to find a similarity solution for the velocity distribution.

The differential formulation of the problem, Eq. (4.1.3) and Eq. (4.1.13) with $dp/dx = 0$ and $f_x = 0$, is

$$\frac{\partial u}{\partial x} + \frac{\partial v}{\partial y} = 0,$$

$$u\frac{\partial u}{\partial x} + v\frac{\partial u}{\partial y} = \nu\frac{\partial^2 u}{\partial y^2}; \tag{5.3.10}$$

$$u(0, y) = U_\infty, \qquad u(x, 0) = 0, \qquad v(x, 0) = 0, \qquad u(x, \infty) = U_\infty.$$

Place x, y, u, and v dimensionless in terms of arbitrarily selected reference lengths x_0, y_0, the inherent characteristic velocity U_∞, and arbitrarily selected velocity V_0, respectively. Then the differential equations of Eq. (5.3.10) may be rearranged as*

$$\left(\frac{U_\infty y_0}{V_0 x_0}\right)\frac{\partial(u/U_\infty)}{\partial(x/x_0)} + \frac{\partial(v/V_0)}{\partial(y/y_0)} = 0,$$

$$\left(\frac{U_\infty y_0^2}{\nu x_0}\right)\left(\frac{u}{U_\infty}\right)\frac{\partial(u/U_\infty)}{\partial(x/x_0)} + \left(\frac{V_0 y_0}{\nu}\right)\left(\frac{v}{V_0}\right)\frac{\partial(u/U_\infty)}{\partial(y/y_0)} = \frac{\partial^2(u/U_\infty)}{\partial(y/y_0)^2}, \tag{5.3.11}$$

which imply that

$$\frac{u}{U_\infty} = f_1\left(\frac{x}{x_0}, \frac{y}{y_0}, \frac{V_0 y_0}{\nu}, \frac{U_\infty y_0}{V_0 x_0}, \frac{U_\infty y_0^2}{\nu x_0}\right),$$

$$\frac{v}{V_0} = g_1\left(\frac{x}{x_0}, \frac{y}{y_0}, \frac{V_0 y_0}{\nu}, \frac{U_\infty y_0}{V_0 x_0}, \frac{U_\infty y_0^2}{\nu x_0}\right). \tag{5.3.12}$$

Since the physics of the problem rejects V_0, x_0, and y_0 as being characteristic properties, let us successively eliminate these reference quantities. Start, for example, with V_0, and transform Eq. (5.3.12) such that only one term remains depending on V_0. This may be done by introducing a new parameter in place of $U_\infty y_0/V_0 x_0$, obtained multiplying $U_\infty y_0/V_0 x_0$ with $V_0 y_0/\nu$. However, Eq. (5.3.12) already contains this parameter. Consequently, we have the mathematical expressions†

$$\frac{u}{U_\infty} = f_2\left(\frac{x}{x_0}, \frac{y}{y_0}, \frac{V_0 y_0}{\nu}, \frac{U_\infty y_0^2}{\nu x_0}\right),$$

$$\frac{v y_0}{\nu} = g_2\left(\frac{x}{x_0}, \frac{y}{y_0}, \frac{V_0 y_0}{\nu}, \frac{U_\infty y_0^2}{\nu x_0}\right), \tag{3.5.13}$$

---

*This arrangement is not the only possible one. However, all arrangements lead to the same answer.

†An arrangement of Eq. (5.3.10) different from Eq. (5.3.11) may directly lead to Eq. (5.3.13), thereby eliminating the step between Eqs. (5.3.12) and (5.3.13).

which assume physical significance in u and v only when independent of V_0. Thus,

$$\frac{u}{U_\infty} = f_2\left(\frac{x}{x_0}, \frac{y}{y_0}, \frac{U_\infty y_0^2}{\nu x_0}\right),$$
$$\frac{v y_0}{\nu} = g_2\left(\frac{x}{x_0}, \frac{y}{y_0}, \frac{U_\infty y_0^2}{\nu x_0}\right). \tag{5.3.14}$$

The elimination of x_0 and y_0, on the other hand, follows identically the elimination process of these variables in the preceding examples. Ultimately, we have

$$\frac{u}{U_\infty} = f_4\left(\frac{U_\infty y^2}{\nu x}\right),$$
$$v = \frac{\nu}{y} g_4\left(\frac{U_\infty y^2}{\nu x}\right), \tag{5.3.15}$$

which may be rearranged as

$$\frac{u}{U_\infty} = f_5(\eta) \tag{5.3.16}$$

and, in terms of $\eta g_4 = g_5$, as

$$v = \left(\frac{\nu U_\infty}{x}\right)^{1/2} g_5(\eta), \tag{5.3.17}$$

where $\eta = y/(\nu x/U_\infty)^{1/2}$ is the similarity variable. (Why does this variable for $\nu = a$ become identical to the variable of Exs. 5.2.1 and 5.3.1?) Next, inserting Eqs. (5.3.16) and (5.3.17) into the conservation of mass, the new dependent variables f_5 and g_5 are found to be related according to

$$\frac{dg_5}{d\eta} = \frac{1}{2}\eta \frac{df_5}{d\eta},$$

whose integration yields, within a constant,

$$g_5 = \frac{1}{2}\left(\eta f_5 - \int f_5 \, d\eta\right).$$

Finally, in terms of a more convenient dependent variable, $f = \int f_5 \, d\eta$, we have

$$\frac{u}{U_\infty} = f', \qquad v = \frac{1}{2}\left(\frac{\nu U_\infty}{x}\right)^{1/2} (\eta f' - f), \tag{5.3.18}$$

originally obtained by Blasius (1908) following a different (physical) argument. Introducing Eq. (5.3.18) into the momentum equation of Eq. (5.3.10), and expressing the boundary conditions in terms of η and f, we get

$$2f''' + ff'' = 0;$$
$$f(0) = 0, \qquad f'(0) = 0, \qquad f'(\infty) = 1. \tag{5.3.19}$$

The original solution of Eq. (5.3.19) obtained by Blasius (1908) rests on a power series expansion about $\eta = 0$ and an asymptotic expansion (a regular perturbation on the potential flow) near $\eta = \infty$, the two solutions being *patched* at a suitable point (for details and other early solution procedures, see Schlichting 1968). A rather accurate tabulation of the functions f, f', and f'' was given by Howarth (1938).

A computational scheme for the solution of Eq. (5.3.19) based on the Runge–Kutta procedure is given in Ex. 7.2.2. The resulting tabulation of functions appears in Table 7.7. In terms of this table, Fig. 5.1(a) shows the variation of the longitudinal velocity $u/U_\infty = f'(\eta)$ versus η. The curvature of this velocity vanishes as $\eta \longrightarrow 0$. (Why?) Also, Fig. 5.1(b) represents the variation of the transversal velocity

$$\frac{v}{U_\infty}\left(\frac{U_\infty x}{\nu}\right)^{1/2} = \frac{1}{2}(\eta f' - f)$$

versus η. Note that at the outer edge of the boundary layer this velocity approaches the *finite* value

$$\frac{v_\infty}{U_\infty} = 0.865\left(\frac{\nu}{U_\infty x}\right)^{1/2}.$$

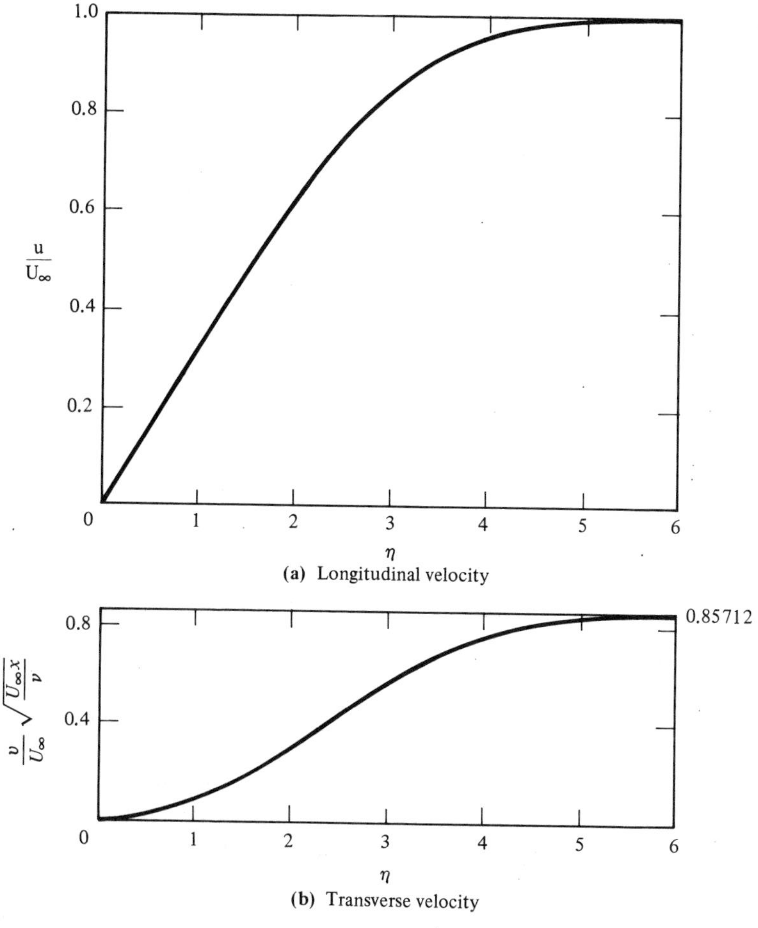

Fig. 5.1 Velocity components for the boundary layer along a flat plate at zero incidence, Ex. 5.3.3

This means that at the outer edge there is an outward flow because the increasing boundary layer thickness causes the fluid to be displaced from the plate walls. However, v_∞ is of order of $R_x^{-1/2}$ and is compatible with the boundary layer approximation (recall Eq. 4.1.15).

Having obtained the velocity distribution, we now proceed to the evaluation of the skin friction (viscous drag).

Skin friction. The shear stress at the plate walls

$$\tau_w = \mu\left(\frac{\partial u}{\partial y}\right)_{y=0},$$

integrated over the length l of the plate, gives the skin friction (force) per unit width,

$$F_w = \int_0^l \tau_w \, dx.$$

Explicitly,

$$\tau_w = \mu U_\infty \left(\frac{U_\infty}{\nu x}\right)^{1/2} f''(0) = 0.332\mu U_\infty \left(\frac{U_\infty}{\nu x}\right)^{1/2}, \tag{5.3.20}$$

where $f''(0) = 0.332$ according to Table 7.7 and the dimensionless shear stress

$$c_x = \frac{\tau_w}{\frac{1}{2}\rho U_\infty^2} = \frac{0.664}{R_x^{1/2}}, \tag{5.3.21}$$

where $R_x = U_\infty x/\nu$ denotes the local Reynolds number. Then for one side of the plate, the skin friction per unit width becomes

$$F_w = 0.664(\rho\mu l)^{1/2} U_x^{3/2}.$$

Furthermore, introducing the dimensionless drag coefficient

$$c_f = \frac{2F_w}{\frac{1}{2}\rho A U_\infty^2},$$

$A = 2l$ denoting the wetted surface area per unit width, we have

$$C_f = \frac{1.328}{R^{1/2}}, \tag{5.3.22}$$

where $R = U_\infty l/\nu$ is the Reynolds number based on the plate length. Clearly, this law of friction, first deduced by Blasius (1908), is valid only in the region of laminar flow, that is, for $R = U_\infty l/\nu < 5 \times 10^5$ to 10^6. For turbulent flow, $R > 10^6$, the drag becomes much greater than the foregoing value.

There remains a number of boundary layers (such as flow to a stagnation point, flow over a wedge, etc.) for which $U = U(x)$. Dimensional analysis applied to these flows leads to $u = U(x)f'(\eta)$, where $\eta \sim y/x^{1/2}$. The momentum equation then places the limitation to forms of $U(x)$ acceptable for a possible similarity solution (see Ch. IX of Schlichting 1968, and Problem 5-16).

EXAMPLE 5.3.4 Reconsider Ex. 4.3.1 or Ex. 5.3.3 with thermal effects added. We wish to find a similarity solution for the temperature distribution.

The formulation of the problem (Eq. 4.1.3, Eq. 4.1.13 with $dp/dx = 0$ and

$f_x = 0$ and Eq. 4.1.21 with $u''' = 0$) is

$$\frac{\partial u}{\partial x} + \frac{\partial v}{\partial y} = 0,$$

$$u\frac{\partial u}{\partial x} + v\frac{\partial u}{\partial y} = \nu\frac{\partial^2 u}{\partial y^2},$$

$$u\frac{\partial T}{\partial x} + v\frac{\partial T}{\partial y} = a\frac{\partial^2 T}{\partial y^2} + \frac{\mu}{\rho c_p}\left(\frac{\partial u}{\partial y}\right)^2, \tag{5.3.23}$$

$$u(0, y) = U_\infty, \qquad u(x, 0) = 0, \qquad v(x, 0) = 0, \qquad u(x, \infty) = U_\infty,$$

$$T(0, y) = T_\infty, \qquad T(x, 0) = T_w \quad \text{or} \quad \frac{\partial T(x, 0)}{\partial y} = 0, \qquad T(x, \infty) = T_\infty.$$

Note that, since we deal with an incompressible flow and assume constant properties, the momentum equation is decoupled from the energy equation. Consequently, similarity considerations of the preceding example apply equally to the velocity distribution of the present example. What remains, then, is to determine whether the energy equation accepts a formulation in terms of (and compatible with) the same similarity variable. This appears to be the case when Eq. (5.3.18) is introduced into the energy of Eq. (5.3.23) and all derivatives with respect to x and y are expressed in terms of $\eta = y/(\nu x/U_\infty)^{1/2}$. (Explain why this is so by comparatively inspecting the transformation steps of momentum and energy equations.) The result is

$$\frac{d^2 T}{d\eta^2} + \tfrac{1}{2}\mathrm{P} f\frac{dT}{d\eta} = -\mathrm{P}\left(\frac{U_\infty^2}{c_p}\right)f''^2, \tag{5.3.24}$$

$\mathrm{P} = \nu/a$ being the Prandtl number. In the literature, the energy equation with viscous dissipation neglected and wall temperature specified is known as the *cooling problem* and that with viscous dissipation and insulated wall as the *thermometer problem*. We now proceed to the solution of these problems.

Cooling problem. Neglecting the viscous dissipation, employing the specified wall temperature, and introducing $\theta_1(\eta, \mathrm{P}) = (T - T_\infty)/(T_w - T_\infty)$, we have, from Eq. (5.3.24)

$$\theta_1'' + \tfrac{1}{2}\mathrm{P} f\theta_1' = 0; \qquad \theta_1(0) = 1, \qquad \theta_1(\infty) = 0. \tag{5.3.25}$$

Rearranging the differential equation of Eq. (5.3.25) as

$$\frac{d\theta_1'}{d\eta} + \tfrac{1}{2}\mathrm{P} f\theta_1' = 0,$$

and integrating it yields

$$\theta_1 = C_2 + C_1\int_\eta^\infty e^{-(1/2)\mathrm{P}\int_0^\eta f\,d\eta}\,d\eta, \tag{5.3.26}$$

where the limits of integrals, in view of two yet to be specified integration constants, remain arbitrary and are selected conveniently. The boundary condition for $\eta \to \infty$ gives $C_2 = 0$. The boundary condition at $\eta = 0$ yields

$$1 = C_1\int_0^\infty e^{-(1/2)\mathrm{P}\int_0^\eta f\,d\eta}\,d\eta,$$

by which Eq. (5.3.24) may be rearranged to give the temperature distribution as

$$\theta_1(\eta, \mathrm{P}) = \frac{\int_\eta^\infty e^{-(1/2)\mathrm{P}\int_0^\eta f\,d\eta}\,d\eta}{\int_0^\infty e^{-(1/2)\mathrm{P}\int_0^\eta f\,d\eta}\,d\eta}. \tag{5.3.27}$$

Furthermore, this distribution may be given a more convenient form in terms of the velocity problem stated by Eq. (5.3.19). Rearranging Eq. (5.3.19) as

$$\frac{df''}{d\eta} = -\frac{1}{2}f''f,$$

and integrating it yields

$$f'' = Ce^{-(1/2)\int_0^\eta f\,d\eta}, \tag{5.3.28}$$

by which Eq. (5.3.27) becomes

$$\theta_1(\eta, \mathrm{P}) = \frac{\int_\eta^\infty [f''(\eta)]^{\mathrm{P}}\,d\eta}{\int_0^\infty [f''(\eta)]^{\mathrm{P}}\,d\eta}. \tag{5.3.29}$$

For P = 1, Eq. (5.3.29) reduces to

$$\theta_1(\eta, 1) = \frac{f'(\eta)|_\eta^\infty}{f'(\eta)|_0^\infty} = \frac{f'(\infty) - f'(\eta)}{f'(\infty) - f'(0)},$$

which, in view of the boundary conditions $f'(0) = 0$ and $f'(\infty) = 1$ of Eq. (5.3.19), and Eq. (5.3.18), may be rearranged as

$$\theta_1(\eta, 1) = 1 - f'(\eta) = 1 - \frac{u(\eta)}{U_\infty}; \tag{5.3.30}$$

that is, the dimensionless velocity and temperature profiles become identical for P = 1 (this fact suggests the Reynolds analogy, already discussed in Section 4.3). The temperature distribution against η calculated from Eq. (5.3.29) is shown plotted in Fig. 5.2. Because of the analogy just mentioned, the curve P = 1 also represents the velocity distribution $u/U_\infty = f(\eta)$. Thus, for P > 1, say for a viscous oil with a Prandtl number P = 1000, Fig. 5.2 indicates to a thermal boundary layer which is only one-tenth of the momentum boundary layer, confirming the ratio $\delta/\Delta \cong \mathrm{P}^{-1/3}$ already obtained in Section 4.3 by the integral approach and approximate profiles. For P < 1 the same figure shows that the thickness of thermal boundary layer exceeds that of momentum boundary layer.

Heat transfer. The local heat flux on plate walls,

$$q_w(x) = -k\left(\frac{\partial T}{\partial y}\right)_{y=0} = -k\left(\frac{U_\infty}{\nu x}\right)^{1/2}\left(\frac{dT}{d\eta}\right)_{\eta=0}, \tag{5.3.31}$$

integrated over the length l of the plate gives the total heat transfer per unit width from two sides of the plate,

$$Q = 2\int_0^l q_w(x)\,dx = -2k\left(\frac{U_\infty}{\nu x}\right)^{1/2}\left(\frac{dT}{d\eta}\right)_{\eta=0}. \tag{5.3.32}$$

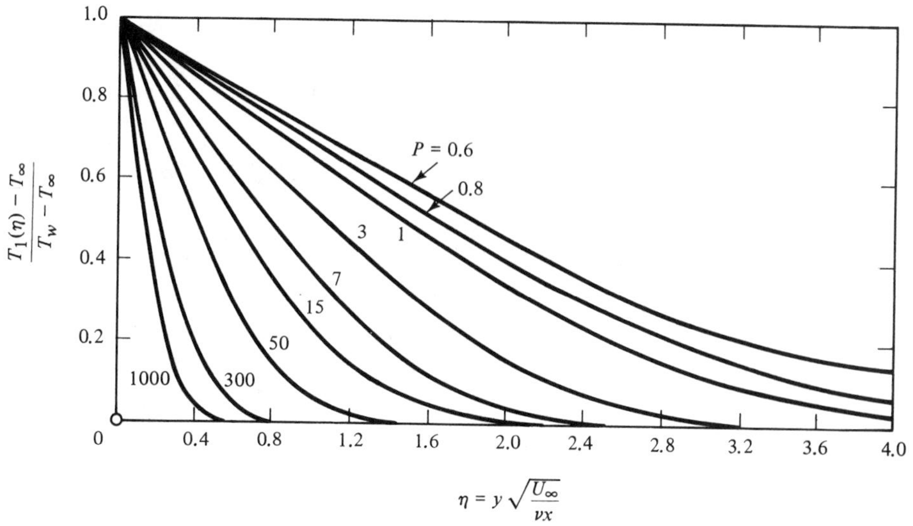

Fig. 5.2 Thermal boundary layer of the cooling problem, Ex. 5.3.4

Here, the wall temperature gradient, calculated from Eq. (5.3.29) with $f''(0) = 0.332$, is

$$\left(\frac{dT}{d\eta}\right)_{\eta=0} = \frac{(T_w - T_\infty)(0.332)^{\mathrm{P}}}{\int_0^\infty [f''(\eta)]^{\mathrm{P}}\, d\eta},$$

which for $0.6 < \mathrm{P} < 10$ can be interpolated with good accuracy by (Pohlhausen 1921)

$$\left(\frac{dT}{d\eta}\right)_{\eta=0} = (T_w - T_\infty)0.332\mathrm{P}^{1/3}. \qquad (5.3.33)$$

Employing this result, Eqs. (5.3.31) and (5.3.32) may be rearranged as

$$q_w(x) = 0.332k\mathrm{P}^{1/3}\mathrm{R}_x^{1/2}\,(T_w - T_\infty),$$

$$Q = 0.664k\mathrm{P}^{1/3}\mathrm{R}^{1/2}\,(T_w - T_\infty).$$

Then, introducing these into the dimensionless local and total heat fluxes defined in terms of the local and averaged Nusselt numbers,

$$q_w(x) = \left(\frac{k}{x}\right)\mathrm{N}_x(T_w - T_\infty) \qquad \text{and} \qquad Q = k\mathrm{N}(T_w - T_\infty),$$

we have, for $0.6 < \mathrm{P} < 10$,

$$\mathrm{N}_x = 0.332\mathrm{P}^{1/3}\mathrm{R}_x^{1/2} \qquad (5.3.34)$$

and

$$\mathrm{N} = 0.664\mathrm{P}^{1/3}\mathrm{R}^{1/2}, \qquad (5.3.35)$$

which for $\mathrm{P} = 1$ is identical with the result derived from the Reynolds analogy (see Section 4.3). The case of turbulent flow (Chapter 11) can be approximated by

$$\mathrm{N}_x = 0.0296\mathrm{P}^{1/3}\mathrm{R}_x^{0.8} \qquad (5.3.36)$$

and

$$N = 0.037 P^{1/3} R^{0.8}, \tag{5.3.37}$$

which are quoted here only for the sake of comparison.

Thermometer problem. Including the viscous dissipation, assuming insulated plate walls, and introducing $T(\eta) - T_\infty = (U_\infty^2/2c_p)\theta_2(\eta, \mathrm{P})$, we have, from Eq. (5.3.24),

$$\theta_2'' + \tfrac{1}{2}\mathrm{P} f \theta_2' = -2\mathrm{P} f''; \qquad \theta_2'(0, \mathrm{P}) = 0, \qquad \theta_2(\infty, \mathrm{P}) = 0. \tag{5.3.38}$$

After rearranging as

$$\frac{d\theta_2'}{d\eta} + \tfrac{1}{2}\mathrm{P} f \theta_2' = 0,$$

the integration of the homogeneous part of the differential equation of Eq. (5.3.38) yields

$$\theta_2' = C_1 e^{-(1/2)\mathrm{P}\int_0^\eta f\, d\eta},$$

or, in view of Eq. (5.3.28),

$$\theta_2' = C_1[f''(\eta)]^{\mathrm{P}}. \tag{5.3.39}$$

Now, for the solution of the nonhomogeneous differential equation of Eq. (5.3.38), assume that $C_1 = C_1(\eta)$ and employ the method of the variation of parameters. This yields

$$C_1 = -2\mathrm{P} \int_0^\eta [f''(\eta)]^{2-\mathrm{P}}\, d\eta + C_2,$$

where $C_2 = 0$ because of the boundary condition $\theta_2'(0) = 0$. Then, Eq. (5.3.39) may be rearranged as

$$\theta_2' = -2\mathrm{P}[f''(\eta)]^{\mathrm{P}} \int_0^\eta [f''(\eta)]^{2-\mathrm{P}}\, d\eta,$$

whose integration over the interval (η, ∞) results in

$$\theta_2 \big|_\eta^\infty = -2\mathrm{P} \int_\eta^\infty [f''(\eta)]^{\mathrm{P}} \left(\int_0^\eta [f''(\eta)]^{2-\mathrm{P}}\, d\eta \right) d\eta + C_3,$$

where $C_3 = 0$ because of the boundary condition $\theta_2(\infty) = 0$. Consequently, we have

$$\theta_2(\eta, \mathrm{P}) = 2\mathrm{P} \int_\eta^\infty [f''(\eta)]^{\mathrm{P}} \left(\int_0^\eta [f''(\eta)]^{2-\mathrm{P}}\, d\eta \right) d\eta, \tag{5.3.40}$$

which, for $\mathrm{P} = 1$, reduces to

$$\theta_2(\eta, 1) = 2 \int_\eta^\infty f' f''\, d\eta = 1 - f'^2(\eta). \tag{5.3.41}$$

The temperature assumed by the adiabatic wall because of the viscous dissipation, the *adiabatic wall temperature*, is

$$T_a - T_\infty = (T_2)_w - T_\infty = \left(\frac{U_\infty^2}{2c_p}\right)\theta_2(0, \mathrm{P}).$$

Figure 5.3 shows this temperature plotted against P. Note from Eq. (5.3.41) that $\theta_2(0, 1) = 1$. Consequently, for a gas with $\mathrm{P} = 1$ and flowing with velocity U_∞ past

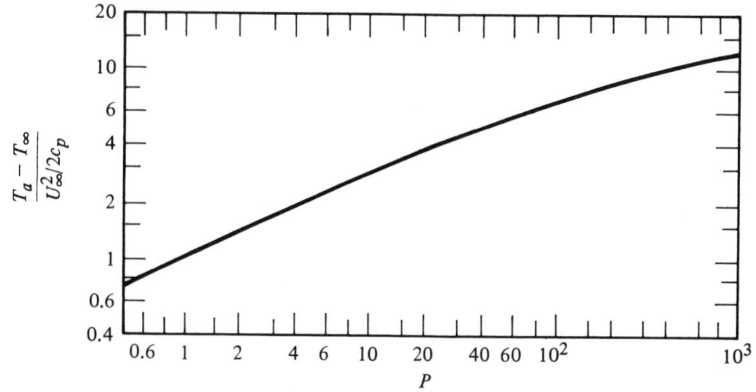

Fig. 5.3 Adiabatic wall temperature for the thermometer problem (Eckert and Drewitz 1940), Ex. 5.3.4 (adapted from Schlichting 1968; used by permission)

an insulated flat plate the temperature rise due to viscous dissipation is equal to the *stagnation temperature.* The temperature distribution made dimensionless by the adiabatic wall temperature,

$$\frac{T_2(\eta, \mathrm{P}) - T_\infty}{T_a - T_\infty} = \frac{\theta_2(\eta, \mathrm{P})}{\theta_2(0, \mathrm{P})},$$

is shown plotted in Fig. 5.4 for various Prandtl numbers.

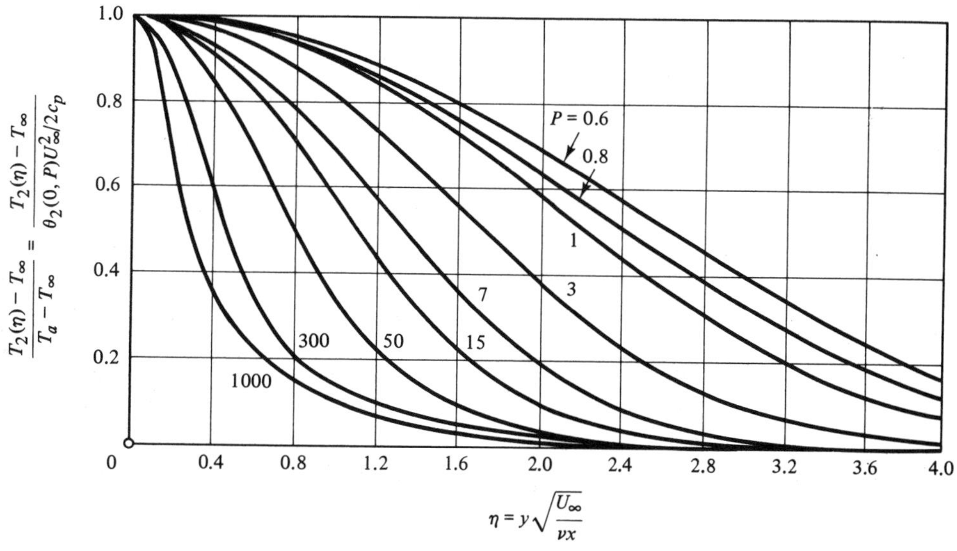

Fig. 5.4 Thermal boundary layer of the thermometer problem, Ex. 5.3.4 (adapted from Schlichting 1968; used by permission)

Since, for the same velocity distribution, two temperatures to be obtained from Eq. (5.3.24) are superposable, the cooling and thermometer problems can be combined to demonstrate the simultaneous effect of specified wall temperature and viscous dissipation. Thus, for a prescribed difference between the wall and free-stream temperatures, $T_w - T_\infty$, we have

$$T(\eta, \mathrm{P}) - T_\infty = [(T_w - T_\infty) - (T_a - T_\infty)]\theta_1(\eta, \mathrm{P}) + \left(\frac{U_\infty^2}{2c_p}\right)\theta_2(\eta, \mathrm{P}), \tag{5.3.42}$$

or, in dimensionless form,

$$\frac{T(\eta, \mathrm{P}) - T_\infty}{T_w - T_\infty} = \left[1 - \frac{1}{2}\mathrm{E}\theta_2(0, \mathrm{P})\right]\theta_1(\eta, \mathrm{P}) + \frac{1}{2}\mathrm{E}\theta_2(\eta, \mathrm{P}), \tag{5.3.43}$$

where $\mathrm{E} = U_\infty^2/c_p(T_w - T_\infty)$ is the Eckert number. Clearly, for $T_w > T_\infty$ but $\mathrm{E}\theta_2(0, \mathrm{P}) > 2$ the boundary layer near the wall is warmer than the wall itself because of the viscous dissipation. In this case, even though the wall is warmer than the free stream, it is being heated (rather than being cooled) by the cold fluid. In Fig. 5.5, Eq. (5.3.43) is shown plotted against η for $\mathrm{E}\theta_2(0, \mathrm{P}) \lesseqgtr 2$ and $\mathrm{P} = 0.7$(air). (Sketch and explain the temperature distributions prescribed by Eq. 5.3.43).

EXAMPLE 5.3.5 Reconsider Ex. 4.3.1, or the cooling problem of Ex. 5.3.4. We wish to know the behavior of thermal boundary layers for vanishingly small and increasingly large Prandtl numbers.

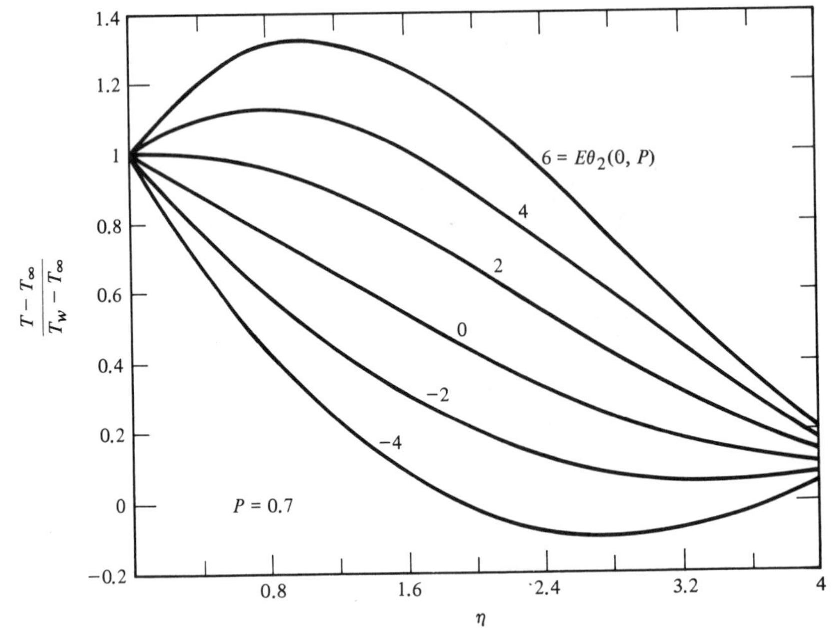

Fig. 5.5 Temperature distribution in a laminar boundary layer on a heated ($E > 0$) and cooled ($E < 0$) flat plate at zero incidence, Ex. 5.3.4 (adapted from Schlichting 1968; used by permission)

In Section 3.5 we learned that as $\mathrm{P} \longrightarrow 0$ for liquid metals, the momentum boundary layer may be ignored when the thermal boundary layer and the Nusselt modulus are being evaluated; in fact, the longitudinal velocity $u(x, y)$ may be approximated by the inviscid stream velocity $U(x)$. Then, the conservation of mass determines the form of the transversal velocity, $v(x, y) = -(dU/dx)y$, and the thermal energy of Eq. (5.3.23) becomes

$$U(x)\frac{\partial T}{\partial x} - y\left(\frac{dU}{dx}\right)\frac{\partial T}{\partial y} = a\frac{\partial^2 T}{\partial y^2}, \qquad \mathrm{P} \longrightarrow 0. \tag{5.3.44}$$

We also learned in the same section that as $\mathrm{P} \longrightarrow \infty$ for viscous oils, the momentum boundary layer may be assumed fully developed (or much thicker than the thermal boundary layer); following Levêque (1928), the longitudinal velocity in the thermal boundary layer may then be approximated by its tangent, $u(x, y) = (\partial u/\partial y)_w y$. Consequently, the conservation of mass gives the transversal velocity,

$$v(x, y) = -\tfrac{1}{2}y^2 \frac{d[(\partial u/\partial y)_w]}{dx},$$

and the thermal energy of Eq. (5.3.44) may be rearranged for a Newtonian fluid to give

$$y\left(\frac{\tau_w}{\mu}\right)\frac{\partial T}{\partial x} - \tfrac{1}{2}y^2 \frac{d}{dx}\left(\frac{\tau_w}{\mu}\right)\frac{\partial T}{\partial y} = a\frac{\partial^2 T}{\partial y^2}, \qquad \mathrm{P} \longrightarrow \infty, \tag{5.3.45}$$

where $\tau_w/\mu = (\partial u/\partial y)_w$, τ_w denoting the wall shear stress. It should be noted that the similarity variable for Ex. 5.3.4 is dictated by the momentum equation, and the energy equation turns out to be compatible with this variable. On the other hand, since velocities of Eqs. (5.3.44) and (5.3.45) are already specified, the similarity variable, if it exists, must be determined by these (energy) equations.

Case of P* $\longrightarrow$ *0. First, we consider Eq. (5.3.44). Let x, y, and T be made dimensionless in terms of arbitrarily selected reference lengths x_0, y_0, and the inherent characteristic temperature T_w. Then, Eq. (5.3.44) may be rearranged to yield

$$\frac{U(x)y_0^2}{ax_0}\frac{\partial \theta}{\partial(x/x_0)} - \frac{y_0 y}{a}\left(\frac{dU}{dx}\right)\frac{\partial \theta}{\partial(y/y_0)} = \frac{\partial^2 \theta}{\partial(y/y_0)^2}, \tag{5.3.46}$$

where $\theta = (T - T_\infty)/(T_w - T_\infty)$. From Eq. (5.3.46) we have, implicitly,

$$\theta = f\left[\frac{x}{x_0}, \frac{y}{y_0}, U(x)\frac{y_0^2}{ax_0}, \frac{y_0 y}{a}\left(\frac{dU}{dx}\right)\right],$$

which, after the elimination of x_0 and y_0 in the usual manner, leads to

$$\theta = f_1\left[U(x)\frac{y^2}{ax}, \left(\frac{dU}{dx}\right)\frac{y^2}{a}\right]. \tag{5.3.47}$$

Dimensional analysis stops here, being unable to provide a similarity variable; yet, it may lead to this variable through a simple reasoning: inspection of Eq. (5.3.47) revels that θ depends explicitly on y^2/a and functionally (implicitly) on $U(x)$. Then, if it exists, the similarity variable must have the form $(y^2/a)G[U(x)]$, where G denotes an arbitrary function of $U(x)$ to be determined by considering the energy equation, Eq. (5.3.44), as a compatibility condition. Interestingly, for the present type of problems, the mathematical steps remaining for dimensional analysis are identical to those to be encountered with separation of variables. Thus, conveniently assuming that

$$\eta = \frac{y}{g(x)}, \tag{5.2.8}$$

and noting that

$$\frac{\partial \theta}{\partial x} = -\left(\frac{y}{g^2}\right)\left(\frac{dg}{dx}\right)\frac{d\theta}{d\eta},$$

$$\frac{\partial^2 \theta}{\partial y^2} = \left(\frac{1}{g^2}\right)\frac{d^2\theta}{d\eta^2},$$

Eq. (5.3.44) may be rearranged to give

$$-\left(\frac{U}{g}\frac{dg}{dx} + \frac{dU}{dx}\right)\eta\frac{d\theta}{d\eta} = \frac{a}{g^2}\frac{d^2\theta}{d\eta^2}. \tag{5.3.48}$$

Equating to $a/2g^2$ the terms of Eq. (5.3.48) in parentheses, we have the ordinary differential equation previously obtained for Ex. 5.2.1,

$$\theta'' + \tfrac{1}{2}\eta\theta' = 0 \tag{5.1.5}$$

and

$$U\frac{dg^2}{dx} + 2\left(\frac{dU}{dx}\right)g^2 = a,$$

which yields, after setting the integration constant to zero for the compatibility of two boundary conditions,

$$g^2 = U^{-2}\left(a\int_0^x U\,dx\right),$$

or, in terms of Eq. (5.2.8),

$$\eta = \frac{yU}{\left(a\int_0^x U\,dx\right)^{1/2}}. \tag{5.3.49}$$

Finally, the wall heat flux*

$$q_w = -k\left(\frac{\partial T}{\partial y}\right)_w = -k(T_w - T_\infty)U\left(a\int_0^x U\,dx\right)^{-1/2}\theta'(0),$$

combined with Eq. (5.1.6) and the definition of the coefficient of heat transfer, gives the local Nusselt number,

$$\mathrm{N_x} = \frac{xU}{\pi^{1/2}\left(a\int_0^x U\,dx\right)^{1/2}}, \qquad \mathrm{P} \longrightarrow 0. \tag{5.3.50}$$

For the flat plate $U(x) = U_\infty$, and Eq. (5.3.50) reduces to

$$\mathrm{N_x} = \pi^{-1/2}(\mathrm{PR_x})^{1/2}, \qquad \mathrm{P} \longrightarrow 0, \tag{5.3.51}$$

where $\mathrm{R_x} = U_\infty x/\nu$ is the local Reynolds number.

---

*Note that θ is defined differently in Eqs. (5.1.1) and (5.3.46).

Case of $P \longrightarrow \infty$. Next, we consider Eq. (5.3.45). Following the same procedure, leaving the details to Problem 5-17, employing $\eta = y/g(x)$, and satisfying

$$\left(\frac{\tau_w}{\mu}\right)\frac{dg^3}{dx} + \frac{3}{2}g^3\frac{d}{dx}\left(\frac{\tau_w}{\mu}\right) = a, \tag{5.3.52}$$

Eq. (5.3.45) may be reduced to the ordinary differential equation obtained previously for Ex. 5.3.2,

$$3\theta'' + \eta^2\theta' = 0. \tag{5.3.8}$$

When inserted into $\eta = y/g(x)$, the solution of Eq. (5.3.52) compatible with boundary conditions gives

$$\eta = \frac{y(\tau_w/\mu)^{1/2}}{\left[\frac{1}{3}a\int_0^x(\tau_w/\mu)^{1/2}\,dx\right]^{1/3}}. \tag{5.3.53}$$

Finally, in terms of Eqs. (5.3.9) and (5.3.53), the local Nusselt number is found to be

$$\mathrm{N_x} = \frac{0.5384x(\tau_w/\mu)^{1/2}}{\left[\frac{1}{3}a\int_0^x(\tau_w/\mu)^{1/2}\,dx\right]^{1/3}}, \qquad \mathrm{P} \longrightarrow \infty. \tag{5.3.54}$$

For the flat plate $U(x) = U_\infty$, and, according to

$$\frac{\tau_w}{\mu} = 0.332U_\infty\left(\frac{U_\infty}{\nu x}\right)^{1/2}, \tag{5.3.20}$$

Eq. (5.3.54) reduces to

$$\mathrm{N_x} = 0.339\mathrm{P}^{1/3}\mathrm{R_x^{1/2}}, \qquad \mathrm{P} \longrightarrow \infty. \tag{5.3.55}$$

It is of some interest to note that, although obtained for $\mathrm{P} \longrightarrow \infty$, Eq. (5.3.55) remains to be a very good approximation even for values of P near unity (recall Fig. 4.11.) In recent years this fact has prompted a concentrated effort to generalize this problem, including pressure correction on the approximation of $u(x, y)$ and arbitrary wall temperature (for a brief account of the subject, see Chap. 6 in Curle 1962).

EXAMPLE 5.3.6 Reconsider Ex. 4.3.3. We wish to find a similarity solution for the natural convection from an isothermal vertical plate.

The formulation of the problem, after neglecting viscous dissipation, is

$$\begin{aligned}
&\frac{\partial u}{\partial x} + \frac{\partial v}{\partial y} = 0,\\
&u\frac{\partial u}{\partial x} + v\frac{\partial u}{\partial y} = \nu\frac{\partial^2 u}{\partial y^2} + g\beta\theta,\\
&u\frac{\partial \theta}{\partial x} + v\frac{\partial \theta}{\partial y} = a\frac{\partial^2 \theta}{\partial y^2},\\
&u(0, y) = 0, \quad u(x, 0) = 0, \quad v(x, 0) = 0, \quad u(x, \infty) = 0,\\
&\theta(0, y) = 0, \quad \theta(x, 0) = \theta_w, \quad \theta(x, \infty) = 0,
\end{aligned} \tag{5.3.56}$$

where $\theta = T - T_\infty$, $\theta_w = T_w - T_\infty$ and $\beta = 1/T_\infty$ (for gases). Placing x, y, u, v, and θ dimensionless in terms of arbitrarily selected lengths x_0, y_0, arbitrarily selected velocities U_0, V_0, and the characteristic temperature θ_w, respectively, Eq. (5.3.56) may be rearranged as

$$\frac{\partial(u/U_0)}{\partial(x/x_0)} + \left(\frac{x_0 V_0}{y_0 U_0}\right)\frac{\partial(v/V_0)}{\partial(y/y_0)} = 0,$$

$$\left(\frac{u}{U_0}\right)\frac{\partial(u/U_0)}{\partial(x/x_0)} + \left(\frac{x_0 V_0}{y_0 U_0}\right)\left(\frac{v}{V_0}\right)\frac{\partial(u/U_0)}{\partial(y/y_0)} = \left(\frac{x_0 \nu}{y_0^2 U_0}\right)\frac{\partial^2(u/U_0)}{\partial(y/y_0)^2} + \left(\frac{x_0 g\beta\theta_w}{U_0^2}\right)\left(\frac{\theta}{\theta_w}\right),$$

$$\left(\frac{u}{U_0}\right)\frac{\partial(\theta/\theta_w)}{\partial(x/x_0)} + \left(\frac{x_0 V_0}{y_0 U_0}\right)\left(\frac{v}{V_0}\right)\frac{\partial(\theta/\theta_w)}{\partial(y/y_0)} = \left(\frac{x_0 a}{y_0^2 U_0}\right)\frac{\partial^2(\theta/\theta_w)}{\partial(y/y_0)^2},$$

which imply that

$$\frac{u}{U_0} = f_1\left(\frac{x}{x_0}, \frac{y}{y_0}, \frac{x_0\nu}{y_0^2 U_0}, \frac{x_0 a}{y_0^2 U_0}, \frac{x_0 g\beta\theta_w}{U_0^2}\right),$$

$$\frac{v}{V_0} = g_1\left(\frac{x}{x_0}, \frac{y}{y_0}, \frac{x_0 V_0}{y_0 U_0}, \frac{x_0\nu}{y_0^2 U_0}, \frac{x_0 a}{y_0^2 U_0}, \frac{x_0 g\beta\theta_w}{U_0^2}\right),$$

$$\frac{\theta}{\theta_w} = \Theta_1\left(\frac{x}{x_0}, \frac{y}{y_0}, \frac{x_0\nu}{y_0^2 U_0}, \frac{x_0 a}{y_0^2 U_0}, \frac{x_0 g\beta\theta_w}{U_0^2}\right).$$

First, note that the parameters $x_0\nu/y_0^2 U_0$ and $x_0 a/y_0^2 U_0$ may be replaced with a pair including one of them, say $x_0\nu/y_0^2 U_0$, and their ratio, the Prandtl number P. Also, eliminating x_0 and y_0 in the usual manner, we have

$$\frac{u}{U_0} = f_2\left(\frac{x\nu}{y^2 U_0}, \frac{x g\beta\theta_w}{U_0^2}, \mathrm{P}\right),$$

$$\frac{v}{V_0} = g_2\left(\frac{x V_0}{y U_0}, \frac{x\nu}{y^2 U_0}, \frac{x g\beta\theta_w}{U_0^2}, \mathrm{P}\right),$$

$$\frac{\theta}{\theta_w} = \Theta_2\left(\frac{x\nu}{y^2 U_0}, \frac{x g\beta\theta_w}{U_0^2}, \mathrm{P}\right).$$

Next, eliminating U_0 from u and θ/θ_w, and, V_0 and U_0 from v, we get

$$u = \frac{\nu x}{y^2} f_3\left(\frac{g\beta\theta_w y^4}{x\nu^2}, \mathrm{P}\right),$$

$$v = \frac{\nu}{y} g_3\left(\frac{g\beta\theta_w y^4}{x\nu^2}, \mathrm{P}\right),$$

$$\frac{\theta}{\theta_w} = \Theta_3\left(\frac{g\beta\theta_w y^4}{x\nu^2}, \mathrm{P}\right),$$

or, introducing $f_4 = f_3/\eta^2$ and $g_4 = g_3/\eta$,

$$u = \nu c^2 x^{1/2} f_4(\eta, \mathrm{P}),$$

$$v = \nu c x^{-1/4} g_4(\eta, \mathrm{P}),$$

$$\frac{\theta}{\theta_w} = \Theta(\eta, \mathrm{P}),$$

where

$$\eta = \frac{cy}{x^{1/4}}, \qquad c = \left(\frac{g\beta\theta_w}{4\nu^2}\right)^{1/4}, \tag{5.3.57}$$

is the similarity variable. Furthermore, in view of the conservation of mass, f_4 and g_4 are related according to

$$\frac{dg_4}{d\eta} = \frac{1}{4}\eta\frac{df_4}{d\eta} - \frac{1}{2}f_4,$$

whose integration readily gives, within a constant,

$$g_4 = \frac{1}{4}\left(\eta f_4 - 3\int f_4\, d\eta\right).$$

Finally, introducing the convenient dependent variable $\zeta = \frac{1}{4}\int f_4\, d\eta$, we find that

$$u = 4\nu c^2 x^{1/2}\zeta', \qquad v = \nu c x^{-1/4}(\eta\zeta' - 3\zeta). \tag{5.3.58}$$

Now in terms of Eqs. (5.3.57) and (5.3.58), and $\theta/\theta_w = \Theta$, Eq. (5.3.56) may be transformed to

$$\begin{gathered} \zeta''' + 3\zeta\zeta'' - 2\zeta'^2 + \Theta = 0, \qquad \Theta'' + 3\mathrm{P}\zeta\Theta' = 0; \\ \zeta(0) = 0, \quad \zeta'(0) = 0, \quad \zeta'(\infty) = 0, \quad \Theta(0) = 1, \quad \Theta(\infty) = 0. \end{gathered} \tag{5.3.59}$$

Again, the split of Eq. (5.3.59) to a set of coupled first-order differential equations permits ready numerical integration. Results illustrated in Fig. 5.6 show the temperature and velocity profiles against η for various values of P. Interestingly, for $\mathrm{P} > 1$, the velocity boundary layer is thicker than its originating thermal boundary layer. (Why? Note that the apparently similar situation in forced convection has a different explanation.)

Heat transfer. The local heat flux

$$q_w(x) = -k\left(\frac{\partial T}{\partial y}\right)_{y=0} = kcx^{-1/4}\left(-\frac{d\theta}{d\eta}\right)_{\eta=0}(T_w - T_\infty)$$

integrated over the length l of the plate gives the total heat flux per unit width

$$Q = \int_0^l q_w(x)\, dx = \frac{4}{3}kcl^{3/4}\left(-\frac{d\theta}{d\eta}\right)_{\eta=0}(T_w - T_\infty).$$

Introducing these fluxes into the local and averaged Nusselt numbers,

$$q_w(x) = \frac{k}{x}\mathrm{N}_x(T_w - T_\infty) \qquad \text{and} \qquad Q = k\mathrm{N}(T_w - T_\infty),$$

we have

$$\mathrm{N}_x = cx^{3/4}\left(-\frac{d\theta}{d\eta}\right)_{\eta=0}, \tag{5.3.60}$$

$$\mathrm{N} = \frac{4}{3}cl^{3/4}\left(-\frac{d\theta}{d\eta}\right)_{\eta=0}. \tag{5.3.61}$$

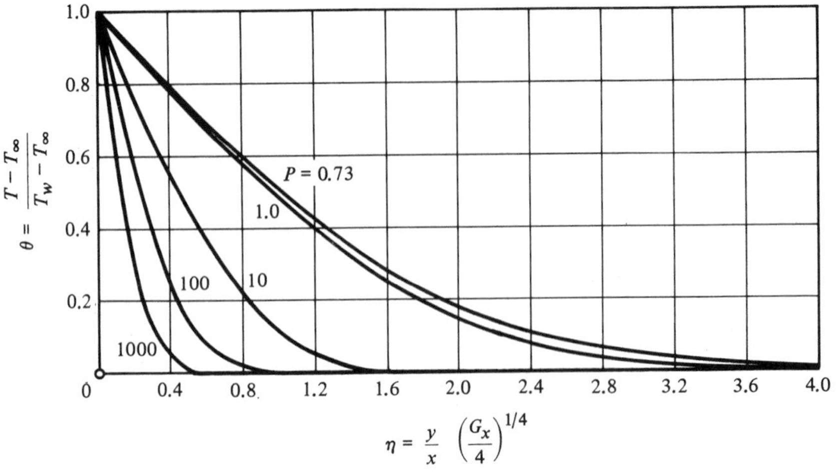

(a) Temperature profiles

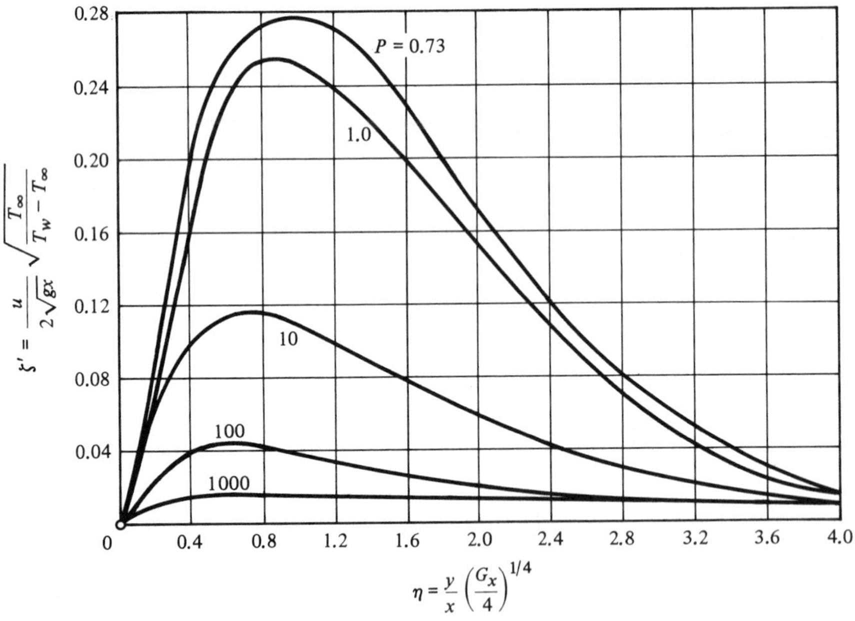

(b) Velocity profiles

Fig. 5.6 Temperature and velocity boundary layers of natural convection near a hot vertical plate (Pohlhausen 1921, Ostrach 1953), Ex. 5.3.6 (adapted from Schlichting 1968; used by permission)

In particular, $(-d\theta/d\eta)_{\eta=0} = 0.508$ for P = 0.733, and Eqs. (5.3.60) and (5.3.61) may be rearranged as

$$\mathrm{N_x} = 0.359\mathrm{G_x^{1/4}}, \tag{5.3.62}$$

$$\mathrm{N} = 0.478\mathrm{G}^{1/4}, \tag{5.3.63}$$

where $\mathrm{G_x} = g\beta(T_w - T_\infty)x^3/\nu^2$ and $\mathrm{G} = \mathrm{G}_{x=l}$ are the local and averaged Grashof numbers, respectively (for comparison of the analytical and experimental values of N, and for measurement techniques, see references cited in Chap. XII of Schlichting 1968).

We terminate our study on the similarity solution of boundary layers by considering a two-domain problem.

EXAMPLE 5.3.7 Consider steady forced convection film boiling of subcooled or saturated liquid over an isothermal flat plate (Cess & Sparrow 1961; also recall Ex. 4.5.1 for an integral approach). Both phases are Fourier–Newton fluids whose properties are assumed constant. We wish to find the local skin friction and heat transfer along the plate.

Neglecting viscous dissipation, we have for the boundary layer of each phase (shown in Fig. 5.7 and prescribed by the governing equations of Eq. 5.3.23),

$$\frac{\partial u}{\partial x} + \frac{\partial v}{\partial y} = 0, \tag{5.3.64}$$

$$u\frac{\partial u}{\partial x} + v\frac{\partial u}{\partial y} = \nu\frac{\partial^2 u}{\partial y^2}, \tag{5.3.65}$$

$$u\frac{\partial \theta}{\partial x} + v\frac{\partial \theta}{\partial y} = a\frac{\partial^2 \theta}{\partial y^2}, \tag{5.3.66}$$

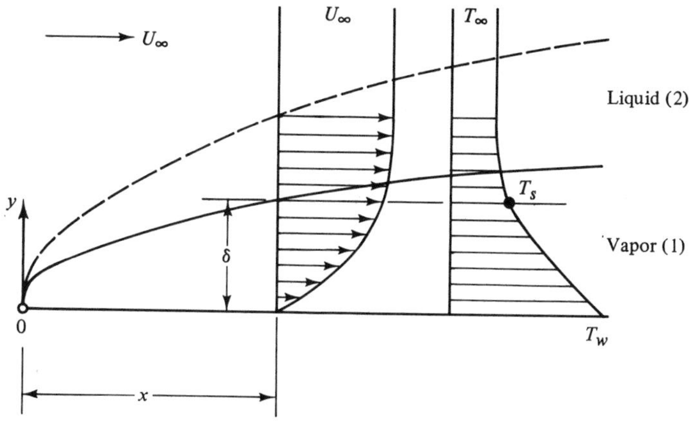

Fig. 5.7 Forced convection film boiling; double boundary layer, Ex. 5.3.7

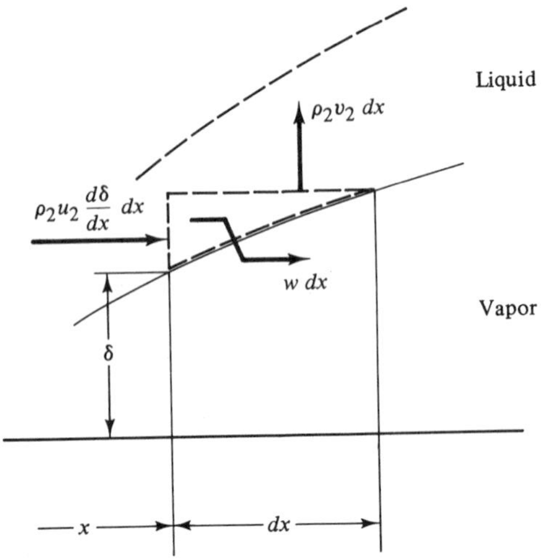

Fig. 5.8 Mass balance along vapor-liquid interface, Ex. 5.3.7

where $\theta = T - T_w$ and subscripts 1 and 2 for vapor and liquid have been left out. Referring to Eqs. (4.4.1) through (4.4.10) for interface conditions and to Fig. 5.8 for interface boundary conditions, we have

$$
\begin{aligned}
x = 0: &\quad u_1 = u_2 = U_\infty; \quad \theta_1 = \theta_2 = \theta_\infty \\
y = 0: &\quad u_1 = v_1 = 0; \quad \theta_1 = 0 \\
y = \delta: &\quad u_1 = u_2; \quad \theta_1 = \theta_2 = \theta_s \\
&w = \rho_1\left(u_1 \frac{d\delta}{dx} - v_1\right) = \rho_2\left(u_2 \frac{d\delta}{dx} - v_2\right) \\
&\mu_1 \frac{\partial u_1}{\partial y} = \mu_2 \frac{\partial u_2}{\partial y} \\
&wh_{21} = -k_1 \frac{\partial \theta_1}{\partial y} + k_2 \frac{\partial \theta_2}{\partial y} \\
y \longrightarrow \infty: &\quad u_2 = U_\infty; \quad \theta_2 = \theta_\infty,
\end{aligned}
\tag{5.3.67}
$$

where w denotes the evaporating mass flow.

Inspection of the foregoing formulation suggests the possibility of a similarity transformation. Let

$$
\begin{aligned}
u_1 &= U_\infty f_1'(\eta); & u_2 &= U_\infty f_2'(\xi) \\
\theta_1 &= \theta_\infty g_1(\eta); & \theta_2 &= \theta_\infty g_2(\xi) \\
\eta &= y\left(\frac{U_\infty}{\nu_1 x}\right)^{1/2}; & \xi &= (y - \delta)\left(\frac{U_\infty}{\nu_2 x}\right)^{1/2}.
\end{aligned}
\tag{5.3.68}
$$

Then Eq. (5.3.64) implies that

$$v_1 = \frac{1}{2}\left(\frac{U_\infty \nu_1}{x}\right)^{1/2}(\eta f'_1 - f_1)$$
$$v_2 = \frac{1}{2}\left(\frac{U_\infty \nu_2}{x}\right)^{1/2}(\xi f'_2 - f_2), \tag{5.3.69}$$

and Eqs. (5.3.65) and (5.3.66) transform into

$$2f'''_1 + f_1 f''_1 = 0; \qquad 2g''_1 + \mathrm{P}_1 f_1 g'_1 = 0 \tag{5.3.70}$$
$$2f'''_2 + f_2 f''_2 = 0; \qquad 2g''_2 + \mathrm{P}_2 f_2 g'_2 = 0, \tag{5.3.71}$$

where $\mathrm{P}_1 = \nu_1/a_1$ and $\mathrm{P}_2 = \nu_2/a_2$ are Prandtl numbers. The boundary conditions stated by Eq. (5.3.67) now become

$$f_1(0) = 0; \qquad f'_1(0) = 0 \tag{5.3.72}$$
$$f'_1(\eta_\delta) = f'_2(0); \qquad f'_2(\infty) = 1 \tag{5.3.73}$$
$$g_1(0) = 0; \qquad g_1(\eta_\delta) = \frac{\theta_s}{\theta_\infty} \tag{5.3.74}$$
$$g_2(0) = \frac{\theta_s}{\theta_\infty}; \qquad g_2(\infty) = 1 \tag{5.3.75}$$
$$f_1(\eta_\delta) = \left(\frac{\rho_2\mu_2}{\rho_1\mu_1}\right)^{1/2} f_2(0); \qquad f''_1(\eta_\delta) = \left(\frac{\rho_2\mu_2}{\rho_1\mu_1}\right)^{1/2} f''_2(0) \tag{5.3.76}$$
$$\frac{\mathrm{P}_1}{2\mathrm{J}_1} f_1(\eta_\delta) = g'_1(\eta_\delta) - \frac{k_2}{k_1}\left(\frac{\nu_1}{\nu_2}\right)^{1/2} g'_2(0), \tag{5.3.77}$$

where $\mathrm{J}_1 = c_1(-\theta_\infty)/h_{12}$ is the Jakob number characteristic of phase change. Note that $\theta = T - T_w$ was defined to be negative for this problem to show the analogy between similarity functions for velocity and temperature.

For a similarity solution to exist, η_δ must be constant, which gives the position of the boundary layer interface

$$\delta(x) = \eta_\delta\left(\frac{\nu_1 x}{U_\infty}\right)^{1/2}. \tag{5.3.78}$$

Employing the foreging 11 boundary conditions, we can determine the constant η_δ and can numerically integrate the two pairs of ordinary differential equations of order three and two, respectively. Then the local skin friction (evaluated from Eq. 5.3.21) is

$$c_{f,x} = \frac{\tau_{xy}(x, 0)}{\frac{1}{2}\rho U_\infty^2} = 2f''_1(0)\mathrm{R}_{1,x}^{-1/2}, \tag{5.3.79}$$

where $\mathrm{R}_{1,x} = U_\infty x/\nu_1$, and the local heat transfer (from Eq. 5.3.31)

$$\mathrm{N}_x = \frac{hx}{k_1} = g'_1(0)\mathrm{R}_{1,x}^{1/2}, \tag{5.3.80}$$

where h is defined by $q_y(x, 0) = h(-\theta_\infty)$.

Clearly, the problem depends on six parameters,

$$\mathrm{P}_1, \quad \mathrm{P}_2, \quad \mathrm{J}_1, \quad \left(\frac{\rho_2\mu_2}{\rho_1\mu_1}\right)^{1/2}, \quad \frac{k_2}{k_1}\left(\frac{\nu_1}{\nu_2}\right)^{1/2}, \quad \frac{\theta_s}{\theta_\infty},$$

and is rather involved for a parametric study. However, various simplifying approximations can be made. Consider, for example, the case of small Jakob number $\mathrm{J}_1 < 1$ and saturated liquid $\theta_s = \theta_\infty$. Following the discussion in Section 4.5 (see also Ex. 4.5.1), we can ignore sensible heat of the vapor compared with latent heat of evaporation whenever the Jakob number is small. This implies a quasi-developed formulation in the balance of thermal energy for the vapor. For a Prandtl number of order unity the same approximation is justified for the balance of vapor momentum. Hence, ignoring the left of Eqs. (5.3.65) and (5.3.66) for the vapor reduces Eq. (5.3.70) to

$$f_1''' = 0; \qquad g_1'' = 0 \tag{5.3.81}$$

which implies linear profiles for velocity and temperature,

$$f_1(\eta) = \tfrac{1}{2}f_1''(0)\eta^2 \qquad \text{or} \qquad f_1'(\eta) = f_1''(0)\eta, \tag{5.3.82}$$

$$g_1(\eta) = \frac{\eta}{\eta_\delta}, \tag{5.3.83}$$

where $\theta_s = \theta_0$ has been used. Evaluating the second of Eq. (5.3.82) for $\eta = \eta_\delta$ gives the relation

$$f_1'(\eta_\delta) = f_1''(0)\eta_\delta. \tag{5.3.84}$$

Then Eq. (5.3.77) and the second of Eq. (5.3.76), noting also that $f_1''(0) = f_1''(\eta_\delta)$, yield

$$\frac{1}{4}f_1''(0)\eta_\delta^3 = \frac{J_1}{P_1}, \tag{5.3.85}$$

$$\eta_\delta = \left(\frac{\rho_1\mu_1}{\rho_2\mu_2}\right)^{1/2}\frac{f_2'(0)}{f_2''(0)}. \tag{5.3.86}$$

Furthermore, since the liquid is saturated, $g_2 = 1$, and from Eq. (5.3.71) we need only to integrate

$$2f_2''' + f_2 f_2'' = 0, \tag{5.3.87}$$

subject to Eqs. (5.3.73) and (5.3.76). Then, employing Eq. (5.3.84), we have

$$f_2(0) = \left(\frac{\rho_1\mu_1}{\rho_2\mu_2}\right)^{1/2} f_1(\eta_\delta) \sim 0,$$

$$f_2'(0) = f_1''(0)\eta_\delta,$$

$$f_2'(\infty) = 1.$$

The first of these boundary conditions may usually be put equal to zero because $\rho_1\mu_1/\rho_2\mu_2 \ll 1$ for vapor–liquid states not close to the thermodynamic critical state.

For specified values of the parameters, $(\rho_1\mu_1/\rho_2\mu_2)^{1/2}$ and $\mathrm{J}_1/\mathrm{P}_1$, the solution need be obtained iteratively. Thus, specifying a value of $f_2'(0)$ integration of Eq. (5.3.87) yields the corresponding value of $f_2''(0)$. Then Eq. (5.3.86) yields η_δ and Eq. (5.3.85) the value of $\mathrm{J}_1/\mathrm{P}_1$ for which the solution is valid. Figure 4.24 shows $f_2''(0) =$

$\tau_w/[\mu U_\infty(U_\infty/\nu x)^{1/2}]$ as a function of $f_2'(0) = u(\eta_\delta)/U_\infty$. A tabulation of these derivatives is given by Cess & Sparrow (1961).

Finally, the skin friction and heat transfer evaluated from Eqs. (5.3.79) and (5.3.80), may be written in terms of η_δ as

$$c_{f,x} = 8\frac{\mathrm{J}_1}{\mathrm{P}_1\eta_\delta^3}\mathrm{R}_{1,\mathrm{x}}^{-1/2}, \tag{5.3.88}$$

$$\mathrm{N}_\mathrm{x} = \frac{1}{\eta_\delta}\mathrm{R}_{1,\mathrm{x}}^{1/2}. \tag{5.3.89}$$

Among the interesting aspects of the problem is the large drag reduction (relative to a liquid flow) encountered in the film boiling.

5.4 *NONSIMILAR BOUNDARY LAYERS*

So far we have been concerned exclusively with similar boundary layers. Yet many flows are inherently nonsimilar, although some of them do, in fact, contain limited regions where the flow is *locally* similar; for example, in flows over blunt bodies, the conditions for similarity are usually satisfied near the forward stagnation point. Because of the simplifications afforded by similarity transformations, many attempts have been made to develop analogous methods which can be applied to nonsimilar boundary layers. In the literature, one of these, the Merk method, appears to have received particular attention in the past. For a concise presentation of the method, we refer to Chao & Fagbenle (1974). A related method suitable to numerical integration of laminar and turbulent boundary layers has been developed by Cebeci & Smith (1974) (see also Cebeci & Bradshaw 1977). We outline this method next.

In principle the method is based on use of a common similarity transformation, say $(x, y) \longrightarrow \eta$. In the absence of similarity the transformed equations are no longer ordinary differential equations in η, but also depend on x. Employing approximations involving known quantities, say upstream finite difference representations, there results ordinary differential equations at any x which are solved in the usual way. The method thus involves a stepwise finite difference integration of the boundary layer development in x.

Let us illustrate the approach by considering the incompressible boundary layer flow resulting from an arbitrary free-stream velocity $U(x)$.

In Ex. 5.3.3 we learned that

$$u = Uf'(\eta), \qquad \eta = \frac{y}{(\nu x/U)^{1/2}} \tag{5.4.1}$$

transforms the boundary layer formulation

$$\frac{\partial u}{\partial x} + \frac{\partial v}{\partial y} = 0$$

$$u\frac{\partial u}{\partial x} + v\frac{\partial u}{\partial y} = U\frac{dU}{dx} + \nu\frac{\partial^2 u}{\partial y^2}, \tag{5.4.2}$$

$$u(x, 0) = 0, \qquad u(x, \infty) = U, \qquad u(0, y) = U \quad \text{(leading edge)},$$

for $U =$ constant into the ordinary differential equation (5.3.19). Furthermore, it may be readily shown (see Problem 5-16) that Eq. (5.4.1) is also successful in achieving a similarity transformation when $U \sim x^m$, yielding the Falkner–Skan class of similarity solutions, governed by

$$f''' + \frac{m+1}{2} ff'' + m(1 - f'^2) = 0$$
$$f(0) = f'(0) = 0, \qquad f'(\infty) = 1, \tag{5.4.3}$$

where

$$m = \frac{x}{U}\frac{dU}{dx}. \tag{5.4.4}$$

The free-stream velocity $U \sim x^m$ is the potential flow solution for uniform flow onto a wedge of half-angle $\beta\pi/2$, where $\beta = 2m/(m+1)$. The wedge reduces to the flat plate for $m = 0$.

Consider now an arbitrary free-stream velocity $U = U(x)$, define $m(x)$ by Eq. (5.4.4), and introduce, in place of Eq. (5.4.1), the transformation

$$u = U(x)f'(\eta, x), \qquad \eta = \frac{y}{(\nu x/U)^{1/2}}. \tag{5.4.5}$$

Noting the differentiation rules

$$\left.\frac{\partial}{\partial x}\right|_y = \left.\frac{\partial}{\partial x}\right|_\eta + \left.\frac{\partial \eta}{\partial x}\frac{\partial}{\partial \eta}\right|_x, \qquad \left.\frac{\partial}{\partial y}\right|_x = \left(\frac{U}{\nu x}\right)^{1/2}\left.\frac{\partial}{\partial \eta}\right|_x,$$

eliminating v by the conservation of mass, yielding

$$v = \frac{1}{2}\left(\frac{\nu U}{x}\right)^{1/2} f' - \frac{\partial}{\partial x}[(U\nu x)f], \tag{5.4.6}$$

Eq. (5.4.2) transforms into

$$f''' + \frac{m+1}{2} ff'' + m(1 - f'^2) = x\left(f'\frac{\partial f'}{\partial x} - f''\frac{\partial f}{\partial x}\right), \tag{5.4.7}$$

subject to

$$f(0, x) = f'(0, x) = 0, \qquad f'(\infty, x) = 1,$$

where $f' = \partial f/\partial \eta$.

Now, if at a given location x we express $\partial f'/\partial x$ and $\partial f/\partial x$ on the right-hand side of Eq. (5.4.7) in terms of known upstream distributions and the local unknown values, Eq. (5.4.7) becomes an ordinary differential equation; hence we can speak of local similarity.

In problems of true similarity, solving one ordinary differential equation for $f(\eta)$, such as Eq. (5.4.3), say by an iterative Runge–Kutta integration over the interval $\eta = 0$ to η_∞, determines the velocity distribution for all values of x, y. Here, however, the solution of Eq. (5.4.7) requires integration of the equation for each value of x as x is incremented in a finite difference scheme, $x = x_1, x_2, \ldots, x_{i-1}, x_i, \ldots, x_n$, along the boundary layer. Given the distributions $f_{i-1}, f'_{i-1}, \ldots$ at x_{i-1}, derivatives with respect to x on the right-hand side of Eq. (5.4.7) are approximated by upstream differences

$$\left(\frac{\partial f}{\partial x}\right)_i \simeq \frac{f_i - f_{i-1}}{x_i - x_{i-1}}, \qquad \left(\frac{\partial f'}{\partial x}\right)_i \simeq \frac{f'_i - f'_{i-1}}{x_i - x_{i-1}}, \tag{5.4.8}$$

and the equation is integrated by a usual iterative Runge–Kutta scheme to yield local values of $f_i, f'_i, f''_i, \ldots$, hence local values of boundary layer thickness δ_1 and δ_2, as well as the coefficient of local skin friction

$$c_f(x_i) = 2f''_i(0, x_i)\mathrm{Re}_{x_i}^{-1/2}, \tag{5.4.9}$$

where $\mathrm{Re}_{x_i} = U(x_i)x_i/\nu$.

Integration in η is extended to a suitably large and constant value η_∞, say on the order of 10 (see Table 7.7). η_∞ is a constant for true similarity and is nearly constant for local similarity. Furthermore, the change in f and f' with x is small, so the foregoing approximations are acceptable even when relatively large steps in x are employed.

Once the velocity distribution has been found, the temperature distribution and the heat transfer may be calculated in a similar manner, now solving Eq. (5.3.24).

REFERENCES

ARPACI, V. S., CLARK, J. A., & LARSEN, P. S., 1965, Proc. R. Soc. A, **283**, 50.

BIRKHOFF, G., 1950, *Hydrodynamics, a Study in Logic, Fact and Similitude*, Princeton University Press, Princeton, N.J.

BLASIUS, H., 1908, Z. Math. Phys., **56**, 1 (see also NACA TM-1256).

CEBECI, T., & BRADSHAW, P., 1977, *Momentum Transfer in Boundary Layers*, Hemisphere/McGraw-Hill, New York.

CEBECI, T., & SMITH, A. M. O., 1974, *Analysis of Turbulent Boundary Layers*, Academic Press, New York.

CESS, R. D., and SPARROW, E. M., 1961, J. Heat Transfer (ASME), **83**, 370, 377.

CHAO, B. T., & FAGBENLE, R. O., 1974, Int. J. Heat Mass Transfer, **17**, 223.

CURLE, N., 1962, *The Laminar Boundary Layer Equations*, Oxford University Press, New York.

ECKERT, E. R. G., & DREWITZ, O., 1940, Forsch. Ing.-Wes., **11**, 116 (or NACA TM-1045, 1943).

FRIEDRICHS, K. O., 1955, Bull. Am. Math. Soc., **61**, 485.

FRÖBERG, C.-E., 1970, *Introduction to Numerical Analysis*, Addison-Wesley, Reading, Mass.

GOLDSTEIN, S., 1939, Proc. Camb. Philos. Soc., **35**, 338.

HANSEN, A. G., 1964, *Similarity Analyses of Boundary Value Problems in Engineering*, Prentice-Hall, Englewood Cliffs, N.J.

HELLUMS, J. D., & CHURCHILL, S. W., 1964, AIChE J., **10**, 110.

HOWARTH, L., 1938, Proc. R. Soc. A, **164**, 551.

KOH, J. C. Y., SPARROW, E. M., & HARTNETT, J. P., 1961, Int. J. Heat Transfer, **2**, 69–82.

LANGLOIS, W. E., 1963, J. Fluid Mech., **15**, 111.

LEVÊQUE, M. A., 1928, Ann. Mines, **13**, 201, 305, 381.

OSTRACH, S., 1953, NACA Rep. 1111.

POHLHAUSEN, E., 1921, Z. Angew. Math. Mech., **1**, 115.

SCHLICHTING, H., 1968, *Boundary Layer Theory*, 6th ed., McGraw-Hill, New York.

SCRIVEN, L. E., 1959, Chem. Eng. Sci., **10**, 1.

SEDOV, L. I., 1959, *Similarity and Dimensional Methods in Mechanics*, Academic Press, New York.

SPARROW, E. M., & GREGG, J. L., 1959, J. Heat Transfer (ASME), **81**, 13.

PROBLEMS

5-1 Consider a semi-infinite stagnant viscous fluid over a flat plate. Let the plate assume suddenly a constant velocity U. Find the unsteady motion of the fluid.

(a) Repeat the problem with a constant force F suddenly applied to the plate.

(b) Describe the physics of the analogous conduction problem leading to the same mathematical formulation.

5-2 Repeat Problem 5-1 for a non-Newtonian fluid governed by $\tau = K(\partial u/\partial y)^n$.

5-3 Repeat Ex. 5.3.2 for q_w = constant.

5-4 Repeat Ex. 5.3.2 for a large Prandtl fluid whose velocity is $(u/U) = (y/l)^n$, approximately.

5-5 Repeat Problem 5-4 for a non-Fourier fluid governed by $q_y = -K_1(\partial T/\partial y)^m$.

5-6 Consider a semi-infinite stagnant viscous fluid bounded by a two-dimensional corner (Fig. 5P-6). Let the corner assume suddenly a constant velocity. Find the unsteady two-dimensional motion of the fluid.

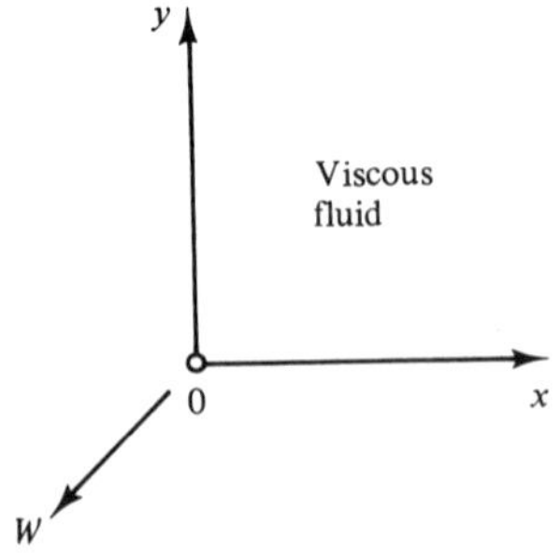

Fig. 5P-6

5-7 Consider the corner of a two-dimensional solid at a uniform initial temperature (Fig. 5P-7). The corner is subjected suddenly to uniform heat flux. Find the unsteady temperature of the fluid.

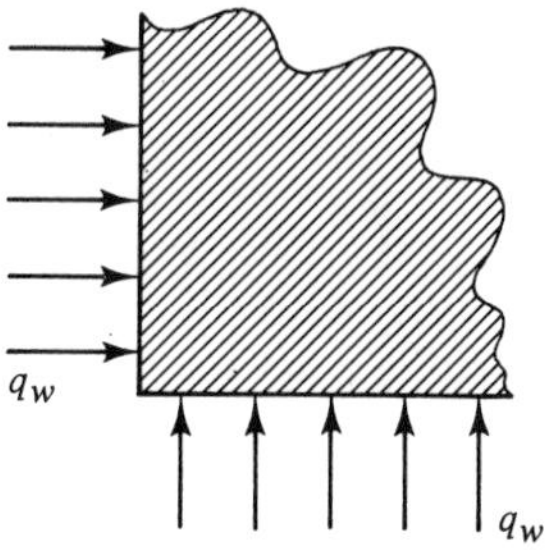

Fig. 5P-7

5-8 Resolve the "cooling problem" of Ex. 5.3.4 for q_w = constant.

5-9 Resolve Ex. 5.3.5 for q_w = constant.

5-10 Resolve Ex. 5.3.6 for q_w = constant.

5-11 Resolve Ex. 5.3.7 for q_w = constant.

5-12 Consider the momentum and thermal boundary layers of an incompressible fluid. Let the potential flow outside of the momentum layer be specified by $U = cx^m$.
(a) Solve the momentum boundary layer in terms of a similarity variable.
(b) Can the same variable be used for the thermal boundary layer with T_w = constant or q_w = constant?

5-13 Consider the steady filmwise condensation of saturated vapor on a vertical isothermal wall (the Nusselt problem).
(a) Ignoring the effects of induced flow of the vapor, determine velocity and temperature distributions of the liquid film, the film thickness, and Nusselt number as function of distance from the start of the plate. Obtain the solution neglecting the effect of liquid inertia (Sparrow & Gregg, 1959).
(b) Resolve the problem, including the effect of vapor flow (Koh, Sparrow & Hartnett, 1961).

5-14 Assume the growth of a spherical vapor bubble be governed by the rate of heat transfer from the surrounding liquid to the interface. Let the liquid be incompressible and uniformly superheated initially. Assume that the vapor is saturated and the initial bubble radius is zero. Neglect the effect of surface tension.

Determine the transient temperature distribution of the liquid and show that the transient bubble radius is governed by $R(t) = 2\lambda(at)^{1/2}$, where $\lambda = f(\rho_v/\rho_l, \text{Ja})$ and $\text{Ja} = c_p\,\Delta T/h_{lv}$ (Scriven 1959, Langlois 1963, Arpaci, Clark & Larsen 1965).

5-15 Resolve Ex. 5.3.1 by the method of separation of variables.

5-16 Discuss the possible forms of the free-stream velocity $U(x)$ that lead to a similarity solution for momentum boundary layers starting from a leading edge and from a stagnation point, respectively.

5-17 Show that Ex. 5.3.5 can be reduced to Eq. (5.3.8) for $\mathbf{P} \longrightarrow \infty$.

5-18 Determine the exact solution to the advancing plane solidification front of Ex. 4.4.1.

5-19 Consider the steady viscous flow between porous parallel plates spaced a distance $2l$ apart and subject to uniform blowing v_0 at both plates. Downstream, the flow may be considered to be developed in the sense that the transverse pressure gradient and momentum change are negligible and that the axial pressure gradient is constant. We seek this pressure gradient, which is the limiting exact solution to the vapor flow in a heat pipe (Ex. 4.5.3).

Chapter 6

PERIODIC CONVECTION

In this chapter we are concerned with the solution of an important class of unsteady convection problems. With respect to their dependence on time, unsteady problems may be classified as *transient* or *periodic*. Also, each transient or periodic problem may involve a *starting*, a *steady*, and an *ending* time interval, as shown in Fig. 6.1. Among these problems, we are interested only in periodic problems, more particularly, in the steady part of these. Clearly, the steady part of the solution of a periodic problem should be simpler than its transient (or complete) solution, which is often involved, if not impossible to obtain. Yet, in many practical applications, such as the flow and heat transfer associated with vibrating components, reciprocating engines, and so on, the steady part of periodic solutions is quite important.

6.1 ORIGIN OF METHOD—LINEAR PROBLEMS

Consider a convection problem involving a *periodic boundary disturbance* or a *periodic volume disturbance*. The most frequent boundary disturbance for convection problems is the fluctuating inlet or upstream velocity or temperature; the most frequent volume disturbance is a fluctuating pressure gradient or body force or a fluctuating internal energy generation.

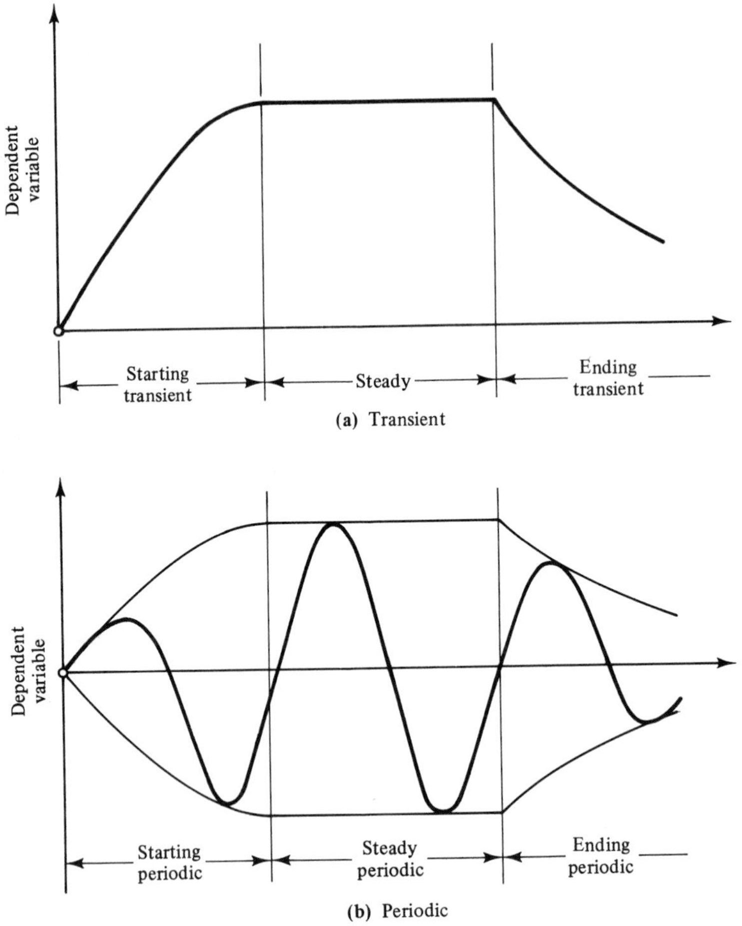

Fig. 6.1 Classification of unsteady problems

Rather than proceeding directly to the solution of an *original periodic problem* with angular frequency ω, say $A(\mathbf{r}, t)$, the method of *complex (dependent) variables** considers also a *complementary periodic problem*, say $A^*(\mathbf{r}, t)$, subject to a fluctuating disturbance of the same angular frequency but differing in phase from that of the original problem by an amount $\pi/2$. Then we have the complex problem in terms of the complex (dependent) variable

$$\boldsymbol{\Psi}(\mathbf{r}, t) = A(\mathbf{r}, t) + iA^*(\mathbf{r}, t), \tag{6.1.1}$$

where i is the usual imaginary unit, $i = (-1)^{1/2}$. The formulation (governing equation plus the initial and boundary conditions) to be satisfied by $\boldsymbol{\Psi}(\mathbf{r}, t)$ is readily obtained

*For application of the method to conduction problems, see Arpaci (1966).

after multiplying by i the problem (original or complementary) subject to the sin ωt disturbance and adding the result to the problem subject to the cos ωt disturbance.

Next, a product solution, $\boldsymbol{\Psi}(r, t) = \boldsymbol{\Phi}(r)\tau(t)$, is assumed. For large values of time, $\boldsymbol{\Psi}(r, t)$ approaches a steady periodic solution with frequency identical to that imposed on the original problem. Consequently,

$$\boldsymbol{\Psi}(r, t) = \boldsymbol{\Phi}(r) \exp(i\omega t). \tag{6.1.2}$$

Strictly speaking, Eq. (6.1.2) applies only to linear problems. Nonlinear problems that require the consideration of higher-order harmonics will be explained in the following section. Note that, being valid only for large values of time, Eq. (6.1.2) can no longer satisfy the initial condition associated with $\boldsymbol{\Psi}(r, t)$. The governing equation and boundary conditions to be satisfied by $\boldsymbol{\Phi}(r)$ are readily obtained by introducing Eq. (6.1.2) into the formulation of $\boldsymbol{\Psi}(r, t)$, and eliminating $\exp(i\omega t)$ from each term. Thus, the method of complex dependent variables reduces the original *unsteady* problem $A(r, t)$ to the *steady* problem $\boldsymbol{\Phi}(r)$. The solution of $\boldsymbol{\Phi}(r)$ may be obtained by conventional methods. Insertion of this solution into Eq. (6.1.2) gives $\boldsymbol{\Psi}(r, t)$. The real part of $\boldsymbol{\Psi}(r, t)$ is the solution of the problem subject to the cos ωt disturbance, and the imaginary part of $\boldsymbol{\Psi}(r, t)$ is the solution of the problem subject to the sin ωt disturbance.

We are ready now to illustrate the application of the method of complex (dependent) variables in terms of a simple thermal problem. Consider for this purpose a fluid flowing steadily with mean velocity V through an infinitely long pipe (Fig. 6.2). The upstream half of the pipe is insulated, the downstream half transfers heat with coefficient h to an ambient at temperature zero. The upstream temperature of the fluid oscillates as $\theta_i(1 + \epsilon \cos \omega t)$, ϵ being the relative amplitude of oscillations. Axial conduction is negligible. Cross-sectional area and periphery of the pipe are A and P, respectively. We wish to determine the radially lumped and axially distributed temperature fluctuations in the downstream flow.

The balance of thermal energy applied to the radially lumped and axially differ-

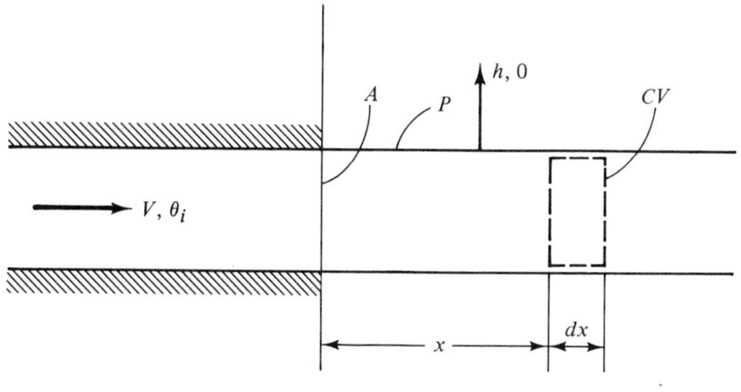

Fig. 6.2 Flow through a radially lumped and axially distributed pipe subject to inlet oscillations in temperature $\theta_i(l + \epsilon \cos \omega t)$ or in velocity $V(1 + \epsilon \cos \omega t)$.

ential control volume shown in Fig. 6.2 yields, after neglecting the effect of axial conduction and introducing the definition of the heat transfer coefficient,

$$\rho c A \frac{\partial \theta}{\partial t} + \rho c A V \frac{\partial \theta}{\partial x} = -hP\theta, \tag{6.1.3}$$

subject to the inlet boundary condition

$$\theta(0, t) = \theta_i(1 + \epsilon \cos \omega t). \tag{6.1.4}$$

Since we are interested only in the steady periodic fluctuations valid for large values of time, the initial condition specifying $\theta(x, 0)$ is not needed here. Introducing $m = hP/\rho c A$, Eq. (6.1.3) may be rearranged as

$$\frac{\partial \theta}{\partial t} + V \frac{\partial \theta}{\partial x} + m\theta = 0. \tag{6.1.5}$$

For large values of time, this formulation, including the inlet condition given by Eq. (6.1.4), can be written as the superposition of a steady problem,

$$V \frac{d\theta_0}{dx} + m\theta_0 = 0, \tag{6.1.6}$$

$$\theta_0(0) = \theta_i, \tag{6.1.7}$$

and an unsteady problem

$$\frac{\partial \theta_1}{\partial t} + V \frac{\partial \theta_1}{\partial x} + m\theta_1 = 0, \tag{6.1.8}$$

$$\theta_1(0, t) = \epsilon \theta_i \cos \omega t. \tag{6.1.9}$$

The solution of the steady problem is readily found to be

$$\theta_0(x) = \theta_i e^{-(m/V)x}. \tag{6.1.10}$$

Now, we apply the method of complex (dependent) variables to the unsteady problem.

The complementary problem, obtained from the unsteady (original) problem by replacing θ with θ^* and $\cos \omega t$ with $\sin \omega t$, is

$$\frac{\partial \theta_1^*}{\partial t} + V \frac{\partial \theta^*}{\partial x} + m\theta^* = 0, \tag{6.1.11}$$

$$\theta_1^*(0, t) = \epsilon \theta_i \sin \omega t. \tag{6.1.12}$$

Multiplying Eqs. (6.1.11) and (6.1.12) by i and respectively adding the results to Eqs. (6.1.8) and (6.1.9) yields the complex problem in terms of complex temperature $\psi = \theta + i\theta^*$,

$$\frac{\partial \psi}{\partial t} + V \frac{\partial \psi}{\partial x} + m\psi = 0, \tag{6.1.13}$$

$$\psi(0, t) = \epsilon \theta_i e^{i\omega t}. \tag{6.1.14}$$

The steady periodic solution has the form

$$\psi(x, t) = \phi(x) e^{i\omega t}. \tag{6.1.15}$$

Inserting Eq. (6.1.15) into Eqs. (6.1.13) and (6.1.14), and eliminating $e^{i\omega t}$ from each term, we have

$$\frac{d\phi}{dx} + \frac{1}{V}(m + i\omega)\phi = 0, \tag{6.1.16}$$

$$\phi(0) = \epsilon\theta_i, \tag{6.1.17}$$

which has the solution

$$\phi(x) = \epsilon\theta_i e^{-(m+i\omega)(x/V)}. \tag{6.1.18}$$

Inserting Eq. (6.1.18) into Eq. (6.1.15) gives the complex temperature,

$$\psi(x, t) = \epsilon\theta_i e^{-(m/V)x} e^{i\omega(t-x/V)}. \tag{6.1.19}$$

The real part of this temperature is the solution of the unsteady part of the original problem,

$$\theta_1(x, t) = \epsilon\theta_i e^{-(m/V)x} \cos\omega\left(t - \frac{x}{V}\right). \tag{6.1.20}$$

Finally, the complete solution of the original problem, obtained from the superposition of Eqs. (6.1.10) and (6.1.20), may be written as

$$\theta(x, t)/\theta_i = e^{-(m/V)x}\left[1 + \epsilon\cos\omega\left(t - \frac{x}{V}\right)\right]. \tag{6.1.21}$$

Because of the linearity of our problem, Eq. (6.1.21) is *exact* and the amplitude of fluctuations ϵ is *arbitrary*.

Actually, there is a shorter way to employ the method of complex (dependent) variables. We may conveniently learn this in terms of the foregoing problem. For this purpose, reconsider Eq. (6.1.5) and the complex form of Eq. (6.1.4),

$$\frac{\partial\theta}{\partial t} + V\frac{\partial\theta}{\partial x} + m\theta = 0, \tag{6.1.5}$$

$$\theta(0, t) = \theta_i[1 + \epsilon\Re(e^{i\omega t})], \tag{6.1.22}$$

where $\Re$ indicates "the real part of." Now, write the solution in the form

$$\theta(x, t) = \theta_0(x) + \epsilon\Re[\theta_1(x)e^{i\omega t}]. \tag{6.1.23}$$

This solution, being the real part of the complex temperature, gives directly the original temperature. Inserting Eq. (6.1.23) into Eqs. (6.1.5) and (6.1.22), and rearranging the result with respect to powers of ϵ, leads to the formulations for θ_0 and θ_1 obtained previously.

Having learned the application of the method of complex (dependent) variables to linear problems, we proceed now to an example which incorporates the effect of oscillatory disturbances to Ex. 3.2.1.

EXAMPLE 6.1.1 Consider a viscous fluid flowing steadily between two parallel plates a distance $2l$ apart. The pressure gradient causing this steady flow starts oscillating as

$$-\frac{dp}{dx}(1 + \epsilon\sin\omega t),$$

where ϵ is the relative amplitude of oscillations. We wish to determine the steady periodic velocity distribution of the fluid.

The differential formulation of the problem, excluding the initial condition, is

$$\frac{1}{\nu}\frac{\partial u}{\partial t} = -\frac{1}{\mu}\left(\frac{dp}{dx}\right)[1 + \epsilon\mathscr{I}(e^{i\omega t})] + \frac{\partial^2 u}{\partial y^2}, \tag{6.1.24}$$

$$\frac{\partial u(0, t)}{\partial y} = 0, \qquad u(l, t) = 0,$$

where $\mathscr{I}$ indicates "the imaginary part of." In view of Eq. (6.1.24), we write a solution in the form

$$u(y, t) = u_0(y) + \epsilon\mathscr{I}[u_1(y)e^{i\omega t}]. \tag{6.1.25}$$

Note that we used the imaginary rather than the real part in Eq. (6.1.25) because $\sin \omega t$ is the imaginary part of

$$e^{i\omega t} = \cos \omega t + i \sin \omega t.$$

Inserting Eq. (6.1.25) into the foregoing formulation and rearranging the result with respect to powers of ϵ leads to the Poiseuille flow for the steady part,

$$u_0(y) = \frac{1}{2\mu}\left(-\frac{dp}{dx}\right)(l^2 - y^2), \tag{6.1.26}$$

and the following formulation for the unsteady part:

$$\frac{d^2 u_1}{dy^2} - \left(\frac{i\omega}{\nu}\right)u_1 = -\frac{\epsilon}{\mu}\left(-\frac{dp}{dx}\right), \tag{6.1.27}$$

$$\frac{du_1(0)}{dy} = 0, \qquad u_1(l) = 0. \tag{6.1.28}$$

The solution of Eq. (6.1.27) subject to Eq. (6.1.28) is

$$u_1(y) = \frac{\epsilon}{i\rho\omega}\left(-\frac{dp}{dx}\right)\left[1 - \frac{\cosh (i\omega/\nu)^{1/2}y}{\cosh (i\omega/\nu)^{1/2}l}\right]. \tag{6.1.29}$$

Noting that $i^{1/2} = (1 + i)/2^{1/2}$, introducing $(2\nu/\omega)^{1/2} = \delta_s$, Eq. (6.1.29) may be rearranged as

$$u_1(y) = \frac{\epsilon}{i\rho\omega}\left(-\frac{dp}{dx}\right)\left[1 - \frac{\cosh (1 + i)(y/\delta_s)}{\cosh (1 + i)(l/\delta_s)}\right]. \tag{6.1.30}$$

Recalling from trigonometry that

$$\cosh (1 + i)\left(\frac{y}{\delta_s}\right) = \cosh \left(\frac{y}{\delta_s}\right) \cos \left(\frac{y}{\delta_s}\right) + i \sinh \left(\frac{y}{\delta_s}\right) \sin \left(\frac{y}{\delta_s}\right)$$

is a complex number, and introducing the shorter notation

$$\cosh (1 + i)\left(\frac{y}{\delta_s}\right) = A_y + iB_y,$$

Eq. (6.1.30) may be reduced to

$$u_1(y) = \frac{\epsilon}{i\rho\omega}\left(-\frac{dp}{dx}\right)\left(1 - \frac{A_y + iB_y}{A_l + iB_l}\right). \tag{6.1.31}$$

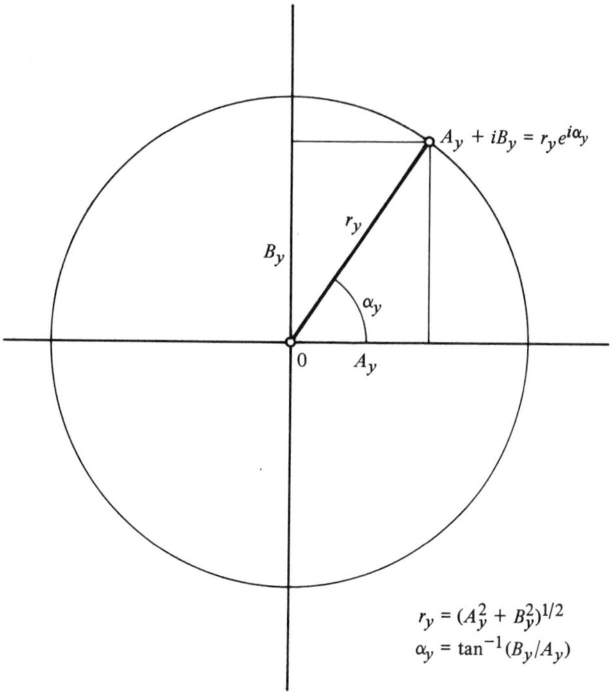

Fig. 6.3 Cartesian and polar representations of a complex number

Furthermore, recalling the polar representation of complex variables, we have from Fig. 6.3,

$$A_y + iB_y = r_y e^{i\alpha_y}, \tag{6.1.32}$$

where $r_y = (A_y^2 + B_y^2)^{1/2}$ and $\tan \alpha_y = B_y/A_y$. Inserting Eq. (6.1.32) into Eq. (6.1.31) gives

$$u_1(y) = \frac{\epsilon}{i\rho\omega}\left(-\frac{dp}{dx}\right)\left[1 - \left(\frac{r_y}{r_l}\right)e^{-i(\alpha_l - \alpha_y)}\right]. \tag{6.1.33}$$

Finally, introducing Eqs. (6.1.26) and (6.1.33) into Eq. (6.1.25) yields the steady periodic velocity fluctuations,

$$\frac{u(y, t)}{(-dp/dx)l^2/2\mu} = 1 - \left(\frac{y}{l}\right)^2 - \epsilon\left(\frac{\delta_s}{l}\right)^2 \left\{\sin \omega t - \left(\frac{r_y}{r_l}\right) \sin\left[\omega t - (\alpha_l - \alpha_y)\right]\right\}. \tag{6.1.34}$$

The quantity $\delta_s = (2\nu/\omega)^{1/2}$, defined following Eq. (6.1.29), is a characteristic length for oscillating flows. It is usually referred to as the *Stokes layer*. For sufficiently large ω, the amplitude of oscillations remains essentially constant in the bulk of flow, except for a thin layer δ_s from walls in which it drops rapidly and vanishes on walls. Note that Eq. (6.1.34) is an exact solution and there is no restriction on ϵ. It repeats (with a phase shift and spatial distribution in amplitude) the externally imposed oscillations. Also, there is no *net effect*, say on the friction, because a temporal average over one period eliminates the contribution of oscillations. These are important features of linear problems.

Now, consider the oscillating flow in a pipe rather than between two parallel plates. Assume that the walls of the pipe are elastic and flexible, and the fluid has a constitution somewhat more complicated than the usual viscous fluids. These conditions describe the blood flow in veins. This problem is an example for a contemporary research area involving both health and engineering sciences. These considerations, however, are beyond the scope of this text (see the report by Brocher 1977).

Having learned the solution of oscillating linear problems, we proceed now to nonlinear problems.

6.2 *NONLINEAR EFFECTS*

Following the philosophy of the preceding section, we utilize an illustrative example in our study of nonlinear effects. For this purpose, reconsider the example of Fig. 6.2. Now, let the pipe flow oscillate as

$$V(1 + \epsilon \cos \omega t).$$

Again, we wish to determine the radially lumped and axially distributed temperature fluctuations in the downstream flow.

The formulation of the problem needs no elaboration, and may be written as

$$\frac{\partial \theta}{\partial t} + V[1 + \epsilon \Re(e^{i\omega t})]\frac{\partial \theta}{\partial x} + m\theta = 0 \tag{6.2.1}$$

subject to

$$\theta(0, t) = \theta_i. \tag{6.2.2}$$

Now, a solution in the form of Eq. (6.1.25) is not good enough because velocity coupled to oscillating temperature gradient in the enthalpy flow gives rise to oscillating temperature terms of higher frequency. This is a feature of *nonlinear* problems. Note that Eq. (6.2.1) is linear with respect to inlet temperature but not with respect to velocity. A series in the form

$$\theta(x, t) = \theta_0(x) + \epsilon\Re[\theta_1(x)e^{i\omega t}] + \epsilon^2\Re[\theta_2(x)e^{2i\omega t}] + \cdots \tag{6.2.3}$$

incorporates higher harmonics into the solution. The amplitude ϵ is no longer arbitrary and must be small enough to ensure the convergence of Eq. (6.2.3). The solution of a nonlinear problem in terms of a series of coupled linear problems such as Eq. (6.2.3) is called the *perturbation solution*. Inserting Eq. (6.2.3) into Eqs. (6.2.1) and (6.2.2), rearranging the result with respect to powers of ϵ, and requiring that terms of each power separately vanish yield for ϵ^0, after eliminating the common operator $\Re$ and $e^{i\omega t}$,

$$V\frac{d\theta_0}{dx} + m\theta_0 = 0, \tag{6.1.6}$$

$$\theta_0(0) = \theta_i, \tag{6.1.7}$$

for ϵ^1, after eliminating $\Re$,

$$V\frac{d\theta_1}{dx} + (m + i\omega)\theta_1 = -V\frac{d\theta_0}{dx}, \tag{6.2.4}$$

$$\theta_1(0) = 0, \tag{6.2.5}$$

and for ϵ^2,

$$\Re(2i\omega\theta_2 e^{2i\omega t}) + V\frac{\partial}{\partial x}\Re(\theta_2 e^{2i\omega t}) + V\Re(e^{i\omega t})\frac{\partial}{\partial x}\Re(\theta_1 e^{i\omega t}) + m\Re(\theta_2 e^{2i\omega t}) = 0, \tag{6.2.6}$$

$$\theta_2(0) = 0. \tag{6.2.7}$$

Equations (6.1.6) and (6.2.4) may be readily solved yielding the first two terms of the solution. These terms, reflecting only the first-order (linear) effect, have no importance for the present discussion and are left to the reader (see Problem 6-3). Equation (6.2.6) involves the second-order (nonlinear) effect. This effect results from the product term

$$V\Re(e^{i\omega t})\frac{\partial}{\partial x}\Re(\theta_1 e^{i\omega t}). \tag{6.2.8}$$

Ignoring $V\,\partial/\partial x$ for the time being, the remainder of Eq. (6.2.8) may be rearranged as

$$\begin{aligned} &\cos\omega t\Re[(\theta_{1R} + i\theta_{1I})(\cos\omega t + i\sin\omega t)], \\ &= \cos\omega t(\theta_{1R}\cos\omega t - \theta_{1I}\sin\omega t), \\ &= \theta_{1R}\cos^2\omega t - \theta_{1I}\sin\omega t\cos\omega t, \\ &= \tfrac{1}{2}[\theta_{1R} + (\theta_{1R}\cos 2\omega t - \theta_{1I}\sin 2\omega t)], \\ &= \tfrac{1}{2}\Re[\theta_1(x) + \theta_1(x)e^{2i\omega t}]. \end{aligned} \tag{6.2.9}$$

Here, we have used the trigonometric relations

$$\cos^2\omega t = \tfrac{1}{2}(1 + \cos 2\omega t), \qquad \sin\omega t\cos\omega t = \tfrac{1}{2}\sin 2\omega t,$$

and, in the last step, the identity

$$\Re(\theta_1 e^{2i\omega t}) = \theta_{1R}\cos 2\omega t - \theta_{1I}\sin 2\omega t.$$

In view of Eq. (6.2.9), Eq. (6.2.6) may be rearranged, after eliminating $\Re$, as

$$2i\omega\theta_2 e^{2i\omega t} + V\frac{\partial}{\partial x}(\theta_2 e^{2i\omega t}) + m\theta_2 e^{2i\omega t} = -\frac{1}{2}V\frac{\partial}{\partial x}(\theta_1 + \theta_1 e^{2i\omega t}). \tag{6.2.10}$$

Since a part of the enthalpy flow is the nonhomogeneous part of the differential equation for θ_2, and since it involves a steady part plus an oscillatory part, the solution of θ_2 may be conveniently written as

$$\theta_{20}(x) + \theta_{22}(x)e^{2i\omega t}. \tag{6.2.11}$$

Here the first term is the solution resulting from the steady part of θ_1 and the second term is the solution resulting from the oscillatory part of θ_1. The series solution of Eq. (6.2.3) may now be rearranged in terms of Eq. (6.2.11) as

$$\theta(x, t) = \theta_0(x) + \epsilon\Re[\theta_1(x)e^{i\omega t}] + \epsilon^2\Re[\theta_{20}(x) + \theta_{22}(x)e^{2i\omega t}] + \cdots. \tag{6.2.12}$$

With Eq. (6.2.12), Eq. (6.2.10) may be split into two equations, one for θ_{20} and the other for θ_{22}. Thus, we have for θ_{20},

$$V\frac{d\theta_{20}}{dx} + m\theta_{20} = -\frac{1}{2}V\frac{d\theta_1}{dx}, \tag{6.2.13}$$

$$\theta_{20}(0) = 0, \tag{6.2.14}$$

and for θ_{22},

$$V\frac{d\theta_{22}}{dx} + (m + 2i\omega)\theta_{22} = -\frac{1}{2}V\frac{d\theta_1}{dx}, \tag{6.2.15}$$

$$\theta_{22}(0) = 0. \tag{6.2.16}$$

Solutions for θ_{20} and θ_{22} are not significant for the present discussion and will not be given here (see Problem 6-4).

However, from the foregoing discussion we learn the important fact that the steady periodic solution of a nonlinear problem resulting from a single harmonic is a series solution. The successive terms of this series involve frequencies which are increasing integer multiples of the input frequency. Furthermore, the average of $\theta(x, t)$ over one period involves no contribution from θ_1 and θ_{22} but adds θ_{20} as a steady net effect to the original steady temperature, $\theta_0(x)$.

Having learned the application of the method of complex (dependent) variables to nonlinear problems, we proceed to an example involving boundary layer flows.

EXAMPLE 6.2.1 Consider a blunt body oscillating with (the real part of) velocity $U_\infty e^{i\omega t}$ in a viscous fluid at rest far from the body, say a sphere or cylinder of diameter D. The problem is identical, in coordinates attached to the body, to a viscous fluid oscillating with $U_\infty e^{i\omega t}$ far from a body at rest. This oscillation gives rise to the free stream (potential flow) velocity $U(x)e^{i\omega t}$ about the body, x being the distance along the surface of the body (Fig. 6.4). We wish to determine the fluid motion neighboring the body.

First, recall from the preceding example that the steady temperature increase θ_{20}, due to nonlinear oscillations resulting from $V(\partial\theta/\partial x)$, depends on $\partial\theta_1/\partial x$ (see Eq. 6.2.13), which in turn depends on $\partial\theta_0/\partial x$ (see Eq. 6.2.4). Similarly, for a boundary layer flow, the nonlinear terms $u(\partial u/\partial x) + v(\partial u/\partial y)$ are the sources of any second-order effects which depend on $\partial U/\partial x$ of the free-stream velocity. Consequently, the effect of any oscillation on boundary layer flows is nill for a flat plate of zero incidence ($\partial p/\partial x = 0$), but it may be appreciable for a blunt body, such as a sphere or cylinder. It may be also shown that the phase changes, as well as amplitude changes, in free-stream velocity induce significant effects (Batchelor 1967).

On dimensional grounds,

$$\text{momentum rate} \sim \text{momentum flux} \tag{6.2.17}$$

along boundaries of the blunt object yields, after noting that $t^{-1} \sim \omega$,

$$\omega U_\infty \sim \frac{\nu U_\infty}{\delta_s^2}, \tag{6.2.18}$$

where δ_s is the penetration depth of oscillations (Stokes layer). Nondimensionalize

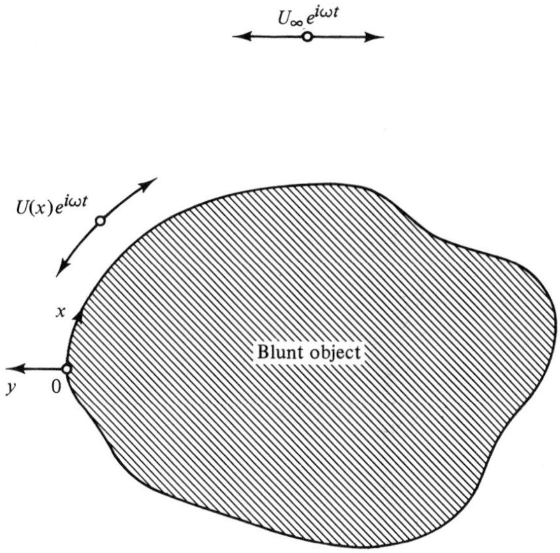

Fig. 6.4 Stationary blunt object in oscillating flow, Ex. 6.2.1

δ_s with respect to a characteristic length D,

$$\frac{\delta_s}{D} \sim \left(\frac{\nu}{\omega D^2}\right)^{1/2}. \tag{6.2.19}$$

For $\nu/\omega D^2 \ll 1$, a boundary layer exists in the usual sense.

The formulation of the boundary layer is

$$\frac{\partial u}{\partial x} + \frac{\partial v}{\partial y} = 0, \tag{4.1.3}$$

$$\frac{\partial u}{\partial t} + u\frac{\partial u}{\partial x} + v\frac{\partial u}{\partial y} = -\frac{1}{\rho}\frac{\partial p}{\partial x} + \nu\frac{\partial^2 u}{\partial y^2}, \tag{4.1.13}$$

$$\frac{\partial U}{\partial t} + U\frac{\partial U}{\partial x} = -\frac{1}{\rho}\frac{\partial p}{\partial x}, \tag{4.1.16}$$

subject to

$$u(x, 0, t) = v(x, 0, t) = 0, \qquad u(x, \infty, t) = U(x, t). \tag{6.2.20}$$

Rearrange Eq. (4.1.13) by eliminating its pressure gradient with Eq. (4.1.16). This gives

$$\frac{\partial u}{\partial t} - \nu\frac{\partial^2 u}{\partial y^2} - \frac{\partial U}{\partial t} = U\frac{\partial U}{\partial x} - u\frac{\partial u}{\partial x} - v\frac{\partial u}{\partial y}. \tag{6.2.21}$$

On dimensional grounds, all terms on the left-hand side are proportional to ωU_∞ and those on the right to U_∞^2/D. For $\omega U_\infty \gg U_\infty^2/D$, or

$$\frac{U_\infty}{\omega D} \ll 1, \tag{6.2.22}$$

the momentum assumes, to first order in $U_\infty/\omega D$, the linear form

$$\frac{\partial u_1}{\partial t} - \nu\frac{\partial^2 u_1}{\partial y^2} - \frac{\partial U}{\partial t} = 0. \tag{6.2.23}$$

Inserting the solution of Eq. (6.2.23) into the right-hand side of Eq. (6.2.21) yields the momentum, to second order in $U_\infty/\omega D$,

$$\frac{\partial u_2}{\partial t} - \nu\frac{\partial^2 u_2}{\partial y^2} - \frac{\partial U}{\partial t} = U\frac{\partial U}{\partial x} - u_1\frac{\partial u_1}{\partial x} - v_1\frac{\partial u_1}{\partial y}, \tag{6.2.24}$$

where v_1 is obtained from Eq. (4.1.3). This scheme of iteration on a base solution, which may be in principle continued to higher orders, is an alternative to the series expansion already introduced as the perturbation solution (see Eqs. 6.2.3 and 6.2.12). This alternative (proposed by Schlichting 1932 for the present problem) is convenient when a small parameter does not appear explicitly in the governing equation.

Before proceeding to a solution, let us comment on Eq. (6.2.22), which may be rearranged as

$$\frac{U_\infty}{\omega D} = \frac{\nu/\omega}{\nu D/U_\infty} \sim \left(\frac{\delta_s}{\delta_\infty}\right)^2 \ll 1,$$

stating that the oscillatory boundary layer δ_s remains well within the boundary layer δ_∞ resulting from a (steady) external velocity U_∞. Also, there is an upper limit for the frequency

$$\frac{U_\infty}{c_\infty} \sim \frac{\omega D}{c_\infty} \ll 1,$$

which ensures the assumption of incompressibility. Here c_∞ is the speed of sound upstream.

Now, we may follow either one of two approaches for a solution. Either solve u_1 from Eq. (6.2.23), and u_2 from Eq. (6.2.24) and obtain v_1, v_2 from Eq. (4.1.3), or replace Eqs. (6.2.23), (6.2.24), and Eq. (4.1.3) with the equation for the stream function whose solution gives u and v (recall Section 2.4). Selecting the latter, Eq. (6.2.23) may be rearranged in terms of $u = \partial\psi/\partial y$, and of $U(x, t) = U(x)e^{i\omega t}$,

$$\frac{\partial^2 \psi_1}{\partial t\,\partial y} - \nu\frac{\partial^3 \psi_1}{\partial y^3} = i\omega U e^{i\omega t}. \tag{6.2.25}$$

Introducing

$$\psi_1 = \left(\frac{2\nu}{\omega}\right)^{1/2} U(x)\zeta_1(\eta)e^{i\omega t}, \tag{6.2.26}$$

$\eta = y(\omega/2\nu)^{1/2}$ being in view of Eq. (6.2.19) a convenient dimensionless distance normal to boundaries, Eq. (6.2.25) may be reduced to

$$\zeta_1''' + 2i(1 - \zeta_1') = 0 \tag{6.2.27}$$

subject to

$$\zeta_1(0) = \zeta_1'(0) = 0, \qquad \zeta_1'(\infty) = 0, \tag{6.2.28}$$

where a prime denotes differentiation with respect to η.

The solution of Eq. (6.2.27) satisfying Eq. (6.2.28) is

$$\zeta_1 = -\frac{1}{2}(1 - i)[1 - e^{-(1+i)\eta}] + \eta. \tag{6.2.29}$$

Inserting Eq. (6.2.29) into Eq. (6.2.26) and employing $u = \partial\psi/\partial y$ gives

$$u_1 = U(x)[\cos \omega t - e^{-\eta} \cos (\omega t - \eta)], \tag{6.2.30}$$

and $v = -\partial\psi/\partial x$ gives

$$v_1 = -\left(\frac{2\nu}{\omega}\right)^{1/2}\left(\frac{dU}{dx}\right)\left[\eta \cos \omega t + \frac{1}{\sqrt{2}} \cos\left(\omega t + \frac{3}{4}\pi\right) + \frac{e^{-\eta}}{\sqrt{2}} \cos\left(\omega t - \eta - \frac{1}{4}\pi\right)\right]. \tag{6.2.31}$$

The third term of Eq. (6.2.31) tends to zero outside the boundary layer. The first term represents the continuity of the main flow and the second term the boundary layer displacement of the external flow (the diffusion of periodic vorticity). The skin friction is

$$\tau_w = \mu\left(\frac{\omega}{2\nu}\right)^{1/2} U(x) \cos\left(\omega t + \frac{1}{4}\pi\right), \tag{6.2.32}$$

which has a phase lead of $\frac{1}{4}\pi$ over the velocity fluctuations.

Now, to account for second-order effects, assume that $\psi = \psi_1 + \psi_2$ and consider Eq. (6.2.24). Since the right-hand side of this equation is of the order of $U_\infty/\omega D$, $\psi_1 \sim U_\infty/\omega D$ and $\psi_2 \sim (U_\infty/\omega D)\psi_1$. Thus

$$\psi_2 \sim \psi_1^2. \tag{6.2.33}$$

Since ψ_1 involves the frequency of the external flow, ψ_2 would involve terms both of zero frequency and of a frequency twice that of the external flow. Consequently, in a manner similar to Eq. (6.2.12),

$$\psi_2 = \left(\frac{2\nu}{\omega}\right)^{1/2} \frac{U}{\omega}\left(\frac{dU}{dx}\right)(\zeta_{20} + \zeta_{22}e^{2i\omega t}). \tag{6.2.34}$$

Inserting Eq. (6.2.34) into the left-hand side and Eqs. (6.2.30) and (6.2.31) into the right-hand side of Eq. (6.2.24), solving the result subject to boundary conditions

$$\begin{aligned} &\zeta_{20} = \zeta'_{20} = \zeta_{22} = \zeta'_{22} = 0 \qquad \text{for } \eta = 0, \\ &\zeta'_{22} \longrightarrow 0 \quad \text{and} \quad \zeta'_{20} \longrightarrow \text{finite} \qquad \text{as } \eta \longrightarrow \infty, \end{aligned} \tag{6.2.35}$$

we obtain

$$\zeta_{20} = \frac{13}{8} - \frac{3}{4}\eta - \frac{1}{8}e^{-2\eta} - \frac{3}{2}e^{-\eta} \cos \eta - e^{-\eta} \sin \eta - \frac{1}{2}\eta e^{-\eta} \sin \eta, \tag{6.2.36}$$

$$\zeta_{22} = \frac{1+i}{4\sqrt{2}}e^{-(1+i)\sqrt{2}\eta} + \frac{1}{2}i\eta e^{-(1+i)\eta} - \frac{1+i}{4\sqrt{2}}. \tag{6.2.37}$$

At large values of η, corresponding to the edge of the boundary layer, $\zeta'_{20} = -\frac{3}{4}$, and the steady flow has a component $-3UU'/4\omega$ parallel to boundaries. The component of this flow normal to boundaries is

$$\frac{1}{\omega}\frac{d}{dx}\left(U\frac{dU}{dx}\right)\left[\frac{3}{4}y - \frac{13}{8}\left(\frac{2\nu}{\omega}\right)^{1/2}\right].$$

Clearly, a small, steady (so-called streaming) motion is generated which continues to persist outside the boundary layer. This motion results from the nonlinearity of momentum flow. Its steady velocity at the edge of the boundary layer has a magnitude of order $U_\infty^2/\omega D$, which may be utilized to introduce a *streaming* Reynolds number, $\text{Rs} = U_\infty^2/\omega\nu$. The behavior of the steady flow outside the boundary layer depends on the magnitude of this number. For sufficiently large values of Rs, there exists a second outer boundary layer at the edge of which u_{20} tends to zero (Fig. 6.5). Further considerations of this problem, including thermal effects, are beyond the scope of this text (see Riley 1966, 1975 for the isothermal flow near an oscillating cylinder and Davidson 1973 for the associated thermal effects).

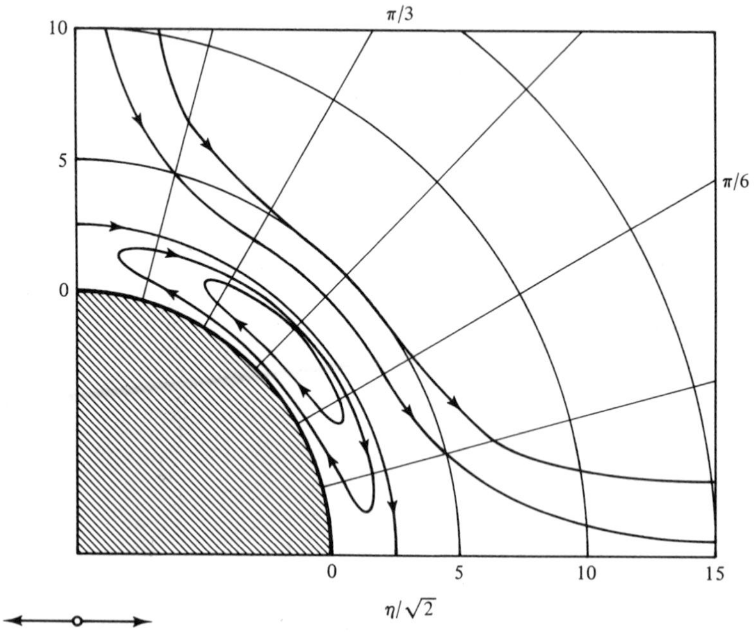

Fig. 6.5 Steady streaming for $\omega D^2/\nu = 400$ (adapted from Andres & Ingard 1953; used by permission)

REFERENCES

ANDRES, J. M., & INGARD, U., 1953, J. acoust. Soc. Amer., **25**, 928–932.

ARPACI, V. S., 1966, *Conduction Heat Transfer*, Addison-Wesley, Reading, Mass.

BATCHELOR, G. K., 1967, *Fluid Dynamics*, Cambridge University Press, New York.

BROCHER, E., 1977, J. Fluid Mech., **79**, 113–126.

DAVIDSON, B. J., 1973, Int. J. Heat Mass Transfer, **16**, 1703–1727.

ESHGHY, S., ARPACI, V. S., & CLARK, J. A., 1965, Trans. ASME, **32**, Ser. E. 183–191.

KAKAC, S., & YENER, Y., 1973, Int. J. Heat Mass Transfer, **16**, 2205–2214.

LARSEN, P. S., & JENSEN, J. W., 1978, Int. J. Heat Mass Transfer, **21**, 511–517.

LIGHTHILL, M. J., 1954, Proc. R. Soc. Lond., A, **224**, 1–23.

LIN, C. C., 1956, Proc. 9th Int. Congr. Appl. Mech., Brussels, Vol. **4**, pp. 155–167.

RICHARDSON, P. D., 1967, Appl. Mech. Rev., **20**, 2011–2017.

RILEY, N., 1966, Matematika, **12**, 161–175.

RILEY, N., 1975, J. Fluid Mech., **68**, 801–812.

ROSENHEAD, L., ed., 1963, *Laminar Boundary Layers*, Oxford University Press, New York.

SCHLICHTING, H., 1932, Phys. Z., **33**, 327–335.

SOUNDALGEKAR, 1973, Proc. R. Soc. A, **333**, 25–36.

STUART, J. T., 1955, Proc. R. Soc. A, **231**, 116–130.

YANG, W. J., 1964, J. Heat Transfer (ASME), **86**, 133–142.

YANG, W. J., CLARK, J. A., & ARPACI, V. S., 1961, J. Heat Transfer (ASME), **83**, 321–338.

PROBLEMS

6-1 A fluid layer of thickness l flows with a lumped velocity V on both sides of a flat fuel plate of thickness $2L$ (Fig. 6P-1). The inlet temperature of the fluid is T_i. Energy generation in the fuel plate oscillates as

$$u''' = u_0'''(1 + \epsilon \sin \omega t).$$

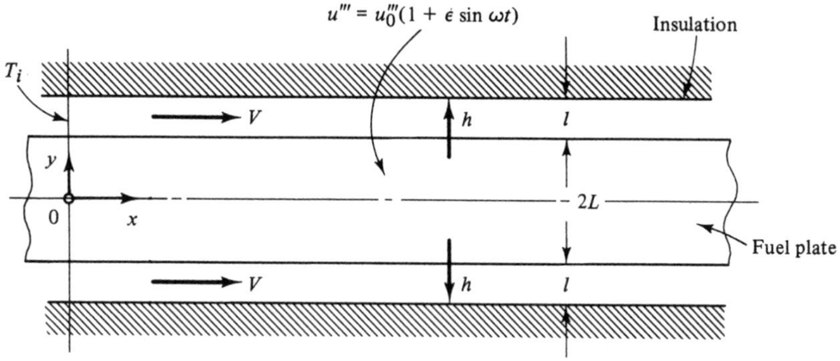

Fig. 6P-1

(a) Find the steady periodic temperature of the system based on a transversly lumped and longitudinally differential analysis.

(b) Resolve the problem after allowing transversal temperature variation in the fuel plate.

6-2 A fluid flows with a velocity

$$V = V_0(1 + \epsilon \cos \omega t)$$

through a thick-walled pipe (Fig. 6P-2). The inlet temperature of the fluid is T_i, and the ambient temperature is T_∞. The inner and outer heat transfer coefficients are h_i and h_o, respec-

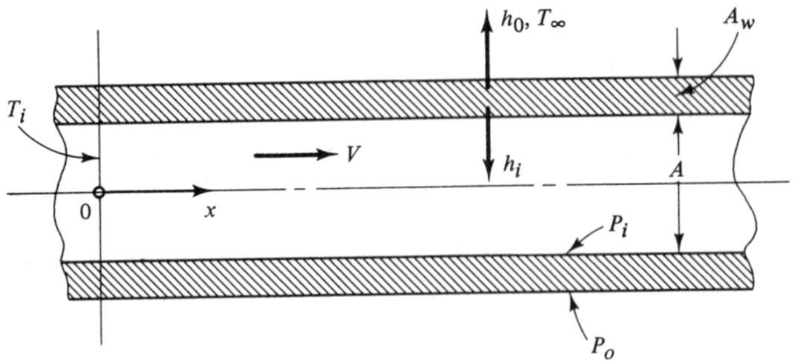

Fig. 6P-2

tively. Find the steady periodic temperature of the system based on a radially lumped and axially differential analysis.

6-3 Obtain the solution of Eq. (6.2.4) subject to Eq. (6.2.5).

6-4 Obtain the solution of Eq. (6.2.13) subject to Eq. (6.2.14) and that of Eq. (6.2.15) subject to Eq. (6.2.16).

6-5 Consider a semi-infinite viscous fluid at rest above an infinite plate which starts oscillating in its own plane. Find the steady periodic motion of the fluid. Determine the penetration depth of oscillating fluid motion. Describe the physics of the conduction problem which has the analogous mathematical formulation.

6-6 Consider a viscous fluid between two parallel plates separated a distance $2l$ apart. One of the plates oscillates in its own plane. Find the steady periodic motion of the fluid. Under what condition does this motion become independent of the stationary plate?

6-7 Reconsider now Problem 6-5, for the linear viscoelastic Maxwell fluid whose constitutive relation for parallel flow is

$$\tau_{yx} + \lambda_\tau \frac{\partial \tau_{yx}}{\partial t} = \mu \frac{\partial u}{\partial y},$$

μ denoting the viscosity and λ_τ the relaxation time for stress. Find the steady periodic velocity of the fluid. Discuss the result in terms of the natural frequency of the fluid and the imposed frequency of the plate.

6-8 Assume that the surface temperature of a semi-infinite solid is oscillating periodically. Thermal conductivity of the solid depends linearly on temperature, $k = k_0(1 + \beta T)$. Find the steady periodic temperature of the solid.

6-9 Reconsider Problem 6-5. Assume the effect of viscous dissipation to be appreciable. Find the steady periodic temperature of the fluid respectively for a plate kept at uniform temperature T_w, and for an insulated plate.

6-10 Reconsider the Graetz problem of Ex. 3.5.5 including an oscillation to its flow. Find the steady periodic Nusselt number.

6-11 Reconsider the natural convection in an enclosure shaped as a vertical slot discussed in Ex. 3.4.2. Let the enclosure oscillate in the vertical direction. Find the steady periodic temperature distribution in the slot.

6-12 Consider a viscous flow, oscillating about a mean velocity, past a flat plate with uniform suction (recall Ex. 4.3.2). Find the steady periodic velocity of the boundary layer (Stuart 1955).

6-13 Consider a viscous flow, oscillating about a mean, over a blunt body. Find the steady periodic velocity and temperature of the boundary layer, and the coefficients of friction and heat transfer (Lighthill 1954, Lin 1956).

6-14 Consider the natural convection boundary layer near a vertical flat plate with uniform suction. The plate temperature is oscillating about a mean value. Find the steady periodic velocity of the boundary layer (Soundalgekar 1973).

6-15 Consider the natural convection boundary layer near a vertical flat plate. Assume that the plate is oscillating in its own plane. Find the steady periodic velocity of and heat transfer across the boundary layer (Eshghy, Arpaci & Clark 1965).

6-16 Survey the literature on Ex. 6.2.1, including thermal effects (Davidson 1973).

Chapter 7

COMPUTATIONAL CONVECTION

As we have learned in Part II, nearly parallel and nonparallel (recirculating) convection problems are inherently nonlinear. Using the integral formulation and approximate profiles, or applying similarity transformations, we were able to reduce the formulation of a number of two-dimensional laminar boundary layer problems from partial differential equations to ordinary but nonlinear differential equations. For more general laminar boundary layers, and for turbulent boundary layers formulated in terms of one- or two-equation models, as well as for two- and three-dimensional recirculating flows, we are faced with the task of solving nonlinear partial differential equations. These equations usually require numerical methods for a solution.

Basically, numerical methods are discretizations of analytical methods. By this discretization the local (differential) formulation leads to finite difference formulations while the global (integral, variational, or any other method of weighted residual) formulation leads to finite element formulations (Fig. 7.1). Both numerical methods lead, after linearization if required, to the solution of systems of linear algebraic equations.

Finite element methods, originally developed for structural analysis, are being used increasingly in fluid mechanics and convection. Although more accurate for a given discretization, but also more elaborate than finite difference methods, the finite element methods are most suited to irregular geometry problems which are not

Discretization of Analytical Method	→	Numerical Method
Local (Differential) Formulation	→	Finite Difference Formulation
Global {Integral / Variational / Weighted residual} Formulation	→	Finite Element Formulation

Fig. 7.1 Foundation of numerical methods

frequently enountered in convection. Finite difference methods require relatively little algebra to set up for a problem having regular boundaries, and these methods are widely used in convection.

The reader is assumed to be familiar with elements of numerical analysis. Certain aspects hereof are included only for completeness, and we refer to Fröberg (1969) or Dahlquist, Björck & Anderson (1974) and to Ames (1977). For an introduction to computational fluid mechanics, we refer to Chow (1979) and Roache (1976) for finite difference methods, and to Gallagher et al. (1975) and Chung (1978) for finite element methods. Boundary layer computations are described by Patankar & Spalding (1970), Cebeci & Bradshaw (1977), and Bradshaw, Cebeci & Whitelaw (1981). An early detailed account of a finite difference computer program for recirculating two-dimensional flows is given by Gosman et al. (1969). Some useful hints on methods of computational convection may be found in Patankar (1980).

The structure of the chapter is as follows. The foundations of finite difference and finite element methods in terms of a one-dimensional illustrative problem are presented in Section 7.1. From then on, only finite difference methods are treated, because these methods are still most widely used in convection. Ordinary differential equations, originating from integral formulation and similarity transformation of boundary layer problems, are discussed in Section 7.2. The partial differential equations associated with nearly parallel (boundary layer) flows are parabolic and the subject matter of Section 7.3, while those of nonparallel (recirculating) flows are elliptical and treated in Section 7.4.

7.1 FINITE DIFFERENCE AND FINITE ELEMENT METHODS

To begin our study of computational convection, reconsider the steady part of the introductory example in Section 6.1, now including axial diffusion of heat. A moving rod (or slug flow) with velocity U_0, temperature $T(x)$, cross-sectional area A, and perimeter P is subject to ambient heat transfer and specified boundary conditions at $x = 0$ and L. The heat transfer coefficient to the ambient at T_∞ is h. We seek the radially lumped and axially distributed temperature $T(x)$.

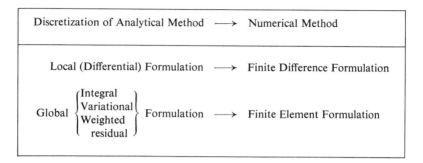

Employing a lumped-differential control volume (see Fig. 6.2 or 7.2) the balance of thermal energy becomes

$$\rho c A U_0 \frac{dT}{dx} = kA \frac{d^2T}{dx^2} - Ph(T - T_\infty),$$

or, in dimensionless form,

$$\frac{d^2\theta}{d\xi^2} - \text{Pe}\frac{d\theta}{d\xi} - M^2\theta = 0, \tag{7.1.1}$$

where $\theta = (T - T_\infty)/(T_1 - T_\infty)$, $\xi = x/L$, $M^2 = PhL^2/(\rho c A)$, and $\text{Pe} = U_0 L/a$ denotes the Péclet number.

For our main objective, which is the illustration of the numerical solution methods, we simplify the problem by neglecting the convective term in Eq. (7.1.1). The simultaneous effects of axial diffusion and convection, being of particular importance for nonparallel flows, are considered separately in Section 7.4.

To explore the relation between analytical and numerical methods shown in Fig. 7.1, we begin by reviewing a number of analytical methods. Accordingly, the following examples related to this problem are considered:

1. Differential formulation and exact solution
2. Integral formulation and solution by approximating function
3. Variational formulation and solution by approximating function
4. Finite difference formulation by discrete control volume (physical approach) and by discretization of differential formulation (mathematical approach)
5. Finite element formulation by integral and variational forms of discrete control volume
6. Numerical solution of discretized formulation

EXAMPLE 7.1.1 Consider a fin of length L, having constant cross-sectional area A, perimeter P, specified base temperature T_0, and insulated tip. The heat transfer coefficient to the ambient at T_∞ is $\bar{h}$. Neglecting the temperature variation over the cross section, we seek the distribution of the lumped temperature $T(x)$ along the fin and the total heat transfer Q from the fin to the ambient.

From Fig. 7.2 and Eq. (7.1.1) the dimensionless formulation becomes

$$\frac{d^2\theta}{d\xi^2} - M^2\theta = 0, \tag{7.1.2}$$

$$\frac{d\theta(0)}{d\xi} = 0, \qquad \theta(1) = 1,$$

whose exact solution is

$$\theta(\xi) = \frac{\cosh M\xi}{\cosh M}. \tag{7.1.3}$$

The total heat transfer from the fin $Q = P\bar{h}\int_0^L (T - T_\infty)\,dx$, which equals the heat transfer by conduction at the base, may be expressed in dimensionless form as

$$\frac{Q}{kA(T_0 - T_\infty)/L} = M^2 \int_0^1 \theta\, d\xi = M \tanh M. \tag{7.1.4}$$

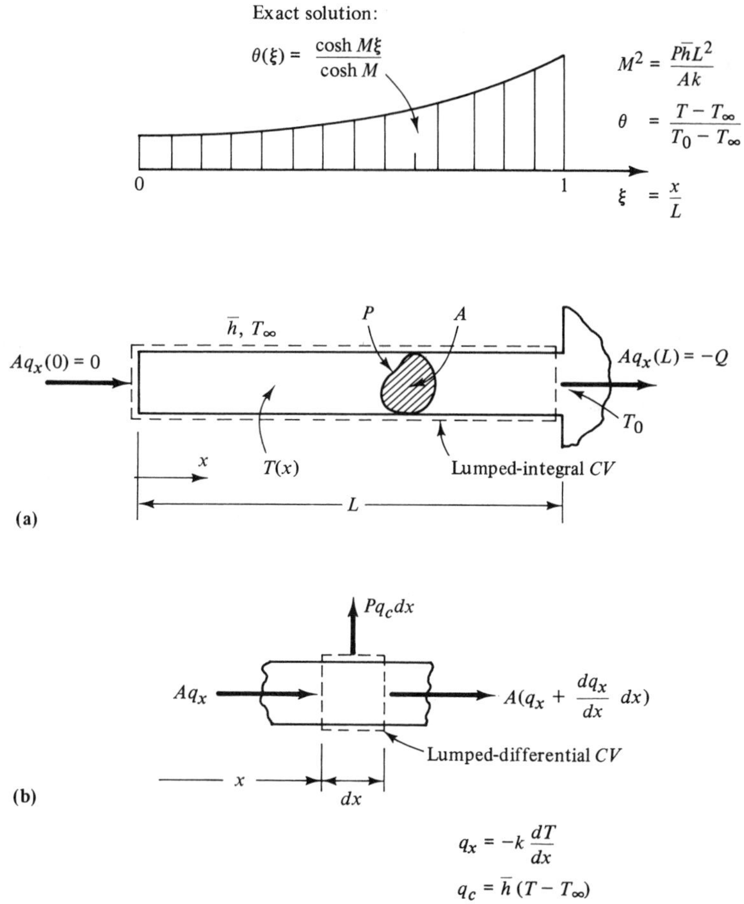

Fig. 7.2 Steady temperature distribution of fin, having specified base temperature and insulated tip; control volume for lumped-differential formulation

Finally, for later use, referring to Fig. 7.4(a), let us derive the exact relation between θ_{i-1}, θ_i, and θ_{i+1} located at $x = (i-1)h$, ih, and $(i+1)h$, respectively, where $h = L/n$. Introducing $m^2 = (Mh/L)^2$ and using Eq. (7.1.3), we obtain

$$\frac{\theta_{i-1}}{\theta_i} = \frac{\cosh[m(i-1)]}{\cosh mi},$$

$$\frac{\theta_{i+1}}{\theta_i} = \frac{\cosh[m(i+1)]}{\cosh mi},$$

whose sum may be expressed as

$$\frac{\theta_{i-1}}{\theta_i} + \frac{\theta_{i+1}}{\theta_i} = 2\cosh m,$$

or as

$$\theta_{i-1} - (2\cosh m)\theta_i + \theta_{i+1} = 0. \tag{7.1.5}$$

We shall compare this result to discrete approximate formulations obtained in later examples.

EXAMPLE 7.1.2 The lumped-integral formulation for the whole fin gives (Fig. 7.2a)

$$0 = Ak\frac{dT}{dx}\bigg|_{x=L} - \int_0^L P\bar{h}(T - T_\infty)\,dx, \tag{7.1.6}$$

or

$$\frac{d\theta}{d\xi}\bigg|_{\xi=1} - M^2\int_0^1 \theta\,d\xi = 0, \tag{7.1.7}$$

which also could have been obtained by integrating Eq. (7.1.2), noting that $d\theta(0)/d\xi = 0$.

For a first-order solution by approximating function (the Rayleigh–Ritz method), we assume a parabolic profile which satisfies the physical boundary conditions of Eq. (7.1.2),

$$\theta = 1 - a_2(1 - \xi^2). \tag{7.1.8}$$

Inserting Eq. (7.1.8) into Eq. (7.1.7) determines the unknown constant

$$a_2 = \frac{M^2}{2 + \frac{2}{3}M^2}, \tag{7.1.9}$$

hence the approximate temperature distribution

$$\theta(\xi) = 1 - \frac{M^2}{2 + \frac{2}{3}M^2}(1 - \xi^2), \tag{7.1.10}$$

and the heat transfer

$$\frac{Q}{kA(T_0 - T_\infty)/L} = \frac{M^2}{1 + \frac{1}{3}M^2}, \tag{7.1.11}$$

which in this case, in view of Eq. (7.1.7), equals $\theta'(1)/\theta_0$.

EXAMPLE 7.1.3 Considering Eq. (7.1.2) to be the *Euler equation* corresponding to a variational problem, the formulation of this problem is obtained as (see, for example, Arpaci 1966)

$$\delta I = \int_0^1 \left(\frac{d^2\theta}{d\xi^2} - M^2\theta\right)\delta\theta\,d\xi = 0, \tag{7.1.12}$$

or, after integrating by parts and noting that $(d\theta/dx)\,\delta\theta\int_0^1 = 0$ in view of the boundary conditions, as

$$\delta I = 0, \qquad I = -\tfrac{1}{2}\int_0^1 \left[\left(\frac{d\theta}{d\xi}\right)^2 + M^2\theta^2\right]d\xi. \tag{7.1.13}$$

For a first-order solution by approximating function, employing again Eq. (7.1.8), we obtain

$$I = -\tfrac{1}{2}[\tfrac{4}{3}a_2^2 + M^2(1 - \tfrac{4}{3}a_2 + \tfrac{8}{15}a_2^2)], \tag{7.1.14}$$

and a_2 is now determined by the condition that I be stationary,

$$\delta I = \frac{\partial I}{\partial a_2}\,\delta a_2 = 0 \qquad \text{or} \qquad \frac{\partial I}{\partial a_2} = 0,$$

which yields

$$a_2 = \frac{M^2}{2 + \frac{4}{5}M^2}. \tag{7.1.15}$$

Thus, the integrated heat transfer becomes

$$\frac{Q}{kA(T_0 - T_\infty)/L} = \frac{M^2(1 + \frac{1}{15}M^2)}{1 + \frac{2}{5}M^2}. \tag{7.1.16}$$

In higher-order approximations $\theta(\xi)$ depends on several unknown parameters a_i which are determined from $\partial I/\partial a_i = 0$, $i = 1, 2, \ldots$.

The total heat transfer from the fin, obtained from the exact and approximate integral and variational solutions given by Eqs. (7.1.4), (7.1.11), and (7.1.16), as a function of the parameter M, are compared in Fig. 7.3. As expected, the variational solution is more accurate than the integral solution employing the same order of approximating function.

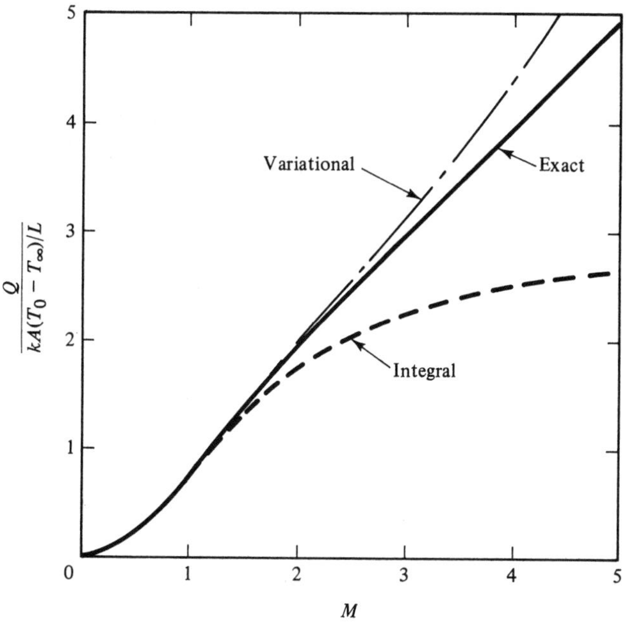

Fig. 7.3 Total dimensionless heat transfer Q versus parameter M

EXAMPLE 7.1.4 For a finite difference formulation of the problem, first subdivide the length of the fin into n intervals of length $\Delta x_i = x_{i+1} - x_i$, $i = 1, 2, \ldots, n$, and introduce the discrete values of the temperature θ_i at nodes $i = 1, 2, \ldots, n + 1$. Confining our attention to equidistant nodes, we denote the length of intervals by $\Delta x_i = h = L/n$.

A numerical solution consists in determining the discrete temperatures θ_i at nodes satisfying the balance of thermal energy, constitutive relations, and the boundary conditions. The discretized formulation may be derived by the (physical) control volume method or by the (mathematical) Taylor series method. We illustrate both methods.

First, for the control volume method, consider $(CV)_{FD}$ at internal node i, $(CV)_i$ of Fig. 7.4(c). The balance of thermal energy gives

$$0 = Aq_{i-1} - Aq_i - hPq_c. \tag{7.1.17}$$

Employing centered difference approximations for the Fourier law of conduction and the definition of the heat transfer coefficient,

$$q_{i-1} \simeq k\frac{T_{i-1} - T_i}{h}, \qquad q_i \simeq k\frac{T_i - T_{i+1}}{h}, \qquad q_c = \bar{h}(T_i - T_\infty), \tag{7.1.18}$$

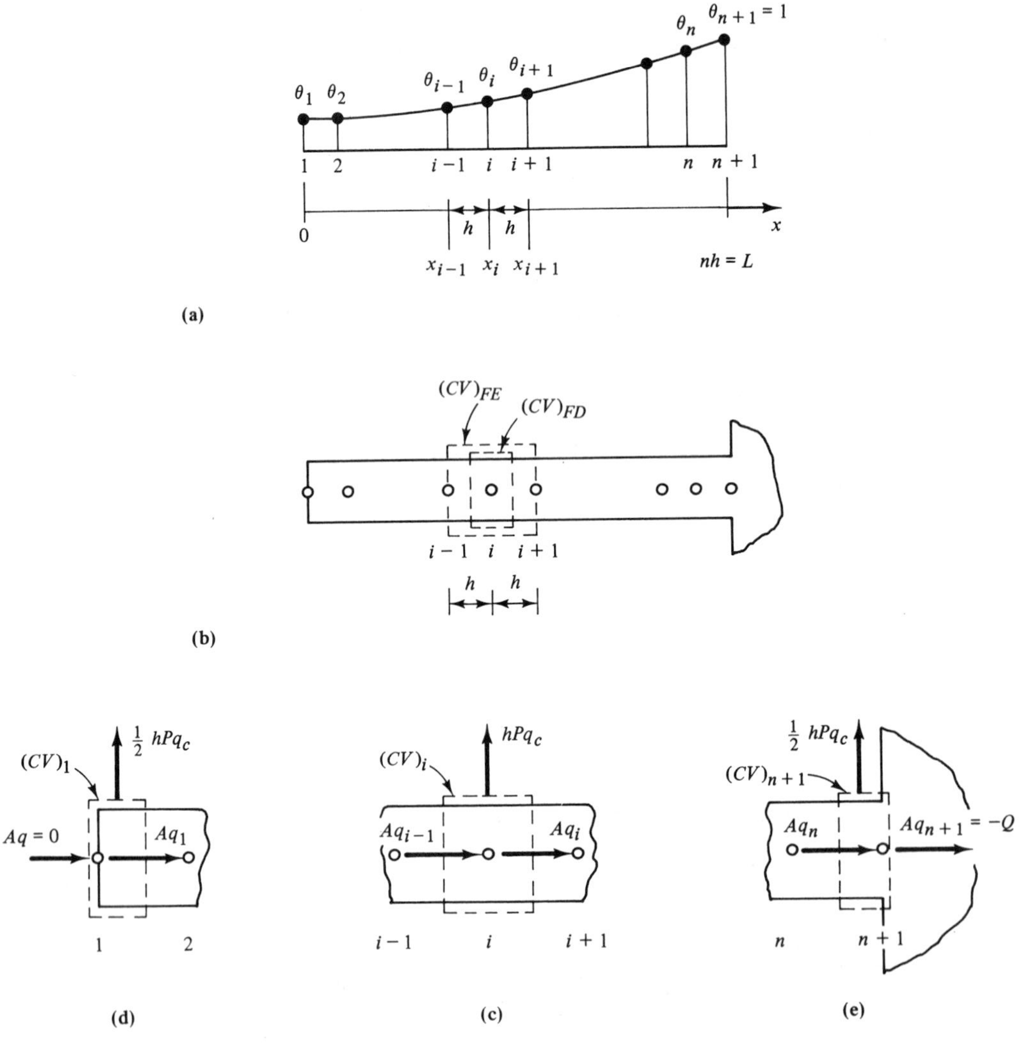

Fig. 7.4 Discrete formulations; notation and control volumes

and introducing $m^2 = P\bar{h}h^2/kA = (Mh/L)^2 = (M/n)^2$, Eq. (7.1.17) may be rearranged as

$$\theta_{i-1} - (2 + m^2)\theta_i + \theta_{i+1} = 0, \tag{7.1.19}$$

which applies to internal nodes, $i = 2, 3, \ldots, n$. At the insulated boundary node $i = 1$, however, we obtain from $(CV)_1$ of Fig. 7.4(d)

$$0 = -Ak\,\frac{T_1 - T_2}{h} - \tfrac{1}{2}hP\bar{h}(T_1 - T_\infty),$$

or

$$-\tfrac{1}{2}(2 + m^2)\theta_1 + \theta_2 = 0, \tag{7.1.20}$$

while boundary node $i = n + 1$ is specified by the base temperature

$$\theta_{n+1} = 1. \tag{7.1.21}$$

Note that $(CV)_{n+1}$ of Fig. 7.4(e) yields

$$0 = Ak\,\frac{T_n - T_{n+1}}{h} - \frac{1}{2}hP\bar{h}(T_{n+1} - T_\infty) - Aq_{n+1},$$

from which the total heat transfer $Q = -Aq_{n+1}$ may be obtained. It is more accurate, however, to employ the integral of the heat transfer to the ambient along the fin, say, approximated by the Simpson formula (Fröberg 1969), for n even,

$$\frac{Q}{kA(T_0 - T_\infty)/L} = M^2 \int_0^1 \frac{\theta(\xi)}{\theta_0}\,d\xi \simeq \frac{M^2}{3n\theta_0}\left(\theta_1 + 4\sum_{j=1}^{n/2}\theta_{2j} + 2\sum_{j=1}^{(n/2)-1}\theta_{2j+1} + \theta_{n+1}\right). \tag{7.1.22}$$

Other faster and more accurate algorithms for numerical integration are available in standard subroutine libraries.

Next, delaying the solution to the discrete formulation, let us consider the alternative (mathematical) method. Starting from the differential formulation Eq. (7.1.2), we obtain the discrete formulation for internal grid points by replacing differential operators by finite difference operators derived from, say, Taylor series expansions of $\theta(x)$ from $\theta_i = \theta(x_i)$ to $x_{i+1} = x_i + h$ and $x_{i-1} = x_i - h$, respectively,

$$\theta(x_i + h) = \theta_{i+1} \simeq \theta_i + \theta_i' h + \frac{1}{2!}\theta_i'' h^2 + \frac{1}{3!}\theta_i''' h^3 + \frac{1}{4!}\theta_i^{(\mathrm{iv})} h^4 + \cdots \tag{7.1.23}$$

$$\theta(x_i - h) = \theta_{i-1} \simeq \theta_i - \theta_i' h + \frac{1}{2!}\theta_i'' h^2 - \frac{1}{3!}\theta_i''' h^3 + \frac{1}{4!}\theta_i^{(\mathrm{iv})} h^4 + \cdots, \tag{7.1.24}$$

where θ_i' denotes $d\theta/dx$ at x_i.

The first-order derivative θ_i' may be approximated, from Eq. (7.1.23), by the *forward-difference* formula

$$\theta_i' \simeq \frac{\theta_{i+1} - \theta_i}{h}, \tag{7.1.25}$$

from Eq. (7.1.24) by the *backward-difference* formula

$$\theta_i' \simeq \frac{\theta_i - \theta_{i-1}}{h}, \tag{7.1.26}$$

or, from the difference between Eqs. (7.1.23) and (7.1.24) by the *central-difference* formula

$$\theta_i' \simeq \frac{1}{2}\frac{\theta_{i+1} - \theta_{i-1}}{h}, \tag{7.1.27}$$

all having a *truncation error* of order $h^2\theta_i'''$.

The second-order derivative may be approximated, from the sum of Eqs. (7.1.23) and (7.1.24), by the *central-difference* formula

$$\theta_i'' \simeq \frac{\theta_{i+1} - 2\theta_i + \theta_{i-1}}{h^2} \tag{7.1.28}$$

with a truncation error of order $h^2\theta_i^{(\text{iv})}$.

Finite difference operators of higher order may be derived recursively from Eqs. (7.1.23) and (7.1.24). Note that use of node spacings that are not equidistant leads to an increase in truncation errors (Problem 7-1).

Then, evaluating Eq. (7.1.2) at x_i, employing Eq. (7.1.28) and noting that $(Mh/L)^2 = m^2$, we obtain

$$\theta_{i-1} - (2 + m^2)\theta_i + \theta_{i+1} = 0, \tag{7.1.29}$$

which is identical to Eq. (7.1.19), applying to internal nodes.

The boundary condition $\theta'(0) = 0$ may be approximated to first or second order using Eq. (7.1.23) for $i = 1$, or equivalently by a polynomial approximation. To first order, this yields

$$\theta_1 - \theta_2 = 0. \tag{7.1.30}$$

To second order, requiring

$$\theta(x) = a_0 + a_1 x + a_2 x^2$$

to satisfy $\theta(0) = \theta_1$, $\theta(h) = \theta_2$, and $\theta(2h) = \theta_3$, and then setting $d\theta(0)/dx = a_1 = 0$, yields

$$\theta_1 - \tfrac{4}{3}\theta_2 + \tfrac{1}{3}\theta_3 = 0. \tag{7.1.31}$$

Note that either of equations Eqs. (7.1.30) or (7.1.31) are different from Eq. (7.1.20) derived from the control volume method, although Eq. (7.1.20) approaches Eq. (7.1.30) for $m \to 0$. Such differences are common in discrete methods. Equation (7.1.31) gives a smooth temperature distribution, but then $q_1 \neq 0$. Strictly speaking, node 1 is only adiabatic if either of Eqs. (7.1.20) or (7.1.30) are used. The specified base temperature again yields Eq. (7.1.21) and the total heat transfer may be evaluated from Eq. (7.1.22).

While the control volume method is often useful because it ensures a proper account of the physics of the problem, the method of Taylor series expansion gives directly the truncation errors associated with replacing differential operators by finite difference operators. Since $h \sim 1/n$ the truncation errors decrease with increasing number of grid points. In a stable and convergent numerical scheme the discrete solution therefore approaches the exact solution as $h \to 0$. However, in addition to the truncation error there will be a minimum numerical *round-off error* associated with the limited number of digits effectively used in the computer employed to perform the numerical calculations. An increasing number of grid points implies an increasing number of algebraic manipulations for a solution. Hence, the accumulation of round-off errors may be expected to set a lower limit for h. For a discussion of round-off errors, see Fröberg (1969), and for stability, convergence, and consistency of numerical solutions, see Roache (1976).

EXAMPLE 7.1.5 For a finite element formulation we employ the subdivision of the fin into n equidistant intervals as before (Fig. 7.4(a)). For reasons to become clear later, we select as *finite element* the control volume $(CV)_{FE}$ of Fig. 7.4(b) for internal node i. Although finite element methods are usually associated with the variational form, one may equally employ the integral form. Here we illustrate both formulations. As indicated in Fig. 7.1, any method of weighted residuals, such as the Galerkin or least-squares methods, may be employed.

First we consider the integral finite element formulation. Recalling the approach in Ex. 7.1.2 and introducing the local coordinate $\zeta = (x - x_i)/h$, the integral formulation for $(CV)_{FE}$ becomes

$$\frac{d\theta}{d\zeta}\bigg|_{-1}^{1} - m^2 \int_{-1}^{1} \theta \, d\zeta = 0. \tag{7.1.32}$$

To proceed, we introduce an approximating function, say, a polynomial of second order in ζ satisfying the discrete values θ_{i-1}, θ_i, and θ_{i+1} at $\zeta = -1$, 0, and 1 to ensure continuity. The approximating function is conveniently expressed as

$$\theta(\zeta) = \sum_{i-1}^{i+1} N_j(\zeta)\theta_j, \tag{7.1.33}$$

where $N_j(\zeta)$, usually called *interpolation functions*, become

$$N_{i-1} = \tfrac{1}{2}\zeta(\zeta - 1), \qquad N_i = 1 - \zeta^2, \qquad N_{i+1} = \tfrac{1}{2}\zeta(\zeta + 1). \tag{7.1.34}$$

Figure 7.5 shows both the present parabolic and the corresponding linear interpolation functions for a one-dimensional three-node element.

Inserting Eq. (7.1.33) into Eq. (7.1.32) yields

$$\theta_{i-1} - 4\frac{3 + m^2}{6 - m^2}\theta_i + \theta_{i+1} = 0, \tag{7.1.35a}$$

which applies to internal nodes, $i = 2, 3, \ldots, n$. For $i = 2$, imposing the condition of insulation $d\theta/d\zeta = 0$ at boundary node $i - 1 = 1$ in Eq. (7.1.33), yields Eq. (7.1.31), by which θ_1 is eliminated from Eq. (7.1.35) for $i = 2$. Similarly, Eq. (7.1.21) applies as boundary condition to eliminate θ_{n+1} from Eq. (7.1.35) for $i = n$.

It may readily be shown that use of the linear interpolation functions of Fig. 7.5 in Eq. (7.1.32) yields

$$\theta_{i-1} - 2\frac{2 + m^2}{2 - m^2}\theta_i + \theta_{i+1} = 0, \tag{7.1.35b}$$

which is severely in error, even for small values of m^2.

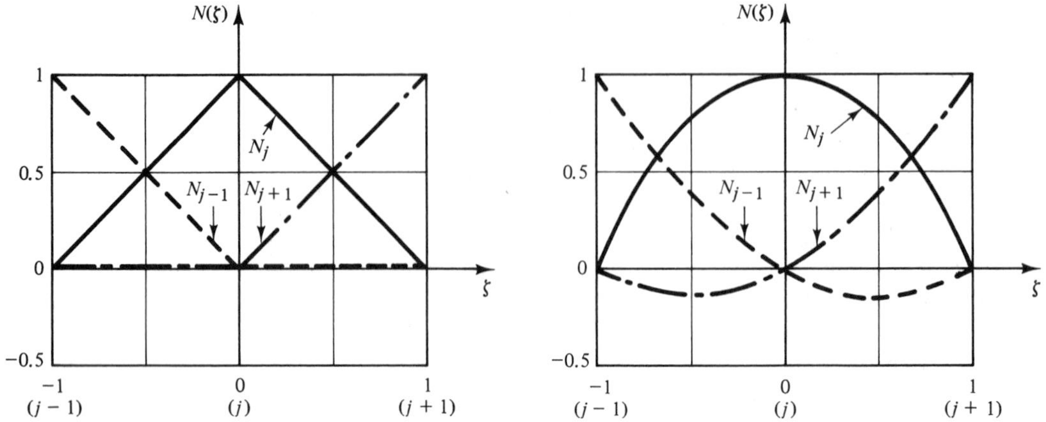

Fig. 7.5 Linear and parabolic interpolation functions for one-dimensional, three-node finite element

Next, we consider the variational finite element formulation. Recalling Ex. 7.1.3, the variational formulation for the element $(CV)_{FE}$ of Fig. 7.4(b) yields

$$I = -\frac{1}{2}\int_{-1}^{1}\left[\left(\frac{d\theta}{d\zeta}\right)^2 + m^2\theta^2\right]d\zeta, \tag{7.1.36}$$

subject to $\delta I = 0$.

Considering θ_i to be the variational parameter to be determined, assuming θ_{i-1} and θ_{i+1} to be fixed boundary conditions for this element (but variational parameters for adjacent elements),* we require that

$$\delta I = \frac{\partial I}{\partial \theta_i}\delta\theta_i \qquad \text{or} \qquad \frac{\partial I}{\partial \theta_i} = 0. \tag{7.1.37}$$

Inserting Eq. (7.1.33) into Eq. (7.1.36) subject to Eq. (7.1.37) yields, after some lengthy but straightforward algebra,

$$\theta_{i-1} - 4\frac{5 + 2m^2}{10 - m^2}\theta_i + \theta_{i+1} = 0, \tag{7.1.38}$$

which applies to internal nodes, $i = 2, \ldots, n$, subject to the boundary conditions, Eqs. (7.1.31) and (7.1.21).

Before proceeding to the solution of discrete formulations, we compare in Table 7.1 the formulations for internal node i of Eqs. (7.1.19), (7.1.35), and (7.1.38) with that of the exact solution, Eq. (7.1.5). The differences between expressions are

*Normally, all nodal values of an element are considered to be variational parameters. This approach, however, would lead to more involved equations precluding ready comparison with the results already derived.

TABLE 7.1 Discretized formulation for internal node

Exact	$\theta_{i-1} - (2\cosh m)\theta_i + \theta_{i+1} = 0$
Finite difference	$\theta_{i-1} - (2 + m^2)\theta_i + \theta_{i+1} = 0$
Finite element—integral (linear profile)	$\theta_{i-1} - 2\frac{2 + m^2}{2 - m^2}\theta_i + \theta_{i+1} = 0$
Finite element—integral (parabolic) and variational (linear)	$\theta_{i-1} - 4\frac{3 + m^2}{6 - m^2}\theta_i + \theta_{i+1} = 0$
Finite element—variational (parabolic)	$\theta_{i-1} - 4\frac{5 + 2m^2}{10 - m^2}\theta_i + \theta_{i+1} = 0$

TABLE 7.2 Series expansion of coefficient to θ_i for small values of m

Exact	$2 + m^2 + \frac{1}{12}m^4 + \frac{1}{360}m^6 + \cdots$
Finite difference	$2 + m^2$
Finite element—integral (linear profile)	$2 + 2m^2 + m^4 + \frac{1}{2}m^6 + \cdots$
Finite element—integral (parabolic) and variational (linear)	$2 + m^2 + \frac{1}{6}m^4 + \frac{1}{36}m^6 + \cdots$
Finite element—variational (parabolic)	$2 + m^2 + \frac{1}{10}m^4 + \frac{1}{100}m^6 + \cdots$

confined to the coefficient of θ_i. These coefficients are shown as function of m^2 in Fig. 7.6, and their series expansions, for small values of m, are shown in Table 7.2. Although we may not reach general conclusions on the basis of an isolated (and simple linear) example, it is clear to see that the finite element variational solution is superior to other comparable solutions. Note, however, that as the number of nodes n is increased, the parameter m, for a given physical problem, decreases and all formulations approach the exact one. This implies that the variational finite element method with fewer elements may be expected to yield the same accuracy as finite difference methods.

Nevertheless, the finite element methods so far have found only limited use for convection problems, particularly those involving turbulent flows, mainly because of the additional labor in the preparatory steps of formulation and the availability of large and fast computers. The absense of exact variational forms for most problems of convection is of little practical consequence since the Galerkin method (or a pseudo-variational form) may be employed with good results. The integral form, giving less accuracy for a given number of elements, is not recommended. Again, the strength of finite element methods lies in their use of fewer elements and their ability to match complex geometrical shapes with great accuracy.

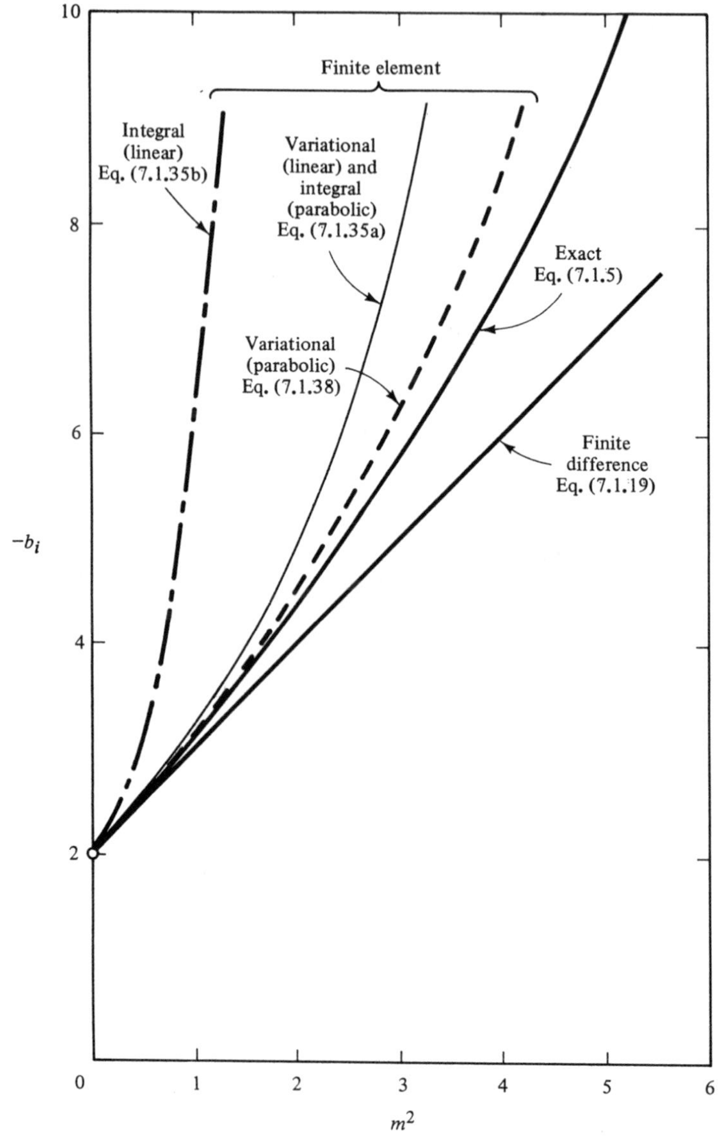

Fig. 7.6 Comparison of finite difference and finite element formulations with exact formulation; coefficient to θ_i

EXAMPLE 7.1.6 Here we consider the solution of the coupled, linear equations arising from a discretized formulation, say from Ex. 7.1.4, Eq. (7.1.29) subject to Eqs. (7.1.31) and (7.1.21). Expressing Eq. (7.1.29) as

$$a_i\theta_{i-1} + b_i\theta_i + c_i\theta_{i+1} = d_i, \tag{7.1.39}$$

the system of linear equations may be written as the *tridiagonal* system

$$\begin{pmatrix} b_2 & c_2 & 0 & 0 & 0 & 0 \\ a_3 & b_3 & c_3 & 0 & 0 & 0 \\ 0 & a_4 & b_4 & c_4 & 0 & 0 \\ \cdot & \cdot & \cdot & \cdot & \cdot & \cdot \\ 0 & \cdots & 0 & a_{n-1} & b_{n-1} & c_{n-1} \\ 0 & \cdots & 0 & 0 & a_n & b_n \end{pmatrix} \begin{pmatrix} \theta_2 \\ \theta_3 \\ \theta_4 \\ \cdot \\ \theta_{n-1} \\ \theta_n \end{pmatrix} = \begin{pmatrix} d_2 \\ d_3 \\ d_4 \\ \cdot \\ d_{n-1} \\ d_n \end{pmatrix}, \tag{7.1.40}$$

or as

$$\boldsymbol{A\theta} = \boldsymbol{D}, \tag{7.1.41}$$

where for $i = 2$ to n,

$$a_i = 1, \qquad b_i = -(2 + m^2), \qquad c_i = 1, \qquad d_i = 0, \tag{7.1.42}$$

subject to subsequent fix-up at boundaries according to Eqs. (7.1.31) and (7.1.21)

$$b_2 \longleftarrow b_2 + \tfrac{4}{3}, \qquad c_2 \longleftarrow c_2 - \tfrac{1}{3}, \qquad d_n \longleftarrow -1. \tag{7.1.43}$$

Because of the local nature of the finite difference (or the finite element) operators, involving only a few terms in the neighborhood of each node, these difference formulations in general lead to systems of equations with banded (sparse) coefficient matrices. To save storage space in the computer while solving such systems, since there is no need to store the whole matrix, each band, such as a_i, b_i, c_i, and d_i of Eq. (7.1.40), is usually stored as a vector.

Equation (7.1.40) may be solved by the *Gauss elimination* method, which consists in first creating zeros below the diagonal (by one forward sweep of linear operations on the equations) and then solving successively the equations (by one backward sweep of back-substitutions).

The FORTRAN listing of subroutine TRI in Table 7.3 shows an algorithm for solving a tridiagonal system of linear equations, numbered from M (= 2) to N (= n). Vectors E, F, and G are dummy vectors of temporary use, and the solution is stored in vector A. Standard subroutine packages are usually available at computing centers for handling a wide variety of sparse matrices more complex than the present one.

TABLE 7.3 FORTRAN listing of subroutine TRI

```
C
      SUBROUTINE TRI(A,B,C,D,M,N)
C     SOLVES TRI-DIAGONAL LINEAR SYSTEM OF EQUATIONS
      DIMENSION A(65),B(65),C(65),D(65),E(65),F(65),G(65)
C     GAUSS ELIMINATION
      E(M)=B(M)
      F(M)=D(M)
      M1=M+1
      DO 10 I=M1,N
      G(I)=A(I)/E(I-1)
      E(I)=B(I)-G(I)*C(I-1)
   10 F(I)=D(I)-G(I)*F(I-1)
C     BACK SUBSTITUTION. ANSWER STORED IN A(I)
      A(N)=F(N)/E(N)
      DO 20 J=M1,N
      I=N+M1-1-J
   20 A(I)=(F(I)-C(I)*A(I+1))/E(I)
      RETURN
      END
```

TABLE 7.4 FORTRAN listing of program calculating temperature distribution and heat transfer from fin, Ex. 7.1.4

```
C
      DIMENSION A(65),B(65),C(65),D(65)
C
      WRITE(6,1001)
 1001 FORMAT('1',
     C 5X,'NUMERICAL SOLUTION TO TEMPERATURE DISTRIBUTION AND            ',/
     C 6X,'HEAT TRASFER FROM FIN, HAVING SPECIFIED BASE TEMPERATURE ',/
     C 6X,'AND INSULATED TIP.                                              ',/
     C 6X,'EXAMPLE 7.1.4  FINITE DIFFERENCE FORMULATION.                   ',/
     C 6X,'                SECOND ORDER BOUNDARY CONDITION AT TIP  ',//,
     C 8X,'CML',5X,'N',5X,'QFD',7X,'QEX',6X,'QINT',6X,'QVAR',7X,'PCT',/)
C
      DATA N,TO,CML/2,1.0,0.0/
C
      DO 30 IC=1,20
      CML=CML+.5
      WRITE(6,1002)
 1002 FORMAT(' ')
C
      DO 30 IN=1,5
      N=2**(IN+1)
C
      CM=CML/FLOAT(N)
      CM2=CM**2
      CML2=CML**2
C     EXACT, INTEGRAL AND VARIATIONAL SOLUTIONS
      QEX=CML*(EXP(CML)-EXP(-CML))/(EXP(CML)+EXP(-CML))
      QINT=CML2/(1.+CML2/3.)
      QVAR=CML2/(1.+2.*CML2/5.)*(1.+CML2/15.)
C
C     NUMERICAL SOLUTION. COEFFICIENT MATRIX - INTERNAL NODES
      DO 10 I=2,N
      A(I)=1.
      B(I)=-(2.+CM2)
      C(I)=1.
   10 D(I)=0.
C     BOUNDARY CONDITIONS - FIX-UP
      B(2)=B(2)+4./3.
      C(2)=C(2)-1./3.
      D(N)=-TO
C
C     SOLVE TRI-DIAGONAL SYSTEM OF LINEAR ALGEBRAIC EQUATIONS
      CALL TRI(A,B,C,D,2,N)
C     COMPUTE BOUNDARY VALUES
      A(1)=4.*A(2)/3.-A(3)/3.
      A(N+1)=TO
C
C     DIMENSIONLESS TOTAL HEAT TRANSFER BY SIMPSON'S FORMULA
      QFD=0.
      DO 20 J=2,N,2
   20 QFD=QFD+4.*A(J)+2.*A(J+1)
      QFD=(QFD+A(1)-A(N+1))/(3.*N*TO)*CML**2
C     PERCENT ERROR IN TOTAL HEAT TRANSFER
      PCT=(QFD-QEX)/QEX*100.
C     PRINT OUT
      WRITE(6,1003)CML,N,QFD,QEX,QINT,QVAR,PCT
 1003 FORMAT(6X,F6.1,I6,5F10.3)
C
   30 CONTINUE
      STOP
      END
```

Table 7.4 shows the FORTRAN listing of a program that calculates the coefficient vectors A, B, C, and D according to the algorithm of Eqs. (7.1.42) and (7.1.43). It then calls the subroutine TRI and calculates the dimensionless heat transfer by Eq. (7.1.22). For comparison the program also calculates the exact solution and the approximate integral and variational solutions.

The results of numerical calculations of total heat transfer for a range of parameters, giving the percent errors compared to the exact solution, are shown in Fig. 7.7.

Here we terminate our study of the foundations for discretized formulations and turn to the solution of convection problems leading to ordinary differential equations.

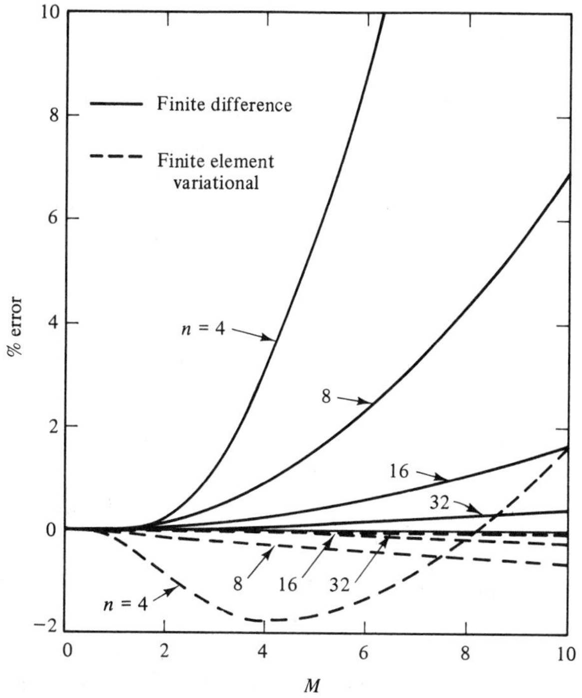

Fig. 7.7 Results of numerical solution by finite difference and finite element formulations; percent error in total heat transfer from fin, Exs. 7.1.4 and 7.1.5, for increasing numbers of elements, $n = 4$, 8, 16, and 32

7.2 ORDINARY DIFFERENTIAL EQUATIONS

In Chapters 3 and 4 we learned that the differential-integral formulation of thermal and momentum boundary layers, subject to approximating functions, leads to the solution of first-order, ordinary differential equations. These equations describe initial value problems in time (Eq. 3.2.11) or in space (Eqs. 3.5.15, 4.2.13, 4.2.39, or 4.3.47). For the equations given, although nonlinear, solutions may be obtained analytically. But for more general problems of momentum boundary layers subject to arbitrary pressure gradients, and for turbulent flows, numerical solution is required.

Furthermore, in Chapter 5, we learned that similarity transformations of boundary layer problems lead to higher-order, nonlinear, ordinary differential equations (Eqs. 5.3.5, 5.3.19, 5.3.24, or 5.3.59). These equations describe two-point boundary value problems. When governed by a single equation, such problems may be solved by a finite difference scheme following Section 7.1. However, difficulties may arise from discretization of derivatives of order greater than 2, from boundary conditions, and the lack of a clearly defined (semi-infinite) solution domain, as well as from nonlinear terms which require linearization, leading to an iterative solution scheme. For these reasons, and in general for problems governed by two or more coupled equations, it is often more convenient to reformulate the problem in terms of sets of coupled first-order ordinary differential equations of an initial value problem. While nonlinearities then present no problem, the solution of this problem now needs an iterative scheme to satisfy the boundary condition at the second boundary.

Let us consider the typical initial value problem in space coordinate x presented by the differential-integral formulation of a developing boundary layer of thickness $\delta(x)$,

$$\frac{d\delta}{dx} = f(x, \delta), \qquad \delta(0) = \delta_0, \tag{7.2.1}$$

where $f(x, \delta)$ is a specified function.

The simplest explicit numerical integration procedure for solving Eq. (7.2.1), employing the forward-difference formula of Eq. (7.1.25), leads to the *Euler* method,

$$\delta(x + h) \simeq \delta(x) + f(x, \delta)h, \tag{7.2.2}$$

where $\delta(x)$, $\delta(x + h)$, . . . is the discrete solution for step size h in x. The truncation error is of order h^2, to which must be added the round-off errors determined by the finite precision employed in the computer. Therefore, even for small values of h, the accumulation of errors makes the method too inaccurate for most work.

The method is improved by a judicious choice of the value f of the derivative to be employed in Eq. (7.2.2). The most common explicit method, having truncation error of order h^5, is the *fourth-order Runge–Kutta* method,

$$\delta(x + h) = \delta(x) + kh, \tag{7.2.3}$$

where the choice of derivative

$$k = \tfrac{1}{6}(k_1 + 2k_2 + 2k_3 + k_4) \tag{7.2.4}$$

is determined in a succession of calculations

$$\begin{aligned} k_1 &= f(x, \delta) \\ k_2 &= f(x + \tfrac{1}{2}h, \delta + \tfrac{1}{2}k_1 h) \\ k_3 &= f(x + \tfrac{1}{2}h, \delta + \tfrac{1}{2}k_2 h) \\ k_4 &= f(x + h, \delta + k_3 h). \end{aligned} \tag{7.2.5}$$

For a discussion of the Runge–Kutta method, its derivation from series expansions, its stability and accuracy, as well as alternative implicit (multistep or predictor–corrector) methods, see, for example, Fröberg (1969). Note that $f(x, \delta)$ must be finite for this procedure. Therefore, integration of a leading-edge boundary layer problem must be started a short distance downstream of the leading-edge discontinuity, for example, from initial conditions determined by an approximate analytical method. We illustrate this point in the first of the following two examples. In the second example we obtain the solution of a third-order equation resulting from a similarity transformation.

EXAMPLE 7.2.1 Reconsider Ex. 10.2.4, the turbulent entrance flow between parallel plates. The approximate differential-integral formulation leads to Eq. (10.2.68), which by the further assumptions given in the example may be reduced to

$$\frac{d\tilde{\delta}}{d\tilde{x}} = f(\tilde{\delta}), \tag{7.2.6}$$

$$f(\tilde{\delta}) = 0.0125\left(\frac{1 - H\tilde{\delta}}{\tilde{\delta}}\right)^{1/4}\left[1 + \frac{(2 + H)H\tilde{\delta}}{1 - H\tilde{\delta}}\right]^{-1}, \tag{7.2.7}$$

subject to $\tilde{\delta}(0) = 0$, where $\tilde{\delta} = \delta_2/h$, $\tilde{x} = (x/h)/\mathrm{Re}_h^{1/4}$, and $H = 1.29$. The entrance length $\tilde{x} = \tilde{L}$ is determined from the condition $\tilde{\delta} = 0.097$.

Since $f(\tilde{\delta})$ is unbounded for $\tilde{x} = 0$, we start the numerical integration from a small positive value, say $\tilde{x}_0$, where $\tilde{\delta}_0 = \tilde{\delta}(\tilde{x}_0)$ is determined analytically by considering Eqs. (7.2.6) and (7.2.7) for $\tilde{\delta} \longrightarrow 0$,

$$\frac{d\tilde{\delta}}{d\tilde{x}} \simeq 0.0125\tilde{\delta}^{-1/4},$$

or, after integration,

$$\tilde{\delta}_0 \simeq 0.0359\tilde{x}_0^{4/5}. \tag{7.2.8}$$

Clearly, $\tilde{x}_0$ may be chosen small enough to yield an accuracy of the initial condition satisfactory for that of the subsequent numerical integration.

Table 7.5 shows the FORTRAN listing of a computer program performing the numerical integration by use of the fourth-order Runge–Kutta method of Eqs. (7.2.3)–(7.2.5). F(D) is conveniently defined in a function subprogram.

The resulting solution, $\tilde{\delta}$ versus $\tilde{x}$, based on use of a constant integration step size DX = 0.01 (= h of Eq. 7.2.3) and a starting point at X0 = 0.01 (= $\tilde{x}_0$ of Eq. 7.2.8), is plotted in Fig. 10.12, indicating that $\tilde{L} \simeq 4.42$. Actually, a 10-fold increase in step size, to DX = 0.1, yields the same result.

EXAMPLE 7.2.2 Reconsider Ex. 5.3.3, the flow over the flat plate for the laminar case, whose differential formulation in the boundary layer approximation accepts the similarity transformation (Eq. 5.3.19)

$$f''' + \tfrac{1}{2}ff'' = 0, \tag{7.2.9}$$

$$f(0) = 0, \qquad f'(0) = 0, \qquad f'(\eta_\infty) = 1,$$

where $u/U_\infty = f'(\eta)$, $\eta = y/(\nu x/U_\infty)$, and η_∞ is a suitably large constant, say, on the order of 10, as shown by experience.

TABLE 7.5 FORTRAN listing of program for Ex. 7.2.1

```
C
      COMMON H
      REAL K,K1,K2,K3,K4
C
      WRITE(6,1001)
 1001 FORMAT('1',
     C 5X,'EXAMPLE 7.2.1                                          ',/
     C 6X,'NUMERICAL INTEGRATION OF EQ.(7.2.6) BY THE             ',/
     C 6X,'FOURTH-ORDER RUNGE-KUTTA METHOD                        ',//
     C 11X,'X',10X,'D',/)
C
      DATA XO,DX,IP/ 0.010, 0.010, 10/
      H=1.29
C
C     APPROXIMATE ANALYTICAL INITIAL CONDITION
      D=(5./4.*0.0125*XO)**0.8
      WRITE(6,1002)XO,D
 1002 FORMAT(6X,2F10.4)
C
      X=XO
      I=0
C     START INTEGRATION LOOP
   10 I=I+1
C     RUNGE-KUTTA PROCEDURE
      K1=F(D)
      K2=F(D+K1*DX/2.)
      K3=F(D+K2*DX/2.)
      K4=F(D+K3*DX)
      K=(K1+2.*K2+2.*K3+K4)/6.
      D=D+K*DX
      X=X+DX
C
      IF(D.LT. 0.09) GOTO 20
      IF((I/IP)*IP .NE. I) GO TO 20
      WRITE(6,1002)X,D
   20 IF(D .LT. 0.100) GO TO 10
C     END OF INTEGRATION LOOP
      STOP
      END
C
      FUNCTION F(D)
      COMMON H
      F=0.0125*(1./D-H)**0.25/(1.+(2.+H)*H*D/(1.-H*D))
      RETURN
      END
```

We wish to obtain a numerical solution to Eq. (7.2.9), in particular to determine $f''(0)$, which enters the expression for the skin friction, Eq. (5.3.20).

Equation (7.2.9) is a two-point boundary value problem and may either be solved as such by a finite difference scheme, or be converted into an initial value problem. In the first case, the required linearization leads to an iterative solution procedure. In the second case, one initial condition at $\eta = 0$, besides the two conditions given, must be guessed and iterated on to satisfy the remaining boundary condition at η_∞. We only outline the first case and then describe the second case in detail and employ it to obtain the solution.

In the first case, following Ex. 7.1.4, we should discretize and linearize Eq. (7.2.9) and then solve implicitly the resulting system of linear, algebraic equations subject to the boundary conditions. Three principal forms of linearizations may be considered. In a *crude linearization*, denoting by an overbar a known solution (to be guessed to start the procedure), Eq. (7.2.9) is replaced by, say,

$$f''' + \tfrac{1}{2}\bar{f}f'' = 0. \tag{7.2.9a}$$

This approach appears to be rather arbitrary and does not immediately apply to terms not composed of products of powers of the dependent variable or its derivatives.

In the *gradient method*, introducing the supposedly small quantities $\tilde{f} = f - \bar{f}$, $\tilde{f}' = f' - \bar{f}', \ldots$, where an overbar denotes a known solution, a nonlinear term $N(f, f', \ldots)$ is formally linearized as

$$N(f, f', \ldots) \simeq \bar{N}(\bar{f}, \bar{f}', \ldots) + \overline{\frac{\partial N}{\partial f}}(f - \bar{f}) + \overline{\frac{\partial N}{\partial f'}}(f' - \bar{f}') + \cdots. \tag{7.2.10}$$

Applying this approach to Eq. (7.2.9) yields

$$f''' + \tfrac{1}{2}\bar{f}f'' + \tfrac{1}{2}f\bar{f}'' = \tfrac{1}{2}\bar{f}\bar{f}''. \tag{7.2.9b}$$

The solution f to this equation is expected to be better than $\bar{f}$, which is then replaced by f, and this iterative scheme is continued until the difference $|f - \bar{f}|$ at all nodes has been reduced to an acceptable small level. It should be noted that the difference $|f - \bar{f}|$ is not necessarily a measure for the absolute error, but only a measure for the change at any iterative step. Also, the discretization of f''' leads to a coefficient matrix having four diagonal bands, requiring two forward sweeps in the Gauss elimination method to create zeros below the diagonal (Problem 7-2).

A more formal approach employs the multidimensional *Newton method*. Again denoting a known approximate solution by $\bar{f}$, we introduce a presumably small perturbation $\tilde{f}$ to be determined such that

$$f = \bar{f} + \tilde{f} \tag{7.2.11}$$

becomes an improved solution. Inserting Eq. (7.2.11) into Eq. (7.2.9) and ignoring terms of order $\tilde{f}^2$, or applying Eq. (7.2.10) to all terms of Eq. (7.2.9), that is,

$$N(f, f', \ldots) \simeq \bar{N}(\bar{f}, \bar{f}', \ldots) + \overline{\frac{\partial N}{\partial f}}\tilde{f} + \overline{\frac{\partial N}{\partial f'}}\tilde{f}' + \cdots, \tag{7.2.12}$$

we obtain the linearized equation governing the perturbation $\tilde{f}$,

$$\tilde{f}''' + \tfrac{1}{2}\bar{f}\tilde{f}'' + \tfrac{1}{2}\bar{f}''\tilde{f} = -\bar{f}''' - \tfrac{1}{2}\bar{f}\bar{f}'', \tag{7.2.9c}$$
$$\tilde{f}(0) = 0, \qquad \tilde{f}'(0) = 0, \qquad \tilde{f}'(\delta_\infty) = 0,$$

where the homogeneous boundary conditions on $\tilde{f}$ follow from Eqs. (7.2.9) and (7.2.11). After discretization and solution of the resulting system of linear equations for nodal values of $\tilde{f}$, the improved solution from Eq. (7.2.11) is introduced into Eq. (7.2.9c) in place of $\bar{f}$ to obtain the next iterate (Problem 7-3). Note that if f converges toward the exact solution, the right-hand side of Eq. (7.2.9c) becomes zero, hence $\tilde{f}$ also approaches zero. From Eq. (7.2.11), $\tilde{f}$ is thus an *absolute error*. Furthermore, if the initial guess $\bar{f}$ lies within the circle of convergence for the problem, the Newton scheme has absolute convergence (see, for example, Dahlquist, Björck & Anderson 1974). Note that a finite difference formulation of the two-point boundary value problem employing one of the foregoing linearizations requires considerable preparation to set up and many nodes for an acceptable accuracy. Therefore, it is more common to employ the second approach, mentioned at the beginning of the section, and convert the problem into an initial value problem, which is now considered.

Equation (7.2.9) is rewritten as the following three, first-order, coupled ordinary differential equations,

$$\begin{aligned} z_1' &= z_2 & & z_1(0) = 0 \\ z_2' &= z_3 & \text{subject to}\quad & z_2(0) = 0 \\ z_3' &= -\tfrac{1}{2}z_1 z_3 & & z_3(0) = f''(0), \end{aligned} \tag{7.2.13}$$

where we have defined $z_1 = f$, $z_2 = f'$, and $z_3 = f''$. The three equations are solved simultaneously by the Runge–Kutta scheme of Eqs. (7.2.3)–(7.2.5) as shown by the algorithm in the FORTRAN listing of the computer program in Table 7.6.

In the first integration we guess the unknown initial value, say, $g = f''(0) = 0.5$ (= Z30). Experience shows that integration to $\eta_\infty = 10$ (= EMAX) more than

TABLE 7.6 FORTRAN listing of program for Ex. 7.2.2

```
C
      REAL   K1,K2,K3,K4,L1,L2,L3,L4,M1,M2,M3,M4
C
      WRITE(6,1001)
 1001 FORMAT('1',
     C 5X,'EXAMPLE 7.2.2                                           ',/
     C 6X,'NUMERICAL INTEGRATION OF THE BLASIUS PROBLEM            ',/
     C 6X,'REWRITTEN AS AN INITIAL VALUE PROBLEM BY USE OF         ',/
     C 6X,'THE FOURTH-ORDER RUNGE-KUTTA METHOD                     ',//
     C 11X,'E',5X,'F=Z1',7X,5HF'=Z2,7X,5HF"=Z3,/)
C
      DATA DE,EMAX,Z30,DZ30,EPS,IP/ 0.050, 10., 0.5,0.20,1.E-06, 4/
C
      IT=0
      IPP=0
C     START ITERATION LOOP IN SHOOTING METHOD
C     INITIAL CONDITIONS
   10 I=0
      IT=IT+1
      Z1=0.
      Z2=0.
      Z3=Z30
      E=0.
C     PRINT WHEN CONVERGED (IPP=1)
      IF(IPP.EQ.0) GO TO 20
      WRITE(6,1002)E,Z1,Z2,Z3
 1002 FORMAT(6X,F7.3,3F11.6)
C     START INTEGRATION LOOP
   20 I=I+1
C     RUNGE-KUTTA PROCEDURE
      K1=Z2
      L1=Z3
      M1=-0.5*Z1*Z3
C
      K2=Z2+L1*DE/2.
      L2=Z3+M1*DE/2.
      M2=-0.5*(Z1+K1*DE/2.)*(Z3+M1*DE/2.)
C
      K3=Z2+L2*DE/2.
      L3=Z3+M2*DE/2.
      M3=-0.5*(Z1+K2*DE/2.)*(Z3+M2*DE/2.)
```

TABLE 7.6 FORTRAN listing of program for Ex. 7.2.2 (cont.)

```
C
      K4=Z2+L3*DE
      L4=Z3+M3*DE
      M4=-0.5*(Z1+K3*DE)*(Z3+M3*DE)
C
      Z1=Z1+(K1+2.*K2+2.*K3+K4)/6.*DE
      Z2=Z2+(L1+2.*L2+2.*L3+L4)/6.*DE
      Z3=Z3+(M1+2.*M2+2.*M3+M4)/6.*DE
      E=E+DE
C
      IF (IPP.EQ.0) GO TO 30
      IF((I/IP)*IP.NE.I) GO TO 30
      WRITE(6,1002)E,Z1,Z2,Z3
   30 IF(E.LT.EMAX) GO TO 20
C     CHECK BOUNDARY CONDITION AT EMAX FOR ERROR=ERR
      ERR=Z2-1.0
      IF(ABS(ERR).LE.EPS) GOTO 40
C     FIRST ITERATION. STORE VALUES FOR NEWTON ITERATION
      IF(IT.GT.1) GOTO 100
      ERR1=ERR
      Z301=Z30
      Z30=Z30+DZ30
      GOTO 10
C     NEWTON STEP
  100 DZ30=  -ERR/(ERR-ERR1)*(Z30-Z301)
C     STORE NEW VALUES AND INTEGRATE ONCE MORE
      ERR1=ERR
      Z301=Z30
      Z30=Z30+DZ30
      GOTO 10
C     END OF ITERATION LOOP. RE-DO INTEGRATION FOR PRINT OUT
  130 CONTINUE
   40 IF(IPP)50,50,60
   50 IPP=1
      GOTO 10
   60 WRITE(6,1003)IT
 1003 FORMAT(/,6X,'NO. OF ITERATIONS = ',I4)
      STOP
      END
```

adequately covers the boundary layer thickness. At this point we check if the remaining boundary condition $f'(\delta_\infty) = 1.0$ (= Z2) is satisfied with desired accuracy (EPS). If not satisfied, we now need a criterion to change the guess (Z30) to improve the agreement in successive iterations.

One simple scheme is the half-interval method. Assuming $f'(\delta_\infty)$ to be a monotoneous function of $f''(0)$, the scheme consists in increasing $f''(0)$ by some initial increment $\delta f''(0)$ (= DZ30), and continue to do so, as long as the error $e = f'(\delta_\infty) - 1.0$ (= Z2 − 1.0 = ERR) is negative. When the error becomes positive, the increment is halved and its sign changed, using the simple FORTRAN algorithm

```
IF ((DZ30 .GT. 0. .AND. ERR .GT. 0.) .OR.
    (DZ30 .LT. 0. .AND. ERR .LT. 0.)) DZ30 = -0.5*DZ30        (7.2.14)
Z30 = Z30 + DZ30
```

before beginning the next integration. This scheme is usually safe, but slow.

Alternatively, we may employ a numerical Newton scheme as done in the program listed. On the assumption that the error e is a monotoneous function of the guess g, a series expansion of $e(g)$ from a first result gives

$$e \simeq e_1 + \left(\frac{\partial e}{\partial g}\right)_1 \Delta g. \tag{7.2.15}$$

Hence, for the error after the subsequent integration to become zero, g_1 should be incremented by

$$\Delta g = -\frac{e_1}{(\partial e/\partial g)_1} \tag{7.2.16}$$

by employing the improved guess

$$g = g_1 + \Delta g. \tag{7.2.17}$$

In the program of Table 7.6 we denote by Z301 and Z30 two successive values of the guess g, and by ERR1 and ERR the corresponding values of the error e in boundary condition at η_∞, and the algorithm of Eqs. (7.2.16) and (7.2.17) becomes

$$\begin{aligned} &\text{DZ30} = -\text{ERR}/(\text{ERR} - \text{ERR1})*(\text{Z30} - \text{Z301}) \\ &\text{Z30} = \text{Z30} + \text{DZ30}. \end{aligned} \tag{7.2.18}$$

Details of the successive updating of storage of previous values appear in the program.

The tabulated results for η, f, f', and f'' are given in Table 7.7. The computation converged (to $|\text{ERR}| \leq \text{EPS} = 10^{-6}$) in seven iterations. For comparison, the half-interval algorithm of Eq. (7.2.14) required 22 iterations. In both cases $f''(0) \simeq 0.33206$, and the dimensionless boundary layer thickness was $\eta(u/U_\infty = 0.99) \simeq 5$.

TABLE 7.7 Results from FORTRAN program listed in Table 7.6, Ex. 7.2.2

$\eta = y\sqrt{\dfrac{U_\infty}{\nu x}}$	f	$f' = \dfrac{u}{U_\infty}$	f''
0.000	0.000000	0.000000	0.332064
0.200	0.006641	0.066409	0.331990
0.400	0.028560	0.132767	0.331476
0.600	0.059736	0.198941	0.330085
0.800	0.108110	0.264714	0.327395
1.000	0.165574	0.329786	0.323013
1.200	0.237953	0.393783	0.318595
1.400	0.322987	0.456270	0.307871
1.600	0.420327	0.516765	0.296668
1.800	0.529526	0.574767	0.282935
2.000	0.650034	0.629776	0.266755
2.200	0.781205	0.681321	0.248354
2.400	0.922304	0.728993	0.228094
2.600	1.072521	0.772466	0.206456
2.800	1.230993	0.811521	0.184008
3.000	1.398824	0.846056	0.161361

TABLE 7.7 Results from FORTRAN program listed in Table 7.6, Ex. 7.2.2 (cont.)

$\eta = y\sqrt{\frac{U_\infty}{\nu x}}$	f	$f' = \frac{u}{U_\infty}$	f''
3.200	1.569111	0.876093	0.139128
3.400	1.748986	0.901773	0.117876
3.600	1.929542	0.923341	0.098086
3.800	2.118048	0.941129	0.080126
4.000	2.305765	0.955529	0.064234
4.200	2.498058	0.966968	0.050519
4.400	2.692380	0.975881	0.038972
4.600	2.888267	0.982694	0.029483
4.800	3.085340	0.987800	0.021871
5.000	3.263294	0.991552	0.015907
5.200	3.481888	0.994255	0.011342
5.400	3.680941	0.996185	0.007927
5.600	3.880312	0.997487	0.005432
5.800	4.079903	0.998385	0.003848
6.000	4.279642	0.998982	0.002402
6.200	4.479479	0.999371	0.001550
6.400	4.679378	0.999620	0.000981
6.600	4.879317	0.999777	0.000808
6.800	5.079281	0.999872	0.000370
7.000	5.279260	0.999930	0.000220
7.200	5.479247	0.999964	0.000129
7.400	5.679240	0.999984	0.000074
7.600	5.879237	0.999995	0.000041
7.800	6.079234	1.000000	0.000023
8.000	6.279231	1.000001	0.000012
8.200	6.479228	1.000001	0.000006
8.400	6.679225	1.000001	0.000003
8.600	6.879222	1.000001	0.000002
8.800	7.079219	1.000001	0.000001
9.000	7.279218	1.000001	0.000000
9.200	7.479213	1.000001	0.000000
9.400	7.679210	1.000001	0.000000
9.600	7.879207	1.000001	0.000000
9.800	8.079204	1.000001	0.000000
10.000	8.279201	1.000001	0.000000

The foregoing method of numerical integration may be readily extended to solve more complex problems, say, Eq. (5.3.59) of Ex. 5.3.6. The two coupled equations of order 2 and 3, respectively, are rewritten as five coupled, first-order equations. Here the numerical Newton–Raphson iteration on two guessed initial conditions requires the solution of two algebraic equations for the corrections to two guesses (Problem 7-5).

As mentioned in Section 5.4, the present method, combined with a finite difference integration scheme in x, may be also employed to solve problems of nonsimilar boundary layer flows governed by Eq. (5.4.7). In the next section we consider other methods for solution of the differential formulation of boundary layer problems.

7.3 PARABOLIC PARTIAL DIFFERENTIAL EQUATIONS

In Section 4.1 we learned that for nearly parallel (boundary layer) flows the streamwise (say x-) diffusion is negligible compared to the streamwise convection of momentum and thermal energy when the Reynolds and Péclet number, respectively, is sufficiently large. Accordingly, the complete differential formulation of the Navier–Stokes equations and the balance of thermal energy reduce to the boundary layer formulation. Both formulations, for two-dimensional flows, are summarized in Table 7.8 for convenience (recall Table 2.4 and Eqs. 4.1.3–4.1.21).

Mathematically, dropping the diffusion in x reduces the highest-order derivative in x from second to first order. Hereby, the governing partial differential equations change from elliptical to parabolic form.

TABLE 7.8 Two-dimensional, incompressible, constant-property convection

(a) Fully two-dimensional (elliptical) formulation

$$\frac{\partial u}{\partial x} + \frac{\partial v}{\partial y} = 0 \tag{7.3.2}$$

$$u\frac{\partial u}{\partial x} + v\frac{\partial u}{\partial y} = f_x - \frac{1}{\rho}\frac{\partial p}{\partial x} + \nu\left(\frac{\partial^2 u}{\partial x^2} + \frac{\partial^2 u}{\partial y^2}\right) \tag{7.3.3}$$

$$u\frac{\partial v}{\partial x} + v\frac{\partial v}{\partial y} = f_y - \frac{1}{\rho}\frac{\partial p}{\partial y} + \nu\left(\frac{\partial^2 v}{\partial x^2} + \frac{\partial^2 v}{\partial y^2}\right) \tag{7.3.4}$$

$$u\frac{\partial T}{\partial x} + v\frac{\partial T}{\partial y} = a\left(\frac{\partial^2 T}{\partial x^2} + \frac{\partial^2 T}{\partial y^2}\right) + \frac{u'''}{\rho c} + \frac{\nu}{c}\Phi_v \tag{7.3.5}$$

(b) Two-dimensional boundary layer (parabolic) formulation

$$\frac{\partial u}{\partial x} + \frac{\partial v}{\partial y} = 0 \tag{4.1.3}$$

$$u\frac{\partial u}{\partial x} + v\frac{\partial u}{\partial y} = f_x - \frac{1}{\rho}\frac{\partial p}{\partial x} + \nu\frac{\partial^2 u}{\partial y^2} \tag{4.1.13}$$

$$0 = -\frac{1}{\rho}\frac{\partial p}{\partial y} \tag{4.1.14}$$

$$U\frac{dU}{dx} = f_x - \frac{1}{\rho}\frac{\partial p}{\partial x} \tag{4.1.16}$$

$$u\frac{\partial T}{\partial x} + v\frac{\partial T}{\partial y} = a\frac{\partial^2 T}{\partial y^2} + \frac{u'''}{\rho c} + \frac{\nu}{c}\left(\frac{\partial u}{\partial y}\right)^2 \tag{4.1.21}$$

The type of a second-order partial differential equation, given by the general form

$$a_{11}\frac{\partial^2 u}{\partial x^2} + 2a_{12}\frac{\partial^2 u}{\partial x \partial y} + a_{22}\frac{\partial^2 u}{\partial y^2} = f\left(x, y; \frac{\partial u}{\partial x}, \frac{\partial u}{\partial y}\right),$$

is determined by the coefficients to the highest-order terms, as

$$a_{12}^2 - a_{11}a_{22} \lesseqgtr 0 \qquad \begin{cases}\text{elliptical}\\ \text{parabolic}\\ \text{hyperbolic.}\end{cases} \tag{7.3.1}$$

Hyperbolic problems arise from fluid compressibility and involve wave propagation, a subject that is outside the scope of this text.

Let us explain the difference between elliptical and parabolic problems in terms of Fig. 7.8. Physically, the solution at point E of the elliptical problem (Fig. 7.8(a)) is determined by the conditions along the complete boundary enclosing the domain. Conversely, a perturbation of the solution at E affects the solution in the whole domain. The solution at point P of the parabolic problem (Fig. 7.8(b)), on the other hand, is determined only by the conditions at the upstream boundary at x_1 and by those along boundaries at y_1 and y_2 for $x \leq x_p$. Conversely, the solution at P, or a perturbation of the solution at P, affects only the solution downstream of x_p, that is, for $x > x_p$.

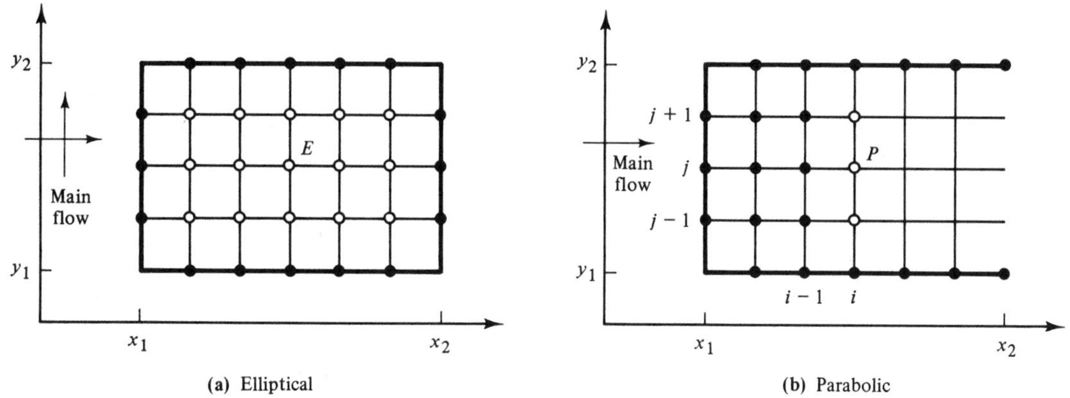

Fig. 7.8 Elliptical and parabolic problems

Note that these physical interpretations are consistent with the number of boundary conditions in x and y, respectively. Also, considering an unsteady diffusion problem, such as that given by the parabolic Eq. (3.2.14), it becomes clear that the solution at any given time depends only on that at earlier times, not on that at later times. In the steady parabolic problems of boundary layer flows, x is a time-like coordinate. Convection in x overshadows the diffusion in x to the extent that the latter may be ignored and information from downstream stations has no effect on the solution at any upstream point. Diffusion in y, on the other hand, is comparable to or overshadows the convection in y. The solution at any x is therefore affected by the conditions at all y, in particular at the boundaries y_1 and y_2.

For these reasons the solution of a parabolic problem may be obtained by a marching procedure in x. Given the solution, say at x_{i-1}, and boundary conditions at $y_1(x_i)$ and $y_2(x_i)$, the solution at interior nodes at x_i is determined (Fig. 7.8(b)).

In the following examples we first illustrate three basic methods for the linear problem of a thermal boundary layer. These methods are the *explicit method*, the implicit *Crank–Nicholson method*, and the implicit *Keller box method*. Then turning to the nonlinear problem of a momentum boundary layer, we illustrate the usefulness of coordinate transformations and discuss methods for handling nonlinearities.

EXAMPLE 7.3.1 Reconsider Ex. 3.5.1, the heat transfer to slug flow associated with a developing thermal boundary layer due to a step change in wall temperature. The differential formulation, preceding the exact solution of Eq. (3.5.4), may be restated as

$$\frac{U}{a}\frac{\partial \theta}{\partial x} = \frac{\partial^2 \theta}{\partial y^2}, \tag{7.3.6}$$

$$\theta(0, y) = 1, \qquad \theta(x, 0) = 0, \qquad \frac{\partial \theta(x, l)}{\partial y} = 0,$$

where $\theta = (T - T_w)/(T_0 - T_w)$, $y = 0$ at the lower plate and $y = l$ at the upper, insulated plate.

For a numerical solution by a finite difference method, the (x, y)-domain is divided into a rectangular grid consisting of, say, $n_j = l/h$ equally sized intervals $\Delta y = h$ from $y = 0$ to l, and $n_i = l/k$ intervals of size $\Delta x = k$ over length l starting from $x = 0$. Then Eq. (7.3.6) is discretized employing finite difference operators discussed in Section 7.1. Since we have a choice between, say, forward- and central-difference formulas, several methods may be considered. In all cases we assume the solution to be known at station x_{i-1} (starting from the upstream boundary condition at $x = 0$) and we seek the solution at x_i.

The *explicit method* employs a first-order forward-difference in x (Eq. 7.1.25) and a second-order central-difference in y (Eq. 7.1.28), evaluated at the known upstream station x_{i-1}, yielding at (i, j) (Fig. 7.9(a))

$$\theta_{i,j} = \theta_{i-1,j} + r(\theta_{i-1,j-1} - 2\theta_{i-1,j} + \theta_{i-1,j+1}), \tag{7.3.7}$$

where the dimensionless parameter $r = ak/Uh^2$ must satisfy the stability criterion

$$r \le \tfrac{1}{2}. \tag{7.3.8}$$

In view of the differences employed in Eq. (7.3.7), the truncation error is of order (k, h^2).

The computational stability of any marching scheme, such as Eq. (7.3.7), may be investigated by introducing a perturbation, say $\epsilon_{i-1,j}$ on $\theta_{i-1,j}$, which then gives rise to a perturbation $\epsilon_{i,j}$ on $\theta_{i,j}$. The scheme is stable provided that

$$|\epsilon_{i,j}| \le |\epsilon_{i-1,j}|,$$

expressing the fact that disturbances do not amplify. Adding the perturbations to $\theta_{i-1,j}$ and $\theta_{i,j}$ in Eq. (7.3.7), and subtracting this equation from the result, yields $\epsilon_{i,j} = -2r\epsilon_{i-1,j}$, and hence, by the foregoing inequality, Eq. (7.3.8).

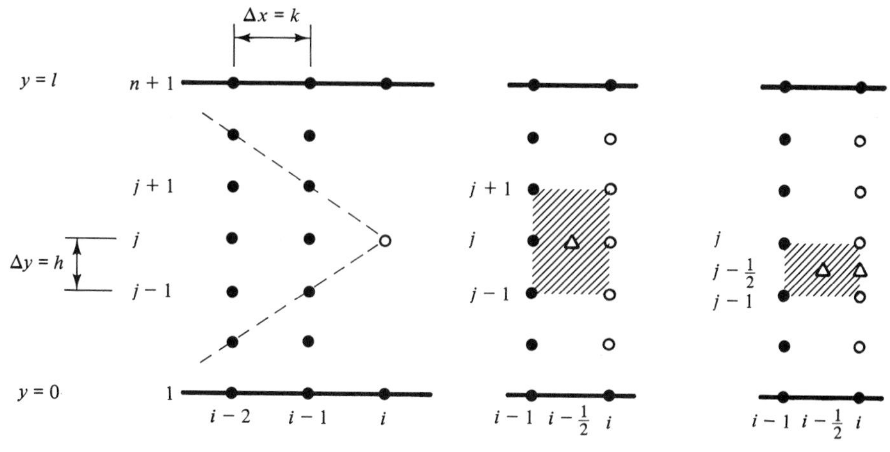

(a) Explicit (b) Crank-Nicholson (c) Keller box

Fig. 7.9 Finite difference grids and some numerical methods for solving parabolic problems (Exs. 7.3.1 and 7.3.2); nodal values: ● known, ○ to be determined, △ centered

Clearly, Eq. (7.3.7) may be solved explicitly for $\theta_{i,j}$, $j = 2, \ldots, n$. Conversely, the solution $\theta_{i,j}$ depends only on $\theta_{i-1,k}$, $k = j-1, j, j+1$, which in turn depend only on $\theta_{i-2,k}$, $k = j-2, \ldots, j+2$, and so on. This limited region of influence (as shown by dashed lines in Fig. 7.9(a)) contradicts the physics and mathematics of the parabolic problem described by Eq. (7.3.6). Although the region of influence may be increased by decreasing the step size k, the explicit method is too inaccurate for most applications and will not be considered further.

The implicit *Crank–Nicholson method* employs a first-order central-difference formula in x and a second-order central-difference formula in y, both evaluated at $(i - \frac{1}{2}, j)$, that is, at $(x_i - \frac{1}{2}h, y_j)$, yielding

$$\frac{U}{a}\frac{\theta_{i,j} - \theta_{i-1,j}}{k} = \frac{1}{2}\left(\frac{\theta_{i-1,j-1} - 2\theta_{i-1,j} + \theta_{i-1,j+1}}{h^2} + \frac{\theta_{i,j-1} - 2\theta_{i,j} + \theta_{i,j+1}}{h^2}\right). \tag{7.3.9}$$

Introducing

$$m = \frac{2Uh^2}{ak} = 2\text{Pe}\,\frac{n_i}{n_j^2}, \qquad n_i = \frac{l}{k}, \qquad n_j = \frac{l}{h}, \tag{7.3.10}$$

where $\text{Pe} = Ul/a$ denotes the Péclet number based on l, Eq. (7.3.9) may be rearranged as

$$\theta_{i,j-1} - (2 + m)\theta_{i,j} + \theta_{i,j+1} = -\theta_{i-1,j-1} + (2 - m)\theta_{i-1,j} - \theta_{i-1,j+1}, \tag{7.3.11}$$

which is of the form Eq. (7.1.39)

$$a_j\theta_{i,j-1} + b_j\theta_{i,j} + c_j\theta_{i,j+1} = d_j, \qquad j = 2, \ldots, n, \tag{7.3.12}$$

implying the solution of a tridiagonal system of $n - 1$ linear equations for interior node values $\theta_{i,j}$, $j = 2, \ldots, n$, in analogy to Eq. (7.1.40).

After computing coefficients in Eq. (7.3.12) according to Eq. (7.3.11), we invoke the boundary conditions of Eq. (7.3.6). At $y = 0$,

$$\theta_{i,1} = 0 \qquad \text{or} \qquad a_2 \longleftarrow 0, \tag{7.3.13}$$

TABLE 7.9 FORTRAN listing of program for Ex. 7.3.1

```
C
      DIMENSION A(65),B(65),C(65),D(65),T(65),T1(65)
      REAL M,NUL,NULI,NULAP,NULB
      WRITE(6,1001)
 1001 FORMAT('1',
     C 5X,'EXAMPLE 7.3.1                                             ',/
     C 6X,'HEAT TRASFER TO SLUG FLOW ASSOCIATED WITH A               ',/
     C 6X,'DEVELOPING THERMAL BOUNDARY LAYER DUE TO A STEP           ',/
     C 6X,'CHANGE IN WALL TEMPERATURE. NUMERICAL SOLUTION            ',/
     C 6X,'USING THE CRANK-NICHOLSON METHOD                          ',//)
C
      DATA NI,NJ,PE,IMAX/  10, 20, 100., 1000/
      IP=1
      IO = FLOAT(NI)*PE/12.
      NJP=NJ+1
      M=PE*2.*FLOAT(NI)/FLOAT(NJ)**2
C     INITIAL CONDITIONS AT  X = 0
      DO 10 J=1,NJP
   10 T1(J)=1.
      WRITE(6,1002) NI,NJ,IMAX,IO,PE,M
 1002 FORMAT(6X,'NI,NJ,IMAX,IO,PE,M = ',4I6,F6.0,F8.3,//,
     C16X, 'XT',11X,'NUL',11X,'NULI',8X,'NULAP',8X,'NULB'/)
C
C     START MARCHING INTEGRATION LOOP
      DO 70 I=1,IMAX
      IF (I.GT.IO/10)IP=NI*2
      XT=FLOAT(I)/FLOAT(NI)/PE
C     COEFFICIENT MATRIX. INTERNAL NODES
      DO 20 J=2,NJ
      A(J)=1.
      B(J)=-(2.+M)
      C(J)=1.
   20 D(J)=-T1(J-1)+(2.-M)*T1(J)-T1(J+1)
C     BOUNDARY CONDITIONS  -  FIX-UP
      A(1)=0.
      A(NJ)=A(NJ)-1./3.
      B(NJ)=B(NJ)+4./3.
C
C     SOLVE TRI-DIAGONAL SYSTEM FOR T(J)
      CALL TRI(A,B,C,D,2,NJ)
      DO 30 J=1,NJ
   30 T(J)=A(J)
C     UPPER-PLATE TEMPERATURE FROM BOUNDARY CONDITION
      T(NJ+1)=T(NJ)*4./3.-T(NJ-1)/3.
C
C     IF((I/10)*10 .EQ. I)WRITE(6,1004) (J,T(J),T1(J),J=1,NJP)
C1004 FORMAT(6X,I4,2F10.6)
C
```

TABLE 7.9 FORTRAN listing of program for Ex. 7.3.1 (cont.)

```
C     COMPUTE HEAT TRANSFER AND PRINT-OUT EVERY IP STEP
      IF((I/IP)*IP .NE. I) GOTO 60
C     DIMENSIONLESS LOCAL HEAT TRANSFER
C     - BY LOCAL GRADIENT
      NUL=0.5*(-3.*T(1)+4.*T(2)-T(3))*FLOAT(NJ)
C     - BY CHANGE IN INTEGRATED TEMPERATURE DISTRIBUTION (SIMPSON)
      NULI=0.
      DO 42 J=2,NJ,2
   42 NULI=NULI+4.*(T1(J)-T(J))+2.*(T1(J+1)-T(J+1))
      NULI= NULI+(T1(1)-T(1)-T1(NJ+1)+T(NJ+1))
      NULI= NULI*PE *FLOAT(NI)/3./FLOAT(NJ)
C     -BY APPROXIMATE INTEGRAL SOLUTION
      IF (I-IO) 44,46,46
C     FIRST DOMAIN
   44 NULAP=1./SQRT(3.*XT)
      GOTO 48
C     SECOND DOMAIN
   46 NULAP=2.*EXP(-3.*FLOAT(I-IO)/FLOAT(NI)/PE)
C     BULK TEMPERATURE. INTEGRATION BY THE SIMPSON FORMULA
   48 TB=0.
      DO 50 J=2,NJ,2
   50 TB=TB+4.*T(J)+2.*T(J+1)
      TB=(TB+T(1)-T(NJP))/3./FLOAT(NJ)
C     LOCAL NUSSELT NUMBER BASED ON L AND TB=T-BULK
      NULB=NUL/TB
      WRITE(6,1003) XT,NUL,NULI,NULAP,NULB
 1003 FORMAT(6X,5F14.3)
C
C     END THIS STEP IN X
C     UPDATE  T1(J)
   60 DO 40 J=1,NJP
   40 T1(J)=T(J)
C
C     END INTEGRATION LOOP
   70 CONTINUE
      STOP
      END
```

which then requires no change of other coefficients. At $y = l$, approximating $\theta_i(y)$ through $\theta_{i,n-1}$, $\theta_{i,n}$, $\theta_{i,n+1}$ by a second-order polynomial and requiring $d\theta_i(l)/dy = 0$, we obtain in analogy to Eq. (7.1.31)

$$\theta_{i,n+1} - \tfrac{4}{3}\theta_{i,n} + \tfrac{1}{3}\theta_{i,n-1} = 0, \tag{7.3.14}$$

which implies the fix-up of coefficients

$$a_n \longleftarrow a_n - \tfrac{1}{3}, \qquad b_n \longleftarrow b_n + \tfrac{4}{3}. \tag{7.3.15}$$

Note that instead of accounting for boundary conditions in the fix-up of coefficients to the first and last of the $n - 1$ equations of interior nodes, we could have included an equation of form Eq. (7.3.12) for $j = 1$ and for $j = n + 1$, respectively, and thus have solved a system of $n + 1$ equations for $\theta_{i,j}, j = 1, 2, \ldots, n + 1$. The equation for $j = 1$, satisfying $\theta_{i,1} = 0$, would require that

$$b_1 = 1, \qquad c_1 = 0, \qquad d_1 = 0. \tag{7.3.16}$$

The equation for $j = n + 1$, employing now the first-order approximation to the condition of zero flux (recall Eq. 7.1.30), $\theta_{i,n} = \theta_{i,n+1}$, would require that

$$a_{n+1} = 1, \qquad b_{n+1} = -1, \qquad d_{n+1} = 0. \tag{7.3.17}$$

The discretization in Eq. (7.3.9), employing central-difference formulas of first and second order, implies a truncation error of order (k^2, h^2). Furthermore, the scheme may be shown to be unconditionally stable.

Table 7.9 shows the FORTRAN listing of a computer program performing the numerical integration of Eq. (7.3.6) by the Crank–Nicholson method. For a choice of values n_i (= NI), n_j (= NJ) and Pe (= PE), given by a DATA-statement, the initial condition $\theta(0, y) = 1.0$ is set in vector T1(J). Note that there is no need to store discrete temperature distributions for all values of x_i. It is sufficient to allocate space for two vectors, one containing the known distribution $\theta_{i-1,j}$ [= T1(J)] and one containing that to be determined $\theta_{i,j}$ [= T(J)]. Coefficients of Eq. (7.3.12) are computed according to Eqs. (7.3.11), (7.3.13), and (7.3.15),* and the tridiagonal system of equations are solved for $\theta_{i,j}, j = 2, \ldots, n$, by subroutine TRI listed in Table 7.3. $\theta_{i,1}$ and $\theta_{i,n+1}$ are then given by Eqs. (7.3.13) and (7.3.14). Before the next step of integration, T(J) is restored in T1(J) and the distribution may be printed out if desired.

The marching integration is performed over IMAX steps, starting from $x = 0$. For comparison with the approximate integral solution of the first and second domains of Ex. 3.5.1, we calculate the penetration distance of Eq. (3.5.12), $x_0 = Ul^2/(12a) = i_0 k$, hence $i_0 = \frac{1}{12} \text{Pe}\, n_i$ (= I0). The program calculates and prints the local dimensionless x-coordinate (XT), $\tilde{x} = (x/l)/\text{Pe} = (i/n_i)/\text{Pe}$, and the local Nusselt number (NUL),

$$\text{Nu}_l = \frac{\bar{h}l}{k} = \frac{\partial\theta(x, 0)}{\partial(y/l)} = \tfrac{1}{2}(-3\theta_{i,1} + 4\theta_{i,2} - \theta_{i,3})n_j. \tag{7.3.18}$$

The latter is derived by approximating $\theta_i(y)$ through $\theta_{i,1}, \theta_{i,2}, \theta_{i,3}$, by a second-order polynomial $\theta_i(y) = a_0 + a_1 y + a_2 y^2$ and then evaluating $\partial\theta_i(0)/\partial y = a_1$.

An alternative and cumulatively more accurate approach is to obtain the heat transfer over length $\Delta x = k$ as the change in integrated thermal energy. From the integral of Eq. (7.3.6), changing sign and using the third boundary condition of Eq. (7.3.6),

$$-\text{Pe}\frac{\partial}{\partial x}\int_0^l \theta(x, y)\, dy = \frac{\partial\theta(x, 0)}{\partial(y/l)} \equiv \text{Nu}_l,$$

or, discretizing in x,

$$\text{Nu}_l = \text{Pe}\frac{1}{k}\int_0^l [\theta_{i-1}(y) - \theta_i(y)]\, dy. \tag{7.3.19}$$

The discretized integral in y, evaluated by the Simpson formula (recall Eq. 7.1.22) is given in the program listing in Table 7.9 (NULI). The program also calculates the approximate integral solution (NULAP), from Eq. (3.5.19) for the first domain and

*Actually, since they are constants, we do not need to recalculate coefficients a_j, b_j, and c_j in every step of the integration. It has been done here for clarity of programming.

from Eq. (3.5.16) for the second domain,

$$\mathrm{Nu}_{l,a} = \begin{cases} (3\tilde{x})^{-1/2}, & \tilde{x} \le \tilde{x}_0 \\ 2\exp[-3(\tilde{x} - \tilde{x}_0)], & \tilde{x} > \tilde{x}_0. \end{cases} \tag{7.3.20}$$

Note that the exact solution for $\tilde{x} \ll 1$, obtained from Eq. (5.1.2), $\mathrm{Nu}_l \simeq (\pi\tilde{x})^{-1/2}$, agrees with the first of Eq. (7.3.20) to within 2%. Far downstream, being based on the constant temperature difference $T_w - T_0$, Nu_l approaches zero. Introducing instead the bulk temperature difference, which for the present slug flow is calculated from

$$\theta_b = \frac{T_b - T_w}{T_0 - T_w} = \frac{1}{l}\int_0^l \theta_i(y)\,dy, \tag{7.3.21}$$

we obtain the Nusselt number usually employed in channel flow (NULB),

$$\mathrm{Nu}_{l,b} = \frac{\mathrm{Nu}_l}{\theta_b}. \tag{7.3.22}$$

$\mathrm{Nu}_{l,b}$ approaches a constant value ($\simeq 2.466$) for fully developed heat transfer as calculated in two iterations on Eqs. (3.6.20) and (3.6.21) for $u/u_m = 1.0$ starting from a cubic profile.

The numerical integration gives $\mathrm{Nu}_{l,b} \simeq 2.469$ for $\tilde{x} \gtrsim 0.4$. Figure 7.10 shows

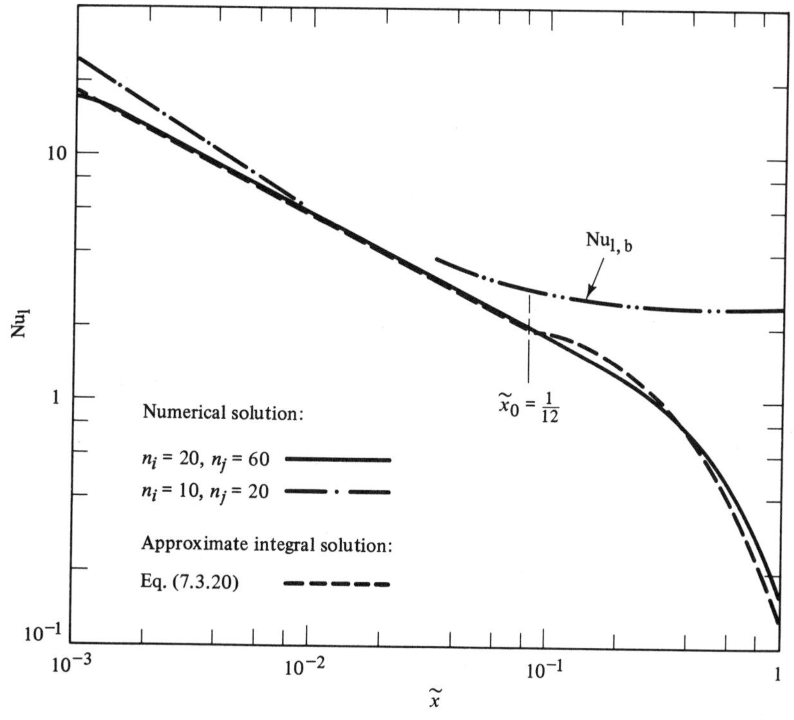

Fig. 7.10 Local heat transfer to slug flow, due to step change in wall temperature, Ex. 7.3.1; Nu_l and $\mathrm{Nu}_{l,b}$ versus $\tilde{x} = (x/l)/\mathrm{Pe}$

Nu_l versus $\tilde{x}$ computed for $n_i = 20$ and $n_j = 60$, compared to the approximate integral solutions of Eq. (7.3.20). The latter appear to be remarkably accurate over the $\tilde{x}$-range shown. Also, the exact solution $(\pi \, \mathrm{Pe})^{-1/2}$ proves to satisfactory for $\tilde{x} \lesssim 0.2$. A comparison of numerical solutions for a range of step sizes in x and y shows that for $\tilde{x} \gtrsim 0.01$ step size has little effect on the solution. The case $n_i = 10$, $n_j = 20$, for example, is significantly in error for $\tilde{x} < 0.01$, as shown in Fig. 7.10. The approach to fully developed heat transfer, $\mathrm{Nu}_{l,b}$ versus $\tilde{x}$, is also shown in the figure for large $\tilde{x}$. Note that $\mathrm{Nu}_{l,b}$ approaches Nu_l for small $\tilde{x}$. Clearly, as seen from Eq. (7.3.6), we could have simplified the algebra of the present example from the onset by employing the dimensionless independent variables y/l and $\tilde{x} = (x/l)/\mathrm{Pe}$.

For the case of a thermal boundary layer developing as a result of a step change in specified heat flux at the wall, instead of wall temperature, at $x \geq 0$, the gradient boundary condition at $y = 0$ now specifies a relation between $\theta_{i,1}$, $\theta_{i,2}$, and $\theta_{i,3}$ to second order (Problem 7-14). For greater accuracy, gradient-type boundary conditions should be handled as described in Ex. 7.1.4 in relation to Fig. 7.4(d) and (e).

Recalling that the thermal boundary layer development is a penetration problem in space, clearly the rectangular grid employed in the present numerical solution is not used effectively for small values of x. Here the accuracy of the solution is limited by the effective number of nodes in the boundary layer. This observation points to the usefulness of an expanding grid whose nodes are all located within the boundary layer. A coordinate transformation, such as one of those derived by similarity transformations, may accomplish this objective; see Section 5.4 and the discussion later in the present section.

EXAMPLE 7.3.2 Reconsider Ex. 7.3.1, now employing a third numerical method, the implicit *Keller box method* (Keller 1970, see also Cebeci & Bradshaw 1977, and Bradshaw, Cebeci & Whitelaw 1981).

This method is based on exclusive use of first-order central differences, ensuring truncation errors of order (k^2, h^2) and unconditional stability. Introducing the auxiliary variable $\theta' = l\,\partial\theta/\partial y$, we rewrite Eq. (7.3.6), which is of second order, as two coupled first-order equations to be discretized at centered nodes (Fig. 7.9(c)), as

$$\left.\begin{aligned} l\frac{\partial\theta}{\partial y} &= \theta' \\ \frac{Ul}{a}\frac{\partial\theta}{\partial x} &= \frac{\partial\theta'}{\partial y} \end{aligned}\right\} \Longrightarrow \left\{\begin{aligned} l\left(\frac{\partial\theta}{\partial y}\right)_{i,j-1/2} &= \theta'_{i,j-1/2} && (7.3.23) \\ \frac{Ul}{a}\left(\frac{\partial\theta}{\partial x}\right)_{i-1/2,j-1/2} &= \left(\frac{\partial\theta'}{\partial y}\right)_{i-1/2,j-1/2}. && (7.3.24) \end{aligned}\right.$$

Recalling Eq. (7.1.27) and defining $\theta'_{i,j-1/2} = \frac{1}{2}(\theta'_{i,j-1} + \theta'_{i,j})$, and so on, Eqs. (7.3.23) and (7.3.24) become

$$\frac{l}{h}(\theta_{i,j} - \theta_{i,j-1}) = \tfrac{1}{2}(\theta'_{i,j-1} + \theta'_{i,j})$$

$$\frac{Ul}{ak}[\tfrac{1}{2}(\theta_{i,j-1} + \theta_{i,j}) - \tfrac{1}{2}(\theta_{i-1,j-1} + \theta_{i-1,j})]$$

$$= \frac{1}{h}[\tfrac{1}{2}(\theta'_{i-1,j} + \theta'_{i,j}) - \tfrac{1}{2}(\theta'_{i-1,j-1} + \theta'_{i,j-1})],$$

or

$$-\theta_{i,j-1} - s\theta'_{i,j-1} + \theta_{i,j} - s\theta'_{i,j} = 0 \tag{7.3.25}$$

$$r\theta_{i,j-1} + \theta'_{i,j-1} + r\theta_{i,j} - \theta'_{i,j} = d_j, \tag{7.3.26}$$

where

$$s = \frac{h}{2l} = \frac{1}{2n_j}, \qquad r = \frac{Ulh}{ak} = \mathrm{Pe}\,\frac{n_i}{n_j}, \qquad d_j = \theta'_{i-1,j} - \theta'_{i-1,j-1} + r(\theta_{i-1,j-1} + \theta_{i-1,j}). \tag{7.3.27}$$

Pairs of Eqs. (7.3.25) and (7.3.26) apply to each node, $j = 2, \ldots, n+1$. Adding a first equation, expressing the wall condition at $y = 0$, $\theta_{i,1} = 0$, and a last equation, expressing the other wall condition at $y = l$, $\theta'_{i,n+1} = 0$, we have the following system of $2(n+1)$ linear equations, suppressing subscript i:

$$\left\{\begin{array}{cccc} \begin{pmatrix} 1 & 0 \\ -1 & -s \end{pmatrix} & \begin{pmatrix} 0 & 0 \\ 1 & -s \end{pmatrix} & \begin{pmatrix} 0 & 0 \\ 0 & 0 \end{pmatrix} & \begin{pmatrix} 0 & 0 \\ 0 & 0 \end{pmatrix} \\ \begin{pmatrix} r & 1 \\ 0 & 0 \end{pmatrix} & \begin{pmatrix} r & -1 \\ -1 & -s \end{pmatrix} & \begin{pmatrix} 0 & 0 \\ 1 & -s \end{pmatrix} & \begin{pmatrix} 0 & 0 \\ 0 & 0 \end{pmatrix} \\ \hline \begin{pmatrix} 0 & 0 \\ 0 & 0 \end{pmatrix} & \begin{pmatrix} r & 1 \\ 0 & 0 \end{pmatrix} & \begin{pmatrix} r & -1 \\ -1 & -s \end{pmatrix} & \begin{pmatrix} 0 & 0 \\ 1 & -s \end{pmatrix} \\ \begin{pmatrix} 0 & 0 \\ 0 & 0 \end{pmatrix} & \begin{pmatrix} 0 & 0 \\ 0 & 0 \end{pmatrix} & \begin{pmatrix} r & 1 \\ 0 & 0 \end{pmatrix} & \begin{pmatrix} r & -1 \\ 0 & 1 \end{pmatrix} \end{array}\right\} \left\{\begin{array}{c} \begin{pmatrix} \theta_1 \\ \theta'_1 \end{pmatrix} \\ \begin{pmatrix} \theta_2 \\ \theta'_2 \end{pmatrix} \\ \hline \begin{pmatrix} \theta_n \\ \theta'_n \end{pmatrix} \\ \begin{pmatrix} \theta_{n+1} \\ \theta'_{n+1} \end{pmatrix} \end{array}\right\} = \left\{\begin{array}{c} \begin{pmatrix} 0 \\ 0 \end{pmatrix} \\ \begin{pmatrix} d_2 \\ 0 \end{pmatrix} \\ \hline \begin{pmatrix} d_n \\ 0 \end{pmatrix} \\ \begin{pmatrix} d_{n+1} \\ 0 \end{pmatrix} \end{array}\right\}. \tag{7.3.28}$$

Equation (7.3.28) may be solved by a standard subroutine for sparse matrices, here having two bands above and two bands below the diagonal.

Alternatively, noting that Eq. (7.3.28) may be written in terms of block matrices

$$\begin{Bmatrix} \mathbf{B}_1 & \mathbf{C}_1 & \mathbf{0} & \mathbf{0} \\ \mathbf{A}_2 & \mathbf{B}_2 & \mathbf{C}_2 & \mathbf{0} \\ \mathbf{0} & \mathbf{A}_n & \mathbf{B}_n & \mathbf{C}_n \\ \mathbf{0} & \mathbf{0} & \mathbf{A}_{n+1} & \mathbf{B}_{n+1} \end{Bmatrix} \begin{Bmatrix} \boldsymbol{\theta}_1 \\ \boldsymbol{\theta}_2 \\ \boldsymbol{\theta}_n \\ \boldsymbol{\theta}_{n+1} \end{Bmatrix} = \begin{Bmatrix} \mathbf{D}_1 \\ \mathbf{D}_2 \\ \mathbf{D}_n \\ \mathbf{D}_{n+1} \end{Bmatrix}, \tag{7.3.29}$$

where definitions of 2×2 matrices $\mathbf{A}_j$, $\mathbf{B}_j$, and $\mathbf{C}_j$ and 2×1 vectors $\boldsymbol{\theta}_j$ and $\mathbf{D}_j$ follow from comparison of Eqs. (7.3.28) and (7.3.29). Equation (7.3.29) is a tridiagonal block-matrix system which may be solved, in analogy to Eq. (7.1.40), by an efficient block elimination method (Keller 1974), comprising the forward sweep

$$\mathbf{E}_1 = \mathbf{B}_1, \qquad \mathbf{F}_1 = \mathbf{D}_1 \tag{7.3.30}$$

$$\left.\begin{aligned} \mathbf{G}_j\mathbf{E}_{j-1} &= \mathbf{A}_j \\ \mathbf{E}_j &= \mathbf{B}_j - \mathbf{G}_j\mathbf{C}_{j-1} \\ \mathbf{F}_j &= \mathbf{D}_j - \mathbf{G}_j\mathbf{F}_{j-1} \end{aligned}\right\} \quad j = 2, \ldots, n+1, \tag{7.3.31}$$

followed by the back-substitution sweep

$$\mathbf{E}_{n+1}\boldsymbol{\theta}_{n+1} = \mathbf{F}_{n+1} \tag{7.3.32}$$

$$\mathbf{E}_j\boldsymbol{\theta}_j = \mathbf{F}_j - \mathbf{C}_j\boldsymbol{\theta}_{j+1}, \qquad j = n, n-1, \ldots, 1. \tag{7.3.33}$$

TABLE 7.10 FORTRAN listing of subroutine TRIBLK

```
      SUBROUTINE TRIBLK(A,B,C,D,NJ,T)
C     SOLVES TRI-DIAGONAL BLOCK-MATRIX SYSTEM  -  2 X 2
      DIMENSION A(65,2,2),B(65,2,2),C(65,2,2),D(65,2),
     C          E(65,2,2),F(65,2),G(65,2,2),T(65,2)
      NJP=NJ+1
C     FORWARD SWEEP
      DO 20 K=1,2
      DO 10 L=1,2
   10 E(1,K,L)=B(1,K,L)
   20 F(1,K)=D(1,K)
      DO 60 J=2,NJP
      J1=J-1
      DET=E(J1,1,1)*E(J1,2,2)-E(J1,1,2)*E(J1,2,1)
      DO 30 K=1,2
      G(J,K,1)=(A(J,K,1)*E(J1,2,2)-A(J,K,2)*E(J1,2,1))/DET
   30 G(J,K,2)=(A(J,K,2)*E(J1,1,1)-A(J,K,1)*E(J1,1,2))/DET
C
      DO 50 K=1,2
      DO 40 L=1,2
   40 E(J,K,L)=B(J,K,L)-G(J,K,1)*C(J1,1,L)-G(J,K,2)*C(J1,2,L)
   50 F(J,K)=D(J,K)-G(J,K,1)*F(J1,1)-G(J,K,2)*F(J1,2)
   60 CONTINUE
C     BACK-SUBSTITUTION SWEEP
      DO 80 I=1,NJP
      J=NJP+1-I
      JP=J+1
      F1=F(J,1)
      F2=F(J,2)
      IF(J.EQ.NJP) GOTO 70
      F1=F1-C(J,1,1)*T(JP,1)-C(J,1,2)*T(JP,2)
      F2=F2-C(J,2,1)*T(JP,1)-C(J,2,2)*T(JP,2)
   70 DET=E(J,1,1)*E(J,2,2)-E(J,1,2)*E(J,2,1)
      T(J,1)=(F1*E(J,2,2)-F2*E(J,1,2))/DET
   80 T(J,2)=(F2*E(J,1,1)-F1*E(J,2,1))/DET
      RETURN
      END
```

Note that auxiliary variables $\boldsymbol{E}_j$, $\boldsymbol{G}_j$ are 2×2 matrices and $\boldsymbol{F}_j$ is a 2×1 vector, implying the solution of systems of four and two algebraic equations, respectively. Table 7.10 gives a FORTRAN listing of subroutine TRIBLK which performs the matrix calculations of Eqs. (7.3.30)–(7.3.33). Note that subroutine TRI of Table 7.3 performs the scalar calculations corresponding to the same equations. We leave to problems (see Problem 7-15) the solution of the present example. Note that Table 7.10 is general and can be further simplified for a given problem, depending on the form of block matrices.

Having introduced the physics of parabolic problems and some numerical methods for their solution, in terms of a linear problem, let us conclude this section by considering briefly the solution of the boundary layer equations of Table 7.8b. The details being too elaborate to mention here, we only comment on the conceptual aspects of the methods and refer readers to special texts on the subject.

First, before we can solve the equations of the boundary layer flow over external or internal surfaces, the *free-stream velocity* $U(x)$, or pressure gradient $\partial p/\partial x$, of Eq. (4.1.16) must be known. If not specified, $U(x)$ is obtained from the inviscid free-stream velocity distribution evaluated at the boundary, assuming the boundary layer to be of negligible thickness. Usually considered to be irrotational (Section 2.4.4), the free-stream flow may be calculated from one of a number of methods for potential flow (Robertson 1965). An introduction to numerical procedures, such as the panel method of distributions of sources over surfaces of arbitrary shape, may be found in Chow (1979).

Next, since the equations of motion are nonlinear and since the numerical methods are ultimately based on the solution of systems of linear equations, any nonlinear terms must be *linearized.* One of the three methods discussed at the beginning of Ex. 7.2.2 may be employed (recall the linearization of Eq. 7.2.9 to one of Eqs. 7.2.9a, b, or c). Often this step is taken in conjunction with the discretization.

Finally, to make effective use of the discretization it is advisable to employ a *coordinate transformation* which ensures that all node points in the direction normal to the main flow lie within the boundary layer. Two approaches that accomplish this objective will be mentioned in the following examples.

EXAMPLE 7.3.3 In the *quasi-similarity method* (Cebeci & Smith 1974) we employ an appropriate similarity transformation. For wall boundary layers this is the Falkner–Skan transformation introduced in Section 5.4.

For the case of variable viscosity, which is required when considering first-order turbulence models (Chapters 10 and 11), the balance of x-momentum, Eq. (5.4.7), may be written as

$$(bf'')' + \frac{m+1}{2}ff'' + m(1-f'^2) = x\left(f'\frac{\partial f'}{\partial x} - f''\frac{\partial f}{\partial x}\right), \tag{7.3.34}$$

$$f(0, x) = f'(0, x) = 0, \qquad f'(\eta_\infty, x) = 1,$$

where

$$\frac{u}{U} = f'(\eta, x); \qquad \eta = \frac{y}{(\nu_0 x/U)^{1/2}}, \qquad m = \frac{x}{U}\frac{dU}{dx}, \tag{7.3.35}$$

$$v = \frac{1}{2}\left(\frac{\nu_0 U}{x}\right)^{1/2} f' - \frac{\partial}{\partial x}[(U\nu_0 x)f], \tag{7.3.36}$$

and where $b = \nu/\nu_0$, ν_0 being a constant reference value, ν the variable viscosity ($\nu + \nu_T$, according to the notation of Chapter 10), and a prime denotes differentiation with respect to η.

Employing the same independent variables, the balance of thermal energy (recall Ex. 5.3.4, but ignoring the dissipation) transforms into

$$(c\theta')' + \frac{1}{2}\Pr f\theta' = x\left(f'\frac{\partial\theta}{\partial x} - \theta'\frac{\partial f}{\partial x}\right), \tag{7.3.37}$$

$$\theta(0, x) = 1, \qquad \theta(\eta_\infty, x) = 0,$$

where

$$\frac{T - T_\infty}{T_w - T_\infty} = \theta(\eta, x), \tag{7.3.38}$$

and where $c = a/a_0$, a_0 being a constant reference value and a the variable diffusivity ($a + a_T$, according to the notation of Chapter 11).

The use of the variable η ensures, for Prandtl numbers not too different from unity, that momentum and thermal boundary layers remain within a constant range, say $\eta_\infty \lesssim 10$ (recall Ex. 7.2.2 for the case of zero pressure gradient) even though the pressure gradient is different from zero. Also, assuming for simplicity $b(\eta)$ and $c(\eta)$ to be specified, the formulation is complete. The discretization and numerical solution of Eqs. (7.3.34) and (7.3.37) may follow several approaches, of which we mention two.

In one approach, after linearizing and discretizing the right-hand sides according to Eq. (5.4.8), the equations are rewritten as sets of three and two coupled first-order ordinary differential equations. These equations are then integrated repeatedly as an initial value problem by the Runge–Kutta procedure while iterating on the missing boundary conditions at $\eta = 0$ to satisfy boundary conditions at η_∞, say by a Newton–Raphson scheme (recall Ex. 7.2.2). During this iteration, nonlinearities are also provided for. When converged at one value of x, we increment x and repeat the procedure, in this way obtaining a stepwise solution while marching in the x-direction.

In another approach, the equations are discretized over the range $\eta = 0$ to η_∞, linearized and solved implicitly by a Crank–Nicholson or Keller box scheme. The latter involves now three dependent variables for Eq. (7.3.34) (say f, f', and f'') and two for Eq. (7.3.37) (say θ and θ'), all discretized in terms of first-order central differences (recall Eqs. 7.3.23 and 7.3.24). The solution of block-matrix systems, of order 3×3 and 2×2, such as Eq. (7.3.29), is required a number of times at a given x while iterating on nonlinear terms before x may be incremented.

A detailed account of the method is given by Cebeci & Smith (1974) (see also Cebeci & Bradshaw 1977 and Bradshaw, Cebeci & Whitelaw 1981). Here, the linearizations are handled by a formal Newton iteration in the manner suggested by Eq. (7.2.9c). Equations for perturbations on the dependent variables, rather than for the variables themselves, are solved, employing the solution from the previous x-position as the first guess for the iteration. The method may be applied to nearly parallel flow in channels as well as be extended to handle axisymmetric flows. The same holds for the other method, which is outlined in the following example and which also makes efficient use of all grid points.

EXAMPLE 7.3.4 In the *method of streamline coordinates* (Patankar & Spalding 1970 and Spalding 1977) use is made of the *von Mises* transformation. Introducing the stream function ψ of two-dimensional plane flow,

$$u = \frac{\partial \psi}{\partial y}, \qquad v = -\frac{\partial \psi}{\partial x}, \tag{2.4.36}$$

we note that the transformation $(x, y) \longrightarrow (\xi, \psi)$ implies, for $\xi = x$, that

$$\begin{aligned}\frac{\partial u}{\partial x} &= \frac{\partial u}{\partial \xi}\frac{\partial \xi}{\partial x} + \frac{\partial u}{\partial \psi}\frac{\partial \psi}{\partial x} = \frac{\partial u}{\partial \xi} - v\frac{\partial u}{\partial \psi} \\ \frac{\partial u}{\partial y} &= \frac{\partial u}{\partial \xi}\frac{\partial \xi}{\partial y} + \frac{\partial u}{\partial \psi}\frac{\partial \psi}{\partial y} = u\frac{\partial u}{\partial \psi}.\end{aligned} \tag{7.3.39}$$

Therefore, the convective terms on the left-hand side of Eq. (4.1.13) transform simply as

$$u\frac{\partial u}{\partial x} + v\frac{\partial u}{\partial y} = u\frac{\partial u}{\partial \xi},$$

and the balance of x-momentum, replacing now ξ by x and dividing by u, becomes

$$\frac{\partial u}{\partial x} = \frac{\partial}{\partial \psi}\left(\nu u \frac{\partial u}{\partial \psi}\right) + \frac{1}{u}\left(f_x - \frac{1}{\rho}\frac{\partial p}{\partial x}\right). \tag{7.3.40}$$

Next, we introduce the dimensionless stream function

$$\omega = \frac{\psi - \psi_I}{\psi_E - \psi_I}, \tag{7.3.41}$$

where $\psi_I(x)$ and $\psi_E(x)$ denote the values of ψ along boundaries of the nearly parallel flow being considered, say wall and free stream. At any x, the physical y-coordinate is related to ω by

$$d\omega = \frac{d\psi}{\psi_E - \psi_I} = u\frac{dy}{\psi_E - \psi_I},$$

or

$$y - y_I = (\psi_E - \psi_I)\int_0^\omega \frac{d\omega}{u}, \tag{7.3.42}$$

from which $y(\omega)$, hence $u(x, y)$, may be determined when $u(x, \omega)$ is known.

Finally, changing variables $(x, \psi) \longrightarrow (x, \omega)$, Eq. (7.3.40) may be written, with $\varphi = u$, in the general form

$$\frac{\partial \varphi}{\partial x} + (a + b\omega)\frac{\partial \varphi}{\partial \omega} = \frac{\partial}{\partial \omega}\left(c\frac{\partial \varphi}{\partial \omega}\right). \tag{7.3.43}$$

Clearly, the balance of thermal energy, with $\varphi = T$, may be written in the same general form. With proper definition of stream function and coefficients, the general form applies to axisymmetrical flows with variable properties.

The discretization and linearization of Eq. (7.3.43) and the numerical solution of systems of such equations by an implicit marching scheme is described in detail by Patankar & Spalding (1970) and Spalding (1977). The method applies to wall bounded or free boundary layers as well as to confined flows in channels. A salient feature of the method is to keep all grid points selected for the interval $\omega = 0$ to 1.0 within the flow considered.

For flow in a channel, having a known geometry, this requirement presents no problem. The grid points are simply spaced across the channel at will, the grid is fixed and—for impermeable walls—ψ_I and ψ_E are constants. However, $\partial p/\partial x$ is not known but is a result of the calculations. The computation at any step in x therefore starts by assuming a value for the pressure gradient $\partial p/\partial x$, which must then be corrected if (after satisfying the balance of momentum) the conservation of mass is not satisfied.

For a boundary layer flow, on the other hand, $\partial p/\partial x$ is known, but the thickness of the boundary layer, hence the position of the grid points in y or ω, are not known. As fluid from the free stream is entrained into the boundary layer, the ω-grid must expand. The computation at any step in x therefore starts by assuming the magnitude of the entrainment flow, which is subsequently checked in the integration step against the conservation of mass to ensure that the outer streamline is just in the free stream. These schemes and other details of the method, which may handle multicomponent, chemically reacting flows, have been tested extensively and form the basis of the computer program GENMIX (Spalding 1977).

Here we terminate our introduction to the numerical solution of boundary layer problems and turn now to those of nonparallel, two-dimensional flows.

7.4 ELLIPTICAL PARTIAL DIFFERENTIAL EQUATIONS

In Section 7.3 we learned the mathematical and physical distinction between parabolic and elliptical problems, and studied the numerical solution of the former. We turn now to problems leading to elliptical partial differential equations.

In the first example, considering a linear problem, we illustrate the discretized finite difference formulation and its solution by point-iteration and block-iteration methods, including the application of relaxation. Next, in a one-dimensional example involving convection and diffusion, we explain the important concept of upstream differences, to be used in convective terms to ensure stability. Turning then to a viscous two-dimensional flow and convection problem, we discuss various formulations resulting from alternative choices of dependent variables, such as u, v, p, T or ψ, ω, T or ψ, T. The further discussion is confined to steady problems of nonparallel flows. However, as noted in Section 4.6, considering flows at higher values of the Reynolds or the Péclet number, there may in fact exist no steady solution to the governing nonlinear equations. Complex and large computational problems should therefore be treated as being unsteady, starting from a state at rest.

EXAMPLE 7.4.1 Reconsider Ex. 3.2.3, the steady, fully developed laminar flow in a channel of rectangular cross section. We wish to compute the velocity distribution and the friction factor numerically.

The differential formulation of Eq. (3.2.32) may be conveniently restated in dimensionless form as the Poisson equation

$$\frac{\partial^2 u}{\partial \eta^2} + m^2 \frac{\partial^2 u}{\partial \zeta^2} = -1, \tag{7.4.1}$$

$$\frac{\partial u(0, \zeta)}{\partial \eta} = 0, \qquad u(1, \zeta) = 0$$

$$\frac{\partial u(\eta, 0)}{\partial \zeta} = 0, \qquad u(\eta, 1) = 0,$$

where u now denotes $u(y, z)/[(-dp/dx)\, l^2/\mu]$, $m = l/L$, $\eta = y/l$, and $\zeta = z/L$ (Fig. 3.8). Equation (7.4.1) describes one symmetric quandrant of the flow, assumed to be parallel and symmetric.

For the numerical solution by a finite difference scheme, the (η, ζ)-domain is divided into a rectangular grid consisting of, say, N equally sized intervals $\Delta\eta = 1/N$ from $\eta = 0$ to 1, and M equally sized intervals $\Delta\zeta = 1/M$ from $\zeta = 0$ to 1 (Fig. 7.11). A square grid results for $N/M = m$.

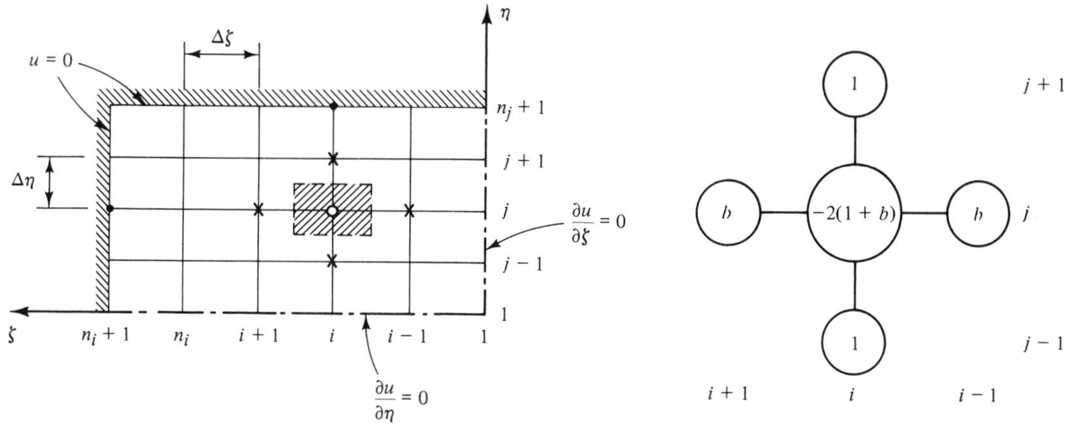

(a) Grid and coordinate system

(b) Five node cell with weight factors for Laplacian

Fig. 7.11 Finite difference formulation, Ex. 7.4.1

Employing the central-difference formula Eq. (7.1.28), the discretized form of Eq. (7.4.1), to order $(\Delta\eta)^2$ and $(\Delta\zeta)^2$, becomes

$$\frac{u_{j-1,i} - 2u_{j,i} + u_{j+1,i}}{(\Delta\eta)^2} + \frac{m^2(u_{j,i-1} - 2u_{j,i} + u_{j,i+1})}{(\Delta\zeta)^2} = -1,$$

or

$$(u_{j-1,i} + u_{j+1,i}) + b(u_{j,i-1} + u_{j,i+1}) - 2(1 + b)u_{j,i} = -\frac{1}{N^2}, \tag{7.4.2}$$

where $b = (mM/N)^2 = (\Delta\eta/\Delta\zeta)^2$. Note that the discretization of the Laplace operator in Eq. (7.4.1), according to Eq. (7.4.2), employs the five-node cell with weight factors shown in Fig. 7.11. Also, for a square grid, $b = 1$, the value at any grid point should apparently equal $\frac{1}{4}$ of the sum of the four adjacent values if $\nabla^2 u = 0$. This statement is the *mean value theorem* for harmonic functions that satisfy the Laplace equation. For a discussion of higher-order finite difference approximations to the Laplace operator, employing up to nine points, we refer to Ames (1977).

Equation (7.4.2) applies to all internal nodes. At $i = M$ and $j = N$ the equation involves known values, such as $u_{j,M} = 0$ and $u_{N,i} = 0$, specified through the no-slip boundary conditions at walls. At $i = 2$ or $j = 2$, the equation involves unknown values, such as $u_{j,1}$ and $u_{1,i}$ along lines of symmetry. Such values may be expressed, to second order, in terms of the two nearest internal node values by approximating the solution by a parabola and requiring the gradient normal to the line of symmetry to be zero, as illustrated by Eq. (7.1.31). Alternatively, also writing Eq. (7.4.2) for nodes along lines of symmetry, noting that $u_{j,2} = u_{j,0}$ and $u_{2,i} = u_{0,i}$ by symmetry, where $i = 0$ refers to $\zeta = -\Delta\zeta$, and $j = 0$ to $\eta = -\Delta\eta$, we obtain, for $i = 1$ and $j \geq 2$,

$$(u_{j-1,1} + u_{j+1,1}) + 2bu_{j,2} - 2(1 + b)u_{j,1} = -\frac{1}{N^2}, \tag{7.4.3}$$

for $j = 1$ and $i \geq 2$,

$$2u_{2,i} + b(u_{1,i-1} + u_{1,i+1}) - 2(1 + b)u_{1,i} = -\frac{1}{N^2}, \tag{7.4.4}$$

and specifically for $i = 1$ and $j = 1$,

$$2u_{2,1} + 2bu_{1,2} - 2(1 + b)u_{1,1} = -\frac{1}{N^2}. \tag{7.4.5}$$

Equations (7.4.2)–(7.4.5) constitute a linear system of $N \times M$ algebraic equations of the form

$$\boldsymbol{A}\boldsymbol{U} = \boldsymbol{B}. \tag{7.4.6}$$

The $N \times M$ square coefficient matrix $\boldsymbol{A}$ has three diagonal bands plus a single side-band N spaces both above and below the diagonal.

A number of sparse-matrix algorithms exist for solving such systems. These methods, which lead to a fully implicit solution, find increasing usage, particularly when boundary conditions are simple. However, the more straightforward algorithms, based on iterative solutions by points or blocks, are still often applied and will be now considered. They require less computer memory and are readily implemented, also for problems involving large numbers of nodes. They are indispensable for nonlinear problems which always need iterative solution procedures.

First, in the explicit, *point-iteration* method we march through the grid from node to node, solving explicitly for $u_{j,i}$ from the appropriate one of Eqs. (7.4.2)–(7.4.5), employing the known (old) values $\bar{u}_{k,l}$ at adjacent nodes available from the previous iteration. In the *Jacobi* method, known values $\bar{u}_{k,l}$ are kept at all nodes until new values $u_{j,i}$ have been computed throughout the grid. Then $\bar{u}_{j,i}$ are updated by $u_{j,i}$. In the *Gauss–Seidel* method, on the other hand, old values $\bar{u}_{j,i}$ are replaced by new ones $u_{j,i}$ as soon as they become available. This latter approach speeds convergence and is preferred.

To further improve convergence, we may employ *successive overrelaxation* (SOR), where, instead of the explicitly computed value $u_{j,i}$, a relaxed value $u^r_{j,i}$ is taken to be the new value at the point,

$$u^r_{j,i} = \bar{u}_{j,i} + \alpha(u_{j,i} - \bar{u}_{j,i}). \tag{7.4.7}$$

The relaxation parameter α controls the rate of convergence. It may be shown (Roache 1976, for example) that $1 \leq \alpha < 2$ for convergence and that the optimum value, for simple problems such as the present one, is given by

$$\alpha_{\text{opt}} = \frac{2(1 - \sqrt{1 - a^2})}{a^2}, \tag{7.4.8}$$

where

$$a = \frac{\cos(\pi/N) + b\sin(\pi/M)}{1 + b}; \qquad b = \left(\frac{\Delta\eta}{\Delta\zeta}\right)^2.$$

For more general problems, α_{opt} cannot be obtained by analytical means and experimentation is needed. For nonlinear problems, it is often necessary to employ *under-relaxation*, that is, $0 < \alpha < 1$, to control convergence.

It may readily be shown that relaxation is equivalent to solving a parabolic time-dependent problem by the explicit method. Iterative methods are therefore often referred to as time-like methods.

Suppose that Eq. (7.4.1) is replaced by the unsteady problem

$$\frac{\partial u}{\partial t} = \frac{\partial^2 u}{\partial \eta^2} + m^2 \frac{\partial^2 u}{\partial \zeta^2} + 1. \tag{7.4.9}$$

Discretizing this equation by use of Eq. (7.3.7) extended to two dimensions in space, denoting by $u^r_{j,i}$ the solution at time $t + \Delta t$, and by an overbar the known one at time t, we have

$$\frac{u^r_{j,i} - \bar{u}_{j,i}}{\Delta t} = \frac{\bar{u}_{j-1,i} - 2\bar{u}_{j,i} + \bar{u}_{j+1,i}}{(\Delta\eta)^2} + \frac{m^2(\bar{u}_{j,i-1} - 2\bar{u}_{j,i} + \bar{u}_{j,i+1})}{(\Delta\zeta)^2} + 1, \tag{7.4.10}$$

or, introducing $b = (m\,\Delta\eta/\Delta\zeta)^2$ and rearranging,

$$u^r_{j,i} = \bar{u}_{j,i} + \frac{2(1 + b)\,\Delta t}{(\Delta\eta)^2}$$

$$\left\{\frac{1}{2(1 + b)}\left[(\bar{u}_{j-1,i} + \bar{u}_{j+1,i}) + b(\bar{u}_{j,i-1} + \bar{u}_{j,i+1}) + \frac{1}{N^2}\right] - \bar{u}_{j,i}\right\}, \tag{7.4.11}$$

which is exactly Eq. (7.4.7) with $\alpha = 2(1 + b)\,\Delta t/(\Delta\eta)^2$, provided that $u_{j,i}$ is obtained from Eq. (7.4.2). The stability criterion of Eq. (7.4.10), in analogy to Eq. (7.3.8), implies $\alpha < 2$ for a square grid. For a discussion of optimum values of α and other iterative schemes, we refer to Roache (1976) and Ames (1977).

To check how well the difference approximation, say Eq. (7.4.2), is satisfied at any stage of the iteration, we may calculate the *residual* at every node

$$r_{j,i} = |\,N^2(u_{j-1,i} - 2u_{j,i} + u_{j+1,i}) + m^2M^2(u_{j,i-1} - 2u_{j,i} + u_{j,i+1}) + 1\,|. \tag{7.4.12}$$

Convergence is usually expressed in terms of the number of iterations n_{10} required to reduce $\max|\,r_{j,i}\,|$ by a factor of 10. However, because it is costly to evaluate Eq. (7.4.12) at all nodes, often the *relative residual*

$$R = \max\left|\frac{u_{j,i} - \bar{u}_{j,i}}{u_{j,i}}\right| \tag{7.4.13}$$

is used instead as an indicator of convergence. It should be noted that the magnitude of R in itself is not an indication of how close the numerical solution is to the exact solution. A small residual merely implies a small change during an iterative step (recall the discussion following Eq. 7.2.12 related to the use of a formal Newton procedure). Ealier relaxation methods selectively reduce only the largest residual in each step. While suitable to calculations by hand for problems involving relatively few nodes, this approach requires a time-consuming search through all nodes at each step and is therefore not employed in computer solutions.

TABLE 7.11 FORTRAN listing of program for Ex. 7.4.1: Solution by point iteration and SOR

```
C
      DIMENSION U(65,65)
      REAL M
      WRITE(6,1001)
 1001 FORMAT('1',
     C 5X,'EXAMPLE 7.4.1                                            ',/
     C 6X,'STEADY, FULLY DEVELOPED LAMINAR FLOW IN A                ',/
     C 6X,'RECTANGULAR CHANNEL                                      ',/
     C 6X,'EXPLICIT POINT-ITERATION BY SOR METHOD OF SOLUTION       ',/
     C 6X,'                                                         ',//)
C
      DATA NJ,NI,M,A/   8,   10,   0.80,   1.00/
      DATA RES,JRES,IRES,IT/ 0., 0, 0, 0/
      NJP=NJ+1
      NIP=NI+1
      B=(M*FLOAT(NI)/FLOAT(NJ))**2
      CNJ2=1./FLOAT(NJ)**2
      PI=3.14159
      A2 =(COS(PI/FLOAT(NJ))+COS(PI/FLOAT(NI)))**2
      AOPT=8.*(1.-SQRT(1.-A2/4.))/A2
C     A=AOPT
      WRITE(6,1002)NJ,NI,M,A,AOPT
 1002 FORMAT(6X,'NI,NJ,M,A,AOPT = ',2I6,3F8.4//,
     C       16X,'IT',9X,'RES',2X,'(JRES,IRES)',5X,'FMRE'/)
C
C     INITIAL GUESS     U-BAR
      DO 10 I=1,NIP
      ZE=FLOAT(I-1)/FLOAT(NI)
      DO 10 J=1,NJP
      ET=FLOAT(J-1)/FLOAT(NJ)
   10 U(J,I)=0.5*(1.-ET**2)*(1.-ZE**2)
C
      IF(IT.EQ.0) GOTO 60
C     START POINT-ITERATION LOOP
   20 CONTINUE
      RES=0.
      IT=IT+1
      DO 50 I=1,NI
      DO 50 J=1,NJ
      IF((I.EQ.1).AND.(J.EQ.1)) GOTO 22
      IF((I.EQ.1).AND.(J.NE.1)) GOTO 24
      IF((I.NE.1).AND.(J.EQ.1)) GOTO 26
      Z=(U(J-1,I)+U(J+1,I) +B*(U(J,I-1)+U(J,I+1))+CNJ2)/2./(1.+B)
      GOTO 30
   22 Z=(U(2,1)+B*U(1,2))/(1.+B)
      GOTO 30
   24 Z=((U(J-1,1)+U(J+1,1))+B*2.*U(J,2)+CNJ2)/2./(1.+B)
      GOTO 30
   26 Z=(2.*U(2,I)+B*(U(1,I-1)+U(1,I+1))+CNJ2)/2./(1.+B)
   30 R1=1.-U(J,I)/Z
C
```

TABLE 7.11 FORTRAN listing of program for Ex. 7.4.1: Solution by point iteration and SOR (cont.)

```
C     CHECK FOR MAX RESIDUAL AND ITS LOCATION
      IF(ABS(R1).LT.RES) GOTO 40
      RES=R1
      JRES=J
      IRES=I
C     NEW OVERRELAXED VALUE
   40 U(J,I)=U(J,I)+A*(Z-U(J,I))
   50 CONTINUE
C
C     MEAN VELOCITY
   60 S=0.25*U(1,1)
      DO 62 J=2,NJ
   62 S=S+0.5*U(J,1)
      DO 64 I=2,NI
   64 S=S+0.5*U(1,I)
      DO 66 I=2,NI
      DO 66 J=2,NJ
   66 S=S+U(J,I)
      UM=S/FLOAT(NJ)/FLOAT(NI)
C
C     FRICTION FACTOR      FMRE = FM*RE
      FMRE=32./(1.+M)**2/UM
C
      WRITE(6,1003)IT,RES,JRES,IRES,FMRE
 1003 FORMAT(14X,I4,E12.3,2I5,F14.4)
      IF(IT.EQ.0) GOTO 20
      IF(ABS(RES) .GT. 0.000100) GOTO 20
C
C     END ITERATION LOOP
      STOP
      END
```

As an alternative, or a supplement, to relaxation one may often speed convergence by periodically computing (say, every fifth or tenth iteration) an improved *extrapolated solution* from consideration of the residuals of two successive iterations. Employing for simplicity a single subscript notation j to refer to all nodes, the system of equations

$$A_{ij}u_j = B_i \tag{7.4.14}$$

in the kth and $(k+1)$th iteration define the residuals r_i^k and r_i^{k+1}

$$A_{ij}^k u_j^k = B_i^k + r_i^k \tag{7.4.15}$$

$$A_{ij}^{k+1} u_j^{k+1} = B_i^{k+1} + r_i^{k+1}. \tag{7.4.16}$$

If A_{ij} and B_i were constants, we could define a constant C_j for node j such that the extrapolated solution

$$u_j^{\text{ext}} = u_j^k + C_j(u_j^{k+1} - u_j^k) \tag{7.4.17}$$

would satisfy Eq. (7.4.14) at node j. Multiplying Eq. (7.4.15) by $1 - C_j$ and Eq. (7.4.16) by C_j, the sum of these results shows that the residual for u_j^{ext} would be zero if

$$C_j = \frac{r_i^k}{r_i^k - r_i^{k+1}}. \tag{7.4.18}$$

Usually, coefficients in Eq. (7.4.14) are not constant but change from iteration to iteration. In a simpler approach one therefore employs a constant value $C_j = C$ in Eq. (7.4.17), obtained from a suitable averaging over all nodes. One choice that seems to work well is based on requiring the normalized square sum of increments in residuals to be zero, yielding

$$C = \frac{\sum_j r_i^{k+1}(r_i^{k+1} - r_i^k)}{\sum_j (r_i^{k+1} - r_i^k)^2}. \tag{7.4.19}$$

Periodic use of C from Eq. (7.4.19) in Eq. (7.4.17) often may reduce n_{10} by a factor of 3 to 4 early in the iterative process. This approach may be justified for nonlinear problems and where evaluation of coefficients is laborious.

Now, to start the iterative solution procedure we need some known approximate solution $\bar{u}(\eta, \zeta)$, for example Eq. (3.2.41) or simply that of second-order polynomials in η and ζ, satisfying the boundary conditions, say

$$\bar{u}(\eta, \zeta) = \tfrac{1}{2}(1 - \eta^2)(1 - \zeta^2). \tag{7.4.20}$$

Table 7.11 shows the listing of a FORTRAN program which, starting from Eq. (7.4.20), computes the velocity distribution by point iteration with the option of SOR by a specified or the optimum value of α given by Eq. (7.4.8). R (= RES) of Eq. (7.4.13) is computed as an indicator of convergence. At the end of each iterative sweep of all nodes the friction factor is calculated according to Eqs. (3.2.55) and (3.2.56), or

$$f_m \,\mathrm{Re} = \frac{32}{(1 + m)^2 u_m}, \tag{7.4.21}$$

where the dimensionless mean velocity is obtained from

$$u_m = \int_0^1 \int_0^1 u(\eta, \zeta)\, d\eta\, d\zeta. \tag{7.4.22}$$

The numerical integration employs a trapesoidal rule for simplicity, implying an area weight factor of 1.0 for internal nodes, 0.5 for nodes along lines of symmetry, and 0.25 at the center node.

Numerical results agree with the exact solution of Fig. 3.11. But the crude numerical integration will underestimate u_m and give too large values of f_m Re when few nodes are used. An improved algorithm for calculating u_m is therefore recommended.

Let us turn now to a second method of solution, that of *block iteration*, where each block of node values is solved implicitly. A column of nodes, $u_{1,i} \cdots u_{N,i}$, for a given value of i is a convenient block (Fig. 7.12(b)), leading to a tridiagonal system of equations, as would a row of nodes, $u_{j,1} \cdots u_{j,M}$, for a given j (Fig. 7.12(c)). Considering blocks to be columns, for example, one iteration then consists in solving implicitly node values for all j at $i = 1, 2, \ldots, M$, successively updating known values $\bar{u}_{j,i}$ by new values $u_{j,i}$ as they become available—or by relaxed values $u^r_{j,i}$ if Eq. (7.4.7) is employed—for a column at a time. This approach is expected to speed convergence because the implicitness makes boundary conditions at $j = 1$ and $j = N + 1$ felt at all internal

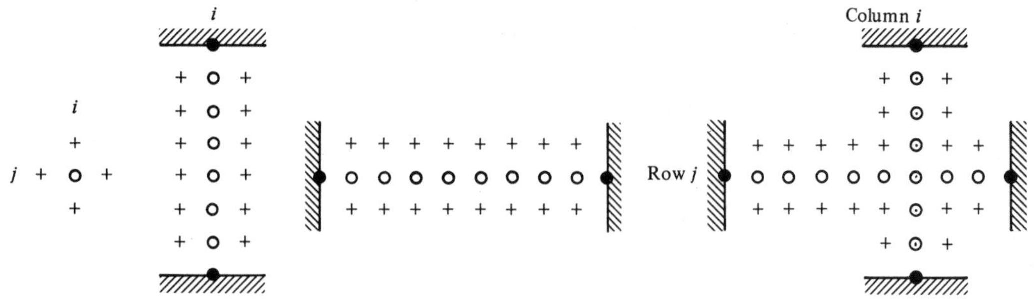

(a) Point iteration (b) Block iteration by column (c) Block iteration by row (d) ADI iteration by column and row

+ known iterate (old value)

o next iterate to be determined (new value)

⊙ intermediate iterate

● known value at boundary

Fig. 7.12 Finite difference grid and some iterative methods for solving elliptical problems, Ex. 7.4.1

nodes along a given column. The old (incorrect) values in adjacent columns of course still slows the convergence.

To determine coefficients of the tridiagonal system of equations associated with column i (≥ 2), we rewrite Eq. (7.4.2) as

$$u_{j-1,i} - 2(1+b)u_{j,i} + u_{j+1,i} = -\frac{1}{N^2} - b(\bar{u}_{j,i-1} + \bar{u}_{j,i+1}),$$

which is of form Eq. (7.1.39), or

$$a_j u_{j-1,i} + b_j u_{j,i} + c_j u_{j+1,i} = d_j. \tag{7.4.23}$$

Hence, for internal nodes, $i \geq 2$ and $j \geq 2$,

$$a_j = 1, \qquad b_j = -2(1+b), \qquad c_j = 1$$
$$d_j = -\frac{1}{N^2} - b(\bar{u}_{j,i-1} + \bar{u}_{j,i+1}), \tag{7.4.24}$$

while for $i \geq 2$ and $j = 1$, in view of Eq. (7.4.4), we correct Eq. (7.4.24) by

$$c_1 = 2. \tag{7.4.25}$$

For the column along the line of symmetry, $i = 1$, Eq. (7.4.24) is modified, according to Eq. (7.4.3) for $j \geq 2$, as

$$d_j = -\frac{1}{N^2} - 2b\bar{u}_{j,2}, \tag{7.4.26}$$

and, according to Eq. (7.4.5), for $j = 1$ as Eq. (7.4.25).

TABLE 7.12 FORTRAN listing for Ex. 7.4.1: Solution by block iteration by columns and SOR

```
C     START BLOCK-ITERATION BY COLUMNS
   20 CONTINUE
      RES=0.
      IT=IT+1
      DO 50 I=1,NI
      DO 30 J=1,NJ
      AI(J)=1.
      BI(J)=-2.*(1.+B)
      CI(J)=1.
      IF(I.EQ.1) GOTO 22
      DI(J)=-CNJ2-B*(U(J,I-1)+U(J,I+1))
      GOTO 24
   22 DI(J)=-CNJ2-2.*B*U(J,I+1)
   24 IF(J.EQ.1) CI(J)=2.
   30 CONTINUE
C     SOLVE TRI-DIAGONAL SYSTEM OF EQUATIONS FOR COLUMN I
      CALL TRI(AI,BI,CI,DI,1,NJ)
      DO 50 J=1,NJ
      Z=AI(J)
      R1=1.-U(J,I)/Z
C
C     CHECK FOR MAX RESIDUAL AND ITS LOCATION
      IF(ABS(R1).LT.RES) GOTO 40
      RES=R1
      JRES=J
      IRES=I
C     NEW OVERRELAXED VALUE
   40 U(J,I)=U(J,I)+A*(Z-U(J,I))
   50 CONTINUE
```

Table 7.12 shows the FORTRAN listing of the foregoing algorithm for coefficients of tridiagonal systems computed for each column, followed by their solution by subroutine TRI of Table 7.3. Table 7.12 replaces the point-iteration portion of the program of Table 7.11. Sample calculations with these two programs show that n_{10} is on the order of 40 for point iteration compared to about 10 for block iteration.

A third method of solution, originally introduced for time-dependent problems by Peaceman & Rachford (1955), is the *alternating direction implicit* (ADI) method. There are several variants of ADI methods. But in principle they all consist of alternatively solving rows and columns implicitly in a block-iteration scheme.

Viewed as a time-like iterative method for a steady problem, the ADI method consists of determining a new iterate $u_{j,i}^{k+1}$ from a known iterate $u_{j,i}^{k}$ in two steps. In the first step the intermediate iterate $u_{j,i}^{k+1/2}$ is found by implicit solution of all node values at column i. Recalling Eq. (7.4.10), the required tridiagonal system may be written

$$\rho_k(u_{j,i}^{k+1/2} - u_{j,i}^{k}) = (u_{j-1,i}^{k+1/2} - 2u_{j,i}^{k+1/2} + u_{j+1,i}^{k+1/2}) + b(u_{j,i-1}^{k} - 2u_{j,i}^{k} + u_{j,i+1}^{k}) + \frac{1}{N^2}. \tag{7.4.27}$$

In the next step the new iterate $u_{j,i}^{k+1}$ is found by implicit solution of all node values at row j, implying the tridiagonal system

$$\rho_k(u_{j,i}^{k+1} - u_{j,i}^{k+1/2}) = (u_{j-1,i}^{k+1/2} - 2u_{j,i}^{k+1/2} + u_{j+1,i}^{k+1/2}) + b(u_{j,i-1}^{k+1} - 2u_{j,i}^{k+1} + u_{j,i+1}^{k+1}) + \frac{1}{N^2}. \tag{7.4.28}$$

In these equations ρ_k is a relaxation parameter having the meaning $(\Delta\eta)^2/(\frac{1}{2}\,\Delta t_k)$ in a time-dependent sense. ρ_k is kept constant while sweeping the whole grid in the two steps comprising one iteration, but it may be changed from one iteration to the next. The implicitness, alternating between columns and rows, making all boundary conditions felt in the interior (Fig. 7.12d), often improves convergence.

Here we terminate our introductory example, involving a linear diffusion problem, and turn next to a one-dimensional convection-diffusion problem.

EXAMPLE 7.4.2 Reconsider the introductory example of Section 7.1. Excluding the heat transfer to the ambient, we now examine the simultaneous effects of axial convection and diffusion in slug flow with velocity U_0, subject to specified temperatures T_0 for $x \leq 0$ and T_1 for $x \geq L$ (Fig. 7.13). We seek the steady temperature $T(x)$ by numerical

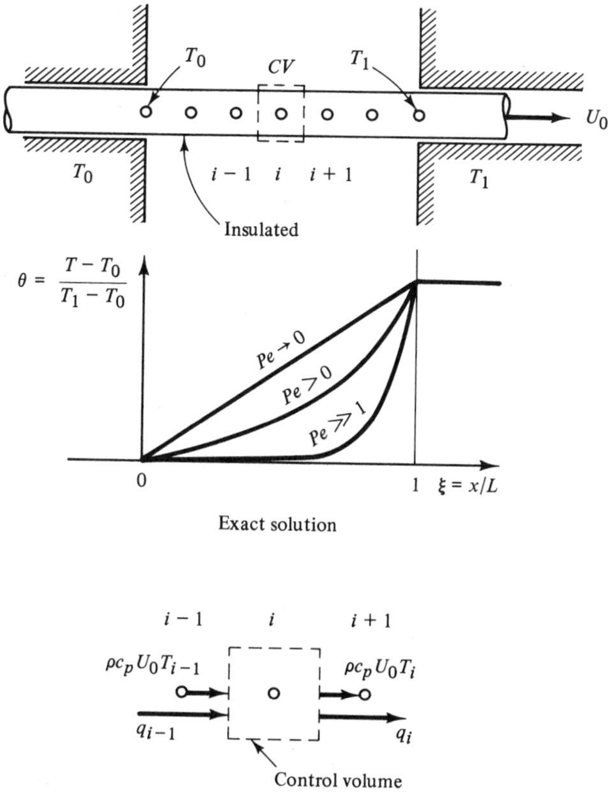

Fig. 7.13 One-dimensional convection and diffusion, Ex. 7.4.2

means. This example explains the need for upstream differencing of convective terms to ensure stability, and it shows the existence of an upper limit of the cell Péclet (or Reynolds) number.

From Eq. (7.1.1) for $M = 0$, the dimensionless differential formulation, $\text{Pe} = U_0 L/a$ denoting the Péclet number, becomes

$$\text{Pe}\frac{d\theta}{d\xi} = \frac{d^2\theta}{d\xi^2}, \tag{7.4.29}$$

$$\theta(0) = 0, \qquad \theta(1) = 1,$$

whose exact solution is

$$\theta(\xi) = \frac{\exp(\text{Pe}\,\xi) - 1}{\exp(\text{Pe}) - 1}. \tag{7.4.30}$$

According to Eq. (2.4.15) the Péclet number is the ratio of enthalpy flow to conduction. When $\text{Pe} \longrightarrow 0$, Eq. (7.4.30) reduces to the conduction solution $\theta = \xi$. When $\text{Pe} \gg 1$, enthalpy flow dominates, and the temperature change from $\theta = 0$ to 1 is confined to a thin region near $\xi = 1$. This implies a steep temperature gradient (Fig. 7.13),

$$\frac{d\theta(1)}{d\xi} = \frac{\text{Pe}\exp(\text{Pe})}{\exp(\text{Pe}) - 1} \simeq \text{Pe} \qquad (\text{Pe} \gg 1). \tag{7.4.31}$$

Now, for a numerical solution, subdividing the interval $\xi = 0$ to 1 into n equidistant intervals of length $\Delta\xi = 1/n$, we discretize Eq. (7.4.29) and solve the resulting tridiagonal system of linear equations by the implicit approach explained in Ex. 7.1.6 (Problem 7-23).

In terms of central-difference formulas, for example, Eqs. (7.1.27) and (7.1.28), Eq. (7.4.29) becomes

$$\tfrac{1}{2}\,\text{Pe}_c\,(\theta_{i+1} - \theta_{i-1}) = \theta_{i-1} - 2\theta_i + \theta_{i+1}, \tag{7.4.32}$$

where

$$\text{Pe}_c = \left(\frac{U_0 L}{a}\right)\Delta\xi = \frac{U_0\,\Delta x}{a} \tag{7.4.33}$$

denotes the *cell Péclet number*. Solving for θ_i,

$$\theta_i = \tfrac{1}{4}[(2 + \text{Pe}_c)\theta_{i-1} + (2 - \text{Pe}_c)\theta_{i+1}]. \tag{7.4.34}$$

Hence, for $\text{Pe}_c \longrightarrow 0$ we find that $\theta_i \longrightarrow \frac{1}{2}(\theta_{i-1} + \theta_{i+1})$, which is in agreement with the conduction limit. For $\text{Pe}_c \gg 1$, on the other hand, $\theta_i \longrightarrow \frac{1}{4}\text{Pe}_c(\theta_{i-1} - \theta_{i+1})$, which for θ increasing with ξ yields absurd and even negative values θ_i. Clearly, Eq. (7.4.32) does not satisfy computational stability unless $\text{Pe}_c < 2$.

Note also, that the discrete solution can at most, for $\theta_i = 0$, $i = 1, 2, \ldots, n$, and $\theta_{n+1} = 1$, represent a temperature gradient at $\xi = 1$ of magnitude

$$\frac{d\theta(1)}{d\xi} \simeq \frac{\theta_{n+1} - \theta_n}{\Delta\xi} = n. \tag{7.4.35}$$

This limit, according to Eq. (7.4.34) for $\theta_{n-1} = 0$, is reached at $\text{Pe}_c = 2$, or $n = \frac{1}{2}\text{Pe}$. If $\text{Pe}_c \geq 2$, the solution shows "wiggles" (Roache 1976), which is a property of the exact solution to the difference equations, a computational instability.

This difficulty may be avoided by the use of *upstream differencing* for convective terms. For $U_0 > 0$, employing the backward-difference formula (see Eq. 7.1.26) for the convective term, we obtain

$$\mathrm{Pe_c}\,(\theta_i - \theta_{i-1}) = \theta_{i-1} - 2\theta_i + \theta_{i+1}, \tag{7.4.36}$$

and solving for θ_i,

$$\theta_i = \frac{(1 + \mathrm{Pe_c})\theta_{i-1} + \theta_{i+1}}{2 + \mathrm{Pe_c}}. \tag{7.4.37}$$

For $\mathrm{Pe_c} \longrightarrow 0$, $\theta_i \longrightarrow \frac{1}{2}(\theta_{i-1} + \theta_{i+1})$ and for $\mathrm{Pe_c} \gg 1$, $\theta_i \longrightarrow \theta_{i-1}$, which is the exact solution. Thus, Eq. (7.4.36) is computationally stable. Also, the steepest gradient at $\xi = 1$, represented by Eq. (7.4.37) for $\theta_{n-1} = 0$, becomes

$$\frac{d\theta(1)}{d\xi} \simeq \frac{1 - 1/(2 + \mathrm{Pe_c})}{\Delta\xi} = \frac{n(n + \mathrm{Pe})}{2n + \mathrm{Pe}}. \tag{7.4.38}$$

Comparing to Eq. (7.4.31), we conclude that $n \geq \frac{1}{2}(1 + \sqrt{5})\,\mathrm{Pe}$ or $\mathrm{Pe_c} \lesssim 0.62$. In practice, such strict requirements are not needed. Upstream differencing ensures computational stability and one often permits local cell Péclet numbers as large as 10. (For a discussion of the relation between upstream differencing and artificial viscosity of time-dependent problems, see Chow 1979.)

For a clarification of the need for upstream differencing of convective terms, consider the control volume approach. Referring to Fig. 7.13, noting that for $U_0 > 0$ the enthalpy of inflow must be based on T_{i-1} and that of outflow on T_i, we immediately obtain

$$\rho c_p U_0 T_i - \rho c_p U_0 T_{i-1} = k\,\frac{T_{i+1} - T_i}{\Delta x} - k\,\frac{T_i - T_{i-1}}{\Delta x}, \tag{7.4.39}$$

which is Eq. (7.4.36). Clearly, it would contradict the physics of the problem if the state of the inflow were determined by a downstream state.

Finally, we may extend the results of the foregoing example to the two-dimensional problem involving the convection of a property b per unit mass. First, the convective terms

$$\rho u\,\frac{\partial b}{\partial x} + \rho v\,\frac{\partial b}{\partial y},$$

whenever possible by use of the conservation of mass, should be written in the *conservation form*

$$\left(\frac{\partial(\rho u b)}{\partial x} + \frac{\partial(\rho v b)}{\partial y}\right)_{i,j} \simeq \frac{\Delta^u(u\rho b)_{i,j}}{\Delta x} + \frac{\Delta^u(v\rho b)_{i,j}}{\Delta y}, \tag{7.4.40}$$

to be discretized by upstream differences Δ^u, as explained below. Equation (7.4.40) is actually the divergence form, representing the net outflow of ρb per unit volume, or outflow minus inflow which appears naturally in the control volume formulation.

Considering the control volume associated with node i, j (Fig. 7.14), we have, for example, the inflow of $(\rho b)_{i-1,j}\,\Delta y$ over the left face provided that the velocity

$$u_{i-1/2,j} = \tfrac{1}{2}(u_{i-1,j} + u_{i,j}),$$

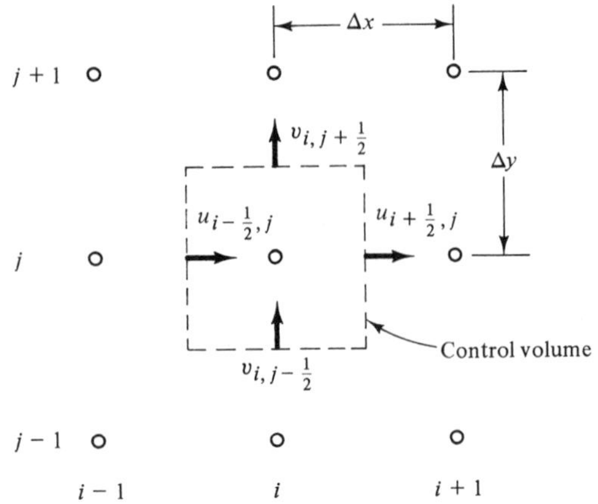

Fig. 7.14 Control volume at node i, j

is positive. Otherwise, we have ouflow of $(\rho b)_{i,j}\,\Delta y$ over this face with the (negative) velocity $u_{i-1/2,j}$.

Employing the absolute-value operator, all contributions to Eq. (7.4.40) may be written in the compact forms suitable to computer programming (Torrance & Rockett 1969),

$$\begin{aligned}\Delta^u(u\rho b) = \tfrac{1}{2}[&(u_{i+1/2,j} - |u_{i+1/2,j}|)(\rho b)_{i+1,j} \\ &+ (u_{i+1/2,j} + |u_{i+1/2,j}| - u_{i-1/2,j} + |u_{i-1/2,j}|)(\rho b)_{i,j} \\ &- (u_{i-1/2,j} + |u_{i-1/2,j}|)(\rho b)_{i-1,j}],\end{aligned} \tag{7.4.41}$$

and

$$\begin{aligned}\Delta^u(v\rho b) = \tfrac{1}{2}[&(v_{i,j+1/2} - |v_{i,j+1/2}|)(\rho b)_{i,j+1} \\ &+ (v_{i,j+1/2} + |v_{i,j+1/2}| - v_{i,j-1/2} + |v_{i,j-1/2}|)(\rho b)_{i,j} \\ &- (v_{i,j-1/2} + |v_{i,j-1/2}|)(\rho b)_{i,j-1}].\end{aligned} \tag{7.4.42}$$

This algorithm ensures the dissappearence of downstream contributions. We shall employ the foregoing differences in the following example involving a two-dimensional convection problem.

EXAMPLE 7.4.3 Reconsider Ex. 4.2.2, the laminar entrance flow between two parallel plates spaced a distance $2l$ apart. Also, let the plates be isothermal at T_w while the fluid upstream is isothermal at T_0. We wish to determine the velocity and temperature distributions for moderate values of the Reynolds number $\mathrm{Re} = U_0 l/\nu$ and the Péclet number $\mathrm{Pe} = U_0 l/a$.

Except close to the entrance [small $\tilde{x} = (x/l)/\mathrm{Re}$] the approximate solution derived in Ex. 4.2.2 based on the boundary layer approximation gives quite satisfactory results when Re is large (Fig. 4.7). For low values of Re, however, the boundary layer approximation is in error, not only near the entrance, but far downstream, as seen for

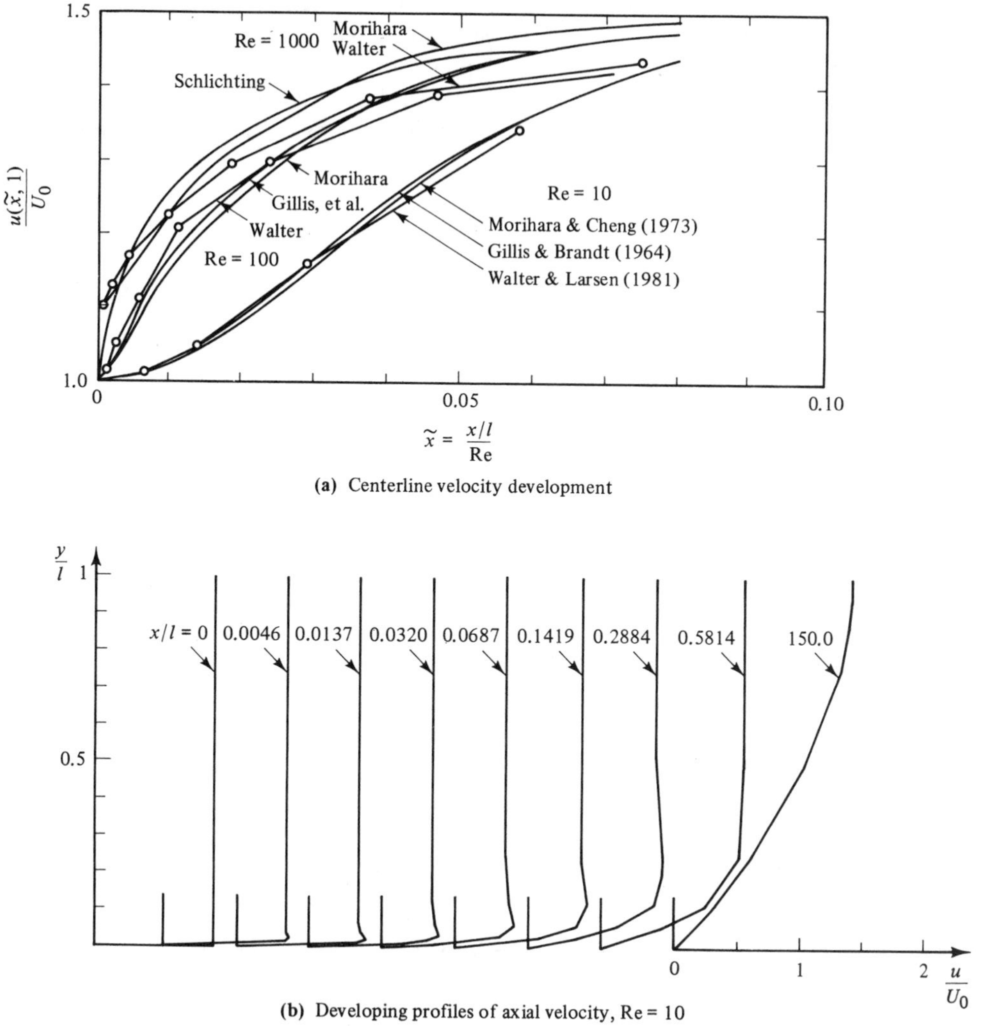

(a) Centerline velocity development

(b) Developing profiles of axial velocity, Re = 10

Fig. 7.15 Entrance flow, Ex. 7.4.3

Re = 10 in Fig. 7.15(a) (Walter & Larsen 1981). The flow is not nearly parallel near the wall, vorticity diffuses upstream, and velocity profiles differ markedly by overshoot near the wall from those employed in the boundary layer model (Fig. 7.15(b)). For a reliable solution at low values of Re, the complete two-dimensional (elliptical) formulation should therefore be employed.

Let us discuss the numerical solution of the present problem by considering in turn each of the three alternative differential formulations of two-dimensional incompressible convection problems given in Section 2.4.3. As part of the formulation we shall in particular stress the boundary conditions in each case.

First, in terms of the basic (primitive) dependent variables u, v, p, T, the differential formulation is given in Table 7.8(a). Figure 7.16(a) shows the corresponding boundary conditions for the rectangular domain, covering one symmetrical half of the channel entrance of length L, to be considered in the computation. The length L is chosen to be long enough that velocity and temperature distributions are fully

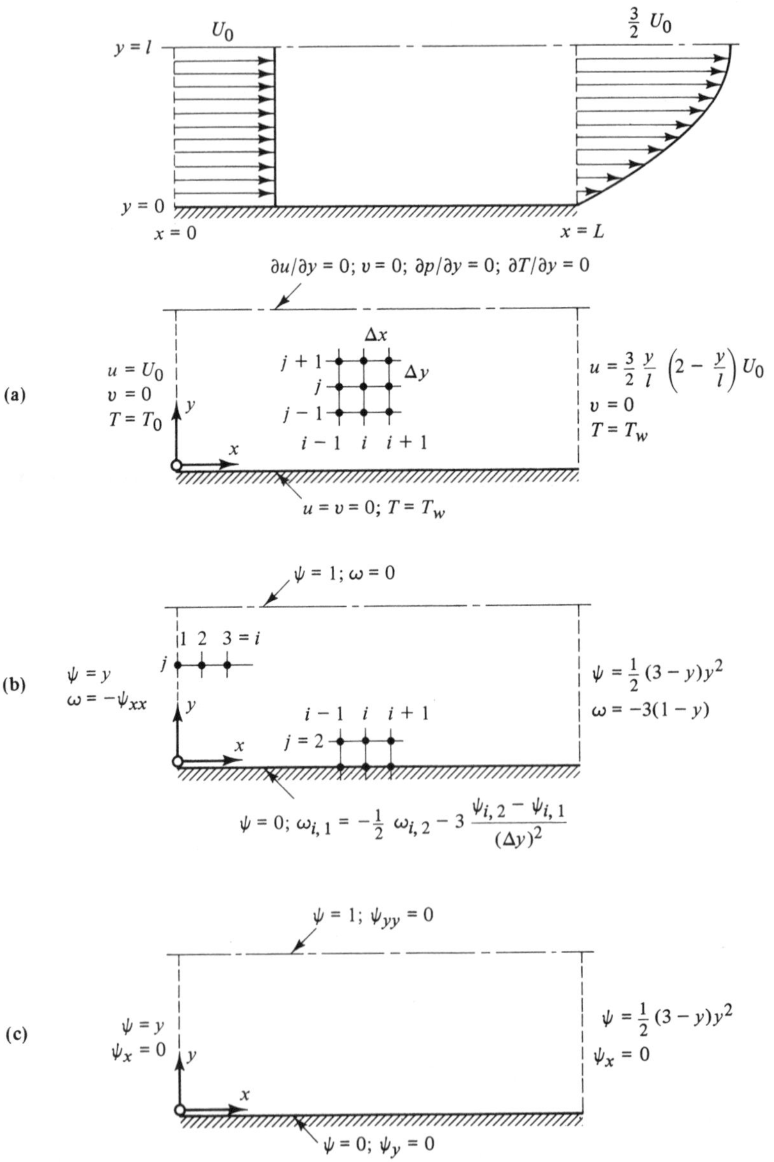

Fig. 7.16 Boundary conditions for entrance flow, Ex. 7.4.3

developed at $x = L$. The boundary conditions of Fig. 7.16(a) follow from the statement of the problem, including the assumptions of symmetry, zero slip at the wall, a fully developed parabolic velocity profile and a uniform temperature at $x = L$.

The discussion of the numerical solution based on use of primitive variables, being beyond the scope of the text, will be confined to a summary of the principal steps. Note that for the case of constant properties, the fluid motion is independent of the temperature distribution and may be solved separately from Eqs. (7.3.2)–(7.3.4) for a steady problem. Then, as will be elaborated later, the temperature distribution may be computed from Eq. (7.3.5).

One way of solving Eqs. (7.3.2)–(7.3.4), by a point- or block-iterative method, consists if the following steps:

a. Assume known distributions $\bar{u}$, $\bar{v}$, $\bar{p}$.
b. Compute the velocity field u^*, v^* from Eqs. (7.3.3)–(7.3.4) based on the pressure field $\bar{p}$.
c. Correct u^*, v^* to obtain the new iterate u, v by satisfying the conservation of mass Eq. (7.3.2) at each node.
d. Compute the pressure field p based on the velocity field u, v from Eq. (2.4.42).

Steps b to d are repeated to attain convergence. Note that unless the correct pressure distribution is employed in step b the velocity field, u^*, v^* will not satisfy the conservation of mass. For various strategies, their details as well as a discussion of the proper discretization of pressure gradients using staggered grids, we refer to Roache (1976) and Patankar (1980) and the references cited therein. While boundary conditions on u, v present no problem, those on p employed in step *d* need particular care.

Involving no actual recirculation, the present problem may be also solved by employing the *parabolized Navier–Stokes equations*, obtained by neglecting the axial diffusion terms $\partial^2 u/\partial x^2$ and $\partial^2 v/\partial x^2$ in Eqs. (7.3.3)–(7.3.4). Computationally, the elliptical nature is preserved through the pressure distribution, which effectively transmits downstream information upstream. For numerical details of this approach, see Chilukuri & Pletcher (1980).

Second, let us turn to the formulation in terms of dependent variables ψ, ω, T given in Table 2.6 as obtained by elimination of the pressure. The steady and dimensionless form of these equations, ignoring buoyancy, dissipation, and energy sources, and employing the conservation of mass to rewrite convection terms in their conservation (divergence) form, becomes

$$\frac{\partial^2 \psi}{\partial x^2} + \frac{\partial^2 \psi}{\partial y^2} = -\omega, \tag{7.4.43}$$

$$\frac{\partial^2 \omega}{\partial x^2} + \frac{\partial^2 \omega}{\partial y^2} = \frac{U_0 l}{\nu}\left[\frac{\partial}{\partial x}\left(\frac{u}{U_0}\omega\right) + \frac{\partial}{\partial y}\left(\frac{v}{U_0}\omega\right)\right], \tag{7.4.44}$$

$$\frac{\partial^2 \theta}{\partial x^2} + \frac{\partial^2 \theta}{\partial y^2} = \frac{U_0 l}{a}\left[\frac{\partial}{\partial x}\left(\frac{u}{U_0}\theta\right) + \frac{\partial}{\partial y}\left(\frac{v}{U_0}\theta\right)\right], \tag{7.4.45}$$

where x and y denote x/l and y/l, respectively, $\theta = (T - T_w)/(T_0 - T_w)$, and

$$\omega = \frac{1}{U_0}\left(\frac{\partial v}{\partial x} - \frac{\partial u}{\partial y}\right); \qquad \frac{u}{U_0} = \frac{\partial \psi}{\partial y}; \qquad \frac{v}{U_0} = -\frac{\partial \psi}{\partial x}. \tag{7.4.46}$$

In view of the elliptical nature of Eqs. (7.4.43)–(7.4.45), one condition for each of the variables ψ, ω, and T has been specified along the closed boundary of the computational domain as shown in Fig. 7.16(b).

Arbitrarily selecting $\psi = 0$ along the wall, Eq. (7.4.46) implies that $\psi = 1$ along the centerline. While other boundary conditions follow readily from those shown in Fig. 7.16(a), those on vorticity at the inlet and along the wall deserve explanation.

At $x = 0$ we cannot put $\omega = 0$, since vorticity should be permitted to diffuse upstream from the wall region for moderate values of Re. Instead, we apply Eq. (7.4.43) to get

$$x = 0: \qquad \omega = -\frac{\partial^2 \psi}{\partial x^2}. \tag{7.4.47}$$

Approximating $\psi_j(x)$ by a parabola through the first three nodes in x, Eq. (7.4.47) is discretized, for an equidistant grid, to

$$x = 0: \qquad \omega_{1,j} = -\frac{(\psi_{1,j} - 2\psi_{2,j} + \psi_{3,j})}{(\Delta x)^2}. \tag{7.4.48}$$

In an iterative scheme of solution, given a known distribution $\bar{\psi}_{i,j}$ over internal nodes, the boundary values $\omega_{1,j}$ are updated from Eq. (7.4.48). Along the wall $y = 0$, assuming the flow to be essentially parallel, $\partial\psi/\partial x \simeq 0$ and $\partial/\partial x \simeq 0$, Eqs. (7.4.43) and (7.4.44) reduce to

$$y = 0: \qquad \frac{\partial^2 \psi}{\partial y^2} \simeq -\omega; \qquad \frac{\partial^2 \omega}{\partial y^2} \simeq 0, \tag{7.4.49}$$

implying a linear variation in $\omega_i(y)$,

$$y = 0: \qquad \omega_i(y) \simeq \omega_{i,1} + (\omega_{i,2} - \omega_{i,1})\frac{y}{\Delta y},$$

hence, after twice integration of the first of Eq. (7.4.49), a cubic variation in $\psi_i(y)$, yielding the relation

$$y = 0: \qquad \omega_{i,1} \simeq -\tfrac{1}{2}\,\omega_{i,2} - \frac{3(\psi_{i,2} - \psi_{i,1})}{(\Delta y)^2}, \tag{7.4.50}$$

from which the boundary value $\omega_{i,1}$ can be updated from known values at internal nodes. Alternative expressions may be found in the literature (Problem 7-24). It may be said that the advantage gained by elimination of pressure from the equations shows up as difficulties in the boundary conditions on the vorticity.

Having established the boundary conditions we next discretize Eqs. (7.4.43)–(7.4.45). Although stretched grids are often used in practice, employing more closely spaced nodes in regions of rapid variation of the dependent variables, such as near the entrance and near the wall, we here consider a constant grid size for simplicity,

$\Delta x = \Delta y = 1/n$, and obtain

$$(\psi_{i-1,j} - 2\psi_{i,j} + \psi_{i+1,j}) + (\psi_{i,j-1} - 2\psi_{i,j} + \psi_{i,j+1}) = -\frac{\omega_{i,j}}{n^2}, \tag{7.4.51}$$

$$(\omega_{i-1,j} - 2\omega_{i,j} + \omega_{i+1,j}) + (\omega_{i,j-1} - 2\omega_{i,j} + \omega_{i,j+1}) = \mathrm{Re_c}[\Delta^u(u\omega)_{i,j} + \Delta^u(v\omega)_{i,j}], \tag{7.4.52}$$

$$(\theta_{i-1,j} - 2\theta_{i,j} + \theta_{i+1,j}) + (\theta_{i,j-1} - 2\theta_{i,j} + \theta_{i,j+1}) = \mathrm{Pe_c}[\Delta^u(u\theta)_{i,j} + \Delta^u(v\theta)_{i,j}], \tag{7.4.53}$$

where upstream differences Δ^u are defined by Eqs. (7.4.41) and (7.4.42), velocities are computed from central-difference formulas

$$u_{i,j} = \frac{\frac{1}{2}(\psi_{i,j+1} - \psi_{i,j-1})}{\Delta y}; \qquad v_{i,j} = -\frac{\frac{1}{2}(\psi_{i+1,j} - \psi_{i-1,j})}{\Delta x}, \tag{7.4.54}$$

and $\mathrm{Re_c} = U_0\,\Delta x/\nu$, $\mathrm{Pe_c} = U_0\,\Delta x/a$ denote the *cell Reynolds number* and *cell Péclet number*, respectively.

The simplest, explicit point-iterative scheme, solving first Eqs. (7.4.51) and (7.4.52), consists of the following steps:

a. Assume known distributions $\bar{\psi}, \bar{\omega}$ (say, uniform ones), and set boundary conditions.
b. Update boundary conditions from Eqs. (7.4.48) and (7.4.50).
c. Compute u, v at all nodes from Eq. (7.4.54).
d. Sweep all nodes, solving for ω_{ij}, considering all other terms to be known in Eq. (7.4.52), and updating ω_{ij}, say by SOR.
e. Sweep all nodes, solving for $\psi_{i,j}$, considering all other terms to be known in Eq. (7.4.51), and updating $\psi_{i,j}$, say by SOR.

Steps b through e are repeated until convergence has been achieved. Note that the nonlinear convective terms on the right-hand side of Eq. (7.4.52) are based on known values in this approach.

Convergence may be speeded only mariginally by employing block iteration by columns in Eqs. (7.4.51) and (7.4.52), separately. Combining the equations, however, and including Eq. (7.4.50) into the implicit scheme, provides a more significant gain (Problem 7-25). A number of fully implicit schemes are discussed by Roache (1975). Employing a Laplacian driver and a Poisson solver, $\psi_{i,j}$ from Eq. (7.4.51) and ω_{ij} from Eq. (7.4.52) are solved at all internal nodes at once. Denoting by an overbar a known iterate, we may summarize the use of a Laplacian driver in a fully implicit approach in the following steps:

a. Assume $\bar{\psi}, \bar{\omega}$.
b. Get new ω from $\nabla^2\omega = \mathrm{Re}(\bar{\psi}_y\bar{\omega}_x - \bar{\psi}_x\bar{\omega}_y)$.
c. Get new ψ from $\nabla^2\psi = -\omega$.
d. Update ω on wall from Eq. (7.4.50).

Steps b through d are repeated to convergence. Here subscripts denote differentiation with respect to x and y.

Similarly, employing a Poisson solver allows the solution of an Oseen-type linearization of Eq. (7.4.44), involving the iterative repetition of steps b to d in the following list:

a. Assume $\bar{\psi}, \bar{\omega}$.
b. Get new ω from $\nabla^2\omega - \text{Re}(\bar{\psi}_y\omega_x - \bar{\psi}_x\omega_y) = 0$.
c. Get new ψ from $\nabla^2\psi = -\omega$.
d. Update ω on walls from Eq. (7.4.50).

However, common to such methods is the need for an iteration on the vorticity at the wall, which will lag the remaining solution and which is most efficient when severely underrelaxed.

Once the ψ, ω, u, v distributions are determined, we may solve the temperature distribution from Eq. (7.4.53). Although linear, the complexity of the right-hand side suggests that it be based on known values from the previous step of point iteration, as other terms in the equation, which is then solved for $\theta_{i,j}$. This step is repeated, and $\theta_{i,j}$ is updated, say by SOR, until convergence has been attained. Also, convergence may be speeded by employing block iteration by columns since no iteration is required on boundary conditions, or better yet by using a Laplacian driver in a fully implicit scheme, or a Poisson solver.

Finally, let us turn to the third formulation, involving only ψ, T as dependent variables, where ψ is given by Eq. (2.4.43), or

$$\nabla^4\psi = \text{Re}\left[\frac{\partial\psi}{\partial y}\nabla^2\left(\frac{\partial\psi}{\partial x}\right) - \frac{\partial\psi}{\partial x}\nabla^2\left(\frac{\partial\psi}{\partial y}\right)\right]. \tag{7.4.55}$$

T, of course, is given by Eq. (7.4.45), the solution of which has already been discussed.

Equation (7.4.55), involving the biharmonic operator of order 4 (recall Eq. 2.4.44), requires two conditions on ψ to be specified along the closed boundary of the computational domain. As seen from Fig. 7.16(c), we specify ψ along the boundaries as well as a zero normal derivative of ψ (for no slip or parallel flow), or a zero normal derivative of second order (for symmetry).

The discretization of Eq. (7.4.55) requires use of a 13-node cell centered at i, j to represent $\nabla^4\psi$ to second order (Ames 1977). The algebraic work prior to programming, as well as the bookkeeping associated with handling near boundary nodes, is quite elaborate and will not be given. Instead, we refer to Roache & Ellis (1975), and for an alternative formulation of third order in terms of variables u, v, to Morihara & Cheng (1973).

Besides using an explicit point-iterative method of solution for $\psi_{i,j}$, it is possible to use a fully implicit approach, involving a biharmonic solver giving at once $\psi_{i,j}$ at all internal nodes. Various schemes of linearization of the right-hand side of Eq. (7.4.55) may be considered (Roache & Ellis 1975), requiring relatively few iterations to attain convergence.

Thus, with the availability of a biharmonic driver, the iterative solution involves the repetition of step b in the following list:

a. Assume $\bar{\psi}$ (for example, from solving $\nabla^4\bar{\psi} = 0$), and
b. Get ψ from $\nabla^4\psi = \text{Re}(\bar{\psi}_y\nabla^2\bar{\psi}_x - \bar{\psi}_x\nabla^2\bar{\psi}_y)$,

until desired convergence. Here $\nabla^4\psi$ need be inverted only once. Use of an Oseen-type linearization of the convective terms calls for a more complex solver and new matrix inversion in each step.

Alternatively, employing a formal Newton method for the linearization, denoting by $\tilde{\psi}$ a presumably small perturbation on the exact solution, this approach involves the repetition of step b in the following list:

a. Assume $\bar{\psi}$ (for example, from solving $\nabla^4\bar{\psi} = 0$) and
b. Get ψ from $\psi = \bar{\psi} + \tilde{\psi}$, where

$$\begin{aligned} \nabla^4\tilde{\psi} &- \text{Re}\,(\bar{\psi}_y\,\nabla^2\tilde{\psi}_x + \tilde{\psi}_y\,\nabla^2\bar{\psi}_x - \bar{\psi}_x\,\nabla^2\tilde{\psi}_y - \tilde{\psi}_x\,\nabla^2\bar{\psi}_y) \\ &= -[\nabla^4\bar{\psi} - \text{Re}(\bar{\psi}_y\,\nabla^2\bar{\psi}_x - \bar{\psi}_x\,\nabla^2\bar{\psi}_y)], \end{aligned} \tag{7.4.56}$$

until convergence $\tilde{\psi} \longrightarrow 0$ (Problem 7-26). This approach requires elaborate preparation, a complicated solver, and matrix inversion in each step. However, convergence is achieved in 2 to 5 steps for Re = 1 to 1000 for the present example (Walter & Larsen 1981).

Let us conclude this example by noting that it is only weakly elliptical, involving no zones of recirculation. Computational experience also shows that downstream boundary conditions at $x = L$ have only a negligible influence on the solution. The influence decreases with increasing Re and Pe, as the problem then becomes convection dominated, the condition for a problem to be parabolic.

For more complex through-flow problems, such as flow over a step (Fig. 7.17(a)), the solution will depend more strongly on the downstream boundary conditions, which may be also difficult to specify if the flow is not fully developed here. An exception is the flow over repeated steps (Fig. 7.17(b)), which may be handled by the use of periodic boundary conditions. As the iteration process proceeds, the state computed at nodes just upstream of the outflow boundary is inserted at the corresponding nodes at the inflow boundary before the next iterative sweep of internal nodes. Convergence requires that the outflow state remains unchanged.

As more complex, nonrectangular geometries of the computational domain are considered, it becomes less attractive to attempt to work out implicit solution schemes, and the simple point iteration is favored for simplicity. For discretization and interpolation formulas for handling nodes near irregular boundaries, we refer to Ames (1977) and Roache (1976).

An alternative approach is to transform the irregular physical domain into a regular, say rectangular, domain In this computational domain we may select a rectangular grid of nodes which maps back to a suitable irregular grid of nodes in the

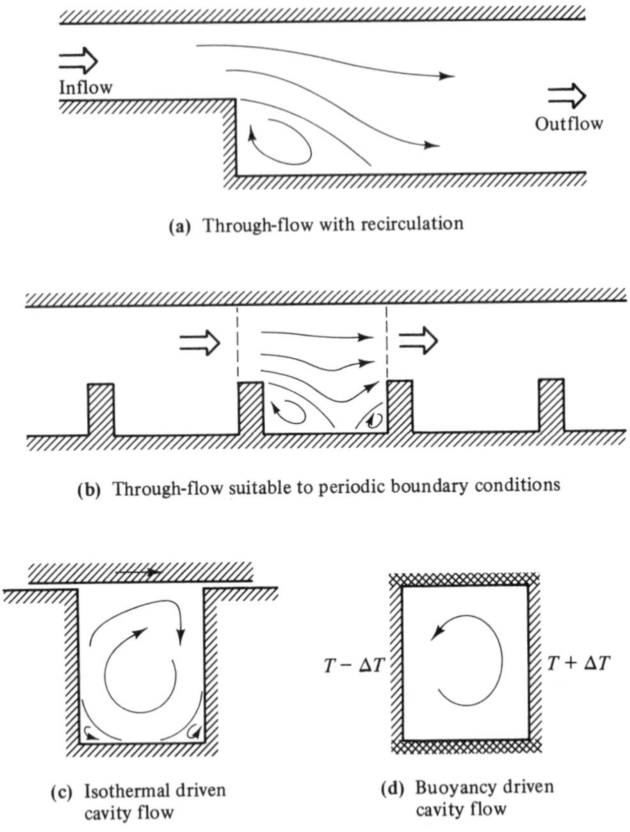
Inflow
Outflow
(a) Through-flow with recirculation
(b) Through-flow suitable to periodic boundary conditions
$T - \Delta T$
$T + \Delta T$
(c) Isothermal driven cavity flow
(d) Buoyancy driven cavity flow

Fig. 7.17 Examples of through-flow and cavity flow

physical domain. For some problems the mapping is conformal and analytical methods apply. In general, however, given an arbitrary curvelinear domain, the mapping requires a numerical method. For a two-dimensional problem this leads to the solution of a Poisson equation (Thames & Thompson 1977), having sufficient free parameters to give a suitable grid in the physical domain. Once the transformation has been established, it is also applied to the governing equations and boundary conditions. The resulting equations are then solved in the rectangular grid and the solution at any node applies to the equivalent physical node.

Flows in closed cavities, say driven by a moving wall (Figs. 7.17(c) and 4.30) or by buoyancy (Figs. 7.17(d) and 4.31), usually present no problem in regard to boundary conditions. For regular geometries, the various implicit methods mentioned in Ex. 7.4.3 may be applied. For all elliptical problems, the upper limit, say 10, on the local cell Reynolds (or Péclet) number (recall Ex. 7.4.2) should be noted. It implies the need for employing successively finer grids for increasing values of Re (or Pe) of the problem. For a limited computer memory, and in view of truncation errors and the accumulation of round-off errors, this requirement sets a practical upper limit for the value of Re (or Pe) that can be handled in a finite difference scheme. This limitation may be also recognized by noting that Re^{-1} (or Pe^{-1}) is a factor multiplying the diffusion terms involving the second-order derivatives, which make the problem elliptical. As Re (or Pe) becomes large, the influence of these terms becomes weak and the system of equations to be solved becomes ill-conditioned. One approach to circumvent these difficulties is to employ a grid-free method, such as the discrete vortex method (Chorin 1978).

So far we have discussed laminar flows with constant properties, but the methods apply in principle equally well to turbulent flows. The essential new feature, common to all problems having variable transport properties, is the discretization of diffusion terms. For an equidistant grid, considering for example the diffusion of x-momentum in y, the control volume approach suggests the discretization

$$\frac{\partial}{\partial y}\left(\nu \frac{\partial u}{\partial y}\right)_{i,j} \simeq \frac{\nu_{i,j+1/2}\dfrac{u_{i,j+1}-u_{i,j}}{\Delta y} - \nu_{i,j-1/2}\dfrac{u_{i,j}-u_{i,j-1}}{\Delta y}}{\Delta y}, \tag{7.4.57}$$

where viscosity ν is a variable to be evaluated between nodes. Instead of the linear approximation $\nu_{i,j+1/2} = \frac{1}{2}(\nu_{i,j} + \nu_{i,j+1})$, which is limited to a slowly varying $\nu_i(y)$, it proves to be more meaningful to employ

$$\nu_{i,j+1/2} = \frac{2}{1/\nu_{i,j} + 1/\nu_{i,j+1}}. \tag{7.4.58}$$

Physically, this approach ensures the correct expression for the flux $\nu\, \partial u/\partial y$, say in the limit of $\nu_{i,j}/\nu_{i,j+1} \ll 1$, as seen by consideration of series-coupled resistances (Patankar 1980). For laminar problems with temperature-dependent ν, equations are coupled and the distribution of temperature must be computed together with those of velocity, steadily updating $\nu(T)$. For turbulent problems, ν is a variable which is computed from algebraic or transport equations as explained in Chapters 10 and 11.

Here we terminate our introduction to computational convection. At the time of this writing, even though there is a wealth of information beyond what has been possible to bring here, the field of computational convection is still undergoing rapid developments. The reader is therefore encouraged to consult new texts in the field, as well as the journal literature.

REFERENCES

AMES, W. F., 1977, *Numerical Methods for Partial Differential Equations*, 2nd ed., Academic Press, New York.

ARPACI, V. S., 1966, *Conduction Heat Transfer*, Addison-Wesley, Reading, Mass.

BRADSHAW, P., CEBECI, T., & WHITELAW, J. H., 1981, *Engineering Calculation Methods of Turbulent Flow*, Academic Press, New York.

CEBECI, T., & BRADSHAW, P., 1977 *Momentum Transfer in Boundary Layers*, Hemisphere/McGraw-Hill, New York.

CEBECI, T., & SMITH, A. M. O., 1974, *Analysis of Turbulent Boundary Layers*, Academic Press, New York.

CHILUKURI, R., & PLETCHER, R. H., 1980, Numer. Heat Transfer, **3**, 169.

CHORIN, A. J., 1978, J. Comput. Phys., **27**, 428.

CHOW, C.-Y., 1979, *An Introduction to Computational Fluid Dynamics*, Wiley, New York.

CHUNG, T. J., 1978, *Finite Element Analysis in Fluid Dynamics*, McGraw-Hill, New York.

DAHLQUIST, G., BJÖRCK, A., & ANDERSON, N., 1974, *Numerical Methods*, Prentice-Hall, Englewood Cliffs, N.J.

FRÖBERG, C.-E., 1969, *Introduction to Numerical Analysis*, Addison-Wesley, Reading, Mass.

GALLAGHER, R. H., ODEN, J. T., TAYLOR, C., & ZIENKIEWICZ, O. C., eds., 1975, *Finite Elements in Fluids*, Wiley, New York.

GILLIS, J., & BRANDT, A., 1964, Air Force European Office of Aerospace Research Scientific Report 63–73.

GOSMAN, A. D., PUN, W. M., RUNCHAL, A. K., SPALDING, D. B., & WOLFSHTEIN, M., 1969, *Heat and Mass Transfer in Recirculating Flows*, Academic Press, New York.

KELLER, H. B., 1970, in *Numerical Solution of Partial Differential Equations* (ed. J. Bramble), Vol. II, Academic Press, New York.

KELLER, H. B., 1974, SIAM J. Numer. Anal. **11**, 305.

MORIHARA, H., & CHENG, R. T.-S., 1973, J. Comput. Phys. **11**, 550.

PATANKAR, S. V., 1980, *Numerical Heat Transfer and Fluid Flow*, Hemisphere/McGraw-Hill, New York.

PATANKAR, S. V., & SPALDING, D. B., 1970, *Heat and Mass Transfer in Boundary Layers*, 2nd ed., Intertext-Books, London.

PEACEMAN, D. W., & RACHFORD, H. H., Jr., 1955, J. Soc. Ind. Appl. Math., **3**, 28.

ROACHE, P. J., 1975, Comput. Fluids, **3**, 179.

ROACHE, P. J., 1976, *Computational Fluid Dynamics*, Hermosa, Alberquerque, N. Mex.

ROACHE, P. J., & ELLIS, M. A., 1975, Comput. Fluids, **3**, 305.

ROBERTSON, J. M., 1965, *Hydrodynamics in Theory and Applications*, Prentice-Hall, Englewood Cliffs, N.J.

SCHLICHTING, H., 1968, *Boundary Layer Theory*, 6th ed., McGraw-Hill, New York.

SPALDING, D. B., 1977, *GENMIX—A General Computer Program for Two-Dimensional Parabolic Phenomena*, Pergamon Press, Elmsford, N.Y.

THAMES, F. C., & THOMPSON, J. F., 1977, J. Comput. Phys. **24**, 245.

TORRANCE, K. E., & ROCKETT, J. A., 1969, J. Fluid Mech., **36**, 33.

WALTER, K. T., & LARSEN, P. S., 1981, Comput. Fluids, **9**, 365.

PROBLEMS

7-1 Derive finite difference operators for derivatives of orders 1 and 2 for the case of node spacings that are not equidistant, say $\Delta x_{i+1}/\Delta x_i = h_{i+1}/h_i = s \neq 1$, by use of Taylor series expansions analogous to Eqs. (7.1.23) and (7.1.24). Discuss the magnitude of truncation errors as function of s.

7-2 Reconsider Ex. 7.2.2. Write a computer program that solves Eq. (7.2.9), linearized respectively as Eq. (7.2.9a) and Eq. (7.2.9b), iteratively by an implicit finite difference scheme, starting for example with the initial guess $\bar{f} = \frac{3}{4}\eta^2 - \frac{1}{8}\eta^4$, corresponding to a cubic velocity profile. Determine the number of nodes and the number of iterations required to obtain the same accuracy as that of Table 7.7.

7-3 Reconsider Problem 7-2, employing now the Newton method, solving Eqs. (7.2.11) and (7.2.9c).

7-4 Resolve Ex. 7.1.1 as an initial value problem using the Runge–Kutta method (see Ex. 7.2.2). Start integration from $\xi = 1$, for example, with an assumed value of the derivative $d\theta(1)/d\xi$. Iterate to satisfy the boundary condition of insulation at $\xi = 0$.

7-5 Write a computer program which solves Eq. (5.3.59) of Ex. 5.3.6 by the fourth-order Runge–Kutta method and using a Newton–Raphson iteration to satisfy the boundary conditions at the free stream.

7-6 Extend the program of Ex. 7.2.2 to also solve the temperature distribution, Eq. (5.3.24). Compare the results to those of Fig. 5.2.

7-7 Solve Eq. (10.2.19) of Ex. 10.2.1 numerically, employing the mixing length of Eqs. (10.2.4) and (10.2.5), respectively.

7-8 Write a computer program that solves Eq. (10.2.44) of Ex. 10.2.2.

7-9 Using a Runge–Kutta scheme, write a program that solves the constant-property, incompressible, laminar momentum boundary layer for arbitrary pressure gradient given by the integral von Karman–Pohlhausen method (see, for example, Schlichting 1968 or Cebeci & Bradshaw 1977).

7-10 Write a computer program which solves a laminar boundary layer flow for arbitrary free-stream velocity $U(x)$, employing the Thwaites method (see Problem 4-27). Consider the solution of Ex. 4.2.2, the laminar entrance flow.

7-11 Write a computer program that solves Ex. 7.2.1 by the Head method (see Problem 10-9 for the formulation).

7-12 Discuss the numerical solution of fully developed turbulent flow between parallel plates spaced a distance $2l$ apart, employing the high-and the low-Reynolds-number (k, ϵ)-model of Tables 10.3 and 10.4, respectively.

7-13 Resolve Ex. 7.3.1, now for fully developed laminar flow (recall Ex. 3.5.5, Eq. 3.5.60).

7-14 Resolve Ex. 7.3.1 for the case of a constant heat flux q_w imposed for $x \geq 0$.

7-15 Write the computer program that solves Eq. (7.3.28) of Ex. 7.3.2, employing the Keller box method.

7-16 Discuss the numerical solution of Eq. (3.2.1) of Ex. 3.2.1.

7-17 Write a computer program that solves Eqs. (3.4.15) and (3.4.16) of Ex. 3.4.2.

7-18 Formulate a finite difference scheme for the solution of the nonlinear parabolic differential equation of Problem 4-28 describing the gravity-driven flow of a thin viscous liquid film on a horizontal plane. (*Suggestion*: Employ δ^4 as dependent variable.)

7-19 Reconsider Problem 4-28, now for an axisymmetric disturbance in a circular shallow container. Formulate the problem approximately and obtain a numerical solution. Present the results, for example, as a plot of rms-film thickness versus time, employing suitable dimensionless variables.

7-20 Write a computer program solving numerically Eq. (7.3.40) for the flat plate boundary layer subject to a zero pressure gradient.

7-21 Reconsider Ex. 7.4.1, solving the finite difference formulation by various schemes. Compare convergence rates for two or more of the methods: point iteration by Jacobi, Gauss–Seidel, or SOR, and block iteration by columns and with SOR, and ADI methods.

7-22 Reconsider Ex. 7.4.1, solving the finite difference formulation by block iteration by columns, one iteration comprising two sweeps, the first one handling the odd-numbered columns, the second one the even-numbered columns.

7-23 Program and compute the numerical solution to Ex. 7.4.2 by the implicit method, employing Eqs. (7.4.32) and (7.4.36), respectively. Study the computational instability of the former equation as the cell Péclet number increases. Compare the computed heat flux at $\xi = 1$ to that obtained from the exact solution, Eq. (7.4.30).

7-24 Verify Eq. (7.4.50). Also, show that the assumption of $\omega_i(y) \simeq$ constant in the first of Eq. (7.4.49) yields the first-order approximation

$$y = 0: \qquad \omega_{i,1} = \frac{-2(\psi_{i,2} - \psi_{i,1})}{(\Delta y)^2},$$

while use of a cubic polynomial of $\psi_i(y)$, and the added no-slip condition $\partial\psi(x, 0)/\partial y = 0$, yields

$$y = 0: \qquad \omega_{i,1} = \frac{\frac{1}{2}(\psi_{i,3} - 8\psi_{i,2} + 7\psi_{i,1})}{(\Delta y)^2}.$$

7-25 Show that writing Eq. (7.4.51) alternating with Eq. (7.4.52) for the nodes of a column (considering the right-hand side of the latter to be known) and incorporating the wall condition on vorticity, Eq. (7.4.50), there results a four-diagonal banded system of linear equations. This system is solved readily in an implicit scheme, leading to a block-iterative solution by columns which incorporates the vorticity boundary condition.

7-26 Verify Eq. (7.4.56), recalling the discussion leading to Eq. (7.2.9c).

7-27 Discuss, based on a literature study, the appropriate boundary conditions for ψ, ω to be used at a convex corner, say in the flow over a step, Fig. 7.17(a).

7-28 Write a computer program and study the numerical solution of the square cavity flow, including buoyancy effects (see Problem 4-40 and Fig. 4.29 for $H = l$). Employ the (ψ, ω, T)-formulation and consider separately the driven flow (say, starting from small values of $\mathrm{Re} = Ul/\nu$), the buoyancy driven flow, and the flow resulting from the combined effects.

7-29 Consider a large rectangular water basin for thermal energy storage (Fig. 7P-29). The top and bottom are insulated, while the sides are isothermal at T_w, which is assumed to remain below the bulk temperature $T(x, t)$, which in turn is stabily stratified [$T(x, t)$ decreases with x]. During charging, hot water at T_0 is fed uniformly over the width $2b$ with a very small velocity U_0 to the top ($x = 0$) at the same rate as cold water is withdrawn at the bottom ($x = H$). We seek the heat loss from the vertical walls and the bulk temperature distribution. Discuss the formulation and numerical solution of the unsteady problem, considering both a full two-dimensional formulation and an approximate coupled boundary layer flow–bulk flow problem. In the latter formulation, the one-dimensional unsteady bulk flow is coupled to quasi-steady buoyancy-driven boundary layer flows along the cooled vertical walls. Note that the bulk flow is governed by the combined effects of charging and entrainment into boundary layer flows.

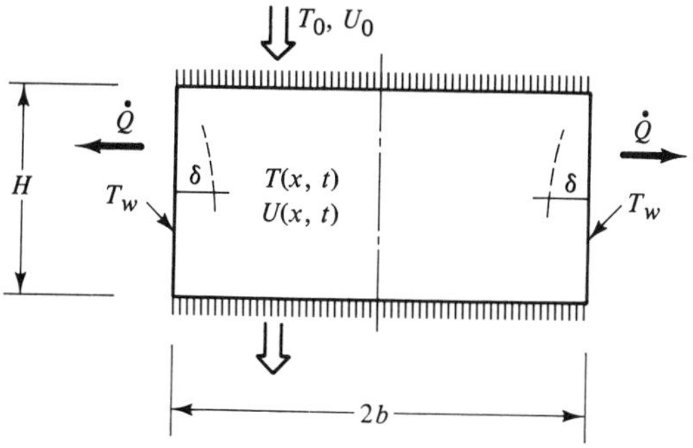

Fig. 7P-29

Part IV

TURBULENCE

In the first three parts, we have learned the foundations of convection and tested our knowledge on these foundations in terms of laminar problems. However, the flow in almost all actual problems is turbulent. Consequently, the rest of this text is devoted to turbulent convection. First, recall the celebrated Reynolds experiment mentioned in Chapter 1 (Fig. 1.2). We learned from this experiment that instability of laminar flow at high Reynolds numbers leads, through a transition, to turbulence. Therefore, some knowledge on the instability of laminar flows is essential for an understanding of turbulence. Accordingly, the next chapter begins with an introduction to the theory of stability. Although important for the separation of stable and unstable flow regimes, this theory is restricted to fluctuations not large enough to handle turbulent flows. The chapter proceeds to the instantaneous and the (temporally) mean aspects of turbulence, and terminates with a statistical interpretation of the mean turbulence. Although the statistical approach enhances our understanding of (homogeneous) turbulence, it is unable to solve problems involving turbulent (shear) flows. Consequently, Chapters 9 and 10 are devoted to models on turbulence and Chapter 11 to the application of these models to convection problems. Although modeling efforts are useful for solution of turbulent problems, because of their inherent nature, they are unable to provide sufficiently universal solutions. They are restricted to a large extent to specific conditions of a problem they are supposed to represent. The ultimate engineering tool for turbulent problems thus remains the correlation of experimental data by means of appropriate dimensionless numbers. Accordingly, Chapter 12 begins with a clear account for the methods of dimensional analysis and proceeds, in terms of the microscales developed in Chapter 9, to demonstrate the basic foundations of the heat transfer correlations, which are assumed to be empirical.

Chapter 8

INSTABILITY AND TURBULENCE

We begin this chapter with a demonstration of the common foundations of the instability theory. In Section 8.1 we state the original energy concept of the stability theory, and apply this concept to a couple of simple problems, one from mechanics the other from conduction. Also, utilizing the conduction problem, we replace the concept with an alternative and often more convenient characteristic-value method for stability problems. In Section 8.2 we apply the method to the Poiseuille flow between two parallel plates, and develop some appreciation for the mathematical complexity of flow instabilities. Also, we begin to realize how futile an approach to turbulence via the stability theory would be. Having failed to approach turbulence by this theory, we search for another tool. In this process, accepting the fact that by existing or projected computers (as well as by analytical means), a long-time direct integration of nonlinear (Navier–Stokes) equations is not feasible, we give up the search for a tool for instantaneous turbulence. Thus, we are forced to settle for a tool which would provide information about the mean turbulence. This leads us, in Sections 8.3 and 8.4, to the Reynolds stress and its vortical interpretation. A natural tool for the investigation of the Reynolds stress is statistics. In Sections 8.5 through 8.7 we deal with a statistical description of this stress and other aspects of turbulence.

8.1 FOUNDATIONS OF INSTABILITY

The original concept of the instability theory rests on the considerations of the change of total energy of a system resulting from an infinitesimal disturbance introduced to the initial state of the system. Let this change of total energy be ΔE. The stability criterion is then

$$\begin{aligned} \Delta E &> 0, \quad \text{stable,} \\ \Delta E &= 0, \quad \text{marginal,} \\ \Delta E &< 0, \quad \text{unstable.} \end{aligned} \tag{8.1.1}$$

The celebrated example is the change of the potential energy of a marble corresponding to its infinitesimal displacement in a bowl or over an inverted bowl, or on a flat horizontal plate. Figure 8.1 shows the stable and unstable cases, and the marginal state separating these cases.

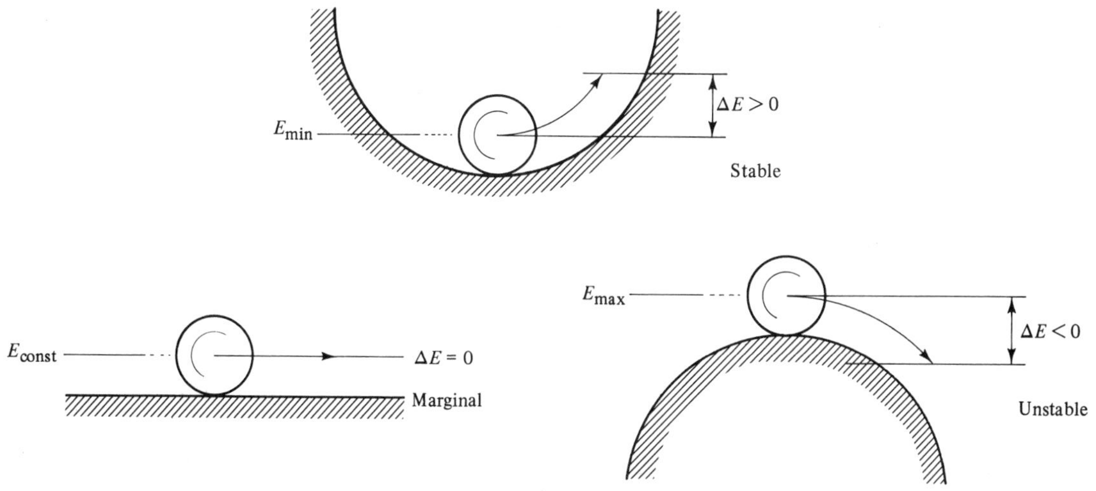

Fig. 8.1 Original stability problem

Two examples from mechanics and conduction elaborate the application of the energy method to instability problems. The first example, depicted in Fig. 8.2, deals with the buckling load P of a rigid vertical beam, hinged at the bottom and attached to a horizontal spring at the top. The beam length is l, the spring constant is k.

For a disturbance α, the sum of the increase in potential energy of the load,

$$-Pl(1 - \cos \alpha) \cong -\tfrac{1}{2}Pl\alpha^2,$$

and the increase in potential energy of the spring,

$$\tfrac{1}{2}kx^2 \cong \tfrac{1}{2}kl^2\alpha^2,$$

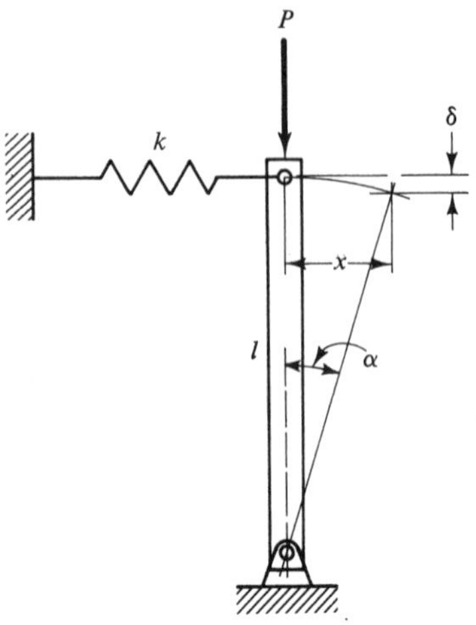

Fig. 8.2 Stability of a rigid beam

gives the change in total potential energy of the system,

$$\Delta E = \tfrac{1}{2}kl^2\alpha^2 - \tfrac{1}{2}Pl\alpha^2 \gtrless 0, \tag{8.1.2}$$

yielding the marginal force

$$P = kl, \tag{8.1.3}$$

above which the beam buckles.

In terms of the foregoing simple problem, we learned the application of the energy method to the instability of mechanics problems. Toward the instability of fluid flow and convection problems, we now consider a transitory problem from conduction.

EXAMPLE 8.1.1 Consider a flat glass plate of thickness l. The plate is heated electrically. The temperature dependence of the electrical conductivity $k_e(T)$ of glass and its model are given in Fig. 8.3. Plate walls are kept at a constant temperature. The temperature instability (thermal buckling) of the plate is required.

Since the electrical conductivity of glass below a critical temperature is very small, the energy generation $u''' = e^2k_e$ below this temperature is negligible. Assume that the temperature distribution in the plate stays above, and the walls at, the critical temperature T_{crit}.

The change of thermal energy,

$$\Delta E = \text{thermal energy [loss(by conduction)} - \text{generation]} \gtrless 0, \tag{8.1.4}$$

gives the stability criterion,

$$\Delta E = -k\left(\frac{d\theta}{dx}\right)\Big|_{x=0}^{x=l} - e^2\beta\int_0^l \theta\,dx \gtrless 0, \tag{8.1.5}$$

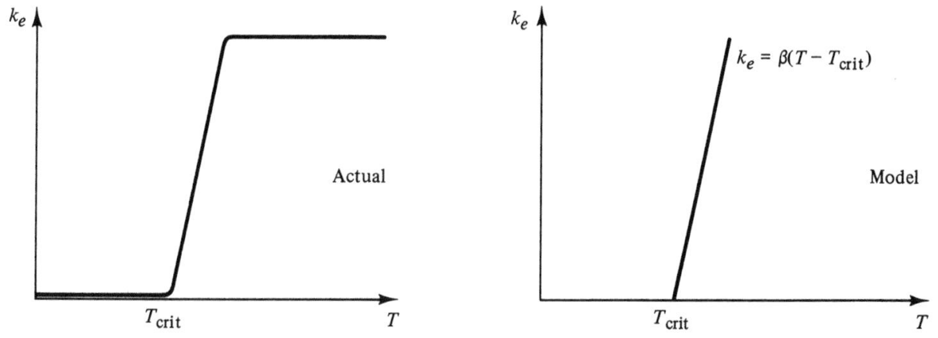

Fig. 8.3 Electrical conductivity of glass, Ex. 8.1.1

or

$$-k\left(\frac{d\theta}{dx}\right)\Bigg|_{x=0}^{x=l} \gtrless e^2\beta\int_0^l \theta\,dx, \tag{8.1.6}$$

$\theta = T - T_{crit}$ and e being the electric potential. Clearly, the energy method applied to a spatially distributed problem requires the selection of an approximate profile for the dependent variable. For the present problem, we need to know the temperature distribution. Assuming, for example, that

$$\theta = A\sin\pi\left(\frac{x}{l}\right), \tag{8.1.7}$$

A being an arbitrary small amplitude, and inserting this profile into Eq. (8.1.6) results in

$$e\left(\frac{\beta}{k}\right)^{1/2} = \frac{\pi}{l} \tag{8.1.8}$$

for the marginal electric potential above which the system is unstable and the temperature of the glass plate would increase until meltdown.

Actually, there exists an alternative method which is usually preferred for instability problems. This method reduces the instability problems to a characteristic-value problem which, for the present case, is

$$k\frac{d^2\theta}{dx^2} + \beta e^2\theta = 0, \tag{8.1.9}$$

$$\theta(0) = \theta(l) = 0. \tag{8.1.10}$$

This leads to the characteristic functions

$$\theta_n = \sin\lambda_n x, \tag{8.1.11}$$

and to the characteristic values

$$\lambda_n l = n\pi, \qquad n = 1, 2, 3, \ldots, \tag{8.1.12}$$

where

$$\lambda_n = e_n\left(\frac{\beta}{k}\right)^{1/2} = n\left(\frac{\pi}{l}\right). \tag{8.1.13}$$

The lowest characteristic value,

$$\lambda_1 = e_1\left(\frac{\beta}{k}\right)^{1/2} = \frac{\pi}{l}, \tag{8.1.14}$$

gives the electric potential at which the glass temperature becomes unstable.

Note that, in the process of selecting an approximate circular profile for the energy method, we apparently picked the exact solution of the problem. Consequently, both methods resulted in the same answer. This does not usually turn out to be the case for more complicated problems.

So far we have ignored the unsteadiness of the foregoing example. Clearly, in the example taken from mechanics (statics) the dynamic effects are negligible. However, the conduction in the glass plate may be unsteady, and it requires the consideration of the unsteady thermal energy,

$$\rho c\frac{\partial\theta}{\partial t} = k\frac{\partial^2\theta}{\partial x^2} + \beta e^2\theta. \tag{8.1.15}$$

We now wish to determine the effect of unsteadiness on the stability of the plate temperature. Let

$$\theta_n = A_n e^{\sigma_n t}\sin\lambda_n x, \tag{8.1.16}$$

where λ_n is as given by Eq. (8.1.13), A_n is the amplitude, and

$$\sigma_n \gtreqless 0 \tag{8.1.17}$$

determines the (unstable) growth, the (marginal) steady state, and the (stable) decay. Equation (8.1.16) inserted into Eq. (8.1.15) yields the same stability conditions,

$$\frac{\sigma_n}{a} = \left[e_n^2\left(\frac{\beta}{k}\right) - n^2\left(\frac{\pi}{l}\right)^2\right] \gtreqless 0. \tag{8.1.18}$$

The marginal state, depending on the growth or damping of small disturbances actually occurs in two distinct ways. In one case, the transition from stability to instability takes place via a *steady* marginal state (illustrated in the foregoing glass plate problem). In the other case, the transition takes place via an *oscillatory* state (illustrated in the following flow problem). In the former case σ_n is real, and there is no need to consider the unsteady effects for the stability problem. In the latter case σ_n is complex, and the stability problem needs to be formulated, including the unsteady effects. Elaborations on these cases, however, are beyond the scope of the text. See, for example, Chandrasekhar (1961) for details.

So far we learned the foundations of the stability theory in terms of two illustrative examples. Actually, our main objective is the instability and eventual turbulence of the fluid flow and convection problems. We proceed now to the instability of isothermal flows.

8.2 *FLOW INSTABILITY*

The simple examples of the preceding section somewhat conceal the complexity of the formal solution procedure for the instability of flow problems, especially for those involving the effect of viscosity. To demonstrate this complexity, we review below the instability of the fully developed laminar flow between two parallel plates.

EXAMPLE 8.2.1 Consider the Poiseuille flow between two parallel plates separated a distance apart of $2l$. We wish to determine the instability of this flow.

The instability of nonlinear fluid mechanics problems is customarily treated by the method of small perturbations. The linearization of the governing equations allows a one-term Fourier representation of the disturbances. In view of the three-dimensional nature of turbulence, it is necessary to assume three-dimensional disturbances. However, the instability of parallel flows usually starts in the form of two-dimensional waves. Accordingly, we consider only two-dimensional disturbances.

Let

$$u = U + u', \quad v = v', \quad w = 0, \quad p = P + p', \tag{8.2.1}$$

where

$$U(y) = \frac{1}{2\mu}\left(-\frac{dP}{dx}\right)(l^2 - y^2) \tag{3.2.3}$$

is the undisturbed Poiseuille flow. Inserting Eq. (8.2.1) into the two-dimensional Navier–Stokes equations, neglecting the nonlinear terms, we have

$$\begin{aligned} \frac{\partial u'}{\partial t} + U\frac{\partial u'}{\partial x} + v'\frac{dU}{dy} &= \frac{1}{\rho}\left(-\frac{\partial p'}{\partial x}\right) + \nu\,\nabla^2 u', \\ \frac{\partial v'}{\partial t} + U\frac{\partial v'}{\partial x} &= \frac{1}{\rho}\left(-\frac{\partial p'}{\partial y}\right) + \nu\,\nabla^2 v', \end{aligned} \tag{8.2.2}$$

and the conservation of mass

$$\frac{\partial u'}{\partial x} + \frac{\partial v'}{\partial y} = 0, \tag{8.2.3}$$

which is linear without any approximation.

After cross-differentiation, eliminating the pressure in Eq. (8.2.2) by employing the stream function

$$u' = \frac{\partial \psi}{\partial y}, \qquad v' = -\frac{\partial \psi}{\partial x} \tag{8.2.4}$$

results in

$$\left(\frac{\partial}{\partial t}+U\frac{\partial}{\partial x}\right)\nabla^2\psi-\left(\frac{d^2U}{dy^2}\right)\frac{\partial\psi}{\partial x}=\nu\,\nabla^4\psi. \tag{8.2.5}$$

Since Eq. (8.2.5) is linear in ψ with coefficients independent of (x, t), assuming that

$$\psi(x, y, t)=\phi(y)e^{i\alpha(x-\sigma t)}, \tag{8.2.6}$$

we obtain for ϕ, after some nondimensionalization,

$$(U-\sigma)(\phi''-\alpha^2\phi)-U''\phi=-\frac{i}{\mathrm{Re}}(\phi^{(\mathrm{iv})}-2\alpha^2\phi''+\alpha^4\phi), \tag{8.2.7}$$

which is the celebrated *Orr–Sommerfeld equation.* Here $\mathrm{Re}=Ul/\nu$ is the Reynolds number. Boundary conditions to be satisfied by Eq. (8.2.7) are

$$\phi=\phi'=0 \qquad \text{for } y=\pm 1, \tag{8.2.8}$$

where y is the transversal distance nondimensionalized with l.

Equation (8.2.7), subject to Eq. (8.2.8) (homogeneous differential equation and homogeneous boundary conditions), constitutes a characteristic-value problem (recall a much simpler characteristic-value problem given by Eqs. 8.1.9 and 8.1.10). Here σ is the required characteristic value for each pair of (α, Re). Let the general solution of Eq. (8.2.7) be

$$\phi(y)=C_1\phi_1+C_2\phi_2+C_3\phi_3+C_4\phi_4. \tag{8.2.9}$$

This solution satisfying Eq. (8.2.8) leads to

$$\begin{vmatrix}\phi_1(-1) & \phi_2(-1) & \phi_3(-1) & \phi_4(-1)\\ \phi_1(1) & \phi_2(1) & \phi_3(1) & \phi_4(1)\\ \phi_1'(-1) & \phi_2'(-1) & \phi_3'(-1) & \phi_4'(-1)\\ \phi_1'(1) & \phi_2'(1) & \phi_3'(1) & \phi_4'(1)\end{vmatrix}=0, \tag{8.2.10}$$

which may be written, implicitly,

$$\sigma=\sigma(\alpha, \mathrm{Re}). \tag{8.2.11}$$

Although α and Re are real and positive, σ is in general complex, say $\sigma=\sigma_R+i\sigma_I$. Here σ_R gives the perturbation velocity and σ_I the exponential decay or amplification factor. A parallel laminar flow $U(y)$ is unstable for these values of (α, Re) for which $\sigma_I>0$. The region of (α, Re)-plane for stable $U(y)$ and that for unstable $U(y)$ are separated by the neutral stability curve along which $\sigma_I=0$ (Fig. 8.4). Details leading to an explicit relation for Eq. (8.2.11) are not important for the present discussion and will not be given here (see, for example, Lin 1955 or Schlichting 1968).

A natural extension of the foregoing example is the instability of natural convection in a slot. In Chapter 3 we discussed the base flow of the vertical slot (see Eq. 3.4.13 of Ex. 3.4.1). Addition of the buoyancy effect to the Navier–Stokes equations given by Eq. (8.2.2) and consideration of the thermal energy equation would complete the for-

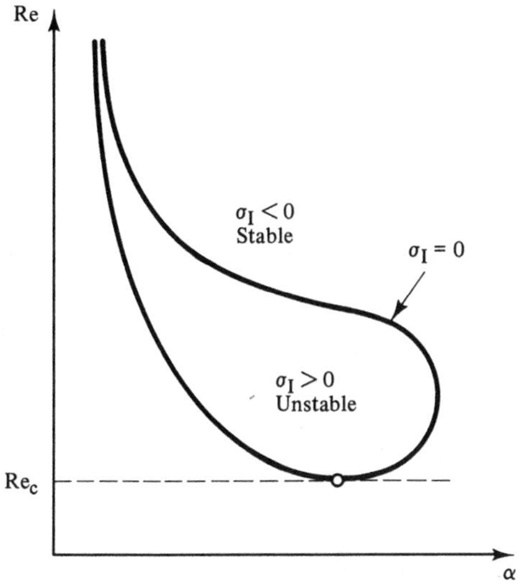

Fig. 8.4 Marginal stability curve, Ex. 8.2.1

mulation of the stability problem of the vertical slot. The solution of this problem will not be discussed here, for reasons to be given in the next paragraph (see, however, Problem 8-15). A photograph of the unstable flow in a vertical slot, taken from Vest (1967) and Vest & Arpaci (1969), is shown in Fig. 8.5. Interested readers are referred to Elder (1965) for an in-depth study, and to Gershuni & Zhukhovitskii (1976) and Monin & Yaglom (1971) for extensive reviews on this problem. Effects of elasticity on the problem may be found in Gözum (1972) and Gözum & Arpaci (1974), that of radiation in Arpaci & Bayazitoglu (1973), and that of rotation in Korpela (1972) and Korpela & Arpaci (1976).

Under conditions of instability determined by the method of small perturbations, disturbances grow exponentially with time. In reality, however, perturbations quickly attain a magnitude such that the neglect of product terms ceases to be valid. Also, recent advances in the instability theory which extend the infinitesimal theory (by carrying the next-order terms) to a finite-amplitude theory, although quite impressive, are not sufficient to handle (because of the neglected higher-order terms) the large fluctuations needed for turbulence (Fig. 8.6). Consequently, there is no hope for an analytical approach to turbulence, at least in a foreseeable future, by a finite-amplitude instability theory. Since our prime objective in this text is turbulent convection, no further elaborations on stability theory are useful for this objective. Accordingly, we look for another approach to our study of turbulence. First, note that a developing (laminar or turbulent) flow or any other field is more involved than its fully developed limit (refer to Chapter 6). Consequently, we proceed directly to a study of fully developed turbulence. Delaying thermal effects to the next chapter, we consider here only isothermal flow.

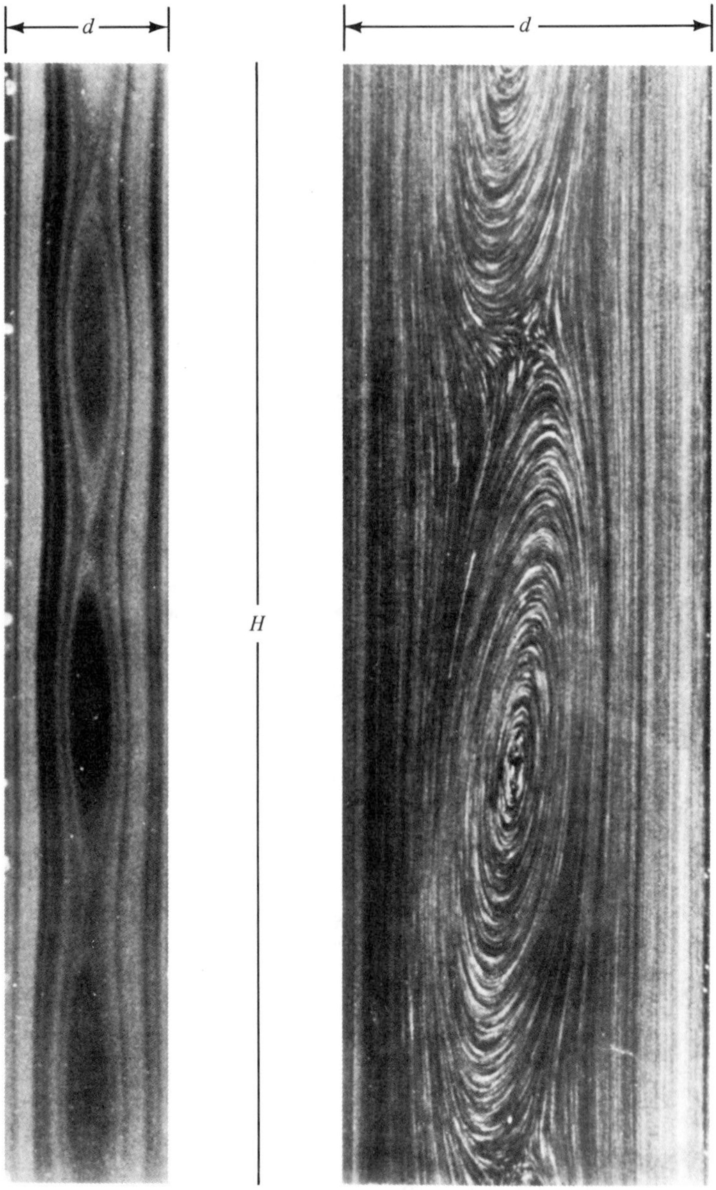

Fig. 8.5 Streak photographs of the secondary flow in a vertical slot. Left: conduction regime, Gr = 9500, $H/d = 33$; Pr = 0.71. Right: boundary-layer regime, Ra $\simeq 4.5 \times 10^5$, $H/d = 20$, Pr $\simeq$ 900

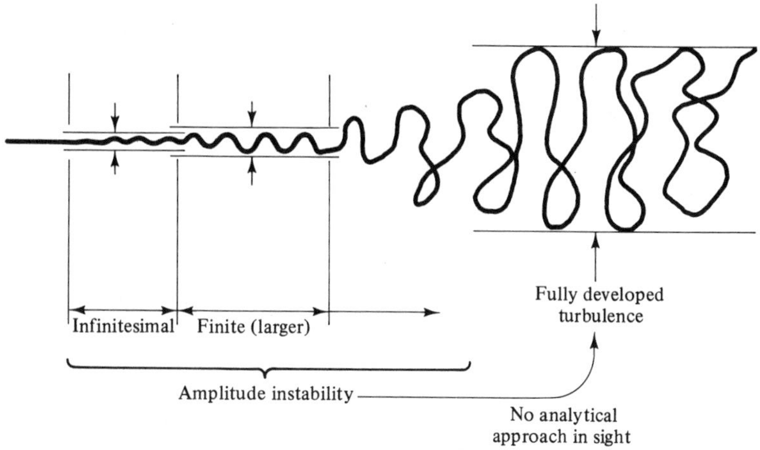

Fig. 8.6 Origin of turbulence

8.3 REYNOLDS STRESS

Assume that the instantaneous velocity of turbulence satisfies the usual momentum equations, which are, for an incompressible fluid,

$$\frac{\partial \tilde{u}_i}{\partial t} + \tilde{u}_j \frac{\partial \tilde{u}_i}{\partial x_j} = \frac{1}{\rho} \frac{\partial}{\partial x_j} \tilde{\sigma}_{ij}, \tag{8.3.1}$$

where $\tilde{\sigma}_{ij}$ is the *stress tensor*. For an incompressible Newtonian fluid, Eqs. (2.3.10) and (2.3.11) become

$$\tilde{\sigma}_{ij} = -\tilde{p}\, \delta_{ij} + 2\mu \tilde{s}_{ij}, \tag{8.3.2}$$

$\tilde{s}_{ij}$ being the *rate of strain tensor*, defined by

$$\tilde{s}_{ij} = \frac{1}{2}\left(\frac{\partial \tilde{u}_i}{\partial x_j} + \frac{\partial \tilde{u}_j}{\partial x_i}\right). \tag{8.3.3}$$

Inserting Eq. (8.3.2) into Eq. (8.3.1) utilizing the conservation of mass, $\partial \tilde{u}_i/\partial x_i = 0$, we have the Navier–Stokes equations for an incompressible fluid,

$$\frac{\partial \tilde{u}_i}{\partial t} + \tilde{u}_j \frac{\partial \tilde{u}_i}{\partial x_j} = -\frac{1}{\rho} \frac{\partial \tilde{p}}{\partial x_i} + \nu\, \nabla^2 \tilde{u}_i. \tag{8.3.4}$$

There are two approaches to the integration of Eq. (8.3.4), the *statistical* and *computational.* The present state of the statistical approach, including recent impressive advances, is far from providing an answer to turbulence. The long-time integration of Eq. (8.3.4) by computational means is also beyond the capacity of presently available as well as projected future computers. If there is no statistical or computational method available for studying instantaneous turbulence, what about its average over a meaningful time interval? Actually, not the instantaneous turbulence, for example, in a pipe, but rather its mean value, which provides the net flow through this pipe, has prime technological importance. Instantaneous turbulence, and its mean value relative to laminar flow near the walls of a pipe, are sketched in Figs. 8.7 and 8.8. (see also Fig. 10.5(a)).

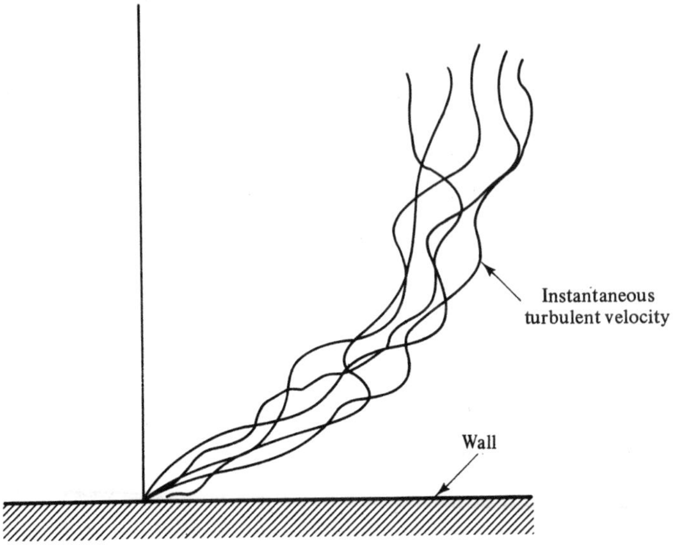

Fig. 8.7 Instantaneous turbulence

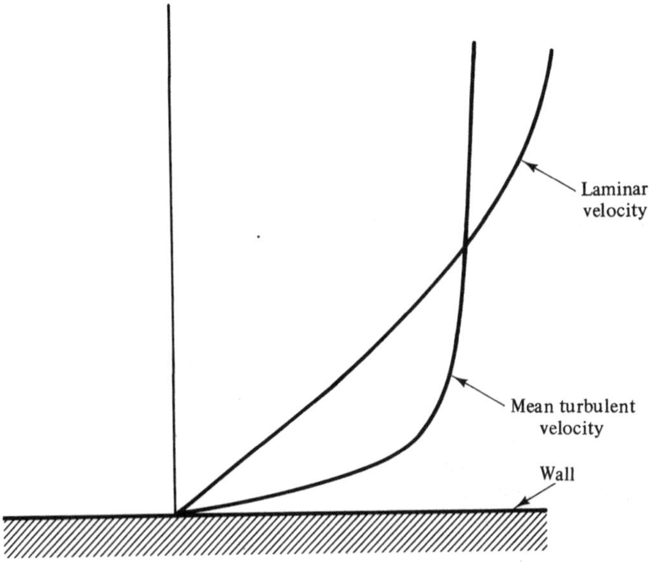

Fig. 8.8 Mean turbulence

What we need now are the Navier–Stokes equations satisfied by the mean turbulence. First, consider the instantaneous velocity $\tilde{u}_i$ expressed as velocity fluctuations u_i superimposed on the mean flow U_i such that

$$\tilde{u}_i = U_i + u_i, \tag{8.3.5}$$

where

$$\bar{\tilde{u}}_i = U_i = \lim_{T\to\infty} \frac{1}{T} \int_{t_0}^{t_0+T} \tilde{u}_i \, dt. \tag{8.3.6}$$

Hereafter, capital letters are used for mean values and lower-case letters for fluctuating values. An overbar denotes averaging, and we confine our interest to turbulence that is steady on the mean, hence $\partial U_i/\partial t = 0$.

Because of the nonlinearity of the Navier–Stokes equations, we need the average of the product of two instantaneous variables. To develop this average, consider variables $\tilde{a}$ and $\tilde{b}$. In terms of means and fluctuations,

$$\tilde{a} = A + a, \qquad \tilde{b} = B + b.$$

By definition,

$$\bar{\tilde{a}} = A, \qquad \bar{\tilde{b}} = B.$$

Consequently, from

$$\bar{\tilde{a}} = \overline{A + a} = A, \qquad \bar{\tilde{b}} = \overline{B + b} = B,$$

we have

$$\bar{a} = 0, \qquad \bar{b} = 0.$$

Then, the average of $\tilde{a}\tilde{b}$,

$$\overline{\tilde{a}\tilde{b}} = \overline{(A + a)(B + b)} = AB + \overline{Ab} + \overline{aB} + \overline{ab},$$

in view of

$$\overline{Ab} = A\bar{b} = 0, \qquad \overline{aB} = \bar{a}B = 0,$$

is

$$\overline{\tilde{a}\tilde{b}} = AB + \overline{ab}. \tag{8.3.7}$$

Now, we are ready for the average of the instantaneous Navier–Stokes equations. In addition to $\tilde{u}_i = U_i + u_i$, introducing

$$\tilde{p} = P + p, \qquad \bar{p} = 0,$$
$$\tilde{\sigma}_{ij} = \Sigma_{ij} + \sigma_{ij}, \qquad \bar{\sigma}_{ij} = 0,$$
$$\Sigma_{ij} = -P\,\delta_{ij} + 2\mu S_{ij},$$

and noting, in view of Eq. (8.3.7) and $\partial u_j/\partial x_j = 0$, that

$$\begin{aligned}\overline{\tilde{u}_j \frac{\partial \tilde{u}_i}{\partial x_j}} &= \overline{(U_j + u_j)\frac{\partial}{\partial x_j}(U_i + u_i)} \\ &= U_j \frac{\partial U_i}{\partial x_j} + \frac{\partial}{\partial x_j}(\overline{u_i u_j}),\end{aligned} \tag{8.3.8}$$

the average of Eq. (8.3.1) may be written as

$$U_j \frac{\partial U_i}{\partial x_j} = \frac{1}{\rho} \frac{\partial}{\partial x_j}\left(\Sigma_{ij} - \rho\overline{u_i u_j}\right), \tag{8.3.9}$$

where

$$T_{ij} = \Sigma_{ij} - \rho\overline{u_i u_j} = -P\delta_{ij} + 2\mu S_{ij} - \rho\overline{u_i u_j} \tag{8.3.10}$$

is the total mean stress in a turbulent flow. The contribution of the turbulent motion to the mean stress,

$$\boxed{\tau_{ij} = -\rho\overline{u_i u_j},} \tag{8.3.11}$$

is called the *Reynolds stress*. This stress is symmetric, $\tau_{ij} = \tau_{ji}$. The diagonal components of τ_{ij} are normal stresses (negative pressures), and off-diagonal components are shear stresses. The latter play an important role in the mean momentum transfer by turbulent motion.

Let us summarize in Table 8.1 what we have learned so far. For instantaneous turbulence, the formulation is complete. We have four equations (conservation of mass and three components of the balance of momentum) and four unknowns ($\tilde{p}, \tilde{u}_i$). However, there is no solution (by any method) resulting from long-time integration of this formulation. Long-time integration of nonlinear equations is still in its stage of infancy. On the other hand, for the mean (steady) turbulence, the formulation becomes

TABLE 8.1

Instantaneous turbulence
$\frac{\partial \tilde{u}_i}{\partial x_i} = 0$ $\frac{\partial \tilde{u}_i}{\partial t} + \tilde{u}_j \frac{\partial \tilde{u}_i}{\partial x_j} = -\frac{1}{\rho}\frac{\partial \tilde{p}}{\partial x_i} + \nu \nabla^2 \tilde{u}_i$
Complete formulation (four equations and four unknowns). No solution (by any method) is in sight.

Mean (steady) turbulence
$\frac{\partial U_i}{\partial x_i} = 0$ $U_j \frac{\partial U_i}{\partial x_j} = -\frac{1}{\rho}\frac{\partial P}{\partial x_i} + \nu \nabla^2 U_i + \frac{1}{\rho}\frac{\partial \tau_{ij}}{\partial x_j}$
Incomplete formulation (because of additional unknowns τ_{ij}). A solution is possible after a model for $\tau_{ij} = -\rho\overline{u_i u_j}$.

incomplete because we continue to have four equations but now have an additional unknown τ_{ij}. However, in terms of a model for τ_{ij} (developed in Chapters 10 and 11), this formulation becomes complete and can be solved. We have then a steady problem which does not require any temporal integration.

The appearence of extra unknowns τ_{ij}, which result from the averaging of the instantaneous formulation, is the *closure problem* of turbulence. However, it should be emphasized that, any time or space averaging of a (linear or nonlinear) formulation generates additional unknowns. This is an important concept, and it may be worth illustrating in terms of a simple example used previously.

Reconsider the introductory example of Section 6.2, which illustrates the thermal fluctuations resulting from the oscillations in the pipe flow velocity (Fig. 6.2 and Eqs. 6.2.1 and 6.2.2). Here the interest lies in the steady mean value of these fluctuations.

Replace Eq. (6.1.5), in terms of the present nomenclature, with

$$\frac{\partial \tilde{\theta}}{\partial t} + \tilde{V}\frac{\partial \tilde{\theta}}{\partial x} + m\tilde{\theta} = 0. \tag{8.3.12}$$

Let

$$\tilde{V} = V + v, \tag{8.3.13}$$

$$\tilde{\theta} = \Theta + \theta, \tag{8.3.14}$$

where V and Θ are the steady mean values, and v and θ are the fluctuations. By definition,

$$\bar{\tilde{V}} = V, \qquad \bar{v} = 0, \tag{8.3.15}$$

$$\bar{\tilde{\theta}} = \Theta, \qquad \bar{\theta} = 0. \tag{8.3.16}$$

Inserting Eqs. (8.3.15) and (8.3.16) into Eq. (8.3.12), and averaging, yields

$$V\frac{d\Theta}{dx} + m\Theta = -\frac{d}{dx}(\overline{v\theta}). \tag{8.3.17}$$

Clearly, the averaging process creates a new unknown, $\overline{v\theta}$, in addition to Θ. Before the solution of Θ, $\overline{v\theta}$ needs to be determined by some means.

The closure problem of turbulence, that is, trying to complete the formulation of the mean steady turbulence by seeking a model for $\tau_{ij} = -\rho\overline{u_i u_j}$, is an unusual and difficult task. Many researchers have tried a relation between τ_{ij} and S_{ij} by imitating the functional form of the viscous stress in the Navier–Stokes equations. Before investigating these models, some remarks on the vorticity dynamics are useful, for reasons explained in the next section.

8.4 *VORTICAL INTERPRETATION OF REYNOLDS STRESS*

Turbulent flows are three-dimensional. Since an important property of three-dimensional flows is vortex stretching, the vorticity dynamics plays an important role in the understanding of turbulence. We shall learn here the relation between the Reynolds stresses and the vorticity.

Rearranging the momentum flow as

$$\tilde{u}_j \frac{\partial \tilde{u}_i}{\partial x_j} = \tilde{u}_j\left(\frac{\partial \tilde{u}_i}{\partial x_j} - \frac{\partial \tilde{u}_j}{\partial x_i}\right) + \tilde{u}_j \frac{\partial \tilde{u}_j}{\partial x_i}$$
$$= 2\tilde{u}_j \tilde{r}_{ij} + \frac{\partial}{\partial x_i}\left(\frac{1}{2}\tilde{u}_j\tilde{u}_j\right)$$
$$= -\epsilon_{ijk}\tilde{u}_j\tilde{\omega}_k + \frac{\partial}{\partial x_i}\left(\frac{1}{2}\tilde{u}_j\tilde{u}_j\right), \tag{8.4.1}$$

and the momentum diffusion as

$$\nu \frac{\partial^2 \tilde{u}_i}{\partial x_j \partial x_j} = \nu \frac{\partial}{\partial x_j}\left(\frac{\partial \tilde{u}_i}{\partial x_j} - \frac{\partial \tilde{u}_j}{\partial x_i}\right) + \nu \frac{\partial}{\partial x_i}\left(\frac{\partial \tilde{u}_j}{\partial x_j}\right)$$
$$= 2\nu \frac{\partial \tilde{r}_{ij}}{\partial x_j} = -\nu\epsilon_{ijk}\frac{\partial \tilde{\omega}_k}{\partial x_j}, \tag{8.4.2}$$

the instantaneous Navier–Stokes equations may be rearranged as

$$\frac{\partial \tilde{u}_i}{\partial t} = -\frac{\partial}{\partial x_i}\left(\frac{\tilde{p}}{\rho} + \frac{1}{2}\tilde{u}_j\tilde{u}_j\right) + \epsilon_{ijk}\tilde{u}_j\tilde{\omega}_k - \nu\epsilon_{ijk}\frac{\partial \tilde{\omega}_k}{\partial x_j}. \tag{8.4.3}$$

For an irrotational flow, $\tilde{\omega}_k = 0$ by definition. Consequently, the viscous term and the vorticity part of the momentum flow vanish. The momentum flow reduces then to the gradient of the dynamic pressure, and Eq. (8.4.3) reduces to the (unsteady) Bernoulli equations (recall Section 2.4.4). In turbulent flows, however, none of these terms vanish.

The cross-product $\epsilon_{ijk}\tilde{u}_j\tilde{\omega}_k$ plays a vital role in our understanding of turbulence. This product is proportional to the Coriolis force involved with the equations of motion in a rotating coordinate system. It is also related to the lift (Magnus) force acting on a vortex line in a velocity field $\tilde{u}_j$ in a fixed coordinate system. The x_1-component of this force is illustrated in Fig. 8.9.

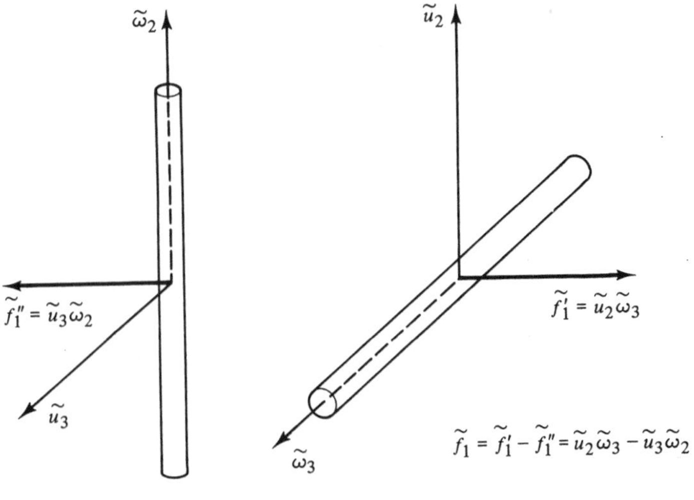

Fig. 8.9 x_1-component of Magnus force

We are ready now to demonstrate the important relation which exists between the Reynolds stress and the Magnus force. For a turbulent flow which is steady in the mean, Eq. (8.4.3) may be written for the mean velocity, after assuming that $\tilde{\omega}_i = \Omega_i + \omega_i$ and noting that $\bar{\omega}_i = 0$, as

$$0 = -\frac{\partial}{\partial x_i}\left(\frac{P}{\rho} + \frac{1}{2}U_j U_j + \frac{1}{2}\overline{u_j u_j}\right) + \epsilon_{ijk}(U_j \Omega_k + \overline{u_j \omega_k}) + \nu \frac{\partial^2 U_i}{\partial x_j \partial x_j}. \qquad (8.4.4)$$

Neglect the contribution of turbulent fluctuations to dynamic pressure (by assuming that $\frac{1}{2}\overline{u_j u_j} \ll \frac{1}{2}U_j U_j$). Then, for a two-dimensional mean flow (of a boundary layer), $U_1 \gg U_2$, $U_3 = 0$, $\partial/\partial x_1 \ll \partial/\partial x_2$, and the only nonzero component of vorticity is

$$\Omega_3 = \frac{\partial U_2}{\partial x_1} - \frac{\partial U_1}{\partial x_2}. \qquad (8.4.5)$$

The contribution of the mean Magnus force to the x_1-momentum,

$$F_1 = U_2 \Omega_3 - U_3 \Omega_2,$$

reduces, in view of $U_3 = 0$ and $\Omega_2 = 0$, to $F_1 = U_2 \Omega_3$. Then, in terms of Eq. (8.4.5),

$$F_1 = U_2 \frac{\partial U_2}{\partial x_1} - U_2 \frac{\partial U_1}{\partial x_2}. \qquad (8.4.6)$$

Also, the contribution of the mean flow to the dynamic pressure gradient in x_1 is

$$-\frac{\partial}{\partial x_1}\left(\frac{1}{2}U_j U_j\right) = -U_1 \frac{\partial U_1}{\partial x_1} - U_2 \frac{\partial U_2}{\partial x_1}. \qquad (8.4.7)$$

Thus, the mean momentum in x_1 may be written as

$$U_1 \frac{\partial U_1}{\partial x_1} + U_2 \frac{\partial U_1}{\partial x_2} = -\frac{1}{\rho}\frac{\partial P}{\partial x_1} + \overline{u_2 \omega_3} - \overline{u_3 \omega_2} + \nu \frac{\partial^2 U_1}{\partial x_2^2}. \qquad (8.4.8)$$

The comparison of Eq. (8.4.8) with the x_1-momentum of the mean flow given in Table 8.1 (after applying the boundary layer approximation) yields the relation between the Reynolds stress and the Magnus force:

$$\frac{1}{\rho}\frac{\partial \tau_{12}}{\partial x_2} = \overline{u_2 \omega_3} - \overline{u_3 \omega_2}, \qquad (8.4.9)$$

or

$$\frac{\partial \tau_{12}}{\partial x_2} = f_1. \qquad (8.4.10)$$

In Chapter 10, one of the models for the Reynolds stress (the vorticity transport theory) rests on Eq. (8.4.9).

8.5 STATISTICAL DESCRIPTION OF TURBULENCE

In Section 8.3, realizing that the instantaneous turbulence is beyond reach, we proceeded to the steady mean turbulence, which involves an additional unknown,

$$\tau_{ij} = -\rho \overline{u_i u_j}, \qquad (8.3.11)$$

the Reynolds stress. We learned also that this unknown is associated with the usual closure problem resulting from the averaging of the instantaneous turbulence equations. Thus, we reduced the mean turbulence to finding an appropriate model for τ_{ij}. Actually, there is an early statistical attempt which provides information for some aspects of turbulence. Before proceeding to the modeling efforts of Chapters 9 through 11, a brief review of the statistical description of turbulence is useful.

Our life experiences based on casual observations of cigarette smoke, the plume of a smokestack, or the formation of clouds indicate the irregularity of all turbulent flows. For example, on a recorder coupled with these observations, a component of instantaneous velocity at a fixed point would appear as shown in Fig. 8.10. The recorder receives the signal from a velocity probe such as the hot wire. Inspection of Fig. 8.10 reveals the randomness of velocity fluctuations and suggests a statistical description of turbulence, which we consider next.

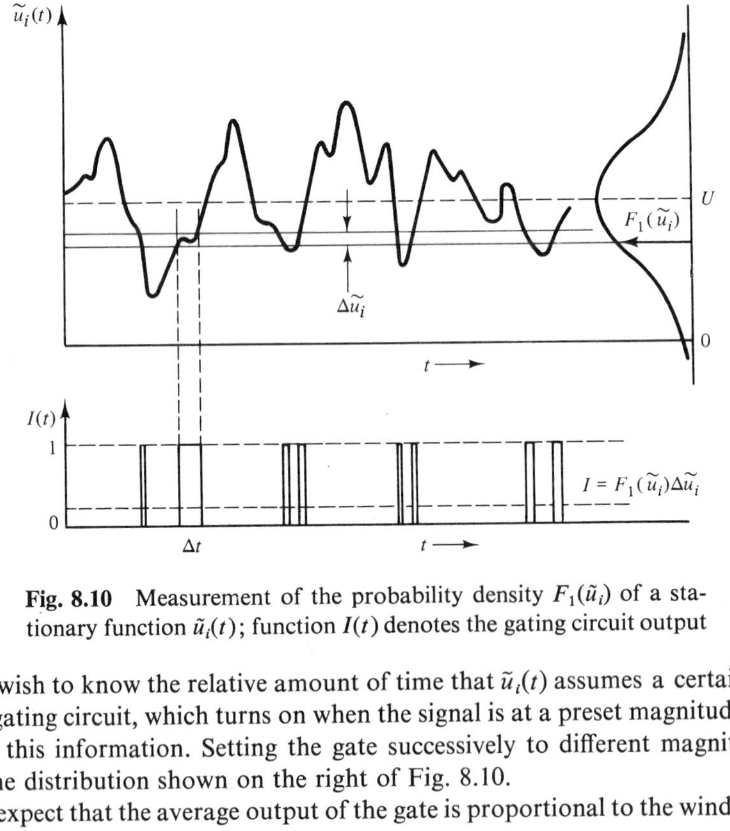

Fig. 8.10 Measurement of the probability density $F_1(\tilde{u}_i)$ of a stationary function $\tilde{u}_i(t)$; function $I(t)$ denotes the gating circuit output

We wish to know the relative amount of time that $\tilde{u}_i(t)$ assumes a certain magnitude. A gating circuit, which turns on when the signal is at a preset magnitude, readily provides this information. Setting the gate successively to different magnitudes, we obtain the distribution shown on the right of Fig. 8.10.

We expect that the average output of the gate is proportional to the window width $\Delta\tilde{u}_i$. Accordingly,

$$F_1(\tilde{u}_i)\,\Delta\tilde{u}_i = \lim_{T\to\infty} \frac{1}{T} \sum (\Delta t), \tag{8.5.1}$$

where the proportionality, $F_1(\tilde{u}_i)$, is called a one-point *probability-density distribution function*, or, for short, the distribution function. In other words, the probability of finding $\tilde{u}_i(t)$ between $\tilde{u}_i$ and $\tilde{u}_i + \Delta\tilde{u}_i$ is proportional to the time spent in that velocity range. Representing a fraction of time, $F_1(\tilde{u}_i)$ is always positive, and the sum of $F_1(\tilde{u}_i)$ over all $\tilde{u}_i$ adds to 1:

$$\int_{-\infty}^{\infty} F_1(\tilde{u}_i)\, d\tilde{u}_i = 1, \qquad F_i(\tilde{u}_i) > 0. \tag{8.5.2}$$

The usual averaging may be alternatively expressed in terms of the distribution function. For example, Eq. (8.3.6) may be rearranged, with the help of Eq. (8.5.1), as

$$U_i = \lim_{T\to\infty} \frac{1}{T} \int_{t_0}^{t_0+T} \tilde{u}_i\, dt = \int_{-\infty}^{\infty} \tilde{u}_i F(\tilde{\boldsymbol{u}})\, d\tilde{\boldsymbol{u}}. \tag{8.5.3}$$

Usually, we are interested in the averaging depending on the fluctuating component of the velocity alone rather than on its instantaneous value. Since $F_1(\tilde{u}_i) = F_1(U_i + u_i)$, the distribution function $F_1(u_i)$ is readily obtained by shifting $F_1(\tilde{u}_i)$ over a distance U_i along the $\tilde{u}_i$-axis. The moments formed with u_i^n and $F_1(u_i)$, n being a positive integer greater than 1, are called *central* moments. For $n = 2$, 3, and 4, for example, these are measures for the standard deviation, skewness, and flatness of the distribution function (see Tennekes & Lumley 1972 for details). Furthermore, because of its relation to the Reynolds stress, here we are primarily interested in the joint moment $\overline{u_i u_j}$, which is defined as

$$\overline{u_i u_j} = \int_{-\infty}^{\infty} u_i u_j F_1(\boldsymbol{u})\, d\boldsymbol{u}, \tag{8.5.4}$$

and called the *correlation* between u_i and u_j. Thus, by this correlation, the additional unknown (Reynolds stress) problem of the steady mean turbulence is reduced to that of finding $F_1(\boldsymbol{u})$.

The equation satisfied by the one-point distribution function $F_1(\boldsymbol{u})$ is in terms of the two-point (joint) distribution function $F_2(\boldsymbol{u}, \boldsymbol{u}')$.* Similarly, the equation for the two-point distribution function depends on the three-point distribution function, so there are never enough equations. This is the *closure problem*, which results from any averaging (and not necessarily turbulent) process.† The closure of the equation for the one-point distribution function, depending on models for the two-point distribution function, may be found in the literature (see, for example, Lundgren 1969). We shall not follow this approach in the evaluation of Reynolds stresses, for the reason given below.

Explicitly, the two-point distribution function $F_2(\boldsymbol{u}, \boldsymbol{u}')$ gives the probability that at the instant t, the velocity at $\boldsymbol{x}$ lies in the range $\boldsymbol{u}, \boldsymbol{u} + d\boldsymbol{u}$, while at $\boldsymbol{x}'$ it lies in the range $\boldsymbol{u}', \boldsymbol{u}' + d\boldsymbol{u}'$. The correlation between velocity components u_i at $\boldsymbol{x}$ and u_j' at $\boldsymbol{x}'$ is given by

$$\overline{u_i u_j'} = \int_{-\infty}^{\infty} \int_{-\infty}^{\infty} u_i u_j' F_2(\boldsymbol{u}, \boldsymbol{u}')\, d\boldsymbol{u}\, d\boldsymbol{u}'. \tag{8.5.5}$$

---

*Readers with any background in kinetic theory and/or radiative transfer are reminded of the conceptual similarity between this equation and the Boltzmann and Fokker–Planck equations of kinetic theory and the transfer equation of radiative transfer.

†See Problem 8-16.

All two-point correlations can be generated from this two-point distribution function. Multipoint correlations involving velocity components at several points in the field may similarly be expressed in terms of multipoint distribution functions. Now, having explained the probabilistic basis for correlations in terms of distribution functions, rather than studying these functions, let us turn to the correlation functions themselves. Actually, statistical turbulence theory is to a large extent concerned with the properties of the *double velocity correlation* of Eq. (8.5.5), called the *correlation tensor*

$$R_{ij}(\boldsymbol{x}, \boldsymbol{x}') = \overline{u_i(\boldsymbol{x})u'_j(\boldsymbol{x}')}. \tag{8.5.6}$$

A complete picture of turbulence structure, however, cannot be obtained from this function alone. In addition, multiple correlations are needed. The importance of the double velocity correlation is that it readily admits a physical interpretation and is easily measured.

To determine R_{ij} from the Navier–Stokes equations, multiply the equations for u_i by u'_j and average, and add the equation obtained by interchanging u_i and u'_j. The result, in the case of isotropic turbulence where

$$\overline{p(\boldsymbol{x})u'_i(\boldsymbol{x}')} = 0, \tag{8.5.7}$$

is

$$\frac{\partial}{\partial t}R_{ij} - \frac{\partial}{\partial r_k}(T_{ik,j} + T_{jk,i}) = 2\nu\,\nabla^2 R_{ij}, \tag{8.5.8}$$

where $\boldsymbol{r} = \boldsymbol{x}' - \boldsymbol{x}$, u^2 is the mean-square velocity fluctuation $u^2 = \overline{u_i u_i}/3$, and $T_{ij,k}$ is a particular case of the triple correlation

$$T_{i,j,k} = \overline{u_i(\boldsymbol{x})u'_j(\boldsymbol{x}')u''_k(\boldsymbol{x}'')}, \tag{8.5.9}$$

namely,

$$T_{ij,k} = \overline{u_i(\boldsymbol{x})u_j(\boldsymbol{x})u'_k(\boldsymbol{x}')}. \tag{8.5.10}$$

Similarly, forming the equations for the third-order correlations introduces fourth-order correlations, and so on. Again, we are faced with a closure problem. The futility of this approach should be clear in view of the involved solution of the illustrative example of Section 6.2. A further perturbation of this example is quite lengthy. What about the cases requiring, because of the size of ϵ, perturbations beyond the third perturbation? Clearly, application of a successive (perturbation) approximation to large-amplitude fluctuations of any nonlinear problem is not productive.

Returning to the solution of Eq. (8.5.8), and the equations for higher-order correlations, we need to replace the resulting infinite system of differential equations for correlation functions of all orders by a finite system. It becomes necessary then to make assumptions enabling higher-order correlations to be expressed in terms of those of lower order. This is usually done by assuming the one-point and two-point distribution functions to be Gaussian, which is approximately true in the isotropic case. Certain kinematic properties of R_{ij} in this case are reviewed below, following closely the concise presentation of Hunt (1964).

8.6 HOMOGENEOUS ISOTROPIC TURBULENCE

Isotropy requires that the correlation between u_i at $\boldsymbol{x}$ and u'_j at $\boldsymbol{x}' = \boldsymbol{x} + \boldsymbol{r}$ depends only on the directions of u_i, u'_j relative to the vector $\boldsymbol{r}$ joining the two points, and not on $\boldsymbol{x}$. When $\boldsymbol{r}$ lies in the x_1-direction, for example, the *longitudinal correlation* $f(r)$ is given by

$$\overline{u_1 u'_1} = u^2 f(r), \tag{8.6.1}$$

and the *transverse correlation* $g(r)$ by

$$\overline{u_2 u'_2} = \overline{u_3 u'_3} = u^2 g(r), \tag{8.6.2}$$

while the mixed correlations $\overline{u_1 u'_2} = \overline{u_2 u'_3} = \cdots = 0$. It follows that the *correlation tensor* $R_{ij}(\boldsymbol{r})$ has elements $\overline{u_i u'_j}$ which are of the form

$$R_{ij}(\boldsymbol{r}) = \overline{u_i u'_j} = u^2\left[\frac{f(r) - g(r)}{r^2} r_i r_j + g(r)\,\delta_{ij}\right], \tag{8.6.3}$$

where r_i is the component of $\boldsymbol{r} = \boldsymbol{x}' - \boldsymbol{x}$ in the x_i-direction and $\delta_{ij} = 1$ when $i = j$, $\delta_{ij} = 0$ when $i \neq j$. Equation (8.6.3) defines the most general isotropic second-order tensor depending only on $\boldsymbol{r}$ and possessing invariance under rigid rotations and reflections. We also require that

$$R_{ij}(\boldsymbol{r}) = R_{ji}(-\boldsymbol{r}), \tag{8.6.4}$$

and from continuity

$$\overline{u_i(\boldsymbol{x}) \frac{\partial}{\partial r_j} u'_j(\boldsymbol{x} + \boldsymbol{r})} = 0 = \frac{\partial}{\partial r_j} R_{ij}(\boldsymbol{r}),$$

and similarly,

$$\frac{\partial}{\partial r_i} R_{ij}(\boldsymbol{r}) = 0. \tag{8.6.5}$$

Subsituting Eq. (8.6.3) into Eq. (8.6.5) shows that the two scalar functions f and g are connected by the relation

$$g = f + \frac{1}{2} r \frac{\partial f}{\partial r}, \tag{8.6.6}$$

which has experimental support. We conclude that in homogeneous isotropic turbulence all second-order velocity correlations depend on a single scalar function $f(r)$. Moreover, the symmetry condition on Eq. (8.6.4) requires that $f(r)$ and $g(r)$ shall be even functions of r.

Vorticity correlations can also be derived from $R_{ij}(\boldsymbol{r})$. Because of the extent to which we wish to get involved with the statistical description of turbulence, however, these correlations are not elaborated here (see, however, Problem 8-18 and, for example, Batchelor 1967).

While the triple velocity correlation $T_{i,j,k}$ given by Eq. (8.5.9) depends on the two vectors $\boldsymbol{r} = \boldsymbol{x}' - \boldsymbol{x}$ and $\boldsymbol{r}' = \boldsymbol{x}'' - \boldsymbol{x}$, the simpler two-point triple correlation

$T_{ij,k}$ occurring in Eq. (8.5.10) depends only on $\mathbf{r}$. Under the conditions of isotropy, $T_{ij,k}$ may be expressed in terms of a single scalar function in a similar manner. For example, with $\mathbf{r}$ in the x_1-direction, we have the purely longitudinal correlation $k(r)$ given by

$$u^3 k(r) = \overline{u_1^2 u_1'} \tag{8.6.7}$$

and two mixed correlations $h(r)$ and $q(r)$ given by

$$u^3 h(r) = \overline{u_2^2 u_1'}, \qquad u^3 q(r) = \overline{u_1 u_2 u_2'}. \tag{8.6.8}$$

The general expression for $T_{ij,k}(\mathbf{r})$ is

$$T_{ij,k} = u^3\left[\frac{k-h-2q}{r^3}r_i r_j r_k + \delta_{ij} r_k \frac{h}{r} + (\delta_{ik} r_j + \delta_{jk} r_i)\frac{q}{r}\right], \tag{8.6.9}$$

where, from continuity, $h(r)$, $k(r)$, and $q(r)$ are linked by the relations

$$k = -2q, \qquad q = -\left(h + \frac{1}{2} r \frac{\partial h}{\partial r}\right). \tag{8.6.10}$$

Expansion of h, k, and q into Taylor series near $r = 0$ shows them to be odd functions of r and their expansion to begin with terms of order r^3.

Substituting R_{ij} from Eq. (8.6.3) and $T_{ij,k}$ from Eq. (8.6.9) in Eq. (8.5.8) leads to the single scalar equation

$$\frac{\partial}{\partial t}(u^2 f) + 2u^3\left(h' + 4\frac{h}{r}\right) = 2\nu u^2\left(f'' + 4\frac{f'}{r}\right), \tag{8.6.11}$$

referred to as the *von Kármán–Howarth equation*. Introducing the expansions

$$f(r) = 1 + \frac{f_0''}{2!}r^2 + \frac{f_0^{(\text{iv})}}{4!}r^4 + \cdots$$

and

$$h(r) = \frac{h_0'''}{3!}r^3 + \cdots$$

in Eq. (8.6.11) and equating powers of r gives equations linking derivatives of f and h. The first equation, due to terms independent of r, is

$$\frac{du^2}{dt} = 10\nu u^2 f_0'' = -10\nu\frac{u^2}{\lambda^2}, \tag{8.6.12}$$

where

$$\lambda^2 = -\frac{1}{f_0''}, \tag{8.6.13}$$

which defines the rate of decrease of turbulent kinetic energy due to viscous dissipation. The second equation, related to the vorticity, is not considered here (see Problem 8-18).

From Eq. (8.6.13), λ must be a measure of the size of the smallest eddies and is accordingly named the *Taylor microscale*. In view of Eq. (8.6.12), λ is also related to a scale of eddies which could be regarded as giving the total viscous dissipation, so that λ is sometimes said to be the dissipation scale. In Chapter 9 we shall give a further interpretation of this scale and utilize it in the construction of a turbulence model. Here we develop the relation between this scale and the (rate of) dissipation. We elaborated on dissipation in Chapter 2 (see Sections 2.1 through 2.4, and recall Eq. 2.4.6 for the definition of the dissipation function, Φ_v). Introducing

$$\epsilon = \nu\Phi_v = \frac{1}{\rho}\tau_{ij}\left(\frac{\partial u_i}{\partial x_j}\right) = \frac{1}{\rho}\boldsymbol{\tau} : \mathbf{s},$$

the dissipation in terms of the mean of the velocity fluctuations may be written as

$$\epsilon = \nu\,\overline{\left(\frac{\partial u_i}{\partial x_j} + \frac{\partial u_j}{\partial x_i}\right)\frac{\partial u_i}{\partial x_j}} = \frac{1}{2}\nu\,\overline{\left(\frac{\partial u_i}{\partial x_j} + \frac{\partial u_j}{\partial x_i}\right)^2}, \tag{8.6.14}$$

which may be rearranged by consideration of the identity

$$\overline{\left(\frac{\partial u_i}{\partial x_k}\right)\left(\frac{\partial u_j}{\partial x_l}\right)'} = \frac{\partial}{\partial x_k}\left(\frac{\partial}{\partial x_l}\right)'\overline{u_i u_j'}, \tag{8.6.15}$$

where the prime denotes the location x_l' of u_j'. It follows that

$$\overline{\frac{\partial u_i}{\partial x_k}\left(\frac{\partial u_j}{\partial x_l}\right)} = -\left(\frac{\partial^2 R_{ij}}{\partial r_k\,\partial r_l}\right)_{r=0}, \tag{8.6.16}$$

where $r_k = x_k' - x_k$. From Eq. (8.6.3),

$$\left(\frac{\partial^2 R_{ij}}{\partial r_k\,\partial r_l}\right)_{r=0} = u^2\left(2\delta_{kl}\,\delta_{ij} - \frac{1}{2}\,\delta_{il}\,\delta_{jk} - \frac{1}{2}\,\delta_{jl}\,\delta_{ik}\right)f_0''. \tag{8.6.17}$$

Equation (8.6.14) can be now expressed in terms of f_0''. For example,

$$\overline{\left(\frac{\partial u_1}{\partial x_1}\right)^2} = \overline{\left(\frac{\partial u_2}{\partial x_2}\right)^2} = \frac{1}{2}\,\overline{\left(\frac{\partial u_1}{\partial x_2}\right)^2} = \cdots = -u^2 f_0'', \tag{8.6.18}$$

and so on. Accordingly,

$$\epsilon = 15\nu\,\overline{\left(\frac{\partial u_1}{\partial x_1}\right)^2} = -15\nu u^2 f_0'' = -15\nu\frac{u^2}{\lambda^2}, \tag{8.6.19}$$

which agrees with Eq. (8.6.12) because, in view of the kinetic energy,

$$E = \tfrac{1}{2}\overline{u_i u_i} = \tfrac{3}{2}u^2, \tag{8.6.20}$$

Eqs. (8.6.12) and (8.6.19) yield

$$\frac{dE}{dt} = -\epsilon. \tag{8.6.21}$$

Here, we conclude our brief review of velocity correlations and proceed to the spectral aspects of statistical turbulence.

8.7 *SPECTRAL DYNAMICS*

The spectral interpretation of turbulence allows us to draw further conclusions, which are not attainable by other means. Spectra are decompositions of nonlinear functions into waves of different wavelengths. The spectrum at a given wavelength gives the mean energy in that wave, and also explains the way in which waves of different scales exchange energy with each other. The energy exchange is the prime objective of the spectral analysis by which we learn that the turbulence receives its energy at large scales, while its energy dissipates at very small scales; there also exist waves within a range of wavelengths (or eddies within a size range) which are not directly affected by the maintenance and dissipation mechanisms of energy.

The correlation tensor $R_{ij}(\boldsymbol{r})$ is related to a spectral tensor $\phi_{ij}(\boldsymbol{\kappa})$, which depends on the complex wave number $\boldsymbol{\kappa}$. The transformation from one to the other is achieved by the three-dimensional Fourier transform

$$R_{ij}(\boldsymbol{r}) = \int_{-\infty}^{\infty} \phi_{ij}(\boldsymbol{\kappa}) e^{i\boldsymbol{\kappa}.\boldsymbol{r}}\, d\boldsymbol{\kappa} \tag{8.7.1}$$

and its inverse

$$\phi_{ij}(\boldsymbol{\kappa}) = \frac{1}{8\pi^3} \int_{-\infty}^{\infty} R_{ij}(\boldsymbol{r}) e^{-i\boldsymbol{\kappa}.\boldsymbol{r}}\, d\boldsymbol{r}. \tag{8.7.2}$$

Of particular significance is the limit of Eq. (8.7.1) as $\boldsymbol{r} \longrightarrow 0$,

$$R_{ij}(0) = \overline{u_i u_j} = \int_{-\infty}^{\infty} \phi_{ij}(\boldsymbol{\kappa})\, d\boldsymbol{\kappa}, \tag{8.7.3}$$

which gives the spectral interpretation of the Reynolds stress,

$$\tau_{ij} = -\rho R_{ij}(0).$$

The correlation components $\overline{u_i u_j}$ determine the energy in the various velocity components. Equation (8.7.3) shows that $\phi_{ij}(\boldsymbol{\kappa})$ gives the division of energy in $\overline{u_i u_j}$ among the eddy sizes or wave numbers present in the motion. Consequently, $\phi_{ij}(\boldsymbol{\kappa})$ is termed the *energy spectrum tensor*.

The sum of the diagonal components of ϕ_{ij} represents the kinetic energy at a given wave number. Clearly, by summation of Eq. (8.7.3) with $i = j$,

$$R_{ii}(0) = \overline{u_i u_i} = 3u^2 = \int_{-\infty}^{\infty} \phi_{ii}(\boldsymbol{\kappa})\, d\boldsymbol{\kappa}, \tag{8.7.4}$$

or

$$E = \tfrac{1}{2}\overline{u_i u_i} = \tfrac{3}{2}u^2 = \tfrac{1}{2}R_{ii}(0), \tag{8.7.5}$$

E being the kinetic energy of turbulent fluctuations per unit mass.

In a manner similar to the correlation tensor, the energy spectrum tensor can be expressed in terms of a single scalar function $A(\kappa)$ which is even in κ,

$$\phi_{ij}(\boldsymbol{\kappa}) = A(\kappa)(\kappa_i \kappa_j - \kappa^2\, \delta_{ij}). \tag{8.7.6}$$

The scalar function $A(\kappa)$ and $f(r)$ are linked by the Fourier transforms given by Eqs. (8.7.1) and (8.7.2). The various wave numbers present in the motion contribute to the total energy according to the energy density distribution

$$\tfrac{1}{2}\phi_{ii}(\boldsymbol{\kappa}) = -\kappa^2 A(\kappa) \tag{8.7.7}$$

obtained by summation of Eq. (8.7.6) with $i = j$. The contribution to the total energy from wavenumbers with magnitudes in the range $(\kappa, \kappa + d\kappa)$ is defined as $E(\kappa)\, d\kappa$, where for isotropic conditions

$$E(\kappa) = 4\pi\kappa^2 \tfrac{1}{2}\phi_{ii}(\boldsymbol{\kappa}) = -4\pi\kappa^4 A(\kappa), \tag{8.7.8}$$

and $E(\kappa)$ is termed the energy spectrum function, or, for short, the energy spectrum. From Eqs. (8.7.1) and (8.7.8), we have

$$\tfrac{1}{2}R_{ii}(r) = \int_0^\infty E(\kappa)\,\frac{\sin \kappa r}{\kappa r}\,d\kappa, \tag{8.7.9}$$

where

$$R_{ii}(r) = u^2(3f + rf'), \tag{8.7.10}$$

obtained by letting $i = j$ in Eq. (8.6.3) and summing.

The limit of Eq. (8.7.9) as $r \longrightarrow 0$, in view of Eq. (8.7.5), gives the relation between the (kinetic) energy and the energy spectrum,

$$E = \int_0^\infty E(\kappa)\,d\kappa. \tag{8.7.11}$$

Also, inserting Eq. (8.7.9) in Eq. (8.7.10) and integrating the latter for $f(r)$ yields

$$u^2 f(r) = 2\int_0^\infty E(\kappa)\left[\frac{\sin \kappa r}{(\kappa r)^3} - \frac{\cos \kappa r}{(\kappa r)^2}\right] d\kappa, \tag{8.7.12}$$

from which

$$u^2 f_0'' = -\frac{2}{15}\int_0^\infty \kappa^2 E(\kappa)\,d\kappa. \tag{8.7.13}$$

Comparison with Eq. (8.6.19) now gives

$$\epsilon = 2\nu \int_0^\infty \kappa^2 E(\kappa)\,d\kappa. \tag{8.7.14}$$

The weighting factor κ^2 in this integral shows the rate of energy dissipation to be greatest for the smallest eddies, that is, at the highest wave numbers.

A balance for the energy spectrum may be written by the help of Fig. 8.11. Thus,

$$\frac{\partial E}{\partial t} = -\frac{\partial T}{\partial \kappa} + P(\kappa) - 2\nu\kappa^2 E(\kappa), \tag{8.7.15}$$

where T is the energy flux, P is the production of energy, and

$$D(\kappa) = 2\nu\kappa^2 E(\kappa) \tag{8.7.16}$$

is the dissipation of energy in wave-number space (for a formal development, see Hinze 1975). A similar balance exists for the thermal spectrum. Further remarks on these spectra, and models for T and P, however, are postponed to Chapter 9.

In this chapter we made an attempt to study the difficult turbulence problem. First, we approached the problem from the view of the instability of laminar flows. Although needed for the separation of the stable and unstable flow regimes, this approach failed to describe the large fluctuations associated with turbulence because its foundations rest on small disturbances. We then turned to another view and, noting the random nature of turbulent fluctuations, proceeded to a statistical approach. Although providing important information for the homogeneous turbulence, this approach also failed to provide an answer for turbulent shear flows because its foundations rest on a closure problem. Accordingly, in the next chapter we change our

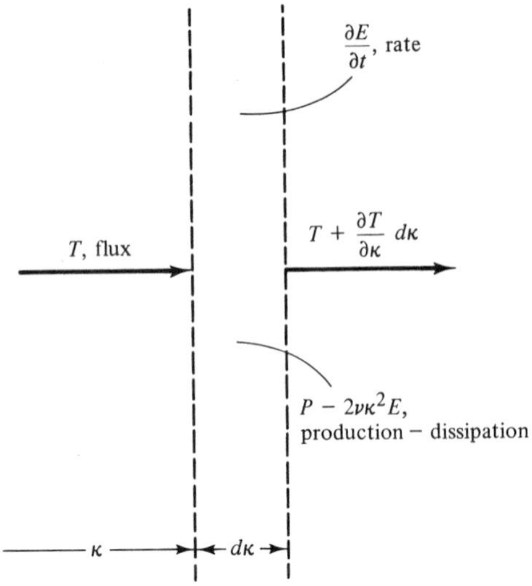

Fig. 8.11 Balance of energy spectrum in the wave-number space

philosophy, and approach turbulence by intuitive models rather than analytical methods.

REFERENCES

ARPACI, V. S., & BAYAZITOGLU, Y., 1973, Phys. Fluids, **16**, 589.

BATCHELOR, G. K., 1967, *The Theory of Homogeneous Turbulence*, Cambridge University Press, New York.

CHANDRASEKHAR, S., 1961, *Hydrodynamic and Hydromagnetic Stability*, Oxford University Press, New York.

ELDER, J. W., 1965, J. Fluid Mech., **23**, 99.

GERSHUNI, G. Z., & ZHUKHOVITSKII, E. M., 1976, *Convective Stability of Incompressible Fluids*, Israel Program for Scientific Translations (available from U.S. Dept. of Commerce).

GÖZUM, D., 1972, Ph. D. thesis, University of Michigan.

GÖZUM, D., & ARPACI, V. S., 1974, J. Fluid Mech., **64**, 439.

HINZE, J. O., 1975, *Turbulence*, McGraw-Hill, New York.

HUNT, J. N., 1964, *Incompressible Fluid Dynamics*, American Elsevier, New York.

KORPELA, S., 1972, Ph. D. thesis, University of Michigan.

KORPELA, S., & ARPACI, V. S., 1976, J. Fluid Mech., **75**, 385.

LAMB, H., 1945, *Hydrodynamics*, Dover, New York.

LIN, C. C., 1955, *The Theory of Hydrodynamic Stability*, Cambridge University Press, Cambridge.

LUNDGREN, T. S., 1969, Phys. Fluids, **12**, 485.

MONIN, A. S., & YAGLOM, A. M., 1971, *Statistical Fluid Mechanics*, Vol. 1, MIT Press, Cambridge, Mass.

SCHLICHTING, H., 1968, *Boundary Layer Theory*, McGraw-Hill, New York.

TENNEKES, H., & LUMLEY, J. L., 1972, *A First Course in Turbulence*, MIT Press, Cambridge, Mass.

VEST, C. M., 1967, Ph. D. thesis, University of Michigan.

VEST, C. M., & ARPACI, V. S., 1969, J. Fluid Mech., **36**, 1.

PROBLEMS

8-1 Interpret the original foundations of the instability theory in terms of the First and Second Laws of thermodynamics.

8-2 Find the buckling load of a simply supported elastic beam (Fig. 8P-2) by following the energy and characteristic-value methods.

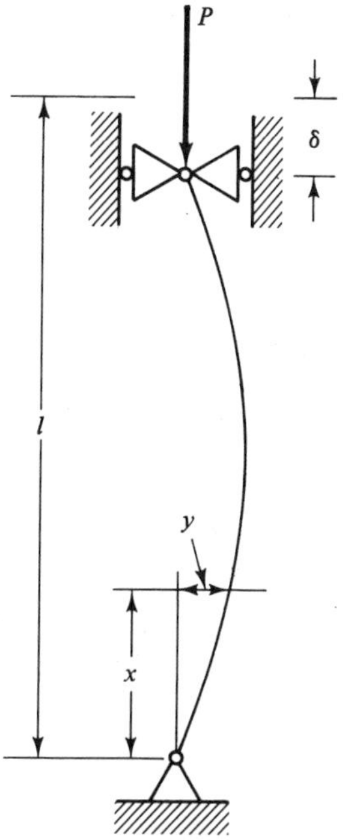

Fig. 8P-2

8-3 Reconsider Ex. 8.1.1. Show the thermal instability of the glass plate by an integral approach employing an approximate temperature profile

$$\theta \sim e^{\sigma t}(l - x)x.$$

8-4 Reconsider Problem 8-3. Show the thermal stability of the glass plate in terms of a dimensional analysis.

8-5 The interfacial behavior of two immiscible fluids under the effect of gravity is the celebrated Rayleigh–Taylor problem. Find the instability of this interface by a dimensional interpretation of the energy approach.

8-6 Following the philosophy of Problem 8-5, add the following to the Rayleigh–Taylor problem:
(a) Surface tension.
(b) Evaporation and superheat (film boiling).
(c) Viscosity.

8-7 The interfacial behavior of two immiscible fluids in uniform relative motion parallel to the interface is the celebrated Kelvin–Helmholtz problem. Find the instability of this interface by the approach followed in Problem 8-5.

8-8 Repeat Problem 8-6 for the Kelvin–Helmholtz problem.

8-9 Study the coupled Rayleigh–Taylor and Kelvin–Helmholtz problems from:
(a) Lamb (1945) for the inviscid case.
(b) Chandrasekhar (1961) for the viscous case.

8-10 Study from Lamb (1945) the stability of an inviscid Cartesian jet of thickness $2l$ in a stagnant fluid of the same density.

8-11 The instability of a stagnant horizontal fluid layer heated from below is the celebrated Rayleigh–Benard problem. Discuss this problem following dimensional arguments and the energy method.

8-12 Study from Chandrasekhar (1961) the exact solution of the Rayleigh–Benard problem for free boundaries.

8-13 Review by chapter titles the types of instability discussed in Chandrasekhar (1961). What is new in Chandrasekhar relative to Lamb (1945)?

8-14 Carry out the development leading to the Orr–Sommerfeld equation given by Eq. (8.2.7).

8-15 Add the effect of buoyancy to the Orr–Sommerfeld equation and develop the energy equation to be coupled to this equation.

8-16 Consider the simple steady conduction problem involving uniform energy generation u''' in a flat plate of conductivity k and thickness $2l$ which transfers heat with a coefficient $h \longrightarrow \infty$ to an ambient at temperature T_∞. The solution of this problem is

$$T - T_\infty = \frac{u'''}{2k}(l^2 - x^2).$$

Discuss this problem in terms of a formulation averaged in x.

8-17 Employing Eq. (8.6.3), or, contracting Eq. (8.6.17), show that

$$\overline{\frac{\partial u_i}{\partial x_j}\frac{\partial u_i}{\partial x_j}} = -\nabla^2 R_{ii}(0) = -15u^2 f_0''.$$

8-18 Show that the second equation to order r^2 resulting from Eq. (8.6.11) is

$$\frac{d}{dt}\overline{\omega^2} = 70u^3h_0''' - 70\nu u^2 f_0^{(\text{iv})}$$

$$= \overline{2\omega_i\omega_j\frac{\partial u_i}{\partial x_j}} - 10\nu\frac{\overline{\omega^2}}{\lambda_\omega^2},$$

where

$$\lambda_\omega^2 = -\frac{15}{7}\frac{f_0''}{f_0^{(\text{iv})}}$$

is the rate of change of mean-square turbulent vorticity.

8-19 Show from

$$u^2 f(r) = \overline{u_1(x)u_1'(x+r)}$$

and

$$u^3 h(r) = \overline{u_2^2(x)u_1(x+r)}$$

that, with isotropy

$$u^2 f_0'' = -\overline{\left(\frac{\partial u_1}{\partial x_1}\right)^2} \qquad \text{and} \qquad u^3 h_0''' = -\frac{1}{2}\overline{\left(\frac{\partial u_1}{\partial x_1}\right)^3},$$

and that

$$\overline{2\omega_i\omega_j\frac{\partial u_i}{\partial x_j}} = 70u^3h_0''' = \frac{7}{3\sqrt{15}}(\overline{\omega^2})^{3/2}S,$$

where skewness is

$$S = -\frac{\overline{(\partial u_1/\partial x_1)^3}}{[\overline{(\partial u_1/\partial x_1)^2}]^{3/2}},$$

and

$$\overline{\omega^2} = 15\overline{\left(\frac{\partial u_1}{\partial x_1}\right)^2}.$$

Chapter 9

KINETIC AND THERMAL SCALES AND SPECTRA

Our approach to turbulence in Chapter 8, as an instability problem or as a statistical problem, has been mathematical. This approach, at present or in the foreseeable future, does not appear to hold any promise for engineering problems involving turbulence. Instead, we must seek another approach which may slowly but steadily increase our understanding of turbulence. Opposed diametrically to the formal nature of the previous approach, the new approach is going to be intuitive. With this approach, we develop in this chapter a qualitative understanding on the physics of turbulence. The new approach is also expected to lead to increasingly realistic future models for the prediction of technologically important turbulence problems.

In Section 9.1 we discuss the kinetic energy* of the mean flow and of fluctuations. In Section 9.2, we introduce some scales, and in Section 9.3 we construct a model based on these scales. Proceeding to thermal aspects of turbulence in Section 9.4, we define the Reynolds flux and introduce some thermal scales. In Section 9.5 we distinguish the thermal scales appropriate for natural convection. Finally, in Section 9.6 we discuss kinetic and thermal equilibrium spectra.

*For customary reasons, the mechanical energy balance will be referred to in this chapter as the kinetic energy balance.

9.1 KINETIC ENERGY OF MEAN FLOW AND OF FLUCTUATIONS

We have already obtained the momentum equations for the steady mean flow of an incompressible fluid,

$$U_j \frac{\partial U_i}{\partial x_j} = \frac{1}{\rho} \frac{\partial T_{ij}}{\partial x_j}, \tag{8.3.9}$$

where

$$T_{ij} = -P\delta_{ij} + 2\mu S_{ij} - \rho\overline{u_i u_j}. \tag{8.3.10}$$

Since the mean momentum $\bar{u}_i$ of turbulent velocity fluctuations is zero, we look for a way other than the mean momentum for a discussion on the interaction between the mean flow and turbulence. We shall do this by studying the kinetic energy of mean flow and the mean kinetic energy of turbulence.

The equation governing the mean flow kinetic energy $\frac{1}{2}U_iU_i$ is obtained by multiplying Eq. (8.3.9) by U_i (recall Eq. 2.1.17). After splitting the stress term in the resulting equation into two components, we have

$$\rho U_j \frac{\partial}{\partial x_j}\left(\frac{1}{2} U_i U_i\right) = \frac{\partial}{\partial x_j}(T_{ij}U_i) - T_{ij}\frac{\partial U_i}{\partial x_j}. \tag{9.1.1}$$

Because T_{ij} is symmetric, the last term of Eq. (9.1.1) may be rearranged as $T_{ij}S_{ij}$, where

$$S_{ij} = \frac{1}{2}\left(\frac{\partial U_i}{\partial x_j} + \frac{\partial U_j}{\partial x_i}\right)$$

is the symmetric part of $\partial U_i/\partial x_j$. Then Eq. (9.1.1) becomes

$$\rho U_j \frac{\partial}{\partial x_j}\left(\frac{1}{2} U_i U_i\right) = \frac{\partial}{\partial x_j}(T_{ij}U_i) - T_{ij}S_{ij}. \tag{9.1.2}$$

We have already explored (see Eq. 2.1.28) the physical significance of the right-hand terms of Eq. (9.1.2). Recall that the dissipation $T_{ij}S_{ij}$ is the deformation part of the work term, and is the irreversible conversion of kinetic (or mechanical) energy into internal (thermal) energy. Consequently, it is a term in the balance of thermal energy. Reasons for the balance of kinetic energy rearranged to involve this term will become clear in the development of and the discussion following Eq. (9.1.14).

Inserting Eq. (8.3.10) into $T_{ij}S_{ij}$ yields

$$T_{ij}S_{ij} = 2\mu S_{ij}S_{ij} - \rho\overline{u_i u_j}S_{ij}. \tag{9.1.3}$$

Note that the contribution of the pressure to deformation work in an incompressible fluid is zero,

$$-P\,\delta_{ij}S_{ij} = -PS_{ii} = -P\frac{\partial U_i}{\partial x_i} = 0,$$

that $2\mu S_{ij}S_{ij}$ is the viscous part and $-\rho\overline{u_i u_j}S_{ij}$ the turbulent part of dissipation in Eq. (9.1.3). Since turbulent stresses provide the latter dissipation, the kinetic energy of turbulence gains from this dissipation. Accordingly, $-\rho\overline{u_i u_j}S_{ij}$ is known as the *turbulent energy production*.

Inserting Eq. (8.3.10) into Eq. (9.1.2) yields, after some rearrangement,

$$U_j \frac{\partial}{\partial x_j}\left(\frac{1}{2}U_iU_i\right) = \frac{\partial}{\partial x_j}\left(-\frac{P}{\rho}U_j + 2\nu U_iS_{ij} - \overline{u_iu_j}U_i\right) - 2\nu S_{ij}S_{ij} + \overline{u_iu_j}S_{ij}, \tag{9.1.4}$$

the first three terms on the right-hand side being the net work flux associated with mean pressure, mean viscous stress, and turbulent stress, and the last two terms being the dissipation related to the mean viscous stress and the turbulent stress, respectively.

Let us introduce now, somewhat loosely for the time being, a length scale l, a velocity scale u, and a time scale $t = l/u$. Here u may be defined, for example, as

$$u^2 = \tfrac{1}{3}\overline{u_iu_i}, \tag{9.1.5}$$

and l as an integral scale proportional to a length characterizing the geometry. Turbulence scales will prove indispensable in modeling turbulence. We shall return to this subject in Section 9.2 and introduce additional scales. In the meantime, the foregoing scales lead to the relations

$$\frac{\partial U_i}{\partial x_i} \sim \frac{u}{l}, \qquad -\overline{u_iu_j} \sim u^2, \qquad S_{ij} \sim \frac{u}{l},$$

and

$$\overline{u_iu_j}S_{ij} \sim ulS_{ij}S_{ij}, \tag{9.1.6}$$

$$-\overline{u_iu_j}U_i \sim ulU_iS_{ij}, \tag{9.1.7}$$

in which all (mean as well as fluctuating) components are scaled relative to the velocity introduced by Eq. (9.1.5). This limitation will also be reflected in the turbulence model to be developed later. Meanwhile, comparing Eqs. (9.1.6) and (9.1.7), respectively, with the first and second viscous terms on the right-hand side of Eq. (9.1.4), we have

$$\frac{-\overline{u_iu_j}U_i}{2\nu S_{ij}U_i} \sim \mathrm{Re}_l$$

and

$$\frac{\overline{u_iu_j}S_{ij}}{2\nu S_{ij}S_{ij}} \sim \mathrm{Re}_l, \tag{9.1.8}$$

where $\mathrm{Re}_l = ul/\nu$ is the Reynolds number based on the integral scales. This result shows that the terms associated with turbulent stress are Re_l times larger than those associated with viscous terms. Since this Reynolds number is usually very large, the viscous terms in Eq. (9.1.4) can ordinarily be neglected. That is, the *gross structure of turbulent flows tends to be independent of viscosity*. This is an important result. However, no further information is apparently available from Eq. (9.1.4).

We turn now to the mean kinetic energy of turbulent fluctuations. The equation governing the mean kinetic energy of turbulent fluctuations is obtained by multiplying Eq. (8.3.1) by $\tilde{u}_i$, taking the time average of all terms, and subtracting Eq. (9.1.4). The details of this exercise are left to the reader (see Problem 9-1). The result is

$$U_j \frac{\partial}{\partial x_j}(\tfrac{1}{2}\overline{u_i u_i}) = \frac{\partial}{\partial x_j}\left(-\overline{\frac{p}{\rho}u_j} + 2\nu\overline{u_i s_{ij}} - \tfrac{1}{2}\overline{u_i u_i u_j}\right) - \overline{u_i u_j}S_{ij} - 2\nu\overline{s_{ij}s_{ij}} \tag{9.1.9}$$

with

$$s_{ij} = \frac{1}{2}\left(\frac{\partial u_i}{\partial x_j} + \frac{\partial u_j}{\partial x_i}\right)$$

for the fluctuating rate of strain. The first three right-hand terms in Eq. (9.1.9) denote the net work flux associated with fluctuating pressure and fluctuating viscous and turbulent stresses, and the last two terms denote the two kinds of dissipation. Note that the turbulent production $-\overline{u_i u_j}S_{ij}$ occurs in Eqs. (9.1.9) and (9.1.4) with opposite signs. As expected, this term apparently serves to exchange kinetic energy between the mean flow and the turbulence. In general, the energy exchange involves a loss to the mean flow and a gain to the turbulence. The last term in Eq. (9.1.9) is the rate at which fluctuating viscous stresses perform deformation work against the fluctuating strain rate. The term denotes a loss of energy, since it is quadratic in s_{ij}, and is called the *viscous dissipation of turbulent fluctuations.* Unlike the dissipation related to mean viscous stress in Eq. (9.1.4), this term is essential to the dynamics of turbulence and cannot usually be neglected. In view of the foregoing discussion, introducing, for the mean kinetic energy of fluctuations,

$$E = \tfrac{1}{2}\overline{u_i u_i}, \tag{9.1.10}$$

for the *turbulent diffusion* (total work flux associated with fluctuations)

$$\mathfrak{D}_j = \overline{\frac{p}{\rho}u_j} - 2\nu\overline{u_i s_{ij}} + \tfrac{1}{2}\overline{u_i u_i u_j}, \tag{9.1.11}$$

for the *turbulent production,*

$$\mathcal{P} = -\overline{u_i u_j}S_{ij}, \tag{9.1.12}$$

and for the *turbulent dissipation,*

$$\epsilon = 2\nu\overline{s_{ij}s_{ij}}, \tag{9.1.13}$$

Eq. (9.1.9) may be given the compact form

$$U_j \frac{\partial E}{\partial x_j} = -\frac{\partial \mathfrak{D}_j}{\partial x_j} + \mathcal{P} - \epsilon, \tag{9.1.14}$$

which states that the mean flow of turbulent kinetic energy is balanced by the diffusion, production, and dissipation of this energy.

In a steady, homogeneous, pure shear flow (all averaged quantities except U_i independent of position and S_{ij} = constant), Eq. (9.1.14) reduces to

$$\mathcal{P} = \epsilon, \tag{9.1.15}$$

the balance between the production rate of turbulent energy by Reynolds stresses and the rate of viscous dissipation of turbulent fluctuations. Generally speaking, in a shear flow, production and dissipation do not balance. Thus, Eq. (9.1.15) may be useful only for scaling of turbulence not related to diffusion.

Employing the scale relations

$$S_{ij} \sim \frac{u}{l}, \qquad -\overline{u_i u_j} \sim u^2$$

and

$$-\overline{u_i u_j} S_{ij} \sim ulS_{ij}S_{ij}$$

in Eq. (9.1.15), we have

$$ulS_{ij}S_{ij} \sim \nu\overline{s_{ij}s_{ij}},$$

or

$$\frac{\overline{s_{ij}s_{ij}}}{S_{ij}S_{ij}} \sim \frac{ul}{\nu} = \mathrm{Re}_l. \tag{9.1.16}$$

Then, in view of the fact that Re_l is usually very large, we conclude that

$$\overline{s_{ij}s_{ij}} \gg S_{ij}S_{ij}. \tag{9.1.17}$$

The fluctuating strain rate s_{ij} is thus very much larger than the mean strain rate S_{ij}. Since strain rates have the dimension of $(\text{time})^{-1}$, the eddies contributing most to the dissipation of energy have very small convection time scales compared to the time scale of the mean flow. Accordingly, the direct interaction between the fluctuating strain rate and the mean strain rate is negligible for large Reynolds numbers.

Foregoing considerations suggest that the scales for fluctuations should be different from those for the mean flow. Accordingly, we may construct a turbulence model to be based on two sets of scales, one set for the mean flow and one set for fluctuations.

9.2 *SCALES OF TURBULENCE*

Let the scales for the mean flow be l, u, and $t = l/u$. These scales, being associated with the gross behavior of the flow, are the larger ones of the two sets. They apply to the production part of Eq. (9.1.15). Let the scales for the fluctuations be η, υ, and $\tau = \eta/\upsilon$. These are the smaller ones of the two sets. They apply to the dissipation part of Eq. (9.1.15). Following Kolmogorov, they are called the (Kolmogorov) *microscales*. In terms of the integral scales and the microscales, Eq. (9.1.15) may be written as

$$\mathcal{P} \sim u^2 \frac{u}{l} \sim \nu \frac{\upsilon^2}{\eta^2} \sim \epsilon, \tag{9.2.1}$$

which illustrates that *viscous dissipation of energy can be estimated from the large-scale inviscid dynamics*. The dissipation may then be interpreted as a passive process proceeding at a rate dictated by the inviscid inertial behavior of the large eddies.

Under the assumption of isotropy, the foregoing two-scale turbulence model reduces to a one-scale model in microscales (Fig. 9.1). Thus, from

$$u^2 \frac{u}{l} \sim \epsilon$$

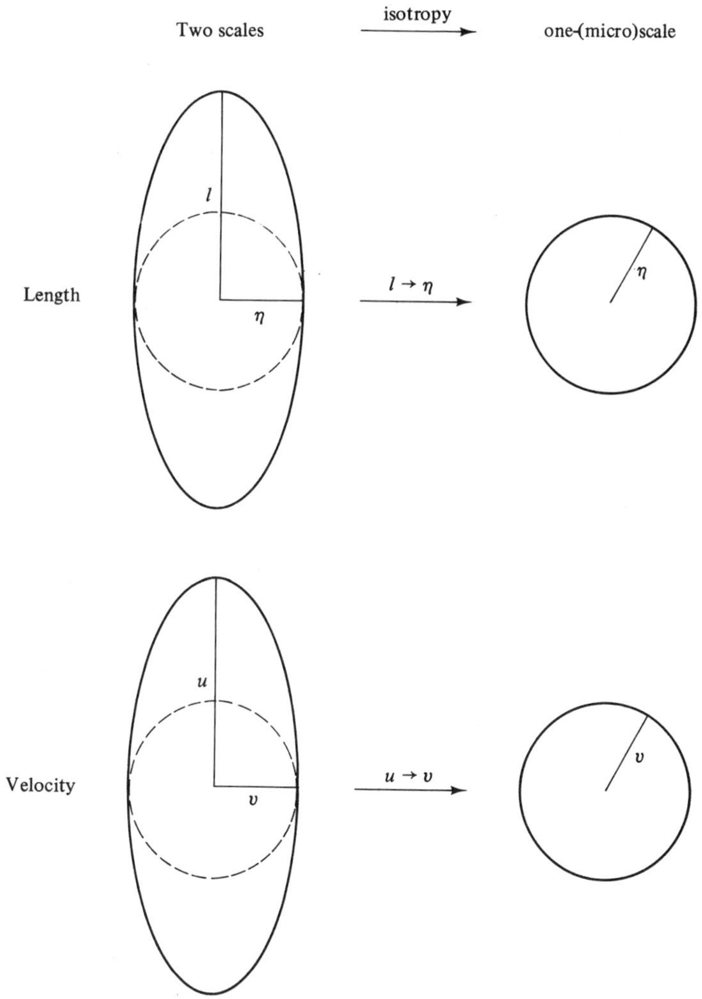

Fig. 9.1 Two-scale turbulence model and its isotropic limit

of Eq. (9.2.1), we obtain as $l \longrightarrow \eta$ and $u \longrightarrow v$,

$$v \sim (\eta\epsilon)^{1/3}, \tag{9.2.2}$$

and from the last proportionality of the same equation,

$$v \sim \eta\left(\frac{\epsilon}{\nu}\right)^{1/2}. \tag{9.2.3}$$

Elimination of v between Eqs. (9.2.2) and (9.2.3) yields the *Kolmogorov microscale*,

$$\boxed{\eta = \left(\frac{\nu^3}{\epsilon}\right)^{1/4},} \tag{9.2.4}$$

and the insertion of this result into Eq. (9.2.2) or Eq. (9.2.3) gives

$$\upsilon \sim (\nu\epsilon)^{1/4}. \tag{9.2.5}$$

Then, for $\tau = \eta/\upsilon$, the ratio of Eq. (9.2.4) into Eq. (9.2.5) yields

$$\tau \sim \left(\frac{\nu}{\epsilon}\right)^{1/2}. \tag{9.2.6}$$

Clearly, Eqs. (9.2.4)–(9.2.6) show that the small (micro) scale motion of turbulence depends only on the momentum diffusivity (kinematic viscosity) and the large-scale energy supply for the dissipation ϵ. The Reynolds number in this scale,

$$\boxed{\mathrm{Re}_K = \frac{\upsilon\eta}{\nu} = 1,} \tag{9.2.7}$$

illustrates that the small-scale motion is slow and viscous. What we learned so far about the two-scale turbulence model is summarized in Table 9.1 for a ready reference.

TABLE 9.1 A two-scale turbulence model

Turbulence—< Mean (large scale) / Fluctuations (small scale)
Large scales: $l, u, t = l/u$, Small scales: $\eta, \upsilon, \tau = \eta/\upsilon$.
Large scale $\equiv$ production scale $\equiv$ integral scale, Small scale $\equiv$ dissipation scale $\equiv$ Kolmogorov microscale
Steady homogeneous pure shear flow, $\mathcal{P} \sim u^2 \dfrac{u}{l} \sim \nu \dfrac{\upsilon^2}{\eta^2} \sim \epsilon$, provides an inviscid estimate for dissipation.
Small scales, $\eta \sim \left(\dfrac{\nu^3}{\epsilon}\right)^{1/4}$, $\upsilon \sim (\nu\epsilon)^{1/4}$, and $\tau \sim \left(\dfrac{\nu}{\epsilon}\right)^{1/2}$, are isotropic.

Actually, a third scale is hidden in the present turbulence model. What would happen to the length η of the Kolmogorov scale if the velocity of this scale were amplified to the velocity u of the integral scale? In other words, what is the length scale associated with

$$\tau^{-1} = \frac{v}{\eta} = \frac{u}{?}. \tag{9.2.8}$$

At first, η is expected to amplify to the length l of the integral scale. However, from the combination of Eqs. (9.2.1) and (9.2.8),

$$u^2 \frac{u}{l} \sim \nu \frac{v^2}{\eta^2} \sim \nu \frac{u^2}{?^2}, \tag{9.2.9}$$

and from the first and last terms of this result, we have

$$\frac{?}{l} \sim \frac{1}{\mathrm{Re}_l^{1/2}} \ll 1, \tag{9.2.10}$$

which shows that the unknown scale must be smaller than the integral scale. Also, according to Eq. (9.2.9), this scale is greater than the Kolmogorov scale. Consequently,

$$\eta < ? < l, \tag{9.2.11}$$

and the unknown scale turns out to be the *Taylor microscale*, λ (recall Eq. 8.6.12). A simple geometric interpretation of Eq. (9.2.8), now written including the definition of λ as

$$\tau^{-1} = \frac{v}{\eta} = \frac{u}{\lambda}, \tag{9.2.12}$$

is shown in Fig. 9.2. From the similarity of two shaded triangles we readily obtain a geometric interpretation of the Taylor scale. Yet a word of caution is in order on this

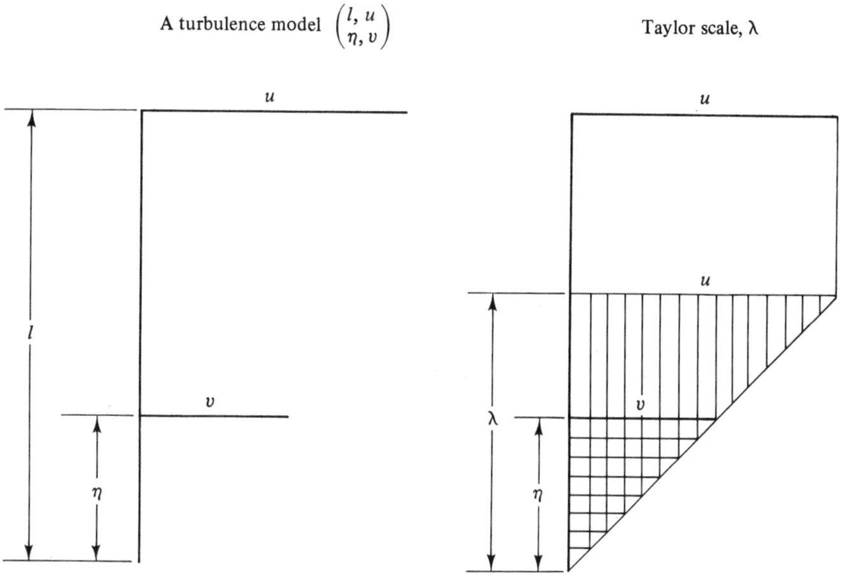

Fig. 9.2 Geometric interpretation of the Taylor scale

scale. Being defined by the velocity of large scales and the time of small scales, the Taylor scale is neither large nor small but rather is a mixed scale. Frequent use of this scale in estimates such as $s_{ij} \sim u/\lambda$ is for reasons of convenience because it involves the measurable velocity of large scales rather than that of small scales. The only scale that can be determined unambiguously is the Kolmogorov time scale of Eq. (9.2.8). The Taylor scale will be interpreted further at the end of the next section.

Note that Eq. (9.2.1) may be rearranged in terms of Eq. (9.2.12) to give

$$u^2 \frac{u}{l} \sim \nu \frac{u^2}{\lambda^2}, \tag{9.2.13}$$

which leads to

$$\frac{\lambda}{l} \sim \frac{1}{\mathrm{Re}_\lambda} \sim \frac{1}{\mathrm{Re}_l^{1/2}}, \tag{9.2.14}$$

where $\mathrm{Re}_\lambda = u\lambda/\nu$ is the Reynolds number based on the Taylor scale. Equation (9.2.1) may also be rearranged in terms of Eq. (9.2.7) to give

$$\frac{\eta}{l} \sim \frac{1}{\mathrm{Re}_\eta^3} \sim \frac{1}{\mathrm{Re}_l^{3/4}}. \tag{9.2.15}$$

Furthermore, from the ratio of Eqs. (9.2.14) and (9.2.15), after utilizing the dependence of these equations on Re_l,

$$\frac{\eta}{\lambda} \sim \frac{1}{\mathrm{Re}_\eta} \sim \frac{1}{\mathrm{Re}_\lambda^{1/2}} \sim \frac{1}{\mathrm{Re}_l^{1/4}}, \tag{9.2.16}$$

or

$$\boxed{\mathrm{Re}_\eta \sim \mathrm{Re}_\lambda^{1/2} \sim \mathrm{Re}_l^{1/4},} \tag{9.2.17}$$

where $\mathrm{Re}_\eta = u\eta/\nu$ is the Reynolds number based on the Kolmogorov scale. Finally, an inspection of Eqs. (9.2.14) and (9.2.15) also yields

$$\left(\frac{\eta}{\lambda}\right)^2 \sim \frac{\lambda}{l}. \tag{9.2.18}$$

The foregoing considerations on the scales of turbulence will be helpful in the construction of a turbulence model, to which we now proceed.

9.3 *A TURBULENCE MODEL*

Following earlier developments on the wave aspects of radiation and the particle aspects of matter, Planck's discovery of the particle aspects of radiation and de Broglie's discovery of the wave aspects of matter have completed the theories resting on the particle–wave duality for both radiation and matter. In a manner similar to this duality, turbulence research appears to employ the wave (spectral) concept and the particle (eddy) concept in attempts for an understanding of turbulence. Although nothing is wrong with an approach loosely involving a spectrum-eddy duality for turbulence, the eddy of turbulence has so far remained a large enough blob without much character. The purpose of this section is to introduce an "eddy with character," hereafter to be referred for short as the "turbulon," and to utilize this concept in the construction of a turbulence model.

A turbulon is assumed to be made of two volumes, the production volume and the dissipation volume, the latter being sheared (with the velocity of the integral scale) in the former, and the ratio of the production volume to the dissipation volume being $(\lambda/\eta)^2$ (Fig. 9.3). Actually, earlier works of Townsend (1951) and Corrsin (1962) recognize the difference between the production and dissipation volumes, which is called the *intermittency* in turbulent flows. The present turbulon model agrees with the Tennekes vortex tube model (1968), and agrees qualitatively with recent experiments of Tavoularis, Bennett & Corrsin (1978) for $Re_\lambda \longrightarrow \infty$.

In terms of the turbulon, the local dissipation per unit production volume is

$$\epsilon \sim \nu \frac{u^2}{\lambda^2} \tag{9.3.1}$$

and per unit dissipation volume is

$$\bar{\epsilon} \sim \epsilon \left(\frac{\lambda}{\eta}\right)^2 \sim \nu \frac{u^2}{\eta^2}. \tag{9.3.2}$$

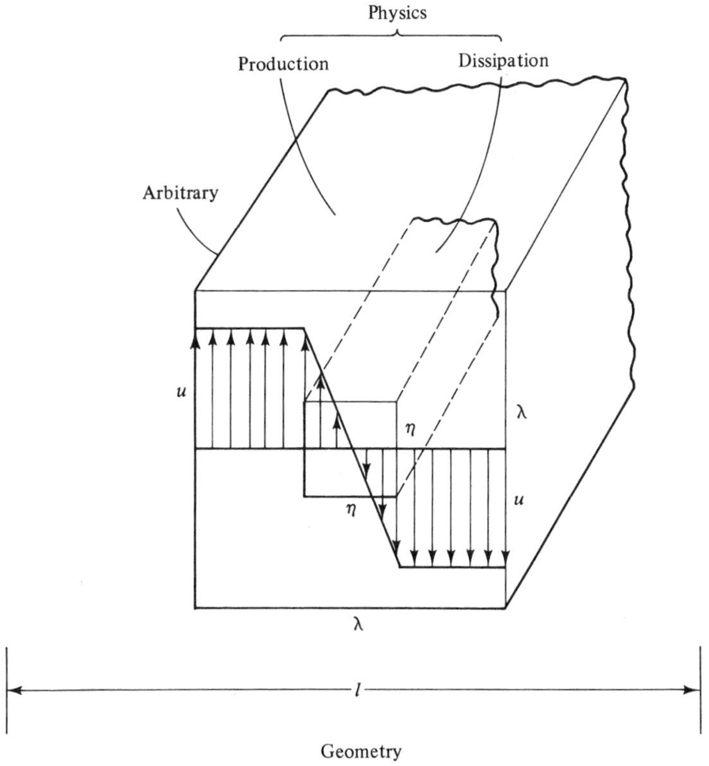

Fig. 9.3 Intermittency of turbulon

Also, the local production per unit production volume is

$$\mathcal{P} \sim u^2 \frac{u}{l}, \tag{9.3.3}$$

and per unit dissipation volume is

$$\bar{\mathcal{P}} \sim \mathcal{P}\left(\frac{\lambda}{\eta}\right)^2 \sim u^2 \frac{u}{l}\left(\frac{\lambda}{\eta}\right)^2,$$

which may be rearranged as

$$\bar{\mathcal{P}} \sim u^2 \frac{u}{\lambda}\left(\frac{\lambda}{l}\right)\left(\frac{\lambda}{\eta}\right)^2,$$

or, in view of Eq. (9.2.18), as

$$\bar{\mathcal{P}} \sim u^2 \frac{u}{\lambda}. \tag{9.3.4}$$

Then the balance between the production and dissipation per unit production volume,

$$\mathcal{P} \sim u^2 \frac{u}{l} \sim \nu \frac{u^2}{\lambda^2} \sim \epsilon, \tag{9.3.5}$$

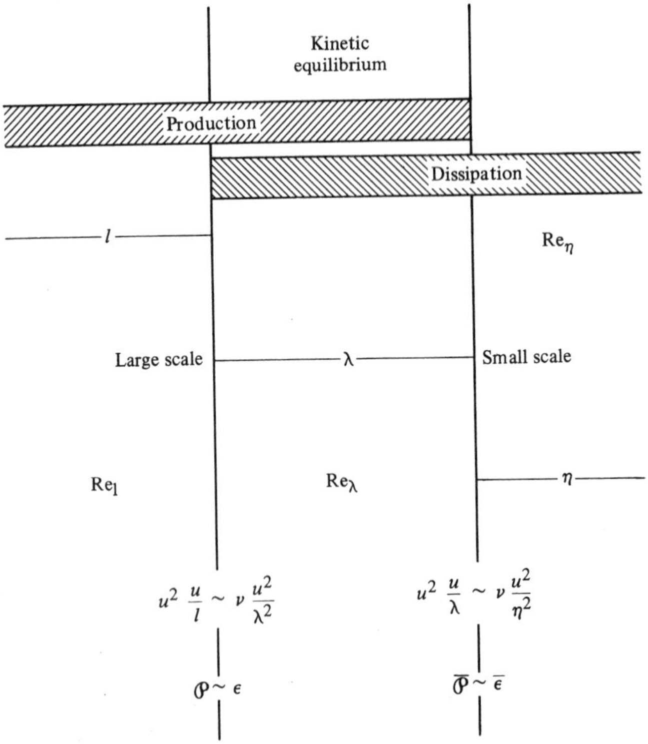

Fig. 9.4 Kinetic equilibrium

and that per unit dissipation volume,

$$\bar{\mathcal{P}} \sim u^2 \frac{u}{\lambda} \sim \nu \frac{u^2}{\eta^2} \sim \bar{\epsilon}, \tag{9.3.6}$$

may be interpreted as the mean kinetic energy balance of turbulent fluctuations near the large-scale tail and the small-scale tail of the equilibrium range, respectively (Fig. 9.4).

Note that by introducing the Taylor scale we replace the turbulence model based on two velocities and two length scales (Fig. 9.2) with an equivalent model based on one measurable velocity and three length scales (Fig. 9.5). For ready reference, we collect in Table 9.2 the various Reynolds numbers and the scale ratios based on the integral, Taylor, and Kolmogorov scales.

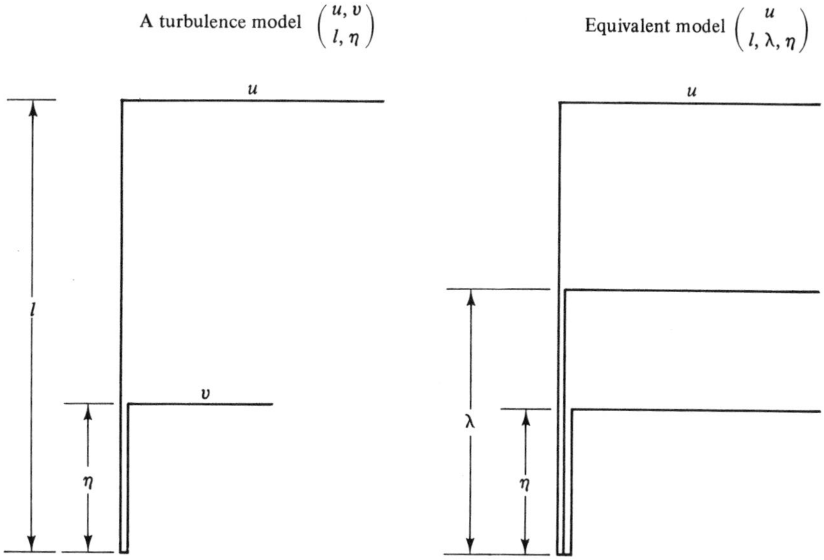

Fig. 9.5 Equivalent representations of turbulence model

Having developed a model for turbulent motion, we proceed to some thermal scales which will be useful in the construction of a model for the transport of thermal energy by this motion.

9.4 *REYNOLDS FLUX AND THERMAL SCALES*

Turbulent flow transports passive contaminants such as chemical species, small particles, and fluid properties such as thermal energy in much the same way as momentum. For use in thermal models, we develop here the equation governing the turbulent transport of thermal energy.

TABLE 9.2 Reynolds numbers and scale ratios

$\mathrm{Re}_K = \dfrac{v\eta}{\nu} = 1$
$\mathrm{Re}_\eta = \dfrac{u\eta}{\nu}$ $\mathrm{Re}_\lambda = \dfrac{u\lambda}{\nu}$ $\mathrm{Re}_l = \dfrac{ul}{\nu}$
$\dfrac{\eta}{l} \sim \mathrm{Re}_\lambda^{-3/2} \sim \mathrm{Re}_l^{-3/4}$ $\dfrac{\lambda}{l} \sim \mathrm{Re}_\lambda^{-1} \sim \mathrm{Re}_l^{-1/2} \sim \left(\dfrac{\eta}{\lambda}\right)^2$ $\dfrac{\eta}{\lambda} \sim \mathrm{Re}_\eta^{-1} \sim \mathrm{Re}_\lambda^{-1/2} \sim \mathrm{Re}_l^{-1/4} \sim \dfrac{v}{u}$
$\mathrm{Re}_\eta \sim \mathrm{Re}_\lambda^{1/2} \sim \mathrm{Re}_l^{1/4}$

The diffusion of thermal energy in an incompressible turbulent flow satisfies, after neglecting the viscous dissipation,

$$\frac{\partial \tilde{\theta}}{\partial t} + \tilde{u}_i \frac{\partial \tilde{\theta}}{\partial x_i} = a \frac{\partial^2 \tilde{\theta}}{\partial x_i \, \partial x_i}, \tag{9.4.1}$$

where a is the thermal diffusivity, which will be assumed constant. Decompose $\tilde{\theta}(x_i, t)$ in a mean value Θ and fluctuation θ as done for $\tilde{u}_i(x_i, t)$,

$$\tilde{\theta} = \Theta + \theta, \tag{9.4.2}$$

where

$$\begin{aligned} \bar{\tilde{\theta}} &= \Theta = \lim_{T\to\infty} \frac{1}{T} \int_{t_0}^{t_0+T} \tilde{\theta} \, dt, \\ \bar{\theta} &= 0, \qquad \frac{\partial \Theta}{\partial t} = 0, \end{aligned} \tag{9.4.3}$$

the last condition confining our interest to the mean steady turbulence.

Inserting Eq. (9.4.2) into Eq. (9.4.1) and taking the average of all terms in the resulting equation yields

$$U_i \frac{\partial \Theta}{\partial x_i} = \frac{\partial}{\partial x_i}\left(a \frac{\partial \Theta}{\partial x_i} - \overline{u_i \theta}\right). \tag{9.4.4}$$

The total mean heat flux q_i in a turbulent flow then becomes

$$q_i = \rho c_p\left(-a \frac{\partial \Theta}{\partial x_i} + \overline{u_i \theta}\right). \tag{9.4.5}$$

This result shows that the turbulent heat flux is a sum involving contributions from molecular motion and turbulent motion. The contribution of turbulent motion to the mean heat flux

$$\boxed{q_i^R = \rho c_p \overline{u_i \theta}} \tag{9.4.6}$$

is called the *Reynolds flux*. The analogy between Eqs. (8.3.11) and (9.4.6) should be noted. This analogy is the foundation of the assumption that turbulence transports thermal energy in the same way as momentum.

Now we proceed to a discussion of thermal fluctuations in an incompressible turbulent flow. A positive measure for these fluctuations is $\overline{\theta^2}$. The equation governing $\overline{\theta^2}$ in a steady flow is obtained in exactly the same way as the equation for $\overline{u_i u_i}$ (see Problem 9-8). The result is

$$\begin{aligned} U_j \frac{\partial}{\partial x_j}\left(\frac{1}{2}\overline{\theta^2}\right) = -\frac{\theta}{\partial x_j}\left[\frac{1}{2}\overline{\theta^2 u_j} - a\frac{\partial}{\partial x_j}\left(\frac{1}{2}\overline{\theta^2}\right)\right] \\ - \overline{u_j\theta}\frac{\partial \Theta}{\partial x_j} - a\overline{\frac{\partial \theta}{\partial x_j}\frac{\partial \theta}{\partial x_j}}, \end{aligned} \tag{9.4.7}$$

or, more compactly,

$$U_j \frac{\partial E_\theta}{\partial x_j} = -\frac{\partial \mathfrak{D}_{\theta,i}}{\partial x_i} + \mathcal{P}_\theta - \epsilon_\theta. \tag{9.4.8}$$

The flow of $\frac{1}{2}\overline{\theta^2}$ is thus controlled by the three right-hand terms, associated with diffusion, production, and dissipation, respectively. Here

$$\mathfrak{D}_{\theta,i} = \frac{1}{2}\overline{\theta^2 u_i} - a\frac{\partial}{\partial x_i}\left(\frac{1}{2}\overline{\theta^2}\right) \tag{9.4.9}$$

may be interpreted as the net flux involving turbulent and molecular contributions. The production and dissipation terms are analogous to those of the turbulent kinetic energy.

In a homogeneous, pure shear flow (all averaged quantities except Θ are independent of position) Eq. (9.4.7) reduces to

$$-\overline{u_j\theta}\frac{\partial \Theta}{\partial x_j} = a\overline{\frac{\partial \theta}{\partial x_j}\frac{\partial \theta}{\partial x_j}}, \tag{9.4.10}$$

or

$$\mathcal{P}_\theta = \epsilon_\theta, \tag{9.4.11}$$

which states that the production and dissipation of E_θ are balanced. Letting $u_i \sim u$ and $\partial\Theta/\partial x_j \sim \theta/l$ (where θ may be assumed as the root mean square (rms) of temperature fluctuations), the production term may be rearranged as

$$\mathcal{P}_\theta \sim ul\frac{\partial \Theta}{\partial x_j}\frac{\partial \Theta}{\partial x_j}. \tag{9.4.12}$$

Combining this result with Eq. (9.4.10) and noting that the Péclet number $\text{Pe}_l = ul/a$

is usually very large, we obtain

$$\frac{\overline{\dfrac{\partial \theta}{\partial x_j}\dfrac{\partial \theta}{\partial x_j}}}{\dfrac{\partial \Theta}{\partial x_j}\dfrac{\partial \Theta}{\partial x_j}} \sim \frac{ul}{a} = \mathrm{Pe}_l \gg 1, \tag{9.4.13}$$

or

$$\overline{\frac{\partial \theta}{\partial x_j}\frac{\partial \theta}{\partial x_j}} \gg \frac{\partial \Theta}{\partial x_j}\frac{\partial \Theta}{\partial x_j}. \tag{9.4.14}$$

That is, the fluctuating temperature gradient is on the average very much larger than the gradient of the mean temperature. This shows, in a manner identical to the discussion following Eq. (9.1.17), that there is little direct interaction between the large-scale production and the small-scale dissipation of E_θ. Actually, the velocities involved with Eq. (9.4.10) depend on the Prandtl number and Eq. (9.4.13) should be corrected accordingly. However, the conclusion reached by Eq. (9.4.14) continues to remain valid after the correction (see Problem 9-9). We proceed now to the effect of Prandtl number on Eq. (9.4.10).

Introducing (l, u_θ, θ) for the large thermal scales and $(\eta_\theta, v_\theta, \vartheta)$ for the small thermal scales, we may express Eq. (9.4.10) as

$$\mathcal{P}_\theta \sim \theta u_\theta \frac{\theta}{l} \sim a\frac{\vartheta^2}{\eta_\theta^2} \sim \epsilon_\theta, \tag{9.4.15}$$

where η_θ denotes the thermal counterpart of the Kolmogorov scale, say the *Batchelor* scale. Also, in analogy to Eq. (9.2.12), introducing the thermal counterpart of the Taylor scale by the relation

$$\frac{\vartheta}{\eta_\theta} = \frac{\theta}{\lambda_\theta}, \tag{9.4.16}$$

Eq. (9.4.15) may be rearanged as

$$\theta u_\theta \frac{\theta}{l} \sim a\frac{\theta^2}{\lambda_\theta^2}. \tag{9.4.17}$$

To proceed further, we need a velocity scale which depends on the Prandtl number. This dependence for fluids with $\mathrm{Pr} > 1$ (water, viscous oils) is different than that for fluids with $\mathrm{Pr} \ll 1$ (liquid metals).

First, let the scales for momentum and E be the same, and those for thermal energy and E_θ be the same. Then, for fluids with $\mathrm{Pr} > 1$, following the knowledge gained on laminar flows (recall Eq. 4.3.18), assume that the analogy between momentum and thermal energy extends to turbulent flows. Accordingly,

$$\left(\frac{\text{flow}}{\text{flux}}\right) \text{ of momentum} \equiv \left(\frac{\text{flow}}{\text{flux}}\right) \text{ of thermal energy}, \tag{9.4.18}$$

which yields, with the help of Fig. 9.6 and $u_\theta = u(\lambda_\theta/\lambda)$,

$$\frac{\rho u(u/l)}{\mu(u/\lambda^2)} \equiv \frac{\rho c_p u(\lambda_\theta/\lambda)(\theta/l)}{k(\theta/\lambda_\theta^2)}, \tag{9.4.19}$$

or

$$\frac{\lambda}{\lambda_\theta} = \text{Pr}^{1/3}, \qquad \text{Pr} > 1. \tag{9.4.20}$$

This is an important result. Scales λ and λ_θ depend on the (turbulent) behavior of flow, but their ratio is a fluid property. Also, the distribution of u and θ need not be linear, as shown in Fig. 9.6. Any distribution, including in part a logarithmic law of the wall (to be discussed in Chapter 11), leads exactly or within a constant to Eq. (9.4.20), depending, respectively, on the analogy or the absence of analogy between u and θ. Departure from the analogy makes the algebraic value of the constant different from unity.

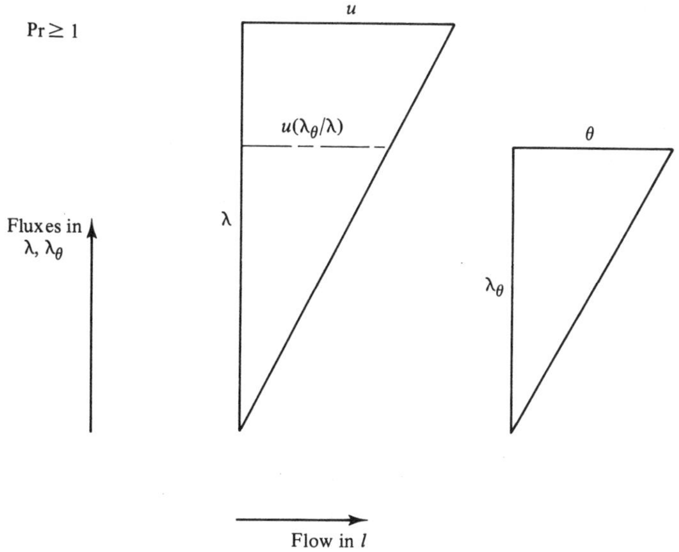

Fig. 9.6 Velocity of enthalpy flow for fluids with Pr > 1

Before proceeding further, let us model, in a manner similar to the earlier turbulon model, the eddy aspects of thermal fluctuations by a "thermal turbulon." Let the thermal turbulon be made also of a production volume and a dissipation volume, the *thermal intermittency* (ratio of these volumes) be $(\lambda_\theta/\eta_\theta)^2$, and this turbulon be attached to the kinetic turbulon (Fig. 9.7). Furthermore, assume the *same* intermittency for kinetic and thermal turbulons, which gives

$$\left(\frac{\lambda}{\eta}\right)^2 = \left(\frac{\lambda_\theta}{\eta_\theta}\right)^2. \tag{9.4.21}$$

Actually, since the analogy between the momentum and thermal energy is expected to extend to the isotropic end of the equilibrium range, Eq. (9.4.20) may be restated as

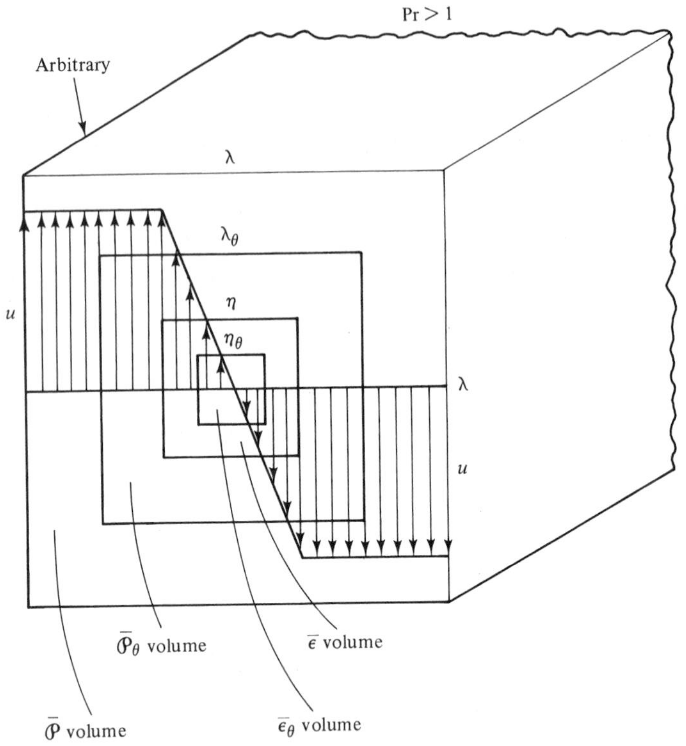

Fig. 9.7 Turbulon and thermal turbulon intermittencies

$$\boxed{\frac{\eta}{\eta_\theta} = \frac{\lambda}{\lambda_\theta} = \mathrm{Pr}^{1/3},} \qquad \mathrm{Pr} > 1, \tag{9.4.22}$$

which agrees with Eq. (9.4.21).

In terms of the thermal intermittency, Eq. (9.4.17) may now be interpreted as the thermal balance per unit thermal production volume,

$$\mathcal{P}_\theta \sim u_\theta \theta \frac{\theta}{l} \sim a \frac{\theta^2}{\lambda_\theta^2} \sim \epsilon_\theta. \tag{9.4.23}$$

Furthermore, multiply each side of Eq. (9.4.23) by $(\lambda_\theta/\eta_\theta)^2$, and rearrange the production term, with the help of Eq. (9.4.21), as

$$\bar{\mathcal{P}}_\theta \sim u_\theta \theta \frac{\theta}{l}\left(\frac{\lambda}{\eta}\right)^2 \sim u_\theta \theta \frac{\theta}{\lambda}\left(\frac{\lambda}{l}\right)\left(\frac{\lambda}{\eta}\right)^2,$$

which reduces, in view of Eq. (9.2.18), to

$$\bar{\mathcal{P}}_\theta \sim u_\theta \theta \frac{\theta}{\lambda}.$$

Accordingly,

$$\bar{\mathcal{P}}_\theta \sim u_\theta \theta \frac{\theta}{\lambda} \sim a \frac{\theta^2}{\eta_\theta^2} \sim \bar{\epsilon}_\theta \tag{9.4.24}$$

may be interpreted as the thermal balance per unit thermal dissipation volume. Thus, Eqs. (9.4.23) and (9.4.24) may be assumed as the balance of the mean-square thermal fluctuations near the large-scale end and the small-scale end of thermal equilibrium, respectively (Fig. 9.8).

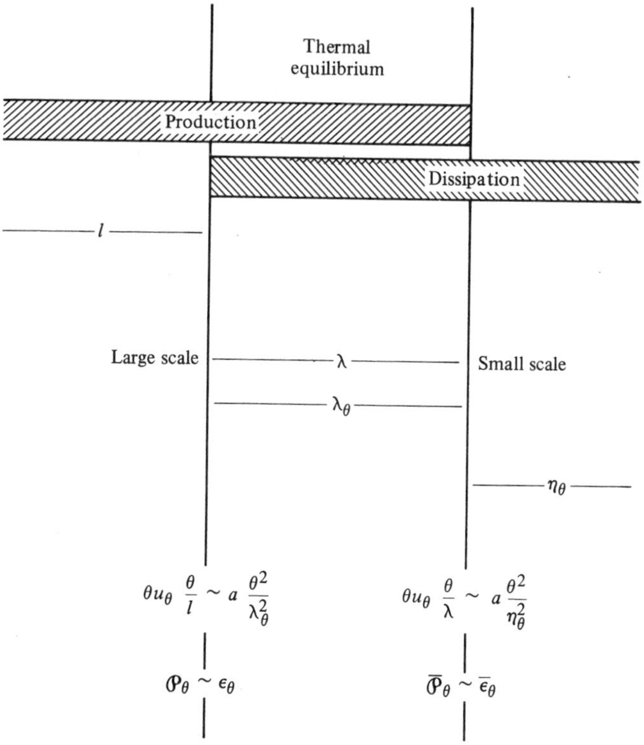

Fig. 9.8 Thermal equilibrium

Finally, we proceed to the thermal counterpart of the condition satisfied by the Reynolds number in Kolmogorov scales,

$$\mathrm{Re}_K = 1. \tag{9.2.7}$$

Letting $\theta \longrightarrow \vartheta$, $u_\theta \longrightarrow v_\theta$, and $l \longrightarrow \eta_\theta$ in Eq. (9.4.15), we get the Péclet number in thermal Kolmogorov scales,

$$\frac{v_\theta \eta_\theta}{a} = 1, \tag{9.4.25}$$

which applies for any Prandtl number. However, since Eq. (9.4.25) is a measure for the thermal energy averaged over η_θ, the natural length scale for this Péclet number is η_θ, but v_θ remains ambiguous until it is related to Prandtl number. To clear this ambiguity, rearrange $u_\theta = u(\lambda_\theta/\lambda)$ by letting $\lambda \longrightarrow \eta$, $u \longrightarrow v$ and $\lambda_\theta \longrightarrow \eta_\theta$, $u_\theta \longrightarrow v_\theta$, as

$$v_\theta = v\left(\frac{\eta_\theta}{\eta}\right). \tag{9.4.26}$$

Elimination of v_θ between Eqs. (9.4.25) and (9.4.26) yields

$$\frac{v\eta_\theta}{a} = \frac{\eta}{\eta_\theta},$$

or, by the consideration of Eq. (9.4.22),

$$\boxed{\mathrm{Pe_K} = \mathrm{Pr}^{1/3},} \qquad \mathrm{Pr} > 1, \tag{9.4.27}$$

where $\mathrm{Pe_K} = v\eta_\theta/a$.

Next, we proceed to fluids with $\mathrm{Pr} \ll 1$ (liquid metals). Noting in view of Fig. 9.9 that $\lambda_\theta \gg \lambda$, neglecting the effect of viscosity, and assuming u to be the characteristic velocity in Eq. (9.4.19), the counterpart of Eq. (9.4.20) is found to be

$$\frac{\lambda}{\lambda_\theta} \sim \mathrm{Pr}^{1/2}, \qquad \mathrm{Pr} \longrightarrow 0. \tag{9.4.28}$$

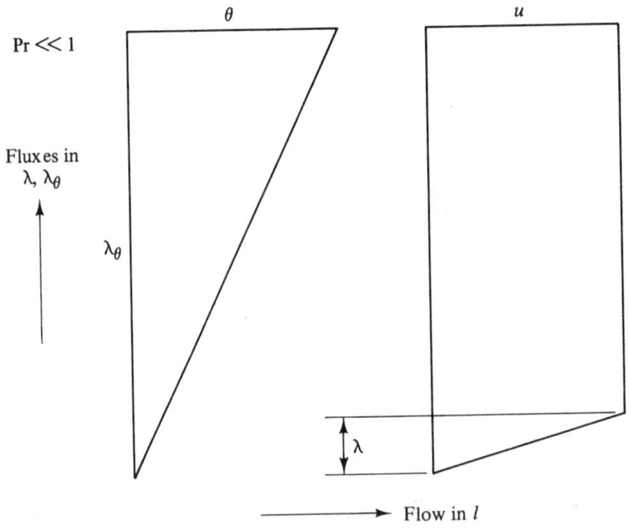

Fig. 9.9 Scales for fluids with $\mathrm{Pr} \ll 1$

Because of the lack of analogy between the momentum and thermal energy in this case, Eq. (9.4.28) is a proportionality rather than being an equality. For the same reason, there is no apparent way to justify the extension of Eq. (9.4.28) to the isotropic end of the equilibrium range, so the isotropy needs a separate treatment.

Assuming that $u_\theta = u$, letting $l \longrightarrow \eta_\theta$ (since $\eta_\theta \gg \eta$ for $\mathrm{Pr} \ll 1$) $u \longrightarrow v$, and $\theta \longrightarrow \vartheta$ in Eq. (9.4.15), we get

$$v \sim \frac{a}{\eta_\theta}. \tag{9.4.29}$$

For these fluids, viscous effects become negligible as $\mathrm{Pr} \longrightarrow 0$. Then from

$$\frac{u^3}{l} \sim \epsilon$$

of Eq. (9.2.1), we have, as $l \longrightarrow \eta_\theta$ and $u \longrightarrow \upsilon$,

$$\upsilon \sim (\eta_\theta \epsilon)^{1/3}. \qquad (9.4.30)$$

Elimination of υ between Eqs. (9.4.29) and (9.4.30) leads to the *Oboukhov–Corrsin microscale*,

$$\boxed{\eta_\theta^c \sim \left(\frac{a^3}{\epsilon}\right)^{1/4},} \qquad \text{Pr} \longrightarrow 0. \qquad (9.4.31)$$

The ratio between the Kolmogorov scale given by Eq. (9.2.4) and the foregoing Oboukhov–Corrsin scale yields

$$\boxed{\frac{\eta}{\eta_\theta^c} \sim \text{Pr}^{3/4},} \qquad \text{Pr} \longrightarrow 0. \qquad (9.4.32)$$

In summary, the equilibrium microscale ratio and the isotropic microscale ratio lead to the same Prandtl dependence for fluids with $\text{Pr} > 1$ because of the assumed analogy between momentum and thermal energy, but to different Prandtl dependence for fluids with $\text{Pr} \ll 1$ because of the lack of this analogy.

For small Prandtl numbers, the Péclet number in Kolmogorov scales, letting $\upsilon_\theta = \upsilon$ in view of Fig. 9.9, is readily found to be

$$\boxed{\text{Pe}_\text{K} = 1,} \qquad \text{Pr} \longrightarrow 0. \qquad (9.4.33)$$

Actually, $\text{Pe}_\text{K} = f(\text{Pr})$ is expected to be a continuous function between Eqs. (9.4.27) and (9.4.33). For $\text{Pr} \ll 1$ but not $\text{Pr} \longrightarrow 0$, Eq. (9.4.33) becomes dependent on Pr, as sketched in Fig. 9.10 (see also Problem 9-11).

All microscales developed so far apply to both forced and natural convection. However, for natural convection, these scales need further interpretation, which will be given in the following section.

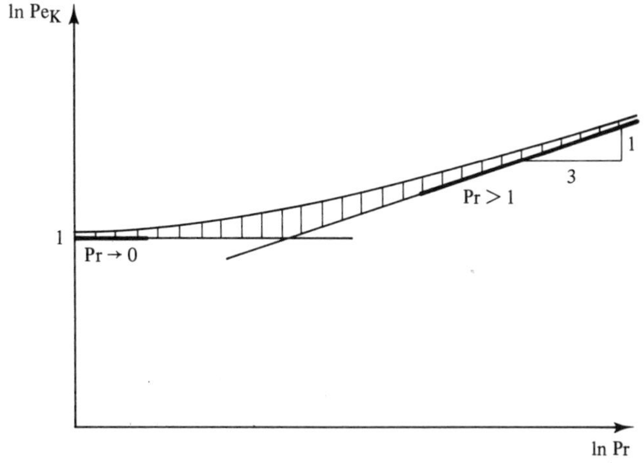

Fig. 9.10 Pe_K versus Pr for forced convection

9.5 SCALES OF NATURAL CONVECTION

All microscales of the preceding sections, expressed in terms of $\epsilon = u^3/l$, are convenient for forced convection. For natural convection, it is more appropriate to express these scales in terms of buoyancy (see Long 1976 for an alternative development without any reference to microscales). Including the buoyant production, the kinetic equilibrium stated by Eq. (9.2.1) may now be rearranged* as

$$\mathcal{P} \sim g\beta\theta u \sim u^2\frac{u}{l} \sim \nu\frac{v^2}{\eta^2} \sim \epsilon, \tag{9.5.1}$$

which shows the coupling between kinetic and thermal equilibria.

For Pr > 1 (involving water and viscous oils), the viscous dissipation is balanced by buoyant production over η_θ part of η, and the viscous dissipation is balanced by the inertial production over $\eta - \eta_\theta$, which may be approximated by η for these fluids (Fig. 9.11). For the isotropic end of kinetic equilibrium, from the balance between buoyant production and viscous dissipation,

$$g\beta\vartheta v_\theta \sim \nu\frac{v_\theta^2}{\eta_\theta^2},$$

which gives

$$v_\theta \sim \frac{g\beta\vartheta\eta_\theta^2}{\nu}. \tag{9.5.2}$$

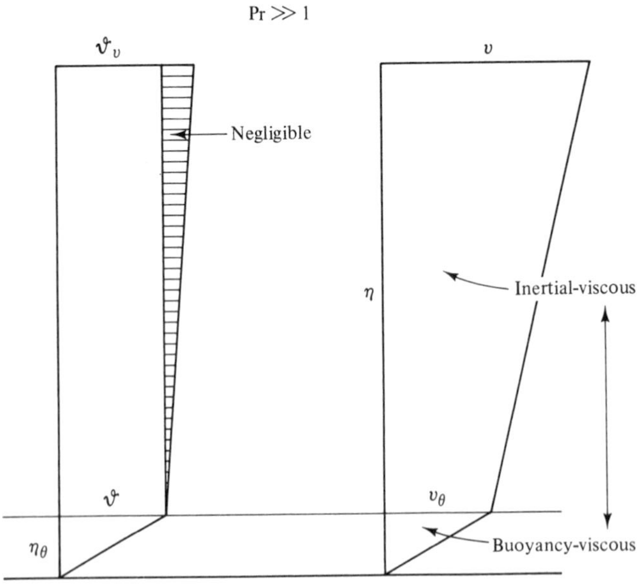

Fig. 9.11 Double layers of kinetic and thermal equilibria for Pr ≫ 1

*See also Eq. (11.4.3).

Also, from the balance between inertial production and viscous dissipation,

$$v^2 \frac{v}{\eta} \sim \nu \frac{v^2}{\eta^2},$$

which yields

$$v \sim \frac{\nu}{\eta}. \tag{9.5.3}$$

In the limit of Pr > 1 for Pr $\longrightarrow$ 1, $\eta_\theta \longrightarrow \eta$ and $v_\theta \longrightarrow v$ and the elimination of velocity between Eqs. (9.5.2) and (9.5.3) results in

$$\boxed{\eta \sim \left(\frac{\nu^2}{g\beta\vartheta}\right)^{1/3},} \qquad \text{Pr} \sim 1, \tag{9.5.4}$$

the Kolmogorov scale appropriate for natural convection (see Problem 9-14 for an alternative development on the same scale).

The thermal equilibrium for an arbitrary Prandtl number is, for both forced and natural convection,

$$\mathcal{P}_\theta \sim \theta u_\theta \frac{\theta}{l} \sim a \frac{\vartheta^2}{\eta_\theta^2} \sim \epsilon_\theta. \tag{9.4.15}$$

For the isotropic end of thermal equilibrium, $l \longrightarrow \eta_\theta$, $u_\theta \longrightarrow v_\theta$, and $\theta \longrightarrow \vartheta$, and Eq. (9.4.15) reduces to

$$\vartheta v_\theta \frac{\vartheta}{\eta_\theta} \sim a \frac{\vartheta^2}{\eta_\theta^2},$$

or

$$v_\theta \sim \frac{a}{\eta_\theta}. \tag{9.5.5}$$

Elimination of v_θ between Eqs. (9.5.2) and (9.5.5) results in

$$\boxed{\eta_\theta^B \sim \left(\frac{\nu a}{g\beta\vartheta}\right)^{1/3},} \qquad \text{Pr} > 1, \tag{9.5.6}$$

the Batchelor scale appropriate for natural convection (see Problem 9-14 for an alternative development, and Bergholtz, Chen & Cheung 1979 for a related development on the same scale). The ratio of Eq. (9.5.4) to Eq. (9.5.6) gives

$$\boxed{\frac{\eta}{\eta_\theta^B} \sim \text{Pr}^{1/3},} \qquad \text{Pr} > 1, \tag{9.5.7}$$

which was also obtained for forced convection (recall Eq. 9.4.22).

Since Pe_K is no longer appropriate, we need to establish another dimensionless number for the behavior of natural convection in microscales. Actually,

$$\frac{v_\theta \eta_\theta}{a} = 1 \tag{9.4.25}$$

continues to apply in form, but v_θ needs to be expressed now in terms of buoyancy.

Inserting Eq. (9.5.2) into Eq. (9.4.25) readily gives

$$\frac{g\beta\vartheta\eta_\theta^3}{\nu a} = 1,$$

or

$$\boxed{\mathrm{Ra}_\mathrm{K} = 1,} \qquad \mathrm{Pr} > 1, \tag{9.5.8}$$

the Rayleigh number in Kolmogorov scales.

For $\mathrm{Pr} \ll 1$ (involving liquid metals), the viscous dissipation is balanced by the inertial production over η, and the buoyant production is convected into inertial production over $\eta_\theta - \eta$, which may be approximated by η_θ for these fluids (Fig. 9.12).

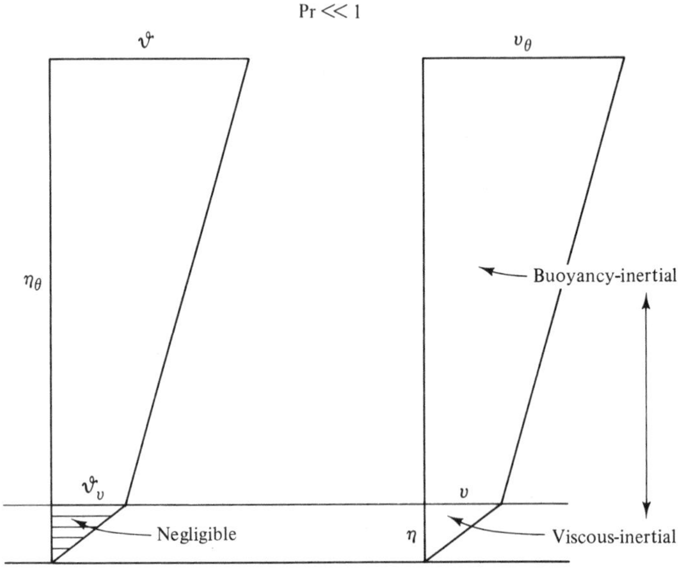

Fig. 9.12 Double layers of kinetic and thermal equilibria for $\mathrm{Pr} \ll 1$

For the isotropic end of the kinetic equilibrium, from the conversion of buoyant production to inertial production,

$$g\beta\vartheta\upsilon_\theta \sim \upsilon_\theta^2 \frac{\upsilon_\theta}{\eta_\theta},$$

or

$$\upsilon_\theta^2 \sim g\beta\vartheta\eta_\theta. \tag{9.5.9}$$

Also, from the balance between inertial production and viscous dissipation,

$$\upsilon \sim \frac{\nu}{\eta}. \tag{9.5.3}$$

In the limit of $\mathrm{Pr} < 1$ for $\mathrm{Pr} \longrightarrow 1$, $\eta_\theta \longrightarrow \eta$ and $\upsilon_\theta \longrightarrow \upsilon$, and the elimination of velocity between Eqs. (9.5.9) and (9.5.3) results in Eq. (9.5.4), as expected.

Recall the isotropic end of thermal equilibrium for an arbitrary Prandtl number,

$$v_\theta \sim \frac{a}{\eta_\theta}. \tag{9.5.5}$$

Elimination of v_θ between Eqs. (9.5.9) and (9.5.5) gives

$$\boxed{\eta_\theta^c \sim \left(\frac{a^2}{g\beta\vartheta}\right)^{1/3},} \qquad \text{Pr} \longrightarrow 0, \tag{9.5.10}$$

the Oboukhov–Corrsin scale for natural convection (see also Problem 9-14). The ratio of Eq. (9.5.4) to Eq. (9.5.10) yields

$$\boxed{\frac{\eta}{\eta_\theta^c} \sim \text{Pr}^{2/3},} \qquad \text{Pr} \longrightarrow 0. \tag{9.5.11}$$

From Eq. (9.4.25), in view of Eq. (9.5.9),

$$\frac{g\beta\vartheta\eta_\theta^3}{a^2} = 1,$$

or

$$\boxed{\text{Ra}_\text{K}\,\text{Pr} = 1,} \qquad \text{Pr} \longrightarrow 0. \tag{9.5.12}$$

Actually, for an arbitrary Prandtl number, $\text{Ra}_\text{K} = f(\text{Pr})$ is expected to be a continuous function between Eqs. (9.5.8) and (9.5.12), as sketched in Fig. 9.13.

In Chapter 12 we shall utilize the kinetic and thermal scales developed in this

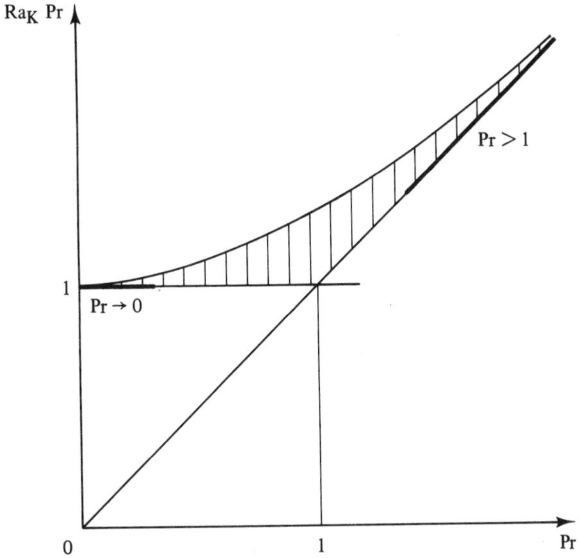

Fig. 9.13 Ra_K Pr versus Pr for natural convection

chapter for the conceptual interpretation of the heat transfer relations hitherto known as the empirical correlations of experimental results. We proceed next to a dimensional interpretation of the equilibrium spectra in terms of these scales. Although these spectra are not directly useful for our later study of heat transfer, they are discussed briefly here for the sake of completeness.

9.6 KINETIC AND THERMAL EQUILIBRIUM SPECTRA

In Section 8.7, we discussed briefly spectral aspects of turbulence, and formally proposed the balance equation for the kinetic (energy) spectrum,

$$\frac{\partial E(\kappa)}{\partial t} = -\frac{\partial T(\kappa)}{\partial \kappa} + P(\kappa) - D(\kappa), \tag{8.7.15}$$

$E(\kappa)$ being the energy spectrum, $T(\kappa)$ the energy flux across κ, $P(\kappa)$ the production, and $D(\kappa)$ the dissipation of energy. The solution of Eq. (8.7.15) requires the knowledge of $T(\kappa)$, $P(\kappa)$, and $D(\kappa)$. As we have showed (recall Eq. 8.7.14), it is easy to establish

$$D(\kappa) = 2\nu\kappa^2 E(\kappa), \tag{8.7.16}$$

but explicit relations for $T(\kappa)$ and $P(\kappa)$ turn out to be rather difficult to obtain for an arbitrary κ. Accordingly, in this section, we propose models for $T(\kappa)$ and $P(\kappa)$, as well as for $E(\kappa)$ and $D(\kappa)$, in a range of large enough wave numbers (or small enough length scales) for which the conditions of equilibrium prevail. First, we introduce three special cases associated with Eq. (8.7.15), equilibrium being one of them.

For steady mean turbulence, Eq. (8.7.15) reduces to

$$0 = -\frac{\partial T(\kappa)}{\partial \kappa} + P(\kappa) - D(\kappa). \tag{9.6.1}$$

Since the dissipation diminishes with decreasing κ, it can be neglected for small enough wave numbers. Then Eq. (9.6.1) reduces to

$$0 = -\frac{\partial T(\kappa)}{\partial \kappa} + P(\kappa), \qquad \kappa \longrightarrow 0. \tag{9.6.2}$$

This case may be interpreted as that of rapid production with no time for dissipation, and is sometimes called the case of *rapid distortion*. In a range of the energy spectrum at intermediate wave numbers, no energy is added by the mean flow and no energy is dissipated by the viscous action. Accordingly, the energy flux across each wave number remains constant, and is equal to ϵ (Fig. 9.14). In this case, Eq. (9.6.1) reduces to

$$P(\kappa) = D(\kappa), \tag{9.6.3}$$

the case of *equilibrium*. Past the equilibrium range, and for large enough wave numbers, $\partial T/\partial \kappa$ may be assumed to remain small, and Eq. (8.7.15) reduces to

$$\frac{\partial E(\kappa)}{\partial t} = -D(\kappa), \qquad \kappa \longrightarrow \infty, \tag{9.6.4}$$

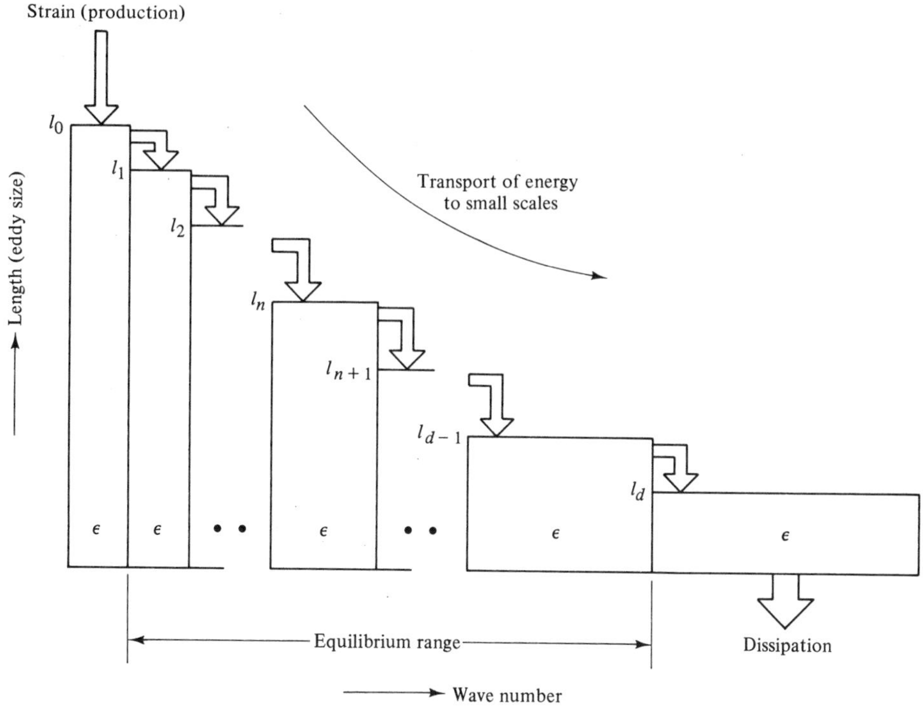

Fig. 9.14 Equilibrium range of energy spectrum

which shows the *decay* of energy. Now we are ready for the main objective of this section, the construction of models for the equilibrium range of $E(\kappa)$, $D(\kappa)$, $T(\kappa)$, and $P(\kappa)$, and their thermal counterparts.

For the equilibrium range of the energy spectrum, an inertial estimate for dissipation, according to Eq. (9.2.1), is

$$\epsilon \sim \frac{u^3}{l}, \tag{9.6.5}$$

which gives

$$u \sim (\epsilon l)^{1/3}. \tag{9.6.6}$$

Since, by definition of the energy,

$$E \sim u^2, \tag{9.6.7}$$

eliminating u between Eqs. (9.6.6) and (9.6.7) yields

$$E \sim (\epsilon l)^{2/3}. \tag{9.6.8}$$

Also, by definition (recall Eq. 8.7.11), the energy and its spectrum are related as

$$E \sim \kappa E(\kappa). \tag{9.6.9}$$

Eliminating E between Eqs. (9.6.8) and (9.6.9), and noting that $l \sim \kappa^{-1}$, we obtain

$$E(\kappa) \sim \epsilon^{2/3}\kappa^{-5/3}, \tag{9.6.10}$$

which is the *Kolmogorov energy spectrum for the equilibrium range.*

For an estimate of the dissipation spectrum $D(\kappa)$, assume from Eq. (9.2.1) a viscous estimate for the dissipation to be

$$\epsilon \sim \nu \frac{\upsilon^2}{\eta^2},$$

which may be rearranged in terms of $\eta \sim \kappa^{-1}$ and $E \sim \upsilon^2$,

$$\epsilon \sim \nu\kappa^3 E(\kappa). \tag{9.6.11}$$

Since, by definition,

$$\epsilon \sim \kappa D(\kappa), \tag{9.6.12}$$

eliminating ϵ between Eqs. (9.6.11) and (9.6.12), we get

$$D(\kappa) \sim \nu\kappa^2 E(\kappa), \tag{9.6.13}$$

or, in terms of Eq. (9.6.10),

$$D(\kappa) \sim \nu\epsilon^{2/3}\kappa^{1/3}. \tag{9.6.14}$$

For an estimate on the energy flux spectrum, interpreting the turbulence production as transfer of energy from larger eddies to smaller ones, we have, from Eq. (9.1.9),

$$T(\kappa) \sim -\overline{u_i u_j} S_{ij},$$

which may be rearranged, in terms of an eddy diffusivity $\nu_T \sim \overline{u_i u_j}/S$, as

$$T(\kappa) \sim \nu_T S^2,$$

or, letting $\nu_T \sim ul$ on dimensional grounds,* as

$$T(\kappa) \sim ulS^2,$$

which may be further rearranged, in terms of Eq. (9.6.6) and $l \sim \kappa^{-1}$, as

$$T(\kappa) \sim S^2\epsilon^{1/3}\kappa^{-4/3}, \tag{9.6.15}$$

or, referring again to Eq. (9.6.10), as

$$T(\kappa) \sim S^2 \frac{\kappa^{1/3}}{\epsilon^{1/3}} E(\kappa). \tag{9.6.16}$$

Since the production and the dissipation are important at different wave numbers, an estimate for the production can now be obtained from Eq. (9.6.2). Thus,

$$P(\kappa) = \frac{dT(\kappa)}{d\kappa}, \tag{9.6.17}$$

which, by the help of Eq. (9.6.15), gives

$$P(\kappa) \sim S^2\epsilon^{1/3}\kappa^{-7/3}, \tag{9.6.18}$$

or, by Eq. (9.6.16),

$$P(\kappa) \sim S^2\epsilon^{-1/3}\kappa^{-2/3}E(\kappa). \tag{9.6.19}$$

An alternative, but somewhat elaborate approach leading to the same estimates for $E(\kappa)$, $D(\kappa)$, $T(\kappa)$, and $P(\kappa)$ may be found in Tennekes and Lumley (1972). Figure 9.15 taken from this reference shows the distribution of some of these functions.

---

*$\nu_T \sim ul$ will be elaborated upon in Chapter 10.

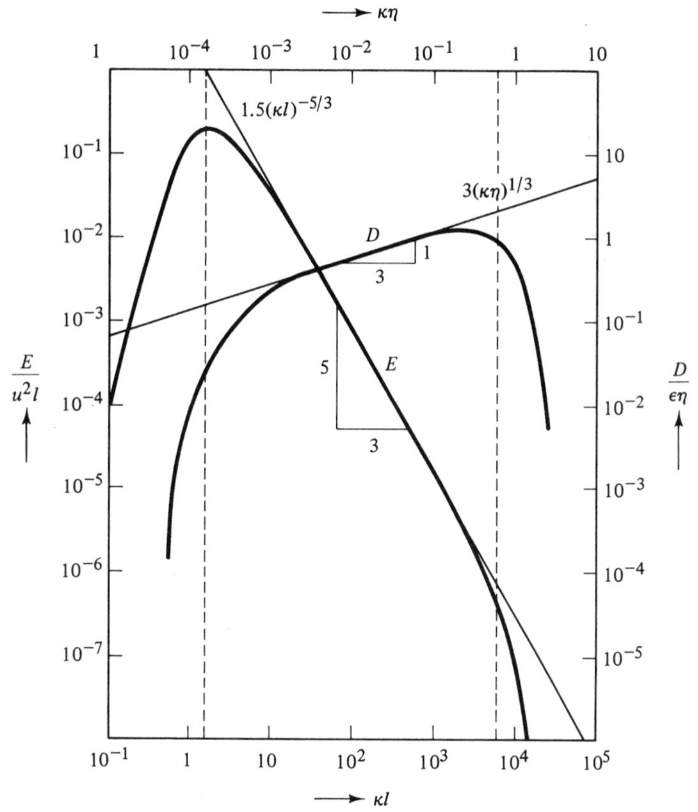

Fig. 9.15 Normalized energy and dissipation spectra for $\mathrm{Re}_l = 2 \times 10^5$ (adapted from Tennekes & Lumley 1972; used by permission)

We proceed now to the thermal spectra. The relation between the mean square of temperature fluctuations and its spectrum is

$$E_\theta = \frac{1}{2}\overline{\theta^2} = \int_0^\infty E_\theta(\kappa)\, d\kappa. \tag{9.6.20}$$

It can be shown, in a manner similar to considerations leading to Eq. (8.7.15), that the balance for the isotropic thermal spectrum satisfies

$$\frac{\partial E_\theta(\kappa)}{\partial t} = -\frac{\partial T_\theta(\kappa)}{\partial \kappa} + P_\theta(\kappa) - D_\theta(\kappa), \tag{9.6.21}$$

$E_\theta(\kappa)$ being the thermal spectrum, $T_\theta(\kappa)$ the thermal flux across κ, $P_\theta(\kappa)$ the production, and $D_\theta(\kappa)$ the dissipation of the thermal spectrum. Although it is easy to establish (recall Eq. 8.7.16) that

$$D_\theta(\kappa) = 2a\kappa^2 E_\theta(\kappa), \tag{9.6.22}$$

explicit relations for $T_\theta(\kappa)$ and $P_\theta(\kappa)$ are not readily available for an arbitrary κ. Accordingly, in the remainder of this section, we propose models only for the equilibrium range of $E_\theta(\kappa)$. Models for $T_\theta(\kappa)$, $P_\theta(\kappa)$, and $D_\theta(\kappa)$ are left to the reader (see Problem 9-18).

For fluids with $\Pr > 1$, an inertial-convection estimate for the thermal dissipation is

$$\epsilon_\theta \sim \theta u\left(\frac{\lambda_\theta}{\lambda}\right)\frac{\theta}{l}. \tag{9.6.23}$$

Eliminating u from this result with the velocity of the inertial estimate given by Eq. (9.6.5) yields

$$\epsilon_\theta \sim \theta^2\left(\frac{\lambda_\theta}{\lambda}\right)\epsilon^{1/3}l^{-2/3}. \tag{9.6.24}$$

Noting the definition of thermal spectrum $E_\theta(\kappa)$,

$$E_\theta \sim \kappa E_\theta(\kappa) \sim \theta^2, \tag{9.6.25}$$

eliminating θ^2 between Eqs. (9.6.24) and (9.6.25), and recalling that $l \sim \kappa^{-1}$ and $\lambda_\theta/\lambda = \Pr^{1/3}$, we obtain the thermal spectrum for the inertial-convective subrange,

$$E_\theta(\kappa) \sim \epsilon_\theta \Pr^{1/3}\epsilon^{-1/3}\kappa^{-5/3}, \qquad \Pr > 1. \tag{9.6.26}$$

There is experimental verification for this spectrum.

For the viscous-convection subrange of the thermal spectrum, let a viscous-convective estimate for the thermal dissipation be

$$\epsilon_\theta \sim a\frac{\vartheta^2}{\eta_\theta^2}. \tag{9.6.27}$$

Noting that the thermal spectrum $E_\theta(\kappa)$ in this subrange is

$$E_\theta \sim \kappa E_\theta(\kappa) \sim \vartheta^2, \tag{9.6.28}$$

eliminating ϑ^2 between Eqs. (9.6.27) and (9.6.28), and employing Eq. (9.4.15) for the elimination of η_θ^2 from the result yields

$$E_\theta(\kappa) \sim \frac{\epsilon_\theta}{a}\eta^2 \Pr^{2/3}\kappa^{-1}. \tag{9.6.29}$$

Note further that in this subrange the Kolmogorov scale,

$$\eta = \left(\frac{\nu^3}{\epsilon}\right)^{1/4} = l_{\min}, \tag{9.6.30}$$

characterizes the size of energy fluctuations which are stopped, by the action of viscosity, from getting smaller. Equation (9.6.29) may be rearranged in terms of Eq. (9.6.30) to give the thermal spectrum for the viscous-convective subrange,

$$E_\theta(\kappa) \sim \epsilon_\theta \Pr^{1/3}\epsilon^{-1/3}l_{\min}^{2/3}\kappa^{-1},$$

or

$$E_\theta(\kappa) \sim \epsilon_\theta \Pr^{1/3}\epsilon^{-1/3}\kappa_{\min}^{-2/3}\kappa^{-1}, \qquad \Pr > 1. \tag{9.6.31}$$

There is no satisfactory verification for this spectrum. Other power laws, such as κ^{-3} and κ^{-7}, have been proposed for the same spectrum (see, for example, Gibson 1968).

Finally, an inspection readily reveals that the thermal spectra discussed so far (given by Eqs. 9.6.26 and 9.6.31) are valid for both forced and natural convection.

Details of the thermal spectra for fluids with $\mathrm{Pr} \ll 1$ will not be given here (see, however, Problem 9-16). The result for the inertial-convection subrange is

$$E_\theta(\kappa) \sim \epsilon_\theta \epsilon^{-1/3} \kappa^{-5/3}, \qquad \mathrm{Pr} \to 0, \tag{9.6.32}$$

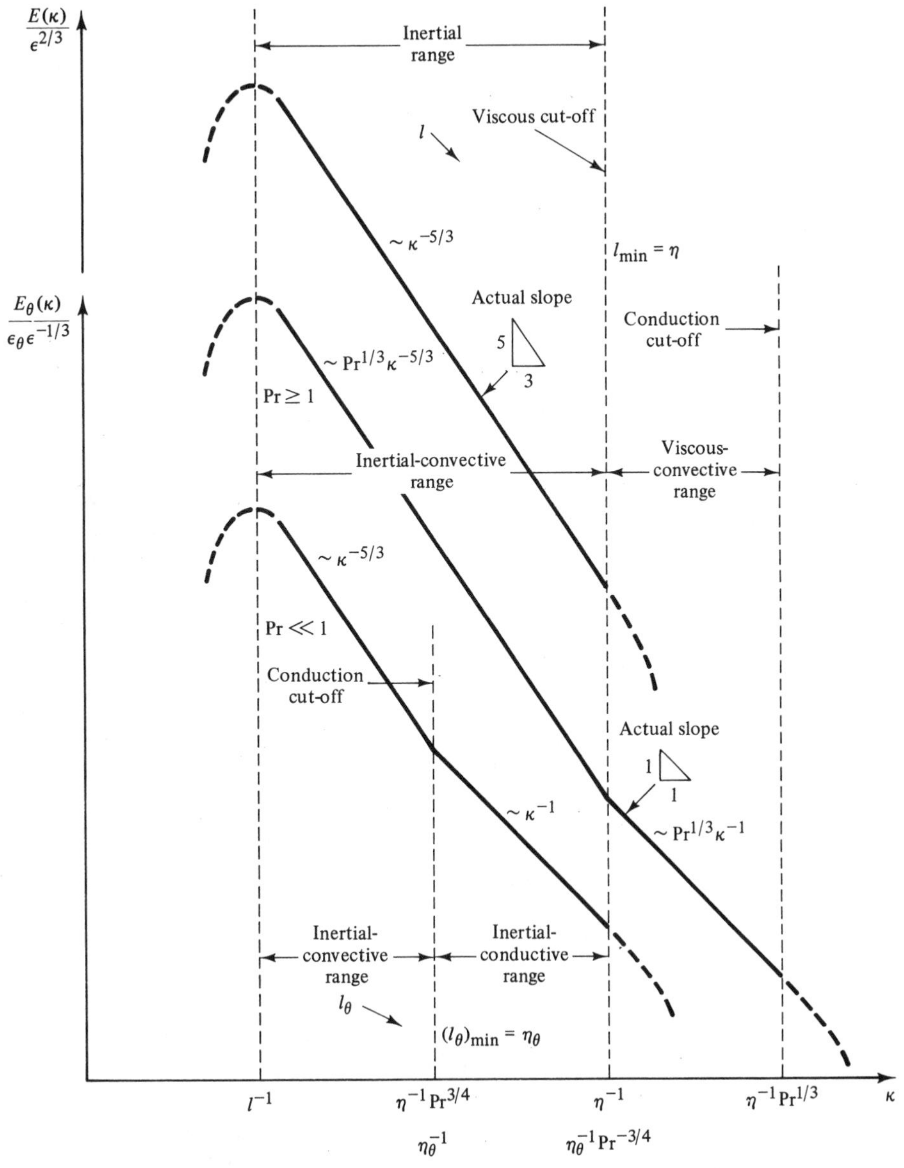

Fig. 9.16 Equilibrium range of kinetic and thermal spectra

and for the inertial-conductive subrange is

$$E_\theta(\kappa) \sim \epsilon_\theta \epsilon^{-1/3}(l_\theta)^{2/3}_{\min}\kappa^{-1},$$

or

$$E_\theta(\kappa) \sim \epsilon_\theta \epsilon^{-1/3}(\kappa_\theta)^{-2/3}_{\min}\kappa^{-1}, \qquad \mathrm{Pr} \longrightarrow 0, \tag{9.6.33}$$

where the Oboukhov–Corrsin scale,

$$\eta^c_\theta = \left(\frac{a^3}{\epsilon}\right)^{1/4} = (l_\theta)_{\min}, \tag{9.6.34}$$

characterizes the size of thermal fluctuations which are stopped, by the effect of conduction, from getting smaller. There is no sufficient experimental support for the thermal spectra for Pr $\longrightarrow 0$.

The kinetic and thermal spectra for Pr > 1 and Pr $\longrightarrow 0$ are sketched in Fig. 9.16. The effect of the Prandtl number on the thermal spectrum is sketched in Fig. 9.17. Inspection of this sketch reveals that improvement about Pr ~ 1 is needed on the thermal spectrum proposed for fluids with Pr > 1 and Pr $\longrightarrow 0$ (see Problem 9-19).

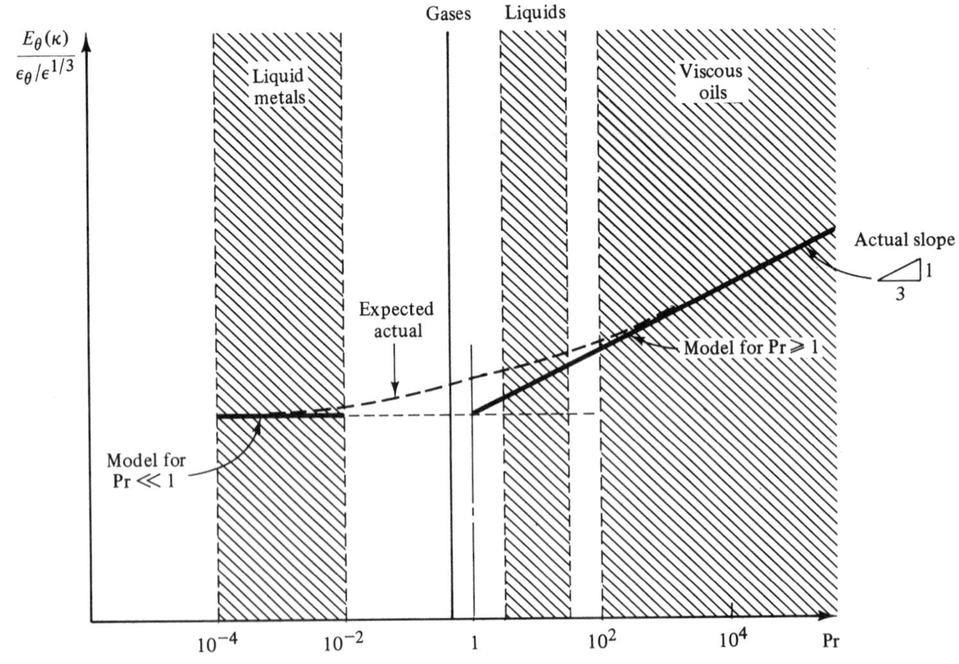

Fig. 9.17 Effect of Prandtl number on thermal spectra

Here we terminate our study on the scales and spectra of turbulence. In Chapter 12, employing these scales, we shall attempt to understand the conceptual foundations of heat transfer relations which are assumed to be a result of experimental correlations. In the meantime, we proceed to more ambitious models which are intended for the prediction of velocity distributions (Chapter 10) and temperature distributions (Chapter 11), as well as the friction and heat transfer coefficients.

REFERENCES

BATCHELOR, G. K., 1959, J. Fluid Mech., **5**, 113.

BATCHELOR, G. K., 1967, *The Theory of Homogeneous Turbulence*, Cambridge University Press, New York.

BERGHOLTZ, R. F., CHEN, M. M., & CHEUNG, F. B., 1979, Int. Heat Mass Transfer, **22**, 763.

CORRSIN, S., 1953, J. Appl. Phys., **22**, 469.

CORRSIN, S., 1959, *Proceedings of the Iowa Thermodynamics Symposium*, State University of Iowa.

CORRSIN, S., 1962, Phys. Fluids, **5**, 1301.

FRISCH, U., SULEM, P.-L., & NELKIN, M., 1978, J. Fluid Mech., **87**, 719.

GIBSON, C. H., 1968, Phys, Fluids **11**, 2316.

GIBSON, C. H., & SCHWARTZ, W. H., 1963, J. Fluid Mech., **16**, 365.

HINZE, J. O., 1975, *Turbulence*, McGraw-Hill, New York.

LONG, R. R., 1976, J. Fluid Mech., **73**, 445.

MONIN, A. S., & YAGLOM, A. M., 1971, *Statistical Fluid Mechanics*, MIT Press, Cambridge, Mass.

OBOUKHOV, A. M., 1949, Izv. Akad. Nauk SSSR, Geogr. Geophys. Ser., **13**, 58.

TAVOULARIS, S., BENNETT, J. C., & CORRSIN, S., 1978, J. Fluid Mech., **88**, 63.

TAYLOR, G. I., 1938, Proc. R. Soc. A, **164**, 476.

TENNEKES, H., 1968, Phys. Fluids, **11**, 669.

TENNEKES, H., & LUMLEY, J. L., 1972, *A First Course in Turbulence*, MIT Press, Cambridge, Mass.

TOWNSEND, A. A., 1951, Proc. R. Soc. A, **208**, 534.

TOWNSEND, A. A., 1976, *The Structure of Turbulent Shear Flow*, Cambridge University Press, New York.

PROBLEMS

9-1 Give details of the development leading to Eq. (9.1.9).

9-2 Assume the small-scale motion to depend on the kinematic viscosity ν and the energy supply (from large scales) for the dissipation ϵ. By dimensional arguments, obtain the small-scale variables (η, υ, τ) in terms of ν and ϵ. Contrast this approach with that used in this chapter for the same scales.

9-3 Show that $\mathrm{Re}_K = v\eta/\nu$ may be interpreted as the ratio of the energy rate of fluctuations to that of the mean flow.

9-4 The equation of the mean-square vorticity fluctuations is obtained by a procedure similar to the one followed for the equation of the turbulent kinetic energy. Show that

$$U_j \frac{\partial}{\partial x_j}\left(\frac{1}{2}\overline{\omega_i\omega_i}\right) = -\overline{u_j\omega_i}\frac{\partial \Omega_i}{\partial x_j} - \frac{1}{2}\frac{\partial}{\partial x_j}(\overline{u_j\omega_i\omega_i}) + \overline{\omega_i\omega_j s_{ij}}$$
$$+ \overline{\omega_i\omega_j}S_{ij} + \Omega_j\overline{\omega_i s_{ij}} + \nu\frac{\partial^2}{\partial x_j\,\partial x_j}\left(\frac{1}{2}\overline{\omega_i\omega_i}\right) - \nu\overline{\frac{\partial\omega_i}{\partial x_j}\frac{\partial\omega_i}{\partial x_j}}.$$

9-5 For sufficiently high Reynolds numbers, the vorticity balance given in Problem 9-4 may be approximated as (Taylor 1938)

$$\overline{\omega_i\omega_j s_{ij}} = \nu\overline{\frac{\partial\omega_i}{\partial x_j}\frac{\partial\omega_i}{\partial x_j}},$$

or, on dimensional grounds, as

$$\omega^2\frac{u}{\lambda} \sim \nu\frac{\omega^2}{\eta^2}.$$

Compare this result with Eq. (9.3.6). What is your conclusion?

9-6 Show from $E = E(\kappa, \epsilon, \nu)$ that the small-scale energy spectrum can be scaled as

$$\frac{E(\kappa)}{\nu^{5/4}\epsilon^{1/4}} = \frac{E(\kappa)}{v^2\eta} = f(\kappa\eta),$$

and from $E = E(\kappa, S, \nu)$ that the large-scale energy spectrum can be scaled as

$$\frac{E(\kappa)}{\epsilon^{3/2}S^{-5/2}} = \frac{E(\kappa)}{u^2 l} = F(\kappa l).$$

9-7 Develop a power-law model (in wave number) for the large-scale energy spectrum.

9-8 Give details of the development leading to Eq. (9.4.7).

9-9 Show the effect of Prandtl number on Eq. (9.4.14) for $\mathrm{Pr} > 1$ and $\mathrm{Pr} \longrightarrow 0$.

9-10 Show for $\mathrm{Pr} \ll 1$, but not $\mathrm{Pr} \longrightarrow 0$, that

$$\frac{\eta}{\eta_\theta} = \frac{\mathrm{Pr}^{3/4}}{1 + \mathrm{Pr}^{3/4}} \quad \text{and} \quad \eta_\theta = \eta_\theta^c(1 + \mathrm{Pr}^{3/4}).$$

9-11 Show the effect of Prandtl number on Eq. (9.4.33) for $\mathrm{Pr} \ll 1$, but not $\mathrm{Pr} \longrightarrow 0$.

9-12 Show that the balance of thermal dissipation (or the mean-square of $\partial\theta/\partial x_i$ fluctuations) satisfies

$$-\overline{\left(\frac{\partial\theta}{\partial x_j}\right)\left(\frac{\partial u_i}{\partial x_j}\frac{\partial\theta}{\partial x_i}\right)} - \frac{\partial}{\partial x_i}\left\{\overline{u_i\frac{1}{2}\left(\frac{\partial\theta}{\partial x_j}\right)\left(\frac{\partial\theta}{\partial x_j}\right)} - a\frac{\partial}{\partial x_i}\left[\frac{1}{2}\overline{\left(\frac{\partial\theta}{\partial x_j}\right)\left(\frac{\partial\theta}{\partial x_j}\right)}\right]\right\}$$
$$= a\overline{\left(\frac{\partial^2\theta}{\partial x_i\,\partial x_j}\right)\left(\frac{\partial^2\theta}{\partial x_i\,\partial x_j}\right)}.$$

9-13 Show that for homogeneous thermal flow, the balance of thermal dissipation given in Problem 9-12 may be approximated as (Corrsin 1953)

$$\overline{\frac{\partial\theta}{\partial x_i}\frac{\partial\theta}{\partial x_j}s_{ij}} = a\overline{\frac{\partial^2\theta}{\partial x_i\,\partial x_j}\frac{\partial^2\theta}{\partial x_i\,\partial x_j}},$$

or, on dimensional grounds, as

$$n^2 \frac{u}{\lambda} \sim a \frac{n^2}{\eta_\theta^2},$$

where $n \sim \partial\theta/\partial x_i$. Compare this result with Eq. (9.4.24). What is the conclusion?

9-14 From the usual definition of Kolmogorov, Oboukhov–Corrsin, and Batchelor scales, utilizing the buoyancy generated production,

$$\mathcal{P} \sim g\beta\theta u \sim \epsilon,$$

obtain the microscales of natural convection given by Eqs. (9.5.4), (9.5.6), and (9.5.10).

9-15 Sketch the production and energy flux spectra on Fig. 9.15.

9-16 Show the steps leading to Eqs. (9.6.32) and (9.6.33).

9-17 Add the effect of Prandtl number to the thermal spectra for $\Pr \ll 1$, but not $\Pr \longrightarrow 0$. Sketch this effect on Fig. 9.17.

9-18 Give estimates on the equilibrium range of $T_\theta(\kappa)$, $P_\theta(\kappa)$, and $D_\theta(\kappa)$ for fluids with $\Pr > 1$ and $\Pr \ll 1$, respectively.

9-19 Show the effect of Prandtl number on thermal spectra for $\Pr \sim 1$.

Chapter 10

FLOW PREDICTION

In Chapter 8 we identified the closure problem. We learned that the equations of the mean motion (Table 8.1) need to be supplemented by a model for the Reynolds stress $\tau_{ij} = -\rho\overline{u_i u_j}$ before any solution can be obtained. In Chapter 9 we developed a one-velocity, three-length flow model (u; l, λ, η), and a one-temperature, three-length thermal model (θ; l, λ_θ, η_θ). Use of these scales in models for turbulent flow and convection problems is new and has to wait further developments. In Chapter 12 we shall make an attempt to *understand*, in terms of these scales, the foundations of turbulent flow and heat transfer correlations which are assumed to be empirical. In this chapter, introducing models resting on an *eddy diffusivity*, we alter and somewhat reduce our objective from attempting to understand turbulence to a *prediction* of turbulence.

Actually, there are a number of models, which are placed in two categories, the *first-order* and the *second-order* models (Table 10.1). In the first-order models, τ_{ij} is related directly to the mean velocity field. In the second-order models, balance equations for τ_{ij} are solved together with the equations of mean motion. The latter approach is still under development (Rotta 1951, Daly & Harlow 1970, Hanjalic & Launder 1972b, Launder, Reece & Rodi 1975, and Lumley 1977) and is beyond the scope of this text. The required transport equations for $\overline{u_i u_j}$, which are generated from the Navier–Stokes equations, are the balance equations for the one-point double

TABLE 10.1 Turbulence Models

FIRST-ORDER (Eddy Diffusivity) MODELS:	
Algebraic model	(algebraic equation for l)
One-equation model	algebraic equation for l, (balance equation for K)
Two-equation model	(balance equations for K and ϵ)
SECOND-ORDER (Reynolds Stress) MODELS:	
Multiequation model	(balance equations for τ_{ij} and ϵ)

correlation tensor $R_{ij}(0)$ of Eq. (8.7.3). To close these equations, the appearing higher-order correlations are expressed in terms of the Reynolds stresses, the mean strain rate, the kinetic energy, and the dissipation.

In the first-order models, the common approach suggested early by Boussinesq (1877) is to imitate the constitutive relation of the isotropic Newton–Stokes fluid, Eq. (2.3.3) for parallel shear flows,

$$\tau_{yx} = -\rho\overline{vu} = \rho\nu_T\frac{\partial U}{\partial y}, \tag{10.1.1}$$

and Eq. (2.3.9) for the general incompressible case,

$$\tau_{ij} - \tfrac{1}{3}\tau_{kk}\,\delta_{ij} = 2\rho\nu_T S_{ij},$$

or

$$\tau_{ij} + \tfrac{2}{3}\rho K\,\delta_{ij} = \rho\nu_T\left(\frac{\partial U_i}{\partial x_j} + \frac{\partial U_j}{\partial x_i}\right). \tag{10.1.2}$$

Here $K = \frac{1}{2}\overline{u_k u_k}$ denotes the turbulent kinetic energy* and ν_T the *turbulent (eddy) diffusivity*. The first-order closures are based on the modeling of this diffusivity. As we shall see, ν_T may be orders of magnitude greater than the molecular diffusivity ν, and it is not a property of the fluid but depends on the turbulent fluid motion, hence varies throughout a flow.

On dimensional grounds, Eq. (10.1.1) may be stated as

$$\rho u^2 \sim \rho\nu_T\frac{u}{l},$$

or

$$\nu_T \sim lu, \tag{10.1.3}$$

where l and u may be assumed, for the time being, denoting the length and velocity scales defined in Chapter 9. These scales will be further interpreted in the following sections of this chapter. Here we exemplify the three first-order models of Table 10.1

*For customary reasons, we use K in this chapter for the turbulent kinetic energy, denoted E in Chapter 9.

by introducing the additional scaling

$$\nu_T \sim \begin{cases} l^2\left|\dfrac{\partial U}{\partial y}\right| \quad \text{or} \quad l\,\Delta U & \text{(algebraic model)} \\ lK^{1/2} & \text{(one-equation model)} \\ \dfrac{K^2}{\epsilon} & \text{(two-equation model).} \end{cases} \qquad (10.1.4)$$

In the *algebraic* model, the mixing length l is described by an algebraic equation, usually determined by the flow configuration, while dU/dy or ΔU are related to the mean flow. In the *one-equation* model, the prescribed mixing length is supplemented by the solution of the balance equation of the turbulent kinetic energy K. In the *two-equation* model, ν_T is obtained through the solution of the balance equations of K and its dissipation ϵ, to be obtained together with the solution to the balance equations of the mean flow.

The first-order models presented in the following sections have been extensively tested and serve well to predict a number of turbulent flows of engineering interest. However, they lack generality because model constants fitted to data for one flow do not in general apply to other flows. Second-order models, although they still require detailed experimental verification, have improved generality but are somewhat cumbersome to use.

Actually, both types of models are basically *local*, being expressed by local values of flow properties. Yet turbulence is a *global* phenomenon, as we learned in Chapter 8 in terms of, for example, the spatial two-point correlation functions. That is, velocity fluctuations entering into the Reynolds stress depend on both small-scale motion near the point and large-scale motion far from the point. The models of Eq. (10.1.4) attempt to bring in the global element through the parameters l, K, and ϵ evaluated at the point. Here l may be the distance to a wall, while K and ϵ, being the solutions to balance equations, depend on the entire flow field and boundary conditions. In recognition of the difficulties associated with the modeling of large-scale motion, current research efforts are directed at another approach, called subgrid modeling (see, for example, Schumann 1975 and Clark, Ferziger & Reynolds 1979). In this approach, the large-scale unsteady motion is solved by direct numerical integration of the Navier–Stokes equations time-averaged over small scales, while the average small-scale motion is modeled. The small-scale motion is assumed to occur at scales smaller than the grid employed in the numerical integration. Although far from practical application, such theoretical studies, as well as experimental studies of large-scale coherent structures, are expected to pave the way for a third generation turbulence models.

In the following sections we first describe quantitatively the characteristics of some turbulent flows. Next, in terms of a few examples, we illustrate classical algebraic models. For a more complete description of these models, see, for example, Hinze (1975), Schlichting (1968), Reynolds (1974), and Cebeci & Smith (1974). The last two sections are devoted to an introduction to one- and two-equation models. For a further study of turbulence models and their applications, we refer to the book by Launder &

Spalding (1972), the review by Reynolds (1976), and the recent text by Bradshaw, Cebeci & Whitelaw (1981).

10.1 SOME CHARACTERISTICS OF TURBULENT FLOWS

In Part II we classified laminar flows, for reasons of formulation, as parallel, nearly parallel, and nonparallel. Turbulent flows need to be divided further into wall-bounded (pipe, channel, boundary layer) flows and free (jet, plume, mixing layer, wake) flows. This further classification is needed because the structure and the transport properties of turbulent flows at rigid and free boundaries are fundamentally different. In addition, the roughness of a wall now becomes a significant parameter because of the importance of the small scales of turbulence.

Consider the three basic flows of Fig. 10.1(a–c). Developed channel flow between two plates is a parallel flow, while developments of the free mixing layer and the flow along a plate at high Reynolds numbers are nearly parallel (boundary layer) flows. In addition to these basic flows we have, of course, the more complex two- or three-dimensional nonparallel (recirculating) flows, such as flow in a cavity, over a step, and downstream of a boundary layer separation, Fig. 10.1(d–f). Although the wall friction and heat transfer associated with nonparallel flows are of considerable technological importance, the prediction of such flows will not be elaborated upon. It requires numerical computations, to which an introduction was given in Chapter 7.

The differences between free and wall-bounded turbulent flows come from the suppression of turbulent fluctuations close to a rigid boundary due to viscous effects which require no slip at the boundary. A free boundary, however, is relatively free to deform according to the large-scale structures of the turbulent flow. Here a thin viscous layer separates the turbulent fluid from nonturbulent fluid. The result is an irregular and unsteady boundary contour, as suggested by Fig. 10.1(b) and (c). A measuring probe placed in this region indicates the fluid to be turbulent only part of the time, that is, only intermittently. Some models account for this intermittency in terms of an actual high turbulence during a fraction of the time. Accordingly, the intermittency varies from unity to zero over the free boundary range. Other models, assuming the usual (unconditional) averaging, simply assume a turbulence that decreases to zero as the free stream is reached.

Let us pursue first the origin of large structures giving free boundary intermittency, and then examine wall-bounded flows to find a different kind of intermittency. Although these characteristics are not explicitly taken into account in models, they provide some understanding for mechanisms at play in the turbulent transport processes. The small-amplitude flow instability, which is the origin of turbulence (Chapter 8), can be extended to large amplitudes to reveal some features of large structures that are found in turbulence. In reality, these features are strongly obscured, however, by secondary instabilities developing on the distorted flow resulting from the primary instabilities. Here we examine only the primary instabilities.

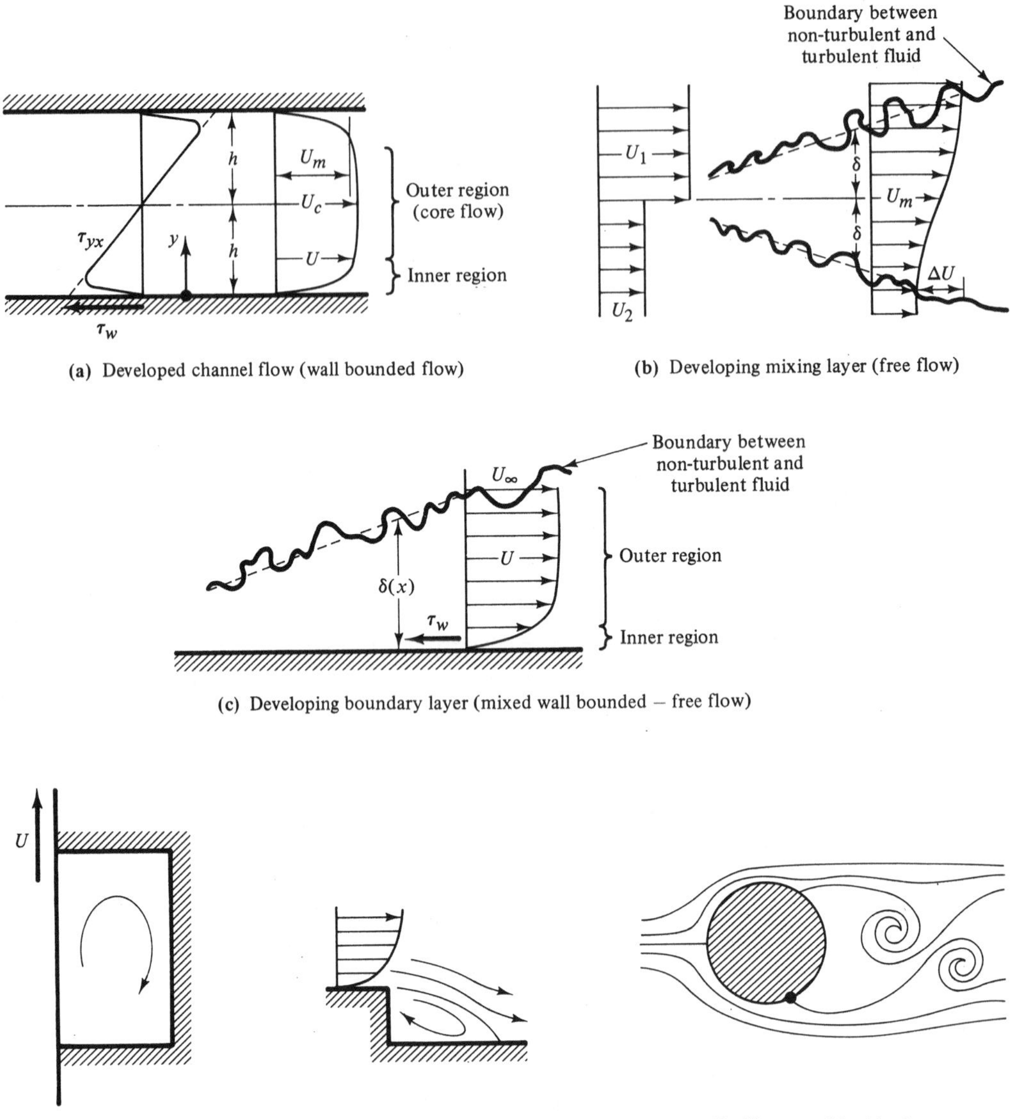

(a) Developed channel flow (wall bounded flow)

(b) Developing mixing layer (free flow)

(c) Developing boundary layer (mixed wall bounded – free flow)

(d) Driven cavity flow

(e) Flow over backward facing step

(f) Flow past blunt body

Fig. 10.1 Basic two-dimensional turbulent flows

Actually, the simplest flow instability is that of the velocity discontinuity in an inviscid fluid (Fig. 10.2(a)). Known as the Kelvin–Helmholtz instability and readily analyzed for small amplitudes by potential flow theory (Lamb 1945), it proves to be unconditionally unstable to any two-dimensional disturbance (Problem 8-7). Furthermore, viewed as a vortex sheet composed of a large number of discrete vortex filaments

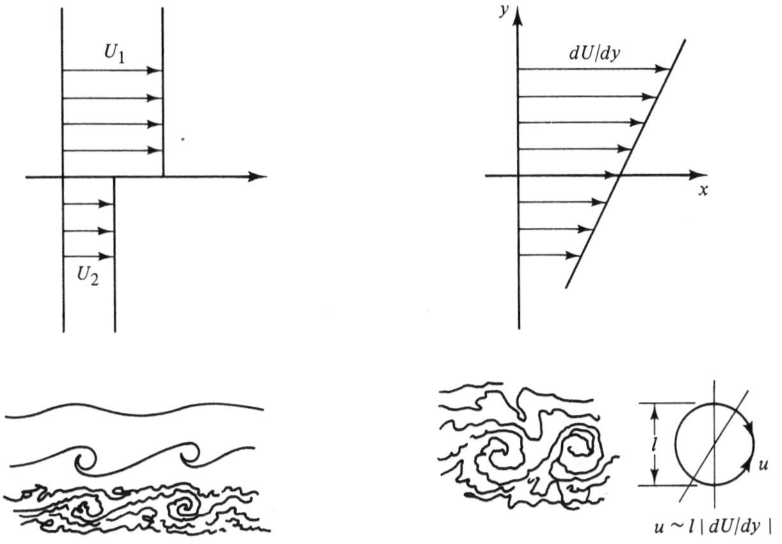

(a) Velocity discontinuity (b) Viscous shear flow

Fig. 10.2 Inviscid Kelvin-Helmholtz instability and viscous shear flow instability; scaling of characteristic velocity of mixing length model

the large-amplitude two-dimensional instability may be readily computed from potential flow theory. The motion of each filament is determined by the induced velocity from all other filaments. This computation is equivalent to the solution of the two-dimensional inviscid Euler equations (for details, see Chow 1979). Figure 10.3 shows some basic features of the solution, such as the concentration of vorticity into large structures (eddies), which in turn move together (vortex pairing).

Now, for a viscous fluid the velocity discontinuity is replaced by a uniform shear flow (Fig. 10.2(b)). Although the viscosity has a stabilizing effect, it does not prevent instability provided that the Reynolds number is sufficiently large. Figure 10.4 shows visualization studies of a developing mixing layer (Fig. 10.1(b)), revealing instability, with the development of large structures having characteristic features similar to those of the inviscid flow. Figure 10.4 also shows the irregular free boundary contour as well as secondary instabilities developing on the large structures and contributing to the small-scale structure of the turbulence, which becomes smaller with increasing Reynolds number. The regular large eddy structures are also well known from the wake of a blunt body, such as a circular cylinder (Fig. 10.1(f)). While easily visualized at low-Reynolds-number flows, they persist (although strongly obscured by structures at smaller scales) in high-Reynolds-number flows. Large-scale structures, limited in size by the width of the flow, are also found in channel flows, as revealed by a camera moving with the flow, or by the experimental technique of conditional sampling and

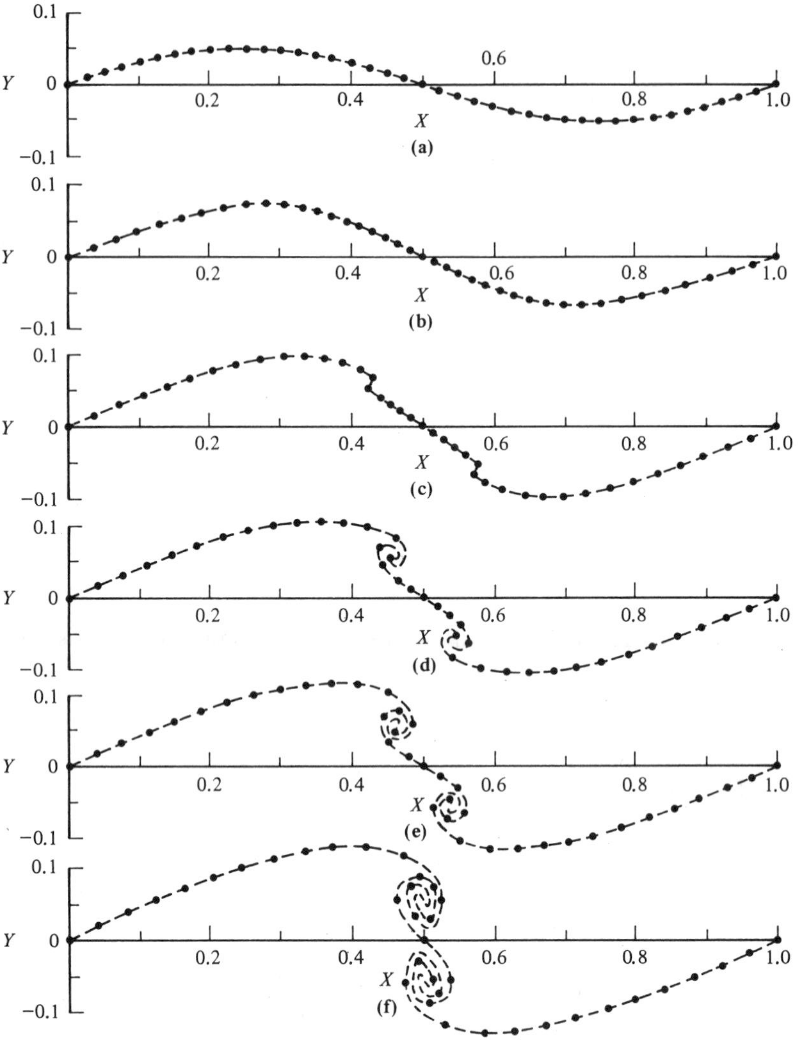

Fig. 10.3 Evolution of large-scale structure in vortex sheet with initial sinusoidal perturbation (adapted from Chow 1979; used by permission)

space–time correlation (see, for example, Cantwell 1981 for a review). These examples illustrate the existence of turbulent structures (eddies) over a range of sizes, limited upward by the geometry of the problem and downward by the viscosity, the smaller eddies becoming smaller with increasing Reynolds number. There is a continuous interaction between these structures, in which kinetic energy is exchanged in a manner expressed by the spectral representation (Chapter 8).

(a) Re_1

(b) $2 \times Re_1$

(c) $4 \times Re_1$

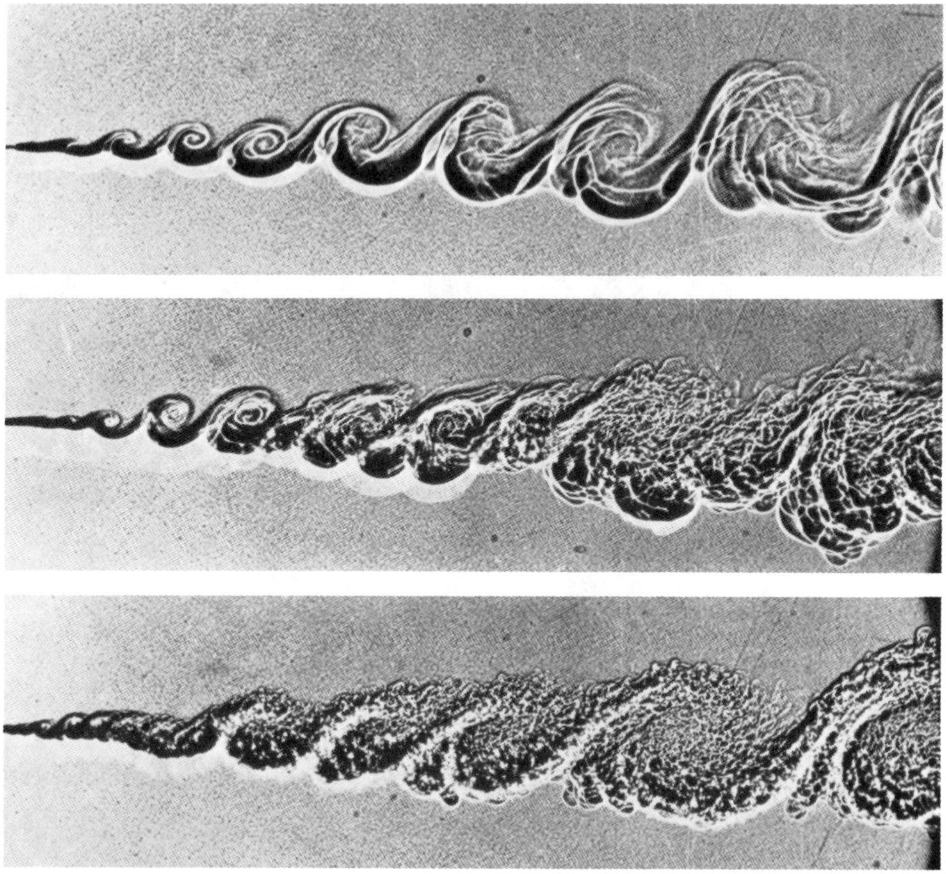

Fig. 10.4 Mixing layer instability developing into turbulence with distinct large structures. Reynolds number doubled from (a) to (b), and from (b) to (c) (adapted from Brown & Roshko 1974; used by permission)

So far we have illustrated the two-dimensional instability of a vortex sheet composed of straight vortex filaments. However, a straight vortex filament is unstable to lateral disturbances, leading to the formation of a "horseshoe" vortex in a shear flow (Fig. 10.5). This process involves vortex stretching and bending and the creation of streamwise vorticity, pointing to the three-dimensional nature of turbulent fluctuations. These large-amplitude instabilities have been observed in wall-bounded shear layers (see Mollo-Christensen 1971 for a discussion of the physics). They give rise to bursts which eject fluid from the wall region followed by sweeps of fluid entering the region (Kim, Kline & Reynolds 1971). These processes occur intermittently at points

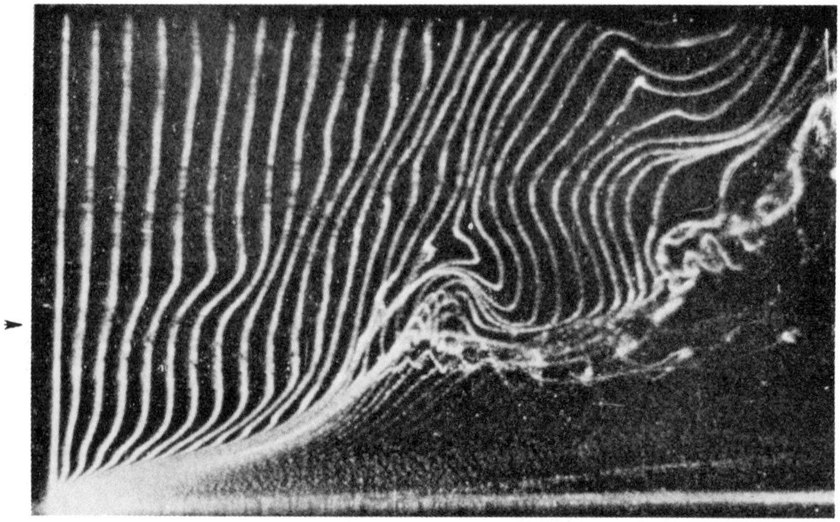

(a) Visualization of bursting process near wall by hydrogen bubble time-lines released from wire perpendicular to wall (at left); formation of transverse vortex; adapted from Kim, Kline & Reynolds 1971 (used by permission)

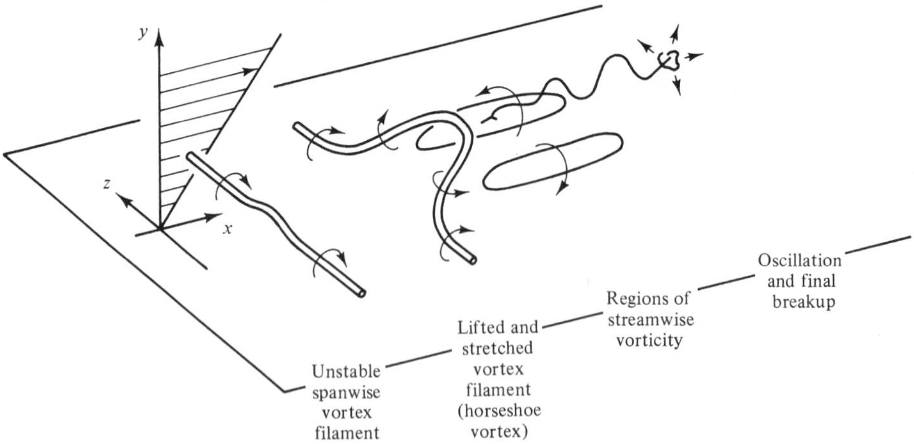

(b) Lateral instability of vortex filament in shear flow near wall (schematic)

Fig. 10.5 Instability and bursting process near wall

along a wall region and constitute the main contribution to the turbulence production and to the Reynolds stresses.

A knowledge of the dynamics of turbulence, briefly outlined above, becomes useful for an appreciation of technical applications. For example, structural vibrations induced by turbulent fluctuations involve both time-dependent stress and pressure loads on structures and the response of turbulent flow to time-dependent boundary conditions on the flow. Flow-generated noise in ducts and free jets originates from stresses and pressure fluctuations, whose spectra determine source distributions and propagation characteristics. The transport, deposition, and reentrainment of particles in turbulent flows of suspensions depend on particle-eddy interaction and turbulent bursting processes at a particle bed boundary. Moreover, in applications of concern to us, such as momentum and heat transfer in single-phase flows, the accumulated knowledge allows the prediction of mean flows, friction, and heat transfer from relatively simple models. At high Reynolds numbers, because of certain similarities among simple flows, these models involve only a few parameters. For example, in wall-bounded flows the main velocity change occurs in a region close to the wall, where experiments show that the velocity distribution follows a universal logarithmic law. We develop this law in Section 10.2. Because the viscosity is important only very close to the wall, these flows are divided into an inner (near-wall) region and an outer (core) region. The logarithmic law joins the solutions of these two regions in the *inertial sublayer* (or equilibrium layer). The Reynolds stress is nearly constant in this layer, most of the turbulent energy production occurs here, and production balances dissipation. Parameters in the inner region include wall distance y, viscosity ν, friction velocity $U_\tau = (\tau_w/\rho)^{1/2}$, and in the outer region, the width of the flow δ and free-stream velocity U_∞ (Fig. 10.1(c)). The complexity of wall-bounded flows is related to the existence of several characteristic length and velocity scales governing these flows. Free turbulent flows, however, are simpler. Characteristic length and velocity scales appear to be the width δ and the velocity difference or ΔU (Fig. 10.1(b) and (c)). In Section 10.2 we study algebraic models based on these few parameters.

10.2 ALGEBRAIC MODELS

Starting with the Boussinesq hypothesis, Eq. (10.1.1), it is our objective to model the turbulent diffusivity ν_T, which is assumed to be a scalar. To relate ν_T to the local parameters of a turbulent flow, Prandtl was inspired by the kinetic theory of gases. According to this theory the *molecular* diffusivity of momentum ν is proportional to the product of the mean free path and the root-mean-square velocity of molecules. Analogously, in his mixing length theory, Prandtl assumed the *turbulent* diffusivity ν_T to be proportional to the product of a mixing length and a characteristic velocity of turbulent fluctuations, Eq. (10.1.3). Actually, the physical processes as well as their scales are fundamentally different. Molecular momentum transfer in a gas is characterized by a large number of identifiable and small particles exchanging momentum through discontinuous interactions. Turbulent momentum transfer, on the other hand, is characterized by a much smaller number of large and poorly defined fluid eddies (say the turbulons of Chapter 9) exchanging momentum through continuous interactions. These interactions include the effect of pressure fluctuations on the momentum transfer, which has no counterpart in the kinetic theory of gases.

Because of these differences we simply employ the scaling introduced in Chapter 9 to obtain Eq. (10.1.3)

$$\nu_T \sim lu, \tag{10.2.1}$$

where l is a length scale characteristic of the size of momentum-transferring eddies and u is a characteristic velocity of transfer. Consider a representative eddy in a turbulent shear flow exemplified by an eddy of the large-amplitude Kelvin–Helmholtz instability (Fig. 10.2). For a mean size l, the velocity of transfer u may be expressed as the difference in mean momentum $l|dU/dy|$ over distance l. This is identical to the previously used scaling, $u/l \sim |dU/dy|$, by which we now eliminate u in Eq. (10.2.1) to obtain the celebrated *mixing length theory* of Prandtl (1925),

$$\nu_T = l^2\left|\frac{dU}{dy}\right|. \tag{10.2.2}$$

Whereas it is physically reasonable that τ_{yx} of Eq. (10.1.1) vanishes at lines of symmetry, where $dU/dy = 0$, there is no reason ν_T should also vanish. Nevertheless, Eq. (10.2.2) has proven to be very useful in the prediction of wall-bounded flows. For free flows, however, Prandtl (1942) suggested $u \sim \Delta U$, where ΔU denotes the momentum difference over the width of the flow $l \sim \delta$ (Fig. 10.1(b)); hence

$$\nu_T = l\,\Delta U. \tag{10.2.3}$$

Consider next the algebraic equations describing the variation of mixing length. For wall-bounded flows, on arguments that the mixing length must vanish at the wall and that eddy sizes are controlled by the distance from the wall, Prandtl introduced the linear relation

$$l = \kappa y, \tag{10.2.4}$$

where $\kappa \simeq 0.40$ denotes the experimentally determined von Kármán *mixing length constant*. Actually, close to the wall the mixing length is smaller than indicated by Eq. (10.2.4) because viscosity dampens the velocity fluctuations. Van Driest (1956) incorporated this effect by considering the Stokes problem of a viscous fluid oscillating parallel to a fixed wall (recall Problem 6-5), finding the amplitude of longitudinal velocity fluctuations to decrease exponentially as the wall is approached. Expecting the fluctuations normal to the wall to behave in a similar way and using the Prandtl mixing length theory, he derived

$$l = \kappa y\left[1 - \exp\left(-\frac{yU_\tau/\nu}{A^+}\right)\right], \tag{10.2.5}$$

where $A^+ = 26$ is the empirically determined van Driest *damping parameter*. Note that Eq. (10.2.5) reduces essentially to Eq. (10.2.4) when the dimensionless wall distance $y^+ = yU_\tau/\nu$ exceeds A^+, which corresponds to the fully turbulent region. Equation (10.2.5) has been found to yield good results for parallel and nearly parallel flows without mass transfer and with moderate pressure gradients. For a discussion and for extensions of this model, see Cebeci & Smith (1974). In the outer region of boundary layers ($y/\delta > 0.2$) l does not continue to increase linearly, but remains approximately constant,

$$l = \lambda\delta, \tag{10.2.6}$$

where $\lambda = 0.075$ to 0.09. In practice, when integrating through a boundary layer, the smaller of Eqs. (10.2.5) and (10.2.6) is therefore selected for use in Eq. (10.2.2).

For free flows, as for the outer region of boundary layers (Eq. 10.2.6), the mixing length is approximately constant and proportional to the width of the layer,

$$l \sim \delta. \tag{10.2.7}$$

This relation is used in Eq. (10.2.3), where ΔU denotes the maximum velocity difference in the layer, for example, the centerline velocity in a free jet. Note that the mixing length l is approximately constant in homogeneous or nearly homogeneous shear flows, but $l = l(y)$ near a wall. In other words,

$$\frac{u}{l} \sim \frac{\Delta U}{\Delta y}$$

appears to hold globally in free flows, but locally in near-wall flows.

Instead of modeling the Reynolds stress itself, Taylor (1932) in his theory of vorticity transport considered the divergence of stress, the term appearing in the balance of momentum. For incompressible flow this term may be expressed as

$$\begin{aligned} \frac{1}{\rho}\frac{\partial \tau_{ij}}{\partial x_i} &= \frac{\partial}{\partial x_i}(-\overline{u_i u_j}) = -\overline{u_i\left(\frac{\partial u_j}{\partial x_i} - \frac{\partial u_i}{\partial x_j} + \frac{\partial u_i}{\partial x_j}\right)} \\ &= -2\overline{u_i r_{ji}} - \frac{\partial}{\partial x_i}\left(\frac{\overline{u_i u_i}}{2}\right) = \epsilon_{jil}\overline{u_i\omega_l} - \frac{\partial K}{\partial x_j}, \end{aligned} \tag{10.2.8}$$

where Eqs. (1.5.4) and (1.5.6) have been used. For parallel or nearly parallel flow, neglecting the last term in Eq. (10.2.8), being the turbulent pressure gradient,

$$\frac{1}{\rho}\frac{d\tau_{yx}}{dy} = \overline{v\omega_z} - \overline{\omega_y w}, \tag{10.2.9}$$

where $\overline{v\omega_z}$ is the turbulent diffusion of z-vorticity and $\overline{\omega_y w}$ is associated with vortex stretching (recall Section 2.4 and Eq. 8.4.9). This interpretation is also demonstrated from Eqs. (10.1.1) and (10.1.3), noting that $\Omega_z \simeq -dU/dy$ and taking $u \sim$ constant,

$$\frac{1}{\rho}\frac{d\tau_{yx}}{dy} \sim \frac{d}{dy}\left(lu\frac{dU}{dy}\right) = \underbrace{-ul\frac{d\Omega_z}{dy}}_{\text{diffusion}} - \underbrace{\Omega_z u\frac{dl}{dy}}_{\text{stretching}}. \tag{10.2.10}$$

Clearly, a hypothesis of vorticity diffusion would be valid only when l is nearly constant. The plane free jet is an example. The round jet, on the other hand, as well as wall-bounded shear layers, involve vortex stretching. Both terms in Eq. (10.2.9) are at play, for example, when vortex filaments are intermittently deformed and displaced in the bursting process near wall (Fig. 10.5). The time average represents the net force due to turbulent momentum transfer.

Having introduced some algebraic models derived from the mixing length concept, we now proceed to illustrative examples. Here we shall reconsider some problems already solved in previous chapters under laminar conditions.

EXAMPLE 10.2.1 Consider steady, turbulent channel flow of an incompressible fluid between smooth parallel plates spaced a distance $2h$ apart (Fig. 10.1(a)). We wish to determine the velocity distribution and the friction factor.

For fully developed flow, $\partial U/\partial x = 0$ and conservation of mass gives $\partial V/\partial y = 0$. Then the balance of momentum in x and y yields

$$0 = -\frac{\partial P}{\partial x} + \frac{d}{dy}\left(\rho\nu\frac{dU}{dy} - \rho\overline{vu}\right) \tag{10.2.11}$$

$$0 = -\frac{\partial P}{\partial y} + \frac{d}{dy}(-\rho\overline{v^2}), \tag{10.2.12}$$

subject to the condition of no slip at the wall, $y = 0$: $U = u = v = 0$ and symmetry at the centerline, $y = h$.

Integration of Eq. (10.2.12) shows that pressure varies over the cross section of the flow,

$$P(x, y) = P_w(x) - \rho\overline{v^2},$$

but $\partial P/\partial x$ equals $\partial P_w/\partial x$, since $\overline{v^2}$ is independent of x, so $\partial P/\partial x$ is independent of x. Integration of Eq. (10.2.11) gives

$$\frac{\tau_{yx}^{\text{tot}}}{\rho} = \nu\frac{dU}{dy} - \overline{vu} = \frac{\tau_w}{\rho}\left(1 - \frac{y}{h}\right), \tag{10.2.13}$$

where

$$\tau_w = -h\frac{\partial P}{\partial x} \quad \text{or} \quad U_\tau^2 = \frac{\tau_w}{\rho} = -\frac{h}{\rho}\frac{\partial P}{\partial x}. \tag{10.2.14}$$

Here U_τ denotes the *friction velocity* employed to define dimensionless parameters

$$u^+ = \frac{U}{U_\tau}; \qquad y^+ = \frac{yU_\tau}{\nu}. \tag{10.2.15}$$

Note that velocity fluctuations, hence Reynolds stresses, vanish at the wall, so $\tau_w = \mu\,\partial U(x, 0)/\partial x$ is purely viscous.

From the definition of the friction factor (recall Eq. 3.2.55)

$$f_m = \frac{(-\partial P/\partial x)D_e}{\frac{1}{2}\rho U_m^2} = 8\left(\frac{U_\tau}{U_m}\right)^2, \tag{10.2.16}$$

where $D_e \equiv 4A/P_w = 4h$ of Eq. (3.2.56) denotes the equivalent diameter, and U_m the mean velocity

$$U_m = \frac{1}{A}\int_A U\,dA = \frac{1}{h}\int_0^h U(y)\,dy. \tag{10.2.17}$$

Note that Eq. (10.2.16) is valid for a channel of any cross section provided that τ_w is the peripheral average.

Employing Eq. (10.1.1) and dimensionless parameters, Eq. (10.2.13) becomes

$$\left(1 + \frac{\nu_T}{\nu}\right)\frac{du^+}{dy^+} = 1 - \frac{y^+}{h^+}, \tag{10.2.18}$$

or, in terms of the Prandtl mixing length model Eq. (10.2.2),

$$\left[1 + (l^+)^2 \left|\frac{du^+}{dy^+}\right|\right]\frac{du^+}{dy^+} = 1 - \frac{y^+}{h^+}. \tag{10.2.19}$$

Solving the resulting quadratic equation in du^+/dy^+ and integrating, subject to $u^+(0) = 0$, gives

$$u^+ = \int_0^{y+} \frac{2(1 - y^+/h^+)\,dy^+}{1 + \sqrt{1 + (l^+)^2(1 - y^+/h^+)}}. \tag{10.2.20}$$

Very close to the wall, $y^+/h^+ \ll 1$, $l^+ \sim 0$, Eq. (10.2.20) yields the viscous, constant shear solution (viscous sublayer)

$$u^+ = y^+. \tag{10.2.21}$$

Moving out from the wall, viscous effects gradually diminish in importance through the transition region as expressed by Eq. (10.2.5). which reduces to Eq. (10.2.4) in the fully turbulent region (inertial sublayer). Ignoring viscous stress, using Eq. (10.2.4) and $y^+/h^+ \ll 1$, which remains valid for large Reynolds numbers, Eq. (10.2.20) yields the fully turbulent, constant shear solution (*logarithmic law-of-the-wall*)

$$u^+ = \frac{1}{\kappa}\ln y^+ + C, \tag{10.2.22}$$

where $\kappa \simeq 0.4$ and $C \simeq 5.5$ according to experiment. Equations (10.2.21) and (10.2.22), which intersect at $y^+ \simeq 11.6$, constitute a simple two-layer velocity distribution to be referred to in Chapter 11.

Carrying out the integration in Eq. (10.2.20) for $y/h \ll 1$, using Eq. (10.2.5), gives a continuous velocity distribution, from the viscous sublayer through the transition region (buffer layer) into the fully turbulent region, yielding $C = 5.24$ (Fig. 10.6). Retaining the factor $1 - y/h$ in Eq. (10.2.20) gives the behavior in the core of the flow. Actually, the constant eddy diffusivity model Eq. (10.2.6) should apply approximately here. Integrating Eq. (10.2.18) from the centerline, ignoring viscous stress, yields

$$\frac{U_C - U}{U_\tau} = \frac{1}{2}\left(\frac{U_\tau h}{\nu_T}\right)\left(1 - \frac{y}{h}\right)^2, \tag{10.2.23}$$

which adequately describes the velocity for $\frac{1}{2} < y/h \leq 1$, with $U_\tau h/\nu_T \sim 13 - 15$.

The analysis leading to Eq. (10.2.22) apparently requires the Reynolds number to be large. In fact, this condition generally ensures the existence of the inertial sublayer of a wall-bounded flow whose velocity distribution then satisfies the universal log law of the wall. Note that the length scales in a channel flow or a boundary layer are the width h and the viscous length ν/u, where u denotes a characteristic fluctuating velocity. When the Reynolds number uh/ν is sufficiently large, there exists a region, $uh/\nu \gg 1$ and $y/h \ll 1$, where the viscous length ν/u is too small, and the geometric length h is too large, for either to influence the dynamics of the turbulence. Hence, the velocity gradient dU/dy, on dimensional grounds, can depend only linearly on u/y, which integrates to the logarithmic law-of-the-wall. The characteristic velocity turns out to be U_τ, but $U_\tau \sim K^{1/2} \sim u$ in the inertial sublayer according to experiment (see Eq. 10.3.8).

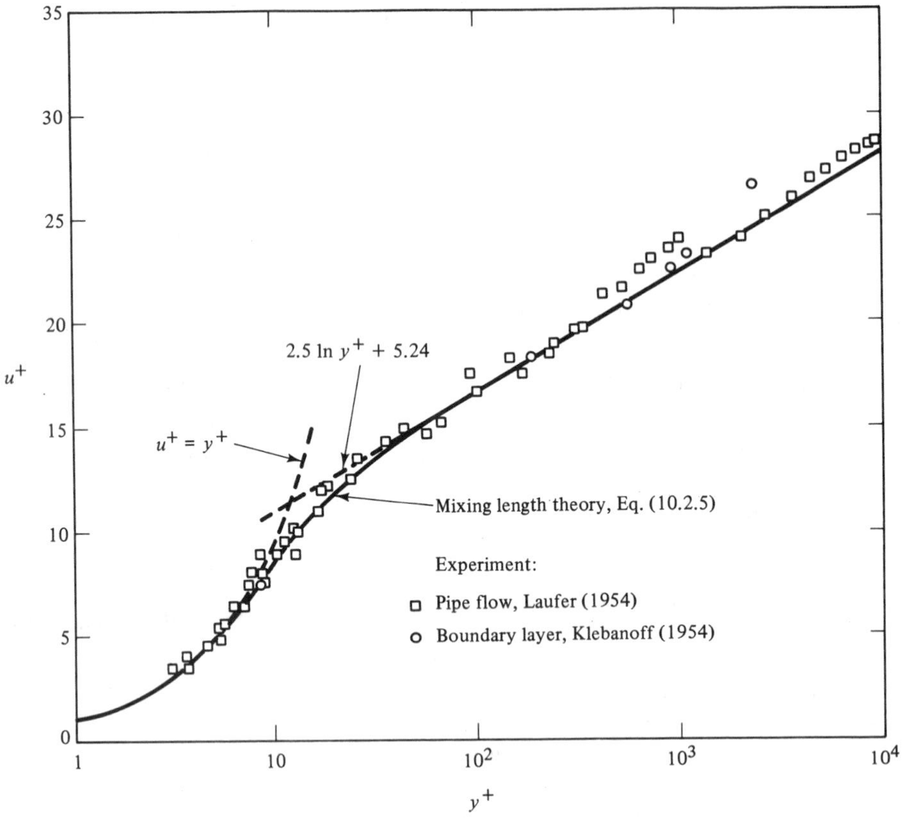

Fig. 10.6 Velocity distribution in the near-wall region (from Cebeci & Smith 1974; used by permission)

There is an analogy between the *inertial sublayer* of near-wall flows (spatially placed between the large-scale inertial-controlled core flow and the small-scale viscous-controlled sublayer), and the *inertial subrange* of energy spectra (spectrally placed between the large-scale energy containing eddies and the small-scale dissipative eddies; recall Chapter 9). Both are universal and appear when the Reynolds number is sufficiently high. This analogy is discussed by Tennekes & Lumley (1972), who also derive Eq. (10.2.22) formally, based on matched asymptotic expansions. The derivation, or a dimensional argument, also shows that the outer solution in the core of the flow must be of the form (*velocity-defect law*)

$$\frac{U_C - U}{U_\tau} = F\left(\frac{y}{h}\right), \tag{10.2.24}$$

where $U_C = U(h)$. Equation (10.2.23) satisfies this law.

Now, to obtain an expression for the friction factor, we merely determine the mean velocity U_m. For this purpose it is an adequate approximation to employ Eq.

(10.2.22) in Eq. (10.2.17), selecting lower limits of integration that give no contributions. Also employing Eq. (10.2.16),

$$\left(\frac{8}{f_m}\right)^{1/2} = \frac{1}{h^+}\int_{y_0^+}^{h^+}\left(\frac{1}{\kappa}\ln y^+ + C\right)dy^+.$$

After integration, and introducing $\mathrm{Re} = D_e U_m/\nu = 4hU_m/\nu = 4(8/f_m)^{1/2}h^+$, we obtain the *logarithmic friction law*

$$\left(\frac{8}{f_m}\right)^{1/2} = 2.5\ln(\mathrm{Re}\, f_m^{1/2}) - 3.07. \qquad (10.2.25)$$

For improved agreement with experimental data of Beavers, Sparrow & Lloyd (1971), shown in Fig. 10.7, the constant 3.07 in Eq. (10.2.25) should be replaced by 2.26.

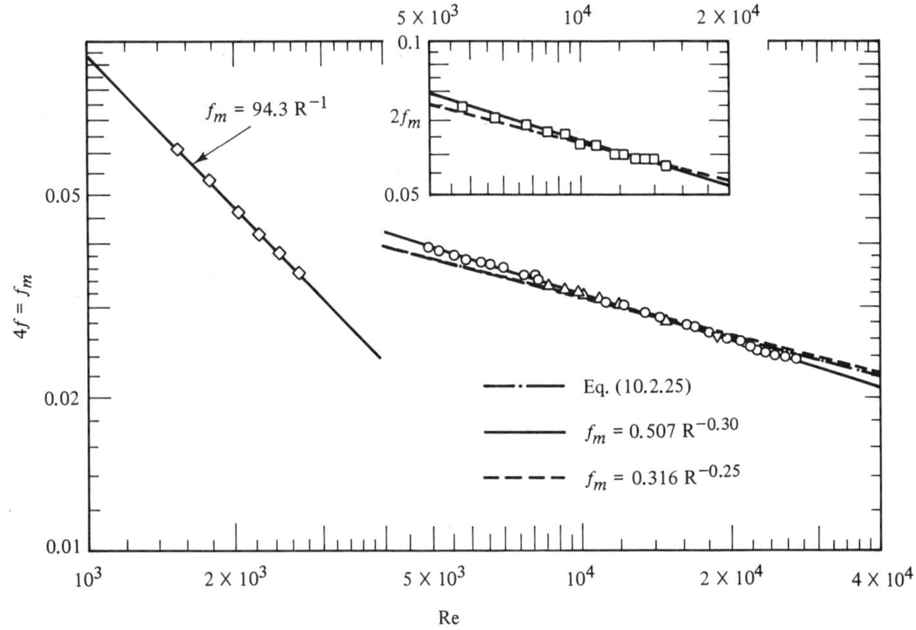

Fig. 10.7 Friction factor f_m versus Reynolds number $\mathrm{Re} = U_m D_e/\nu$ for large aspect ratio rectangular ducts (adapted from Beavers et al. 1971; used by permission)

The foregoing analysis may be applied to flow in smooth channels of different cross sections. For a pipe of circular cross section (see Problem 10-1), we obtain

$$\left(\frac{8}{f_m}\right)^{1/2} = 2.5\ln(\mathrm{Re}\, f_m^{1/2}) - 2.58. \qquad (10.2.26)$$

A value of 2.26 instead of 2.58 gives good agreement with data. Comparing Eqs. (10.2.25) and (10.2.26) shows (as mentioned at the end of Ex. 3.2.4) that the lack of geometric similarity between cross sections has little effect on results for turbulent flows. In place of Eq. (10.2.26) it may be convenient to use an empirical power law, as first suggested by Blasius (1913),

$$f_m = 0.3164\mathrm{Re}_D^{-1/4}, \qquad 5 \times 10^3 < \mathrm{Re}_D < 3 \times 10^4, \tag{10.2.27}$$

$$f_m = 0.184\mathrm{Re}_D^{-1/5}, \qquad 3 \times 10^4 < \mathrm{Re}_D < 10^6. \tag{10.2.28}$$

Also, experimental data for pipe flows are conveniently represented by the smooth three-layer velocity distribution

$$\begin{aligned} 0 \le y^+ < 5: &\qquad u^+ = y^+ \\ 5 \le y^+ < 30: &\qquad u^+ = 5 \ln y^+ - 3.05 \\ 30 \le y^+: &\qquad u^+ = 2.5 \ln y^+ + 5.5, \end{aligned} \tag{10.2.29}$$

which becomes useful in discussing the associated heat transfer (Section 11.1).

So far we have considered a smooth wall. That is, the rms value of the height k of roughness elements was small compared to the thickness of the viscous wall layer, $k^+ = U_\tau k/\nu < 1$. Here, as for fully laminar flow, the roughness has no influence on the velocity distribution. However, when roughness elements protrude into the transition region, Reynolds stress contributes to τ_w. Both the viscous length ν/U_τ and k enter the problem and $u^+ = u^+(y^+, k^+)$ takes the experimentally determined form

$$u^+ = \frac{1}{\kappa} \ln \frac{y}{k} + B, \tag{10.2.30}$$

where $B = B(k^+)$ (Fig. 10.8). When roughness elements protrude into the fully

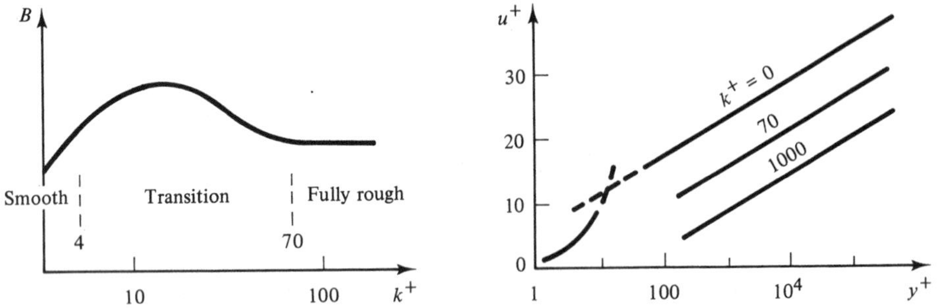

Fig. 10.8 Velocity distribution in near-wall region, Eq. (10.2.30); effect of wall roughness

turbulent region, $k^+ > 70$, viscous effects are eliminated altogether, $B = 8.5$, and the friction factor becomes independent of the Reynolds number. In all cases we assume that $k/h \ll 1$ and the velocity-defect law holds. It is interesting to note that the mixing length constant κ, hence the slope of the log law, remains unchanged.

The Moody (1944) diagram, given by Fig. 10.9, shows empirically determined friction factors for pipe flow over a range of values of the relative roughness k/D_e.

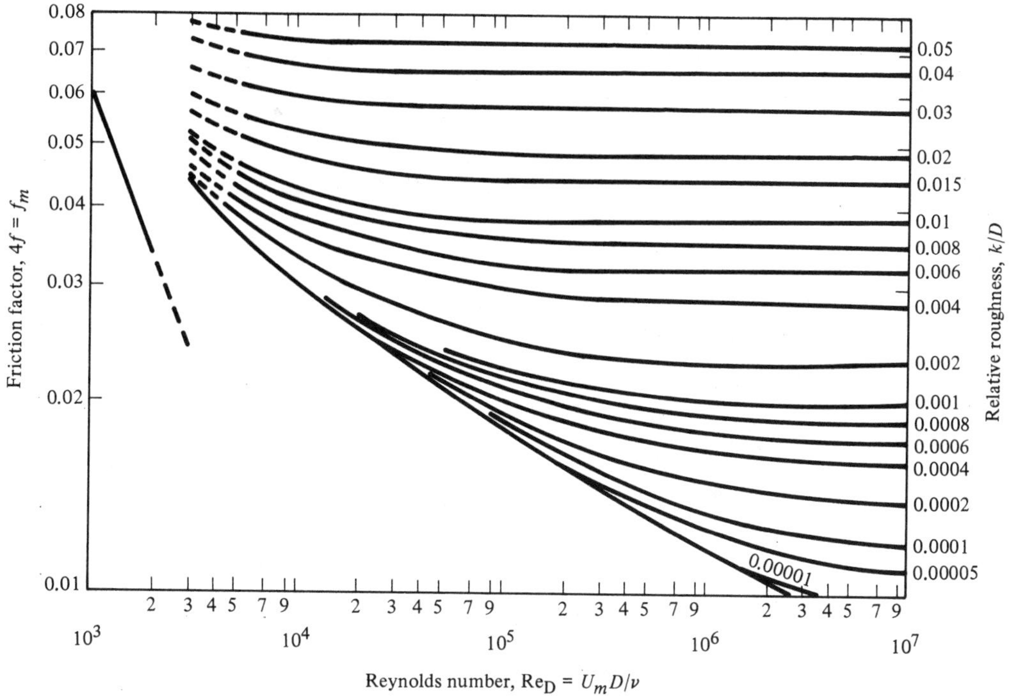

Fig. 10.9 Friction factors for pipe flows (adapted from Shames 1962; used by permission)

For a further discussion of channel flow, including other geometries, one may refer to Schlichting (1968).

Let us finally examine the distribution of eddy diffusivity. Employing Eq. (10.2.29) in Eq. (10.2.18), for example, for pipe flow at Re $= 5 \times 10^4$, gives the result compared to data by Laufer (1954) shown in Fig. 10.10. Evidently, the approximation $\nu_T/\nu = \kappa R^+(y/R)(1 - y/R)$ for the core flow, where $R^+ = U_\tau R/\nu$, is in error. Here, as also shown in the figure, a constant value of eddy diffusivity $\nu_T/\nu \sim R^+/15$, as used in Eq. (10.2.23), is more appropriate. For a continuous variation in the near-wall region, Eq. (10.2.5) may be used in Eq. (10.2.19). Alternatively, we may use the empirical relation proposed by Reichardt (1951),

$$\frac{\nu_T}{\nu} = \frac{1}{6}\kappa R^+\eta(2 - \eta)(3 - 4\eta + 2\eta^2), \tag{10.2.31}$$

where $\eta = y/R$. This relation reduces to the mixing length theory $\nu_T/\nu = \kappa y^+$ near the wall, while being nearly constant in the core, approaching $R^+/15$ at the centerline. Near a fully rough wall ν_T/ν may be estimated from Eq. (10.2.30) for $y/R \lesssim 0.5$.

Having completed our discussion on developed channel flow, let us turn to the velocity discontinuity examined for stability in Section 10.1, now studying its development into a free shear flow (Fig. 10.4).

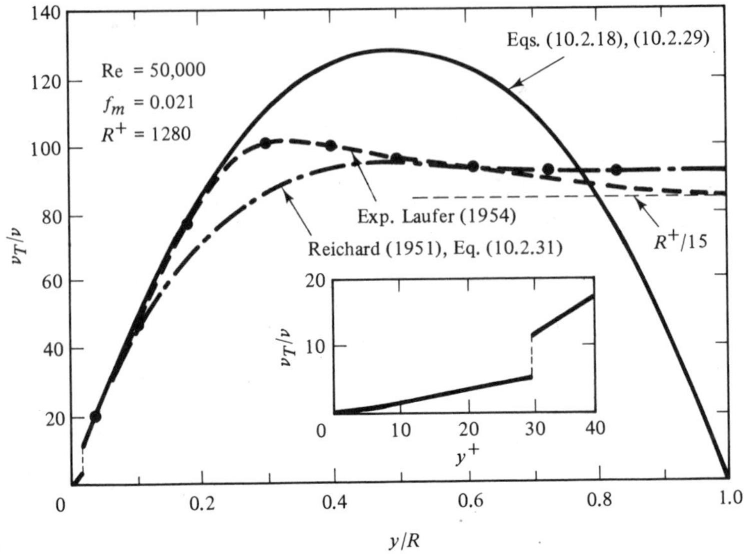

Fig. 10.10 Eddy diffusivity distribution for pipe flow (experimental data by Laufer 1954; used by permission)

EXAMPLE 10.2.2 Two flows with uniform velocities U_1 and U_2, respectively, are brought into contact at $x = 0$ (Fig. 10.1(b)). We wish to determine the distributions of velocity and eddy diffusivity. The mixing layer developing downstream of the velocity discontinuity forms a free flow with zero pressure gradient. The case $U_2 = 0$ corresponds to flow over a step or to the flow in one-half of a free jet in its range of a potential core.

If initially laminar, the mixing layer becomes unstable to small disturbances and turns turbulent when the Reynolds number (based on velocity difference $U_1 - U_2$ and width of the layer 2δ) is greater than about 300 (Townsend 1976). This corresponds to $(U_1 - U_2)x/\nu \gtrsim 7 \times 10^4$. Knowledge of velocity and turbulence properties are needed for calculating heat and mass transfer between two parallel streams brought into contact. Also, it may be of interest to evaluate the mass flow within the mixing layer, entrained from the two streams.

The resulting flow is nearly parallel and we need the boundary layer formulation, which may be derived from the equations of Table 8.1 following the procedure of Section 4.1. However, if the normal components of Reynolds stress cannot be ignored (streamwise turbulent diffusion is not negligible and the pressure gradient through the layer is not zero) the two-dimensional boundary layer formulation is modified to

$$U\frac{\partial U}{\partial x} + V\frac{\partial U}{\partial y} = -\frac{1}{\rho}\frac{\partial P}{\partial x} + \frac{\partial(-\overline{u^2})}{\partial x} + \frac{\partial}{\partial y}\left(\nu\frac{\partial U}{\partial y} - \overline{vu}\right) \tag{10.2.32}$$

$$0 = -\frac{1}{\rho}\frac{\partial P}{\partial y} + \frac{\partial(-\overline{v^2})}{\partial y}, \tag{10.2.33}$$

where the pressure gradient, now at the free stream U_∞, is given by Eq. (4.1.16),

$$U_\infty \frac{dU_\infty}{dx} = -\frac{1}{\rho}\frac{dP(x, \delta)}{dx}. \tag{10.2.34}$$

Integrating Eq. (10.2.33) from y to δ, assuming that $\overline{v^2}(x, \delta) = 0$, differentiating the result with respect to x, and also using Eq. (10.2.34), we eliminate $\partial P/\partial x$ from Eq. (10.2.32) to obtain

$$U\frac{\partial U}{\partial x} + V\frac{\partial U}{\partial y} = U_\infty \frac{dU_\infty}{dx} + \frac{\partial}{\partial x}(\overline{v^2} - \overline{u^2}) + \frac{\partial}{\partial y}\left(\nu\frac{\partial U}{\partial y} - \overline{vu}\right). \tag{10.2.35}$$

Note that streamwise turbulent diffusion is important only for highly turbulent and anisotropic flows.

Considering now low turbulence intensity, the equations of Table 8.1 reduce to the usual boundary layer formulation of Table 10.2. Furthermore, U_1 and U_2 are

TABLE 10.2 Two-dimensional boundary layer formulation (low-intensity turbulent flow)

$$\frac{\partial U}{\partial x} + \frac{\partial V}{\partial y} = 0 \tag{10.2.36}$$

$$U\frac{\partial U}{\partial x} + V\frac{\partial U}{\partial y} = F_x - \frac{1}{\rho}\frac{\partial P}{\partial x} + \frac{\partial}{\partial y}\left(\nu\frac{\partial U}{\partial y} - \overline{vu}\right) \tag{10.2.37}$$

$$0 = -\frac{1}{\rho}\frac{\partial P}{\partial y} \tag{10.2.38}$$

$$U_\infty\frac{dU_\infty}{dx} = F_x - \frac{1}{\rho}\frac{\partial P}{\partial x} \tag{10.2.39}$$

constant, $\partial P/\partial x = 0$ for the present problem, and assuming fully turbulent flow, Eq. (10.2.37) may be reduced to

$$U\frac{\partial U}{\partial x} + V\frac{\partial U}{\partial y} = \frac{\partial(-\overline{vu})}{\partial y}. \tag{10.2.40}$$

Referring to Eqs. (10.1.1) and (10.2.3) for free flows, we write (see Problem 10-7 for an alternative solution based on Eq. 10.2.2)

$$\frac{\tau_{yx}}{\rho} = -\overline{vu} = \nu_T\frac{\partial U}{\partial y}; \qquad \nu_T = l(x)\,\Delta U, \tag{10.2.41}$$

where $\Delta U = \frac{1}{2}(U_1 - U_2)$ is one-half the velocity difference and l a mixing length which is independent of y. Then Eq. (10.2.40) takes the form

$$U\frac{\partial U}{\partial x} + V\frac{\partial U}{\partial y} = l\,\Delta U\frac{\partial^2 U}{\partial y^2}, \tag{10.2.42}$$

subject to $U(x, \infty) = U_1$ and $U(x, -\infty) = U_2$.

Dimensional analysis (recall Section 5.2) shows that a similarity solution exists provided that $l = b_1 x$ where b_1 is a constant. Now let

$$U(x, y) = U_m + \Delta U f'(\eta); \qquad \eta = \frac{y}{x}; \qquad U_m = \tfrac{1}{2}(U_1 + U_2),$$

and consider the antisymmetric upper half of the mixing layer. Integration of Eq. (10.2.36) subject to $V(x, 0) = 0$ gives

$$V = \Delta U(\eta f' - f) \tag{10.2.43}$$

and Eq. (10.2.42) transforms into

$$b_1 f''' + f f'' + \left(\frac{U_m}{\Delta U}\right)\eta f'' = 0, \tag{10.2.44}$$

subject to $f(0) = f'(0) = 0, f'(\infty) = 1$. Solution of Eq. (10.2.44) requires numerical integration (see Section 7.2).

An approximate solution, valid far downstream, is obtained by linearizing Eq. (10.2.42) to

$$U_m \frac{\partial U}{\partial x} = b_1 x \, \Delta U \frac{\partial^2 U}{\partial y^2}. \tag{10.2.45}$$

In effect we consider the flow to be essentially parallel and $\Delta U / U_m$ to be small. Using the foregoing transformation, Eq. (10.2.44) may be reduced to

$$f''' + \left(\frac{b_1 \, \Delta U}{U_m}\right)^{-1} \eta f'' = 0,$$

whose solution is

$$U(x, y) = U_m + \Delta U \operatorname{erf}\left[\left(\frac{2b_1 \, \Delta U}{U_m}\right)^{-1/2} \frac{y}{x}\right]. \tag{10.2.46}$$

The constant b_1 must be determined from experiment, which in agreement with Eq. (10.2.3) shows $\delta \, \Delta U / \nu_T \sim 15.5$, where δ is defined as the y-value at which the velocity is 99% of the free-stream value. This yields

$$b_1 \simeq 0.055 \frac{\Delta U}{U_m}.$$

Suppose, as for laminar problems, that we consider the integral formulation. Rearranging Eq. (4.1.27), we obtain for the upper half of the mixing layer,

$$\frac{d}{dx} \int_0^\delta \rho U(U - U_1) \, dy = \tau_{yx}(x, \delta) - \tau_{yx}(x, 0). \tag{10.2.47}$$

Inserting the approximating cubic profile

$$U = U_m + \Delta U \left[\frac{3}{2}\left(\frac{y}{\delta}\right) - \frac{1}{2}\left(\frac{y}{\delta}\right)^3\right],$$

satisfying $U(x, 0) = U_m$ and $U(x, \delta) = U_m + \Delta U$, and $\tau_{yx}(x, \delta) = 0$ according to Eq. (10.2.41), we obtain

$$\delta(x) = 2x\left(\frac{b_1 \, \Delta U / U_m}{1 + \frac{13}{35} \Delta U / U_m}\right)^{1/2}, \tag{10.2.48}$$

and again using the empirical relation $\delta/l \sim 15.5$,

$$b_1 = \frac{0.017\ \Delta U/U_m}{1 + \frac{13}{35}\ \Delta U/U_m}.$$

For small values of $\Delta U/U_m$, this result agrees with that found above, except that the numerical factor is smaller. This difference is a consequence of differences in profiles.

One aspect of occasional interest for free flows is their entrainment of undisturbed fluid. The mass flow affected by the mixing process, relative to U_2, for example, may be calculated from

$$\rho \int_{-\delta}^{\delta} (U - U_2)\, dy, \tag{10.2.49}$$

and is seen to increase linearly with x.

EXAMPLE 10.2.3 Consider the steady, uniform flow over a thin flat plate of length L, Fig. 10.1(c). We wish to find the drag force per unit width of the plate.

For a smooth wall and low turbulence intensity of the free stream, the initial boundary layer development is laminar (Ex. 4.2.1) and $\delta(x) \sim x^{1/2}$. Instability and transition to turbulence occur from a critical Reynolds number $\mathrm{Re}_{x,c} = (U_\infty x/\nu)_c \sim 3 \times 10^5$. Farther downstream the boundary layer is turbulent and $\delta(x) \sim x^{4/5}$, Fig. 10.11.

The integral balance of momentum Eq. (4.1.28), with $\partial P/\partial x = 0$ from Eq. (10.2.39), may be written

$$\frac{d}{dx}\int_0^{\delta} \rho U(U_\infty - U)\, dy = \tau_w, \tag{10.2.50}$$

or

$$\frac{d\delta_2}{dx} = \tfrac{1}{2} f, \tag{10.2.51}$$

where we have used the definitions of *momentum thickness* δ_2 and *local coefficient of skin friction* f,

$$\delta_2 = \delta \int_0^1 \frac{U}{U_\infty}\left(1 - \frac{U}{U_\infty}\right) d\left(\frac{y}{\delta}\right), \tag{10.2.52}$$

$$f = \frac{\tau_w}{\frac{1}{2}\rho U_\infty^2}. \tag{10.2.53}$$

On the assumption that velocity profiles are similar, $U(x, y)/U_\infty = f[y/\delta(x)]$, and that the Blasius law of friction may be written as

$$f = C_2 \mathrm{Re}_\delta^{-m}, \tag{10.2.54}$$

where $C_2 = 0.045$, $m = 0.25$, and $\mathrm{Re}_\delta = U_\infty \delta/\nu$, Eq. (10.2.51) may be rearranged as

$$\delta^m \frac{d\delta}{dx} = \frac{C_2}{2}\frac{\delta}{\delta_2}\left(\frac{U_\infty}{\nu}\right)^{-m},$$

and integrated, subject to $\delta(0) = 0$,

$$\frac{\delta}{x} = \left(\frac{m+1}{2} C_2 \frac{\delta}{\delta_2}\right)^{1/(1+m)} \left(\frac{U_\infty x}{\nu}\right)^{-m/(1+m)} \tag{10.2.55}$$

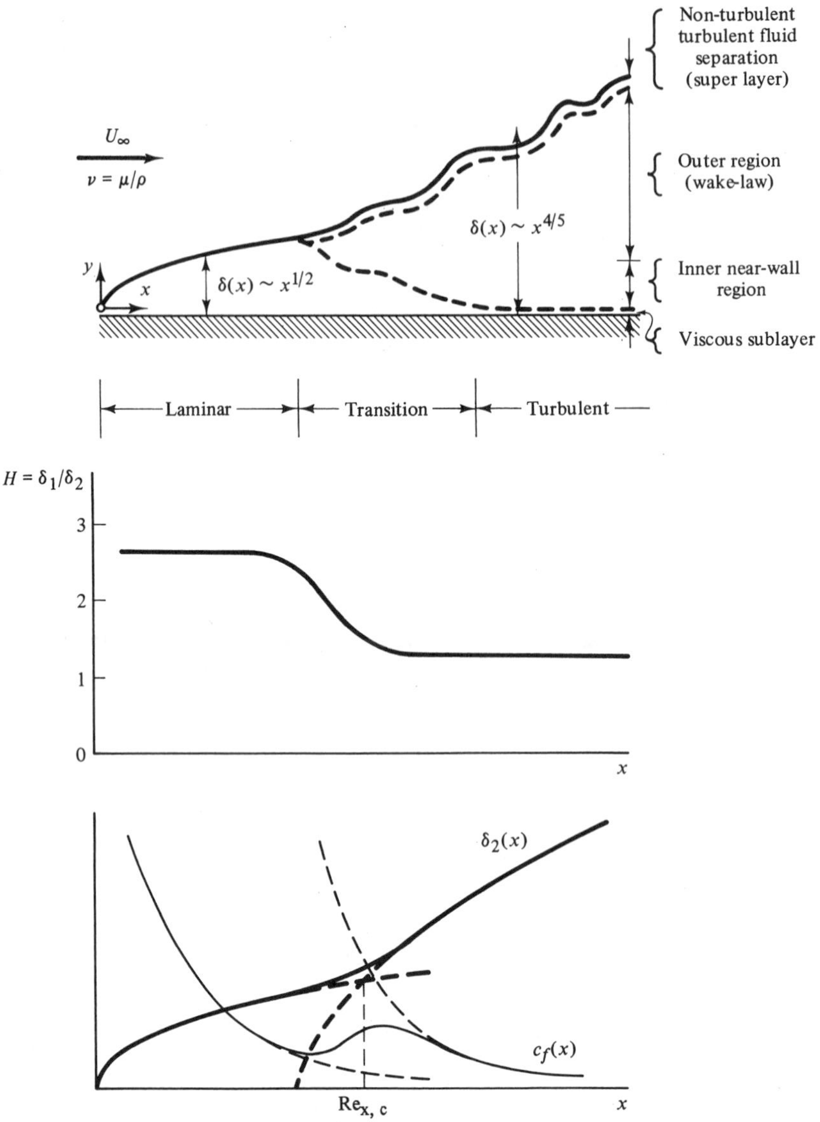

Fig. 10.11 Flat-plate boundary layer; development of shape factor and skin friction

Recall from Exs. 4.2.1 and 5.3.3 that the flat-plate laminar boundary layer satisfies the foregoing assumptions with $m = 1$. Experiments for the turbulent boundary layer show $m = 1/4$ for $3 \times 10^4 < \mathrm{Re}_x < 10^6$ and far downstream the assumption $\delta(0) = 0$ introduces little error. Hence,

$$\delta(x) \sim \begin{cases} x^{1/2} & \text{(laminar)} \\ x^{4/5} & \text{(turbulent).} \end{cases} \tag{10.2.56}$$

The turbulent boundary layer is thicker and grows more rapidly than the laminar one because of the more effective momentum transfer from the wall and because of the increased entrainment of free-stream fluid along the irregular edge.

An approximate solution based on a power-law velocity profile $U/U_\infty = (y/\delta)^{1/n}$, $n = 7$ (see Problem 10-6), leads to

$$\frac{\delta}{x} \simeq 0.37\ \mathrm{Re}_x^{-1/5}; \qquad \frac{\delta_2}{x} \simeq 0.036\ \mathrm{Re}_x^{-1/5}, \tag{10.2.57}$$

and, with a slight adjustment of the coefficient to better fit data,

$$f \simeq 0.0576\ \mathrm{Re}_x^{-1/5}. \tag{10.2.58}$$

The drag force is determined by integration of Eq. (10.2.58). If the flow over a significant part of the plate is laminar, this should be taken into account (Fig. 10.11). For a long plate the integrated mean value of skin friction is approximately

$$f_L = \frac{1}{L}\int_0^L f\,dx \simeq 0.072\ \mathrm{Re}_L^{-1/5}, \tag{10.2.59}$$

and the drag force (on one side, per unit width) is $F_D = \frac{1}{2}\rho U_\infty^2 f_L L$. Alternatively, using the logarithmic law of the wall, Eq. (10.2.22), following the steps leading to Eq. (10.2.25), we here derive

$$\left(\frac{2}{f}\right)^{1/2} = \frac{1}{\kappa}\ln\left[\left(\frac{2}{f}\right)^{1/2}\mathrm{Re}_\delta\right] + C. \tag{10.2.60}$$

For the mean value, Schlichting (1968) recommends the explicit expression which fits data

$$f_L = 0.455(\log \mathrm{Re}_L)^{-2.58}. \tag{10.2.61}$$

Actually, the log law is only valid in the inner 10% of the boundary layer and the empirical three-layer velocity distribution, Eq. (10.2.29), is modified to

$$\begin{aligned} 0 \le y \le 5: &\qquad u^+ = y^+ \\ 30 \le y \le \delta^+/10: &\qquad u^+ = \frac{1}{\kappa}\ln y^+ + C \\ \delta^+/10 \le y \le \delta^+: &\qquad u^+ = \frac{1}{\kappa}\ln y^+ + C + C_3 W\left(\frac{y}{\delta}\right), \end{aligned} \tag{10.2.62}$$

where $\kappa = 0.41$, $C = 5$, $C_3 = 1.35$, and the Coles wake law involves the function

$$W\left(\frac{y}{\delta}\right) = 1 - \cos\left(\frac{\pi y}{\delta}\right).$$

Velocity distributions for boundary layers with arbitrary pressure gradients are very complex and cannot be prescribed locally, but must be determined from a detailed numerical calculation carrying the history of boundary layer development. Among the general features, the inner region appears to be remarkably insensitive to the intermittency of the free edge and to the free-stream pressure gradient. Therefore, it can be usually treated as the wall region of channel flow for high Reynolds number (recall the discussion following Eq. 10.2.23). The outer region, on the other hand, has features common to free flows. It is sensitive, and responds slowly, to changes in pressure gradient, hence carries the upstream history of the boundary layer.

As for laminar flows, a class of turbulent flows are characterized by velocity profiles that have similarity properties. The flat plate with zero pressure gradient is one example. To identify other cases, consider the integral balance of momentum Eq. (4.1.28), including the pressure gradient, which may be rearranged as

$$\frac{1}{U_\infty^2}\frac{d(U_\infty^2\,\delta_2)}{dx} = \frac{1}{2}f(1-\Pi), \tag{10.2.63}$$

where

$$\Pi = \frac{\delta_1}{\tau_w}\left(-\frac{dP}{dx}\right); \qquad \delta_1 = \delta\int_0^1\left(1-\frac{U}{U_\infty}\right)d\left(\frac{y}{\delta}\right). \tag{10.2.64}$$

Clauser has shown empirically (see, for example, Cebeci & Smith 1974) that velocity profiles are similar (self-preserving flows) when the acceleration parameter Π is constant. Empirical relations have been developed for these profiles.

Because of the importance of the turbulent boundary layers with an arbitrary pressure gradient (aerofoils, turbomachinery, entrance flows, etc.), a large number of empirical relations have been developed between 1930 and 1950 based on use of the integral balance of momentum. Although modern turbulence models are increasingly replacing these classical methods, they are still used to a considerable extent. They are fast and comparatively accurate, due to their substantial experimental basis. Introducing the shape parameter $H = \delta_1/\delta_2$, where δ_1 and δ_2 are given by Eqs. (10.2.64) and (10.2.52), Eq. (10.2.63) may be written as (see Problem 10-10)

$$\frac{d\delta_2}{dx} + (2+H)\frac{\delta_2}{U_\infty}\frac{dU_\infty}{dx} = \frac{1}{2}f. \tag{10.2.65}$$

Specifying two additional empirical relations, say,

$$f = f_1(\mathrm{Re}_{\delta_2}, H); \qquad H = f_2(\mathrm{Re}_{\delta_2}, f), \tag{10.2.66}$$

and an upstream boundary condition, Eq. (10.2.65) may be integrated to yield δ_2, H, and f along the wall. Note that this method involves no approximate velocity profiles or any turbulence model. For a summary of these efficient methods, including the more recent entrainment procedure of Head (1958), see Schlichting (1968), Cebeci & Smith (1974), and Problem 10-9. Some recent integral methods (see, for example, Kuhn & Nielsen 1974) are based on explicit use of velocity profiles having the necessary flexibility to reflect change due to nonzero pressure gradients (Problem 10-11). The following example, however, illustrates the application of Eq. (10.2.65) to an accelerating flow, assuming H to be constant.

EXAMPLE 10.2.4 Reconsider Ex. 4.2.2, the entrance flow between parallel plates spaced a distance $2h$ apart (Fig. 4.5), now for fully turbulent flow. We wish to find the pressure drop and estimate the length to achieve fully developed flow.

The conservation of mass integrated from 0 to x, Eq. (4.2.10),

$$\int_0^\delta U\,dy + U_\infty(h-\delta) - U_0 h = 0,$$

may be rearranged, introducing $H = \delta_1/\delta_2$ and δ_1 from Eq. (10.2.64), as

$$U_\infty = U_0 \frac{h}{h - H\delta_2}, \tag{10.2.67}$$

where U_0 is the upstream uniform velocity. Hence

$$\frac{\delta_2}{U_\infty}\frac{dU_\infty}{dx} = \frac{H\delta_2}{h - H\delta_2}\frac{d\delta_2}{dx}.$$

Employing this result and Eq. (10.2.54), the integral balance of momentum for the developing boundary layer, Eq. (10.2.65), may be rearranged as

$$\left(\frac{\delta_2}{h - H\delta_2}\right)^m \left[1 + \frac{(2 + H)H\delta_2}{h - H\delta_2}\right]\frac{d\delta_2}{dx} = \frac{C_2}{2}\left(\frac{\delta_2}{\delta}\right)^m \mathrm{Re}_h^{-m}, \tag{10.2.68}$$

where $\mathrm{Re}_h = U_0 h/\nu$.

Even though velocity distributions are not expected to be similar, let us for simplicity use the power-law approximation $U/U_\infty = (y/\delta)^{1/n}$, $n = 7$, to obtain

$$\frac{\delta_1}{\delta} = 0.125, \qquad \frac{\delta_2}{\delta} = 0.097, \qquad H = \frac{\delta_1}{\delta_2} = 1.29, \tag{10.2.69}$$

and employ the Blasius values $C_2 = 0.045$ and $m = 1/4$.

Numerical integration of Eq. (10.2.68), subject to $\delta_2(0) = 0$, assuming no initial laminar boundary layer, gives $\delta_2(x)$ shown in Fig. 10.12 (see Ex. 7.2.1). Accordingly, the entry length L, determined from $\delta = h$, or $\delta_2(L) = h\delta_2/\delta$, becomes

$$\frac{L}{h} \simeq 4.42\,\mathrm{Re}_h^{1/4}. \tag{10.2.70}$$

The local skin friction may then be calculated from

$$f = 0.025\,\mathrm{Re}_h^{-1/4}\left(\frac{\delta_2}{h - H\delta_2}\right)^{-1/4} \tag{10.2.71}$$

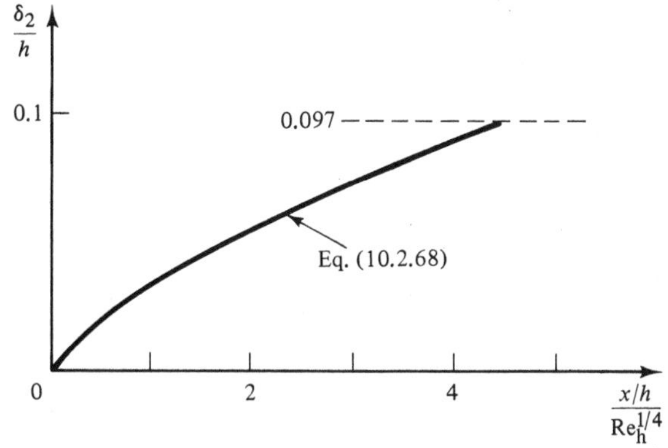

Fig. 10.12 Momentum thickness $\delta_2(x)$ for entrance flow between parallel plates, Ex. 10.2.4

and the pressure gradient, including friction and momentum changes, from

$$-\frac{dP}{dx} = \rho U_\infty \frac{dU_\infty}{dx} = \rho U_0^2 \frac{Hh^2}{(h - H\delta_2)^3} \frac{d\delta_2}{dx}, \tag{10.2.72}$$

which may be further integrated to yield the total pressure drop over the entry length.

Experience shows that the skin friction becomes essentially constant after 10 to 20 diameters, which is predicted approximately by the entry length of Eq. (10.2.70). Recall from Ex. 4.2.2 that the entrance length of laminar flows may be considerably longer.

The present approximate integral solution does not show the complexity of the problem as reflected in nonsimilar velocity profiles and overshoot of centerline velocity. For a discussion of formal solution procedures and comparison with experiment, see Bradshaw, Cebeci, & Whitelaw (1981) and Kline (1981). Here we terminate our study of algebraic models and proceed to one-equation turbulence models.

10.3 ONE-EQUATION MODELS

The models discussed so far in effect attempt to describe turbulent transport processes in terms of only one parameter, the mixing length l, which is provided algebraically. It is not surprising that these models are silent to the level of turbulence, and therefore at best are valid for one (very high) Reynolds number. To improve the models, it seems reasonable to first include the turbulent kinetic energy $K = \frac{1}{2}\overline{u_i u_i}$, being the most basic local parameter characterizing the turbulence.

In the (K, l)-model, starting again with Eq. (10.1.1), using Eq. (10.1.3) and noting that $u \sim K^{1/2}$, we write

$$\nu_T = l_1 K^{1/2}, \tag{10.3.1}$$

prescribe l_1 by an algebraic equation, and determine the distribution of K by solving the balance equation of turbulent kinetic energy together with the equations of the mean flow. Here l_1 is a length scale representative of the large eddies whose energy constitutes the main part of K. In this way the turbulent diffusivity becomes dependent on the turbulent energy transport, hence to some extent on the history of the flow. This idea was first proposed by Kolmogorov (1942), but had to await the advance of computers for practical use. Models based on transport equations, require numerical integration because they lead to systems of highly nonlinear and coupled differential equations.

Assuming l_1 to be prescribed, we need the balance of turbulent kinetic energy (recall Eq. 9.1.9)

$$\begin{aligned} U_j \frac{\partial K}{\partial x_j} &= \frac{\partial}{\partial x_j}\left(-\frac{\overline{pu_j}}{\rho} + 2\nu\overline{u_i s_{ij}} - \frac{1}{2}\overline{u_i u_i u_j}\right) \\ &\quad - \overline{u_i u_j}\frac{\partial U_i}{\partial x_j} - 2\nu\overline{s_{ij}s_{ij}}. \end{aligned} \tag{10.3.2}$$

Closure of Eq. (10.3.2) consists in modeling terms to involve only the variables U_j, K, and l_1. The common approach is to interpret the equation phenomenologically as (see, for example, Ng & Spalding 1972)

$$U_j \frac{\partial K}{\partial x_j} = -\frac{\partial \mathfrak{D}_j}{\partial x_j} + \mathcal{P} - \epsilon,$$

change = diffusion + production − dissipation

and write

$$U_j \frac{\partial K}{\partial x_j} = \frac{\partial}{\partial x_j}\left[\left(\nu + \frac{\nu_T}{\sigma_K}\right)\frac{\partial K}{\partial x_j}\right] + \nu_T\left(\frac{\partial U_i}{\partial x_j} + \frac{\partial U_j}{\partial x_i}\right)\frac{\partial U_i}{\partial x_j} - C_D \frac{K^{3/2}}{l_1}, \qquad (10.3.3)$$

which for nearly parallel (boundary layer) flows reduces to

$$U\frac{\partial K}{\partial x} + V\frac{\partial K}{\partial y} = \frac{\partial}{\partial y}\left[\left(\nu + \frac{\nu_T}{\sigma_K}\right)\frac{\partial K}{\partial y}\right] + \nu_T\left(\frac{\partial U}{\partial y}\right)^2 - C_D\frac{K^{3/2}}{l_1}. \qquad (10.3.4)$$

In the first term on the right of Eq. (10.3.3), the net flux of work associated with viscous stress, pressure fluctuations, and turbulent stress is modeled by gradient diffusion, where $\sigma_K \simeq 1.0$ denotes the turbulent diffusivity ratio of momentum and energy.

The formal derivation of the viscous diffusion of K is based on combining the viscous stress diffusion and dissipation terms, using $s_{ij} = \frac{1}{2}(\partial u_i/\partial x_j + \partial u_j/\partial x_i)$ and $\partial u_j/\partial x_j = 0$,

$$\begin{aligned}\frac{\partial}{\partial x_j}(2\nu\overline{u_i s_{ij}}) - 2\nu\overline{s_{ij}s_{ij}} &= \frac{\partial}{\partial x_j}\left[\nu\overline{u_i\left(\frac{\partial u_i}{\partial x_j} + \frac{\partial u_j}{\partial x_i}\right)}\right] - \frac{1}{2}\nu\overline{\left(\frac{\partial u_i}{\partial x_j}\right)^2} - \frac{1}{2}\nu\overline{\left(\frac{\partial u_j}{\partial x_i}\right)^2} \\ &\quad - \nu\overline{\left(\frac{\partial u_i}{\partial x_j}\right)\left(\frac{\partial u_j}{\partial x_i}\right)} \qquad (10.3.5) \\ &= \frac{\partial}{\partial x_j}\left(\nu\frac{\partial K}{\partial x_j}\right) - \nu\overline{\left(\frac{\partial u_i}{\partial x_j}\right)^2}.\end{aligned}$$

Therefore, ϵ denotes the isotropic viscous dissipation $\nu\overline{(\partial u_i/\partial x_j)^2}$. Note that small scales tend to isotropy at large Reynolds number, and that the viscous diffusion is negligible outside the near-wall region of viscous influence.

The second term on the right of Eq. (10.3.3) expresses the work by turbulent stress associated with the deformation of the mean flow. It is interpreted as the transfer of energy from the mean motion to the turbulent motion through the action of the Reynolds stress. Normally, $-\overline{u_i u_j} > 0$ for $\partial U_i/\partial x_j > 0$, so the production is positive. In view of Eq. (10.1.2), this term needs no modeling.

The third term is the viscous dissipation of turbulent motion. For equilibrium turbulence, from Eqs. (9.2.1) and (9.2.13),

$$\mathcal{P} \sim \frac{u^3}{l} \sim \nu\frac{u^2}{\lambda^2} \sim \epsilon,$$

or in view of $u^2 \sim K$, $\epsilon \sim K^{3/2}/l$ or

$$\epsilon = C_D\frac{K^{3/2}}{l_1}, \qquad (10.3.6)$$

which has been used in the balance of K because of the small and isotropic nature of the dissipation.

In addition to closure, the free constants of any turbulence model must be determined from experiments. Here one looks for flows or flow regions which are universal in some sense and for which the model equations reduce to simple forms. As mentioned in Section 9.6, three cases are commonly referred to and for which the balance of turbulent kinetic energy reduces as follows

$$\text{equilibrium:} \qquad 0 = \mathcal{P} - \epsilon$$

$$\text{decay:} \qquad U_j \frac{\partial K}{\partial x_j} = -\epsilon$$

$$\text{rapid distortion:} \qquad U_j \frac{\partial K}{\partial x_j} = \mathcal{P}.$$

Equilibrium exists in homogeneous shear flow and, more important, in the inertial sublayer of wall-bounded flows. It is therefore the most important condition for shear flow models to satisfy and it will be needed for both one- and two-equation models. Decay of homogeneous isotropic turbulence occurs downstream of a grid, for example, and is readily established in wind tunnels and has been widely studied. Data on decay is satisfied by the two-equation models. Rapid distortion occurs to some extent in very thin shear layers, but is better defined in wind tunnels having rapid changes in cross section. Data on rapid distortion are included into second-order models.

Now the coefficient C_D is determined from experimental data on equilibrium turbulence. Ignoring convective and diffusive terms in Eq. (10.3.4) gives

$$\nu_T \left(\frac{\partial U}{\partial y}\right)^2 = C_D \frac{K^{3/2}}{l_1}, \tag{10.3.7}$$

or, introducing $\partial U/\partial y = \tau_{yx}/(\rho \nu_T)$, where $\tau_{yx}/\rho \simeq \tau_w/\rho = U_\tau^2$ because of the proximity of the wall,

$$K^+ = \frac{K}{U_\tau^2} = C_D^{-1/2} \qquad \text{(equilibrium)}. \tag{10.3.8}$$

Although some data suggest a smaller value, $C_D = 0.09$ is the generally accepted model value for large Reynolds number. Equation (10.3.8) is assumed to be valid at a wall distance $y_0^+ \simeq 30$ to 50, as seen from Fig. 10.13, showing the energy budget for developed pipe flow. Here we may also compared l_1 to the Prandtl mixing length l. Introducing $\partial U/\partial y = (\tau_{yx}/\rho)^{1/2}/l$ from Eqs. (10.1.1) and (10.2.2) on the left of Eq. (10.3.7) gives, with Eq. (10.3.8),

$$\frac{l_1}{l} = C_D^{1/4} \qquad (= 0.548). \tag{10.3.9}$$

In addition to Eq. (10.3.4) the (K, l)-model requires l_1 prescribed algebraically, for example by Eqs. (10.2.4)–(10.2.6). Modifications to the van Driest hypothesis (see Eq. 10.2.5) are suggested by the availability of a local turbulent Reynolds number

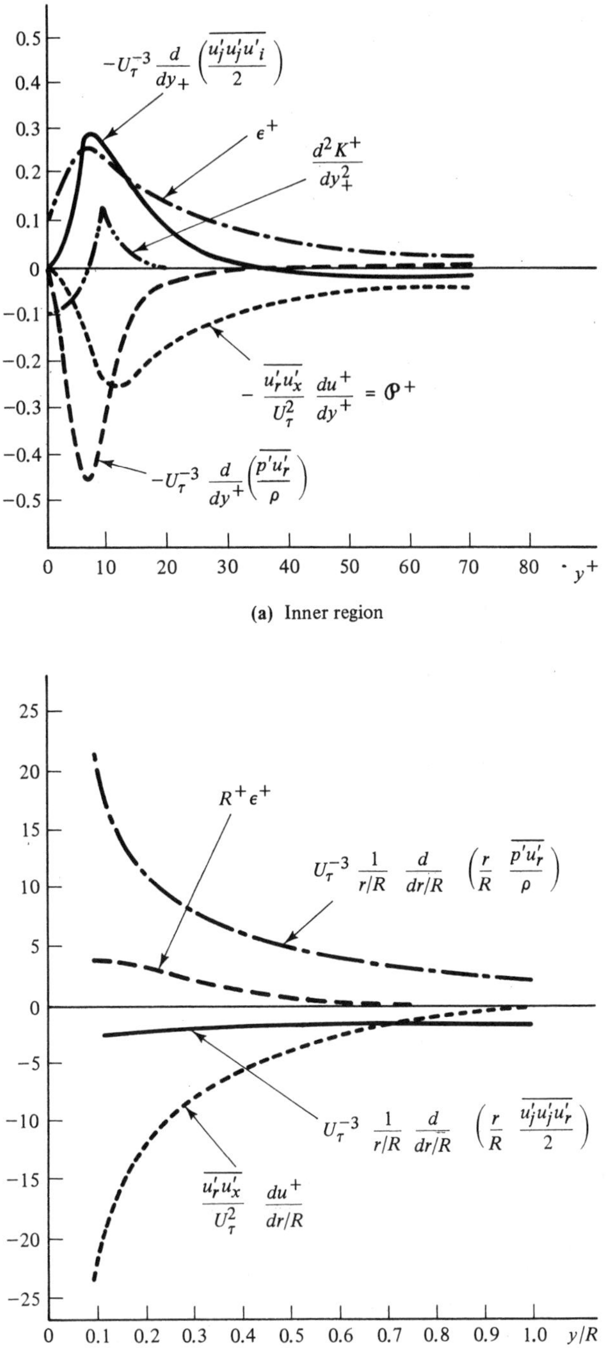

(a) Inner region

(b) Outer region

Fig. 10.13 Energy budget for pipe flow (adapted from Laufer 1954; used by permission)

near a wall

$$R_T = \frac{K^{1/2}y}{\nu}. \tag{10.3.10}$$

Thus, Wolfshtein (1969) proposed that

$$l_{1\nu} = C_D^{1/4}\kappa y[1 - \exp(-A_\nu R_T)]; \qquad A_\nu = 0.016 \tag{10.3.11}$$

to be used in the diffusivity, Eq. (10.3.1), and

$$l_{1D} = C_D^{1/4}\kappa y[1 - \exp(-A_D R_T)]; \qquad A_D = 0.26 \tag{10.3.12}$$

to be used in the dissipation, Eq. (10.3.6). In Eq. (10.2.6) the maximum value of either $l_{1\nu}$ or l_{1D} in boundary layers is replaced by

$$l_{1,\max} = C_D^{5/4}\delta \qquad (= 0.049\delta). \tag{10.3.13}$$

Other one-equation models follow in principle the (K, l)-model described above, with one exception, the Bradshaw stress model for wall boundary layers (Bradshaw, Ferriss & Atwell 1967), which builds on the experimentally supported hypothesis that at any point the quantity $(\tau_{yx}/\rho)/(2K)$ is approximately constant ($\sim$ 0.15). The balance of turbulent kinetic energy is then recast as a balance equation for τ_{yx}, where the modeling of diffusion and dissipation now involves two empirically determined algebraic functions of y/δ.

In regard to boundary conditions it is a common practice to apply the (K, l)-model in two ways. Either the equations are integrated formally from the wall, subject to

$$y = 0: \qquad U = 0, \qquad K = 0, \tag{10.3.14}$$

or—at a considerable saving of computational effort—the integration is started from the equilibrium region (inertial sublayer), say at $y = y_0$, subject to approximate boundary conditions. These so-called wall functions translate the physical conditions at the wall to the equilibrium layer. Treating the sublayer as a Couette flow with negligible pressure gradient corresponds to a matching to the logarithmic law of the wall. For the smooth impermeable wall this may be expressed as (Ng & Spalding 1972)

$$y_0^+ = \frac{U_\tau y_0}{\nu}(\sim 30\text{–}50): \qquad u^+ = \frac{1}{\kappa}\ln(Ey^+)$$
$$K^+ = C_D^{-1/2}; \qquad l_1 = C_D^{1/4}\kappa y, \tag{10.3.15}$$

where $E = 9.0$ and $\kappa = 0.435$. More elaborate wall functions, including pressure gradient and turbulence augmentation (Wolfshtein 1969) and wall suction (Pantankar & Spalding 1970), are available. At a rough wall, Eq. (10.3.15) is replaced by matching to Eq. (10.2.30). Boundary conditions at a line of symmetry present no problem. Similarly, at the free stream of a boundary layer, $y = \delta$, assuming the gradients of K to vanish, Eq. (10.3.3) implies that $U\,\partial K/\partial x = -C_D K^{3/2}/l_1$. Recall that a viscous superlayer here separates turbulent fluid from nonturbulent fluid.

As mentioned in Sections 2.4.3 and 7.4 the numerical solution of incompressible two-dimensional laminar flows is sometimes facilitated by using the (ψ, ω)-formulation of Table 2.6. To employ the corresponding (Ψ, Ω)-formulation to two-dimensional turbulent flows, introduce $\tilde{\psi} = \Psi + \psi$ and $\tilde{\omega} = \Omega + \omega$. Equation (2.4.39), presents no problem, and its average is

$$\nabla^2\Psi + \Omega = 0. \tag{10.3.16}$$

The average vorticity transport equation, however, depending on the way it is derived, is more involved. In one approach, the average of Eq. (2.4.32) leads, for three-dimensional flow, to

$$\underbrace{U_k\frac{\partial\Omega_l}{\partial x_k}}_{\text{change}} = \underbrace{\underbrace{\Omega_k\frac{\partial U_l}{\partial x_k}}_{\text{mean}} + \underbrace{\overline{\omega_k\frac{\partial u_l}{\partial x_k}}}_{\text{fluctuating}}}_{\text{stretching}} + \underbrace{\nu\frac{\partial^2\Omega_l}{\partial x_k\partial x_k} + \frac{\partial}{\partial x_k}(-\overline{u_k\omega_l})}_{\text{diffusion}} + \underbrace{\epsilon_{lij}\frac{\partial}{\partial x_i}[\beta g_j(T - T_0)]}_{\text{buoyancy source}}. \tag{10.3.17}$$

For two-dimensional flow, say parallel to the (x, y)-plane with $\Omega_z = \Omega$, the vortex stretching associated with mean flow vanishes, and we need to model both the Reynolds fluxes of vorticity and the vortex stretching associated with fluctuations. In the other approach, taking the curl of the average of the balance of momentum, Eq. (8.3.9) with τ_{ij} from Eq. (10.1.2), and following the steps leading to Eq. (2.4.32), noting that the curl of a gradient is zero, yields

$$\underbrace{U_k\frac{\partial\Omega_l}{\partial x_k}}_{\text{change}} = \underbrace{\Omega_k\frac{\partial U_l}{\partial x_k}}_{\text{stretching}} + \underbrace{\nu\frac{\partial^2\Omega_l}{\partial x_k\partial x_k} + \epsilon_{lij}\frac{\partial}{\partial x_i}\frac{\partial}{\partial x_k}(2\nu_T S_{kj})}_{\text{diffusion}} + \underbrace{\epsilon_{lij}\frac{\partial}{\partial x_i}[\beta g_j(T - T_0)]}_{\text{buoyancy source}}. \tag{10.3.18}$$

Note that there is no stretching associated with fluctuating motion. For two-dimensional flow all stretching vanishes and the curl of the turbulent diffusion may be conveniently rewritten as

$$\epsilon_{3ij}\frac{\partial}{\partial x_i}\frac{\partial}{\partial x_k}(2\nu S_{kj}) = \frac{\partial^2}{\partial x_k\partial x_k}(\nu_T\Omega) + \mathcal{P}_\omega$$

such that the term

$$\mathcal{P}_\omega = 2\left[\frac{\partial U}{\partial y}\frac{\partial^2\nu_T}{\partial x^2} - \left(\frac{\partial U}{\partial x} - \frac{\partial V}{\partial y}\right)\frac{\partial^2\nu_T}{\partial x\partial y} - \frac{\partial V}{\partial x}\frac{\partial^2\nu_T}{\partial y^2}\right] \tag{10.3.19}$$

depends only on second derivatives of ν_T and may be neglected. Finally, introducing $U = \partial\Psi/\partial y$ and $V = -\partial\Psi/\partial x$, we obtain

$$\begin{aligned}\frac{\partial\Psi}{\partial y}\frac{\partial\Omega}{\partial x} - \frac{\partial\Psi}{\partial x}\frac{\partial\Omega}{\partial y} &= \frac{\partial^2}{\partial x^2}[(\nu + \nu_T)\Omega] + \frac{\partial^2}{\partial y^2}[(\nu + \nu_T)\Omega] + \mathcal{P}_\omega \\ &\quad + \beta\left[\frac{\partial g_y(T - T_0)}{\partial x} - \frac{\partial g_x(T - T_0)}{\partial y}\right].\end{aligned} \tag{10.3.20}$$

Boundary conditions become, replacing Eq. (10.3.14) or Eq. (10.3.15) by

$$y = 0: \qquad \Psi = \Psi_w, \qquad \Omega = -\frac{U_\tau^2}{\nu}, \tag{10.3.21}$$

or, approximately,

$$y = y_0: \qquad \Psi \simeq \Psi_w + \tfrac{1}{2}y_0^2\frac{U_\tau^2}{\nu}, \qquad \Omega = -\frac{U_\tau^2}{\nu}. \tag{10.3.22}$$

Because of the complexity of the (K, l)-model, requiring numerical integration, the following example is confined to a discussion of the behavior of solutions in certain regions of a flow.

EXAMPLE 10.3.1 Reconsider Ex. 10.2.1 of developed channel flow between parallel plates, Fig. 10.1(a). We wish to formulate the (K, l)-model equations for this problem and investigate their behavior.

In terms of dimensionless variables Eqs. (10.2.18), (10.3.1), and (10.3.4) and, say, a single-length-scale van Driest expression of form Eq. (10.3.11), the formulation reduces to

$$(1 + \nu^+)\frac{du^+}{dy^+} = 1 - \frac{y^+}{h^+} \tag{10.2.18}$$

$$\frac{d}{dy^+}\left[\left(1 + \frac{\nu^+}{\sigma_k}\right)\frac{dK^+}{dy^+}\right] + \nu^+\left(\frac{du^+}{dy^+}\right)^2 - C_D\frac{(K^+)^{3/2}}{l_1^+} = 0 \tag{10.3.23}$$

$$\nu^+ = l_1^+(K^+)^{1/2} \tag{10.3.24}$$

$$l_1^+ = C_D^{1/4}\kappa y^+[1 - \exp(-A_1(K^+)^{1/2}y^+)], \tag{10.3.25}$$

subject to $u^+(0) = K^+(0) = 0$, $dK^+(h^+)/dy^+ = 0$.

This highly nonlinear formulation clearly requires numerical integration. The rather complicated nature of the energy budget shown in Fig. 10.13 suggests that solution by approximating functions is not a realistic approach. However, we may study the gross behavior of K by considering three regions of the flow.

First, close to the wall, ignoring turbulent diffusion and production, and for $y/h \ll 1$, Eq. (10.2.18) gives $u^+ = y^+$, and Eq. (10.3.23)

$$\frac{d^2K^+}{dy_+^2} = \frac{C_D(K^+)^{3/2}}{l_1^+} = C_D^{3/4}\frac{K^+}{\kappa A_1(y^+)^2}$$

has the solution

$$K^+ = a_1(y^+)^{a_2}; \qquad a_2 = \frac{1}{2}\left[1 + \left(1 + \frac{4C_D^{3/4}}{\kappa A_1}\right)^{1/2}\right], \tag{10.3.26}$$

where the integration constant a_1 depends on the remaining solution.

Second, in the equilibrium region ($y^+ \sim 30$ to 50) production equals dissipation and Eq. (10.3.23) reduces to Eq. (10.3.7), yielding Eq. (10.3.8). Finally, in the core of the flow, ignoring production, taking the eddy diffusivity to be constant $\nu^+ \simeq h^+/15$, and using Eq. (10.3.24), Eq. (10.3.23) reduces to

$$\frac{d^2K^+}{d(y^+)^2} - \frac{\sigma_K C_D}{(\nu^+)^2}(K^+)^2 = 0. \tag{10.3.27}$$

A power series solution, satisfying symmetry and $K^+(h^+) = K_C^+$,

$$K^+ \simeq K_c^+ + \frac{\sigma_K C_D}{2(\nu^+)^2}(K_c^+)^2\left(1 - \frac{y^+}{h^+}\right)^2, \tag{10.3.28}$$

shows the behavior of K^+, to second order, near the centerline. Note that the constant $\frac{1}{2}\sigma_K C_D/(\nu^+)^2$ is very small, making K^+ nearly uniform near the centerline. The significant

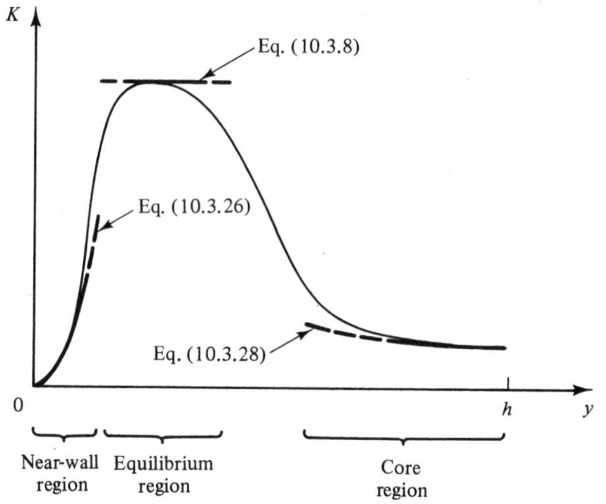

Fig. 10.14 The distribution of turbulent kinetic energy in channel flow (schematic)

increase in K^+ as the equilibrium region is approached (Fig. 10.14) is due to the production that was left out in deriving both of Eqs. (10.3.28) and (10.3.26). For a further study of this model, see Wolfshtein (1969). Let us proceed now to models involving two parameters.

10.4 TWO-EQUATION MODELS

Although the one-equation (K, l)-model represents an improvement over algebraic models, it works satisfactorily only for simple parallel or nearly parallel flows. The model does not have sufficient generality to handle accelerating or decelerating boundary layer flows with strong pressure gradient, nor is it suitable to nonparallel (recirculating) flows. Evidently, the algebraic l-equations rigidly limit the model to apply to flows from which they were derived.

In two-equation models the algebraic equation is therefore replaced by a transport equation, and transport equations for two turbulence parameters are solved together with the equations of the mean flow.

Starting again from Eq. (10.1.1) and $\nu_T \sim lu$ with $u \sim K^{1/2}$, which leads to $\nu_T \sim lK^{1/2}$, we employ Eq. (10.3.6), or $l \sim K^{3/2}/\epsilon$, to obtain

$$\nu_T = C_\mu \frac{K^2}{\epsilon}, \tag{10.4.1}$$

which is the (K, ϵ)-model, C_μ being a constant at high Reynolds number. The choice of ϵ as the second parameter is not unique (we return to this question at the end of this section). However, ϵ appears directly as the sink term in the balance equation of K

and has a clear physical meaning. A balance equation for ϵ can be readily derived from the Navier–Stokes equations, although its closure is subject to some speculation. Nevertheless, the (K, ϵ)-model has been explored and tested on engineering problems more than other two-equation models.

While the mixing length introduced by Prandtl is phenomenological, spectral theory provides a formal definition, for example, that proposed by Rotta (1951):

$$l_R = \frac{\int_0^\infty \frac{1}{\kappa} E(\kappa)\, d\kappa}{\int_0^\infty E(\kappa)\, d\kappa}. \tag{10.4.2}$$

Recall also from Eqs. (8.7.11) and (8.7.14) that

$$K = \int_0^\infty E(\kappa)\, d\kappa, \qquad \epsilon = 2\nu \int_0^\infty \kappa^2 E(\kappa)\, d\kappa. \tag{10.4.3}$$

In this sense, a (K, l)-model as well as the (K, ϵ)-model may be interpreted as involving two moments of the turbulent kinetic energy spectrum, K being weighted toward large scales and ϵ toward small scales. In principle, models involving several moments could be developed. Launder & Schiestel (1978) followed the different approach of dividing the spectral energy range into large and small scales to account for the differences in the dynamics of these scales. This approach involves two sets of (K, ϵ)-variables, requiring the solution of four transport equations. Since the labor of solution then becomes comparable to that of second-order models, one may well favor these.

Now, a transport equation for the isotropic dissipation $\epsilon = \nu\overline{(\partial u_i/\partial x_j)^2}$ is obtained by the operation $\nu\overline{(\partial u_i/\partial x_j)\partial/\partial x_j}$ on the momentum balance (Problem 10-19). The result,

$$\begin{aligned} U_i \frac{\partial \epsilon}{\partial x_i} = &-\frac{\partial}{\partial x_k}\left(\overline{u_k \nu \frac{\partial u_i}{\partial x_j}\frac{\partial u_i}{\partial x_j}} + \frac{\nu}{\rho}\overline{\frac{\partial p}{\partial x_i}\frac{\partial u_k}{\partial x_i}}\right) \\ &- 2\nu\left(\overline{\frac{\partial u_i}{\partial x_j}\frac{\partial u_k}{\partial x_j}} + \overline{\frac{\partial u_j}{\partial x_i}\frac{\partial u_j}{\partial x_k}}\right)\frac{\partial U_i}{\partial x_k} \\ &- 2\nu\overline{\frac{\partial u_i}{\partial x_k}\frac{\partial u_i}{\partial x_j}\frac{\partial u_k}{\partial x_j}} - 2\overline{\left(\nu\frac{\partial^2 u_i}{\partial x_k \partial x_j}\right)^2}, \end{aligned} \tag{10.4.4}$$

may be modeled into the standard form of a balance equation, interpreting the three groups of terms on the right-hand side as the diffusion, the production, and the loss of isotropic dissipation (see Harlow & Nakayama 1967, Hanjalic & Launder 1972a, and Jones & Launder 1972a),

$$U_i \frac{\partial \epsilon}{\partial x_i} = \frac{\partial}{\partial x_k}\left(\frac{\nu_T}{\sigma_\epsilon}\frac{\partial \epsilon}{\partial x_k}\right) + C_{\epsilon 1}\frac{\epsilon}{K}\left(-\overline{u_i u_k} - \frac{2}{3}K\delta_{ik}\right)\frac{\partial U_i}{\partial x_k} - C_{\epsilon 2}\frac{\epsilon^2}{K}. \tag{10.4.5}$$

Note that the production of ϵ is modeled simply as the production of K multiplied (scaled) by ϵ/K. The modeling of the last term of Eq. (10.4.4) merely employs the scaling $\epsilon \sim \nu u^2/\lambda^2$ and $u^2 \sim K$ from Chapter 9, hence

$$\left(\nu\frac{u}{\lambda^2}\right)^2 \sim \frac{\epsilon^2}{u^2} \sim \frac{\epsilon^2}{K}.$$

In Eq. (10.4.5), σ_ϵ denotes the turbulent diffusivity ratio of momentum and dissipation, and $C_{\epsilon 1}$ and $C_{\epsilon 2}$ adjustable coefficients, all to be determined from experiment. At high Reynolds numbers, now conveniently defined as

$$R_T = \frac{K^2}{\nu \epsilon}, \tag{10.4.6}$$

these coefficients are approximately constant, and Eq. (10.4.5) and Eq. (10.3.3) (writing the last term as ϵ) constitute the basic (K, ϵ)-model with $-\overline{u_i u_k}$ given by Eq. (10.1.2) and ν_T by Eq. (10.4.1).

The resulting formulation for nearly parallel flow is summarized in Table 10.3. The numerical values of constants are determined from experimental data, favoring basic universal flows to attain greatest generality of the model.

TABLE 10.3 Boundary layer formulation of the (K, ϵ)-model

$$\frac{\partial U}{\partial x} + \frac{\partial V}{\partial y} = 0 \tag{10.4.7}$$

$$U\frac{\partial U}{\partial x} + V\frac{\partial U}{\partial y} = F_x - \frac{1}{\rho}\frac{\partial P}{\partial x} + \frac{\partial}{\partial y}\left[(\nu + \nu_T)\frac{\partial U}{\partial y}\right] \tag{10.4.8}$$

$$U\frac{\partial K}{\partial x} + V\frac{\partial K}{\partial y} = \frac{\partial}{\partial y}\left(\frac{\nu_T}{\sigma_K}\frac{\partial K}{\partial y}\right) + \nu_T\left(\frac{\partial U}{\partial y}\right)^2 - \epsilon \tag{10.4.9}$$

$$U\frac{\partial \epsilon}{\partial x} + V\frac{\partial \epsilon}{\partial y} = \frac{\partial}{\partial y}\left(\frac{\nu_T}{\sigma_\epsilon}\frac{\partial \epsilon}{\partial y}\right) + C_{\epsilon 1}\frac{\nu_T \epsilon}{K}\left(\frac{\partial U}{\partial y}\right)^2 - C_{\epsilon 2}\frac{\epsilon^2}{K} \tag{10.4.10}$$

$$\nu_T = C_\mu \frac{K^2}{\epsilon}$$

$C_\mu = 0.09$, $\sigma_K = 1.0$, $\sigma_\epsilon = 1.3$

$C_{\epsilon 1} = 1.45$, $C_{\epsilon 2} = 2.0$

First, C_μ is found by considering equilibrium turbulence, production equal to dissipation, in Eq. (10.4.9). Following the steps leading to Eq. (10.3.8) gives

$$\nu_T\left(\frac{dU}{dy}\right)^2 = \epsilon,$$

or

$$C_\mu = C_D = \left(\frac{K}{U_\tau^2}\right)^{-2} = 0.09. \tag{10.4.11}$$

In Eq. (10.4.10) production is not equal to loss of ϵ (which would give $C_{\epsilon 1} = C_{\epsilon 2}$), but diffusion needs to be included in the equilibrium layer. Using Eqs. (10.3.6), (10.3.9), and $l = \kappa y$, yielding $\epsilon \simeq C_\mu^{3/4} K^{1/2}/(\kappa y)$, taking $K \sim$ constant and ignoring convection and viscous effects in Eq. (10.4.10), we obtain

$$C_{\epsilon 1} = C_{\epsilon 2} - \frac{K^2}{C_D^{1/2}\sigma_\epsilon}. \tag{10.4.12}$$

An additional information about $C_{\epsilon 2}$ is obtained by next considering the decay of homogeneous turbulence in the absence of production and diffusion, say behind a grid. Equations (10.4.9) and (10.4.10) then reduce to

$$U\frac{dK}{dx} = -\epsilon$$

$$U\frac{d\epsilon}{dx} = -C_{\epsilon 2}\frac{\epsilon^2}{K},$$

which have the solution

$$\epsilon \sim K^{C_{\epsilon 2}}; \qquad K \sim x^{1-1/C_{\epsilon 2}}. \tag{10.4.13}$$

Experiments indicate that grid turbulence decays as $K \sim x^{-n}$ where during the initial stage (large R_T)

$$1.0 < n < 1.4; \qquad 2.0 > C_{\epsilon 2} > 1.71, \tag{10.4.14}$$

and during the late stage of decay ($R_T \longrightarrow 0$)

$$n \simeq 2.5; \qquad C_{\epsilon 2} \simeq 1.4. \tag{10.4.15}$$

The suggested model constants at large R_T are $C_{\epsilon 2} = 1.92$, $\sigma_\epsilon = 1.3$, and Eq. (10.4.12) with $\kappa = 0.43$ gives $C_{\epsilon 1} = 1.45$. Note that $C_{\epsilon 2} = 2$ and Eq. (10.4.13) lead to $\nu_T =$ constant, which is a good approximation for free flows with negligible production and diffusion (recall Ex. 10.2.2).

To apply this high-Reynolds-number (K, ϵ)-model to wall-bounded flows, we start the integration from the fully turbulent equilibrium layer, using in analogy to Eq. (10.3.15),

$$y_0^+(\sim 30 \text{ to } 50): \qquad u^+ = \frac{1}{\kappa}\ln(Ey^+)$$
$$K^+ = C_D^{-1/2}; \qquad \epsilon^+ \equiv \frac{\nu\epsilon}{U_\tau^4} = (\kappa y^+)^{-1}. \tag{10.4.16}$$

Another iterative practice, avoiding to fix K^+ at the value $C_D^{-1/2}$, consists of specifying (Launder & Spalding 1974)

$$y_0^+(\sim 30 \text{ to } 50): \qquad u^+ C_D^{1/4}(K^+)^{1/2} = \frac{1}{\kappa}\ln[Ey^+ C_D^{1/4}(K^+)^{1/2}]$$
$$\frac{\partial K^+}{\partial y^+} = 0; \qquad \epsilon^+ = (\kappa y^+)^{-1}. \tag{10.4.17}$$

At lines of symmetry and at the free stream of boundary layers, gradients of U, K, and ϵ in Eqs. (10.4.8)–(10.4.10) are assumed to vanish.*

The foregoing (K, ϵ)-model, employing additional empirical information, has been extended to low Reynolds numbers (Jones & Launder 1972a) to permit integration to a wall and prediction of flows that turn laminar. The boundary layer formulation of this low-Reynolds-number (K, ϵ)-model is summarized in Table 10.4. Note that a term, $-2\nu(\partial K^{1/2}/\partial y)^2$ representing the dissipation in the immediate vicinity

*For a rough wall, boundary conditions similar to those of Eq. (10.4.16) may be derived by matching to Eq. (10.2.30).

TABLE 10.4 Boundary layer formulation of low-Reynolds-number (K, ϵ)-model

$$U\frac{\partial K}{\partial x} + V\frac{\partial K}{\partial y} = \frac{\partial}{\partial y}\left[\left(\nu + \frac{\nu_T}{\sigma_K}\right)\frac{\partial K}{\partial y}\right] + \nu_T\left(\frac{\partial U}{\partial y}\right)^2 - \epsilon - 2\nu\left(\frac{\partial K^{1/2}}{\partial y}\right)^2 \quad (10.4.18)$$

$$U\frac{\partial \epsilon}{\partial x} + V\frac{\partial \epsilon}{\partial y} = \frac{\partial}{\partial y}\left[\left(\nu + \frac{\nu_T}{\sigma_\epsilon}\right)\frac{\partial \epsilon}{\partial y}\right] + C_{\epsilon 1}\frac{\nu_T \epsilon}{K}\left(\frac{\partial U}{\partial y}\right)^2 + 2\nu\nu_T\left(\frac{\partial^2 U}{\partial y^2}\right)^2 - C_{\epsilon 2}\frac{\epsilon^2}{K} \quad (10.4.19)$$

$$\nu_T = C_\mu \frac{K^2}{\epsilon}; \qquad R_T = \frac{K^2}{\nu\epsilon}$$

$$C_\mu = 0.09 \exp\left(\frac{-2.5}{1 + R_T/50}\right);$$

$$C_{\epsilon 1} = 1.44; \qquad \sigma_K = 1.0; \qquad \sigma_\epsilon = 1.30$$

$$C_{\epsilon 2} = 2.0[1 - 0.3 \exp(-R_T^2)]$$

of a wall, has been now added to Eq. (10.4.18) to permit use of the more convenient (but fricticious) boundary condition $\epsilon = 0$ at $y = 0$. A production term, $2\nu\nu_T\,(\partial^2 U/\partial y^2)^2$, improving the agreement with data near a wall, has been added to Eq. (10.4.19). In addition, the coefficients C_μ and $C_{\epsilon 2}$ are made to depend on R_T to fit data at low Reynolds numbers, in particular Eq. (11.4.14).

Predictions by Jones & Launder (1972a, b) based on the low-Reynolds-number (K, ϵ)-model are shown in Fig. 10.15 for the flat-plate boundary layer, and in Fig. 10.16 for developed pipe flow. The friction factors predicted for parallel-plate channel flow essentially agree with the data of Beavers, Sparrow & Lloyd (1971) shown in Fig. 10.7. For a comparative study employing revised forms of the low-Reynolds-number (K, ϵ)-model, see Hoffman (1975) and Chien (1982).

Inspection of the literature also shows that different authors suggest slightly different sets of constants for a given model in order to match different experimental data. Further suggestions are expected to appear as more accurate data become available.

Let us return to the choice of variables in two-equation models. Actually, by the free use of the scaling

$$\nu_T \sim lu, \qquad u^2 \sim K, \qquad l \sim \frac{K^{3/2}}{\epsilon},$$

the parameter K may be supplemented by any new parameter of form $K^m l^n$. Although several such models have been proposed, $K^{3/2}l^{-1} \sim \epsilon$ has a clear physical basis and has been most widely studied. However, a different approach is suggested by introducing a characteristic time parameter of the turbulence. Noting that $\omega \sim u/l$ and $K \sim u^2$, hence

$$\nu_T \sim lu \sim \frac{u^2}{\omega} \sim \frac{K}{\omega}, \qquad (10.4.20)$$

or

$$\nu_T = C_\omega \frac{K}{\omega}, \qquad (10.4.21)$$

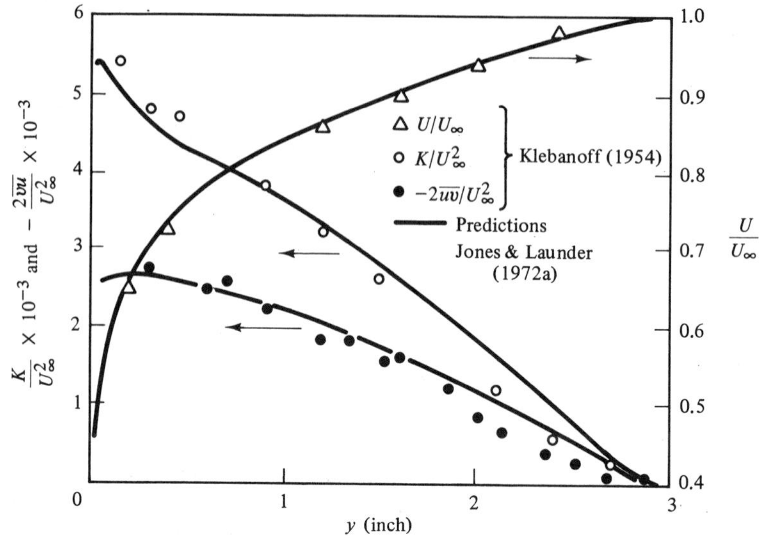

(a) Profiles of mean velocity, turbulent energy and shear stress.

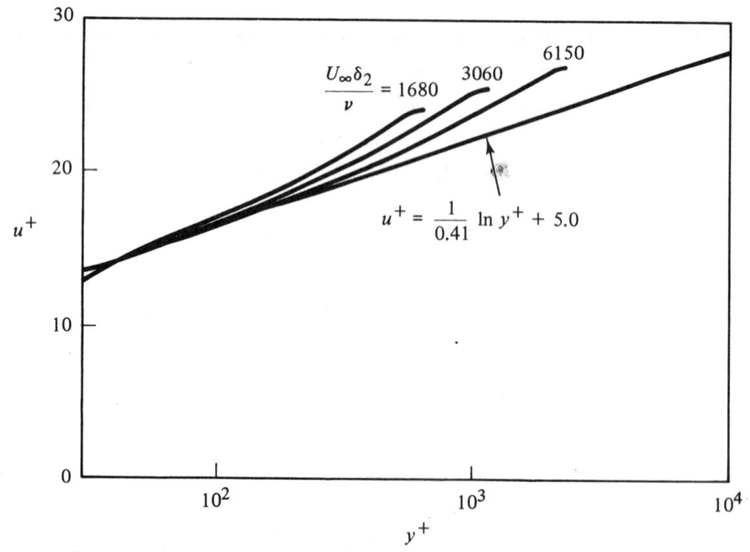

(b) Mean velocity profiles in universal coordinates.

Fig. 10.15 Low Reynolds number (K, ϵ)-model predictions of flat-plate boundary layer flow (adapted from Jones & Launder 1972a; used by permission)

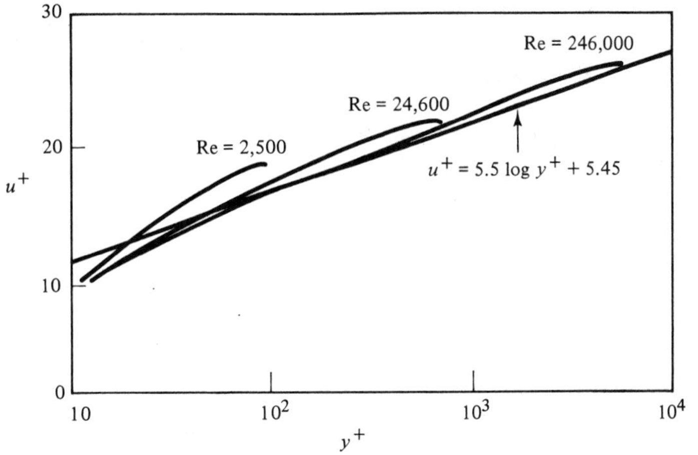

(a) Dependence on Reynolds number

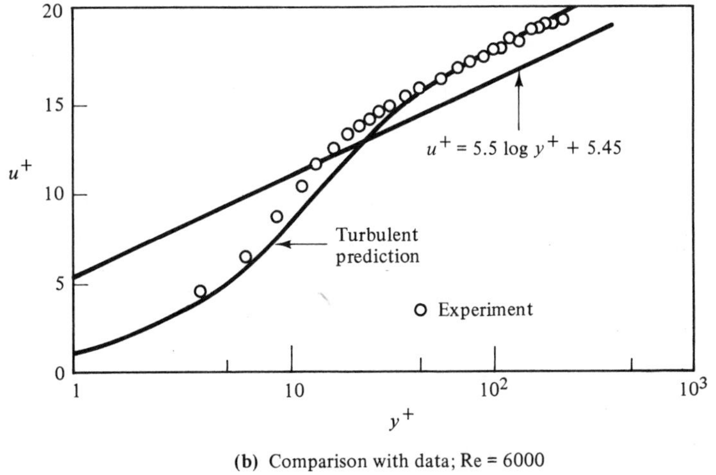

(b) Comparison with data; Re = 6000

Fig. 10.16 Low Reynolds number (K, ϵ)-model predictions for channel flow (adapted from Jones & Launder 1972b; used by permission)

which is the (K, ω)-model. Kolmogorov (1942) first suggested this approach and derived a transport equation for ω, while Saffman (1970) (see also Saffman & Wilcox 1974) proposed a transport equation for ω^2. The (K, ω)-model has been developed to also handle compressible flows. In addition, its boundary conditions at a rough wall has been studied in detail. While ω is unbounded at a smooth wall it approaches a finite value, depending on a universal function of roughness parameter k^+, at a rough wall. Despite the differences in the modeling of terms in transport equations, all two-equation models remain part of one family of models having rather similar characteristics and limitations. Significant improvement requires a different approach. The Reynolds stress closure, still under development, offers improvement at increased computational cost.

Having developed some models for the prediction of turbulent flows, we proceed to the prediction of turbulent heat transfer.

REFERENCES

BEAVERS, G. S., SPARROW, E. M., & LLOYD, J. R., 1971, J. Basic Eng., **93**, 296.

BLASIUS, H., 1913, Forsch. Arb. Ing. Wes., No. 131.

BOUSSINESQ, J., 1877, Mem. Pres. Acad. Sci. XXIII, **46**, Paris.

BRADSHAW, P., FERRISS, D. H., & ATWELL, N. P., 1967, J. Fluid Mech., **28**, 593.

BRADSHAW, P., CEBECI, T., & WHITELAW, J. H., 1981, *Engineering Calculation Methods for Turbulent Flow*, Academic Press, New York.

BROWN, G. T., & ROSHKO, A., 1974, J. Fluid Mech., **64**, 775.

CANTWELL, B. J., 1981, Annu. Rev. Fluid Mech., **13**, 457.

CEBECI, T., & SMITH, A. M. O., 1974, *Analysis of Turbulent Boundary Layers*, Academic Press, New York.

CHIEN, K.-Y., 1982, AIAA J., **20**, 33.

CHOW, C.-Y., 1979, *An Introduction to Computational Fluid Mechanics*, Wiley, New York.

CLARK, R. A., FERZIGER, J. H., & REYNOLDS, W. C., 1979, J. Fluid Mech., **91**, 1.

DALY, B. J., & HARLOW, F. H., 1970, Phys. Fluids, **13**, 2634.

HARLOW, F. H., & NAKAYAMA, P. I., 1967, Phys. Fluids, **10**, 2323.

HANJALIC, K., & LAUNDER, B. E., 1972a, J. Fluid Mech., **51**, 301.

HANJALIC, K., & LAUNDER, B. E., 1972b, J. Fluid Mech., **54**, 609.

HEAD, M. R., 1958, ARCR and M.3152.

HINZE, J. O., 1975, *Turbulence*, McGraw-Hill, New York.

HOFFMAN, G. H., 1975, Phys. Fluids, **18**, 309.

JONES, W. P., & LAUNDER, B. E., 1972a, Int. J. Heat Mass Transfer, **15**, 301.

JONES, W. P., & LAUNDER, B. E., 1972b, ASME Paper 72-HT-20 (or Int. J. Heat Mass Transfer, **16**, 1119, 1973).

KIM, H. T., KLINE, S. J., & REYNOLDS, W. C., 1971, J. Fluid Mech., **50**, 133.

KLEBANOFF, P. S., 1954, NACA TN-3178.

KLINE, A., 1981, J. Fluids Eng., **103**, 243.

KOLMOGOROV, A. N., 1942, IV. Akad. Nauk. SSSR, Ser. Fiz., **56**, 2 (see also Imperial College, Mech. Eng. Dept. Rep. ON/6, 1968).

KUHN, G. D., & NIELSEN, J. N., 1974, AIAA J., **12**, 881.

LAMB, H., 1945, *Hydrodynamics*, Dover, New York.

LAUFER, J., 1954, NACA Rep. 1174.

LAUNDER, B. E., & SCHIESTEL, R., 1978, C. R. Acad. Sci., Paris, **286*A***, 709.

LAUNDER, B. E., & SPALDING, D. B., 1972, *Mathematical Models of Turbulence*, Academic Press, New York.

LAUNDER, B. E., & SPALDING, D. B., 1974, Comput. Methods Appl. Mech. Eng., **3**, 269.

LAUNDER, B. E., REECE, G. J., & RODI, W., 1975, J. Fluid Mech., **68**, 537.

LUDWIEG, H., & TILLMANN, W., 1949, Ing. Arch., **17**, 288.

LUMLEY, J. L., 1977, J. Fluid Mech., **82**, 161.

MOLLO-CHRISTENSEN, E., 1971, AIAA J., **9**, 1217.

MOODY, L. F., 1944, Trans. ASME, **66**, 671.

NG, K. H., & SPALDING, D. B., 1972, Phys. Fluids, **15**, 20.

PATANKAR, S. V., & SPALDING, D. B., 1970, *Heat and Mass Transfer in Boundary Layers*, 2nd ed., International Textbook Company, Scranton, Pa.

PRANDTL, L., 1925, Z. Angew. Math. Mech., **5**, 136.

PRANDTL, L., 1942, Z. Angew. Math. Mech., **22**, 241.

REICHARDT, H., 1951, Arch. Ges. Wärmetechnik, **6**, 129.

REYNOLDS, A. J., 1974, *Turbulent Flows in Engineering*, Wiley, New York.

REYNOLDS, W. C., 1976, Annu. Rev. Fluid Mech., **8**, 183.

ROTTA, J. C., 1951, Z. Phys., **129**, 547; **131**, 51.

SAFFMAN, P. G., 1970, Proc. R. Soc. Lond. A, **317**, 417.

SAFFMAN, P. G., & WILCOX, D. C., 1974, AIAA J., **12**, 541.

SCHLICHTING, H., 1968, *Boundary Layer Theory*, 6th ed., McGraw-Hill, New York.

SCHUMANN, U., 1975, J. Comput. Phys., **18**, 376.

SHAMES, I. H., 1962, *Mechanics of Fluids*, McGraw-Hill, New York.

TAYLOR, G. I., 1932, Proc. R. Soc. Lond. A, **135**, 685.

TENNEKES, H., & LUMLEY, J. L., 1972, *A First Course in Turbulence*, MIT Press, Cambridge, Mass.

TOWNSEND, A. A., 1976, *The Structure of Turbulent Shear Flow*, 2nd ed., Cambridge University Press, New York.

VAN DRIEST, E. R., 1956, J. Aerosp. Sci., **23**, 1007.

WOLFSHTEIN, M., 1969, Int. J. Heat Mass Transfer, **12**, 301.

PROBLEMS

10-1 Derive Eq. (10.2.26) for the friction factor of flow in a smooth pipe of circular cross section.

10-2 Consider developed parallel flow along a smooth porous wall with uniform suction velocity $-V_w$. Show that the fully turbulent velocity distribution in the near-wall region takes the form

$$\frac{2}{v_w^+}[(1 + v_w^+ u^+)^{1/2} - 1] = \frac{1}{\kappa} \ln y^+ + C.$$

10-3 Determine the velocity distribution, flow rate, and friction factor for a fully developed turbulent liquid film of smooth thickness δ flowing down a smooth plane wall inclined the angle α with the vertical.

10-4 Reconsider Problem 10-3, now for a fully rough wall.

10-5 Derive an expression for the friction factor for developed channel flow with fully rough walls. Discuss also the case of flow between parallel plates, one being rough and the other one smooth.

10-6 Show that the classical power law approximation $U/U_c = (y/R)^{1/n}$ is compatible with the empirical Blasius expression for the friction factor for smooth pipe flow, $f_m = C_1 \text{ Re}^{-m}$, with $C_1 = 0.3164$ and $m = \frac{1}{4}$, yielding $n = 7$. Applying the foregoing approach to the flat-plate boundary layer, replacing R by δ, derive Eq. (10.2.55) and $f = 0.045 \text{ Re}_\delta^{-1/4}$, taking $U_m/U_\infty = 0.8$.

10-7 Resolve Ex. 10.2.2, now using Eq. (10.2.2).

10-8 Determine the velocity distribution of turbulent Couette flow for smooth and rough walls, respectively.

10-9 Reconsider Ex. 10.2.3, now using the entrainment procedure of Head. This integral method for a steady incompressible turbulent boundary layer with arbitrary pressure gradient involves the simultaneous solution of two equations,

$$\frac{d}{dx}(U_\infty \delta_2 H_1) = U_\infty F, \tag{a}$$

$$\frac{d\delta_2}{dx} + (H + 2)\frac{\delta_2}{U_\infty}\frac{dU_\infty}{dx} = \frac{1}{2} f, \tag{b}$$

where f, δ_1, and δ_2 are given by Eqs. (10.2.53), (10.2.64), and (10.2.52), shape factors by $H = \delta_1/\delta_2$ and $H_1 = (\delta - \delta_1)/\delta_2$, and empirical functions by

$$F = 0.0306(H_1 - 3.0)^{-0.6169}, \tag{c}$$

$$H_1 = \begin{cases} 0.8234(H - 1.1)^{-0.1287} + 3.3; & H \leq 1.6 \\ 1.5501(H - 0.6778)^{-3.064} + 3.3; & H \geq 1.6, \end{cases} \tag{d}$$

and the friction factor by the Ludwieg & Tillmann (1949) correlation

$$f \equiv \frac{2\tau_w}{\rho U_\infty^2} = 0.246 \times 10^{-0.678H} \times (U_\infty \delta_2/\nu)^{-0.268}. \tag{e}$$

Equation (a) is the proposed empirical law for the velocity V_e of entrainment of free-stream fluid into the boundary layer. Show that $V_e = d(U_\infty \delta_2 H_1)/dx$. Equation (b) is the usual integral balance of momentum. Formulate Ex. 10.2.3 in terms of the Head method, and study Section 7.2 for its numerical solution by the Runge–Kutta procedure.

10-10 Derive Eqs. (10.2.65) and (10.2.63).

10-11 One integral procedure for turbulent boundary layers based on the explicit use of profiles employs (see Kuhn & Nielsen 1974)

$$u^+ = \frac{1}{\kappa}\ln(1+y^+) + 5.1 - (3.39y^+ + 5.1)\exp(-0.37y^+) + \frac{1}{2}u_\beta^+\left[1-\cos\left(\pi\frac{y}{\delta}\right)\right], \tag{a}$$

where $\kappa = 0.4$ and u_β^+ equals u^+ at $y^+ = \delta^+ = \delta U_\tau/\nu$. Equation (a) is employed in Eq. (10.2.65) with f given, for example, by Eq. (e) of Problem 10-9. Discuss the tasks of employing the method to the solution of Ex. 10.2.4, and of comparing it to the Head procedure of Problem 10-9, say obtaining the shape factor relation Eq. (d) of that problem.

10-12 Reconsider the entrance flow of Ex. 10.2.4, now for a channel that diverges slightly so as to keep the core velocity constant and equal to its value at the inlet. This is a normal design practice for a wind tunnel. Estimate the angle of divergence for a mean air velocity of 30 m/s into a 3-m-long by 0.3-m-wide channel, assuming the boundary layer to start fully turbulent from the inlet.

10-13 Derive Eq. (10.3.2).

10-14 Derive Eq. (10.3.20).

10-15 Using the (K, ϵ)-model, explain what happens to the parameters K, ϵ, l, ν_T, and R_T in a homogeneous isotropic turbulent flow of velocity U_0 downstream of a grid. Discuss also the unsteady problem resulting from quickly sweeping a volume of fluid by a grid.

10-16 Determine the asymptotic behavior of K and ϵ near a smooth wall by equating molecular diffusion to dissipation in the equations of K and ϵ, and by assuming a power-law variation $a_i y^{\alpha_i}$ for each variable. Consider both cases, with and without the additional term $2\nu(\partial K^{1/2}/\partial y)^2$.

10-17 Consider a homogeneous shear flow in x, dU/dy = constant. Write the balance equations of Reynolds stresses $\overline{u_1^2}$, $\overline{u_2^2}$, and $\overline{u_3^2}$ to see that the production of turbulent kinetic energy only contributes to $\overline{u_1^2}$, that energy is transferred to $\overline{u_2^2}$ and $\overline{u_3^2}$ by the terms associated with pressure-velocity and with triple-velocity correlations, hence that total energy K is not evenly distributed on the three fluctuating velocity components, as for isotropic turbulence. Experiments show, in fact, that the energy on $\overline{u_1^2}$ is approximately twice that on either one of the other components (Tennekes & Lumley 1972).

10-18 Show that the production term in Eq. (10.3.2) becomes $\frac{2}{3}(K/\rho)\,d\rho/dt$ for isotropic turbulence subject to a volume change (compression or expansion). Discuss the implication for an internal combustion engine, given some turbulence level prior to compression or expansion cycles.

10-19 Derive Eq. (10.4.4).

10-20 Discuss the derivation of Eqs. (10.4.9) and (10.4.10) of the (K, ϵ)-model from the transfer equation of spectral theory (Chapter 9) for the case of homogeneous turbulence.

Chapter 11

HEAT TRANSFER PREDICTION

Because of the importance of fluid motion to convective heat transfer and because of the inherent complexity of turbulent flows, so far we have concentrated on the flow aspects of turbulence. We proceed now to the transport of thermal energy (heat) by these flows.

To determine the temperature distribution and heat transfer from the balance of mean thermal energy, Eq. (9.4.4), the Reynolds fluxes of heat $q = \rho c_p \overline{u_i \theta}$ now need to be modeled. Following the ideas of Table 10.1, we may distinguish between first- and second-order models for these fluxes. In *first-order* models, q_i is expressed by the mean temperature field. In *second-order* models, balance equations for $\overline{u_i \theta}$, generated from the equations of instantaneous motion and temperature, are modeled and solved together with the equations of mean fields. The latter approach is still under development (see, for example, the survey by Launder in Bradshaw 1976), and is beyond the scope of this text.

The common approach of first-order models is to imitate the constitutive relation of an isotropic Fourier fluid, Eq. (2.3.17),

$$\frac{q_i}{\rho c_p} = \overline{u_i \theta} = -a_T \frac{\partial T}{\partial x_i}, \tag{11.1.1}$$

where a_T denotes the *turbulent* (*eddy*) *diffusivity* of heat. What remains is the modeling of this diffusivity. Our present understanding of the thermal energy transport, however, does not allow us to do more than crudely relate the diffusivity of heat a_T to that of momentum ν_T. The most frequently used relation involves the *turbulent Prandtl number*

$$\sigma_T = \frac{\nu_T}{a_T}, \tag{11.1.2}$$

which is assumed to be constant. The success of this approach is limited by the inherent oversimplification embedded in the eddy diffusivity models. The approach also assumes the existence of a similarity between the transfer of energy and momentum. However, using this similarity, we shall obtain the heat transfer indirectly from the available information on momentum transfer.

The foregoing similarity (for boundary conditions as governing equations, as well) is the *analogy* betwen the transfer of heat and momentum, which is introduced in Section 11.1. Examples include classical analogies which make only qualitative reference to distributions of velocity and temperature. Next, in Sections 11.2 and 11.3, we discuss algebraic models and one- and two-equation models. These models also employ analogies but provide temperature distributions as well as heat transfer. Section 11.4 is devoted to buoyancy-driven flows.

11.1 ANALOGY BETWEEN HEAT AND MOMENTUM

We begin with the original analogy due to Reynolds (1874), and proceed to the improvements on this analogy by Prandtl (1910) and Taylor (1919), and by von Kármán (1939) and Martinelli (1947). These analogies have been extensively applied to pipe flows because of the technological importance of such flows. Actually, the pressure gradient in a pipe flow destroys the analogy (see Problem 11-1). Nevertheless, departures from the analogy due to a typical pressure gradient appear to remain well within the limits of experimental error involved with heat transfer measurements.

The starting point for the analogies is the definitions of total fluxes, comprising molecular and turbulent diffusion,

$$\tau_{yx} = \rho(\nu + \nu_T)\frac{dU}{dy}, \tag{11.1.3}$$

$$q_y = -\rho c_p(a + a_T)\frac{dT}{dy}, \tag{11.1.4}$$

where ν_T and a_T are defined by Eqs. (10.1.1) and (11.1.1).

The Reynolds analogy. Assume that the developed flow in a pipe or a channel is composed of a wall layer of thickness δ and a turbulent core (Fig. 11.1). The ratio of Eqs. (11.1.3) and (11.1.4) integrated over the wall layer, assuming the ratios of both fluxes and diffusivities to be constant, yields

$$\frac{\tau_w/\rho}{q_w/\rho c_p} = \frac{\nu + \nu_T}{a + a_T}\frac{U_m}{T_w - T_m}. \tag{11.1.5}$$

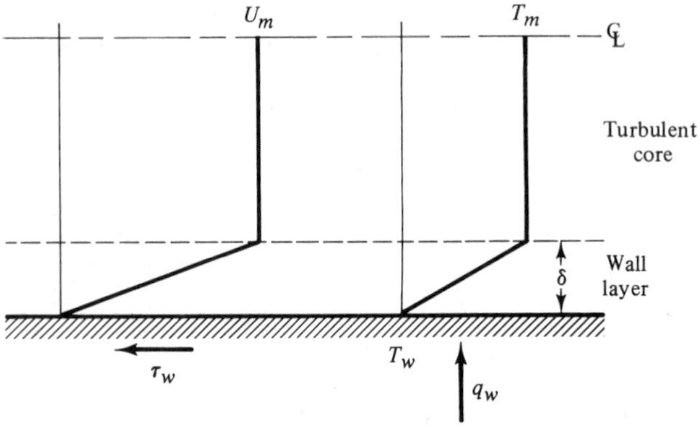

Fig. 11.1 Turbulent pipe flow; the Reynolds analogy

This equation may be rearranged by the definition of the heat transfer coefficient and that of the friction factor,

$$q_w = \bar{h}(T_w - T_m), \qquad f = \frac{\tau_w}{\frac{1}{2}\rho U_m^2}, \tag{11.1.6}$$

as

$$\text{St} = \frac{1}{2} f \frac{a + a_T}{\nu + \nu_T}, \tag{11.1.7}$$

where $\text{St} = \bar{h}/\rho c_p U_m$ is the Stanton number. Equation (11.1.7) reduces to the *Reynolds analogy*,

$$\boxed{\text{St} = \tfrac{1}{2} f,} \tag{11.1.8}$$

provided that the turbulent diffusivities are orders of magnitude larger than molecular diffusivities,

$$\nu_T \gg \nu, \qquad a_T \gg a, \tag{11.1.9}$$

and the turbulent Prandtl number is of the order of unity,

$$\sigma_T = \frac{\nu_T}{a_T} \sim 1. \tag{11.1.10}$$

The first one of these assumptions neglects the wall layer relative to the turbulent core; the second one neglects the differences between velocity and temperature fluctuations in transferring momentum and heat, respectively.

Equation (11.1.8) is an important result, providing information about heat transfer from friction measurements only. That is, when analogy prevails, any analytical, computational, or experimental information on the fluid mechanics of a problem can be related, through this analogy, to the heat transfer of the problem.

Although in reasonable agreement with experimental data for fluids with Prandtl numbers near unity, Eq. (11.1.8) fails for fluids with Prandtl numbers significantly different from unity (recall the Prandtl number ranges of Fig. 3.19). Other analogies have been developed for the purpose of including the effect of Prandtl number to the

Reynolds analogy. We expect these analogies to have the form

$$\mathrm{St} = \tfrac{1}{2} f\, C(\mathrm{Pr}), \tag{11.1.11}$$

where $C(\mathrm{Pr})$ is the Prandtl number correction.

The Prandtl–Taylor analogy. It begins, in a manner identical to that of Reynolds, by assuming that the turbulent flow comprises a viscous wall region (sublayer) and a turbulent core. However, rather than neglecting the sublayer later, the analogy is integrated over each layer separately and the results are then coupled at their interface.

For the sublayer, neglecting the turbulent diffusion of momentum and heat, we have from Eq. (11.1.5) and Fig. 11.2,

$$\frac{c_p \tau_w}{q_w} = \mathrm{Pr}\,\frac{U_\delta}{T_w - T_\delta}. \tag{11.1.12}$$

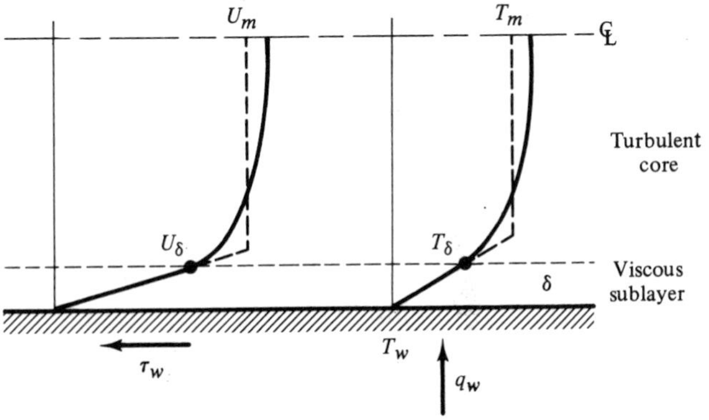

Fig. 11.2 Turbulent pipe flow; the Prandtl–Taylor analogy

For the turbulent core, repeating the steps leading to Eq. (11.1.5), this time neglecting the molecular diffusion of momentum and heat, we have from Fig. 11.2,

$$\frac{c_p \tau_w}{q_w} = \sigma_T \frac{U_m - U_\delta}{T_m - T_\delta}. \tag{11.1.13}$$

Elimination of T_δ between Eqs. (11.1.12) and (11.1.13) leads to

$$T_m - T_w = \frac{q_w}{c_p \tau_w}[U_\delta \mathrm{Pr} + \sigma_T (U_m - U_\delta)],$$

or

$$T_m - T_w = \frac{q_w U_m \sigma_T}{c_p \tau_w}\left[1 + \frac{U_\delta}{U_m}\left(\frac{\mathrm{Pr}}{\sigma_T} - 1\right)\right],$$

or, employing Eq. (11.1.6), to

$$\mathrm{St} = \frac{1}{2} f \left[\frac{1/\sigma_T}{1 + \dfrac{U_\delta}{U_m}\left(\dfrac{\mathrm{Pr}}{\sigma_T} - 1\right)}\right], \tag{11.1.14}$$

where the ratio in brackets is the Prandtl–Taylor correction of the Reynolds analogy for fluids with Prandtl numbers greater than unity. (Why $\mathrm{Pr} \geq 1$?) In Eq. (11.1.14), $\frac{1}{2}f$ and U_δ/U_m may be obtained from velocity data, while σ_T also requires temperature data. Earlier σ_T was assumed to be unity, but a value of 0.9 is now recommended. According to Eqs. (10.2.15), (10.2.16), and (10.2.21), $U_\delta/U_m = (U_\delta/U_\tau)(U_\tau/U_m) = \delta^+(\frac{1}{2}f)^{1/2}$, where $\delta^+ \simeq 11.6$ for a two-layer velocity profile. Consequently, $\mathrm{St}/(\frac{1}{2}f)$ now depends on Re, Pr, and σ_T.

Other analogies. The foregoing analogies make no distinction between centerline values and mean (bulk) values of velocity and temperature in Eq. (11.1.6). This distinction is made in the two following analogies. Using the three-layer velocity distribution of Eq. (10.2.29), *von Kármán* (1939) for $\mathrm{Pr} \gg 1$, assuming $a_T = \nu_T$ and ignoring a and ν in the turbulent core, derived

$$\mathrm{St} = \frac{1}{2}f\frac{c_1}{c_2}\left\{1 + c_1\left(\frac{1}{2}f\right)^{1/2}\left[5(\mathrm{Pr} - 1) + 5\ln\left(\frac{5\mathrm{Pr} + 1}{6}\right)\right]\right\}^{-1}, \qquad (11.1.15)$$

where $c_1 = U_m/U_c$ and $c_2 = (T_w - T_m)/(T_w - T_c)$, the subscripts c, m, and w referring to centerline, bulk, and wall values, respectively. *Martinelli* (1947) for $\mathrm{Pr} \ll 1$, ignoring only ν in the turbulent core, derived

$$\mathrm{St} = \frac{(\frac{1}{2}f)^{1/2}}{5\sigma_T c_2}\left\{\frac{\mathrm{Pr}}{\sigma_T} + \ln\left(1 + 5\frac{\mathrm{Pr}}{\sigma_T}\right) + \frac{1}{2}F\ln\left[\frac{\mathrm{Re}}{60}\left(\frac{1}{2}f\right)^{1/2}\right]\right\}^{-1}, \qquad (11.1.16)$$

where the factor F depends on σ_T, Re, and Pr, as shown in Fig. 11.3 for $\sigma_T = 1$. These analogies in effect employ distributions of momentum diffusivity ν_T derived from

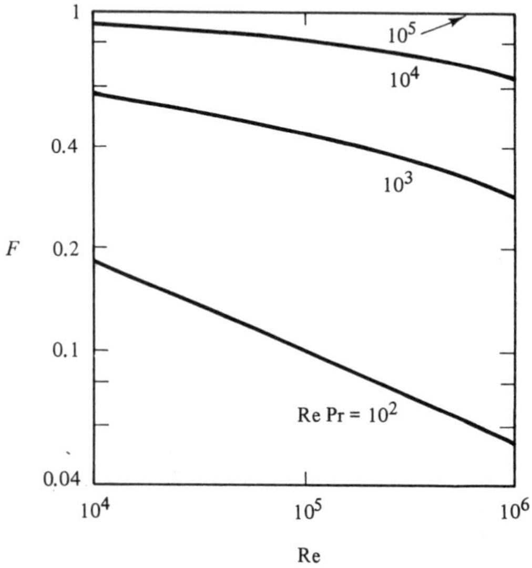

Fig. 11.3 Factor F in Eq. (11.1.16) for $\sigma_T = 1$ (adapted from Rohsenow & Choi 1961; used by permission)

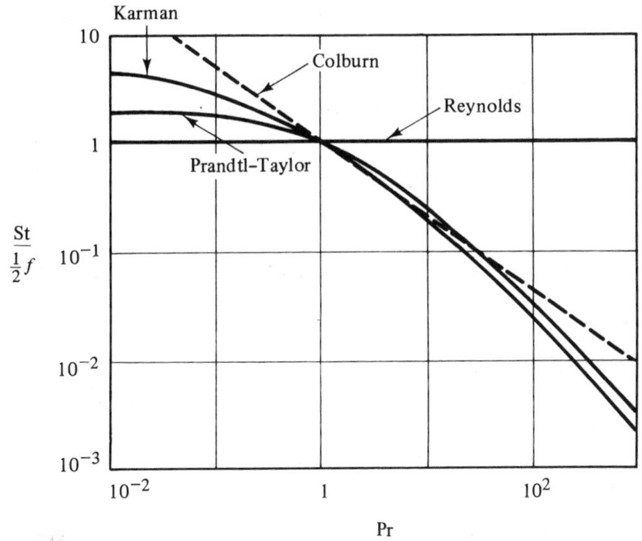

Fig. 11.4 Prandtl number dependency of various analogies

specific velocity distributions in a manner to be discussed in the following section, dealing with algebraic models (Problems 11-2 and 11-3). Figure 11.4 compares three of the foregoing analogies and the experimentally determined Colburn correlation (1933), valid for $\text{Pr} \geq 1$,

$$\text{St} = \tfrac{1}{2} f \, \text{Pr}^{-2/3}, \tag{11.1.17}$$

in terms of their Prandtl number dependence.

Having studied the concept of analogy between heat and momentum transfer, we proceed to the algebraic models for the prediction of heat transfer. However, in Chapter 12 we shall return to these analogies and, rather than using turbulent diffusivities and the turbulent Prandtl number, introduce another analogy based on the microscales of turbulence. By this approach we shall derive Eq. (11.1.17).

11.2 ALGEBRAIC MODELS

As stated in Chapter 10, algebraic models are based on specifying the turbulent diffusivity by algebraic equations. For thermal problems, rather than attempting to derive a special mixing length theory for a_T (see, for example, Cebeci 1973), we employ the analogy $a_T = \nu_T/\sigma_T$, assume σ_T to be constant, and borrow ν_T from velocity distributions, using Eq. (10.1.1). By this approach we may study temperature distributions as well as heat transfer.

Actually, the turbulent Prandtl number σ_T is generally not constant throughout a flow, but depends on position and Reynolds number as well as on the Prandtl number. However, despite many theoretical and experimental investigations (see surveys by Reynolds 1975 and 1976), there remains considerable uncertainty about the variation of σ_T. The existence of a logarithmic law of temperature distributions in near-wall flows suggests that here σ_T is independent of position. Furthermore, experiments indicate the slope of the logarithmic part to be the same for a wide range of Prandtl numbers, implying a constant value $\sigma_T \sim 0.85$ according to Kader & Yaglom (1972). Near the wall ($y^+ < 15$), data suggest that σ_T increases well above unity (see Kays & Crawford 1980). Most studies employ $\sigma_T \sim 0.9$ for wall-bounded flows and $\sigma_T \sim 0.7$ for free flows.

In the following example we study the effect of Prandtl number on the temperature distribution of a simple parallel flow.

EXAMPLE 11.2.1 Consider the steady, developed Couette flow between parallel plates spaced a distance $2h$ apart and kept at temperatures T_w and $T_w + \Delta T$ (Fig. 11.5). We wish to find the temperature distribution and the heat transfer between plates.

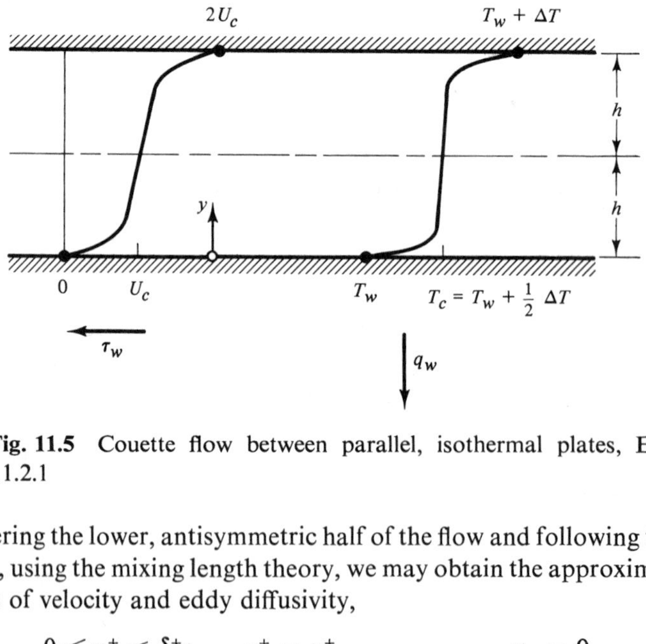

Fig. 11.5 Couette flow between parallel, isothermal plates, Ex. 11.2.1

Considering the lower, antisymmetric half of the flow and following the procedure of Ex. 10.2.1, using the mixing length theory, we may obtain the approximate two-layer distributions of velocity and eddy diffusivity,

$$0 \le y^+ < \delta^+: \qquad u^+ = y^+, \qquad \nu_T = 0, \tag{11.2.1}$$

$$\delta^+ < y^+: \qquad u^+ = \frac{1}{\kappa} \ln y^+ + C, \qquad \frac{\nu_T}{\nu} = \kappa y^+. \tag{11.2.2}$$

Evaluating Eq. (11.2.2) at $h^+ = hU_\tau/\nu = \text{Re}\, U_\tau/U_c$, where $\text{Re} = hU_c/\nu$, we obtain the skin friction $\frac{1}{2}f = (U_\tau/U_c)^2$ from

$$\left(\frac{2}{f}\right)^{1/2} = \frac{1}{\kappa} \ln [\text{Re}\,(\tfrac{1}{2}f)^{1/2}] + C. \tag{11.2.3}$$

Next, the balance of thermal energy, reducing to Eq. (11.1.4) with $q_y = -q_w =$ constant, may be rearranged as

$$1 = \left(\frac{a}{\nu} + \frac{a_T}{\nu}\right)\frac{dT^+}{dy^+} \tag{11.2.4}$$

in terms of the dimensionless variables

$$T^+ = \frac{T - T_w}{T_q}, \qquad T_q = \frac{q_w}{\rho c_p U_\tau}, \qquad y^+ = \frac{yU_\tau}{\nu}, \qquad U_\tau = \left(\frac{\tau_w}{\rho}\right)^{1/2}, \tag{11.2.5}$$

where T_q denotes the *heat flux temperature*, in analogy with the friction velocity U_τ. Note that since both total shear stress and total heat flux, Eqs. (11.1.3) and (11.1.4), are constant across the layer, and since velocity and temperature are constant on walls, this problem satisfies exactly the conditions of similarity in governing equations and boundary conditions. We then employ the local analogy, assuming $\sigma_T = \nu_T/a_T$ to be constant.

Considering the case Pr $\sim$ 1 and $\sigma_T \sim 1$, we expect the diffusion of heat to be dominated by the molecular contribution in the wall layer, Eq. (11.2.1), and by the turbulent contribution in the core, Eq. (11.2.2). Employing ν_T from these equations and integrating Eq. (11.2.5) yields the two-layer temperature distribution,

$$0 \le y^+ < \Delta^+: \qquad T^+ = \mathrm{Pr}\, y^+, \tag{11.2.6}$$

$$\Delta^+ < y^+: \qquad T^+ = \frac{\sigma_T}{\kappa}\ln y^+ + \mathrm{Pr}\,\Delta^+ - \frac{\sigma_T}{\kappa}\ln \Delta^+, \tag{11.2.7}$$

where $\Delta^+ \simeq \delta^+$ denotes the dimensionless thermal sublayer thickness. If we use $\kappa = 0.4$ and $C = 5.5$ the intersection of Eqs. (11.2.1) and (11.2.2) gives $\delta^+ = 11.6$. Kays & Crawford (1980) employ $\kappa = 0.41$ and $C = 5.0$, yielding $\delta^+ = 10.8$, but use $\Delta^+ = 13.2$ to compensate for $\sigma_T > 0.9$ when $y^+ < 15$.

Evaluating Eq. (11.2.7) at the centerline $y^+ = h^+$,

$$T_c^+ = \frac{\sigma_T}{\kappa}\ln h^+ + \mathrm{Pr}\,\Delta^+ - \frac{\sigma_T}{\kappa}\ln \Delta^+, \tag{11.2.8}$$

inserting $(1/\kappa)\ln h^+$ from Eq. (11.2.3), using Eq. (11.2.5) and the definition of the Stanton number, and noting that $T_c^+ = (\frac{1}{2}f)^{1/2}/\mathrm{St}$ for $q_w = \bar{h}[T(h) - T_w]$, we simply recover Eq. (11.1.14), the result of the Prandtl–Taylor analogy.

Now, examining experimental temperature distributions, such as those of Fig. 11.6 covering a range of Prandtl numbers, we make three important observations. The thickness of the thermal sublayer is not constant and equal to δ^+, but depends on Prandtl number. The Prandtl number dependence of Eq. (11.2.7) is stronger for Pr > 1 than is indicated by the data. Data at high Reynolds and Péclet numbers support the existence of a universal logarithmic law of the wall for temperature, such that $\sigma_T \simeq 0.85$ for $\kappa = 0.4$. In fact, similar to the derivation of the logarithmic law of the wall for velocity at high Reynolds numbers, the corresponding law of temperature may be derived from dimensional analysis and matching of laws for regions near and far from the wall. Such an analysis suggests that $T^+ = \varphi(y^+, \mathrm{Pr})$ near the wall,

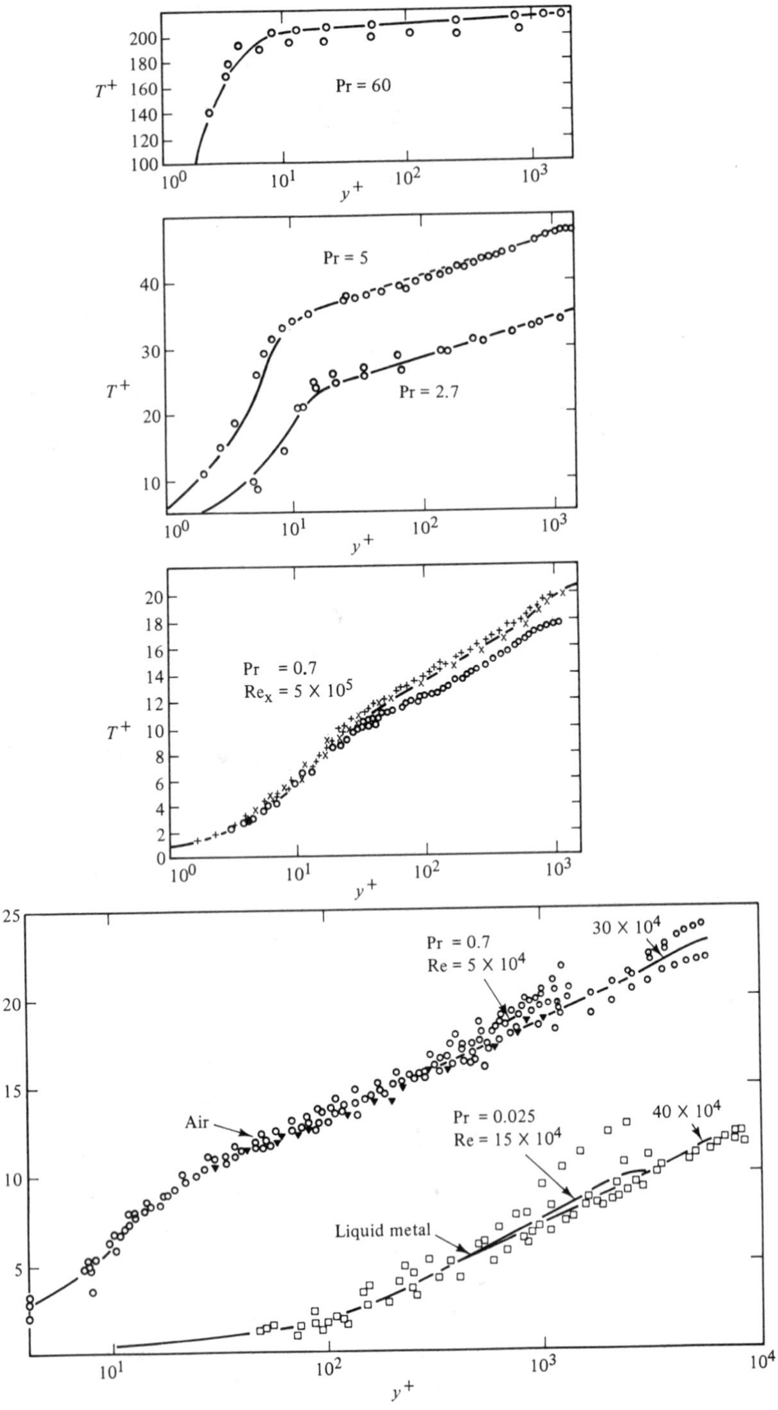

Fig. 11.6 Turbulent temperature profiles and typical data for liquid metals, air, water, and oils (adapted from Kader 1981; used by permission)

while $T_c^+ - T^+ = \Phi_1(y/h, \text{Pr})$ far from the wall near the centerline (or boundary layer edge). For large values of Re and Pe = Re Pr, there exists a region (the thermal inertial sublayer) in which both laws are valid, hence $T^+(y^+, \text{Pr}) = \varphi(y^+, \text{Pr})$. Also, the total temperature difference between wall and centerline has the form $T_c^+ = \Phi_2(h^+, \text{Pr})$. Thus

$$\varphi(y^+, \text{Pr}) + \Phi_1\left(\frac{y}{h}, \text{Pr}\right) = \Phi_2(h^+, \text{Pr}). \tag{11.2.9}$$

By successive differentiations with respect to y and h, it may be shown that

$$\varphi = \alpha \ln y^+ + \beta \tag{11.2.10}$$

$$\Phi_1 = -\alpha \ln\left(\frac{y}{h}\right) + \beta_1 \tag{11.2.11}$$

$$\Phi_2 = \alpha \ln h^+ + \beta + \beta_1, \tag{11.2.12}$$

where β and β_1 depend on Pr. Clearly, Eq. (11.2.10) is the logarithmic law of the wall, Eq. (11.2.11) is the temperature defect law, and Eq. (11.2.12) is the heat transfer law, recalling $T_c^+ = (\frac{1}{2}f)^{1/2}/\text{St}$. The foregoing derivation may be found in Kader & Yaglom (1972).

Let us return to the question of the thickness Δ^+ of the near-wall thermal sublayer (conduction layer), which is dominated by molecular diffusion. When $\text{Pr} > 1$ the influence of molecular diffusion, hence the thickness of the conduction layer, is expected to be less than that of the viscous layer. Conversely, for $\text{Pr} < 1$, the conduction layer is expected to be thicker than the viscous layer. Based on our experience from laminar boundary layers [recall Eq. (4.1.19)], we expect that $\Delta/\delta \sim \text{Pr}^{-1/n}$. This reasoning may require further refinement, however, because the eddy diffusivities a_T and ν_T are nowhere zero, but merely decrease toward zero as the wall is approached. We therefore need improved estimates of the variation of a_T with distance.

For a smooth, explicit distribution of a_T we follow the suggestion of Levich (1962), dividing the near-wall flow into three layers,

$$\frac{a}{\nu} + \frac{a_T}{\nu} = \begin{cases} \text{Pr}^{-1}, & 0 \leq y^+ \leq y_1^+ \quad (11.2.13\text{a}) \\ \text{Pr}^{-1} + \sigma_T^{-1} a_1 (y^+)^m, & y_1^+ \leq y^+ \leq y_2^+ \quad (11.2.13\text{b}) \\ \sigma_T^{-1} \kappa y^+, & y_2^+ \leq y^+. \quad (11.2.13\text{c}) \end{cases}$$

In the thin conduction layer of thickness y_1^+, eddy diffusivity is ignored; in the transition layer which extends to y_2^+, ν_T varies, for $m = 3$, approximately as the van Driest law, Eq. (10.2.5) (see Problem 11-5); in the fully turbulent layer, ν_T varies as the mixing length theory, Eq. (10.2.2).

***The case* Pr $\gtrsim$ 1.** Inserting Eq. (11.2.13) into Eq. (11.2.4), neglecting Pr^{-1} in Eq. (11.2.13b), and integrating yields

$$T^+ = \begin{cases} \text{Pr}\, y^+, & 0 \leq y^+ \leq y_1^+ \quad (11.2.14\text{a}) \\ a_2 - \dfrac{\sigma_T}{(m-1) a_1 (y^+)^{m-1}}, & y_1^+ \leq y^+ \leq y_2^+ \quad (11.2.14\text{b}) \\ \dfrac{\sigma_T}{\kappa} \ln y^+ + \beta, & y_2^+ \leq y^+. \quad (11.2.14\text{c}) \end{cases}$$

It remains to determine y_1^+ and y_2^+ and integration constants a_2 and β. Equating diffusivities at boundaries between adjacent layers in Eq. (11.2.13) gives

$$y_1^+ = \left(\frac{\sigma_T}{a_1 \, \mathrm{Pr}}\right)^{1/m}, \tag{11.2.15}$$

$$y_2^+ = \left(\frac{\kappa}{a_1}\right)^{1/(m-1)} \tag{11.2.16}$$

According to Kader & Yaglom (1972), experiments for channel, pipe, and boundary layer flows suggest that $\kappa = 0.4$, $m = 3$, $\sigma_T^{-1} a_1 \simeq 10^{-3}$, and $\sigma_T \simeq 0.85$, hence $y_1^+ \simeq 10 \, \mathrm{Pr}^{-1/3}$ and $y_2^+ \simeq 21.7$. Figure 11.7 shows the variation of diffusivity implied by Eq. (11.2.13) for $\mathrm{Pr} \gtrsim 1$ [actually for $\mathrm{Pr} > \sigma_T/(\kappa y_2^+) \sim 0.1$].

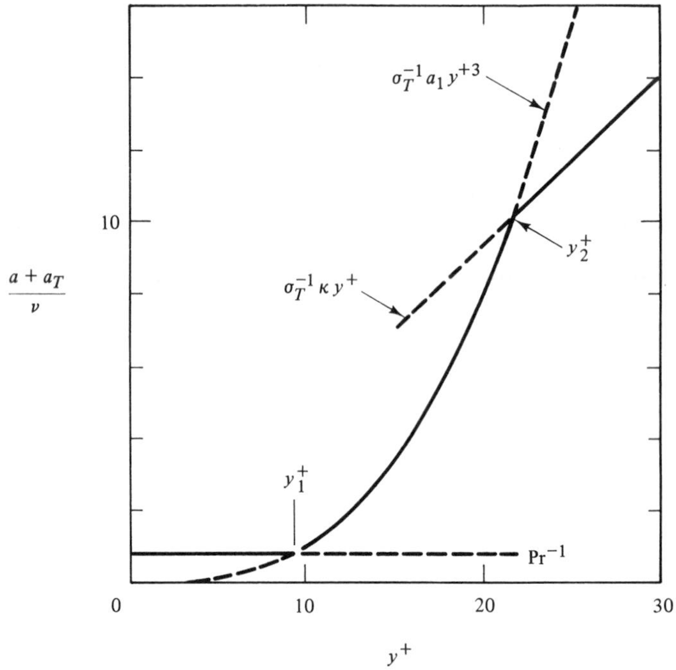

Fig. 11.7 Distribution of effective diffusivity, $(a + a_T)/\nu$ versus y^+, Eq. (11.2.13)

Equating also temperatures from Eq. (11.2.14) at y_1^+ and y_2^+ yields

$$a_2 = a_3 \, \mathrm{Pr}^{(m-1)/m}, \tag{11.2.17}$$

$$\beta = a_3 \, \mathrm{Pr}^{(m-1)/m} - \frac{\sigma_T}{\kappa} \ln y_2^+ - \frac{\sigma_T}{(m-1) a_1 (y_2^+)^{m-1}}, \tag{11.2.18}$$

where

$$a_3 = \frac{m}{m-1}\left(\frac{\sigma_T}{a_1}\right)^{1/m}. \tag{11.2.19}$$

With the foregoing constants we have $a_2 = 15 \, \mathrm{Pr}^{2/3}$ and $\beta \simeq 15 \, \mathrm{Pr}^{2/3} - 7.60$.

A slightly different analysis, including the molecular diffusion of heat in the transition layer, yields

$$a_3 \simeq \frac{m^2+1}{m^2-1}\left(\frac{\sigma_T}{a_1}\right)^{1/m}, \tag{11.2.20}$$

and Kader & Yaglom (1972) furthermore suggest that

$$\beta = 12.5\,\mathrm{Pr}^{2/3} + \frac{\sigma_T}{\kappa}\ln \mathrm{Pr} - 5.3, \tag{11.2.21}$$

which is needed for the heat transfer.

Evaluating Eq. (11.2.14c) at the centerline $y^+ = h^+ = \mathrm{Re}\,(\frac{1}{2}f)^{1/2}$ gives the heat transfer correlation which may be written either to show the correction to the Reynolds analogy, as

$$\mathrm{St} = \frac{1}{2}f\frac{\sigma_T^{-1}}{1+(\frac{1}{2}f)^{1/2}(\beta - C)}, \tag{11.2.22}$$

or, without use of Eq. (11.2.3), as

$$\mathrm{St} = \frac{(\frac{1}{2}f)^{1/2}}{(\sigma_T/\kappa)\ln[\mathrm{Re}\,(\frac{1}{2}f)^{1/2}] + \beta}. \tag{11.2.23}$$

Note that for $\mathrm{Pr} \gg 1$, $\beta \sim 12.5\,\mathrm{Pr}^{2/3} \gg C$ and the denominator of Eq. (11.2.22) is approximately equal to $12.5(\frac{1}{2}f)^{1/2}\,\mathrm{Pr}^{2/3}$, so we obtain the $\mathrm{Pr}^{1/3}$ dependence of the Nusselt number, Eq. (11.1.17).

***The case* Pr $\ll$ 1.** Here, molecular diffusion contributes well into the turbulent region and Fig. 11.7 suggests a two-layer model, matched at y_3^+. Equations (11.2.13a) and (11.2.13b) intersect at $y_3^+ = \sigma_T/(\kappa\,\mathrm{Pr}) = 2.12/\mathrm{Pr}$. Data, however, suggest that $y_3^+ \simeq 5/\mathrm{Pr}$, which yields

$$\beta = \frac{\sigma_T}{\kappa}\ln \mathrm{Pr} - 1.5. \tag{11.2.24}$$

To summarize, temperature distributions calculated from Eq. (11.2.14) are shown in Fig. 11.8 for a number of Prandtl numbers, and are compared to data in Fig. 11.6.

The foregoing analysis also applies to a boundary layer or channel flow, but in channel and pipe flows friction data is usually referred to Reynolds number based on the mean (bulk) velocity $U_m = (1/A)\int_A U\,dA$ rather than on the centerline or free-stream velocity U_c. Similarly, the Nusselt number is referred to the bulk temperature $T_b = (1/AU_m)\int_A UT\,dA$. These changes, affecting only constants in Eqs. (11.2.3) and (11.2.22), are readily made (recall Ex. 10.2.1 and see Problem 11-6). For correlations for pipe flow, see Kader & Yaglom (1972), and for further refinements of temperature distributions, see Kader (1981).

In another approach, Jayatilleke (1959) expressed the integral of Eq. (11.2.4) for the fully turbulent region as

$$T^+ = \sigma_T(u^+ + P_f),$$

or, employing Eq. (11.2.2), as

$$T^+ = \frac{\sigma_T}{\kappa}\ln y^+ + \beta, \qquad \beta = \sigma_T(C + P_f), \tag{11.2.25}$$

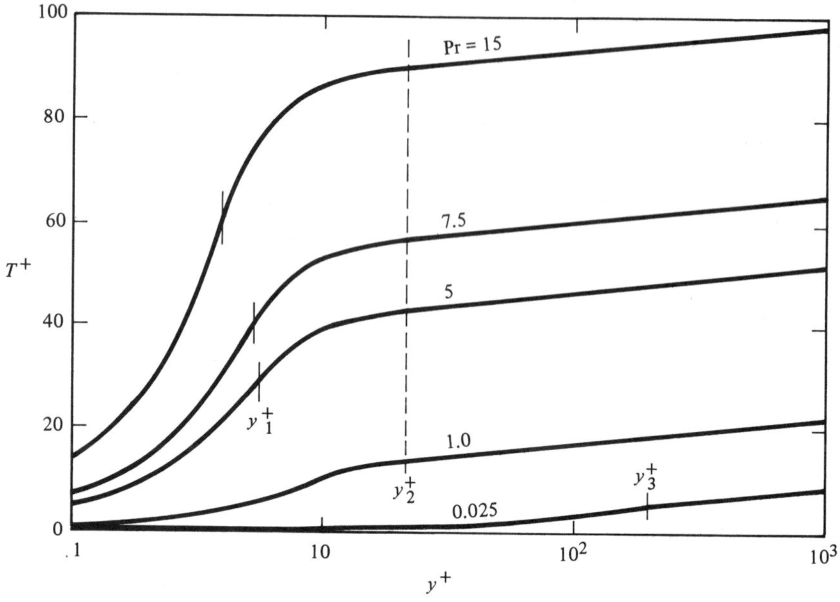

Fig. 11.8 Effect of Prandtl number on temperature distributions, Eq. (11.2.14)

where the empirically based function P_f for smooth surfaces is given by

$$P_f = 9.0\left[\left(\frac{\text{Pr}}{\sigma_T}\right)^{3/4} - 1\right]\left[1 + 0.28 \exp\left(-0.007\,\frac{\text{Pr}}{\sigma_T}\right)\right]. \qquad (11.2.26)$$

These relations have been used extensively for engineering calculations, including those at low Reynolds numbers. As shown in Fig. 11.9, the Prandtl number dependence of β described by Eqs. (11.2.25) and (11.2.21) is quite similar at low Pr, but it deviates increasingly at high Pr such that $\beta \sim \text{Pr}^{3/4}$ according to Eq. (11.2.25) and $\beta \sim \text{Pr}^{2/3}$ according to Eq. (11.2.21). Now, evaluating Eq. (11.2.25) at h^+ yields

$$\text{St} = \tfrac{1}{2}f\{\sigma_T[1 + (\tfrac{1}{2}f)^{1/2}(P_f - C)]\}^{-1}. \qquad (11.2.27)$$

It is interesting to note that the right-hand side of Eq. (11.2.27), for Re $\sim 10^4 - 10^5$ and Pr ~ 1, reflects approximately a $\text{Pr}^{-2/3}$ dependence (recall Eq. 11.1.17).

So far we considered a smooth wall. As the roughness parameter $k^+ = kU_\tau/\nu$ increases through the transition range ($4 \lesssim k^+ \lesssim 70$) into the fully rough range ($k^+ \gtrsim 70$; recall Fig. 10.8) the increase in heat transfer is not as pronounced as that of friction. Empirical heat transfer correlations for these ranges (see, for example, Schlichting 1968) show a continued importance of the Prandtl number. This implies that molecular diffusion of heat is important in the wall region, even for a fully rough wall flow where Eq. (10.2.30) suggests that $\nu_T \simeq \kappa y U_\tau$ is valid right to the wall. It appears that shear stress is transmitted from the rough wall to the fully turbulent fluid through the intervention of pressure fluctuations, a mechanism that has no

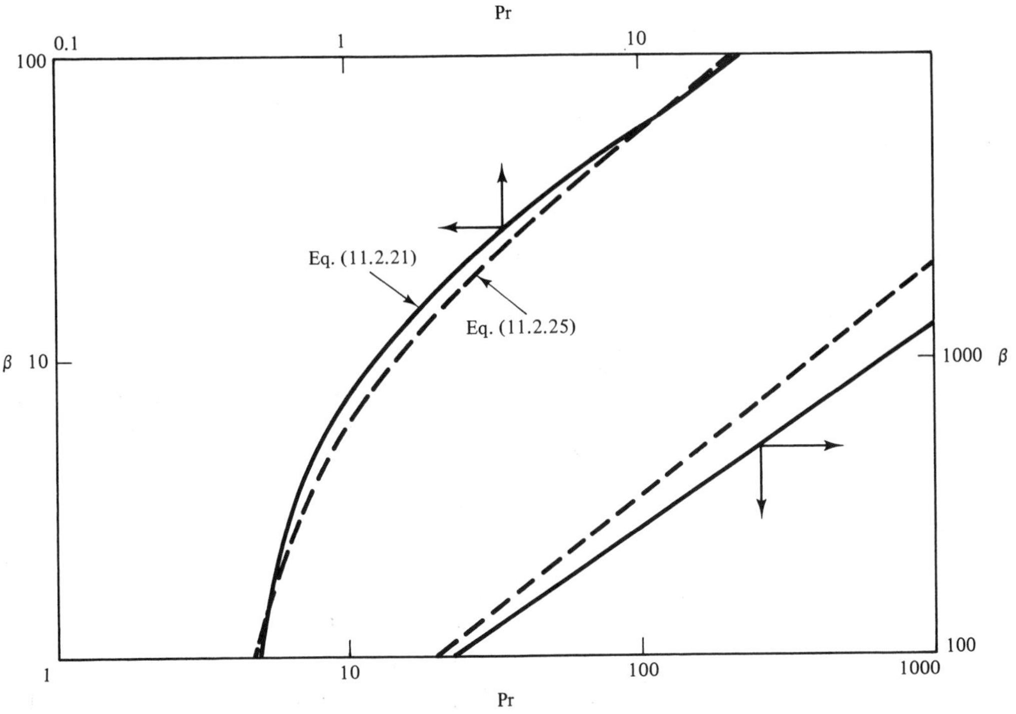

Fig. 11.9 β versus Pr according to Eqs. (11.2.21) and (11.2.25)

counterpart in energy transfer. The analogy between the transfer of momentum and heat therefore fails near the wall and a thermal sublayer need to be included. For a theoretical treatment, see Owen & Thomson (1963) and for correlation of data, see Dipprey & Sabersky (1963).

EXAMPLE 11.2.2 Reconsider Ex. 10.2.2, the mixing layer developing after two parallel flows of velocities U_1 and U_2, and of temperatures T_1 and T_2, are brought into contact (Fig. 10.1b).

The resulting flow being nearly parallel, we employ the formulation of Table 10.2, supplemented by the boundary layer approximation to the balance of thermal energy

$$U\frac{\partial T}{\partial x} + V\frac{\partial T}{\partial y} = \frac{\partial}{\partial y}\left[(a + a_T)\frac{\partial T}{\partial y}\right]. \tag{11.2.28}$$

Here we have used Eq. (11.1.1) and ignored both the axial diffusion and the dissipation.

Following the development in Ex. 10.2.2, ignoring molecular diffusion, introducing $a_T = \nu_T/\sigma_T$, $\sigma_T =$ constant, and ν_T from Eq. (10.2.41), Eq. (11.2.28) becomes

$$U\frac{\partial T}{\partial x} + V\frac{\partial T}{\partial y} = \frac{l\,\Delta U}{\sigma_T}\frac{\partial^2 T}{\partial y^2}, \tag{11.2.29}$$

subject to $T(x, \infty) = T_1$ and $T(x, -\infty) = T_2$. Recalling Eq. (10.2.42), we expect Eq. (11.2.29) to accept a similarity solution provided that $l = b_1 x$, where $b_1 =$ constant.

Introducing

$$T(x, y) = T_m + \Delta T\, g(\eta); \qquad \eta = \frac{y}{x};$$

$$T_m = \tfrac{1}{2}(T_1 + T_2); \qquad \Delta T = \tfrac{1}{2}(T_1 - T_2),$$

considering the upper half of the mixing layer, and employing the velocity field from Ex. 10.2.2, Eq. (10.2.29) becomes

$$\left(\frac{b_1}{\sigma_T}\right) g'' + f g' + \left(\frac{U_m}{\Delta U}\right) \eta g' = 0, \qquad (11.2.30)$$

subject to $g(0) = 0$, $g(\infty) = 1$.

For the special case $\sigma_T = 1$, Eqs. (11.2.30) and (10.2.44) are identical, implying that $g = f$; hence dimensionless temperature and velocity distributions are identical. Otherwise, the two equations require numerical integration for a solution (see Section 7.2). However, a closed-form approximate solution, valid far downstream by the assumption of parallel flow, $\Delta U/U_m \ll 1$, so that Eq. (11.2.30) may be linearized to

$$g'' + \left(\frac{b_1\, \Delta U}{U_m \sigma_T}\right)^{-1} \eta g' = 0,$$

becomes

$$T(x, y) = T_m + \Delta T \operatorname{erf}\left[\left(\frac{2b_1\, \Delta U}{U_m \sigma_T}\right)^{-1/2} \frac{y}{x}\right]. \qquad (11.2.31)$$

This result predicts, for $\sigma_T < 1$, a wider thermal mixing layer than the momentum mixing layer. However, since the latter sustains the eddy diffusivity, we conclude that both layers, according to the present simple model, have the same width.

EXAMPLE 11.2.3 Reconsider Ex. 10.2.3, the flat-plate boundary layer flow. The fluid is at uniform temperature T_∞ and the plate is isothermal at T_w. We wish to find the heat transfer from the plate and the thermal boundary layer thickness $\Delta(x)$.

The integral balance of thermal energy, Eq. (4.1.30), ignoring the dissipation, may be written

$$\frac{d}{dx} \int_0^\Delta \rho c_p U (T - T_\infty)\, dy = q_w, \qquad (11.2.32)$$

or, for constant U_∞ and $T_w - T_\infty$, as

$$\frac{d\Delta_2}{dx} = \mathrm{St}_x, \qquad (11.2.33)$$

where we have introduced the definitions of thermal energy thickness Δ_2 and local Stanton number St_x,

$$\Delta_2 = \int_0^\Delta \frac{U}{U_\infty} \frac{T - T_\infty}{T_w - T_\infty}\, dy, \qquad (11.2.34)$$

$$\mathrm{St}_x = \frac{\bar{h}}{\rho c_p U_\infty}, \qquad \bar{h} = \frac{q_w}{T_w - T_\infty}. \qquad (11.2.35)$$

Clearly, given a relation between St_x and Δ_2 the solution of Eq. (11.2.33) yields $\Delta_2(x)$ and St_x. We expect to find such a relation by use of the analogy on the assumption that velocity and temperature profiles prossess similarity properties,

$$\frac{U}{U_\infty} = f_U\left[\frac{y}{\delta(x)}\right] \quad \text{and} \quad \frac{T - T_\infty}{T_w - T_\infty} = f_T\left[\frac{y}{\Delta(x)}\right].$$

This is the case for the present problem if momentum and thermal boundary layers both start at $x = 0$.

Following the development in Ex. 10.2.3 (recall also Ex. 4.3.1) and employing Eqs. (10.2.54) and (11.1.17), we may rearrange Eq. (11.2.33) as

$$\Delta^m \frac{d}{dx}\left(\frac{\Delta_2}{\Delta}\Delta\right) = \frac{C_2}{2}\mathrm{Pr}^{-2/3}\left(\frac{U_\infty}{\nu}\right)^{-m}\left(\frac{\Delta}{\delta}\right)^m. \tag{11.2.36}$$

Now, on the assumption that Δ_2/Δ and Δ/δ are constants, integration of Eq. (11.2.36) shows that

$$\frac{\Delta}{x} \sim \mathrm{Re}_x^{-m/(1+m)} = \mathrm{Re}^{-1/5}, \tag{11.2.37}$$

which, compared to Eq. (10.2.55), confirms the foregoing assumption, and furthermore yields

$$\frac{\Delta}{\delta} = \mathrm{Pr}^{-2/3}\frac{\Delta}{\Delta_2}\frac{\delta_2}{\delta}. \tag{11.2.38}$$

For an approximate solution, let us employ the power-law profiles

$$\frac{U}{U_\infty} = \left(\frac{y}{\delta}\right)^{1/n}, \qquad \frac{T - T_\infty}{T_w - T_\infty} = 1 - \left(\frac{y}{\Delta}\right)^{1/n}, \qquad n \simeq 7 \tag{11.2.39}$$

in Eqs. (10.2.52) and (11.2.34); hence

$$\frac{\Delta_2}{\Delta} = \left(\frac{\Delta}{\delta}\right)^{1/n}\frac{\delta_2}{\delta}, \qquad \frac{\delta_2}{\delta} = \frac{n}{(n+1)(n+2)} \tag{11.2.40}$$

and Eq. (11.2.38) becomes

$$\frac{\Delta}{\delta} = \mathrm{Pr}^{-2n/[3(n+1)]} \qquad (= \mathrm{Pr}^{-0.58}). \tag{11.2.41}$$

The foregoing development has assumed that the turbulent momentum and thermal boundary layers both start at $x = 0$. Suppose that the wall is only heated to T_w downstream of $x = L$, while it is kept at T_∞ upstream of this point. Then Eq. (11.1.17) no longer applies. Instead, we may use a local analogy, obtaining ν_T from the velocity distribution, and integrate Eq. (11.1.4) to yield the temperature distribution and the local Stanton number expressed by Δ_2. Finally, Eq. (11.2.33) may be integrated. An approximate solution to this problem, using the profiles of Eq. (11.2.39), yields (see Kays & Crawford 1980)

$$\mathrm{St}_x = \frac{1}{2}f\,\mathrm{Pr}^{-2/3}\left[1 - \left(\frac{L}{x}\right)^{9/10}\right]^{-1/9}. \tag{11.2.42}$$

(Actually, following the procedure discussed for channel flow at the end of the next example, one should solve Eq. (11.2.28) and not Eq. (11.1.4) to obtain the temperature distribution.) The case of an arbitrary wall temperature distribution may be treated by the principle of superposition, employing Eq. (11.2.42).

Let us conclude this section by considering the fully developed heat transfer of channel flow.

EXAMPLE 11.2.4 Reconsider Ex. 3.6.1, the fully developed heat transfer in channel flow between parallel plates spaced a distance $2l$ apart and subject to a constant wall heat flux. We wish to determine the temperature distribution and the heat transfer.

Note that the general discussion of Section 3.6 is equally valid for turbulent flow. The balance of thermal energy becomes

$$Ul^2 \frac{\partial T}{\partial x} = \frac{\partial}{\partial \eta}\left[(a + a_T)\frac{\partial T}{\partial \eta}\right], \tag{11.2.43}$$

where $\partial T/\partial x = \partial T_b/\partial x$, according to Eq. (3.6.6), is a constant and $\eta = 1 - y/l$, y being measured from the wall.

Repeating the steps leading to Eqs. (3.6.11) and (3.6.12), but replacing a by $a + a_T$, which is assumed to be a known function of position, we readily obtain the dimensionless temperature distribution

$$\frac{T_w - T}{T_w - T_b} = \frac{\displaystyle\int_\eta^1 \frac{a}{a + a_T}\int_0^\eta \frac{U}{U_m}\,d\eta'\,d\eta}{\displaystyle\int_0^1 \frac{U}{U_m}\left[\int_\eta^1 \frac{a}{a + a_T}\int_0^{\eta'} \frac{U}{U_m}\,d\eta''\,d\eta'\right]d\eta}, \tag{11.2.44}$$

and the heat transfer

$$\mathrm{Nu}_l = \left\{\int_0^1 \frac{U}{U_m}\left[\int_\eta^1 \frac{a}{a + a_T}\int_0^{\eta'} \frac{U}{U_m}\,d\eta''\,d\eta'\right]d\eta\right\}^{-1}, \tag{11.2.45}$$

where $\mathrm{Nu}_l = \bar{h}l/k$ denotes the Nusselt number and U_m the mean (bulk) velocity.

Given the distributions of U/U_m and $a + a_T$, the integrations in Eqs. (11.2.44) and (11.2.45) may be carried out. The calculations are involved and require numerical methods. Note that the increase in complexity, compared to that of Ex. 11.2.1, stems from the fact that the heat flux now varies with position in proportion to $\int_0^\eta U\,d\eta'$ (recall Eq. 3.6.9).

The case of constant wall temperature leads, as in Ex. 3.6.2, to an integral equation for the dimensionless temperature distribution. The solution of this equation requires an iterative numerical procedure.

The case of the developing temperature distribution and heat transfer in fully developed flow (turbulent Graetz problem) is governed by Eq. (11.2.43), where $\partial T/\partial x$ now is not a constant or a known function of position. The solution of this equation may be obtained either by direct numerical integration (see Chapter 7 and the references therein) or, following past practice, by a procedure similar to that of Ex. 3.5.5. Using separation of variables, eigenvalues and constants in the series solution are now computed numerically. Results of this and related cases may be found in Kays & Crawford (1980).

Here we terminate our discussion of algebraic models and proceed to one- and two-equation models.

11.3 ONE- AND TWO-EQUATION MODELS

In our discussion of algebraic models in Section 11.2 we adopted the local analogy $a_T = \nu_T/\sigma_T$, using the previously developed models for ν_T. Actually, from Eq. (11.1.1), employing the large thermal scales of turbulence from Section 9.4,

$$u_\theta \theta \sim a_T \frac{\theta}{l},$$

or

$$a_T \sim l u_\theta, \tag{11.3.1}$$

where l and u_θ denote characteristic scales of length and velocity associated with the turbulent transfer of energy. The velocity scale may also be expressed by the scaling of equilibrium (Eq. 9.4.15)

$$u_\theta \theta \frac{\theta}{l} \sim \epsilon_\theta$$

as

$$u_\theta \sim \frac{\epsilon_\theta l}{K_\theta}, \tag{11.3.2}$$

where $K_\theta = \frac{1}{2}\overline{\theta^2}$ of temperature fluctuations* is the counterpart to $K = \frac{1}{2}\overline{u_i u_i}$ of velocity fluctuations. The dissipation ϵ_θ is the similar counterpart to ϵ. Employing ϵ to scale l according to the relation of equilibrium turbulence (Eq. 9.3.5)

$$u^2 \frac{u}{l} \sim \epsilon$$

gives (recall Eq. 10.3.6)

$$l \sim \frac{K^{3/2}}{\epsilon}. \tag{11.3.3}$$

Inserting Eqs. (11.3.2) and (11.3.3) into Eq. (11.3.1) gives

$$a_T \sim \frac{\epsilon_\theta}{K_\theta} \frac{K^3}{\epsilon^2} = \left(\frac{\epsilon_\theta}{\epsilon} \frac{K}{K_\theta}\right) \frac{K^2}{\epsilon}. \tag{11.3.4}$$

Similar models, such as $a_T \sim K_\theta K/\epsilon_\theta$ (see Plumb & Kennedy 1977), may be found in the literature. For models such as these, we need, in addition to the transport equations of K and ϵ required to obtain ν_T, those of K_θ and ϵ_θ.

First, the K_θ-equation (recall Eq. 9.4.7),

$$U_i \frac{\partial}{\partial x_i}(\tfrac{1}{2}\overline{\theta^2}) = \frac{\partial}{\partial x_j}\left(a \frac{\partial}{\partial x_j} \tfrac{1}{2}\overline{\theta^2} - \tfrac{1}{2}\overline{u_j \theta^2}\right) - \overline{u_j \theta} \frac{\partial T}{\partial x_j} - a\overline{\left(\frac{\partial \theta}{\partial x_j}\right)^2}, \tag{11.3.5}$$

interpreted in the usual way as

$$\begin{array}{ccccccc} U_i \dfrac{\partial K_\theta}{\partial x_i} & = & -\dfrac{\partial \mathfrak{D}_{\theta,i}}{\partial x_i} & + & \mathcal{P}_\theta & - & \epsilon_\theta, \\ \text{change} & = & \text{diffusion} & + & \text{production} & - & \text{dissipation} \end{array}$$

---

*Denoted E_θ in Chapter 9.

is readily modeled (see Corrsin 1952) following the approach used for the K-equation, yielding

$$U_i \frac{\partial K_\theta}{\partial x_i} = \frac{\partial}{\partial x_j}\left[\left(a + \frac{a_T}{\sigma_{K_\theta}}\right)\frac{\partial K_\theta}{\partial x_j}\right] + a_T\left(\frac{\partial T}{\partial x_j}\right)^2 - \epsilon_\theta, \tag{11.3.6}$$

where σ_{K_θ} is to be determined by experiment. The ϵ_θ-equation, which would be similar to Eq. (11.4.5), has not yet been modeled, however. ϵ_θ may be estimated from experimental values for the ratio of the decay of temperature and velocity fluctuations, which so far shows values in the range $(\epsilon_\theta/\epsilon)(K/K_\theta) \sim 1 - 2$.

Presently, while progress is being made on the development of first- and second-order models for heat transfer (see, for example, Launder, p. 231 in Bradshaw 1976, for a review), it is a common practice to put $(\epsilon_\theta/\epsilon)(K/K_\theta) \sim 1$, implying a return to the local analogy. Hence, employing

$$a_T = \frac{C_\mu}{\sigma_T}\frac{K^2}{\epsilon} \tag{11.3.7}$$

in Eq. (11.1.1), we only need the usual two-equation (K, ϵ)-model for either high- or low-Reynolds-number flows, Table 10.3 or 10.4 (Jones & Launder 1973). Similarly, the one-equation model, Eq. (10.3.1), is sufficient if we employ

$$a_T = \frac{l_1 K^{1/2}}{\sigma_T}, \tag{11.3.8}$$

where l_1 is specified algebraically.

Boundary conditions at a wall are handled by starting the integration either from the wall ($y = 0$), where $T = T_w$ or $q_w = -k\,\partial T/\partial y$, or from the equilibrium layer ($y^+ \sim 30 - 50$ for $\mathrm{Pr} \sim 1$), where T is matched to the logarithmic law, Eqs. (11.2.14c) or (11.2.25), in analogy with conditions given by Eq. (10.3.15), (10.4.16), or (10.4.17). At lines of symmetry and at the free stream of boundary layers, gradients of T in Eq. (11.2.28) are assumed to vanish. More elaborate wall functions have been derived by Wolfshtein (1969) based on a detailed numerical study of the (K, l)-model for Couette flows with and without pressure gradient and at various turbulence levels.

During the past decade Eqs. (11.3.7) and (11.3.8) have been used extensively in the numerical computation of convection problems involving boundary layers as well as two- and three-dimensional recirculating flows. Here we refer the reader to the references of Chapter 7 and to the current journal literature on computational fluid dynamics and heat transfer. For a study of the application to channel flow, the reader may extend Ex. 10.3.1 by including the temperature field. As a supplement to this well-behaved flow we examine, in the next example, the application of the (K, l)-model to a one-dimensional approximation of the heat transfer in a separated flow.

EXAMPLE 11.3.1 Consider the separation region behind a blunt body having characteristic length D and temperature T_w. The free-stream velocity and temperature are U_∞ and T_∞, respectively. We wish to estimate the local heat transfer to the fluid in the separated region near the rear stagnation point.

The following analysis, due to Spalding (1967), illustrates some characteristics of the one-equation (K, l)-model and demonstrates how a complicated problem may be studied by a simple model.

As is well known from fluid mechanics, and discussed in Section 4.6, separation from a wall, followed by backflow, occurs in a boundary layer subjected to an excessive adverse (positive) pressure gradient. Examples are: the rear of a blunt body, such as a circular cylinder (Fig. 10.1f); flow past a step; or flow normal to a disk or plate. The shear stress is zero at the point of separation and is small near the surface in the recirculating region. At the outer edge of this region the main flow creates a thin shear layer.

The separation region is modeled as a parallel wall flow. It has essentially zero shear stress in a thin layer $y < y_0$ next to the wall. Outside this layer, for $y_0 < y < y_1$ $(y_1 \gg y_0)$, there exists a layer of constant stress. According to this model, turbulent energy is produced in the outer layer and affects the heat transfer in the inner layer only to an extent determined by its ability to diffuse toward the wall while being dissipated.

The temperature is determined by integrating the balance of thermal energy, Eq. (11.1.4), assuming the heat flux q_w to be constant,

$$\frac{\rho c_p(T_w - T)}{q_w} = \int_0^y \frac{dy}{\mathrm{Pr}^{-1} + \sigma_T^{-1}\nu_T/\nu},$$

subject to $T(y_1) = T_\infty$. Hence, introducing

$$q_w = \bar{h}(T_w - T_\infty), \qquad \mathrm{St} = \frac{\bar{h}}{\rho c_p U_\infty},$$

the heat transfer may be expressed as

$$\mathrm{St} = \frac{\nu}{y_0 U_\infty}\left(\int_0^{y_1/y_0} \frac{d(y/y_0)}{\mathrm{Pr}^{-1} + \sigma_T^{-1}\nu_T/\nu}\right)^{-1}. \tag{11.3.9}$$

Let us now examine the Reynolds number dependence embedded in the factor $y_0 U_\infty/\nu$, assuming the integral of Eq. (11.3.9) to yield the Prandtl number dependence.

Employing $\nu_T = l_1 K^{1/2}$ from Eq. (10.3.1) and a mixing length hypothesis $l_1 = by$, Eq. (10.3.4) for parallel flow becomes

$$b\frac{d}{dy}\left(K^{1/2}y\frac{dK}{dy}\right) - l_1 K^{1/2}\left(\frac{dU}{dy}\right)^2 - a\frac{K^{3/2}}{y} = 0. \tag{11.3.10}$$

In the fully turbulent, zero shear layer $y < y_0$, ignoring the production, Eq. (11.3.10) may be integrated to yield

$$K^{3/2} = C_1 y^m + C_2 y^{-m},$$

or, with $C_2 = 0$ to bound K,

$$\left(\frac{K}{K_0}\right)^{3/2} = \left(\frac{y}{y_0}\right)^m, \tag{11.3.11}$$

where $m = (\frac{3}{2}a/b)^{1/2}$. It is now assumed that the Reynolds number of turbulence $R_0 = y_0 K_0^{1/2}/\nu$ is a constant (on the order of 80 based on near-wall shear flows). Employing Eq. (11.3.11) evaluated at y_1, where $K = K_1$, and the definition of R_0 to eliminate y_0 and K_0 from the factor $y_0 U_\infty/\nu$ and introducing the characteristic length D, we obtain

$$\frac{\nu}{y_0 U_\infty} = R_0^{-3/(m+3)} \left(\frac{K_1^{1/2}}{U_\infty}\right)^{3/(m+3)} \left(\frac{D}{y_1}\right)^{m/(m+3)} \left(\frac{U_\infty D}{\nu}\right)^{-m/(m+3)}. \tag{11.3.12}$$

Omitting the remaining analysis and comparison with data, which suggests that $m \simeq 2$, Eq. (11.3.9) may be written as

$$\text{St}\,\text{Re}_\text{D}^{0.4} = \text{const.} \left(\frac{K_1^{1/2}}{U_\infty}\right)^{0.6} \left(\frac{D}{y_1}\right)^{0.4} f(\text{Pr}), \tag{11.3.13}$$

implying that

$$\text{Nu}_\text{D} \sim \text{Re}_\text{D}^{0.6}, \tag{11.3.14}$$

in essential agreement with heat transfer data. .

The value of $K^{1/2}/U_\infty$ depends on the nature of the turbulence generating shear layer away from the wall. It is on the order of 0.1 for boundary layer separation, but greater (~ 0.25) for free shear layers existing in the flows over a step or normal to a plate. D/y_1 depends on the flow pattern of the separation region, but is typically on the order of 10.

Here we terminate our study of one- and two-equation models of forced convection and turn to buoyancy-driven flows.

11.4 *BUOYANCY-DRIVEN CONVECTION*

Recalling the general formulation of buoyancy-driven flow given in Section 2.4.2, let us formulate the governing equations for the turbulent case. Our starting point is Eq. (2.4.23), which governs the instantaneous motion

$$\frac{\partial \tilde{u}_j}{\partial t} + \tilde{u}_i \frac{\partial \tilde{u}_j}{\partial x_i} = \nu \frac{\partial^2 \tilde{u}_j}{\partial x_i \partial x_i} - \frac{1}{\rho_0} \frac{\partial}{\partial x_j}(\tilde{p} - p_0) + \beta g_j(\tilde{T} - T_0), \tag{11.4.1}$$

where $g_j = -f_j$ is the negative body force per unit mass and T_0 is the reference temperature of neutral buoyancy. The difference between actual and hydrostatic pressures, $\tilde{p} - p_0$, is negligible, except for combined (forced and buoyancy-driven) flows. Inserting $\tilde{T} = T + \theta$ and averaging, Eq. (11.4.1) becomes, for the steady case,

$$U_i \frac{\partial U_j}{\partial x_i} = \frac{\partial}{\partial x_i}\left(\nu \frac{\partial U_j}{\partial x_i} - \overline{u_i u_j}\right) - \frac{1}{\rho_0} \frac{\partial}{\partial x_j}(P - p_0) + \beta g_j(T - T_0). \tag{11.4.2}$$

Note that, within the Boussinesq approximation, the buoyancy contributes to the mean motion through the body force associated with the distribution of mean temperature.

The balance of thermal energy is unchanged, say, Eq. (11.2.28) for nearly parallel flow. However, let us examine the effect of buoyancy on the turbulence fields, specifically its effect on Reynolds stresses and Reynolds fluxes of heat in the context of first-order models.

First, repeating the steps leading to the balance of the dissipation shows that the ϵ-equation should be modified only by the addition of the term $\beta g_j \overline{(\partial u_j/\partial x_k)(\partial \theta/\partial x_k)}$. This term, however, is considered to be negligible and we retain Eq. (10.4.10) for nearly parallel flow.* Next, repeating the steps leading to the balance of turbulent kinetic energy, $K = \frac{1}{2}\overline{u_i u_i}$ and the balance of $K_\theta = \frac{1}{2}\overline{\theta^2}$ yields

$$U_i \frac{\partial K}{\partial x_i} = -\frac{\partial \mathfrak{D}_{K,i}}{\partial x_i} - \overline{u_i u_j}\frac{\partial U_j}{\partial x_i} + \beta g_j \overline{u_j \theta} - \epsilon, \tag{11.4.3}$$

$$U_i \frac{\partial K_\theta}{\partial x_i} = -\frac{\partial \mathfrak{D}_{\theta,i}}{\partial x_i} - \overline{u_j \theta}\frac{\partial T}{\partial x_j} - \epsilon_\theta, \tag{11.4.4}$$

where the diffusion terms are given in Eqs. (10.3.3) and (11.3.6), respectively. Note that the K_θ-equation is unchanged, but that the K-equation has been modified now by the addition of the production term $\beta g_j \overline{u_j \theta} = \beta g_j q_j/\rho c_p$. This term is positive for a positive turbulent heat flux opposing the direction of the body force, while it is negative when the heat flux is in the direction of the body force.

The implications of the foregoing added effect are perhaps best known from the atmospheric boundary layer of horizontal flow subject to a vertical thermal stratification. The buoyant production of K is positive for a decreasing temperature with height. This situation corresponds to an unstable stratification, and turbulence may be maintained even in the absence of shear production. Conversely, production of K is negative for an increasing temperature with height, which represents a stable stratification. Here, although the production of K_θ is never negative, buoyancy may significantly suppress the turbulence. For most technical flows, because of their small size compared to the atmosphere, buoyancy effects on the turbulence structure are believed to be small. Yet, while turbulent eddies caused by buoyancy contribute relatively little to momentum exchange, they may be effective for heat exchange.

Now, consider a vertically driven flow resulting from a horizontal thermal stratification, say the fluid adjacent to the vertical isothermal plate, or the fluid between two vertical plates at unequal temperatures (Fig. 11.10). Here, for parallel or nearly parallel flow, the production terms in Eqs. (11.4.3) and (11.4.4) are $-\overline{vu}\,\partial U/\partial y + \beta g \overline{u\theta}$ and $-\overline{v\theta}\,\partial T/\partial y$, respectively. Since the heat flux in y, $q_y/\rho c_p = \overline{v\theta}$, is usually positive for $\partial T/\partial y$ negative, there is a positive production of K_θ once a turbulent flow has been established. Similarly, $-\overline{vu}\,\partial U/\partial y \simeq \nu_T(\partial U/\partial y)^2$ is positive. At first glance, the buoyancy production $\beta g \overline{u\theta}$ to K should be zero for such flows. Yet the term is usually modeled as a positive production

$$\beta g \overline{v\theta} \simeq -\frac{\nu_T}{\sigma_T}\beta g \frac{\partial T}{\partial y}$$

(see, for example, Lin & Churchill 1978) on the argument that due to near isotropy in the fully turbulent region, $\overline{u\theta}$ is approximately equal to $\overline{v\theta}$. For a further study of

*Some investigators (Plumb & Kennedy 1977) add the production term $C_{\epsilon 3}\beta g_j(\epsilon/K)\overline{u_j\theta}$, where $\overline{u_j\theta} = C_4(K_\theta K)^{1/2}$, and solve the K_θ-equation, with $\epsilon_\theta \sim \epsilon K_\theta/K$.

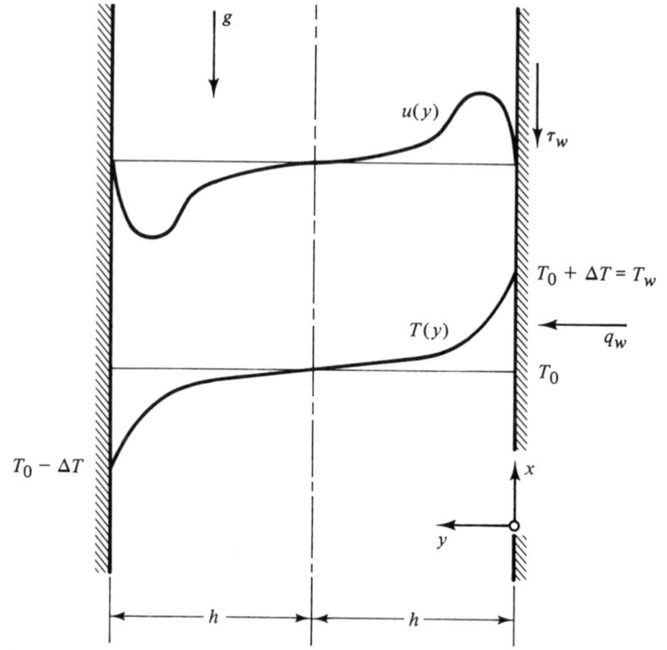

Fig. 11.10 Natural convection in vertical slot with isothermal walls at unequal temperatures, Ex. 11.4.1

buoyancy-driven convection, we refer the reader to Townsend (1976) and for its second-order modeling to Launder (p. 231 in Bradshaw 1976).

EXAMPLE 11.4.1 Reconsider the steady developed part of Ex. 3.4.2, the buoyancy-driven flow between vertical isothermal plates spaced a distance $2h$ apart and kept at temperatures $T_0 + \Delta T$ and $T_0 - \Delta T$, respectively (Fig. 11.10). We seek the velocity and temperature distributions and the heat transfer between plates.

The differential formulation of the parallel, developed flow is

$$0 = \frac{d}{dy}\left[(\nu + \nu_T)\frac{dU}{dy}\right] + \beta g(T - T_0) \tag{11.4.5}$$

$$q_w = -(a + a_T)\frac{dT}{dy}, \tag{11.4.6}$$

subject to

$$U(0) = U(h) = 0, \qquad T(0) = T_w = T_0 + \Delta T, \qquad T(h) = T_0, \tag{11.4.7}$$

where conditions of antisymmetry follow from the conservation of mass

$$\int_0^{2h} U(y)\, dy = 0.$$

Introducing the usual dimensionless variables according to Eq. (11.2.5) and the Grashof number $G = \beta g \, \Delta T \, h^3/\nu^2$, Eqs. (11.4.5) and (11.4.6) become

$$0 = \frac{d}{dy^+}\left[(1 + \nu^+)\frac{du^+}{dy^+}\right] + \frac{G}{h_+^3}\left(1 - \frac{T^+}{T_0^+}\right) \tag{11.4.8}$$

$$1 = (\mathrm{Pr}^{-1} + \sigma_T^{-1}\nu^+)\frac{dT^+}{dy^+}, \tag{11.4.9}$$

subject to

$$u^+(0) = u^+(h^+) = 0, \qquad T^+(0) = 0, \qquad T^+(h^+) = T_0^+.$$

To proceed, we need an appropriate turbulence model for ν^+. So far there is no well-established and generally applicable mixing length theory for buoyancy-driven flows. With the one-equation (K, l)-model one may have to resort to intuitive estimates for the spatial variation of l, as done, for example, by Kaviany & Seban (1981) for an unsteady problem. In this approach, or if the two-equation (K, ϵ)-model is used, solution requires numerical methods.

An algebraic solution may be obtained by employing ν^+ from a two-layer model, following the procedure of Ex. 10.2.1 (see Problem 11-16). This approach may be justified for a confined buoyancy driven flow because of its resemblance to forced channel flow, although the buoyancy force, unlike the pressure gradient of channel flow, is not uniform over the flow. The parameters h^+ and T_0^+ are determined as part of the solution, and defining $\mathrm{Nu} = q_w h/(k \, \Delta T)$, we obtain the heat transfer as

$$\mathrm{Nu} = \mathrm{Pr}\frac{h^+}{T_0^+}. \tag{11.4.10}$$

Actually, the convection in vertical slots of finite height L depends on the ratio $L/2h$. The problem is more complex than the one described above since there will be a vertical temperature gradient along the centerline (see, for example, the recent data by Yin, Wung & Chen 1978 for moderate Grashof numbers). The related problem of the effect of buoyancy on upwards forced flow in a slot has been studied by Nakajima & Fukui (1980).

EXAMPLE 11.4.2 Reconsider Ex. 4.3.3, the steady natural convection from a vertical, isothermal flat plate at temperature T_w in a quiescent fluid at T_∞ (Fig. 4.14); now for turbulent flow, $\mathrm{Ra}_L > 10^9$. We wish to find the heat transfer from the plate.

Lacking reliable algebraic models, let us consider the integral formulation of Ex. 4.4.3, but use an empirical friction relation and the Reynolds analogy to compensate for the fact that approximate profiles for velocity and temperature usually cannot be related to shear stress and heat flux at the wall. (For a discussion of velocity and temperature distributions of this problem based on scaling arguments, see George & Capp 1979.)

Following Eckert & Jackson (1951), considering $\mathrm{Pr} \sim 1$, so that $\delta \sim \Delta$, we introduce the profiles

$$U = U_1 \eta^{1/7}(1 - \eta)^4, \qquad T - T_\infty = (T_w - T_\infty)(1 - \eta^{1/7}), \tag{11.4.11}$$

where $\eta = y/\delta$, into Eqs. (4.3.43) and (4.3.44), replacing their last terms by wall fluxes, and obtain

$$\frac{d}{dx}(U_1^2\delta)\int_0^1 [\eta^{1/7}(1-\eta)^4]^2\,d\eta = \beta g(T_w - T_\infty)\delta\int_0^1 (1-\eta^{1/7})\,d\eta - \frac{\tau_w}{\rho} \tag{11.4.12}$$

$$(T_w - T_\infty)\frac{d}{dx}(U_1\delta)\int_0^1 \eta^{1/7}(1-\eta)^4(1-\eta^{1/7})\,d\eta = \frac{q_w}{\rho c_p}. \tag{11.4.13}$$

To proceed, we assume the validity of the Blasius law of friction, Eq. (10.2.54), written in terms of the reference velocity U_1,

$$f = \frac{\tau_w}{\frac{1}{2}\rho U_1^2} = 0.045\left(\frac{\delta U_1}{\nu}\right)^{-1/4}, \tag{11.4.14}$$

and the Colburn correlation, Eq. (11.1.17),

$$\text{St Pr}^{2/3} = \tfrac{1}{2}f. \tag{11.4.15}$$

Then Eqs. (11.4.12) and (11.4.13) prove to be satisfied by solutions of the form

$$U_1 = C_1 x^m, \qquad \delta = C_2 x^n,$$

with $m = 0.5$ and $n = 0.7$, yielding

$$U_1 = 1.185\left(\frac{\text{Gr}_x}{1 + 0.494\,\text{Pr}^{2/3}}\right)^{1/2}\frac{\nu}{x} \tag{11.4.16}$$

$$\frac{\delta}{x} = 0.565\left(\frac{1 + 0.494\,\text{Pr}^{2/3}}{\text{Gr}_x\,\text{Pr}^{16/3}}\right)^{1/10}, \tag{11.4.17}$$

where $\text{Gr}_x = \beta g(T_w - T_\infty)x^3/\nu^2$ denotes the local Grashof number. Recalling Eq. (4.3.50), we have

$$\delta \sim \begin{cases} x^{1/4} & \text{laminar} \\ x^{7/10} & \text{turbulent.} \end{cases} \tag{11.4.18}$$

Finally, employing Eqs. (11.4.16) and (11.4.17) in Eq. (11.4.15) gives the local heat transfer

$$\text{Nu}_x = \frac{\bar{h}x}{k} = 0.0295\,\text{Ra}_x^{2/5}\left(\frac{\text{Pr}^{1/6}}{1 + 0.494\,\text{Pr}^{2/3}}\right)^{2/5}. \tag{11.4.19}$$

Ignoring an initial laminar boundary layer, Eq. (11.4.19) may be integrated over length L to give the average Nusselt number, $\text{Nu}_L = \frac{5}{6}\text{Nu}_{x=L}$. For Prandtl numbers near unity, inserting $\text{Pr} = 1.0$ in the parenthetical term in Eq. (11.4.19), we obtain

$$\text{Nu}_L \simeq 0.021\,\text{Ra}_L^{2/5}, \tag{11.4.20}$$

where $\text{Ra} = \text{Gr Pr}$ denotes the Rayleigh number. This correlation agrees reasonably well with experimental data for $10^9 < \text{Ra}_L < 5 \times 10^{10}$, as shown in Fig. 11.11. As discussed in Chapter 12, empirical correlations can be put into the form

$$\text{Nu}_L = \text{const. Ra}_L^m\,F(\text{Pr}), \tag{11.4.21}$$

where m ranges from $\frac{1}{3}$ (giving constant $\bar{h}$) to $\frac{2}{5}$.

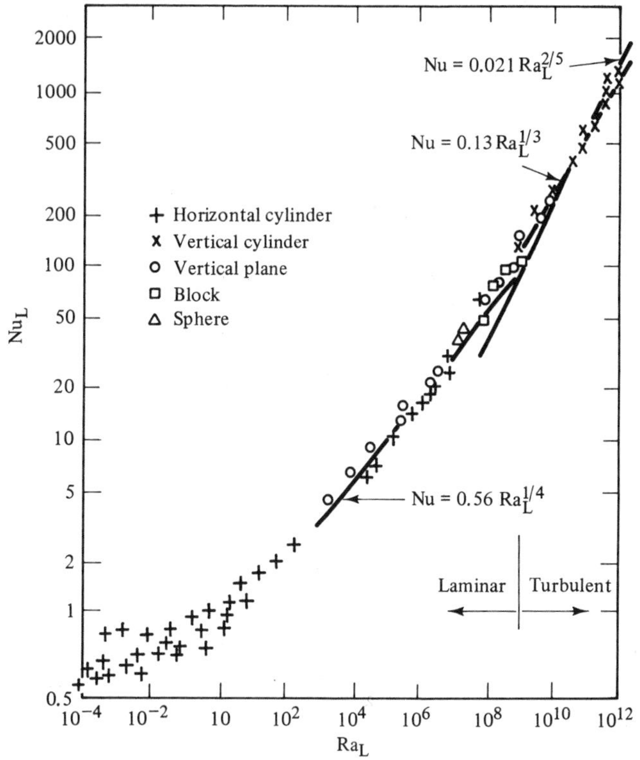

Fig. 11.11 Average Nusselt number versus Rayleigh number for natural convection from various surfaces to air and water (adapted from Jakob 1949; used by permission)

Although modeling of buoyancy-driven flows is still in a stage of development, it is worthwhile to mention the results for the present problem obtained by Lin & Churchill (1978). They used the low-Reynolds-number (K, ϵ)-model with the added production term due to buoyancy in the K-equation, as discussed in the beginning of this section. Their computational results for local heat transfer shown in Fig. 11.12 agree well with data and may be expressed closely by the correlation

$$\mathrm{Nu_x} = 0.0495\ \mathrm{Ra_x^{0.367}}, \qquad 5 \times 10^9 < \mathrm{Ra_x} < 5.8 \times 10^{11}. \tag{11.4.22}$$

Similar results are given in a numerical study by Plumb & Kennedy (1977), who employed a (K, ϵ, K_θ)-model, accounting for low-Reynolds-number effects according to Hoffman (1975) but modified to include the effects of buoyancy. These as well as other studies show promising results: for example, the study of a nonparallel flow of room convection by Gosman et al. (1980) using the high-Reynolds-number (K, ϵ)-model with the usual wall functions. Yet further research remains to be done to develop more reliable first- and second-order models.

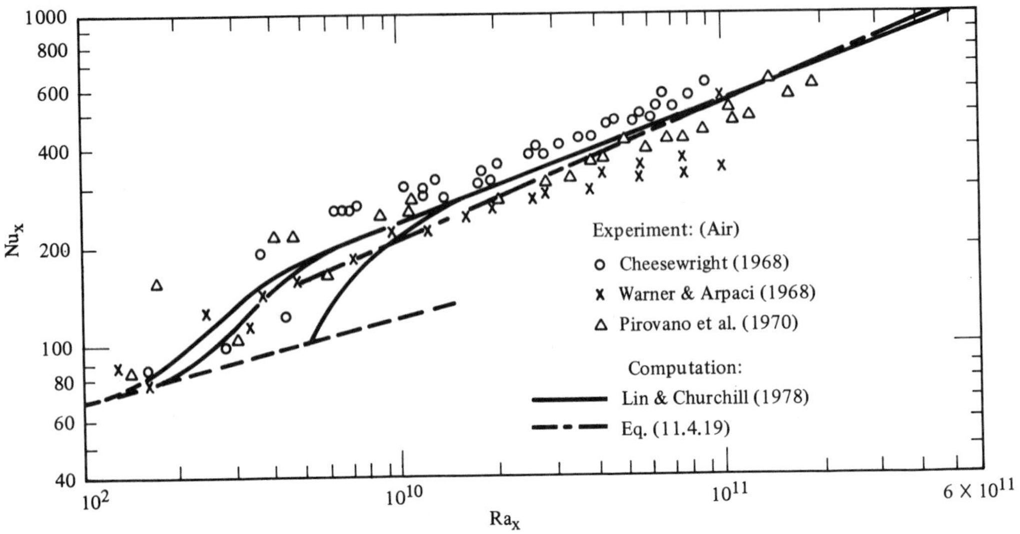

Fig. 11.12 Local Nusselt number versus Rayleigh number for natural convection from vertical isothermal plate, Ex. 11.4.2

Here we terminate our study of heat transfer prediction by turbulence models and proceed to Chapter 12 for a further study of heat transfer correlations, which often prove to be indispensable for the interpretation of experimental data.

REFERENCES

BRADSHAW, P., ed., 1976, *Turbulence*, Topics in Applied Physics, Springer-Verlag, New York.

CEBECI, T., 1973, J. Heat Transfer, **95C**, 227.

CHEESEWRIGHT, R., 1968, J. Heat Transfer, **90C**, 1.

COLBURN, A. P., 1933, Trans. Am. Inst. Chem. Eng., **29**, 174.

CORRSIN, S., 1952, J. Appl. Phys., **23**, 113.

DIPPREY, D. F., & SABERSKY, D. H., 1963, Int. J. Heat Mass Transfer, **6**, 329.

ECKERT, E. R. G., & JACKSON, T. W., 1951, NACA Rep. 1015.

GEORGE JR., W. K., & CAPP, S. P., 1979, Int. J. Heat Mass Transfer, **22**, 813.

GOSMAN, A. D., NIELSEN, P. E., RESTIVO, A., & WHITELAW, J. H., 1980, J. Fluids Eng., **102**, 316.

HOFFMAN, G. H., 1975, Phys. Fluids, **18**, 309.

JAKOB, M., 1949, *Heat Transfer,* Vol. I, John Wiley, New York.

JAYATILLEKE, C. L. V., 1959, Prog. Heat Mass Transfer, **1**, 193.

JONES, W. P., & LAUNDER, B. E., 1973, Int. J. Heat Mass Transfer, **16**, 1119.

KADER, B. A., 1981, Int. J. Heat Mass Transfer, **24**, 1541.

KADER, B. A., & YAGLOM, A. M., 1972, Int. J. Heat Mass Transfer, **15**, 2329.

KAVIANY, M., & SEBAN, R. A., 1981, Int. J. Heat Mass Transfer, **24**, 1742.

KAYS, W. M., & CRAWFORD, M. E., 1980, *Convection Heat and Mass Transfer*, McGraw-Hill, New York.

LEVICH, V. G., 1962, *Physicochemical Hydrodynamics*, Prentice-Hall, Englewood Cliffs, N.J.

LIN, S.-J., & CHURCHILL, S. W., 1978, Numer. Heat Transfer, **1**, 129.

MARTINELLI, R. C., 1947, Trans. ASME, **69**, 947.

NAKAJIMA, M., & FUKUI, K., 1980, Int. J. Heat Mass Transfer, **23**, 1325.

OWEN, P. R., & THOMSON, W. R., 1963, J. Fluid Mech., **15**, 321.

PIROVANO, A., VIANNAY, S., & JANNOT, M., 1970, in *Heat Transfer 1970*, Vol. 4, Paper No. 1.8, Elsevier, Amsterdam.

PLUMB, O. A., & KENNEDY, L. A., 1977, J. Heat Transfer, **99C**, 79.

PRANDTL, L., 1910, Phys. Z., **11**, 1072.

REYNOLDS, A. J., 1975, Int. J. Heat Mass Transfer, **18**, 1055.

REYNOLDS, A. J., 1976, Int. J. Heat Mass Transfer, **19**, 757.

REYNOLDS, O., 1874, Proc. Manchester Lit. Philos. Soc., **14**, 7.

ROHSENOW, W. M., & CHOI, H. Y., 1961, *Heat, Mass and Momentum Transfer*, Prentice-Hall, Englewood Cliffs, N.J.

SCHLICHTING, H., 1968, *Boundary Layer Theory*, 6th ed., McGraw-Hill, New York.

SPALDING, D. B., 1967, J. Fluid Mech., **27**, 97.

TAYLOR, G. I., 1919, ARCR and M. 272.

TOWNSEND, A. A., 1976, *The Structure of Turbulent Shear Flow*, 2nd ed., Cambridge University Press, New York.

VON KÁRMÁN, TH., 1939, Trans. Am. Soc. Mech. Eng., **61**, 705.

WARNER, C. Y., & ARPACI, V. S., 1968, Int. J. Heat Mass Transfer, **11**, 397.

WOLFSHTEIN, M., 1969, Int. J. Heat Mass Transfer, **12**, 301.

YIN, S. H., WUNG, T. Y., & CHEN, K., 1978, Int. J. Heat Mass Transfer, **21**, 307.

PROBLEMS

11-1 Based on the differential formulation of fully developed heat transfer in a pipe, and for developing heat transfer along the isothermal flat plate in uniform flow, discuss the similarity and the basis for the analogy of heat and momentum transfer (recall Sections 3.6 and 4.3 for the laminar case).

11-2 Derive the analogy of von Kármán (1939), Eq. (11.1.15).

11-3 Study the analogy of Martinelli (1947), Eq. (11.1.16) (see, for example, Rohsenow & Choi 1961).

11-4 Derive Eqs. (11.2.10)–(11.2.12).

11-5 Derive $\nu_T/\nu \sim (y^+)^3$ near a wall, using Eq. (10.2.5).

11-6 Derive the heat transfer law for pipe flow corresponding to Eq. (11.2.22).

11-7 Discuss algebraic models for heat transfer from a fully rough wall.

11-8 Consider the fully developed turbulent flow between parallel plates spaced $2h$ apart and being kept at uniform temperatures T_1 and T_2. Employing an algebraic model, determine the temperature distribution and the heat flux from each wall.

11-9 Recalling Ex. 11.2.4, determine the Nusselt number for fully developed heat transfer at constant wall flux and constant wall temperature, respectively.

11-10 Reconsider Ex. 11.2.2 and obtain an integral solution to the temperature distribution (recall Ex. 10.2.2).

11-11 Consider the homogeneous isotropic turbulent flow with uniform velocity U_0 downstream of a grid, one-half of which is heated to impose a temperature discontinuity on the flow. Ignoring buoyancy, determine the temperature distribution downstream of the grid.

11-12 Repeat Problem 11-11 for the case of a line source of heat placed at the grid.

11-13 Reconsider Ex. 4.3.2, the asymptotic solutions for the isothermal flat plate in uniform flow with uniform suction, now for turbulent flow. Ignore the dissipation.

11-14 Reconsider Ex. 4.4.2 of filmwise condensation of vapor (at the saturation temperature T_s) on an inclined isothermal wall (at $T_w < T_s$), Fig. 4.21. Determine the thickness $\delta(x)$ of the film, now assuming the flow to be turbulent, and determine the Nusselt number.

11-15 Derive the buoyancy production term of Eq. (11.4.3).

11-16 Obtain an approximate solution to Ex. 11.4.1 using the two-layer model of ν_T given by Eqs. (10.2.21) and (10.2.22). Employing h as the characteristic length, plot Nu versus Ra and compare the result to Eq. (11.4.20). Examine the resulting two-layer velocity profile.

11-17 Reconsider Ex. 4.3.4, the natural convection from the vertical flat isothermal plate subject to uniform suction, now for turbulent flow.

11-18 Study the second-order models and their reduced forms, denoted by algebraic Reynolds stress closure and algebraic Reynolds flux closure.

11-19 Compare Eqs. (11.3.4) and (11.3.7), and use the scaling of Chapter 9 to evaluate the Prandtl number effect on σ_T.

Chapter 12

DIMENSIONAL ANALYSIS

In Chapters 10 and 11, we have developed some models for the prediction of turbulent flow and turbulent heat transfer. For the verification of these models, we need experimental support. A rational way of investigating a problem experimentally is to describe the problem first in terms of appropriate *dimensionless numbers*. The development leading to a dimensionless description of a problem is called *dimensional analysis*. Interpretation of experimental results in terms of some dimensionless numbers is called *correlation of experimental data.*

Actually, in Chapters 9 through 11, we have already utilized a considerable amount of dimensional reasoning. Because of its vital importance to turbulent flow and turbulent heat transfer, we here return to dimensional analysis and explore a number of available methods. In Section 12.1 we introduce and develop three different methods for dimensional analysis. In Section 12.2 we work with one characteristic length which leads to an implicit relation describing (laminar or turbulent) convection. In Section 12.3 we consider two characteristic lengths (one for flow terms and one for flux terms) which lead to an explicit relation describing laminar convection. In Section 12.4 we employ the three-length model developed in Chapter 9. This model leads to an *explicit* relation describing turbulent convection.

12.1 FOUNDATIONS OF DIMENSIONAL ANALYSIS

When we have a complete understanding of physics and have no difficulty in formulation but are stuck mathematically on a solution, we refer to dimensional analysis for a functional (implicit) form of solution in terms of appropriate dimensionless numbers. Actually, there exist three distinct methods for dimensional analysis:

1. *Formulation* (*nondimensionalized*). Whenever a formulation in terms of governing equations is readily available, term-by-term nondimensionalization of this formulation leads directly to the related dimensionless numbers. The procedure is not suitable to problems that cannot be readily formulated.
2. Π-*theorem*.* If a formulation is not easily accessible but *all* physical and geometric quantities that prescribe a physical situation are *clearly* known, we write an implicit relation among these quantities,

$$f(Q_1, Q_2, \ldots, Q_n) = 0. \tag{12.1.1}$$

   Expressing these quantities in terms of appropriate fundamental units, and making Eq. (12.1.1) independent of the fundamental units by an appropriate combination of Q's yields the dimensionless numbers.
3. *Physical similitude*. Ratios established from the individual terms of appropriate general principles give the dimensionless numbers. The great convenience of this method is that there is no need to worry about an explicit formulation (required for the first method), except for a clear understanding of terms comprising a general principle. Also, there is no need to go through a nondimensionalization process (required for the second method) since a ratio between any two terms of a general principle is automatically dimensionless.

Let us demonstrate the application of the foregoing methods in terms of an illustrative example based on the simple pendulum (in vacuum). We wish to determine the period of this pendulum by dimensional analysis.

Clearly, the tangential acceleration of the pendulum results from the tangential component of the gravitational force. Also, from the tangential component of Newton's law of motion, we have the governing equation (Fig. 12.1)

$$\frac{d^2\varphi}{dt^2} + \frac{g}{l}\sin\varphi = 0. \tag{12.1.2}$$

Following the first method, we have from the nondimensionalization of Eq. (12.1.2) in terms of period T,

$$\frac{d^2\varphi}{d(t/T)^2} + \left(T^2\frac{g}{l}\right)\sin\varphi = 0,$$

*Dimensionless numbers obtained by this method are called $\Pi_1, \Pi_2, \ldots, \Pi_n$.

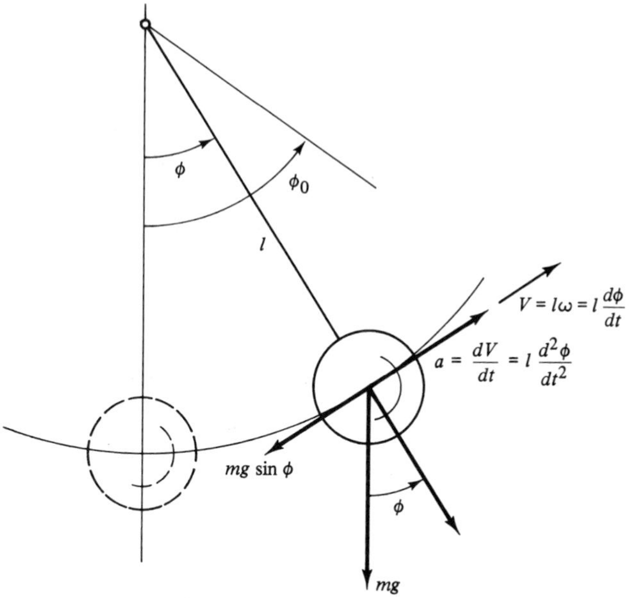

Fig. 12.1 Simple pendulum

which suggests the functional (implicit) relationship

$$\varphi = f\left(\frac{t}{T}, T^2\frac{g}{l}\right). \tag{12.1.3}$$

However, we are not interested in the instantaneous position φ (of the pendulum) but rather in its extrema φ_0, for which t/T assumes integer values, 1, 2, 3, ... Consequently,

$$\varphi_0 = f\left(T^2\frac{g}{l}\right).$$

Inverting this functional relationship, and expressing the result in terms of the period rather than its square, we have

$$T\sqrt{\frac{g}{l}} = f(\varphi_0). \tag{12.1.4}$$

For the second method (the Π-theorem), the approach to be followed is identical to that followed in the construction of similarity variables in Chapter 5. Here, however, the approach will be somewhat more concise.

Recalling the fact that the tangential momentum is balanced by the tangential component of the gravitational body force, and from the inspection of this balance, we conclude that

$$T = f(m, g, l, \varphi_0), \tag{12.1.5}$$

where m, g, l, and φ_0 all are independent quantities. In terms of three fundamental units of mechanics $[M]$, $[L]$, $[T]$,* Eq. (12.1.5) may be expressed as

$$[T] \equiv f\left[M, \frac{L}{T^2}, L, 0\right]. \tag{12.1.6}$$

Now, we begin rearranging Eq. (12.1.5) in such a way that, with each arrangement, it becomes independent of one fundamental unit. First of all, the dimensional homogeneity in $[M]$ suggests that

$$T = f_1(g, l, \varphi_0),$$

or, in terms of the fundamental units,

$$[T] \equiv f_1\left[\frac{L}{T^2}, L, 0\right].$$

Eliminating $[L]$, for example, by the ratio g/l, yields

$$T = f_2\left(\frac{g}{l}, \varphi_0\right),$$

or, in terms of the fundamental units,

$$[T] \equiv f_2\left[\frac{1}{T^2}, 0\right].$$

Finally, eliminating $[T]$ by the product $T\sqrt{g/l}$ gives the dimensionless relation

$$T\sqrt{\frac{g}{l}} = f_3(\varphi_0),$$

which is identical to Eq. (12.1.4). Note that the number of steps in the foregoing nondimensionalization procedure is equal to the number of fundamental units. Consequently, the number of dimensionless numbers is equal to the difference between the number of dimensional quantities in the original statement of a problem and the number of fundamental units. That is, Eq. (12.1.5) is in terms of five quantities, and since there are three fundamental units, the result involves $5 - 3 = 2$ dimensionless numbers.

For the third method (physical similitude), consider the tangential balance between the inertial and gravitational forces, $F_i + F_g = 0$, or the ratio

$$\frac{F_i}{F_g} + 1 = 0. \tag{12.1.7}$$

An order-of-magnitude consideration of this ratio in terms of φ_0 and T reveals that

$$\frac{F_i}{F_g} \sim \frac{ml\varphi_0/T^2}{mgF(\varphi_0)} \sim 1,$$

$F(\varphi_0)$ indicating the variation in the tangential component of the gravitational body force as a function of φ_0, or

$$T^2\frac{g}{l} \sim \frac{\varphi_0}{F(\varphi_0)},$$

*Or $[F]$, $[L]$, $[T]$.

or

$$T\sqrt{\frac{g}{l}} = f(\varphi_0). \tag{12.1.4}$$

Thus, we are able to show by three distinct methods that the dimensionless period of a simple pendulum in vacuum depends only on its initial displacement.

Now, combining Eq. (12.1.4) with a simple experiment to be performed by one pendulum with a number of φ_0's (Fig. 12.2), we can determine the explicit form of

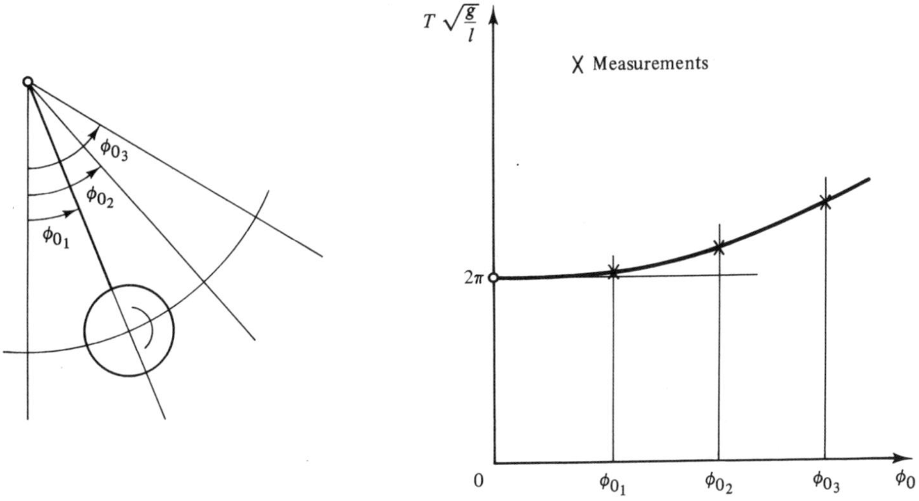

Fig. 12.2 Experiments with a simple pendulum

Eq. (12.1.4). As is well known, $\sin\varphi \sim \varphi$ for small displacements from the equilibrium, and Eq. (12.1.2) reduces to

$$\frac{d^2\varphi}{dt^2} + \frac{g}{l}\varphi = 0, \tag{12.1.8}$$

which characterizes the harmonic motion with angular frequency $\omega = \sqrt{g/l}$. Consequently,

$$T = \frac{2\pi}{\omega} = 2\pi\sqrt{\frac{l}{g}} \qquad \text{as } \varphi_0 \longrightarrow 0. \tag{12.1.9}$$

Having learned the foundations of dimensional analysis in terms of an illustrative example, we proceed to examples more relevant to our convection studies. Consider first an isothermal flow problem which will be useful for the enthalpy term of convection problems.

EXAMPLE 12.1.1 To illustrate problems involving forced flows, let a solid sphere of diameter D be immersed and held in an incompressible fluid streaming steadily with uniform velocity V (Fig. 12.3). The density and viscosity of the fluid are ρ and μ, respectively. We wish to determine force F on the sphere exerted by the moving fluid.

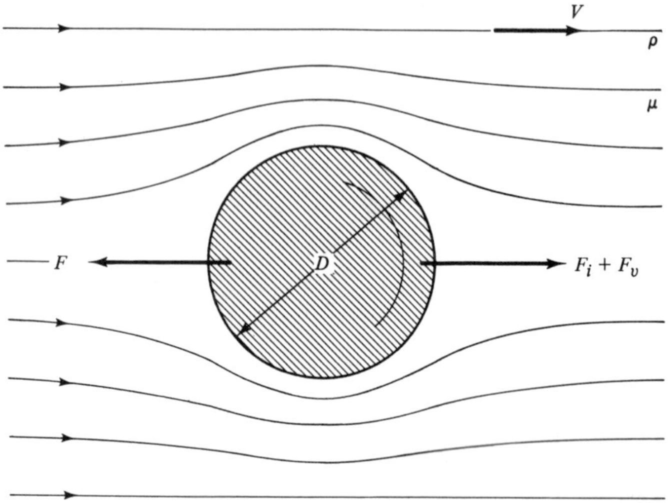

Fig. 12.3 Flow past a sphere

Since the differential formulation of the viscous flow near a sphere is somewhat involved, we proceed directly to the Π-theorem. In view of the fact that force F is balanced by the inertial and viscous forces, we assume that

$$F = f(V, D, \rho, \mu), \tag{12.1.10}$$

which may be expressed in terms of the fundamental units as

$$\left[\frac{ML}{T^2}\right] \equiv f\left[\frac{L}{T}, L, \frac{M}{L^3}, \frac{M}{LT}\right].$$

Now, we begin rearranging Eq. (12.1.10) by making it independent of one fundamental unit at a time. Since the mass dependence is the simplest one, we start with mass. To eliminate $[M]$ we pick any one of the mass-dependent terms of the right-hand side, ρ or μ. For example, we pick μ (later we comment on what would happen if we picked ρ instead) and combine it with F and ρ in such a way that the combination becomes independent of mass. Thus

$$\frac{F}{\mu} = f_1\left(V, D, \frac{\rho}{\mu}\right), \tag{12.1.11}$$

which may be expressed in terms of the remaining fundamental units as

$$\left[\frac{L^2}{T}\right] \equiv f_1\left[\frac{L}{T}, L, \frac{T}{L^2}\right].$$

Clearly, the time dependence of Eq. (12.1.11) is simpler than its length dependence. To eliminate $[T]$, we pick any one of the time-dependent terms on the right-hand side, V or ρ/μ. Since the final dimensionless numbers will ultimately involve all quantities describing Eq. (12.1.10) and since we already manipulated with μ, this time we pick V and combine it with F/μ and ρ/μ in such a way that the combinations become independent of time. Thus

$$\frac{F}{\mu V} = f_2\left(D, \frac{\rho V}{\mu}\right), \tag{12.1.12}$$

which, in terms of the length unit may be expressed as

$$[L] \equiv f_2\left[L, \frac{1}{L}\right].$$

Finally, eliminating $[L]$ by combining D with other terms of Eq. (12.1.12), we get

$$\frac{F}{\mu VD} = f_3\left(\frac{\rho VD}{\mu}\right) = f_3(\text{Re}), \tag{12.1.13}$$

where $\rho VD/\mu = \text{Re}$ is the *Reynolds number.*

Now, let us go back to Eq. (12.1.10) and start all over again. This time we make the equation independent of $[M]$ by manipulating the mass-dependent terms with ρ rather than μ, and we end up with

$$\frac{F}{\rho V^2 D^2} = f_4\left(\frac{\mu}{\rho VD}\right) = f_4(\text{Re}). \tag{12.1.14}$$

Since the dimensional analysis can provide only a functional (implicit) relationship between dimensionless numbers, Eqs. (12.1.13) and (12.1.14) are synonymous dimensionless results. Furthermore, since Eq. (12.1.14) is a functional relationship, it can be rearranged as

$$\left(\frac{F}{\rho V^2 D^2}\right) \text{Re} = f_3(\text{Re}),$$

which is identical to Eq. (12.1.13). That is, by suitable transformations, a dimensionless relation can be converted to another dimensionless relation equivalent, from the standpoint of dimensional analysis, to the original relation.

Dimensional analysis offers no clue as to which one of Eqs. (12.1.13) and (12.1.14) may be most convenient for our problem. Aside from the obvious fact that dependent variable F should be included in only one dimensionless number, it is necessary to rely on past experience and physical insight in the selection of one of these equations. For example, according to the experimental evidence, $F/\rho V^2 D^2$ changes less than $F/\mu VD$ in the practical range of Re, and Eq. (12.1.14) should be preferred.

Next, we proceed to dimensional analysis of the same problem by the method of physical similitude. Since force F on the sphere exerted by the moving fluid is balanced with the inertial and viscous forces,

$$F + F_i + F_v = 0,$$

from which we may establish ratios

$$\frac{F}{F_v} \sim \frac{F}{D^2 \mu(V/D)} = \frac{F}{\mu VD}, \tag{12.1.15}$$

$$\frac{F_i}{F_v} \sim \frac{\rho D^3(V^2/D)}{D^2 \mu(V/D)} = \frac{\rho VD}{\mu} = \text{Re}, \tag{12.1.16}$$

and

$$\frac{F}{F_i} \sim \frac{F}{\rho D^3(V^2/D)} = \frac{F}{\rho V^2 D^2}. \tag{12.1.17}$$

Clearly, from $F_i = m\,(dV/dt)$ we have $F_i \sim m(V/t)$ and (in view of $m \sim \rho D^3$ and $t \sim D/V$) $F_i \sim \rho D^3(V^2/D)$. Also, for a Newtonian fluid, from $F = A\tau_w = A\mu\,(du/dy)$, we have $F_v \sim D^2\mu(V/D)$. Equations (12.1.15) and (12.1.16) are the dimensionless numbers associated with Eq. (12.1.13), and Eqs. (12.1.16) and (12.1.17) are those associated with Eq. (12.1.14).

The concept of physical similitude becomes quite useful for experiments to be conducted with scaled models rather than the actual prototype. Physical similitude is said to exist between two systems if corresponding dimensionless numbers have the same value. Geometric similitude is a prerequisite for the physical similitude. Further elaborations on the similitude, however, belong to texts on fluid mechanics.

Note that so far we have used the same length scale in the nondimensionalization of the inertial and viscous forces. It suffices to consider a one-length scale whenever interest lies in the implicit dimensionless relations. We know, however, from the momentum boundary layers of Chapter 4, that the lengths involved with the flow and flux of momentum are not the same. In Section 12.3, utilizing two-length scales, we shall obtain explicit dimensionless relations. Meanwhile, we proceed to the dimensionless numbers associated with buoyancy-driven flows.

EXAMPLE 12.1.2 To illustrate problems involving buoyancy effects, let the solid sphere of Ex. 12.1.1 now fall, under the effect of gravity, in a fluid of density ρ and viscosity μ (Fig. 12.4). The difference between the density of the sphere and that of the fluid is $\Delta\rho$. We wish to determine the terminal velocity of the sphere.

In a manner similar to the preceding problem, we start with the Π-theorem. First, replacing F of Eq. (12.1.10) with the buoyant force per unit volume, $g\,\Delta\rho$, we write

$$g\,\Delta\rho = f(V, D, \rho, \mu),$$

or, because of the fact that V is now the dependent variable,

$$V = f(g\,\Delta\rho, D, \rho, \mu). \tag{12.1.18}$$

In terms of the fundamental units, Eq. (12.1.18) is equivalent to

$$\left[\frac{L}{T}\right] \equiv f\left[\frac{M}{L^2T^2}, L, \frac{M}{L^3}, \frac{M}{LT}\right].$$

Next, we begin rearranging Eq. (12.1.18) so that it becomes suitable to the elimination of one fundamental unit at a time. To eliminate $[M]$, we pick ρ, for example, and combine it with $g\,\Delta\rho$ and μ in such a way that the combinations become independent of mass. Thus,

$$V = f_1\left(g\frac{\Delta\rho}{\rho}, D, \frac{\mu}{\rho}\right), \tag{12.1.19}$$

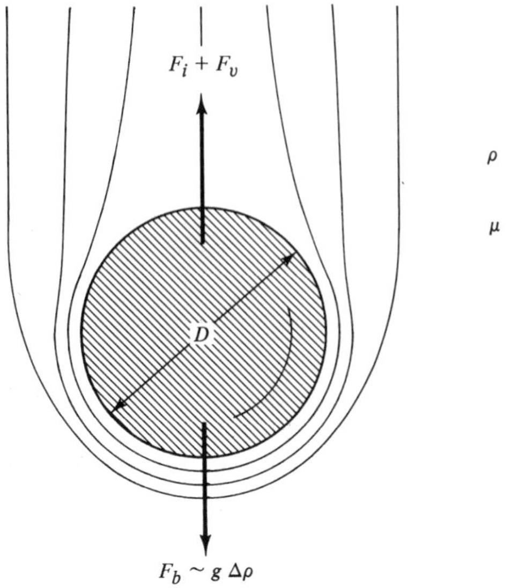

Fig. 12.4 Free fall of a sphere

which, in terms of the remaining fundamental units, is equivalent to

$$\left[\frac{L}{T}\right] \equiv f_1\left[\frac{L}{T^2}, L, \frac{L^2}{T}\right].$$

To eliminate $[T]$, we pick $\mu/\rho = \nu$, for example, and combine it with V and $g\,(\Delta\rho/\rho)$ in such a way that the combinations become independent of time. Thus,

$$\frac{\rho V}{\mu} = f_2\left(\frac{g}{\nu^2}\left(\frac{\Delta\rho}{\rho}\right), D\right), \tag{12.1.20}$$

which, in terms of the length unit, is equivalent to

$$\left[\frac{1}{L}\right] \equiv f_2\left[\frac{1}{L^3}, L\right].$$

Finally, eliminating $[L]$ by combining D with other terms of Eq. (12.1.20), we get

$$\frac{\rho VD}{\mu} = f_3\left(\frac{g}{\nu^2}\left(\frac{\Delta\rho}{\rho}\right)D^3\right), \tag{12.1.21}$$

where

$$\frac{g}{\nu^2}\left(\frac{\Delta\rho}{\rho}\right)D^3 = \mathrm{Gr} \tag{12.1.22}$$

is the *Grashof number*. Thus, the terminal velocity of buoyancy driven bodies is found to be governed by the dimensionless relation

$$\mathrm{Re} = f(\mathrm{Gr}). \tag{12.1.23}$$

Next, we proceed to dimensional analysis of the same problem by the method of physical similitude. Noting that the buoyant force is balanced with the inertial and viscous forces,

$$F_b + F_i + F_v = 0,$$

we may establish ratios

$$\frac{F_i}{F_v} \sim \frac{\rho V D}{\mu} = \text{Re}, \tag{12.1.16}$$

$$\frac{F_b}{F_v} \sim \frac{g\,\Delta\rho\, D^3}{\mu(V/D)D^2} = \frac{g\,\Delta\rho\, D^2}{\mu V}. \tag{12.1.24}$$

Here, we refer to the fact that the physics of any problem may be described by one dimensionless number depending on other dimensionless numbers composed only of independent quantities. Since the velocity of our problem depends on the buoyancy, and the Reynolds number in terms of this velocity is the dependent dimensionless number, the dimensionless number associated with buoyancy must be independent of the velocity. For example, the following combination, obtained from the product of Eqs. (12.1.16) and (12.1.24),

$$\frac{F_b}{F_v} \times \frac{F_i}{F_v} \sim \frac{g\,\Delta\rho\, D^2}{\mu V} \times \frac{\rho V D}{\mu} = \frac{g}{\nu^2}\left(\frac{\Delta\rho}{\rho}\right) D^3 = \text{Gr},$$

is independent of velocity, and it may be considered as a dimensionless number characterizing the buoyancy. Thus, the physical similitude leads us to the relation among force ratios

$$\frac{F_i}{F_v} = f\left(\frac{F_b}{F_v} \times \frac{F_i}{F_v}\right), \tag{12.1.25}$$

which is identical to Eq. (12.1.23), the result already obtained by employing the Π-theorem.

After the foregoing isothermal problems, we proceed to dimensional analysis of convection problems.

12.2 ONE-LENGTH CONVECTION

Delaying the consideration of multiple-length scales to the next two sections, we continue to use a one-length scale in the nondimensionalization process, and begin with forced convection because of its relative simplicity.

Forced convection. Ignoring the method of nondimensionalized governing equations, we proceed directly to the Π-theorem. Consider, for example, a steady two-dimensional problem with temperature distribution

$$\frac{T_w - T}{T_w - T_\infty} = f(\underbrace{\underbrace{x, y, V, \rho, \mu}_{\text{flow}}, c_p}_{\text{enthalpy flow}}, \underbrace{k}_{\text{conduction}}).$$

The wall gradient of this temperature gives the *local heat transfer coefficient*,

$$h_x = f(x, V, \rho, \mu, c_p, k),$$

whose average over distance D (or l) gives the (average) heat transfer coefficient,

$$h = f(D, V, \rho, \mu, c_p, k), \tag{12.2.1}$$

where the right-hand side is composed only of independent quantities. For thermal problems, a fourth fundamental unit is needed in addition to the three fundamental units of mechanics. This unit is usually assumed to be temperature $[\theta]$.* In terms of four fundamental units, Eq. (12.2.1) may be expressed as

$$\left[\frac{M}{T^3\theta}\right] \equiv f\left[L, \frac{L}{T}, \frac{M}{L^3}, \frac{M}{LT}, \frac{L^2}{T^2\theta}, \frac{ML}{T^3\theta}\right].$$

(For the units of c_p, recall that $h^0 = h + V^2/2 + gz$ and employ $h \sim c_pT \sim V^2/2$, or $c_p[\theta] \sim [L^2/T^2]$; for the units of h, $q \sim$ power/area = force × velocity/area $\sim [ML/T^2][L/T]/[L^2] \sim h[\theta]$; and for the units of k, note that $h \sim k/[L]$.) We proceed now to dimensionless numbers of forced convection by successively eliminating the fundamental units from Eq. (12.2.1).

Again we start the elimination process with the mass.† Combining ρ, for example, with other mass-dependent quantities, we have from Eq. (12.2.1),

$$\frac{h}{\rho} = f_1\left(D, V, \nu, c_p, \frac{k}{\rho}\right), \tag{12.2.2}$$

where $\nu = \mu/\rho$ is the kinematic viscosity (or momentum diffusivity). In terms of the fundamental units, Eq. (12.2.2) is equivalent to

$$\left[\frac{L^3}{T^3\theta}\right] \equiv f_1\left[L, \frac{L}{T}, \frac{L^2}{T}, \frac{L^2}{T^2\theta}, \frac{L^4}{T^3\theta}\right].$$

Now, the temperature dependence appears to be the simplest one. Elimination of $[\theta]$ from Eq. (12.2.2) by combining c_p, for example, with other temperature-dependent quantities yields

$$\frac{h}{\rho c_p} = f_2(D, V, \nu, a), \tag{12.2.3}$$

where $a = k/\rho c_p$ is the thermal diffusivity. Equation (12.2.3) is equivalent to

$$\left[\frac{L}{T}\right] \equiv f_2\left[L, \frac{L}{T}, \frac{L^2}{T}, \frac{L^2}{T}\right].$$

The time dependence of Eq. (12.2.3) is somewhat easier than its length dependence. Eliminating $[T]$ by combining V, for example, with other time-dependent quantities‡ gives

$$\frac{h}{\rho c_p V} = f_3\left(D, \frac{V}{\nu}, \frac{V}{a}\right), \tag{12.2.4}$$

*Or heat flux $[Q]$.

†See Problem 12-9 for other possibilities.

‡$[T]$ may also be eliminated with either a or ν.

which is equivalent to

$$[0] \equiv f_3\left[L, \frac{1}{L}, \frac{1}{L}\right].$$

Finally, eliminating $[L]$, we have

$$\frac{h}{\rho c_p V} = f_4\left(\frac{VD}{\nu}, \frac{VD}{a}\right),$$

or

$$\text{St} = f(\text{Re}, \text{Pe}), \tag{12.2.5}$$

where $\text{St} = h/\rho c_p V$ is the *Stanton number* and $\text{Pe} = VD/a$ is the *Péclet number* (or thermal Reynolds number). Noting that

$$\text{Pe} = \text{Re Pr},$$

$$\text{St} = \frac{\text{Nu}}{\text{Pe}} = \frac{\text{Nu}}{\text{Ke Pr}},$$

$\text{Pr} = \rho c_p/k = \nu/a$ being the *Prandtl number*, Eq. (12.2.5) may be written alternatively as

$$\boxed{\text{Nu} = f(\text{Re}, \text{Pr}),} \tag{12.2.6}$$

which is the form used most frequently. Explicit forms of Eq. (12.2.6) associated with laminar boundary layers may be found in Chapter 4, and those resulting from models on turbulent convection have been given in Chapter 11.

Next, we proceed to dimensional analysis of the same problem by the method of physical similitude. From the definition of

$$\text{convection} \equiv \text{conduction in } \underbrace{\text{moving media}}_{\substack{\text{inertial force + viscous force} \\ \text{flow} \\ \downarrow \\ \text{enthalpy flow}}}$$

we may establish ratios

$$\frac{\text{convection}}{\text{conduction}} \sim \frac{hD^2\theta}{kD^2(\theta/D)} = \frac{hD}{k} = \text{Nu}, \tag{12.2.7}$$

$$\frac{\text{inertial force}}{\text{viscous force}} \sim \frac{VD}{\nu} = \text{Re}, \tag{12.2.8}$$

$$\frac{\text{enthalpy flow}}{\text{conduction}} \sim \frac{\rho c_p V D^2\theta}{kD^2(\theta/D)} = \frac{VD}{a} = \text{Pe}, \tag{12.2.9}$$

where Nu is the heat transfer coefficient nondimensionalized relative to conduction, and Re and Pe are obtained from the nondimensionalization of momentum and thermal energy, respectively. For incompressible flows the momentum is decoupled from the energy, and is characterized by Re. However, the energy is coupled to the momentum through enthalpy flow, and the Pe number is silent to this coupling.

Eliminating the velocity between Eqs. (12.2.8) and (12.2.9), we obtain

$$\frac{\text{enthalpy flow}}{\text{conduction}} \times \frac{\text{viscous force}}{\text{inertial force}} \sim \frac{\nu}{a} = \text{Pr}, \tag{12.2.10}$$

which describes the coupling of energy to momentum. Also, Pr characterizes the diffusion of momentum relative to the diffusion of heat, and is the dimensionless number involving only physical properties of the fluid.

Experimental data associated with gases, water, and viscous oils may be correlated with Eq. (12.2.6) as shown in Fig. 12.5, where Re_c denotes the critical Reynolds

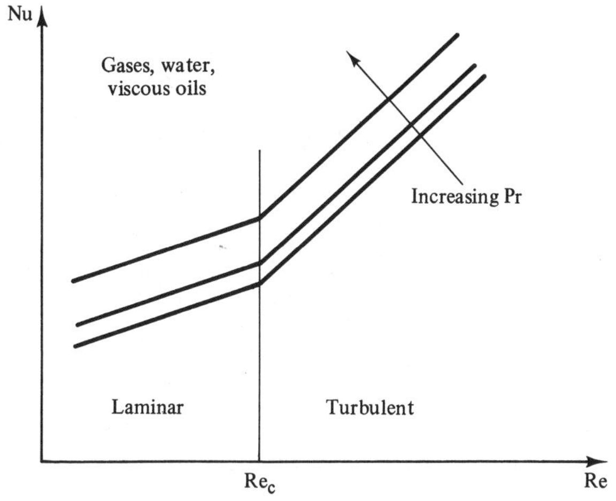

Fig. 12.5 Correlation of data on forced convection in gases, water, and viscous oils

number at which the laminar flow becomes unstable. Beyond Re_c the forced convection is turbulent. Equation (12.2.6) does not correlate the liquid metal data. For liquid metals, viscous forces are small and the momentum equation degenerates to a uniform velocity, and the importance of the Reynolds number diminishes. Consequently,

$$\boxed{\text{Nu} = f(\text{Pe})} \tag{12.2.11}$$

correlates the data on liquid metals. Explicit forms of Eq. (12.2.11) associated with slug flow may be found in Chapter 3, and those resulting from models on turbulent convection or from correlation of experimental data have been given in Chapter 11.

Having learned the dimensionless numbers of forced convection, we proceed to those of natural convection.

Natural convection. So far we learned that the correlation of forced convection starts with

$$h = f(D, V, \rho, \mu, c_p, k), \tag{12.2.1}$$

where the right-hand terms are made only of independent quantities. Since the velocity of natural convection depends on the buoyancy, we may utilize Eq. (12.2.1) for natural convection after replacing V with a buoyancy term.

Buoyancy force per unit volume is

$$g\,\Delta\rho.$$

For small temperature differences, we assume that

$$\Delta\rho = \left(\frac{d\rho}{dT}\right)\Delta T,$$

which may be rearranged in terms of the coefficient of thermal expansion (recall Section 2.4.2), $\beta = -\rho^{-1}(d\rho/dT)$, as

$$\Delta\rho \sim \beta\rho\,\Delta T.$$

Consequently, the buoyancy force becomes

$$g\beta\rho\,\Delta T.$$

Noting that Eq. (12.2.1) already involves ρ because of the inertial force, and replacing V of this equation with $g\beta\,\Delta T$, we have

$$h = f(D, g\beta\,\Delta T, \rho, \mu, c_p, k). \tag{12.2.12}$$

Consider first the method of the Π-theorem, and accordingly, express Eq. (12.2.12) in terms of our fundamental units as

$$\left[\frac{M}{T^3\theta}\right] \equiv f\left[L, \frac{L}{T^2}, \frac{M}{L^3}, \frac{M}{LT}, \frac{L^2}{T^2\theta}, \frac{ML}{T^3\theta}\right].$$

Begin the elimination process with the mass-dependent terms. We have already used μ (in connection with the flow around a sphere) and ρ (in connection with the forced convection) for this elimination. Let us see what happens if we now eliminate $[M]$ by combining k with other mass-dependent quantities:

$$\frac{h}{k} = f_1\left(D, g\beta\,\Delta T, \frac{\rho}{k}, \frac{\mu}{k}, c_p\right), \tag{12.2.13}$$

which, in terms of fundamental units, is equivalent to

$$\left[\frac{1}{L}\right] \equiv f_1\left[L, \frac{L}{T^2}, \frac{T^3\theta}{L^4}, \frac{T^2\theta}{L^2}, \frac{L^2}{T^2\theta}\right].$$

At this stage, the temperature dependence appears to be simplest. Elimination of $[\theta]$ from Eq. (12.2.13) combining c_p, for example, with other temperature-dependent quantities yields

$$\frac{h}{k} = f_2(D, g\beta\,\Delta T, a, \mathrm{Pr}), \tag{12.2.14}$$

where $a = k/\rho c_p$ is the thermal diffusivity. Equation (12.2.14) is equivalent to

$$\left[\frac{1}{L}\right] \equiv f_2\left[L, \frac{L}{T^2}, \frac{T}{L^2}, 0\right].$$

Eliminating the time dependence from Eq. (12.2.14) by combining a with $g\beta\,\Delta T$, for example, we get

$$\frac{h}{k} = f_3\left(D, \frac{g\beta\,\Delta T}{a^2}, \mathrm{Pr}\right), \tag{12.2.15}$$

which is equivalent to

$$\left[\frac{1}{L}\right] \equiv f_3\left[L, \frac{1}{L^3}, 0\right].$$

Finally, eliminating $[L]$ yields

$$\frac{hD}{k} = f_4\left(\frac{g\beta\,\Delta T\,D^3}{a^2}, \mathrm{Pr}\right),$$

or

$$\mathrm{Nu} = f(\Pi, \mathrm{Pr}), \tag{12.2.16}$$

where $\Pi = g\beta\,\Delta T\,D^3/a^2$ is a dimensionless number. Noting that

$$\frac{\Pi}{\mathrm{Pr}} = \frac{g\beta\,\Delta T\,D^3}{\nu a} = \mathrm{Ra} = \mathrm{Gr\,Pr},$$

Eq. (12.2.16) may be given in equivalent dimensionless form as

$$\mathrm{Nu} = f(\mathrm{Ra}, \mathrm{Pr}), \tag{12.2.17}$$

where Ra is the *Rayleigh number*.

Next, we proceed to dimensional analysis of the same problem by the method of physical similitude. From the definition of

$$\text{natural convection} = \text{conduction in } \underbrace{\text{moving media}}_{\underbrace{\left.\begin{matrix}\text{intertial}\\ \text{viscous}\\ \text{buoyancy}\end{matrix}\right\}\text{forces}}_{\substack{\text{flow}\\ \downarrow\\ \text{enthalpy flow}}}}$$

we may establish the ratios

$$\frac{\text{convection}}{\text{conduction}} \sim \frac{hD}{k} = \mathrm{Nu}, \tag{12.2.7}$$

$$\frac{\text{inertial force}}{\text{viscous force}} \sim \frac{VD}{\nu} = \mathrm{Re}, \tag{12.2.8}$$

$$\frac{\text{inertial force}}{\text{buoyancy force}} \sim \frac{mV^2/D}{mg\beta\,\Delta T} = \frac{V^2}{g\beta\,\Delta T\,D}, \tag{12.2.18}$$

$$\frac{\text{enthalpy flow}}{\text{conduction}} \sim \frac{VD}{a} = \mathrm{Pe}. \tag{12.2.9}$$

Here, we recall the fact that dimensionless numbers are composed only of independent quantities, and note that V is *not* an independent quantity for natural convection.

Consequently, Eqs. (12.2.8), (12.2.9), and (12.2.18) may prescribe the natural convection after they are combined and made independent of V. For example, from Eqs. (12.2.8) and (12.2.9),

$$\frac{\text{enthalpy flow}}{\text{conduction}} \times \frac{\text{viscous force}}{\text{inertial force}} \sim \frac{\nu}{a} = \text{Pr}, \tag{12.2.10}$$

which describes the coupling of energy to momentum. From Eqs. (12.1.24) and (12.2.18),

$$\frac{\text{buoyancy force}}{\text{inertial force}} \times \left(\frac{\text{enthalpy flow}}{\text{conduction}}\right)^2 \sim \frac{g\beta\,\Delta T\,D^3}{a^2} = \Pi, \tag{12.2.19}$$

which describes the coupling of momentum to energy. Thus, we have

$$\text{Nu} = f(\Pi, \text{Pr}), \tag{12.2.16}$$

already obtained by the method of the Π-theorem. Furthermore, in natural convection associated with gases, water, and viscous oils, the effect of inertial forces is quite small and may be neglected. The dimensionless number characterizing this situation, obtained by the elimination of the inertial force between Eqs. (12.2.10) and (12.2.19), for example, yields the Rayleigh number,

$$\frac{\text{buoyancy force}}{\text{viscous force}} \times \frac{\text{enthalpy flow}}{\text{conduction}} \sim \frac{g\beta\,\Delta T\,D^3}{\nu a} = \text{Ra}.$$

Accordingly, the experimental data on gases, water, and viscous oils can be correlated with

$$\boxed{\text{Nu} = f(\text{Ra})} \tag{12.2.20}$$

as shown in Fig. 12.6. Here Ra_c denotes the critical Rayleigh number at which the

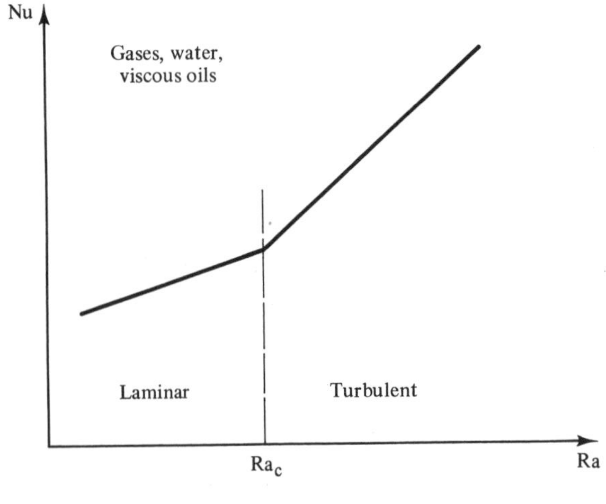

Fig. 12.6 Correlation of data on natural convection in gases, water, and viscous oils

laminar flow becomes unstable. Beyond Ra_c the natural convection is turbulent. Explicit forms of Eq. (12.2.20) associated with laminar boundary layers may be found in Chapter 4, and those resulting from models on turbulent convection or from the correlation of turbulent experimental data have been given in Chapter 11.

Equation (12.2.20) does not correlate the liquid metal data. For liquid metals, the effect of inertial forces becomes important, while the effect of viscous forces diminishes. In this case, ignoring Eq. (12.2.8), and eliminating V between Eqs. (12.2.9) and (12.2.18), we have

$$\frac{\text{buoyancy force}}{\text{inertial force}} \times \left(\frac{\text{enthalpy flow}}{\text{conduction}}\right)^2 = \Pi$$

with

$$\Pi = \mathrm{Ra\,Pr}.$$

Accordingly, the liquid metal data can be correlated with

$$\boxed{\mathrm{Nu} = f(\mathrm{Ra\,Pr}).} \tag{12.2.21}$$

Here it is worth noting the conceptual difference between the Re number of forced convection and the Ra number of natural convection. The Re number results from the nondimensionalized momentum which is uncoupled from the energy of incompressible (and constant property) fluids. On the other hand, the Ra number characterizes the coupling through buoyancy of momentum to energy. Furthermore, since the Gr number results from nondimensionalized momentum only, it cannot alone characterize natural convection. For both forced convection and natural convection the Pr number characterizes the coupling of energy to momentum.

Combined convection. So far we have studied separately forced convection and natural convection. In the case of forced convection, we neglected the buoyancy force, and in the case of natural convection, we neglected the inertial force. The importance of natural convection relative to forced convection may then be determined by the ratio

$$\frac{\text{buoyancy force}}{\text{inertial force}} = \mathrm{Ri},$$

which is called the *Richardson number*. Explicitly,

$$\frac{g\,\Delta\rho\, D^3}{\rho D^3 V^2/D} = \frac{g}{V^2}\left(\frac{\Delta\rho}{\rho}\right) D = \mathrm{Ri}, \tag{12.2.22}$$

which is often rearranged as

$$\frac{(g/\nu^2)(\Delta\rho/\rho)\,D^3}{(VD/\nu)^2} = \frac{\mathrm{Gr}}{\mathrm{Re}^2} = \mathrm{Ri}. \tag{12.2.23}$$

A combined convection problem influenced primarily by inertia, with small contribution from buoyancy, may then be represented as

$$\mathrm{Nu}_{\text{combined}} \sim (1 + \mathrm{Ri})\mathrm{Nu}_{\text{forced}}, \tag{12.2.24}$$

and that influenced primarily by buoyancy, with a small contribution from inertia, may be represented as

$$\mathrm{Nu}_{\text{combined}} \sim (1 + \mathrm{Ri}^{-1})\mathrm{Nu}_{\text{natural}}. \tag{12.2.25}$$

Having studied *one-length dimensional analysis*, we proceed now to *two-length dimensional analysis*.

12.3 TWO-LENGTH LAMINAR CONVECTION

First, a conceptual difference among the three methods of dimensional analysis stated in Section 12.1 should be recalled; the Π-theorem searches for dimensionless numbers from an appropriate product of the components of forces (or terms) involved with general principles; physical similitude establishes dimensionless numbers from ratios of these forces; the formulation method, in the process of nondimensionalizing governing equations, deals directly with general principles and constitutive relations, and provides not only dimensionless numbers but also a relationship among them through these principles. The two-length dimensional analysis is based on this latter method, now illustrated in terms of a number of problems already considered in this chapter.

Boundary layer flows. Reconsider the laminar flow (boundary layer) around a sphere (Fig. 12.7). An order-of-magnitude representation of the momentum gives, in terms of the flow and flux scales shown on Fig. 12.7,

$$U\frac{U}{x} \sim \nu\frac{U}{\delta^2}, \tag{12.3.1}$$

or

$$\frac{\delta}{x} \sim \left(\frac{\nu}{Ux}\right)^{1/2} = \frac{1}{\mathrm{Re}_x^{1/2}}. \tag{12.3.2}$$

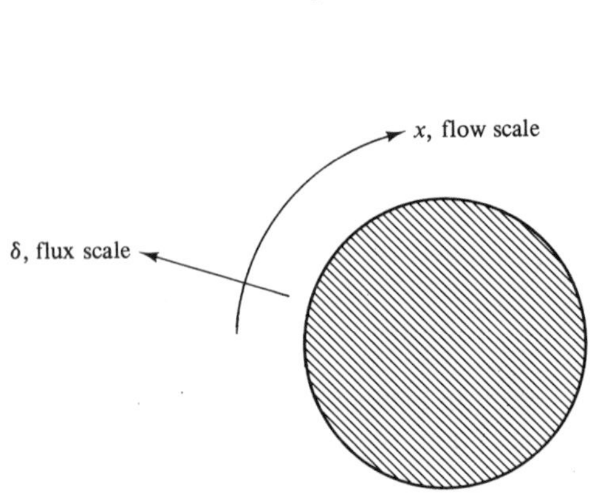

Fig. 12.7 Scales of flow past a sphere

Also, the local friction coefficient is

$$f_x = \frac{\tau_w}{\frac{1}{2}\rho U^2} \sim \frac{\mu\, U/\delta}{\frac{1}{2}\rho U^2}, \tag{12.3.3}$$

or

$$\frac{1}{2} f_x \sim \frac{\nu}{U\delta} = \left(\frac{\nu}{Ux}\right)\frac{x}{\delta},$$

or, in terms of Eq. (12.3.2),

$$\boxed{\frac{1}{2} f_x \sim \frac{1}{\mathrm{Re}_x^{1/2}}.} \tag{12.3.4}$$

For the local coefficient skin friction, following a laminar boundary layer analysis, we obtained in Chapter 4,

$$\frac{1}{2} f_x = \frac{0.323}{\mathrm{Re}_x^{1/2}}. \tag{4.2.6}$$

The difference between Eqs. (12.3.4) and (4.2.6) is the proportionality constant missing from the former. Apparently, a two-length dimensional analysis gives, except for a constant, the answer obtained by analytical means. Thus, the *analytical approach* of Chapter 4 *may be replaced by a two-length dimensional analysis to be completed by a one-data-point experiment for the missing constant.* This is an important conclusion. Actually, there is no need to solve laminar flow problems by dimensional analysis because of the simplicity of these problems. However, dimensional analysis becomes indispensable for turbulent flow problems, which, at present or in the foreseeable future, are not expected to be solved analytically.

Also, following a one-length dimensional analysis, we obtained

$$\frac{F}{\rho V^2 D^2} = f_4(\mathrm{Re}), \tag{12.1.14}$$

which is equivalent to a mean coefficient of friction. Explicitly, Eq. (12.1.14) may be written as

$$\frac{F}{\rho V^2 D^2} = \frac{C}{\mathrm{Re}^n}, \tag{12.3.5}$$

C and n being unknown constants to be determined experimentally. Thus, for the same problem, a *one-length dimensional analysis requires two data-point experiments.*

Boundary layer heat transfer. Now, we proceed to the heat transfer associated with the foregoing problem. First, consider the case of $\mathrm{Pr} > 1$ (water and viscous oils). The nondimensionalization of the energy equation gives

$$U\left(\frac{\Delta}{\delta}\right)\frac{\theta}{x} \sim a\frac{\theta}{\Delta^2}, \tag{12.3.6}$$

where Δ is the flux scale of the energy equation and, in view of Fig. 12.8, the velocity of the enthalpy flow is $U(\Delta/\delta)$. Then, from the ratio of Eqs. (12.3.1) and (12.3.6)

$$\boxed{\frac{\delta}{\Delta} \sim \mathrm{Pr}^{1/3}.} \tag{12.3.7}$$

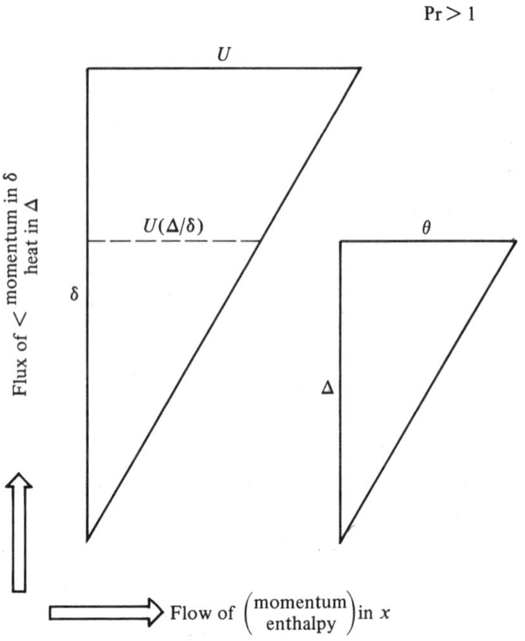

Fig. 12.8 Velocity of the enthalpy flow

Actually, the same result is also available from the use of the analogy between momentum and thermal energy defined as

$$\left(\frac{\text{flow}}{\text{flux}}\right) \text{ of momentum} \equiv \left(\frac{\text{flow}}{\text{flux}}\right) \text{ of thermal energy,} \qquad (9.4.18)$$

which gives, with the help of Fig. 12.8,

$$\frac{\rho U \dfrac{U}{x}}{\mu \dfrac{U^2}{\delta^2}} \equiv \frac{\rho c_p U \left(\dfrac{\Delta}{\delta}\right) \dfrac{\theta}{x}}{k \dfrac{\theta}{\Delta^2}},$$

or the *equality*

$$\frac{\delta}{\Delta} = \text{Pr}^{1/3}.$$

This result is universal and applies to both laminar and turbulent flows of fluids with $\text{Pr} > 1$. It may be used as the base of an analogy alternative to those developed in Section 11.1.

From the ratio of the definitions of the heat transfer coefficient and the friction coefficient, we already know that

$$\text{Nu}_x \sim \left(\frac{1}{2} f_x\right) \text{Re}_x \left(\frac{\delta}{\Delta}\right). \qquad (12.3.8)$$

Insertion of Eqs. (12.3.3) and (12.3.7) into Eq. (12.3.8) gives

$$\mathrm{Nu}_x \sim \mathrm{Re}_x^{1/2}\,\mathrm{Pr}^{1/3}, \qquad \mathrm{Pr} > 1. \tag{12.3.9}$$

Also, the one-scale dimensional analysis leads to an implicit relationship among the dimensionless numbers given by Eq. (12.2.6), or, explicitly,

$$\mathrm{Nu}_D = C\,\mathrm{Re}_D^m\,\mathrm{Pr}^n. \tag{12.3.10}$$

Clearly, the two-length dimensional analysis provides exponents of the dimensionless numbers and again needs a one-point experiment, while the one-length dimensional analysis now requires a three-point experiment.

Next, we proceed to the case of $\mathrm{Pr} \ll 1$ (liquid metals). The dimensional representation of the momentum given by Eq. (12.3.1) remains the same. However, neglecting the contribution of the enthalpy flow within the momentum boundary layer (Fig. 12.9), the energy equation may be written as

$$U\frac{\theta}{x} \sim a\frac{\theta}{(\Delta - \delta)^2}. \tag{12.3.11}$$

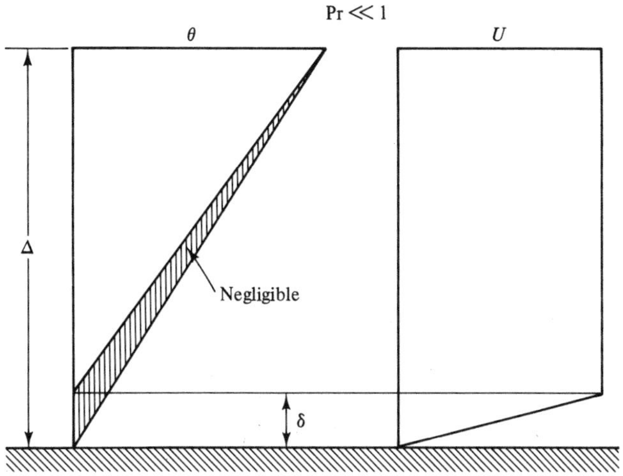

Fig. 12.9 Thermal and momentum boundary layers for fluids with $\mathrm{Pr} \ll 1$

From the ratio of Eqs. (12.3.1) and (12.3.11) we get

$$\boxed{\frac{\delta}{\Delta} \sim \frac{\mathrm{Pr}^{1/2}}{1 + \mathrm{Pr}^{1/2}}.} \tag{12.3.12}$$

Inserting Eqs. (12.3.3) and (12.3.12) into Eq. (12.3.8) gives the local Nusselt number for the laminar flow of liquid metals,

$$\mathrm{Nu}_x \sim \left(\frac{\mathrm{Pr}^{1/2}}{1 + \mathrm{Pr}^{1/2}}\right)\mathrm{Re}_x^{1/2},$$

or, explicitly,

$$\mathrm{Nu}_x = C_1\left(\frac{\mathrm{Pr}^{1/2}}{C_0 + \mathrm{Pr}^{1/2}}\right)\mathrm{Re}_x^{1/2}, \qquad \mathrm{Pr} \ll 1. \tag{12.3.13}$$

Note that two-length dimensional analysis of the present problem, leading to Eq. (12.3.13), requires a two-point experiment. As $\mathrm{Pr} \longrightarrow 0$, in the denominator neglecting $\mathrm{Pr}^{1/2}$ relative to C_0, Eq. (12.3.13) may be reduced to

$$\mathrm{Nu}_x = C_2\,\mathrm{Pe}_x^{1/2}, \qquad \mathrm{Pr} \longrightarrow 0, \tag{12.3.14}$$

where $\mathrm{Pe}_x = \mathrm{Re}_x\,\mathrm{Pr} = Ux/a$ is the local Péclet number. A one-point experiment is needed now for the proportionality constant $C_2 = C_1/C_0$.

Natural convection. The two-length dimensional analysis, so far applied to forced convection, may readily be extended to natural convection. The momentum balance, now including the buoyancy, becomes

$$U\frac{U}{x} + \nu\frac{U}{\delta^2} \sim g\left(\frac{\Delta\rho}{\rho}\right). \tag{12.3.15}$$

Assuming that

$$\delta \sim \Delta \tag{12.3.16}$$

for natural convection, the energy balance may be written as

$$U\frac{\theta}{x} \sim a\frac{\theta}{\delta^2}. \tag{12.3.17}$$

Solving U from Eq. (12.3.17),

$$U \sim \frac{ax}{\delta^2}, \tag{12.3.18}$$

and inserting this velocity into Eq. (12.3.15) yields

$$\frac{a^2x}{\delta^4}(1 + \mathrm{Pr}) \sim g\left(\frac{\Delta\rho}{\rho}\right), \tag{12.3.19}$$

which may be rearranged as

$$\frac{\delta}{x} \sim \left(\frac{1 + \mathrm{Pr}}{\mathrm{Pr}}\right)^{1/4}\mathrm{Ra}_x^{-1/4}, \tag{12.3.20}$$

where

$$\mathrm{Ra}_x = \frac{g}{\nu a}\left(\frac{\Delta\rho}{\rho}\right)x^3 \tag{12.3.21}$$

is the local Rayleigh number.

The local Nusselt number, in view of $\delta \sim \Delta$, is

$$\mathrm{Nu}_x \sim \frac{x}{\delta}. \tag{12.3.22}$$

Inserting Eq. (12.3.20) into Eq. (12.3.22) gives

$$\mathrm{Nu}_x \sim \left(\frac{\mathrm{Pr}}{1 + \mathrm{Pr}}\right)^{1/4}\mathrm{Ra}_x^{1/4}. \tag{12.3.23}$$

For large Prandtl number fluids, Eq. (12.3.23) reduces to

$$\mathrm{Nu}_x \sim \mathrm{Ra}^{1/4}, \qquad \mathrm{Pr} \longrightarrow \infty, \tag{12.3.24}$$

and for small Prandtl number fluids to

$$\mathrm{Nu}_x \sim (\mathrm{Pr}\,\mathrm{Ra}_x)^{1/4}, \qquad \mathrm{Pr} \longrightarrow 0. \tag{12.3.25}$$

Clearly, two-point experiments are needed for Eq. (12.3.23) and a one-point experiment is needed for Eqs. (12.3.24) and (12.3.25). So far, employing the two-length dimensional analysis, we have obtained explicit heat transfer relations for laminar forced and natural convection. This approach is not suitable to turbulent convection. In the next section, employing a three-length dimensional analysis, we obtain relations for turbulent convection.

12.4 *THREE-LENGTH TURBULENT CONVECTION*

In Chapter 9, we introduced a turbulence model involving two-velocity and two-length scales $(u, v; l, \eta)$. Later, we replaced this model with an equivalent model involving one-velocity and three-length scales $(u; l, \lambda, \eta)$. Similarly, we began with a two-temperature and two-length thermal model $(\theta, \vartheta; l_\theta, \eta_\theta)$ and replaced it with an equivalent one-temperature and three-length model $(\theta; l_\theta, \lambda_\theta, \eta_\theta)$. Now, employing the latter models, we attempt to understand turbulent heat transfer relations which are known as correlations of experimental data.

For turbulent flows, assuming that the coefficients of heat transfer and friction are defined in terms of the *dissipation scales*, we introduce

$$q_w \sim h\theta \sim k\left(\frac{\theta}{\eta_\theta}\right) \tag{12.4.1}$$

and

$$\frac{1}{2}f \sim \frac{\tau_w}{\rho u^2} \sim \frac{\mu(u/\eta)}{\rho u^2} \sim \frac{\nu}{u\eta}, \tag{12.4.2}$$

which may be rearranged as

$$\frac{h}{k} \sim \eta_\theta^{-1} \tag{12.4.3}$$

and

$$\frac{u}{\nu} \sim (\tfrac{1}{2}f)^{-1}\eta^{-1}. \tag{12.4.4}$$

The ratio of Eqs. (12.4.3) and (12.4.4) readily gives

$$\frac{h/k}{u/\nu} \sim \left(\frac{1}{2}f\right)\frac{\eta}{\eta_\theta}. \tag{12.4.5}$$

Noting that the left-hand side of Eq. (12.4.5) is independent of any length, multiplying and dividing this side by a convenient length, we get

$$\frac{\mathrm{Nu}}{\mathrm{Re}} \sim \left(\frac{1}{2}f\right)\frac{\eta}{\eta_\theta}. \tag{12.4.6}$$

Now, consider the case of fluids with $\mathrm{Pr} > 1$ and $\mathrm{Pr} \ll 1$ separately.

For Pr > 1, recalling that

$$\frac{\eta}{\eta_\theta} = \mathrm{Pr}^{1/3}, \qquad \mathrm{Pr} > 1, \tag{9.4.22}$$

Eq. (12.4.6) may be rearranged as

$$\frac{\mathrm{Nu}}{\mathrm{Re}} \sim \left(\frac{1}{2}f\right)\mathrm{Pr}^{1/3}, \tag{12.4.7}$$

or, in terms of the Stanton number,

$$\mathrm{St} = \frac{\mathrm{Nu}}{\mathrm{Re\,Pr}}, \tag{12.4.8}$$

as

$$\boxed{\mathrm{St\,Pr}^{2/3} = \tfrac{1}{2}f,} \qquad \mathrm{Pr} > 1. \tag{12.4.9}$$

This relation is called the *Colburn correlation* because of its discovery through the interpretation of turbulent experimental data (recall Eq. 11.1.17). Note that Eq. (12.4.9) is written as an equality rather than proportionality because of the well-known analogy between momentum and heat for fluids with Pr > 1. Furthermore, as we learned from an approximate solution of the integral formulation (Chapter 4) or from the similarity solution of differential formulation (Chapter 5), Eq. (12.4.9) also applies to laminar boundary layers. Consequently, Eq. (12.4.9) turns out to be a *universal relation* valid for turbulent as well as laminar forced convection. Only the friction coefficient involved with Eq. (12.4.9) shows the effect of flow behavior on this equation.

Usually, the friction coefficient is available from the correlation of turbulent experimental data on pressure drop or drag. Here, however, we get this coefficient by referring to microscales of turbulence.

From Eq. (12.4.2), we have

$$\tfrac{1}{2}f \sim \mathrm{Re}_\eta^{-1}, \tag{12.4.10}$$

which may be rearranged, in terms of Eq. (9.2.17), as

$$\tfrac{1}{2}f \sim \mathrm{Re}_l^{-1/4}. \tag{12.4.11}$$

Combining Eq. (12.4.7) or Eq. (12.4.9) with Eq. (12.4.11), we obtain

$$\mathrm{Nu}_l \sim \mathrm{Re}_l^{3/4}\,\mathrm{Pr}^{1/3}. \tag{12.4.12}$$

Experimental data confirm this relation for turbulent heat transfer over a flat plate and in a pipe, provided that the Reynolds number remains moderate in the latter case. Equation (12.4.12) correlates the boundary layer flow better than the pipe flow because the pressure gradient involved with the latter was not taken into account. The pipe flow is usually correlated by

$$\mathrm{Nu}_l \sim \mathrm{Re}_l^{4/5}\,\mathrm{Pr}^{1/3}. \tag{12.4.13}$$

The small quantitative difference between Eqs. (12.4.12) and (12.4.13) results from the pressure gradient driving the pipe flow and from the pressure fluctuations which are apparently of secondary importance for heat transfer.

For $\Pr \ll 1$, employing

$$\frac{\eta}{\eta_\theta^c} \sim \Pr^{3/4}, \qquad \Pr \longrightarrow 0, \tag{9.4.32}$$

Eq. (12.4.6) may be rearranged as

$$\frac{\text{Nu}}{\text{Re}} \sim \left(\frac{1}{2}f\right)\Pr^{3/4} \tag{12.4.14}$$

which, in terms of Eq. (12.4.11), leads to

$$\text{Nu}_l \sim \text{Pe}_l^{3/4}, \tag{12.4.15}$$

an expected result for liquid metals whose heat transfer depends only on the Péclet number. However, since the friction coefficient has been obtained separately, this result could not be guaranteed from the start, but shows now the consistency between the kinetic and thermal models developed in Chapter 9. Experimental data on liquid metals are somewhat scattered and do not appear to provide conclusive support for Eq. (12.4.15).

Next, we proceed to heat transfer relations for natural convection. For $\Pr > 1$, inserting Eq. (9.5.6) into Eq. (12.4.3) gives

$$\frac{h}{k} \sim \left(\frac{g\beta\vartheta}{\nu a}\right)^{1/3}. \tag{12.4.16}$$

Assuming that $\vartheta \sim \theta \sim \Delta T$ (a shortcoming of the present model), and introducing a convenient length scale, say l, Eq. (12.4.16) may be rearranged as

$$\text{Nu}_l \sim \text{Ra}_l^{1/3}. \tag{12.4.17}$$

There is strong experimental support for this result for $\Pr > 1$ (see, for example, Warner & Arpaci 1968, Bergholz, Chen & Cheung 1979, and Goldstein & Tokuda 1980). As Pr gets smaller, because of the increasing effect of inertial forces, Eq. (12.4.17) is replaced with a somewhat more complex relation (see, for example, Long 1976 and Cheung 1980).

For $\Pr \ll 1$, inserting Eq. (9.5.10) into Eq. (12.4.3) gives

$$\frac{h}{k} \sim \left(\frac{g\beta\vartheta}{a^2}\right)^{1/3}, \tag{12.4.18}$$

which may be rearranged, in a manner similar to Eq. (12.4.16), as

$$\text{Nu}_l \sim (\text{Ra}_l \Pr)^{1/3}. \tag{12.4.19}$$

Experimental data on liquid metals are lacking.

In this section, using the microscales of turbulence, we made an attempt to understand the foundations of heat transfer correlations. This is a promising area for turbulence research. Further efforts, in terms of improved models, are needed for a better understanding of turbulence.

REFERENCES

BERGHOLZ, R. F., CHEN, M. M., & CHEUNG, F. B., 1979, Int. J. Heat Mass Transfer **22**, 763.

CHEUNG, F. B., 1980, J. Fluid Mech., **97**, 743.

CHU, T. Y., & GOLDSTEIN, R. J., 1973, J. Fluid Mech., **60**, 141.

GARON, A. M., & GOLDSTEIN, R. J., 1973, Phys. Fluids, **16**, 1818.

GLOBE, S., & DROPKIN, D., 1959, J. Heat Transfer (ASME), **81**, 24.

GOLDSTEIN, E., & TOKUDA, S., 1980, Int. J. Heat Mass Transfer, **23**, 738.

GOLITSYN, G. S., 1979, J. Fluid Mech., **95**, 567.

HOLLANDS, K. G. T., RAITHBY, G. D., & KOENICEK, L., 1975, Int. J. Heat Mass Transfer, **18**, 879.

HUNSAKER, J. C., & RIGHTMIRE, B. G., 1947, *Engineering Applications of Mechanics*, McGraw-Hill, New York.

KRAICHNAN, R. H., 1962, Phys. Fluids, **5**, 1374.

KRISHNAMURTI, R., 1970, J. Fluid Mech.. **42**, 295.

LANGHAAR, H. L., 1951, *Dimensional Analysis and Theory of Models*, Wiley, New York.

LONG, R. R., 1976, J. Fluid Mech., **73**, 445.

PRIESTLEY, C. H. B., 1954, Aust. J. Phys., **7**, 176.

SEDOV, L. I., 1959, *Similarity and Dimensional Methods in Mechanics*, Academic Press, New York.

THRELFALL, D. C., 1975, J. Fluid Mech., **67**, 17.

WARNER, C. Y., & ARPACI, V. S., 1968, Int. J. Heat Mass Transfer. **11**, 397.

PROBLEMS

12-1 Reconsider the isothermal flow problem of Ex. 12.1.1. Including the effect of gravity g and surface tension σ (force/length), write

$$F = f(V, D, \rho, \mu, g, \sigma).$$

Find dimensionless numbers by the Π-theorem and physical similitude. Show that the result may be arranged as

$$\frac{F}{\rho V^2 D^2} = f(\text{Re}, \text{Fr}, \text{We}),$$

where $\text{Fr} = V/\sqrt{gD}$ is the *Froude number* and $\text{We} = \rho V^2 D/\sigma$ is the *Weber number*. Explain clearly the physics characterized by these numbers.

12-2 Including the effect of surface tension σ to Ex. 12.1.2, write for the terminal velocity of a gas bubble rising in a liquid

$$V = f(g\,\Delta\rho, D, \rho, \mu, \sigma).$$

Find the dimensionless numbers associated with this problem by the Π-theorem and physical similitude. Show that the result may be expressed as

$$\text{Re} = f(\text{Gr}, \text{Bo}),$$

where $\text{Bo} = g\,\Delta\rho\, D^2/\sigma$ is the *Bond number.*

12-3 A wooden sphere is held by a string in a water stream (Fig. 12P-3). Determine the string force by means of dimensional analysis.

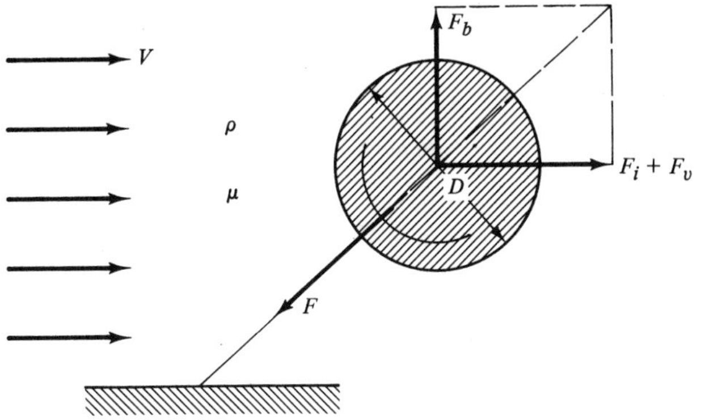

Fig. 12P-3

12-4 Express in terms of appropriate dimensionless numbers the diameter of droplets formed by a liquid jet discharging with a specified velocity from a horizontal tube.

12-5 Express in terms of appropriate dimensionless numbers the diameter of droplets formed by a liquid jet discharging under the effect of gravity from a vertical tube.

12-6 Consider the motion of a viscous fluid between two concentric cylinders in relative motion (Fig. 12P-6). In a manner conceptually identical to the definition of the Grashof

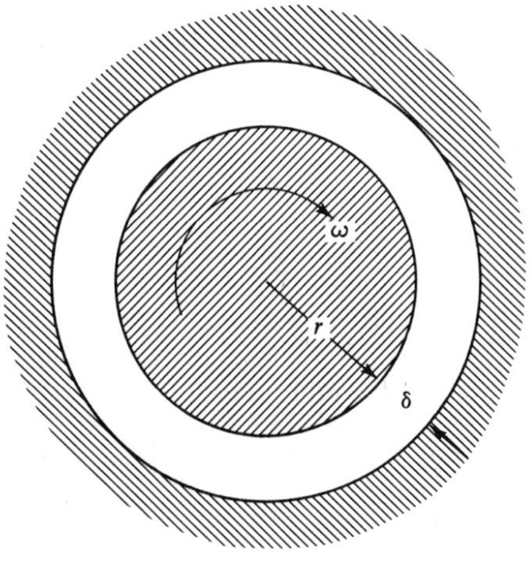

Fig. 12P-6

number, define the Taylor number as

$$\text{Ta} = \left(\frac{\text{centrifugal}}{\text{viscous}} \times \frac{\text{inertial}}{\text{viscous}}\right) \text{forces}$$

and show that

$$\text{Ta} = \frac{r\omega^2 d^3}{\nu^2}.$$

12-7 Consider the boundary layer flow along a concave surface (Fig. 12P-7). Rearrange the statement of the Taylor number in terms of the free stream velocity U and the boundary layer thickness δ. The resulting dimensionless number,

$$\text{Gö} = \left(\frac{U\delta}{\nu}\right)^2 \frac{\delta}{r} = \text{Re}_\delta^2\left(\frac{\delta}{r}\right),$$

is known to be the *Görtler number.*

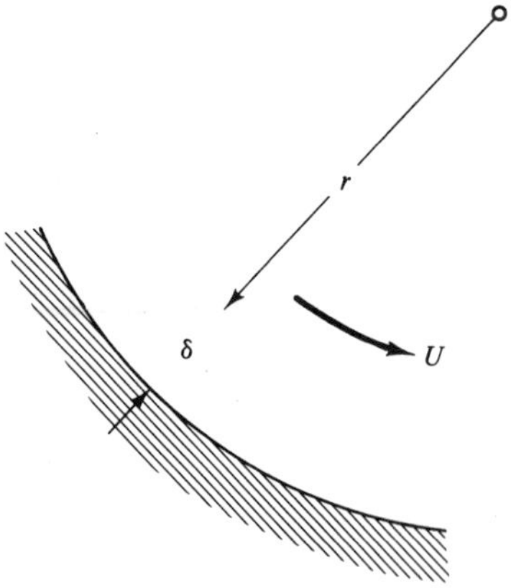

Fig. 12P-7

12-8 A cylinder of diameter D rotates steadily with angular velocity ω in a fluid streaming with velocity U (Fig. 12P-8). Show by means of a dimensional analysis that

$$\text{torque} = f(\text{Re}, \text{Ta}),$$

where Re is the Reynolds number and Ta is the Taylor number.

12-9 Reconsider forced convection

$$h = f(D, V, \rho, \mu, c_p, k). \tag{12.2.1}$$

(a) Start the elimination process by eliminating $[M]$, first with μ, and second with k.
(b) Eliminating $[\theta]$, $[M]$, and $[T]$, respectively, show that the dynamic, kinematic, and geometric equivalents of Eq. (12.2.1) are:

$$\frac{h}{k} = f_1\left(V, D, \rho, \mu, \frac{c_p}{k}\right), \quad \text{dynamic equivalent}$$

$$\frac{h}{k} = f_2(V, D, \nu, a), \quad \text{kinematic equivalent}$$

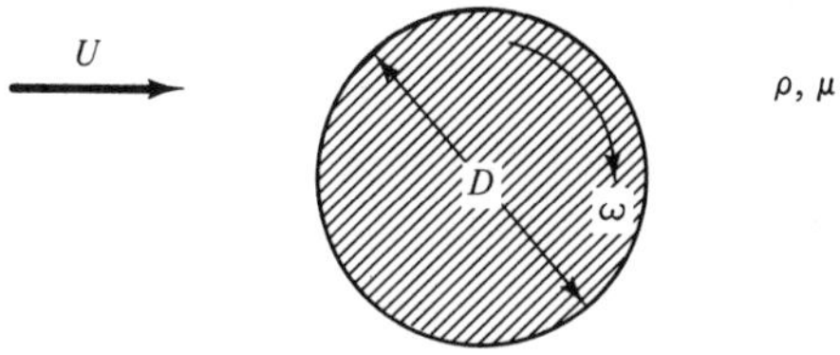

Fig. 12P-8

$$\frac{h}{k} = f_3\left(\frac{V}{\nu}, \frac{V}{a}, D\right), \qquad \text{geometric equivalent}$$

$$\frac{hD}{k} = f_4\left(\frac{VD}{\nu}, \frac{VD}{a}\right), \qquad \text{dimensionless.}$$

12-10 Reconsider natural convection

$$h = f(D, g\beta\,\Delta T, \rho, \mu, c_p, k). \tag{12.2.12}$$

(a) Start the elimination process by eliminating $[M]$, first with ρ and second with μ.
(b) Elimiting $[\theta]$, $[M]$, and $[T]$, respectively, show the dynamic, kinematic, and geometric equivalents of Eq. (12.2.12).

12-11 Consider the natural convection from a horizontal cylinder rotating with angular velocity ω (Fig. 12P-11). The peripheral surface temperature of the cylinder is T_w and the ambient temperature is T_∞. The diameter of the cylinder is D. Assuming that natural convection resulting from rotation and that from gravity can be superimposed, express the Nusselt number in terms of appropriate dimensionless numbers.

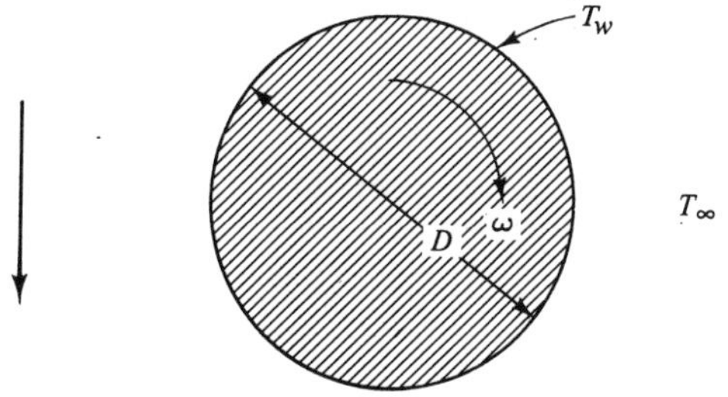

Fig. 12P-11

12-12 Consider the correlations

$$\text{Nu} \sim \text{Re}^{3/4-4/5}\,\text{Pr}^{1/3-2/5}$$

for forced convection, and

$$\text{Nu} \sim \text{Ra}^{1/3-2/5}$$

for natural convection.

Show that the correlation for combined convection may be written as

$$\text{Nu} \sim (1 + \text{Ri}^{2/5})(\text{Re}^2\,\text{Pr})^{2/5}$$

for the case of dominant forced convection, and as

$$\mathrm{Nu} \sim (1 + \mathrm{Ri}^{-2/5})\mathrm{Ra}^{2/5}$$

for the case of dominant natural convection.

12-13 Discuss the physics of a hot-air balloon in terms of appropriate dimensionless numbers.

12-14 Investigate the dimensionless numbers correlating heat transfer data on film boiling.

12-15 Find the terminal velocity of the falling sphere (Fig. 12.4) by means of two-length and three-length dimensional analyses.

12-16 Employing two-length and three-length dimensional analysis, find the Nusselt number of a combined convection boundary layer for the case of dominant
(a) Forced convection.
(b) Natural convection.

APPENDIXES

APPENDIX A. SOME FORMULAS OF VECTOR AND CARTESIAN TENSOR CALCULUS

A.1 Notation and Basic Expressions

Range convention: Each suffix takes values 1, 2, 3, referring to Cartesian axes.

Summation convention: Twice-repeated suffix in product is called a dummy index and held to be summed over the range of its values.

Suffix in parentheses: Suffix enclosed in parentheses does not denote component, but refers to a surface normal. For example, $\boldsymbol{e}_{(i)}$ is unit vector in x_i.

Notation: Scalar (lightface italic), p Vector (boldface italic), $\boldsymbol{V}$ Tensor (boldface Greek), $\boldsymbol{\sigma}$

Unit vectors:

$$\boldsymbol{e}_{(1)} = (1, 0, 0); \qquad \boldsymbol{e}_{(2)} = (0, 1, 0); \qquad \boldsymbol{e}_{(3)} = (0, 0, 1)$$

Unit normal to surface:

$$\boldsymbol{n} = \boldsymbol{e}_{(1)}n_1 + \boldsymbol{e}_{(2)}n_2 + \boldsymbol{e}_{(3)}n_3,$$

where n_i denotes direction cosines.

Kronecker delta:

$$\boldsymbol{\delta} \equiv \delta_{ij} = \begin{cases} 1 & \text{if } i = j \\ 0 & \text{if } i \neq j \end{cases}$$

Permutation symbol:

$$\epsilon_{ijk} = \begin{cases} 1 & \text{if } ijk \text{ is an even permutation (123, 231, 312)} \\ 0 & \text{if any two of } i, j, k \text{ are the same} \\ -1 & \text{if } ijk \text{ is an odd permutation (321, 213, 132)} \end{cases}$$

A.2 Expressions Involving the Nabla Operator

Nabla (del) operator:

$$\nabla = \boldsymbol{e}_{(1)} \frac{\partial}{\partial x_1} + \boldsymbol{e}_{(2)} \frac{\partial}{\partial x_2} + \boldsymbol{e}_{(3)} \frac{\partial}{\partial x_3}$$

Gradient of scalar (a vector):

$$\nabla p = \boldsymbol{e}_{(1)} \frac{\partial p}{\partial x_1} + \boldsymbol{e}_{(2)} \frac{\partial p}{\partial x_2} + \boldsymbol{e}_{(3)} \frac{\partial p}{\partial x_3}$$

Divergence of vector (a scalar):

$$\nabla \cdot \boldsymbol{V} = \frac{\partial V_1}{\partial x_1} + \frac{\partial V_2}{\partial x_2} + \frac{\partial V_3}{\partial x_3}$$

Divergence of gradient, Laplacian:

$$\nabla \cdot \nabla = \nabla^2 = \frac{\partial^2}{\partial x_1^2} + \frac{\partial^2}{\partial x_2^2} + \frac{\partial^2}{\partial x_3^2}$$

Curl of vector (a vector):

$$\nabla \times \boldsymbol{V} = \begin{vmatrix} \boldsymbol{e}_{(1)} & \boldsymbol{e}_{(2)} & \boldsymbol{e}_{(3)} \\ \dfrac{\partial}{\partial x_1} & \dfrac{\partial}{\partial x_2} & \dfrac{\partial}{\partial x_3} \\ V_1 & V_2 & V_3 \end{vmatrix}$$

$$= \boldsymbol{e}_{(1)}\left(\frac{\partial V_3}{\partial x_2} - \frac{\partial V_2}{\partial x_3}\right) + \boldsymbol{e}_{(2)}\left(\frac{\partial V_1}{\partial x_3} - \frac{\partial V_3}{\partial x_1}\right) + \boldsymbol{e}_{(3)}\left(\frac{\partial V_2}{\partial x_1} - \frac{\partial V_1}{\partial x_2}\right)$$

Gradient of vector (a tensor):

$$\nabla \boldsymbol{V} = \begin{vmatrix} \dfrac{\partial V_1}{\partial x_1} & \dfrac{\partial V_2}{\partial x_1} & \dfrac{\partial V_3}{\partial x_1} \\ \dfrac{\partial V_1}{\partial x_2} & \dfrac{\partial V_2}{\partial x_2} & \dfrac{\partial V_3}{\partial x_2} \\ \dfrac{\partial V_1}{\partial x_3} & \dfrac{\partial V_2}{\partial x_3} & \dfrac{\partial V_3}{\partial x_3} \end{vmatrix}$$

Divergence of tensor (a vector):

$$\nabla \cdot \boldsymbol{\sigma} = \boldsymbol{e}_{(1)}\left(\frac{\partial \sigma_{11}}{\partial x_1} + \frac{\partial \sigma_{21}}{\partial x_2} + \frac{\partial \sigma_{31}}{\partial x_3}\right) + \boldsymbol{e}_{(2)}\left(\frac{\partial \sigma_{12}}{\partial \sigma_1} + \frac{\partial \sigma_{22}}{\partial x_2} + \frac{\partial \sigma_{32}}{\partial x_3}\right)$$
$$+ \boldsymbol{e}_{(3)}\left(\frac{\partial \sigma_{13}}{\partial x_1} + \frac{\partial \sigma_{23}}{\partial x_2} + \frac{\partial \sigma_{33}}{\partial x_3}\right)$$

A.3 Other Products and Operations

Regarding scalars as zero-order tensors and vectors (including the nabla operator) as first-order tensors, the tensor order of a product equals the sum of orders of factors less the "order" of the multiplication sign given in the following table:

Multiplication Sign		"Order" of Sign
	(none)	0
$\times$	(cross)	1
$\cdot$	(dot)	2
:	(double dot)	4

Scalar (dot) product of vectors:

$$\boldsymbol{U} \cdot \boldsymbol{V} = U_1 V_1 + U_2 V_2 + U_3 V_3$$

Vector (cross) product of vectors:

$$\boldsymbol{U} \times \boldsymbol{V} = \begin{vmatrix} \boldsymbol{e}_{(1)} & \boldsymbol{e}_{(2)} & \boldsymbol{e}_{(3)} \\ U_1 & U_2 & U_3 \\ V_1 & V_2 & V_3 \end{vmatrix}$$
$$= \boldsymbol{e}_{(1)}(U_2 V_3 - U_3 V_2) + \boldsymbol{e}_{(2)}(U_3 V_1 - U_1 V_3) + \boldsymbol{e}_{(3)}(U_1 V_2 - U_2 V_1)$$

Tensor product of vectors (dyad):

$$\boldsymbol{U}\boldsymbol{V} = \begin{vmatrix} U_1 V_1 & U_1 V_2 & U_1 V_3 \\ U_2 V_1 & U_2 V_2 & U_2 V_3 \\ U_3 V_1 & U_3 V_2 & U_3 V_3 \end{vmatrix}$$

Vector–tensor (dot) product:

$$\boldsymbol{\sigma} \cdot \boldsymbol{V} = \boldsymbol{e}_{(1)}(\sigma_{11} V_1 + \sigma_{12} V_2 + \sigma_{13} V_3) + \boldsymbol{e}_{(2)}(\sigma_{21} V_1 + \sigma_{22} V_2 + \sigma_{23} V_3)$$
$$+ \boldsymbol{e}_{(3)}(\sigma_{31} V_1 + \sigma_{32} V_2 + \sigma_{33} V_3)$$

Note that $\boldsymbol{\sigma} \cdot \boldsymbol{V} = \boldsymbol{V} \cdot \boldsymbol{\sigma}^T$, where $\boldsymbol{\sigma}^T$ denotes the transpose of $\boldsymbol{\sigma}$. When $\boldsymbol{\sigma}$ is symmetric, $\boldsymbol{\sigma} \cdot \boldsymbol{V} = \boldsymbol{V} \cdot \boldsymbol{\sigma}$. Also, $\boldsymbol{V} \cdot \nabla \boldsymbol{V} = (\boldsymbol{V} \cdot \nabla)\boldsymbol{V} = \boldsymbol{V} \cdot (\nabla \boldsymbol{V})$.

Vector–tensor (cross) product:

Note that $\boldsymbol{\nu} = \boldsymbol{r} \times \boldsymbol{\sigma}$ implies that $\nu_{lj} = \epsilon_{lmi} x_m \sigma_{ij}$, and that $\boldsymbol{\pi} = \boldsymbol{\sigma} \times \boldsymbol{r}$ implies that $\pi_{il} = \epsilon_{ljm} \sigma_{ij} x_m = -\epsilon_{lmj} x_m \sigma_{ij}$; hence

$$-\boldsymbol{\sigma} \times \boldsymbol{r} = \begin{vmatrix} x_2\sigma_{13} - x_3\sigma_{12} & x_3\sigma_{11} - x_1\sigma_{13} & x_1\sigma_{12} - x_2\sigma_{11} \\ x_2\sigma_{23} - x_3\sigma_{22} & x_3\sigma_{21} - x_1\sigma_{23} & x_1\sigma_{22} - x_2\sigma_{21} \\ x_2\sigma_{33} - x_3\sigma_{32} & x_3\sigma_{31} - x_1\sigma_{33} & x_1\sigma_{32} - x_2\sigma_{31} \end{vmatrix}$$

Tensor–tensor (double dot) product:

$$\boldsymbol{\sigma} : \nabla \boldsymbol{V} = \sigma_{11}\frac{\partial V_1}{\partial x_1} + \sigma_{12}\frac{\partial V_2}{\partial x_1} + \sigma_{13}\frac{\partial V_3}{\partial x_1} + \sigma_{21}\frac{\partial V_1}{\partial x_2} + \sigma_{22}\frac{\partial V_2}{\partial x_2} + \sigma_{23}\frac{\partial V_3}{\partial x_2} + \sigma_{31}\frac{\partial V_1}{\partial x_3} + \sigma_{32}\frac{\partial V_2}{\partial x_3} + \sigma_{33}\frac{\partial V_3}{\partial x_3}$$

Other vector–tensor products:

$$(\nabla \cdot \boldsymbol{\sigma}) \cdot \boldsymbol{V} = \left(\frac{\partial \sigma_{11}}{\partial x_1} + \frac{\partial \sigma_{21}}{\partial x_2} + \frac{\partial \sigma_{31}}{\partial x_3}\right) V_1 + \left(\frac{\partial \sigma_{12}}{\partial x_1} + \frac{\partial \sigma_{22}}{\partial x_2} + \frac{\partial \sigma_{32}}{\partial x_3}\right) V_2 + \left(\frac{\partial \sigma_{13}}{\partial x_1} + \frac{\partial \sigma_{23}}{\partial x_2} + \frac{\partial \sigma_{33}}{\partial x_3}\right) V_3$$

$$\nabla \cdot (\boldsymbol{\sigma} \cdot \boldsymbol{V}) = \frac{\partial}{\partial x_1}(\sigma_{11}V_1 + \sigma_{12}V_2 + \sigma_{13}V_3) + \frac{\partial}{\partial x_2}(\sigma_{21}V_1 + \sigma_{22}V_2 + \sigma_{23}V_3) + \frac{\partial}{\partial x_3}(\sigma_{31}V_1 + \sigma_{32}V_2 + \sigma_{33}V_3)$$

APPENDIX B. FLOW AND FLUX REPRESENTATIONS

GEOMETRIC INTERPRETATION

Differential volume (point)

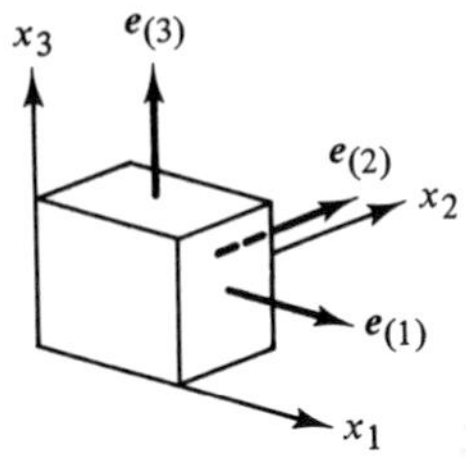

Differential surface

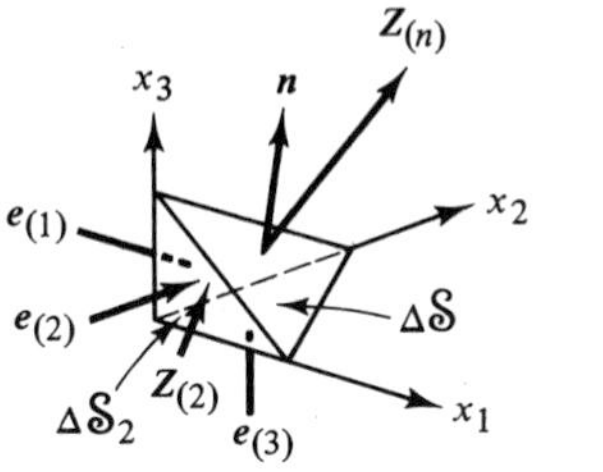

$n_i = \Delta S_i / \Delta S$ (direction cosine)

Change over differential volume

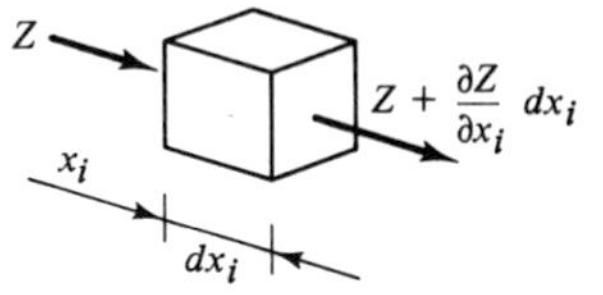

MATHEMATICAL REPRESENTATION

Definition		Symbolic	Indicial	Symbolic	Indicial
Mass (flow)	ρV_i	$\boldsymbol{n} \cdot (\rho \boldsymbol{V})$	$n_i \rho V_i$	$\nabla \cdot (\rho \boldsymbol{V})$	$\frac{\partial}{\partial x_i}(\rho V_i)$
Momentum (flow)	$\rho V_i \boldsymbol{V}$	$\boldsymbol{n} \cdot (\rho \boldsymbol{V}\boldsymbol{V})$	$n_i \rho V_i V_j$	$\nabla \cdot (\rho \boldsymbol{V}\boldsymbol{V})$	$\frac{\partial}{\partial x_i}(\rho V_i V_j)$
Stress (flux)	$\boldsymbol{S}_{(i)}$	$\boldsymbol{S}_{(n)} = \boldsymbol{n} \cdot \boldsymbol{\sigma}$	$n_i \sigma_{ij}$	$\nabla \cdot \boldsymbol{\sigma}$	$\frac{\partial \sigma_{ij}}{\partial x_i}$
Moment of stress (flux)	$\boldsymbol{r} \times \boldsymbol{S}_{(i)}$	$\boldsymbol{r} \times \boldsymbol{S}_{(n)} = \boldsymbol{r} \times (\boldsymbol{n} \cdot \boldsymbol{\sigma})$	$n_i \epsilon_{lmj} x_m \sigma_{ij}$	$\nabla \cdot (-\boldsymbol{\sigma} \times \boldsymbol{r})$	$\frac{\partial}{\partial x_i}(\epsilon_{lmj} x_m \sigma_{ij})$
Power by stress (flux)	$\boldsymbol{S}_{(i)} \cdot \boldsymbol{V}$	$\boldsymbol{S}_{(n)} \cdot \boldsymbol{V} = \boldsymbol{n} \cdot (\boldsymbol{\sigma} \cdot \boldsymbol{V})$	$n_i \sigma_{ij} V_j$	$\nabla \cdot (\boldsymbol{\sigma} \cdot \boldsymbol{V})$	$\frac{\partial}{\partial x_i}(\sigma_{ij} V_j)$
Heat (flux)	q_i	$\boldsymbol{n} \cdot \boldsymbol{q}$	$n_i q_i$	$\nabla \cdot \boldsymbol{q}$	$\frac{\partial q_i}{\partial x_i}$

APPENDIX C. CYLINDRICAL AND SPHERICAL COORDINATES

C.1 Cylindrical Coordinates

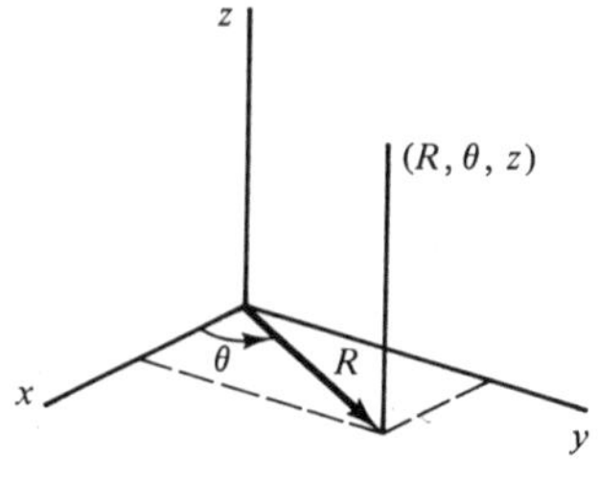

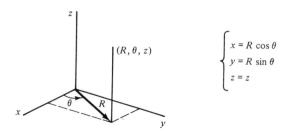

$$\begin{cases} x = R\cos\theta \\ y = R\sin\theta \\ z = z \end{cases}$$

Velocity vector:

$$\boldsymbol{V} = (V_R, V_\theta, V_z)$$

Gradient of scalar:

$$\nabla T = \left(\frac{\partial T}{\partial R}, \frac{1}{R}\frac{\partial T}{\partial \theta}, \frac{\partial T}{\partial z}\right)$$

Divergence of vector:

$$\nabla \cdot \boldsymbol{V} = \frac{1}{R}\frac{\partial (RV_R)}{\partial R} + \frac{1}{R}\frac{\partial V_\theta}{\partial \theta} + \frac{\partial V_z}{\partial z}$$

Curl of vector:

$$\nabla \times \boldsymbol{V} = \left(\frac{1}{R}\frac{\partial V_z}{\partial \theta} - \frac{\partial V_\theta}{\partial z}, \frac{\partial V_R}{\partial z} - \frac{\partial V_z}{\partial R}, \frac{1}{R}\frac{\partial (RV_\theta)}{\partial R} - \frac{1}{R}\frac{\partial V_R}{\partial \theta}\right)$$

Laplacian:

$$\nabla^2 T = \frac{1}{R}\frac{\partial}{\partial R}\left(R\frac{\partial T}{\partial R}\right) + \frac{1}{R^2}\frac{\partial^2 T}{\partial \theta^2} + \frac{\partial^2 T}{\partial z^2}$$

Substantial time derivative of scalar T:

$$\frac{DT}{Dt} = \frac{\partial T}{\partial t} + \boldsymbol{V} \cdot \nabla T$$

Substantial time derivative of vector $\boldsymbol{V}$:

$$\left(\frac{D\boldsymbol{V}}{Dt}\right)_R = \frac{\partial V_R}{\partial t} + V_R\frac{\partial V_R}{\partial R} + \frac{V_\theta}{R}\frac{\partial V_R}{\partial \theta} - \frac{V_\theta^2}{R} + V_z\frac{\partial V_R}{\partial z}$$

$$\left(\frac{D\boldsymbol{V}}{Dt}\right)_\theta = \frac{\partial V_\theta}{\partial t} + V_R\frac{\partial V_\theta}{\partial R} + \frac{V_\theta}{R}\frac{\partial V_\theta}{\partial \theta} + \frac{V_R V_\theta}{R} + V_z\frac{\partial V_\theta}{\partial z}$$

$$\left(\frac{D\boldsymbol{V}}{Dt}\right)_z = \frac{\partial V_z}{\partial t} + V_R\frac{\partial V_z}{\partial R} + \frac{V_\theta}{R}\frac{\partial V_z}{\partial \theta} + V_z\frac{\partial V_z}{\partial z}$$

Divergence of tensor $\boldsymbol{\tau}$:

$$(\nabla \cdot \boldsymbol{\tau})_R = \frac{1}{R}\frac{\partial}{\partial R}(R\tau_{RR}) + \frac{1}{R}\frac{\partial \tau_{\theta R}}{\partial \theta} - \frac{\tau_{\theta\theta}}{R} + \frac{\partial \tau_{zR}}{\partial z}$$

$$(\nabla \cdot \tau)_\theta = \frac{1}{R^2}\frac{\partial}{\partial R}(R^2\tau_{R\theta}) + \frac{1}{R}\frac{\partial \tau_{\theta\theta}}{\partial \theta} + \frac{\partial \tau_{z\theta}}{\partial z}$$

$$(\nabla \cdot \tau)_z = \frac{1}{R}\frac{\partial}{\partial R}(R\tau_{Rz}) + \frac{1}{R}\frac{\partial \tau_{\theta z}}{\partial \theta} + \frac{\partial \tau_{zz}}{\partial z}$$

Components of stress tensor of Newton fluid:

$$\tau_{RR} = \mu\left(2\frac{\partial V_R}{\partial R} - \frac{2}{3}\nabla \cdot V\right)$$

$$\tau_{\theta\theta} = \mu\left[2\left(\frac{1}{R}\frac{\partial V_\theta}{\partial \theta} + \frac{V_R}{R}\right) - \frac{2}{3}\nabla \cdot V\right]$$

$$\tau_{zz} = \mu\left(2\frac{\partial V_z}{\partial z} - \frac{2}{3}\nabla \cdot V\right)$$

$$\tau_{R\theta} = \tau_{\theta R} = \mu\left[R\frac{\partial}{\partial R}\left(\frac{V_\theta}{R}\right) + \frac{1}{R}\frac{\partial V_R}{\partial \theta}\right]$$

$$\tau_{Rz} = \tau_{zR} = \mu\left(\frac{\partial V_z}{\partial R} + \frac{\partial V_R}{\partial z}\right)$$

$$\tau_{\theta z} = \tau_{z\theta} = \mu\left(\frac{\partial V_\theta}{\partial z} + \frac{1}{R}\frac{\partial V_z}{\partial \theta}\right)$$

Viscous dissipation:

$$\mu\Phi_v = \tau_{RR}\frac{\partial V_R}{\partial R} + \tau_{\theta\theta}\left(\frac{1}{R}\frac{\partial V_\theta}{\partial \theta} + \frac{V_R}{R}\right) + \tau_{zz}\frac{\partial V_z}{\partial z}$$
$$+ \tau_{R\theta}\left[R\frac{\partial}{\partial R}\left(\frac{V_\theta}{R}\right) + \frac{1}{R}\frac{\partial V_R}{\partial \theta}\right] + \tau_{Rz}\left(\frac{\partial V_z}{\partial R} + \frac{\partial V_R}{\partial z}\right) + \tau_{\theta z}\left(\frac{1}{R}\frac{\partial V_z}{\partial \theta} + \frac{\partial V_\theta}{\partial z}\right)$$

GOVERNING EQUATIONS OF CONSTANT-PROPERTY, INCOMPRESSIBLE CONVECTION:

Conservation of mass:

$$\frac{1}{R}\frac{\partial}{\partial R}(RV_R) + \frac{1}{R}\frac{\partial V_\theta}{\partial \theta} + \frac{\partial V_z}{\partial z} = 0$$

Balance of momentum:

$$\frac{\partial V_R}{\partial t} + V_R\frac{\partial V_R}{\partial R} + \frac{V_\theta}{R}\frac{\partial V_R}{\partial \theta} - \frac{V_\theta^2}{R} + V_z\frac{\partial V_R}{\partial z}$$
$$= f_R - \frac{1}{\rho}\frac{\partial p}{\partial R} + \nu\left[\frac{\partial}{\partial R}\left(\frac{1}{R}\frac{\partial(RV_R)}{\partial R}\right) + \frac{1}{R^2}\frac{\partial^2 V_R}{\partial \theta^2} - \frac{2}{R^2}\frac{\partial V_\theta}{\partial \theta} + \frac{\partial^2 V_R}{\partial z^2}\right]$$

$$\frac{\partial V_\theta}{\partial t} + V_R\frac{\partial V_\theta}{\partial R} + \frac{V_\theta}{R}\frac{\partial V_\theta}{\partial \theta} + \frac{V_R V_\theta}{R} + V_z\frac{\partial V_\theta}{\partial z}$$
$$= f_\theta - \frac{1}{\rho R}\frac{\partial p}{\partial \theta} + \nu\left[\frac{\partial}{\partial R}\left(\frac{1}{R}\frac{\partial(RV_\theta)}{\partial R}\right) + \frac{1}{R^2}\frac{\partial^2 V_\theta}{\partial \theta^2} + \frac{2}{R^2}\frac{\partial V_R}{\partial \theta} + \frac{\partial^2 V_\theta}{\partial z^2}\right]$$

$$\frac{\partial V_z}{\partial t} + V_R\frac{\partial V_z}{\partial R} + \frac{V_\theta}{R}\frac{\partial V_z}{\partial \theta} + V_z\frac{\partial V_z}{\partial z}$$

$$= f_z - \frac{1}{\rho}\frac{\partial p}{\partial z} + \nu\left[\frac{1}{R}\frac{\partial}{\partial R}\left(R\frac{\partial V_z}{\partial R}\right) + \frac{1}{R^2}\frac{\partial^2 V_z}{\partial \theta^2} + \frac{\partial^2 V_z}{\partial z^2}\right]$$

Balance of thermal energy:

$$\frac{\partial T}{\partial t} + V_R\frac{\partial T}{\partial R} + \frac{V_\theta}{R}\frac{\partial T}{\partial \theta} + V_z\frac{\partial T}{\partial z}$$

$$= a\left[\frac{1}{R}\frac{\partial}{\partial R}\left(R\frac{\partial T}{\partial R}\right) + \frac{1}{R^2}\frac{\partial^2 T}{\partial \theta^2} + \frac{\partial^2 T}{\partial z^2}\right] + \frac{\nu}{c}\Phi_v + \frac{u'''}{\rho c}$$

Dissipation function:

$$\Phi_v = \left(\frac{\partial V_\theta}{\partial z} + \frac{1}{R}\frac{\partial V_z}{\partial \theta}\right)^2 + \left(\frac{\partial V_z}{\partial R} + \frac{\partial V_R}{\partial z}\right)^2 + \left[\frac{1}{R}\frac{\partial V_R}{\partial \theta} + R\frac{\partial}{\partial R}\left(\frac{V_\theta}{R}\right)\right]^2$$

$$+ 2\left\{\left(\frac{\partial V_R}{\partial R}\right)^2 + \left[\frac{1}{R}\left(\frac{\partial V_\theta}{\partial \theta} + V_R\right)\right]^2 + \left(\frac{\partial V_z}{\partial z}\right)^2\right\}$$

C.2 Spherical Polar Coordinates

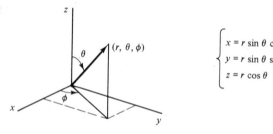

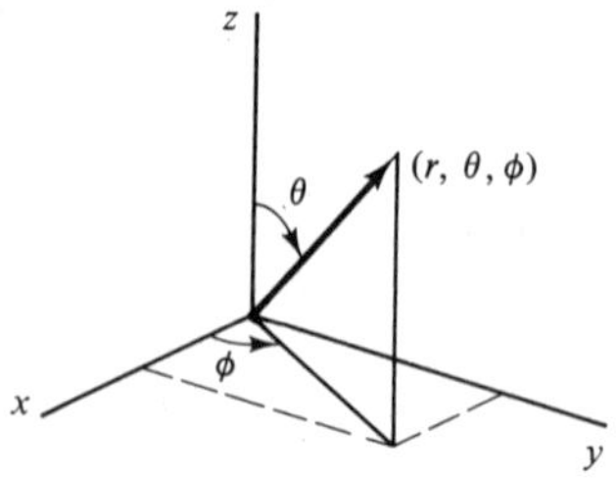

$$\begin{cases} x = r\sin\theta\cos\phi \\ y = r\sin\theta\sin\phi \\ z = r\cos\theta \end{cases}$$

Velocity vector:

$$\boldsymbol{V} = (V_r, V_\theta, V_\varphi)$$

Gradient of scalar:

$$\nabla T = \left(\frac{\partial T}{\partial r}, \frac{1}{r}\frac{\partial T}{\partial \theta}, \frac{1}{r\sin\theta}\frac{\partial T}{\partial \varphi}\right)$$

Divergence of vector:

$$\nabla \cdot \boldsymbol{V} = \frac{1}{r^2}\frac{\partial}{\partial r}(r^2 V_r) + \frac{1}{r\sin\theta}\frac{\partial}{\partial \theta}(V_\theta \sin\theta) + \frac{1}{r\sin\theta}\frac{\partial V_\varphi}{\partial \varphi}$$

Curl of vector:

$$\nabla \times \boldsymbol{V} = \left\{\frac{1}{r\sin\theta}\frac{\partial}{\partial \theta}(V_\varphi \sin\theta) - \frac{1}{r\sin\theta}\frac{\partial V_\theta}{\partial \varphi},\right.$$

$$\left.\frac{1}{r\sin\theta}\frac{\partial V_r}{\partial \varphi} - \frac{1}{r}\frac{\partial(rV_\varphi)}{\partial r}, \frac{1}{r}\frac{\partial(rV_\theta)}{\partial r} - \frac{1}{r}\frac{\partial V_r}{\partial \theta}\right\}$$

Laplacian:

$$\nabla^2 T = \frac{1}{r^2}\frac{\partial}{\partial r}\left(r^2 \frac{\partial T}{\partial r}\right) + \frac{1}{r^2 \sin\theta}\frac{\partial}{\partial\theta}\left(\sin\theta \frac{\partial T}{\partial\theta}\right) + \frac{1}{r^2 \sin^2\theta}\frac{\partial^2 T}{\partial\varphi^2}$$

Substantial time derivative of scalar T:

$$\frac{DT}{Dt} = \frac{\partial T}{\partial t} + \boldsymbol{V} \cdot \nabla T$$

GOVERNING EQUATIONS OF INCOMPRESSIBLE, CONSTANT-PROPERTY CONVECTION:

Conservation of mass:

$$\frac{1}{r^2}\frac{\partial}{\partial r}(r^2 V_r) + \frac{1}{r\sin\theta}\frac{\partial}{\partial\theta}(V_\theta \sin\theta) + \frac{1}{r\sin\theta}\frac{\partial V_\varphi}{\partial\varphi} = 0$$

Balance of momentum:

$$\frac{\partial V_r}{\partial t} + V_r\frac{\partial V_r}{\partial r} + \frac{V_\theta}{r}\frac{\partial V_r}{\partial\theta} + \frac{V_\varphi}{r\sin\theta}\frac{\partial V_r}{\partial\theta} - \frac{V_\theta^2 + V_\varphi^2}{r}$$

$$= f_r - \frac{1}{\rho}\frac{\partial p}{\partial r} + \nu\left(\nabla^2 V_r - \frac{2V_r}{r^2} - \frac{2}{r^2}\frac{\partial V_\theta}{\partial\theta} - \frac{2}{r^2}V_\theta\cot\theta - \frac{2}{r^2\sin\theta}\frac{\partial V_\varphi}{\partial\varphi}\right)$$

$$\frac{\partial V_\theta}{\partial t} + V_r\frac{\partial V_\theta}{\partial r} + \frac{V_\theta}{r}\frac{\partial V_\theta}{\partial\theta} + \frac{V_\varphi}{r\sin\theta}\frac{\partial V_\theta}{\partial\varphi} + \frac{V_r V_\theta}{r} - \frac{V_\varphi^2\cot\theta}{r}$$

$$= f_\theta - \frac{1}{\rho r}\frac{\partial p}{\partial\theta} + \nu\left(\nabla^2 V_\theta + \frac{2}{r^2}\frac{\partial V_r}{\partial\theta} - \frac{V_\theta}{r^2\sin^2\theta} - \frac{2\cos\theta}{r^2\sin^2\theta}\frac{\partial V_\varphi}{\partial\varphi}\right)$$

$$\frac{\partial V_\varphi}{\partial t} + V_r\frac{\partial V_\varphi}{\partial r} + \frac{V_\theta}{r}\frac{\partial V_\varphi}{\partial\theta} + \frac{V_\varphi}{r\sin\theta}\frac{\partial V_\varphi}{\partial\varphi} + \frac{V_\varphi V_r}{r} + \frac{V_\theta V_r}{r}\cot\theta$$

$$= f_\varphi - \frac{1}{\rho r\sin\theta}\frac{\partial p}{\partial\varphi} + \nu\left(\nabla^2 V_\varphi - \frac{V_\varphi}{r^2\sin^2\theta} + \frac{2}{r^2\sin\theta}\frac{\partial V_r}{\partial\varphi} + \frac{2\cos\theta}{r^2\sin^2\theta}\frac{\partial V_\theta}{\partial\varphi}\right)$$

Balance of thermal energy:

$$\frac{\partial T}{\partial t} + V_r\frac{\partial T}{\partial r} + \frac{V_\theta}{r}\frac{\partial T}{\partial\theta} + \frac{V_\varphi}{r\sin\theta}\frac{\partial T}{\partial\varphi} = a\,\nabla^2 T + \frac{\nu}{c}\Phi_v + \frac{u'''}{\rho c}$$

Dissipation function:

$$\Phi_v = \left[r\frac{\partial}{\partial r}\left(\frac{V_\theta}{r}\right) + \frac{1}{r}\frac{\partial V_r}{\partial\theta}\right]^2 + \left[\frac{1}{r\sin\theta}\frac{\partial V_r}{\partial\varphi} + r\frac{\partial}{\partial r}\left(\frac{V_\varphi}{r}\right)\right]^2$$

$$+ \left[\frac{\sin\theta}{r}\frac{\partial}{\partial\theta}\left(\frac{V_\varphi}{\sin\theta}\right) + \frac{1}{r\sin\theta}\frac{\partial V_\theta}{\partial\varphi}\right]^2$$

$$+ 2\left[\left(\frac{\partial V_r}{\partial r}\right)^2 + \left(\frac{1}{r}\frac{\partial V_\theta}{\partial\theta} + \frac{V_r}{r}\right)^2 + \left(\frac{1}{r\sin\theta}\frac{\partial V_\varphi}{\partial\varphi} + \frac{V_r}{r} + \frac{V_\theta\cot\theta}{r}\right)^2\right]$$

APPENDIX D. SOME DIMENSIONLESS NUMBERS

Bingham:	$\mathrm{Bg} = \dfrac{\tau_0 L}{\mu_1 U}$
Biot:	$\mathrm{Bi} = \dfrac{h l_s}{k_s}$
Bond:	$\mathrm{Bo} = \dfrac{\Delta \rho g l^2}{\sigma_s}$
Colburn factor:	$C_H = \dfrac{\mathrm{Nu}}{\mathrm{Re}\,\mathrm{Pr}^{1/3}}$
Eckert:	$\mathrm{Ec} = \dfrac{U^2}{c\,\Delta T}$
Friction factor:	$c_f = \dfrac{\tau_w}{\frac{1}{2}\rho U^2}$
Friction factor:	$f_m = \dfrac{(-dp/dx)D_e}{\frac{1}{2}\rho U^2} \qquad (= \tfrac{1}{4}c_f)$
Froude:	$\mathrm{Fr} = \dfrac{U}{\sqrt{gD}}$
Görtler:	$\mathrm{G\ddot{o}} = 2\left(\dfrac{u_0 \delta}{\nu}\right)^2 \left(\dfrac{\delta}{r}\right)$
Graetz:	$\mathrm{Gz} = \dfrac{aU/l}{x/l}$
Grashof:	$\mathrm{Gr} = \dfrac{g\beta\,\Delta T\, l^3}{\nu^2}$
Jakob:	$\mathrm{Ja} = \dfrac{c\,\Delta T}{h_{lv}}$
Nusselt:	$\mathrm{Nu} = \dfrac{hl}{k}$
Péclet:	$\mathrm{Pe} = \dfrac{Ul}{a}$
Prandtl:	$\mathrm{Pr} = \dfrac{\nu}{a}$
Rayleigh:	$\mathrm{Ra} = \dfrac{g\beta\,\Delta T\, l^3}{\nu a} \qquad (= \mathrm{Gr}\,\mathrm{Pr})$
Reynolds:	$\mathrm{Re} = \dfrac{Ul}{\nu}$
Richardson:	$\mathrm{Ri} = \dfrac{g\beta\,\Delta T\, l}{U^2} \qquad \left(= \dfrac{\mathrm{Gr}}{\mathrm{Re}^2}\right)$
Stanton:	$\mathrm{St} = \dfrac{h}{\rho c U} \qquad \left(= \dfrac{\mathrm{Nu}}{\mathrm{Re}\,\mathrm{Pr}}\right)$

APPENDIX D. SOME DIMENSIONLESS NUMBERS (cont.)

Streaming Reynolds:	$\mathrm{Rs} = \dfrac{U^2}{\omega \nu}$
Taylor:	$\mathrm{Ta} = \dfrac{r d^3 (\Omega_1^2 - \Omega_2^2)}{\nu^2}$
Weber:	$\mathrm{We} = \dfrac{\rho l U^2}{\sigma_s}$

APPENDIX E. THERMOPHYSICAL PROPERTIES OF FOUR FLUIDS

Fluid			Air	Light oil	Water	Mercury
Temperature	T	(°C)	27	60	20	100
Density	ρ	(kg m^{-3})	1.18	860	1000	13,400
Dynamic viscosity	μ	$\times 10^3(\text{kg m}^{-1}\ \text{s}^{-1})$	0.0185	72.2	1.0	1.25
Kinematic viscosity	ν	$\times 10^6(\text{m}^2\ \text{s}^{-1})$	15.7	84	1.006	0.093
Thermal conductivity	k	$(\text{W m}^{-1}\ \text{K}^{-1})$	0.0262	0.14	0.60	10.5
Specific heat	c_p	$\times 10^{-3}(\text{J kg}^{-1}\ \text{K}^{-1})$	1.006	2.05	4.18	0.137
Thermal diffusivity	$a = \dfrac{k}{\rho c_p}$	$\times 10^6(\text{m}^2\ \text{s}^{-1})$	22.1	0.080	0.144	5.72
Prandtl number	$\Pr = \dfrac{\nu}{a}$	(Dimensionless)	0.711	1058	7.01	0.0163
Thermal expansivity	β	$\times 10^3(\text{K}^{-1})$	3.33	0.70	0.18	0.182
Buoyancy parameter	$\dfrac{g\beta}{\nu^2}$	$\times 10^9(\text{K}^{-1}\ \text{m}^{-3})$	0.133	0.00097	1.75	206
Buoyancy parameter	$\dfrac{g\beta}{\nu a}$	$\times 10^9(\text{K}^{-1}\ \text{m}^{-3})$	0.095	1.03	12.2	3.36

INDEX TO EXAMPLES

INSTABILITY

TURBULENT FLOW PREDICTION

Algebraic models

One-equation Models

TURBULENT HEAT TRANSFER PREDICTION

Algebraic models

One- and two-equation models

DIMENSIONAL ANALYSIS

INDEX